PROGRAMMABLE LOGIC CONTROLLERS

Second Edition

James A. Rehg • Glenn J. Sartori

Upper Saddle River, New Jersey
Columbus, Ohio

Library of Congress Cataloging-in-Publication Data

Rehg, James A.

Programmable logic controllers / James A. Rehg, Glenn J. Sartori. -- 2nd ed.

p. cm.

ISBN-13: 978-0-13-504881-8

ISBN-10: 0-13-504881-8

1. Programmable controllers. I. Sartori, Glenn J. II. Title.

TJ223.P76R44 2009

629.8'9--dc22

2008025785

Editor-in-Chief: Vernon Anthony
Acquisitions Editor: Wyatt Morris
Editorial Assistant: Christopher Reed
Project Manager: Rex Davidson
Senior Operations Supervisor: Pat Tonneman
Operations Specialist: Laura Weaver
Art Director: Candace Rowley
Cover Designer: Diane Lorenzo
Cover Photo: iStock
Director of Marketing: David Gesell
Marketing Assistant: Les Roberts

This book was set by Aptara® and was printed and bound by Courier Kendallville, Inc. The cover was printed by Phoenix Color Corp.

Pearson Education Ltd., London
Pearson Education Singapore Pte. Ltd.
Pearson Education Canada, Inc.
Pearson Education—Japan
Pearson Education Australia Pty. Limited
Pearson Education North Asia Ltd., Hong Kong
Pearson Educación de Mexico, S.A. de C.V.
Pearson Education Malaysia Pte. Ltd.

10 9 8 7 6 5 4 3 2 1
ISBN-13: 978-0-13-504881-8
ISBN-10: 0-13-504881-8

Preface

INTRODUCTION

The 1970s witnessed the birth of two types of computers that changed the world and the way business is conducted. The Apple II, introduced in 1976, was the world's first widely used microcomputer. Today's multibillion-dollar personal computer industry is an outgrowth of this small computer company started by two young entrepreneurs in a garage.

The second computer, created in 1972 by Richard Morley and now called a *programmable logic controller* (PLC), does not have the instant name recognition of the personal computer, but it has had an equally significant impact in manufacturing. The PLC is often called the personal computer for the factory floor.

PLCs are the de facto standard used to control automation systems in every industry across the globe. Control applications range from a single machine to an entire production facility with processes that have both analog and discrete control requirements. This textbook addresses the application, operation, programming, and troubleshooting of PLCs used in automation and control applications.

IMPORTANT FEATURES

The text has the following salient features:

- The text is divided into two parts. Part 1, Chapters 1 through 9, introduces the reader to fundamental PLC concepts and the basic operation and programming format for the commonly used instructions in most PLC applications. If you learn the instructions in Part 1 you will be able to program and interpret 90 percent of the ladder rungs used in automation control. Part 2, Chapters 10 through 17, addresses the advanced instructions, covers four of the languages (Ladder Diagram, Function Block, Sequential Function Chart, and Structured Text) in the IEC 61131 PLC standard, and provides a practical introduction to industrial networks. Goals and learning objectives are provided at the beginning of every chapter.
- The presentation format includes a description of programming and PLC instructions for all three Allen-Bradley PLCs, PLC 5, SLC 500, and ControlLogix, with the SLC 500 system used most often in the example problems.

- Standard ladder logic building blocks are developed for PLC instructions in Chapters 4 through 11, 13, 15, and 16. The standards start with a description of the automation control requirement and then present the ladder rung options used for a solution. Example problems show how these standard rungs are grouped for a total automation solution.
- The operation and programming for two generations of PLC software—rack/slot-based addressing and variable- or tag-based addressing systems—is discussed.
- Text content is sequenced to support a laboratory with a class lecture.
- All end-of-chapter questions and problems, including references to Allen-Bradley manuals and chapter illustrations, are conveniently located in Appendix D.
- Troubleshooting is integrated into every chapter.
- The text is written in a direct, clear, and easy-to-read style that is designed for students with no prior PLC experience.
- Real-world control problems are used in illustrated programming examples.
- A pneumatic robot material handler and a process tank control problem are used in Chapters 3, 4, 7, 11, 13, 15, and 16 to illustrate PLC control of a sequential machine and process system. The control solution for each problem changes as new PLC instructions are introduced in the chapters.
- The text includes a generous number of example problems at varying levels of difficulty and a large number of descriptive figures.
- A CD-ROM with reference material from Allen-Bradley is provided with the text.
- The text content and organization permits teachers to adjust the chapter sequence to fit a current syllabus.
- A glossary of terms is provided in an appendix.
- A description of the five IEC 61131 programming languages with detailed coverage of the four supported in Allen-Bradley PLCs is given.

FOR THE STUDENTS

An increasing number of graduates of engineering and engineering technology programs are working in manufacturing automation because production systems have become increasingly complex and highly automated. As a result, students need to understand the theory and operation of PLCs used in the control of production systems. Our primary goal for this book is to create a clear and comprehensive text for students to use to learn programmable logic controllers. Every effort is made to present the material in a logical order, to express the concepts in a writing style that a first-time user of PLCs can understand, and to keep the needs of the student foremost in every part of the text development. Texts often include technical terms when describing new concepts that have not been previously defined or that are not common knowledge for the students. A special effort is made in this text not to use any term or technical language that is not introduced or defined earlier in the text.

In addition, the text is written as both a learning tool and as a future reference resource. If you work in automation, you will have to use PLCs or PLC-like controllers. The information presented in the text describes PLCs clearly for the student learner, and the broad coverage of topics serves as an ideal resource when the student graduate is working with industrial controls.

STUDENT CD-ROM INCLUDED WITH TEXT

The text comes with a student CD-ROM with valuable resources for learning to operate and program programmable logic controllers. Reference material from Allen-Bradley is included on the CD-ROM. The material provides quick access to technical data related to Allen-Bradley PLCs. The chapter questions and problems make use of the PLC data presented on the CD-ROM.

LEVEL, AUDIENCE, AND USE

The primary audiences for the text are students in two- and four-year technology curriculums and engineering students studying PLCs. Part 1 of the text assumes only that the reader has a good understanding of DC and AC circuits and a basic understanding of digital circuits; therefore, the text could be used for an introduction PLC course as early as the third semester/quarter. The second half of the text, Part 2, introduces process control. The differential equation format for the control algorithms is given, but the PLC implementation is reduced to an algebra formulation. In addition, the text-based programming language, Structured Text, is covered in Chapter 15, which uses the structures normally found in a structured programming language. Classes for students without any previous programming courses or experience may require more time for this chapter to introduce programming structures and their operation.

PLCs are taught in one-, two-, and three-course sequences, with courses in the semester and quarter systems. Although the majority of institutions teach one PLC course, the move toward a two-course sequence is growing. Because the courses at each institution are slightly different, the text is presented in a flexible format that can be adapted to each institution's syllabus. The following table lists suggestions for chapters that could be used for the multiple-course sequences in both the semester and quarter systems. The content suggestions are for 3-credit courses with 2 hours of lecture and 2 hours of laboratory per week. If other lecture hour times are used, then adjustments to the content are possible.

The text integrates all three of the Allen-Bradley PLC processors (PLC 5, SLC 500, and ControlLogix) throughout the text. The SLC 500, however, is used for the majority of the examples and is the processor described when an instruction operates in the same way in all three processors. When operation differs for the three processors or where different instructions are present, each process is covered in a separate section. Because the operation of the PLC 5 is similar to the SLC 500 in many cases, these two are frequently covered in the same section. However, the chapter organization permits an instructor to

Single Course – semester system	Chapters 1 through 9 (Part 1) plus introduction to shift registers (Chapter 10), sequencers (Chapter 11), and PLC networks (Chapter 17). The focus is on the SLC 500 with PLC 5 and ControlLogix content added as time permits. Chapter 3 is covered completely since it introduces all three PLC systems' addressing modes.
Single Course – quarter system	Chapters 1 through 9 (Part 1) with the focus on the SLC 500. Chapter 3 is covered completely since it introduces all three PLC systems' addressing modes.
First Course – semester system *Or*	Chapters 1 through 9 (Part 1) with the focus on either the SLC 500 or the PLC 5 plus the ControlLogix instruction format. Chapter 3 is covered completely since it introduces all three PLC systems' addressing modes. Chapters 1 through 12 with the focus on the SLC 500 and PLC 5 PLCs.
Second Course – semester system *Or*	Review Chapter 3 for the addressing format of the ControlLogix PLC system. Then cover Chapters 10 and 11 with the focus on either the SLC 500 or the PLC 5 plus the ControlLogix instruction format. Finally, cover Chapters 12 through 17. Chapters 3 through 11 and 13 through 17 with the focus on the ControlLogix PLC.
First Course – quarter system	Chapters 1 through 9 (Part 1) with the focus in on the SLC 500.
Second Course – quarter system	ControlLogix content in Chapter 3 is covered and then Chapters 10 through 17 for all PLC systems.

separate the SLC 500 and PLC 5 processors from the ControlLogix if the course requires just one type of PLC content.

The text is organized in two parts. Part 1 introduces PLCs and covers in detail the programming instructions used to write a large number of the PLC programs used in industry. Part 2 is written to support additional topics in an introductory course, a second course in PLCs, or an advanced PLC offering as illustrated in the previous table.

VENDOR RESOURCES OR A PLC TEXTBOOK?

A review of PLC texts indicates that many add little new information from what is available online directly from the vendors that manufacture the devices. Vendor reference and resource material is a valuable aid to the PLC application engineer or technician when device-specific information is needed. This text is designed to complement vendor resources by providing concepts and content not provided by the vendors. For example, thirteen of the chapters have standard application solutions for instructions with comments on how an instruction can be used most effectively. In addition, information on troubleshooting and programming not provided by the vendors is included.

The writing style of this text differs from the industry material as well. Industry resources are written for an industry audience and assume a certain minimum knowledge base on the subject. This text makes no such assumption and describes the technology so that students can learn PLCs with no previous experience in PLCs or discrete and analog system control. The text does make good use of vendor resources so that students learn how to use the material, which will be their source of PLC information in the future.

CHAPTER CONTENT

Part 1, titled *Programmable Logic Controller—Fundamental Concepts*, includes nine chapters written to support a first course in programmable logic controllers. Chapter 1 and all subsequent chapters start with chapter goals and objectives. In addition, this chapter defines a PLC and covers a brief history of PLCs, a description of the system and components, an introduction to vendor systems, a description of PLC types and types of input and output modules, and a comparison between relay ladder logic and PLC ladder logic. Every chapter ends with general chapter questions, Web and data sheet questions, and problems divided into general, PLC 5, SLC 500, ControlLogix, and challenge groups.

Chapter 2 focuses on discrete input devices and output actuators and describes the operation of the most frequently used input devices and output actuators in automation control systems. Devices covered include manual and mechanically operated switches, transducers and sensors including proximity and photoelectric devices, interfacing switches and sensors, input wiring, field device current sourcing and current sinking concepts, electromagnetic output and solenoid-controlled devices, control relays, contactors, motor starters, pilot lights, alarms, interfacing output devices, and output wiring.

Chapter 3 covers an introduction to PLC programming with the following topics: decimal, octal, and binary number systems, ladder logic fundamentals, addressing rack/slot and tag-based systems, examine if closed and examine if open inputs selection and applications, retentive and non-retentive coil outputs, virtual or internal relays, scan time, multiple inputs, standard input logic, sealing contacts, multiple outputs, latched outputs, and using internal memory bits. In addition, empirical program design is introduced along with programming devices and software. Troubleshooting techniques for PLC-controlled systems and program design are also covered.

Chapter 4 covers programming timers, and the standard ladder logic used for timers. Topics include mechanical and electronic timing relays, and PLC timer instructions, such as on-delay timers, off-delay timers, retentive timers, and the reset instruction. In addition, cascaded timers, empirical design of timer ladders, and troubleshooting of timer ladders and input/output modules are addressed. The use of timers in pneumatic robot control is also described.

Chapter 5 describes the counter function present in PLCs, and the standard ladder logic used for counters. Topics include counter instructions, such as up counters, down counters, up/down counters, and cascade counters; using counter output bits; programming counter instructions; and counter applications. In addition, the reset instruction and one-shot function are covered along with programming and application issues.

Chapter 6 focuses on arithmetic and move instructions available in the PLC instruction set, and the standard ladder logic used for these instructions. Topics include instructions and format for addition, subtraction, multiplication, division, square root, and move instructions; programming these instructions; and application of math and move instructions.

Chapter 7 describes the binary-coded decimal (BCD) and the hexadecimal numbering systems and covers the conversion and comparison instructions. In addition, the standard ladder logic used for these instructions is covered. The instructions for converting to and from BCD numbers are presented. The comparison instructions covered include equal to, not equal to, less than, greater than, less than or equal to, and greater than or equal to. In addition, programming and application of the comparison instructions are addressed.

Chapter 8 focuses on instructions that change the flow of the program execution and covers some special-purpose instructions. In addition, the standard ladder logic used for these instructions is covered. The program flow instructions described include master control and zone control instructions, jump instructions, subroutines, and immediate input and output instructions. In addition, the clear instruction is covered along with programming and application issues.

Chapter 9 is the last chapter in Part 1 of the text. The addressing modes discussed include direct, indirect, indexed, and indexed indirect. Typical applications for the addressing modes are included to indicate how each mode is used.

Part 2 of the text, titled *Advanced PLC Instructions and Applications*, includes Chapters 10 through 17. The content of Part 2 could be used selectively to enhance a first course in PLCs or it could be used for a second, more advanced PLC offering.

Chapter 10 covers instructions related to data handling and shift register applications. In addition, the standard ladder logic used for these instructions is covered. Topics include the copy and fill instructions, the FIFO, LIFO, and FAL functions, bit patterns in a register, changing a register bit status, shift register functions, programming FIFO, LIFO, and shift register instructions.

Chapter 11 addresses the programming and operation of PLC sequencers, and the standard ladder logic used for sequencers. Topics include electromechanical sequencing, the basic PLC sequencer function, PLC sequencer with timing, cascading sequencers, and programming applications for the sequencer function.

Chapter 12 covers analog PLC applications. The concepts covered include analog sensors and actuators, types of PLC analog modules and systems, PLC analog input and output data, programming analog instructions, and analog applications.

Chapter 13 introduces the first of the new IEC 61131 programming languages, Function Block Diagram (FBD). The Allen-Bradley FBD instruction format is used in the description, which includes programming examples and application information. In addition, the standard ladder logic used for these instructions is covered.

Chapter 14 describes how the analog principles from Chapter 12 and FBD instructions from Chapter 13 are applied to the control of on-off and continuous processes. The topics include on-off control, two-position control, floating control, PID principles, fuzzy logic, and programming the PID function.

Chapter 15 introduces the second of the new IEC 61131 programming languages, Structured Text (ST). The Allen-Bradley ST instruction format is used in the description, which includes programming examples and application information. In addition, the standard ladder logic used for these instructions is covered.

Chapter 16 introduces the third of the new IEC 61131 programming languages, Sequential Function Chart (SFC). The Allen-Bradley instruction format is used in the description, which includes programming examples and application information.

Chapter 17 addresses industrial networks and distributive control. The network topics include PLC network architecture, Ethernet/IP, DeviceNet,

ControlNet, remote I/O, Data Highway Plus, DH 485, Modbus, and Profibus. In addition, wireless networks and the human-machine interface (HMI) are covered.

SUPPLEMENTS

- Online Instructor's Resource Manual with PowerPoints
- Online TestGen
- Laboratory Exercises

To access supplementary materials online, instructors need to request an instructor access code. Go to **www.pearsonhighered.com/irc**, where you can register for an instructor access code. Within 48 hours after registering, you will receive a confirming e-mail, including an instructor access code. Once you have received your code, go to the site and log on for full instructions on downloading the materials you wish to use.

ACKNOWLEDGMENTS

The authors would like to thank the many people that helped to make this text possible. First we want to thank the following individuals from industry who supplied valuable content oversight and application information. The industry participants included Don Cox and Dave Mayewski.

We would also like to acknowledge the support received from Allen-Bradley for software and systems that permitted us to verify the code presented in the text. A special thanks to John Sjolander for assistance above and beyond the call of duty.

We would like to thank Wyatt Morris, Christopher Reed, Lois Porter, and Rex Davidson at Prentice Hall and Jodi Dowling at Aptara. Thanks to Marci Rehg for copy editing. A special thanks to the reviewers that provided valuable feedback and suggestions to improve the content, chapter sequences, and concepts that should be modified and added. The reviewers include:

- Tom Cunningham, Baker College, Owosso
- Daniel Green, Sinclair Community College
- Sam Guccione, Eastern Illinois University
- Gregory Harstine, Stark State College of Technology
- Jude Pearse, University of Maine
- Eduard Plett, Kansas State University
- Dave Setser, Johnson County Community College
- Cree Stout, York Technical College
- Ken Swayne, Pelissippi State Technical Community College
- Marc Timmerman, Oregon Institute of Technology
- Edward Troyan, Lehigh Carbon Community College

In addition, we want to thank the teachers, staff, and students at Penn State Altoona for helping us understand how PLCs should be taught and learned.

Finally, we would like to thank you for adopting the text for your class and for providing us feedback on what you see as beneficial changes to the content and sequencing of the material. We invite you to email us with your comments and suggestions.

James A. Rehg
(jamesrehg@plcteacher.com)
Glenn J. Sartori
(glennsartori@plcteacher.com)

Contents

PART 1 Programmable Logic Controllers—Fundamental Concepts 1

Goal 1
Objectives 1
Career Insight 1

CHAPTER 1 Introduction to Programmable Logic Controllers 3

1-1 GOALS AND OBJECTIVES 3

1-2 THE PLC INDUSTRY TODAY 3
1-2-1 PLC Definitions 4
1-2-2 PC versus PLC 5

1-3 RELAY LADDER LOGIC 5
1-3-1 Electromagnetic Relay 7
1-3-2 Relay Control Systems 8

1-4 PLC SYSTEM AND COMPONENTS 10
1-4-1 Backplane 11
1-4-2 Processor and Power Supply 11
1-4-3 Programming Device 11
1-4-4 Input and Output Interface 12
1-4-5 Special Communications Modules and Network Connections 18
1-4-6 PLC Special-Purpose Modules 20

1-5 PLC TYPES 20
1-5-1 Rack/Slot Address Based 20
1-5-2 Tag Based PLCs 20
1-5-3 Soft PLCs or PC-Based Control 21

1-6 PLC LADDER LOGIC PROGRAMMING 21
1-6-1 PLC Solution 22
1-6-2 Ladder Logic Operation 24

1-6-3 An Alternate Solution 24
1-6-4 PLC Advantages 25

1-7 ELECTRICAL AND PLC SAFETY 28
1-7-1 Electrical Shock—How the Body Reacts 28
1-7-2 The Nature of Electrical Shock 28
1-7-3 Safe Electrical Practices 30
1-7-4 Response to Shock Victims 32

1-8 WEB SITES FOR PLC MANUFACTURERS 32

CHAPTER 2 Input Devices and Output Actuators 33

2-1 GOALS AND OBJECTIVES 33

2-2 MANUALLY OPERATED INDUSTRIAL SWITCHES 33
2-2-1 Toggle Switches 34
2-2-2 Push Button Switches 36
2-2-3 Selector Switches 38

2-3 MECHANICALLY OPERATED INDUSTRIAL SWITCHES 39
2-3-1 Limit Switches 40
2-3-2 Flow Switches 41
2-3-3 Level Switches 41
2-3-4 Pressure Switches 42
2-3-5 Temperature Switches 42
2-3-6 Control Diagrams 43

2-4 INDUSTRIAL SENSORS 45
2-4-1 Proximity Sensors 47
2-4-2 Photoelectric Sensors 51

2-5 INTERFACING INPUT FIELD DEVICES 58
2-5-1 Powering Input Field Devices 58
2-5-2 Input Wiring 59
2-5-3 Current Sinking and Current Sourcing Devices 60

2-6 ELECTROMAGNETIC OUTPUT ACTUATORS 61
2-6-1 Solenoid-Controlled Devices 61
2-6-2 Control Relays 63
2-6-3 Latching Relays 67
2-6-4 Contactors 67
2-6-5 Motor Starters 68

2-7 VISUAL AND AUDIO OUTPUT DEVICES 69
2-7-1 Pilot Lamps 69
2-7-2 Horns and Alarms 70

2-8 INTERFACING OUTPUT FIELD DEVICES 71
2-8-1 Powering Output Field Devices 71
2-8-2 Output Wiring 71
2-8-3 Current Sinking and Current Sourcing Devices 73

2-9 TROUBLESHOOTING INPUT AND OUTPUT DEVICES 73
2-9-1 Troubleshooting Switches 74
2-9-2 Troubleshooting Relays 74
2-9-3 Troubleshooting Proximity Sensors 75
2-9-4 Troubleshooting Photoelectric Sensors 75

CHAPTER 3 Introduction to PLC Programming 77

3-1 GOALS AND OBJECTIVES 77

3-2 NUMBER SYSTEMS 77
3-2-1 Number System Basics 78
3-2-2 Binary System 78
3-2-3 Octal Number System 80

3-3 HEXADECIMAL NUMBER SYSTEM 81
3-3-1 Introduction 81
3-3-2 Comparison of Numbering Systems 82
3-3-3 Binary to Hex Conversion 82
3-3-4 Hex to Binary Conversion 82
3-3-5 Hex to Decimal Conversion 83
3-3-6 Decimal to Hex Conversion 83
3-3-7 Repeated Division Method 83

3-4 BITS, BYTES, WORDS, AND MEMORY 83

3-5 PLC MEMORY AND REGISTER STRUCTURE 84
3-5-1 Allen-Bradley Memory Organization 85
3-5-2 Allen-Bradley PLC 5 Memory Organization 86
3-5-3 Allen-Bradley SLC 500 Memory Organization 86
3-5-4 Allen-Bradley Logix System Memory Organization 89

3-6 INPUT AND OUTPUT ADDRESSING 92
3-6-1 PLC 5 Rack/Group-Based Addressing 92
3-6-2 SLC 500 Rack/Slot-Based Addressing 94
3-6-3 Other Vendors' Rack/Slot PLC Addressing 99
3-6-4 Tag-Based Addressing 100

3-7 INTERNAL CONTROL RELAY BIT ADDRESSING 104
3-7-1 PLC 5 and SLC 500 Binary Bit Addressing 104
3-7-2 ControlLogix Binary Bit Addressing 107
3-7-3 Retentive and Non-retentive Memory 107

3-8 STATUS DATA ADDRESSING 108
3-8-1 PLC 5 and SLC 500 Status Data Addressing 108
3-8-2 Logix System Status 109

3-9 ALLEN-BRADLEY INPUT INSTRUCTIONS AND OUTPUT COILS 110
3-9-1 Examine if Closed and Examine if Open Instructions 110
3-9-2 Output Energize, Output Latch, and Output Unlatch Instructions 112

3-10 INPUTS, OUTPUTS, AND SCAN TIME 116
3-10-1 Scan Time 116
3-10-2 Linking Inputs and Outputs 117
3-10-3 Process Tank Application 119

3-11 PLC PROGRAM DESIGN AND RELAY LADDER LOGIC CONVERSION 121
3-11-1 Examine if Closed and Examine if Open Selection 121
3-11-2 Multiple Inputs 123
3-11-3 Multiple Outputs 126
3-11-4 Empirical Program Design 126
3-11-5 Converting Relay Logic to PLC Solutions 132

3-12 TROUBLESHOOTING LADDER LOGIC CONTROL SYSTEMS 134
3-12-1 System Troubleshooting Tools 134
3-12-2 Troubleshooting Sequence 138
3-12-3 Troubleshooting Input and Output Modules 138

CHAPTER 4 Programming Timers 141

4-1 GOALS AND OBJECTIVES 141

4-2 MECHANICAL TIMING RELAYS 141
4-2-1 Timed Contacts 142
4-2-2 Instantaneous Contacts 143
4-2-3 Timing Relay Operation 144
4-2-4 Selecting Timing Relays 145

4-3 ELECTRONIC TIMING RELAYS 145

4-4 PLC TIMER INSTRUCTIONS 146

4-5 ALLEN-BRADLEY TIMER INSTRUCTIONS 146
4-5-1 Allen-Bradley Timer Symbol and Parameters 146
4-5-2 Allen-Bradley Timer Bits 149
4-5-3 Allen-Bradley TON, TOF, and RTO Instructions 149

4-6 ALLEN-BRADLEY TIMER PARAMETER AND BIT ADDRESSING 153
4-6-1 PLC 5 and SLC 500 Timer Memory Map 153
4-6-2 ControlLogix Timer Addressing 154
4-6-3 Timer Contacts versus PLC Instructions 155

4-7 PROGRAMMING ALLEN-BRADLEY TON AND TOF TIMER LADDER LOGIC 155
4-7-1 Standard Ladder Logic for Allen-Bradley TON Timers 155
4-7-2 Standard Ladder Logic for Allen-Bradley TOF Timers 158
4-7-3 Allen-Bradley TON and TOF Timer Applications 158

4-8 ALLEN-BRADLEY RETENTIVE TIMERS 164
4-8-1 Reset Instruction for RTO Timer and Other Allen-Bradley Instructions 164

4-9 CASCADED TIMERS 167

4-10 EMPIRICAL DESIGN PROCESS WITH PLC TIMERS 169
4-10-1 Adding Timers to the Process 169

4-11 CONVERSION OF RELAY LOGIC TIMER LADDERS TO PLC LOGIC 174

4-12 TROUBLESHOOTING LADDER RUNGS WITH TIMERS 175
4-12-1 Troubleshooting Timer Ladder Logic 175
4-12-2 Temporary End Instruction 177

4-13 LOCATION OF THE INSTRUCTIONS 177

CHAPTER 5 Programming Counters 179

5-1 GOALS AND OBJECTIVES 179

5-2 MECHANICAL AND ELECTRONIC COUNTERS 179

5-3 INTRODUCTION TO ALLEN-BRADLEY COUNTERS 180
5-3-1 Counter Output Bits 181

5-4 ALLEN-BRADLEY COUNTER AND RESET INSTRUCTIONS 183
5-4-1 PLC 5 and SLC 500 Counter and Reset Addressing 183
5-4-2 Logix Counter Instructions 186
5-4-3 Standard Ladder Logic for Counters 188
5-4-4 Allen-Bradley Up Counters 191
5-4-5 Allen-Bradley Down Counters 197
5-4-6 Allen-Bradley Up/Down Counters 197
5-4-7 Allen-Bradley One-Shot Instructions 198

5-5 CASCADED COUNTERS 203

5-6 EMPIRICAL DESIGN PROCESS WITH PLC COUNTERS 203
5-6-1 Adding Counters to the Process 203

5-7 CONVERSION OF RELAY LOGIC COUNTER LADDERS TO PLC LOGIC 207

5-8 TROUBLESHOOTING COUNTER LADDER LOGIC 207
5-8-1 Suspend Instruction 208
5-8-2 Process Speed versus Scan Time 209

5-9 LOCATION OF THE INSTRUCTIONS 209

CHAPTER 6 Arithmetic and Move Instructions 211

6-1 GOALS AND OBJECTIVES 211

6-2 BINARY ARITHMETIC 211

6-3 SIGNED BINARY NUMBERS 213

6-4 ALLEN-BRADLEY ARITHMETIC INSTRUCTIONS 215
6-4-1 Structure for Arithmetic Instructions 215

6-5 OPERATION OF ALLEN-BRADLEY ARITHMETIC AND MOVE INSTRUCTIONS 216
6-5-1 Addition Instruction 217
6-5-2 Subtraction Instruction 218
6-5-3 Multiplication Instruction 219
6-5-4 Division Instruction 219
6-5-5 Square Root Instruction 222
6-5-6 Move Instructions 222

6-6 STANDARD LADDER LOGIC FOR ALLEN-BRADLEY MATH AND MOVE INSTRUCTIONS 223

6-7 EMPIRICAL DESIGN PROCESS WITH MATH AND MOVE INSTRUCTIONS 230
6-7-1 Adding Math and Move Instructions to the Process 230

6-8 TROUBLESHOOTING MATH AND MOVE LADDER LOGIC 235
6-8-1 SLC 500 Test Modes 235

6-9 LOCATION OF THE INSTRUCTIONS 238

CHAPTER 7 Conversion and Comparison Instructions 241

7-1 GOALS AND OBJECTIVES 241

7-2 BINARY CODED DECIMAL SYSTEM 241
7-2-1 Allen-Bradley BCD Instructions and Standard Ladder Logic 243

7-3 HEXADECIMAL SYSTEM 250

7-4 COMPARISON INSTRUCTION STRUCTURE 250

7-5 ALLEN-BRADLEY COMPARISON INSTRUCTIONS 251
7-5-1 Standard Ladder Logic for EQU, NEQ, LES, and GRT Comparison Instructions 251
7-5-2 Standard Ladder Logic for LEQ, GEQ, MEQ, and LIM Comparison Instructions 253
7-5-3 Standard Ladder Logic for Multiple Instructions and Hysteresis 255

7-6 EMPIRICAL DESIGN PROCESS WITH BCD CONVERSION AND COMPARISON INSTRUCTIONS 258
7-6-1 Adding BCD Conversion Instructions to the Process 258
7-6-2 Adding Comparison Instructions to the Process 259
7-6-3 Process Tank Design 259
7-6-4 Pneumatic Robot Design 265

7-7 TROUBLESHOOTING BCD CONVERSION AND COMPARISON LADDER LOGIC 268
7-7-1 Troubleshooting with the Module Indicators 268

7-8 LOCATION OF THE INSTRUCTIONS 272

CHAPTER 8 Program Control Instructions 273

8-1 GOALS AND OBJECTIVES 273

8-2 PROGRAM CONTROL INSTRUCTIONS 273

8-3 ALLEN-BRADLEY PROGRAM CONTROL INSTRUCTIONS 274
8-3-1 Master Control Reset Instructions 274
8-3-2 Jump and Label Zone Control Instructions 277
8-3-3 Subroutine Instructions 280
8-3-4 PLC 5 and SLC 500 Subroutine Instructions 282
8-3-5 PLC 5 and ControlLogix Options for Subroutine Instructions 287

8-4 ALLEN-BRADLEY IMMEDIATE INPUT AND OUTPUT INSTRUCTIONS 296
8-4-1 PLC 5 Immediate Input and Output Instructions 296
8-4-2 SLC 500 Immediate Input and Output Instructions 296
8-4-3 ControlLogix Immediate Output Instruction 298

8-5 EMPIRICAL DESIGN PROCESS WITH PROGRAM CONTROL INSTRUCTIONS 299
8-5-1 Adding Program Control Instructions to the Process 299

8-6 TROUBLESHOOTING PROGRAM CONTROL INSTRUCTIONS IN LADDER LOGIC 301

8-7 LOCATION OF THE INSTRUCTIONS 301

CHAPTER 9 Indirect and Indexed Addressing 303

9-1 GOALS AND OBJECTIVES 303

9-2 ALLEN-BRADLEY ADDRESSING MODES 303
9-2-1 Direct Addressing 303
9-2-2 Indirect Addressing 304
9-2-3 Indexed Addressing 304
9-2-4 Indexed Indirect Addressing 305
9-2-5 PLC 5, SLC 500, and Logix Systems Syntax 305

9-3 EMPIRICAL DESIGN PROCESS WITH INDIRECT AND INDEXED ADDRESSSING 310
9-3-1 Adding Indirect and Indexed Addressing to the Process 310

9-4 TROUBLESHOOTING INDIRECT AND INDEXED ADDRESSING IN LADDER LOGIC 314

PART 2 Advanced PLC Instructions and Applications 315

Goals 315
Objectives 315
Career Insights 315

CHAPTER 10 Data Handling Instructions and Shift Registers 317

10-1 GOALS AND OBJECTIVES 317

10-2 DATA HANDLING 317
10-2-1 Bit Patterns in Words 317
10-2-2 Word Patterns in Files 318

10-3 ALLEN-BRADLEY DATA TRANSFER AND MANIPULATION INSTRUCTIONS 318
10-3-1 AND, OR, and XOR Instructions 319
10-3-2 File-Arithmetic-Logic (FAL) Instruction 320
10-3-3 Shift Registers 326
10-3-4 First In, First Out (FIFO) Function 333
10-3-5 Last In, First Out (LIFO) Function 336
10-3-6 Copy and Fill Instructions 338

10-4 EMPIRICAL DESIGN PROCESS WITH BIT AND WORD OPERATION INSTRUCTIONS 341

10-5 TROUBLESHOOTING DATA HANDLING INSTRUCTIONS AND SHIFT REGISTERS IN LADDER LOGIC 344

10-6 LOCATION OF THE INSTRUCTIONS 345

CHAPTER 11 PLC Sequencer Functions 347

11-1 GOALS AND OBJECTIVES 347

11-2 ELECTROMECHANICAL SEQUENCING 347

11-3 BASIC PLC SEQUENCER FUNCTION 349

11-4 ALLEN-BRADLEY SEQUENCER INSTRUCTIONS 349
11-4-1 PLC 5 and SLC 500 SQO and SQC Sequencer Instruction Structure 349
11-4-2 PLC 5 and SLC 500 SQO Instruction Operation 350
11-4-3 ControlLogix SQO Sequencer Instruction 352
11-4-4 PLC 5 and ControlLogix Sequencer Input (SQI) Instruction 359
11-4-5 Sequencer Compare (SQC) Instruction 366
11-4-6 Sequencer Load (SQL) Instruction 367

11-5 CASCADING SEQUENCERS 369

11-6 EMPIRICAL DESIGN PROCESS WITH SEQUENCER INSTRUCTIONS 369
11-6-1 Adding Sequential Instructions to the Process 370

11-7 TROUBLESHOOTING SEQUENCER INSTRUCTIONS 371

11-8 LOCATION OF THE INSTRUCTIONS 371

CHAPTER 12 Analog Sensors and Control Systems 373

12-1 GOALS AND OBJECTIVES 373

12-2 ANALOG SENSORS 373
12-2-1 Temperature Sensors 374
12-2-2 Pressure Sensors 377
12-2-3 Flow Sensors 380

12-2-4 Position Sensors 389
12-2-5 Vision Systems 390
12-2-6 Troubleshooting Analog Sensors 396

12-3 ANALOG MODULES AND FIELD DEVICES INTERFACING 399
12-3-1 Analog Input and Output Data 400
12-3-2 PLC 5, SLC 500, and Logix Options 401

12-4 CLOSED-LOOP CONTROL SYSTEMS 401
12-4-1 Direct-Acting and Reverse-Acting Controllers 401
12-4-2 Analysis of Closed-Loop Systems 403
12-4-3 Load Change—Process Disturbance 405

12-5 ATTRIBUTES OF AN EFFECTIVE CONTROL SYSTEM 406
12-5-1 Transient Response 406
12-5-2 Response to Change 406
12-5-3 Controller Response and Damping 407
12-5-4 Transient Response Options 408
12-5-5 Steady-State Response 410
12-5-6 Understanding Steady-State Error 410
12-5-7 Correction for Steady-State Error 413
12-5-8 Controller Gain Side Effects 417
12-5-9 Steady-State Error Correction with Bias 418
12-5-10 Stability 418

12-6 TROUBLESHOOTING THE PROPORTIONAL GAIN CONTROLLER 421

CHAPTER 13 PLC Standard IEC 61131-3 Function Block Diagrams 423

13-1 GOALS AND OBJECTIVES 423

13-2 PLC STANDARDS 423
13-2-1 IEC 61131-3 Standard Languages 424

13-3 FUNCTION BLOCK DIAGRAM (FBD) 425
13-3-1 Signal Flow Types, Execution Order, and Data Latching 427
13-3-2 Feedback Loops 428
13-3-3 Function Block Diagram Program Development Sequence 429
13-3-4 Allen-Bradley RSLogix 5000 FBD Programming 431

13-4 EMPIRICAL DESIGN WITH FUNCTION BLOCK DIAGRAMS 433
13-4-1 Standard Function Block Control Solutions 433

13-5 SITES FOR ALLEN-BRADLEY PRODUCTS AND DEMO SOFTWARE 447

CHAPTER 14 Intermittent and Continuous Process Control 449

14-1 GOALS AND OBJECTIVES 449

14-2 PROCESS CONTROL 449

14-3 INTERMITTENT CONTROLLERS 450
14-3-1 On-Off Control 450
14-3-2 Two-Position Control 451
14-3-3 Floating Control 455

14-4 CONTINUOUS CONTROLLERS 457
14-4-1 Proportional Control 457
14-4-2 Proportional Integral (PI) Control 460
14-4-3 Proportional Derivative (PD) Control 462
14-4-4 Proportional Integral and Derivative (PID) Control 465
14-4-5 Fuzzy Control 467

14-5 DIGITAL CONTROL 470
14-5-1 Digital Sample and Hold 471
14-5-2 Proportional Control Mode 472
14-5-3 Integral Control Mode 472
14-5-4 Derivative Control Mode 474
14-6 SCALING IN PROCESS CONTROL 475
14-7 MANUAL CONTROL MODE AND BUMPLESS TRANSFER 476
14-8 LOCATION OF THE INSTRUCTIONS 476

CHAPTER 15 PLC Standard IEC 61131-3—Structured Text Language 477

15-1 GOALS AND OBJECTIVES 477
15-2 OVERVIEW OF IEC 61131-3 TEXT LANGUAGES 477
15-3 ALLEN-BRADLEY IEC 61131 STRUCTURED TEXT IMPLEMENTATION 477
15-4 STRUCTURE TEXT PROGRAMMING 478
15-4-1 Assignment Statements 478
15-4-2 Expressions 478
15-4-3 Operators and Functions 479
15-4-4 Relational Operators 480
15-4-5 Logical Operators and Bitwise Operators 480
15-4-6 Constructs 481
15-5 EMPIRICAL DESIGN WITH STRUCTURED TEXT 487
15-5-1 Standard Structured Text Control Solutions 487
15-5-2 Discrete and Process Implementation 490

CHAPTER 16 PLC Standard IEC 61131-3—Sequential Function Chart 495

16-1 GOALS AND OBJECTIVES 495
16-2 IEC 61131-3 STANDARD LANGUAGES 495
16-3 SEQUENTIAL FUNCTION CHART (SFC) 495
16-3-1 Standard SFC Sequences 496
16-3-2 SFC Sequences 497
16-3-3 SFC Step Programming 498
16-4 SITES FOR ALLEN-BRADLEY PRODUCTS AND DEMO SOFTWARE 509

CHAPTER 17 Industrial Networks and Distributive Control 511

17-1 GOALS AND OBJECTIVES 511
17-2 PLC NETWORK ARCHITECTURE 511
17-3 ETHERNET 512
17-3-1 Ethernet Operation 513
17-3-2 Ethernet Industry Protocol 513
17-4 CONTROLNET 514
17-4-1 ControlNet Operation 514
17-4-2 ControlNet Features 514
17-5 DEVICENET 515
17-5-1 DeviceNet Operation 515
17-5-2 DeviceNet Features 515

17-6 SPECIAL NETWORK INTERFACES 516
17-6-1 SERCOS Interfaces 516
17-6-2 Smart I/O Interfaces 516
17-6-3 Remote I/O Interfaces 517
17-6-4 Serial Communication Interfaces 517
17-6-5 Wireless Interfaces 519
17-6-6 Human-Machine Interfaces 520

17-7 NETWORK APPLICATIONS 520
17-7-1 Profibus Network 520
17-7-2 Data Highway Networks 521
17-7-3 Modbus Network 523

17-8 TROUBLESHOOTING NETWORK SYSTEMS 524

17-9 DISTRIBUTIVE CONTROL 524

17-10 DISTRIBUTED I/O 526
17-10-1 In-cabinet I/O 526
17-10-2 On-machine I/O 526

17-11 SELECTING AND DESIGNING NETWORKS 526

17-12 WEB SITES FOR INDUSTRIAL NETWORKS 529

APPENDIX A Glossary 531

APPENDIX B PLC Module Interface Circuits 543

APPENDIX C Programmable Logic Controller History 549

APPENDIX D Questions and Problems 551

Index 595

Dedicated to my wonderful wife Marci, my sons James and Richard, their loving wives Dorothy and Lorri, and my delightful grandchildren. Also to the thousands of students whose insightful questions have taught me so much during my 39 years in the classroom.

—James Rehg

Dedicated to my family: my loving wife Rosanne and her sister Chris, who are two of my biggest cheerleaders. And to my two sons Michael and Jeffrey, who with their thriving families offer me loving support.

—Glenn Sartori

PART 1

Programmable Logic Controllers—Fundamental Concepts

GOAL

The goal for the first half of the text is to introduce you to the fundamental concepts associated with the operation and programming of programmable logic controllers (PLCs). In every programming language a large percentage of the programs developed use a small subset of the language instructions and functions available in the language. Therefore, the goal of Part 1, Chapters 1 through 9, is for you to master the instructions and formats used most frequently in PLC program development, thus preparing you to write control programs for a large number of automation control applications.

OBJECTIVES

After completing Part 1 you should be able to:

- Identify a PLC and have an appreciation for the history behind the technology.
- Identify and interface standard automation control input and output devices to PLC input and output modules.
- Use data in and convert data between the following number systems: binary, octal, decimal, binary-coded decimal, and hexadecimal.
- Write automation control programs using instructions from the following categories: inputs, outputs, timers, counters, arithmetic and move operations, comparison operations, program control, and data addressing modes.

CAREER INSIGHT

The following career insight provides some information on the careers available to students who master the content of Part 1 of the text. Read the career insight to get a general feel for the career opportunities in PLCs. If you enjoy the technology and problems covered in Part 1, then there are numerous career paths that you can follow in the PLC area.

There are many opportunities associated with programmable logic controllers (PLC). The PLC has become the de facto standard for control of discrete industrial processes, and is moving into the process control areas as well. On the factory floor the PLC plays the same role that the microcomputer has in the office, namely a program-controlled device for problem solution. The PLC has both a hardware and software component, so both aspects of the machine must be addressed for a successful solution. Therefore, there are opportunities on both the hardware and software sides of this technology.

From the hardware viewpoint, you could work for PLC vendors in jobs ranging from designing new hardware to interfacing with customers in a technical sales capacity. You could work for automation vendors who represent the PLC manufacturers and provide design and technical sales assistance to the end users. Numerous additional opportunities exist at system houses (companies that design automation systems for end users) where the complete design of an automation system is performed, including the integration of a PLC for control.

Every hardware solution demands a software solution as well. PLC programmers are required from PLC vendors to end users. The PLC vendors, automation vendors, system houses, and end users all must employ PLC programmers to either develop new programs or maintain existing programs. At present the programming is primarily ladder logic, but as the new programming languages defined in the IEC 61131 PLC standard are embraced, the programming will include five different programming options. The skills needed for this new standard are similar to the skills needed to learn and use high-level text-based languages, such as C+, or graphic languages, such as LabView.

The opportunities are numerous for individuals who have mastered the PLC technology. Students who like the programming aspect of the device can find careers that are focused primarily on programming. Students with an interest in hardware can find jobs that are focused primarily on that area of PLCs. Finally, there are opportunities that require an integration of both the hardware and software sides. If you enjoy this technology, it is a wonderful and rewarding career area.

CHAPTER

1

Introduction to Programmable Logic Controllers

1-1 GOALS AND OBJECTIVES

Chapters 1, 2, and 3 introduce the reader to the programmable logic controller (PLC) technology and the industrial control devices used in automation. As a result, you will find that these chapters cover numerous concepts that prepare the reader for the specific PLC instructions that begin in Chapter 4. The first three chapter's topics are varied, but they all focus on an understanding of the hardware and software in a PLC system. Therefore, the primary goal of this chapter is to introduce programmable logic controllers (PLC) to the student who has no knowledge of this technology. A secondary goal is to show how PLCs fit into the general control needs present in automation.

After completing this chapter you should be able to:

- Write the definition of a PLC.
- Describe the similarities and differences between PLC ladder logic and relay ladder logic.
- Describe the function of all of the component parts of a PLC system.
- Describe the difference between programmable logic controllers and personal computers.
- Define the three types of PLCs currently available: rack/slot address-based, tag-based, and soft PLCs.
- Draw a PLC input and output interface for a typical application.
- Create a simple ladder logic program.
- Describe the electrical safety issues associated with working on PLC systems and the actions to take for an electrical shock victim.

1-2 THE PLC INDUSTRY TODAY

Dick Morley conceived the programmable controller on January 1, 1968. When his new company, Modicon, installed the first model 084 PLC at the Oldsmobile Division of General Motors Corporation and the Landis Company in Landis, Pennsylvania, in 1970, the PLC evolution began. Today the 6.5 billion dollar PLC business is growing at 20 percent per year; however, few people beyond those working in manufacturing automation know it exists. The PLC has been a strong silent partner in promoting manufacturing automation around the globe. A search of the Thomas Register, an online automation manufacturers data base, found 110 listings under the heading of programmable controller vendors, and over 1000 vendors offering PLC add-in boards and other peripherals. In addition, over 2000 companies are developing PLC solutions. This industrial strength microcomputer controls a wide variety of industrial processes, from automobile assembly to stamping out Oreo cookies. The study of PLCs begins with definitions of a programmable logic controller.

1-2-1 PLC Definitions

PLCs are defined as follows:

> *PLCs are special-purpose industrial computers designed for use in the control of a wide variety of manufacturing machines and systems.*
>
> *Or*
>
> *A PLC is a specialized electronic device based on one or more microprocessors that is used to control industrial machinery.*

The definitions state that PLCs are industrial computers, where the term "industrial" implies that PLCs are computers designed to operate in the harsh physical and electrical noise environments present in production plants. They are also specialized electronic devices, so they are not just personal computers that have been moved to the factory floor. Figure 1-1 provides a look at several configurations of Allen-Bradley's Logix systems.

PLC control applications vary from the on/off control of a pump motor using a liquid level switch to control of a conveyor system used to sort packages based upon destination zip codes. Because PLCs are computers, they must be programmed using a programming language. Although most PLCs in the United States use a vendor-specific programming language, called *ladder logic*, there are five standard programming languages that are available for programming new applications.

In 1979 the International Electrotechnical Commission (IEC) established a working group to look at the standardization of PLCs. The PLC standard, called IEC 1131 (changed later to IEC 61131), has six parts. Part 3 is of most interest for this text and is the one that defines the following new languages present in the standard.

- Ladder Diagrams (LD)
- Function Block Diagrams (FBD)
- Structured Text (ST)
- Instruction List (IL)
- Sequential Function Charts (SFC)

The PLC language preferences in specific countries is quite varied. Developers in Europe and Asia, for example, embrace a variety of PLC

FIGURE 1-1: Allen-Bradley Logix systems.

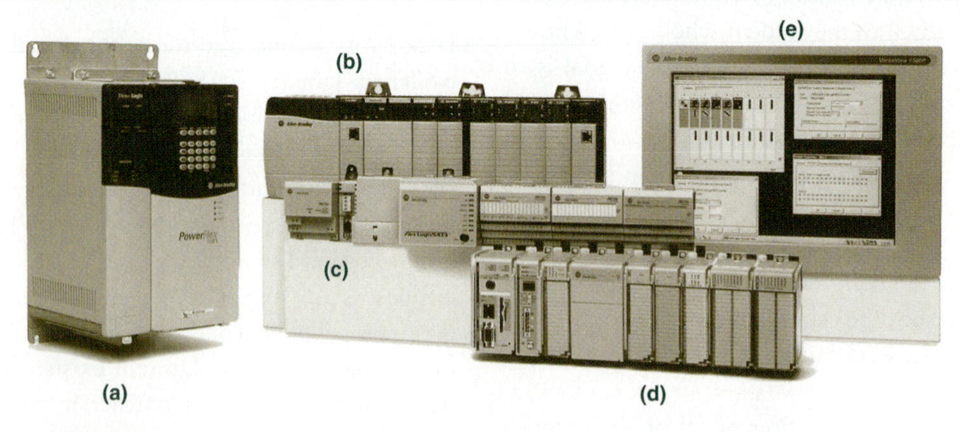

(a) DriveLogix
(b) ControlLogix
(c) FlexLogix
(d) CompactLogix
(e) SoftLogix

Courtesy of Rockwell Automation, Inc.

languages compared to the U.S. market, where ladder logic is used most often. In some applications, such as process and motion control, function block diagrams are used, and in some sequential machine control the SFC language is used. Allen-Bradley, the most commonly used PLC vendor in the United States, offers ladder logic, function block diagrams, sequential function charts, and structured text. These new languages are addressed fully in Part 2 of the text. However, in Part 1, ladder logic is used for the introduction to PLCs.

The reader may notice a similarity between the definition of a PLC and the operation of the personal computer used in offices and homes. The differences between these two technologies are addressed in the next section.

1-2-2 PC versus PLC

The original design for the programmable logic controller was called a *programmable controller*, or PC. The PC abbreviation caused no confusion until the personal computer became widely used and also adopted the PC abbreviation. To avoid confusion, the programmable controller industry added the word *logic* in the title, producing the new term *programmable logic controller*, or PLC. To avoid confusion, this text uses the term programmable logic controller, or PLC. The abbreviation PC will refer to a personal computer.

The PC and PLC have some things in common and many things that make them different. The architecture of the PC and PLC systems are similar, with both featuring a *motherboard*, *processor*, *memory*, and *expansion slots*. Figure 1-2 illustrates the PLC's *central processing unit* (CPU) composed of a microprocessor, often an 8051 integrated circuit, and the computer-type architecture. The PLC processor has a *microprocessor* chip linked to *memory* and *I/O* (input/output) chips through parallel *address*, *data*, and *control* buses. Generally, PLCs do not have removable or fixed storage media such as floppy and hard disk drives, but they do have solid-state memory to store programs. PLCs do not have a monitor, but a human machine interface (HMI) flat screen display is often used to show process or production machine status. PCs do many jobs in homes and offices, but PLCs

FIGURE 1-2: Processor architecture for PLC.

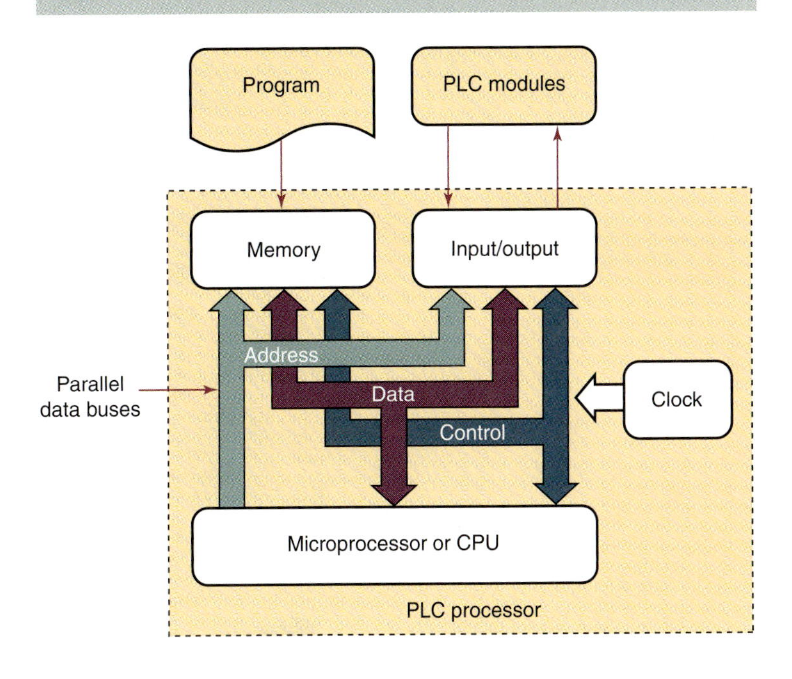

perform only one task, the control of manufacturing machines and processes.

Will PC and PLC technologies converge? According to vendors such as Allen-Bradley and Siemens the answer is someday, but the PC will never replace the PLC. PCs are performing PLC-type control in some applications using software, such as SoftLogic, that allows the PC to simulate the actions of a PLC. Technical differences notwithstanding, the PC and PLC industries are beginning to look more alike; and in manufacturing automation, both are replacing relay ladder logic.

1-3 RELAY LADDER LOGIC

Industrial automation began with relays used to control the sequence of operations in machines. These sequential control systems, called *relay ladder logic*, were the control standard for industry. The early PLCs were designed to eliminate the relay logic used for sequential control applications. To understand how PLCs accomplished this task, it is important to understand the operation of relays and relay ladder logic. Figure 1-3 illustrates five types of relays and identifies all of the parts of this electromechanical device.

FIGURE 1-3: Electromechanical relays.

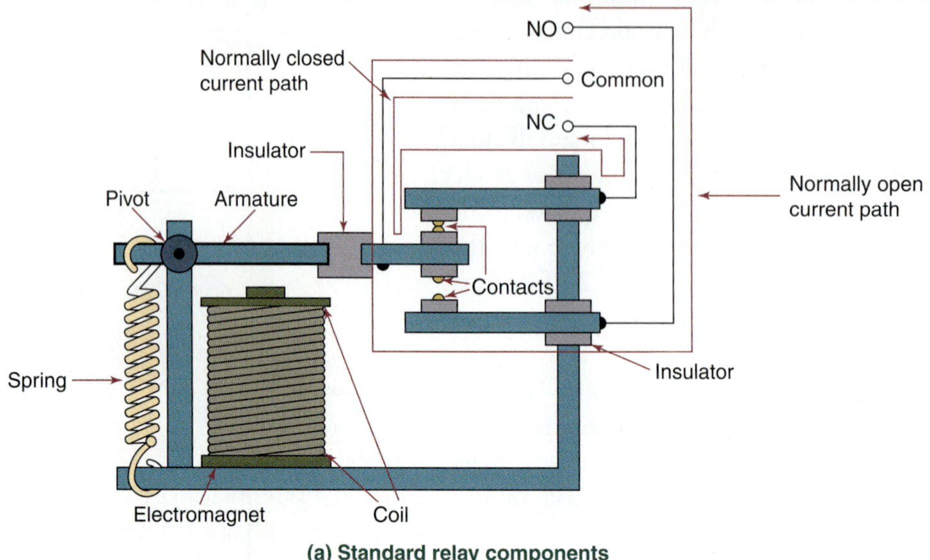

(a) Standard relay components

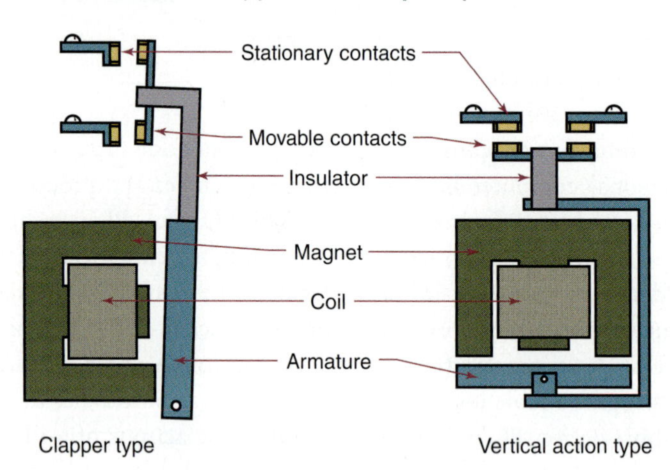

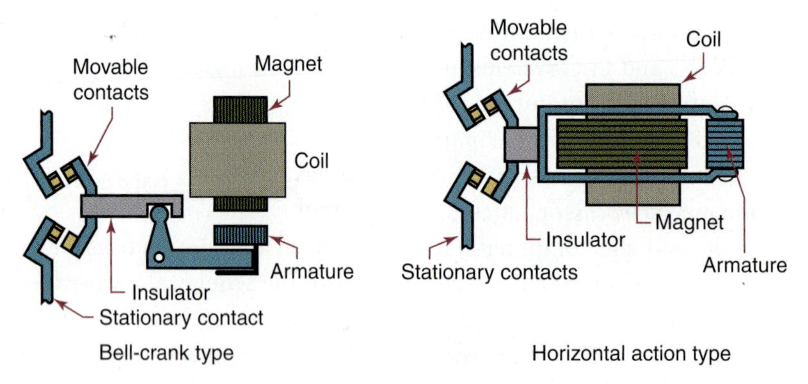

(b) Relay configurations

Source: Kraebber and Rehg, Computer Integrated Manufacturing, *Third Edition, © 2005, p. 503, reprint by permission of Prentice Hall, Upper Saddle River, NJ; and Rehg and Sartori,* Industrial Electronics, *First Edition, © 2006, p. 69, reprinted by permission of Prentice Hall, Upper Saddle River, NJ.*

FIGURE 1-4: Relay.

Courtesy of Square D/Schneider Electric.

FIGURE 1-5: Standard symbols for single pole double throw relay contacts.

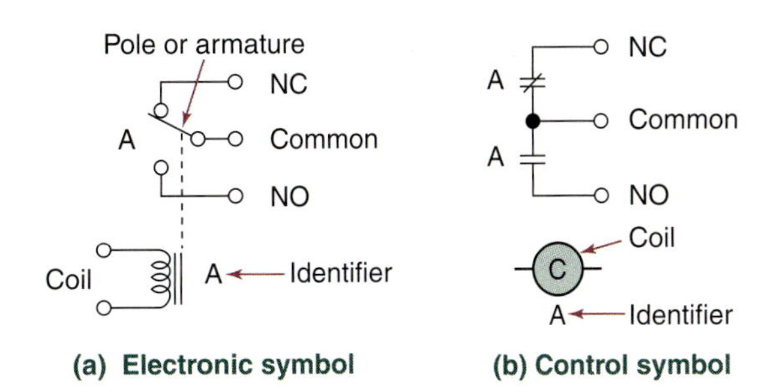

(a) Electronic symbol **(b) Control symbol**

1-3-1 Electromagnetic Relay

Joseph Henry, developer of the electromagnet in 1831, built the first relay-type device in 1836. After nearly 170 years of service, the relay illustrated in Figure 1-3(a) still has the same three components:

1. The *electromagnet* is a magnet, which is created by passing a current through wire wound around a steel core.
2. The *armature*, called a *clapper*, is a hinged metal plate that is pulled toward the coil by the electromagnet when the coil is energized and pulled away from the coil by the spring when the coil is de-energized.
3. The *contacts*, which create one electrical path through the *normally closed* (NC) contacts when the coil is not energized (armature up) and a second path through the *normally open* (NO) contacts when the coil is energized (armature down).

This *single pole double throw* (SPDT) configuration in Figure 1-3(a) has one common contact (*single pole* or *armature*) and two positions (NC and NO) called *throws*. When the coil is not energized, the spring holds the armature in the up position (pulls down on the opposite side of the pivot point). In this position a near zero resistance connection is established between the common armature contact and the NC contact. When the coil is energized, the armature pivots down so contact with the NC contact is broken. In this position, a near zero resistance connection is established between the common armature contact and the NO contact. *Insulators* are used in the armature to isolate the electrical switching contacts of the relay from the rest of the relay components. Figure 1-3(b) shows four other relays with a *single pole single throw* (SPST) relay configuration.

Figure 1-4 shows a Square D relay with contacts like the vertical action in Figure 1-3. Figure 1-5 illustrates two different schematic representations for the relays in Figures 1-3 and 1-4. The relay contact symbols used for electronic circuit schematics are often different from those used in control-type schematics.

Compare the NO and NC conduction paths in Figure 1-3(a) with the electronic and control schematics in Figure 1-5(b). The NO contact symbol has two parallel lines, indicating an open circuit, and the NC symbol has the same two parallel lines with a line across them to indicate closed contacts.

Relays are available in a variety of sizes with a number of contact configurations. Figure 1-6(a) illustrates a SPST relay with the electronic (left) and control symbol (right). Figures 1-6(b) and (c) illustrate symbols for a *two pole double throw* (2PDT) and a *three pole double throw* (3PDT) device. The relay symbols in Figure 1-6(b) identify a device with double break contacts that does not have a common pole for the NC and NO contacts. All four of the relays in Figure 1-3(b) have this *double break* contact configuration. The terms *double*

FIGURE 1-6: Relays with multiple poles and throws.

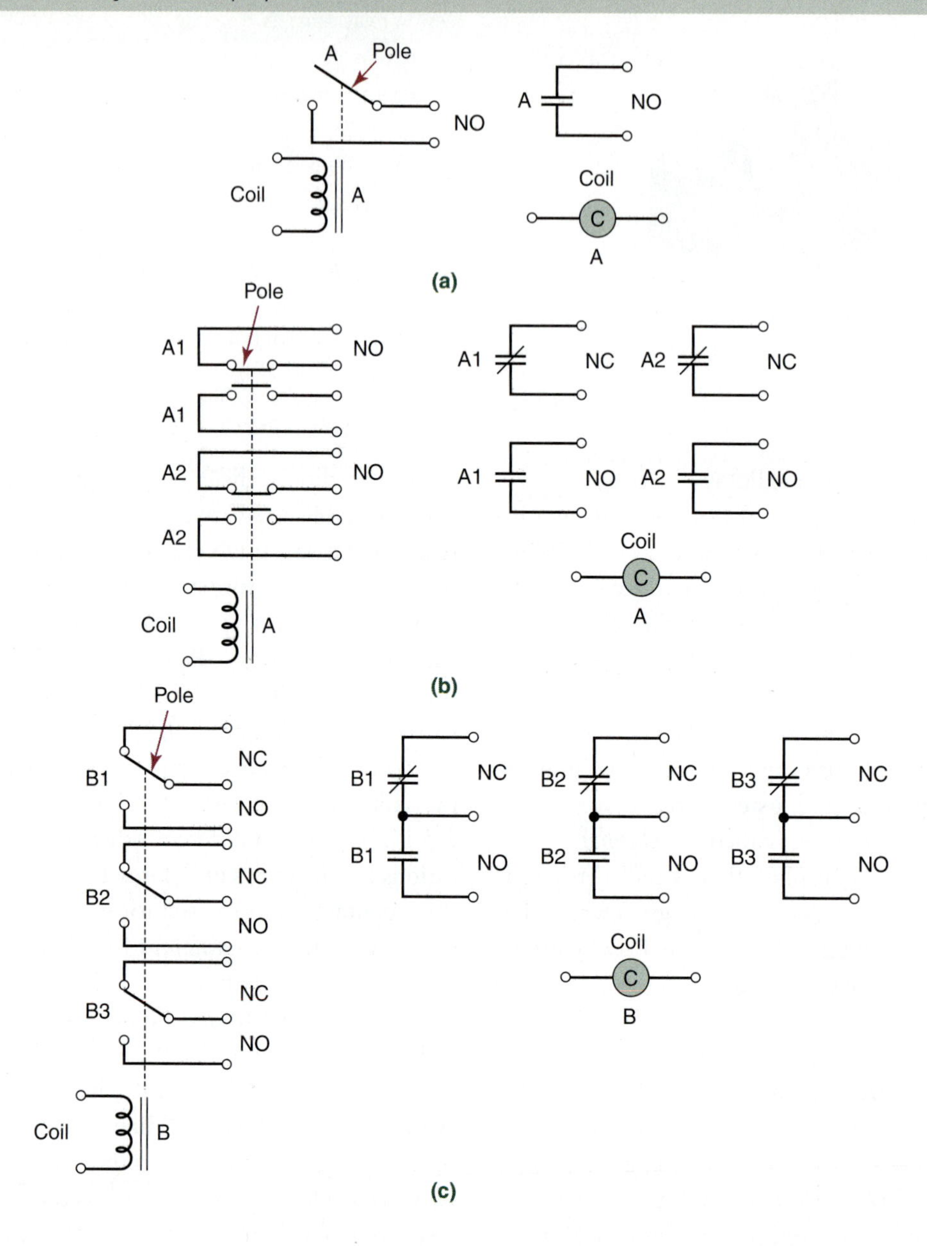

pole and *triple pole* are often used instead of two pole and three pole, respectively. The dashed lines indicate that the relay coil activates *all* sets of NO and NC contacts in unison. With the operation of the relay established, a relay control circuit example is used to introduce basic sequential logic control.

1-3-2 Relay Control Systems

To illustrate how relays are used in machine control, consider the following simple control problem. A tank, illustrated in Figure 1-7(a), is filled through an electrically operated valve and emptied by a motor-driven pump. Control of the valve and pump must satisfy the following logic:

1. The pump can operate only when the input valve to the tank is open.
2. The input valve can be opened when the pump is either operating or not operating.

The *electronic schematic* in Figure 1-7(b) illustrates a solution to the control problem. The same solution is represented in Figure 1-7(c)

FIGURE 1-7: Process tank control systems.

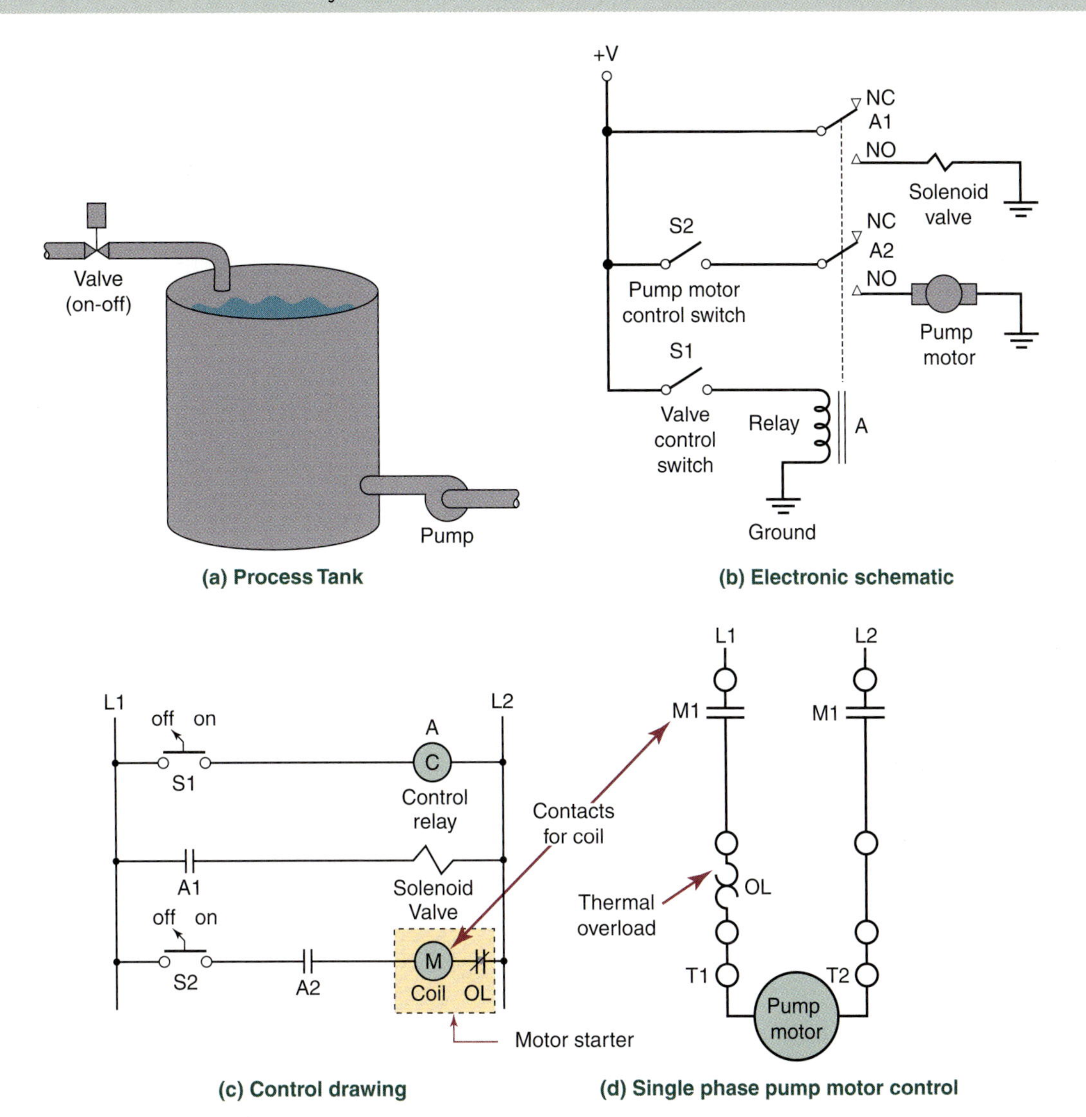

using control schematic symbols in a control drawing. The control drawing is also called a *relay ladder logic diagram* or a *two-wire diagram*. Let S1 close and verify that the outputs in both schematics satisfy the logic required for this control problem. Next, change S2 from open to close and note how the outputs react.

The circuits work as follows:

1. Switch S1 is closed manually and causes the electromagnetic relay A to be energized.
2. When the relay is energized in Figure 1-7(b), the A1 and A2 poles move from the normally closed (NC) positions to the normally open (NO) positions. In Figure 1-7(c), A1 and A2 contacts are closed and power flow results.
3. In Figure 1-7(c), the change in contact A1 energizes the input valve and allows liquid to flow into the tank.
4. The change in contact A2 causes no immediate action.
5. Switch S2 is closed manually and causes the pump to operate.

For the pump to operate when the pump switch is activated, the NO relay contact A2 has to close. If the valve switch S1 is opened manually while the pump is operating, the pump motor stops because the relay is not energized (The NO contacts no longer have power flow).

The control diagram in Figure 1-7(c) provides the same control as the circuit in Figure 1-7(b) but it uses control-type symbols for the components in a configuration called a *relay ladder logic diagram*. It is called this because it uses relays, looks like a ladder, and satisfies the logic control requirements specified for control of the output device. Standard control drawing symbols are used to represent the different input and output devices, such as mechanical switches, sensors, magnetic contactors and relays, and electrical contacts.

These diagrams have a vertical line at the left (marked L1) and right (marked L2) sides. The left vertical line, sometimes called the *left power rail*, usually represents the positive, hot, or high side of the power source; the right vertical line, called the *right power rail*, represents the power return, neutral, or ground. All the circuits containing the switches, sensors, and output actuators used to operate a machine are drawn between these two vertical lines. These types of drawings are sometimes called *two-wire* or *three-wire diagrams*; however, this is an older description and the relay ladder logic identification is a better choice.

In Figure 1-7(b) the motor is connected directly to the switch and relay contacts, but this is only done in industrial control applications when a small factional horsepower motor is used. The control schematic, Figure 1-7(c), indicates the preferred solution, which is to use a specialpurpose motor relay (identified with an M), called a *motor starter*, to switch the power for the pump motor. Starters can switch the high voltage and current associated with large motors. Figure 1-7(d) shows the starter's normally open M1 contacts in the motor power lines. There is one set of *overload*, OL, contacts to protect the motor. The thermal overloads are drawn between the contacts and the motor in Figure 1-7(d), and the starter overloads are shown in the ladder rung between the coil (M) and the L2 power rail. By convention, only one OL contact is shown in the rung, but there would be a contact for every overload relay. Later in this chapter this relay ladder is converted to a PLC solution; however, some PLC terminology and basic operation must be covered first. This preparation for PLC program development starts in the next section with the elements of a PLC system and components used.

1-4 PLC SYSTEM AND COMPONENTS

Figure 1-8 illustrates and names the many modules in the block diagram for a PLC system. These components and modules are described in Sections 1-4-1 through 1-4-6, and the PLC system diagram in Figure 1-8 is referenced often in these sections. Put a bookmark at this figure's page for ease in flipping back to the PLC block diagram.

The heart of the PLC is the PLC processor. The processor is surrounded by input modules on the left, output modules on the right, and a power supply above. Programming is performed by either a handheld programmer, by a directly connected personal computer, or through a computer with a network connection. The processor communicates with input and output devices through input and output modules. Note the variety of input and output devices that interface with the input and output modules (see Figure 1-12).

The processor frequently interfaces with a variety of local area networks (LANs), with the most commonly used networks and subnetworks listed in the upper left and lower right of the figure. In most cases, *network communications* and *sub-network scanner* modules are used to build a data exchange with external systems. Seven commonly used data interfaces (Ethernet/IP to SERCOS interface) are shown and each is described in detail in the following sections.

For the larger systems, the PLC blocks in Figure 1-8 are mounted in a *rack* as illustrated in Figure 1-9(a). The rack provides mechanical support and all the electrical interconnections plus the data interface between all of the PLC modules using the backplane bus structure. In smaller PLC systems, the component modules are integrated into a single unit like in the Allen-Bradley Pico model. The rack illustrated in Figure 1-9(b) is for the Allen-Bradley SLC 500 series PLCs. The PLC in Figure 1-9(a) is a ControlLogix system. The pictured sensors or actuators, called *field devices*, are covered in Chapter 2.

FIGURE 1-8: PLC system block diagram.

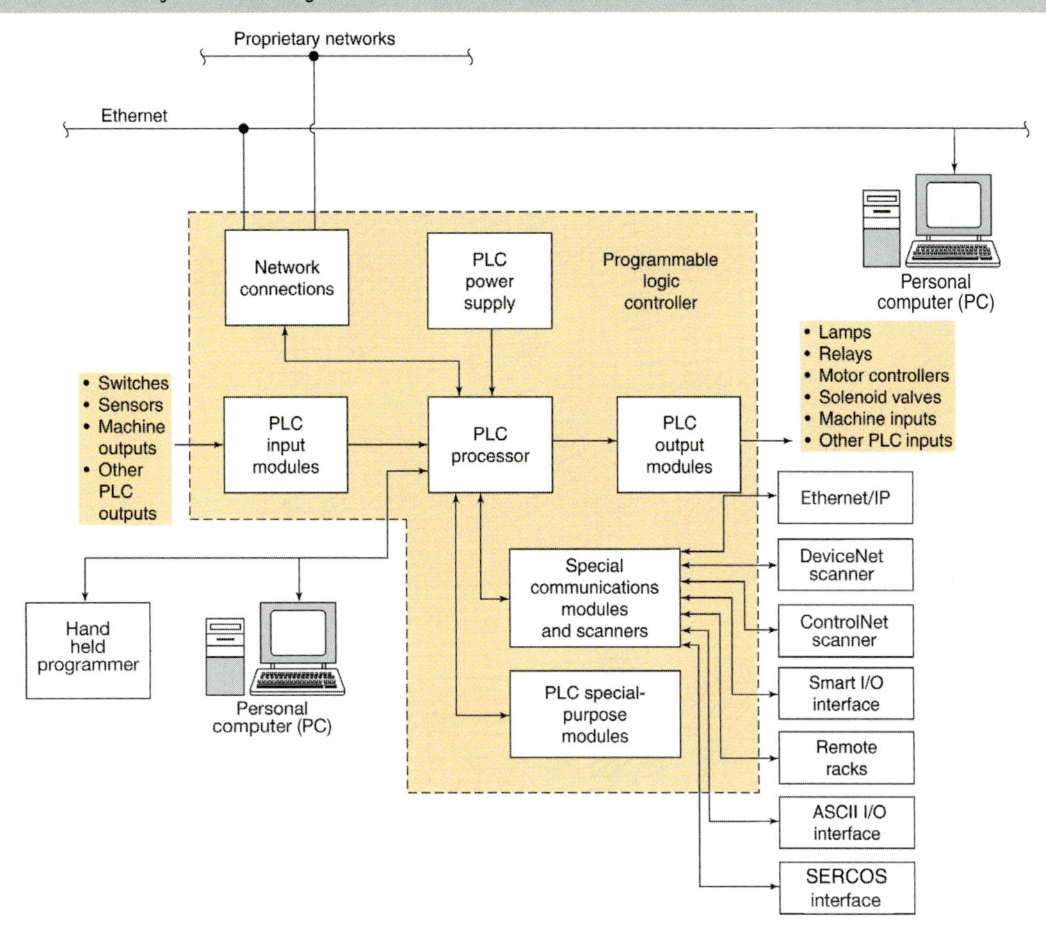

1-4-1 Backplane

The power and data interface for the modules is provided by the *backplane* in the rack in Figure 1-9(b). The backplane has copper conductors, called *lands*, that deliver power to the modules and also provide a data bus to exchange data between the modules and the processor. Modules slide into the rack and engage connectors on the backplane to access the backplane's power and data buses. The number of slots in the rack is determined by the number and type of modules required for the control application. The SLC 500 and the ControlLogix series PLC racks are available in 4, 7, 10, and 13 slot models and the ControlLogix series also has a 17 slot rack available.

1-4-2 Processor and Power Supply

The center box in Figure 1-8, called the *PLC processor*, is the central processing unit (CPU) that handles all logical operations and performs all the mathematical computations. The processor occupies the fourth slot position in the ControlLogix rack illustration in Figure 1-9(a) and is in slot zero for the SLC 500 system in Figure 1-9(b). In older systems the processor must be in slot zero, but in the new PLC models, one or more processors can be in any slot in the rack. In addition, multiple processors can be used in a single rack to enhance performance in models like the Logix family of PLCs. A picture of an Allen-Bradley ControlLogix processor is shown in Figure 1-10.

Figure 1-9(a and b) illustrates a *PLC power supply* in the left-most box in the rack. The power supply provides power to the processor and to the modules plugged into the rack.

1-4-3 Programming Device

The programming devices connected to the processor in Figure 1-8 are used to enter and download programs or to edit existing programs in the PLC. PCs and a handheld programmer are

FIGURE 1-9: PLC racks and view of the backplane.

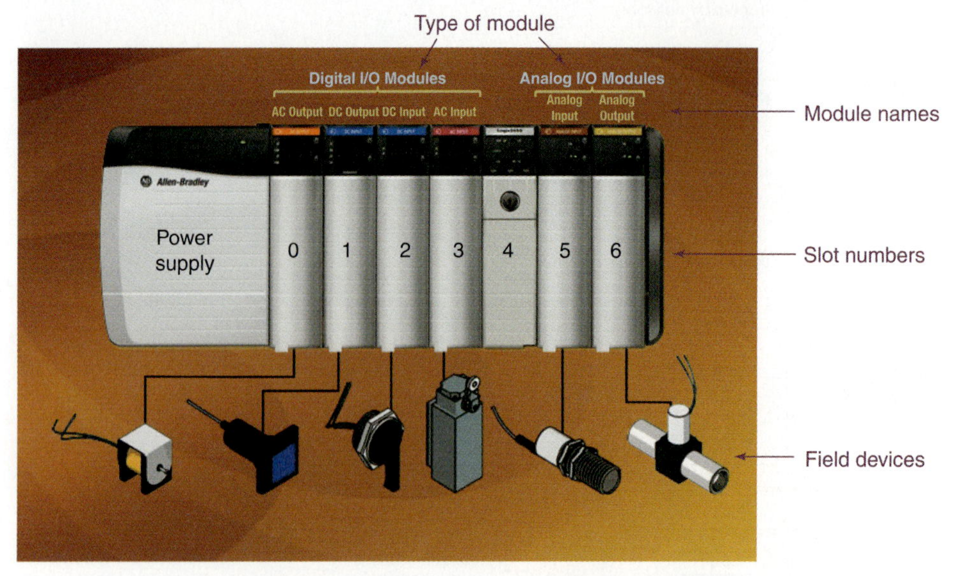

(a) Front panel and rack with seven modules for ControlLogix PLC

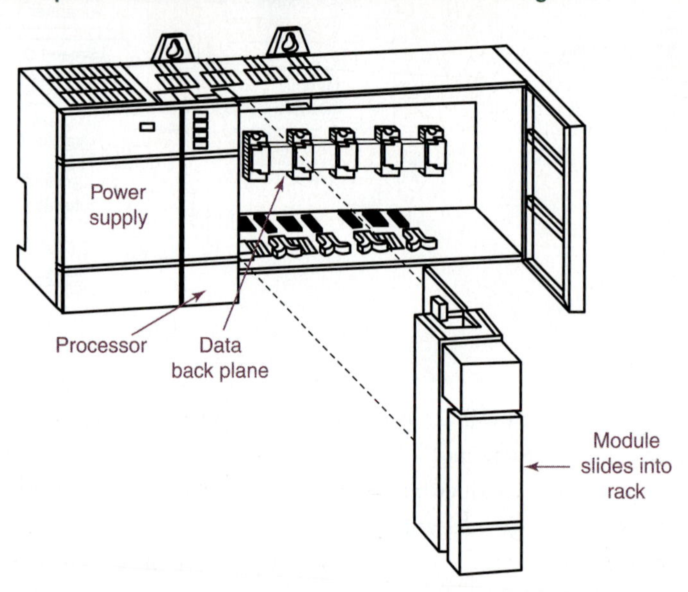

(b) Module and rack interface for SLC 500 PLC

Courtesy of Rockwell Automation, Inc.; and Rehg and Sartori, Industrial Electronics, *First Edition, © 2006, p. 568, reprint by permission of Prentice Hall, Upper Saddle River, NJ.*

shown in the block diagram. Figure 1-11 shows a handheld programmer for the SLC 500 system.

1-4-4 Input and Output Interface

The input and output (I/O) interface used in PLCs can take two forms: *fixed* or *modular*. The fixed type is associated with the small or micro PLC systems where all of the features are integrated into a single unit. The number of I/O ports is fixed within each model and cannot be changed. The modular types, like Figure 1-9, use a rack to hold the I/O modules so the number and type of I/O modules can be varied.

The input interface provides the link between the PLC processor and the external devices that measure the conditions in the production area. The input devices used most often include

FIGURE 1-10: ControlLogix processor.

switches, sensors, machine outputs, or other PLC outputs. These input devices are often called *field devices*, indicating that they are not a part of the PLC hardware. The PLC input modules are both a *physical interface* for the connection of wires and an *electrical/data interface* to determine the on/off state or voltage level from the attached field device. In addition, they act as *signal conditioners* changing the many different types of input voltages to the 0 to 5 volt DC voltage levels used in the PLC processor.

The fixed and modular output modules provide the interface between the PLC processor and the external devices or actuators, such as lamps, relays, motor and heater contactors, solenoid valves, and machine inputs. The term *field device* is also used to address the wide range of output devices attached to a PLC system. The output module is both a location for termination of wiring and a signal conditioner to provide the proper voltage and output drive power required by the output field devices.

As a *signal conditioner* for the PLC, the many different input and output modules match the large variety of field devices to the processor's input and output. Figure 1-12 lists some of Allen-Bradley's input and output modules

FIGURE 1-11: Handheld programmer for SLC 500.

(a) Handheld programmer

(b) Programmer keyboard

available—note the range of I/O points (points is another term for input terminals) and the supported input and output signal types.

Selecting input and output modules requires a through analysis of the input and output field devices. The voltage and current characteristics of the modules must be matched to the specifications of the field device. Voltage type (AC or DC) and levels must be considered. Also, system surge and continuous current levels at turn on and minimum current levels for turn off must be matched to a module with equal capability. There are numerous input and output modules because of the variety of input and output field devices and the voltage and current requirements present.

DC modules can be either *current sourcing* (current flows out from the module when active) or *current sinking* (current flows into the module when active). This important concept of sourcing and sinking current for modules is covered in this chapter for the PLC modules and in Chapter 2 for the field devices. Figure 1-9(a) shows several I/O modules in the ControlLogix's rack.

Interface modules are available for PLCs with a combination of inputs and outputs on the same module. Figure 1-12 illustrates some of the combination modules for the SLC 500 system.

Input current sinking and sourcing circuits. DC input modules are current sourcing, current sinking, or not dependent on the current orientation. The current flow for input sinking and sourcing modules is illustrated and described in

FIGURE 1-12: Allen-Bradley SLC 500 discrete I/O modules.

	ID Code	Voltage Category	Category Number	Input/Output	I/O Points	Module Description
AC Modules	100	100/120V AC	1746-IA4	Input	4	120V AC Input
	300	100/120V AV	1746-IA8	Input	8	120V AC Input
	500	100/120V AC	1746-IA16	Input	16	120V AC Input
	101	200/240V AC	1746-IM4	Input	4	240V AC Input
	301	200/240V AC	1746-IM8	Input	8	240V AC Input
	501	200/240V AC	1746-IM16	Input	16	240V AC Input
	2703	100/120V AC	1746-OA8	Output	8	120/240V AC Output
	2903	100/120V AC	1746-OA16	Output	16	120/240V AC Output
	2803	120/240V AC	1746-OAP12(1)	Output	12	**High Current** 120/240V AC Output
DC Modules	306	24V DC	1746-IB8	Input	8	Current **Sinking** DC Input
	506	24V DC	1746-IB16	Input	16	Current Sinking DC Input
	706	24V DC	1746-IB32(1)	Input	32	Current Sinking DC Input
	519	24V DC	1746-ITB16	Input	16	Fast Response DC Sinking Input
	509	48V DC	1746-IC16	Input	16	Current Sinking DC Input
	507	125V DC	1746-IH16	Input	16	Current Sinking DC Input
	320	24V DC	1746-IV8	Input	8	Current Sinking DC Input
	520	24V DC	1746-IV16	Input	16	Current Sinking DC Input
	720	24V DC	1746-IV32(1)	Input	32	Current Sinking DC Input
	518	24V DC	1746-ITV16	Input	16	Fast Response DC Sourcing Input
	515	5V DC/TTL	1746-IG16(2)	Input	16	Current **Sourcing** TTL Input
	2619	24V DC	1746-OB6EI	Output	6	Isolated Sourcing DC Output
	2713	24V DC	1746-OB8	Output	8	Current Sourcing DC Output
	2913	24V DC	1746-OB16	Output	16	Current Sourcing DC Output
	2920	24V DC	1746-OB16E(1)(3)	Output	16	Current Sourcing DC Output
	3113	24V DC	1746-OB32(1)	Output	32	Current Sourcing DC Output
	3120	24V DC	1746-OB32E(1)	Output	32	Current **Sourcing** DC Output
	2721	24V DC	1746-OBP8(3)	Output	8	High Current Sinking DC Output
	2921	24V DC	1746-OBP16(1)	Output	16	High Current Sinking DC Output
	2714	24V DC	1746-OV8	Output	8	Current Sinking DC Output
	2914	24V DC	1746-OV16	Output	16	Current **Sinking** DC Output
	3114	24V DC	1746-OV32(1)	Output	32	Current Sinking DC Output
	2922	24V DC	1746-OVP16(1)	Output	16	High Current Sinking DC Output
	2915	5V DC/TTL	1746-OG16(2)	Output	16	Current Sinking TTL Output
AC/DC Modules	510	24V AC/DC	1746-IN16	Input	16	24V AC/DC Input
	2500	AC/DC Relay	1746-OW4(1)	Output	4	Relay (Hard Contact) Output
	2700	AC/DC Relay	1746-OW8(1)	Output	8	Relay (Hard Contact) Output
	2900	AC/DC Relay	1746-OW16(1)	Output	16	Relay (Hard Contact) **Output**
	2701	AC/DC Relay	1746-OX8(1)	Output	8	Isolated Relay Output
	800	In-120V AC, Out-Relay	1746-IO4(1)	Input/Output	2 In, 2 Out	Combination Input/Output
	1100	In-120V AC, Out-Relay	1746-IO8(1)	Input/Output	4 In, 4 Out	Combination Input/Output
	1500	In-120V AC, Out-Relay	1746-IO12(1)	Input/Output	6 In, 6 Out	Combination Input/Output
	1512	In-24V DC, Out-Relay	1746-IO12DC(3)	Input/Output	6 In, 6 Out	Combination Input/Output

(1) Certifed for Class 1, Division 2 hazardous location by CSA only.
(2) Not CE marked.
(3) These modules carry the C-UL mark and are certified by UL per CSA only.

EXAMPLE 1-1

An application has the following input field devices that must be interfaced to the SLC 500 PLC. For each combination of field devices select the appropriate combination of input modules from Figure 1-12 to interface them with a PLC.

a. Seventeen 120 VAC inputs, five 5 VDC inputs are sinking, which require a sourcing module, and three 24 VDC inputs are sinking, which require a sourcing module.

SOLUTION

Use one 1746-IA16 AC module and one 1746-IA4 AC module with 16 and 4 AC inputs, respectively, to handle the seventeen inputs. Use one 1746-IG16 5V TTL sourcing module with 16 I/O points for the five DC inputs, and one 1746-IB8 sinking module with 8 I/O points for the three 24 VDC inputs.

EXAMPLE 1-2

For each combination of the following field devices, select the appropriate combination of output modules from Figure 1-12 to interface them with a PLC.

a. Ten 90 VAC output field devices that require isolation and twenty-one 12 to 20 VDC current sinking output field devices.
b. Twenty-eight pneumatic valves that require 28 VDC, ten 24 VDC control signals for a CNC machine with sourcing inputs, and three 120 VAC motors with continuous currents less than 1 amp and starting surge currents less than 10 amps.

SOLUTION

a. Use one 1746-OW16 (16 outputs) relay output module to isolate loads and one 1746-OB32 (32 outputs) 24 VDC sourcing module for the 12 to 20 VDC signals. Whenever inductive loads, like the solenoid valves, are switched in control applications, surge suppressive circuits should be used to reduce the voltage spikes produced.
b. Use two 1746-OW16 (16 relay type outputs) modules for 28 VDC loads. Use one 1746-OV16 (16 output) 24 VDC sinking module for the 24 VDC CNC signals. Finally, use a 1746-OAP12 (12 outputs) 120 VAC high current output module for the three motors.

Figures 1-13(a) and (b). Read the description in these figures before continuing. Note the direction of conventional current flow for each type in the drawings.

The current flows *into* the input terminals of sinking DC input modules [Figure 1-13(a)] and flows *out of* the terminals of sourcing DC input modules [Figure 1-13(b)]. When you wire the interface, you must be sure that the power supply polarity orientation supports the current direction required by the module and field device. Verify in the figure that the current flow is out of the positive battery terminal (conventional current flow) and into the sinking module. The current is out of the sourcing module. Note in the figures that a sinking DC input module requires that the field device has a sourcing type of output. If the input module is sourcing, then the field device must be the opposite.

The interface in Figure 1-13(c) illustrates that AC input modules can interface to an AC field device or to either a sinking or sourcing DC field device. Descriptions of the operation are provided in the figure.

Output sinking and sourcing circuits. DC output modules are current sourcing, current sinking, or not dependent on the current orientation. The current flow for DC output sinking and sourcing modules is illustrated and described in Figures 1-14(a) and (b).

The current flows *into* the output terminals of sinking DC output modules [Figure 1-14(a)] and flows *out of* the terminals of sourcing DC modules [Figure 1-14(b)]. You must verify that the power supply polarity orientation supports the current direction required by the module and field device. Note in the figures that a sinking

FIGURE 1-13: Input module interfaces.

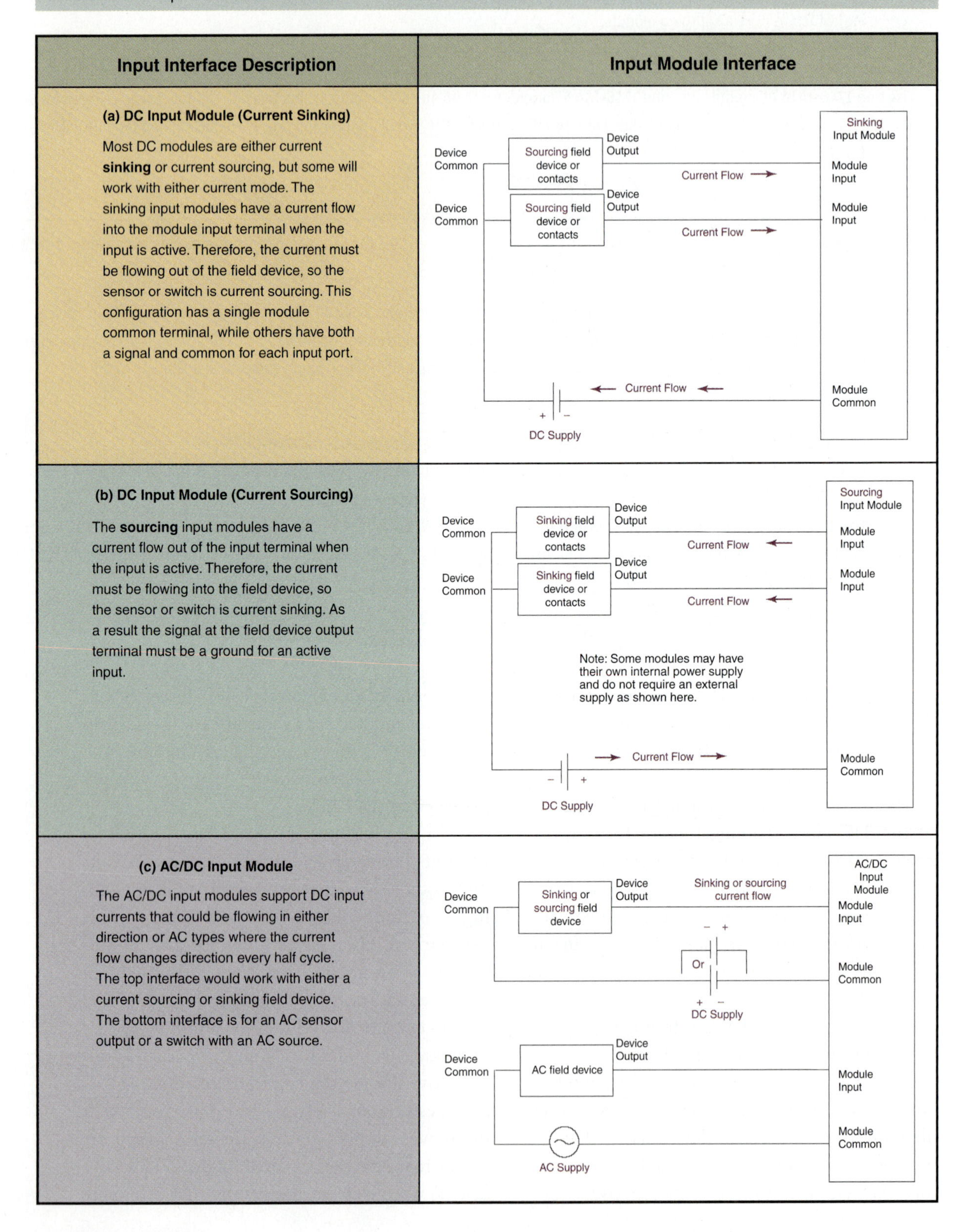

FIGURE 1-14: Output module interfaces.

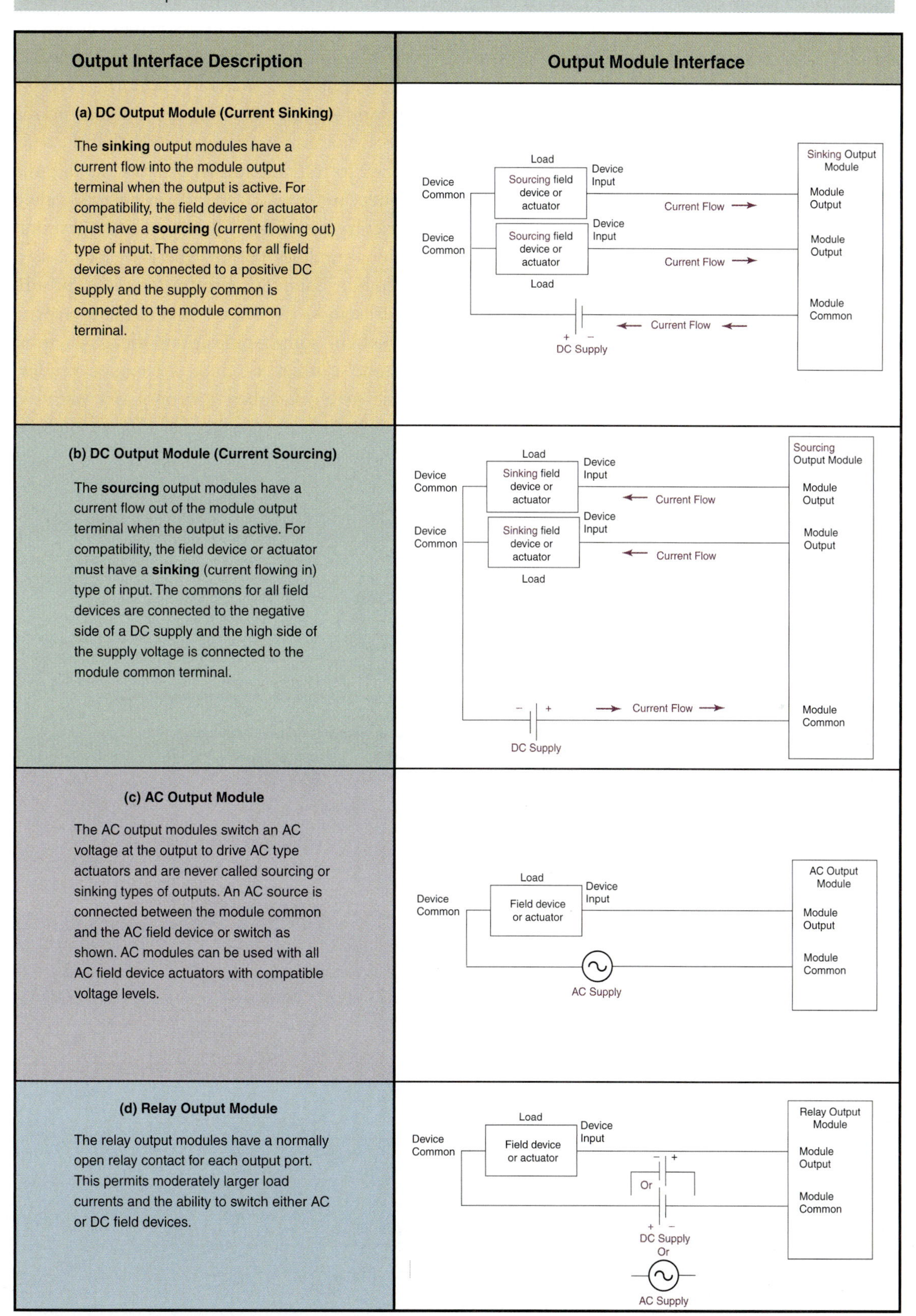

Output Interface Description	Output Module Interface
(a) DC Output Module (Current Sinking) The **sinking** output modules have a current flow into the module output terminal when the output is active. For compatibility, the field device or actuator must have a **sourcing** (current flowing out) type of input. The commons for all field devices are connected to a positive DC supply and the supply common is connected to the module common terminal.	
(b) DC Output Module (Current Sourcing) The **sourcing** output modules have a current flow out of the module output terminal when the output is active. For compatibility, the field device or actuator must have a **sinking** (current flowing in) type of input. The commons for all field devices are connected to the negative side of a DC supply and the high side of the supply voltage is connected to the module common terminal.	
(c) AC Output Module The AC output modules switch an AC voltage at the output to drive AC type actuators and are never called sourcing or sinking types of outputs. An AC source is connected between the module common and the AC field device or switch as shown. AC modules can be used with all AC field device actuators with compatible voltage levels.	
(d) Relay Output Module The relay output modules have a normally open relay contact for each output port. This permits moderately larger load currents and the ability to switch either AC or DC field devices.	

output module requires that the field device is a sourcing type. If the output module is sourcing, then the field device must be the opposite.

The interface in Figures 1-14(c) and (d) illustrates that AC output modules can interface to an AC field device, and that a relay output device can interface to either type of output field device, sinking or sourcing. Descriptions of the operation are provided in the figure.

1-4-5 Special Communications Modules and Network Connections

Communications modules and networks are introduced here and are covered in greater detail in Chapter 17. While Figure 1-15 includes many different networks, PLC systems rarely have all network options present. The figure is used to show you what is possible but not to imply that it is a typical implementation.

The special communication modules listed in Figure 1-8 provide a link for the PLC processor to other computer-controlled machines and devices that must share data and control requirements with the PLC. Seven examples are listed: *DeviceNet*, *ControlNet*, *Ethernet/IP*, *SERCOS interface*, *Smart I/O interface*, *Remote racks*, and *ASCII I/O interface*. In addition, the network connections box indicates that the PLC is linked to information level networks like the Ethernet and vendor-specific networks called proprietary networks. The use of these communications modules permits the PLC to act as a data hub or data concentrator for these six subnetworks and a gateway or link to the enterprise Ethernet. In some applications the PLCs from the same vendor are linked using a vendor-specific network called a *proprietary network* (Figure 1-8). The Allen-Bradley *data highway* is

FIGURE 1-15: Communication network options for PLC control systems.

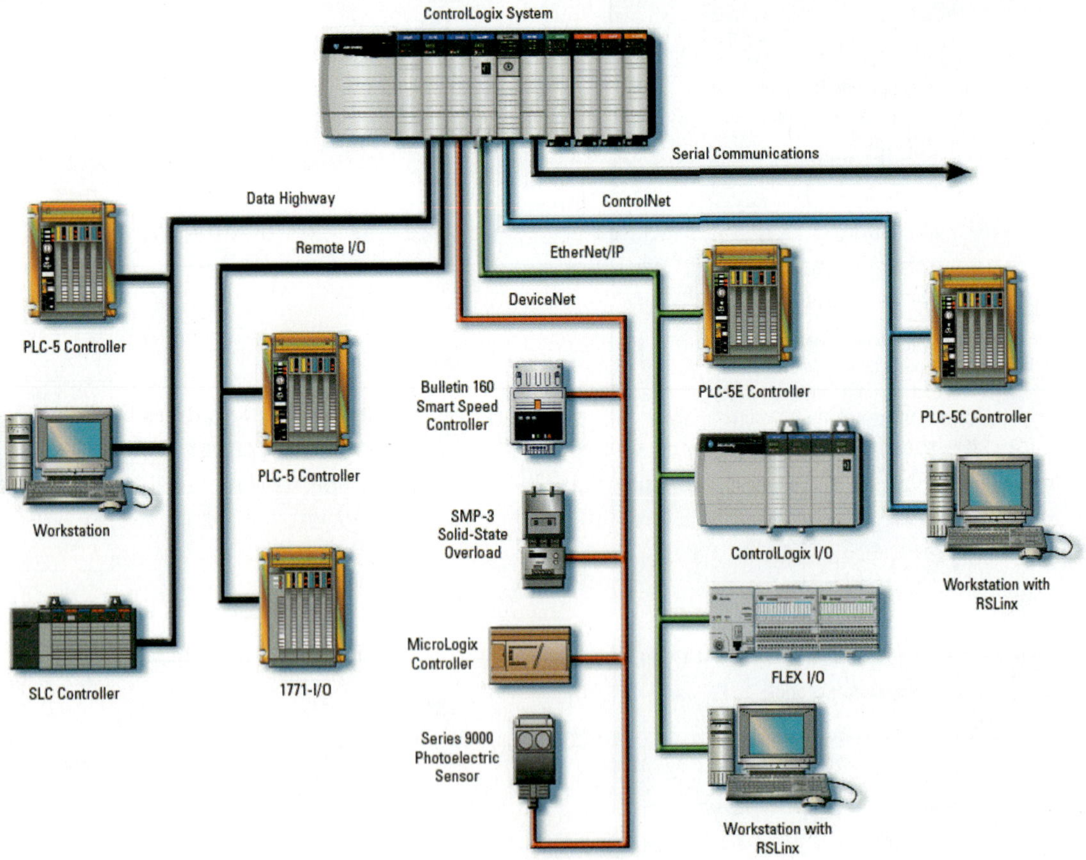

Courtesy of Rockwell Automation, Inc.

an example of this type of vendor-specific network. A brief overview of these network technologies is provided here so you understand the control and network capability of current PLC systems.

- **DeviceNet:** DeviceNet, Figure 1-15, is a low-cost communications network that connects *smart* or intelligent input and output field devices, such as sensors and actuators, to the PLC. A *smart* field device is one that has an embedded microprocessor so that it can communicate over a network, thus eliminating costly hardwiring. Find the DeviceNet in Figure 1-15 and review the range of device types connected to the network.
- **ControlNet:** ControlNet is another open network standard that is one level above DeviceNet in the control hierarchy. While the primary function of DeviceNet is the networking of input and output devices, ControlNet uses the *producer/consumer* network model to efficiently exchange time-critical application information in control systems. Find the ControlNet in Figure 1-15 and look at the type of systems attached.
- **Ethernet/IP:** Ethernet/IP (IP stands for industrial protocol) is an open industrial networking standard that takes advantage of commercial off-the-shelf Ethernet communication devices and physical media. Find the Ethernet/IP in Figure 1-15 and look at the type of systems attached.

Study the three levels of network control illustrated in Figure 1-15. One can see that plantfloor architecture has flattened into three layers: Ethernet/IP, ControlNet, and DeviceNet. The highest level, Ethernet/IP, provides the information layer for plant-wide data collection and program maintenance. The middle level, ControlNet, supports the automation and control layer with input and output control where data cannot be lost and with message passing between systems. The lowest level, DeviceNet, supports a device layer for cost-effective integration of individual input and output field devices. Some overlap in the network layers exists since sensors and actuators are interfaced to each depending upon the application.

PLCs located in this network architecture often use *smart I/O interfaces*, *remote racks*, and *serial communications* to further distribute data communications to remote control locations. A description of the last four special communication modules follows.

- **SERCOS Interfaces:** This special communications module is a Serial Realtime Communications System, or SERCOS for short. SERCOS is a digital motion control network that interfaces the motion control module in the PLC with the servo motor drive through a fiber optic cable.
- **Smart I/O Interfaces:** PLC vendors have proprietary network protocols to allow all their devices to communicate. One common application for a proprietary network is the use of *smart I/O devices*. The Allen-Bradley Data Highway in Figure 1-15 is an example of a proprietary network used to place Allen-Bradley devices on a network. In the figure it is used to connect two PLC models and a programming station. Find the Data Highway in Figure 1-15 and compare the type of systems attached to it and to the other networks.
- **Remote Racks:** In an effort to distribute the control capability across a large automation system, the PLC vendors provide *remote rack* capability. The rack, similar to the rack illustrated in Figure 1-9, uses the standard I/0 modules for control of machines and processes; however, the processor module is replaced with a remote rack communications module. The processor in the main PLC rack sends control instructions over the single network cable to the communication module in the remote rack and then to the I/O modules included in the remote rack.
- **ASCII I/O Interface:** The last special communications module is the ASCII I/O interface. This serial communication resource is either built into the processor module or comes as a separate module. In both cases, the ASCII interface permits serial data

communication using several standard interfaces, such as RS-232 and RS-422.

1-4-6 PLC Special-Purpose Modules

The final element in the PLC system block diagram, Figure 1-8, is labeled *PLC special-purpose modules*. This term represents a broad collection of modules developed by PLC vendors for PLC control of a variety of automation devices. Examples of Allen Bradley modules include analog input and output, temperature measurement and control (thermocouple and resistance temperature device), multiple PID loop control, servo motor control, stepper motor control, high-speed counter, and hydraulic ram control.

Systems from most PLC vendors conform to the block diagram in Figure 1-8 and have similar operational modes. With the introduction of the basic PLC system completed, a detailed study of the three types of PLC systems is presented in the next section.

1-5 PLC TYPES

PLCs are grouped into three operational classifications, namely *rack* or *address-based systems*, *tag-based systems*, and *soft PLCs* or *PC-based control*. The first type, rack or address-based control, was implemented in the initial PLC systems and is still the type used most frequently. The Allen-Bradley PLC-5 and SLC 500 systems use versions of this system.

1-5-1 Rack/Slot Address Based

The PLC system illustrated in Figure 1-9(b) is a rack/slot address-based system because the slot location of the input and output (I/O) modules in the rack establishes the PLC address for the input or output signal attached to the module. The modules placed in the rack to the right of the processor, Figure 1-9(b), are most often some type of input or output module. The type of field device that is connected to the PLC and the type of signal, such as AC, DC, discrete, analog, voltage, or current, dictate the type of I/O card required. While numerous types of I/O modules are used (see Figure 1-12 for some from Allen-Bradley), each has two functions:

1. They provide the interface terminals to which field device wires are attached.
2. They provide an electronic signal conditioning circuit that interfaces the type of signal present with the PLC.

The signal level presented at each input is represented inside the PLC by a variable, which is the letter I followed by an address reference number. While each PLC vendor of rack/slot address-based systems has a different addressing scheme, the address is determined by the type of module present (input or output), the rack/slot number occupied by the module, and the terminal number used for the connection. For example, the following syntax is used to address discrete inputs on an Allen-Bradley SLC 500 system.

I : (rack slot number) / (terminal number)

The slot number in the PLC rack in which the input module is placed and the terminal number on the input module to which the field device wiring is attached determines the input address for the SLC 500 system. The letter I indicates that it is an input; the colon (:) is a delimiter separating the module type letter (I or O) and the module slot number. The back slash (/) is a delimiter between the module slot number and terminal number. Examples 1-3 and 1-4 illustrate this addressing concept.

1-5-2 Tag Based PLCs

Tag-based systems are used in all of the new models of PLCs, such as the Allen-Bradley's ControlLogix family and PLCs from Telemecanique and Siemens. The tag name used in these systems is the same as a variable declared in high-level programming languages such as BASIC and C. In this type of addressing system, field device inputs and outputs are assigned *variable names* at the time the control system design is performed. Allen-Bradley uses the term *tag* instead of the term *variable*. Later the variable name or tag is assigned to an I/O module and a specific terminal number. The tag is the only reference used when the program is developed with

EXAMPLE 1-3

An Allen-Bradley SLC 500 system uses a rack like the one illustrated in Figure 1-9(b). Determine the address for a discrete input signal attached to terminal 5 of the DC input module in slot 2.

SOLUTION

The address for the discrete DC input would be

I:2/5

The letter I indicates an input, the 2 indicates that the DC input module is in slot 2, and 5 indicates that the discrete field device input signal wire is connected to terminal 5.

The addressing for discrete output modules would be similar to the input module except that the letter I would be replaced by the letter O. The following example illustrates the rack/slot address scheme for outputs.

EXAMPLE 1-4

An Allen-Bradley SLC system uses a rack like the one illustrated in Figure 1-9(b). Determine the address for a discrete output field device attached to terminal 12 of the AC output module in slot 5.

SOLUTION

The address for the discrete AC output would be

O:5/12

The letter O indicates that it is an output, the 5 indicates that the AC output module is in slot 5, and the 12 indicates that the output field device wire is connected to terminal 12.

the PLC's ladder logic program. When the tags or variables are defined, the type of data to be represented by the tag is declared.

1-5-3 Soft PLCs or PC-Based Control

A PC-based control system, called Soft PLC by Allen Bradley, is an emulation of a PLC using software on a PC. This implementation uses an industrial PC, an input/output card in the PC for an interface to the field devices, and application software that makes the PC operate like a PLC.

A second implementation uses a standard PLC with an industrial PC module placed in one of the PLC rack slots. This version puts the PC on the PLC backplane and gives the soft PLC application running in the PC access to all of the I/O modules in the PLC rack. Soft PLC solutions use one of these two implementations.

When is a Soft PLC the optimum solution? In general, on and off control of machine outputs with few numerical calculation requirements would be an application for a rack-type PLC solution. However, if large data storage and extensive mathematical manipulation is required along with sequential or process control, then a Soft PLC implementation is the better choice. If the power of the PC can be utilized, such as in data storage and numerical processing or in displaying graphics of the process, then the higher cost associated with a Soft PLC solution is justified.

1-6 PLC LADDER LOGIC PROGRAMMING

Ladder logic is a PLC graphical programming technique that was introduced with the first PLCs more than 35 years ago. The relay logic solution to the pump control problem in Figures 1-16(a) and (b) is similar to the PLC ladder logic program for the same problem. However, the subtle differences between the two systems often create confusion for the new user of PLCs. After developing the PLC solution those differences are explored.

FIGURE 1-16: Process tank control systems.

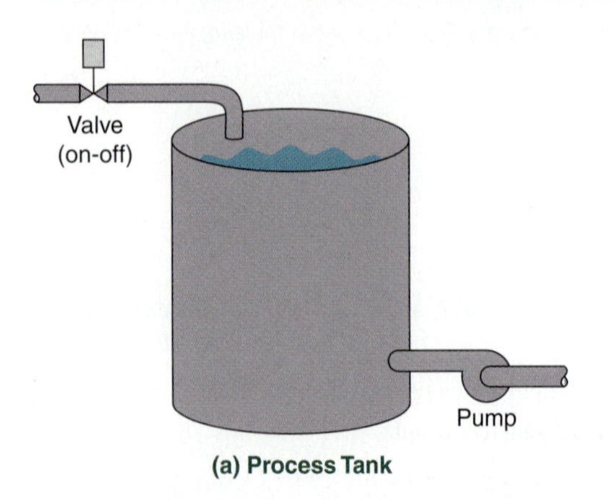

(a) Process Tank

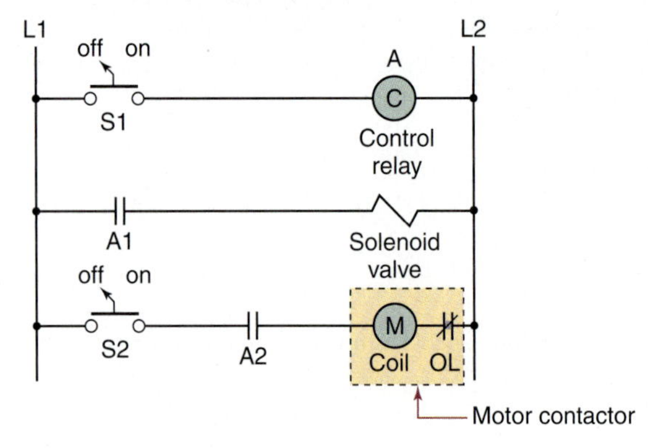

(b) Control drawing

1-6-1 PLC Solution

When the ladder logic solution for the tank control in Figure 1-16(a) is replaced by a PLC program, the field devices remain but the mechanical relay is eliminated. A *field device* is a general term for all input devices and all output actuators. Output actuators in PLC systems, Figure 1-17(a), are mechanical, pneumatic, hydraulic, or electrical devices that produce a mechanical motion in response to an electrical input signal. Note that the input switches (illustrated using the standard symbol for a selector switch) are connected to the PLC input module, and the actuators are wired to the output module. The terminations at the input and output modules are identified by terminal numbers. For example, the switches are connected to terminals 1 and 2 on the input module and the valve and pump are connected to similar terminal numbers on the output module. Any available terminals on the modules, compatable with input and output signals could be used. The illustration indicates that the PLC processor and program are located between the input and output modules.

The PLC ladder logic program that would provide the same logical control as the circuit in Figure 1-16(b) is illustrated in Figure 1-17(b). Compare Figures 1-16(b) and 1-17(b) to identify similarities and differences. They include:

- The control circuit in Figure 1-16(b) exists in the form of physical components and wire, but the ladder logic program in Figure 1-17(b) exists only as a set of instructions and logical statements in the PLC memory.
- The mechanical relay in the relay logic has been replaced by a software or virtual relay (CR1) in the PLC ladder logic. The software or virtual relay exists only in the PLC memory.
- The PLC rung output [-( )-] is called a coil in some literature but should not be used when referring to PLC ladder outputs. A better term is *discrete output instruction*, which is a bit that exists only in memory.
- Each rung of the PLC ladder logic represents a logical statement executed in software with inputs on the left and outputs on the right. If the inputs are true, then the output is true or active. For example, if input instruction I1 is true (continuity is present), then output CR1 will be active. Keep in mind that this is virtual power flow, no actual current flow is present. If CR1 is active then the input instruction with address CR1 in rung 01 is true and output O1 will be active. Input instruction I1 will be true if switch S1 is closed and a voltage is present at terminal 1 of the input module. If output O1 is active, then terminal 1 of the output module has a ground present to turn on the solenoid valve.
- The number of virtual relays, output instructions, and referenced input instructions in the PLC ladder logic is only limited by the size of the PLC memory, while the number of contacts for the mechanical relay is limited to the number of poles present on the relay selected.

- The input and output instructions in the PLC ladder logic do not represent the switches and actuators directly. The PLC input instructions are logical symbols associated with the input signals (voltages) at the input module terminals. The output symbol is associated with the signal (voltage) that will be presented to the actuator connected to the output module.
- The input and output devices have separate power sources that are isolated from the power for the PLC processor.

The solution in Figure 1-17 is used to emphasize that a PLC solution has field devices interfaced to PLC input and output modules and the PLC program logically connects the input devices to the output actuators through the ladder logic program. In the future, only the field device wiring and the ladder logic program are presented for a solution. The PLC ladder logic inputs are called *instructions* and not *contacts* because the input instructions only represent the value of a binary bit in the PLC memory. In summary, all the symbols in the relay logic diagram represent actual components and contacts present in the control system, but the input (-| |-) and output (- ( ) -) instruction symbols in the PLC ladder logic represent only data values stored in the PLC memory.

FIGURE 1-17: PLC solution to the pump control problem in Figure 1-16.

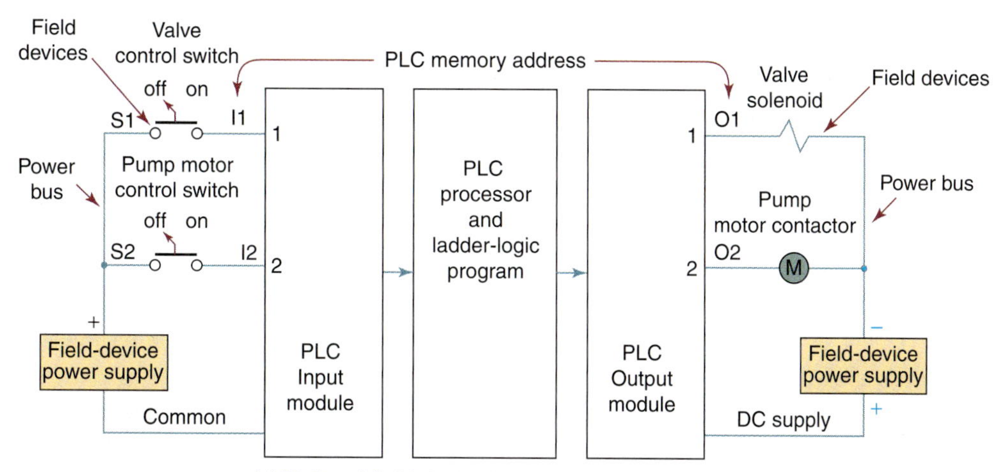

(a) PLC and field device interface for one pump

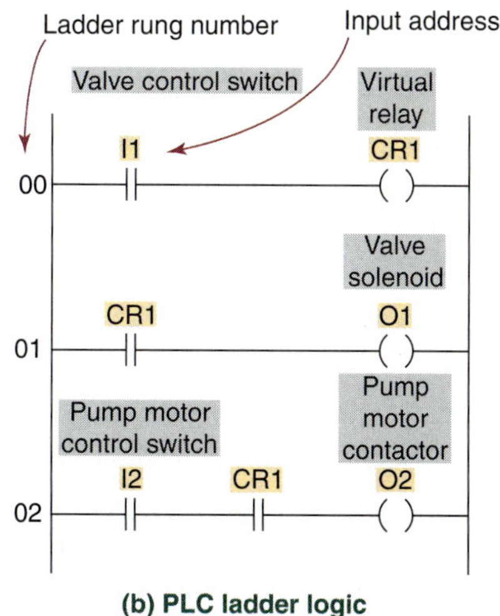

(b) PLC ladder logic

1-6-2 Ladder Logic Operation

The PLC solution in Figure 1-17(b) has three rungs with input instructions on the left and output instructions on the right. The input instructions in rungs 00 and 02 have data addresses of I1 and I2, so the voltage at input terminals 1 and 2 determines if these instruction states are true (continuity) or false (no continuity). The input terminal voltages are set by the position of external switches S1 and S2 in Figure 1-17(a). The input instruction in rung 01 is controlled by the condition of the virtual relay, CR1.

The output instructions in rungs 01 and 02 have data addresses of O1 and O2, so their state, true or false, determines the state, on and off, of the output devices connected to output terminals 1 and 2. For example, if O1 output is active, then the valve solenoid is turned on. The output in rung 00, CR1, is a virtual relay with input instructions referenced to it in rungs 01 and 02. If the output CR1 is active then all the instructions associated with that virtual relay are true. The operation of the ladder logic is summarized as follows.

- Rung 00: Output instruction CR1 is active because input instruction I1 is true (input terminal 1 has a voltage present because switch S1 is closed). If CR1 is active then both of the input instructions associated with CR1 in rungs 01 and 02 are true.
- Rung 01: Output instruction O1 is active because input instruction CR1 is true. If O1 is active then the valve solenoid is on and the inflow valve is open.
- Rung 02: Output instruction O2 is active only if input instructions CR1 and I2 are both true. If output O2 is active then the pump motor contactor connected to output terminal 2 is on and the pump is running. Thus the pump operation requires that output CR1 is true (switch S1 is closed and the input valve is open) AND switch S2 is closed (pump control switch is on).

Study the operation of the program and PLC interface in Figure 1-17 until you are familiar with the notation and operation. In summary, a PLC solution requires that:

- Input field devices, switches and sensors, are wired to terminals on the input module.
- Output field devices are connected to terminals on the output module.
- Input instructions with addresses for each input field device are placed on the left side of the ladder logic rung, and output instructions with addresses for each output field device are placed on the right side of the rung. Outputs can be paralleled on the same rung.
- Virtual relays and combinations of input instructions are placed on the ladder rungs to provide the desired control of the outputs.

FIGURE 1-18: Alternate ladder logic solution to the single-pump problem.

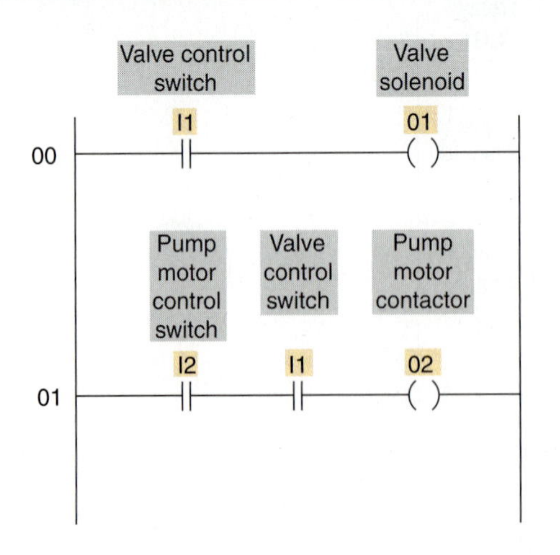

1-6-3 An Alternate Solution

Figure 1-18 illustrates an alternate solution to the ladder logic in Figure 1-17(b). Note that the virtual relay, CR1, has been removed and multiple input instructions with the address I1 have been used to achieve the same logical control. In the relay ladder logic solution, Figure 1-16(b), the field device selector switch for the valve had only one set of contacts. This single contact had to control two different devices so a relay was necessary with two switching contacts (2PST). In contrast, the I1 instruction in the PLC ladder logic is not a physical contact but a virtual one created in memory. The single contact of valve control switch, S1, establishes the input condition at terminal 1 (voltage present or no voltage present). A nearly unlimited number of I1 instructions can be used in the PLC ladder logic, and each will be true if the S1 switch is on. The number of these instructions is limited by PLC memory size. This alternate solution illustrates that multiple solutions to control problems exist, but only one solution is optimum. With only two rungs in the alternate solution, less memory is used and the program runs faster. Verify that the solutions in Figure 1-18, and Figure 1-17 work equally well.

1-6-4 PLC Advantages

In the PLC solution, the only physical wires in the system are the interfaces between the input and output field devices and the PLC input and output modules. All the elements on the ladder program rungs, namely the virtual relays, virtual relay instructions, field device input instructions, and output instructions, exist only in software in the PLC memory. As a result, the PLC ladder logic has many advantages when compared to conventional relay logic. Understanding these advantages is best illustrated by Example 1-5.

Although both systems satisfy the control requirements, the new relay logic requires extensive changes to the physical system compared to the PLC implementation. Both diagrams in Figure 1-19 are notated with an *M* to indicate modified wiring and an *N* for new wiring in the relay logic. The relay ladder logic requires five modifications and five new wires. In contrast, the new PLC solution has no modifications and just new wiring for the second pump motor contactor and selector switch. The relay logic solution has the wiring for the new components plus extensive wiring changes because the original relay and pump switch had to be replaced. The replacement was necessary because the number of poles on the original relay was not sufficient for the new control requirement. There are extensive changes in the ladder logic program for the PLC solution; however, these are implemented in software so it is both faster and performed at much lower cost.

Figure 1-20(b) is an alternate solution to the ladder in Figure 1-20(a) and does not use any virtual control relays. Verify that the ladder logic in Figure 1-20(a and b) would provide the same control for the tank problem.

It is clear from this example why PLCs are the choice for sequential control over relay logic in industrial automation. In addition, the similarity between the PLC ladder logic program and the control diagram [Figure 1-7(c)] used for relay ladder logic provides an easy transition to PLCs for electricians, technicians, and design engineers who must work with both.

Other advantages of PLCs include:

- **Reliability:** Relays are electro-mechanical devices, and physical wear in relay logic controls occurs every time the devices are turned on. PLCs have reliability inherent in all

EXAMPLE 1-5

The tank control system illustrated in Figures 1-7 and 1-17 has a second pump added to drain the tank. This second pump should be on if the following conditions are true: (1) the inlet valve is open, (2) pump 1 is on, and (3) the new pump 2 selector switch is closed. Determine the changes required in the relay logic solution in Figure 1-7 and in the PLC solution in Figure 1-17.

SOLUTION

The second pump and pump selector switch must be added in both the PLC and relay logic solution. No other changes to the physical system are necessary for the PLC solution shown in Figure 1-19(a). However, Figure 1-19(b) shows the three major changes required in the relay logic solution: (1) a new relay with three double throw poles, (2) replacement of the original single pole pump number one control switch with a double pole type, and (3) extensive modifications to the relay control wiring. Compare the new relay ladder logic solution illustrated in Figure 1-19(b) and the PLC ladder logic solution shown in Figure 1-20(a). Review both solutions and then read the analysis of the changes in each.

FIGURE 1-19: Two-pump solution using PLC and relay ladder logic.

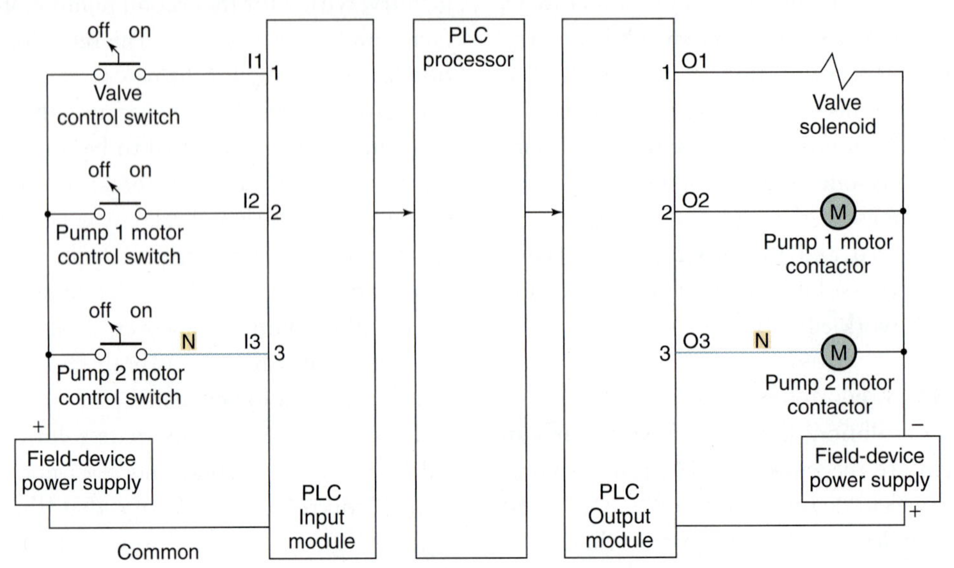

(a) PLC and field device interface for two pumps

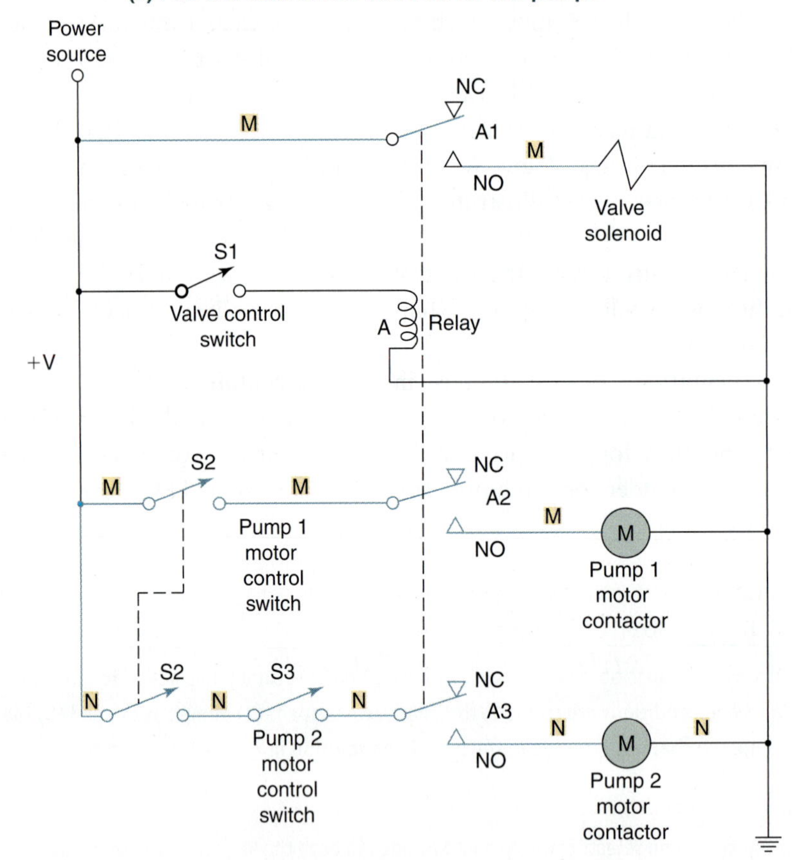

(b) Relay ladder logic solution for two pumps

electronic devices. The mean time between failure (MTBF), a statistical analysis of failure potential, is typically very large for PLCs.

- **Improved maintenance and troubleshooting:** If a problem occurs in any PLC module or in the processor, the module or processor can be changed in a matter of minutes without any changes in wiring. The PLC also makes troubleshooting the entire control system easier because a technician or engineer can check the status of each input or use software to force a change in outputs to identify the input or output device causing the problem.
- **Off-line programming:** In the past, PLCs could be programmed only with special programming terminals supplied by the vendor; however, the present systems use microcomputers (PCs) as programming terminals. The new programming software allows PLC program development on the PC to be tested with emulator software to find problems before the software is used in the control system. The new programming software allows downloads directly to the PLC, through a serial connection, or over the Internet. In most situations, the application programming is performed using only PC hardware and software resources while the PLC is running the process. This process, called *off-line programming*, allows new program development and current program modifications without taking the PLCs out of the production process. In contrast, installing and modifying relay logic circuits often takes days or weeks with considerable lost production as the control circuits are interfaced to the production system.
- **On-line programming:** On-line programming allows the programmer to edit ladder logic rungs while the PLC is executing a production program. The changes are made in a special on-line mode and when change is complete the new ladder logic is made an active part of the current ladder program. This is a large dollar savings in production line industries, such as automotive assembly, where any time the production line is halted, thousands of dollars are lost. Also, changes to variable values and set points are performed while the processor is in the production mode.
- **Broad application base:** PLC software supports a broad range of discrete and analog applications in numerous industries. With only program and module changes, a PLC can be moved from control of an assembly

FIGURE 1-20: Two-pump PLC ladder logic solutions.

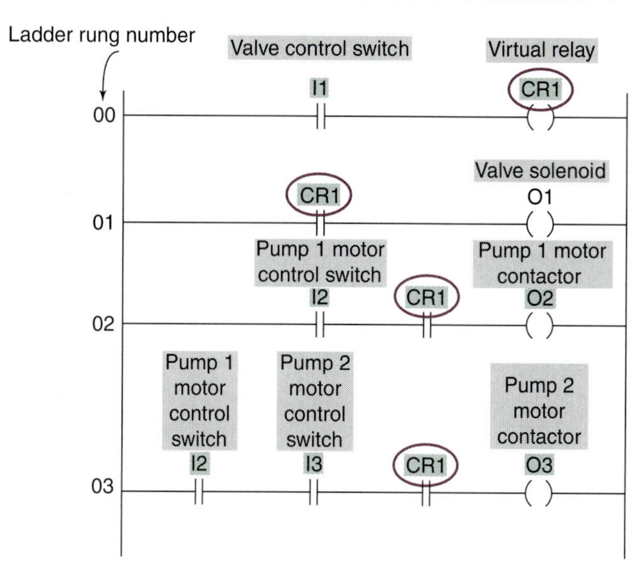

(a) PLC ladder logic solution for two pumps

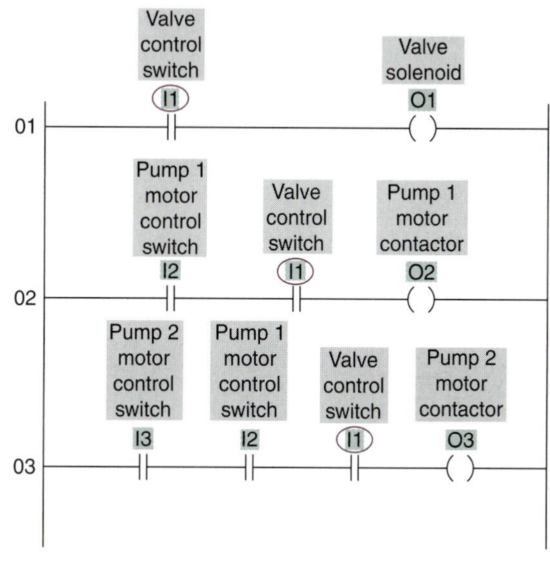

(b) Alternate solution for two pumps

cell to control of liquid level and temperature in a process control system.

- **Low cost and small footprint:** The cost and size of PLCs have dropped significantly in the last 10 years. For example, a microPLC, which would fit in the palm of your hand, offers powerful machine control for less than $300.
- **High-end control grows exponentially:** Although cost and size are dropping on the low end, the capability of large PLC systems expand as well. The ability to network and distribute the control using numerous proprietary and international network standards permits PLCs to take control of entire manufacturing systems and production plants.

The comparison between relay ladder logic and PLC programs illustrates the enormous advantage to using PLC in automation control. However, the move from relay logic to PLC control does not diminish the need for good and safe practices when working on electrical systems. The next section reviews the requirements for electrical safety.

1-7 ELECTRICAL AND PLC SAFETY

It is important to have a healthy respect for the energy sources encountered everyday, like electrical power, but there is no need to fear them. The key is to respect energy sources and know how to work with them safely; this section introduces safety related to electrical energy. The important components of electrical safety include:

- Electrical shock—how the body reacts
- The nature of electrical shock
- Safe electrical practices
- Response to shock victims

Each component is covered in the following sections.

1-7-1 Electrical Shock—How the Body Reacts

When electric current flows through the body, the resistance in the tissue converts most of the energy into heat. When the current is high, the amount of heat generated is sufficient to burn the body tissue. Although the effect is similar to a burn caused by an open flame, it is more serious because the damaged tissue is often beneath the skin and may include internal organs. In addition, electrical shock may cause significant damage to the central nervous system. The primary effect is to overload the electrical signals in the nervous system and take control of the muscles away from the brain. For example, all forearm muscles contract when a large shock or current flows through the forearm muscle. Muscles that both close and open the fingers are present in the forearm, and the muscles that close the fingers are stronger than those used to open them. As a result, the fingers will close strongly and the brain will not be able to command them to open. If the shock or electrical current is a result of the hand touching an electrical energy source, the hand will close on the source and be unable to release it. Involuntary muscle contraction, called *tetanus*, is only stopped by removing the current. The victim may still need to be physically pulled from the shock source, after the electrical source is de-energized, because muscle control is not immediately restored. Internal muscles, such as the diaphragm and heart, can also be immobilized by a DC current so that breathing and/or blood pumping is stopped.

A second problem, called *fibrillation*, produced by a small AC current, causes the heart to flutter rather than beat so blood flow through the body ceases. Direct current (DC) has the tendency to induce *muscular tetanus*, whereas alternating current (AC) causes fibrillation. In either case, electric currents high enough to cause involuntary muscle action are dangerous and should be avoided.

1-7-2 The Nature of Electrical Shock

Electrical shock occurs when some part of the body becomes a current carrying part of an electrical circuit. In Figure 1-21 the path from one hand to the other is passing current because the right hand is in contact with a voltage source and the left hand is touching a ground for the voltage source. In this case the left hand is holding a water pipe with a portion buried in the ground and the energy source has its negative terminal connected to a grounding rod driven into the earth. A similar shock would occur if the victim were to accidentally grab the wires on each side of the load. A shock of this type is dangerous because the current is passing through the chest where the diaphragm (used for lung action) and heart muscles are located. A shock occurs whenever any part of the body becomes a conductor between a point of higher electrical potential and a point of lower electrical potential (i.e., one hand on each side of the load). If the potential is high enough to overcome the contact resistance of the skin, then the current flowing through the body will cause damage to nerves and/or tissue. Table 1-1 indicates the level of current (all values are in milliamps) necessary to stimulate different bodily responses. Note that the current levels are different for men and women, and AC current is dangerous at much lower current levels than DC.

FIGURE 1-21: Electrical shock across the chest.

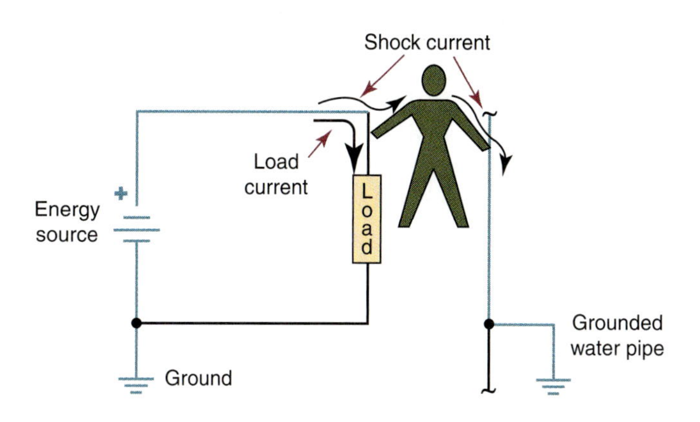

The level of current that flows under shock conditions is dependent on the resistance that the body presents to the electrical source. Research conducted on contact resistance between parts of the human body and points of

TABLE 1-1: Bodily response to electrical shock.

Bodily Effect	Gender	Direct Current dc	Alternating Current 60 Hz ac	Alternating Current 10K Hz ac
Slight sensation felt at hands	Men	1.0	0.4	7.0
	Women	0.6	0.3	5.0
Threshold of perception	Men	5.2	1.1	12
	Women	3.5	0.7	8.0
Painful, but maintained voluntary muscle control	Men	62	9.0	55
	Women	41	6.0	37
Painful, unable to let go of wires	Men	76	16	75
	Women	51	10.5	50
Severe pain, difficulty breathing	Men	90	23	94
	Women	60	15	63
Possible heart fibrillation after 3 seconds	Men	500	100	
	Women	500	100	

contact with electrical circuits provides the following data:

- Hand or foot contact, insulated with rubber: 20 MΩ typical
- Foot contact through leather shoe sole: 100 kΩ to 500 kΩ dry, 5 kΩ to 20 kΩ wet
- Contact between wire and finger: 40 kΩ to 1 mΩ dry, 4 kΩ to 15 kΩ wet
- Wire held by hand: 15 kΩ to 50 kΩ dr, 3 kΩ to 5 kΩ wet
- Metal pliers held by hand: 5 kΩ to 10 kΩ dry, 1 kΩ to 3 kΩ wet
- Contact with palm of hand: 3 kΩ to 8 kΩ dry, 1 kΩ to 2 kΩ wet
- 1.5-inch metal pipe grasped by one hand: 1 kΩ to 3 kΩ dry, 500 Ω to 1.5 kΩ wet
- Hand immersed in conductive liquid: 200 Ω to 500 Ω
- Foot immersed in conductive liquid: 100 Ω to 300 Ω

IMPORTANT: The resistance values between parts of the body and an electrical conductor are presented as examples obtained in one research study. They may not be the same for every person. You should **never** intentionally make contact between some part of your body and dangerous levels of electrical energy based on the assumed contact resistances listed here. Approved safety devices are required when handling dangerous levels of electrical energy.

Safety data for bodily response to current was obtained from tests conducted by The Massachusetts Institute of Technology and the nominal body resistance values were determined by tests conducted by the Lawrence Livermore National Laboratory.

1-7-3 Safe Electrical Practices

Whenever possible, work on all industrial systems should be performed when the system is in a *zero energy state*. In this state all sources of energy are removed to minimize the possibility of injury for persons working on the system. This concept applies to all energy sources. Some examples of energy sources include:

- Voltage sources
- Compressed springs
- High pressure fluids
- High pressure air
- Potential energy from suspended weight
- Chemical energy (flammable and reactive substances)
- Nuclear energy (radioactivity)

Systems that are controlled by a PLC often have multiple power sources. The PLC is usually powered from 110 volts AC, and the input and output modules can have AC and DC sources with a wide range of voltages from 5 volts DC to 440 volts AC. In addition, the system often controls valves that switch high pressure air and fluid. It is often not possible to work or troubleshoot a PLC system in the zero energy state, so good practices should be used to avoid electrical shock. Contact with high pressure air and fluid should also be avoided, and attention paid to the pneumatic and hydraulic devices powered by these energy sources. Deep cuts can be produced by small streams of high pressure air and liquid.

The primary safety process used throughout industry is a *lock-out/tag-out* procedure. This technique is normally not used in instructional laboratories but is quite common in industry. In the PLC laboratory is it good practice to check for the presence of voltage with a meter before actually touching any conductors in the circuit. This is especially important when AC voltage is present or when a DC voltage greater than 5 volts is used. Many industry safety manuals include the following three-step procedure when measuring a voltage:

1. Verify that that the meter is working by measuring a known voltage source.
2. Use the meter to test the circuit you plan to touch for the presence of a voltage.
3. Verify again that that the meter is working by measuring a known voltage source.

This may seem excessive, but avoiding accidental contact with dangerous voltage levels is important, and this is a proven technique for preventing electrical shock.

EXAMPLE 1-6

Determine the electrical DC potential necessary to produce pain with loss of muscle control in men when a wire is held in the dry hand. Minimum resistance values should be used.

SOLUTION

Ohms law is used to calculate the electrical DC source value as follows:

$$\begin{aligned} \text{DC source} &= \text{contact resistance x current level} \\ &= 15\ \text{k}\Omega \times 76\ \text{mA} \\ &= 1140\ \text{volts} \end{aligned}$$

This level of DC voltage is not common in most cases.

EXAMPLE 1-7

Determine the electrical AC potential necessary to produce fibrillation in women when a wire is held in the wet hand. Minimum resistance values should be used.

SOLUTION

Ohms law is used to calculate the electrical AC source value as follows:

$$\begin{aligned} \text{DC source} &= \text{contact resistance x current level} \\ &= 3\ \text{k}\Omega \times 100\ \text{mA} \\ &= 300\ \text{volts} \end{aligned}$$

This level of voltage is common in motor control circuits. Also, if the skin is penetrated by an arc or wire, then the resistance drops sharply and the voltage necessary for a dangerous shock is reduced accordingly.

EXAMPLE 1-8

Determine the level of bodily response if a 28 VAC 60 Hz source is touched by pliers held in the wet hands. Use minimum resistance values.

SOLUTION

Ohms law is used to calculate the electrical AC current value as follows:

$$\begin{aligned} \text{DC source} &= \frac{\text{AC source}}{\text{contact resistance}} \\ &= \frac{28\ \text{VAC}}{1\text{k}\Omega} \\ &= 28\ \text{mA} \end{aligned}$$

This level of discomfort falls between severe pain/difficulty in breathing and fibrillation for both men and women. A large number of control circuits in PLC systems have 28 VAC sources, so the possibility of shock should be an incentive for safe electrical practices.

A final precaution is to make initial contact with the conductor(s) with the back of one hand or fingers before grasping it between the fingers. If voltage is present when the back of the hands or fingers contact the conductor, then the natural muscle reaction will throw the fingers away from the conductor. This is a final precaution and should never be done to determine if a conductor has voltage present. Another suggestion often stated when working around high power circuits is to work with one of your hands in your pocket. This is just a way of emphasizing that you never want to permit shock current to pass through the chest region. If you follow good practices electrical energy is safe to work with, and you never need to fear it or experience an electrical shock.

1-7-4 Response to Shock Victims

Electrical shock need not occur, but if it does you should know what actions must be taken to reduce the likelihood of a serious injury. When a person is in contact with an electrical conductor and is unconscious or unable to release the electrical circuit, the first step is to quickly remove power from the circuit. If you make direct contact with victim, then it is possible that you will also receive the same level of electrical shock. After the energy source has been removed, the breathing and pulse of the victim may have to be restored, so CPR is often necessary while waiting for the arrival of 911 emergency medical teams.

1-8 WEB SITES FOR PLC MANUFACTURERS

A search in Google using the words *PLC* or *programmable logic controller* will produce hundreds of responses. The list will include companies that make PLCs, ones that design automation systems using PLCs, others that make peripheral devices to support PLCs, and organizations that support the PLC industry. The following list of PLC manufacturers and their URLs represents some of the more frequently used PLCs in North American industrial applications. The list also includes some produced in Europe and Japan, since many machines used in U.S. industries are manufactured there. The URLs are often links to the company's home page; links to the PLC products should be selected. Also, URLs frequently change, so use a search engine with the company name if the URL provided is no longer active.

- Allen-Bradley: *http://www.ab.com*
- AutomationDirect: *http://web2.automationdirect.com/adc/Home/Home*
- GE—Fanuc: *http://www.geindustrial.com/cwc/gefanuc/ctrlio.html*
- Mitsubishi Electric: *http://www.mitsubishi-automation.de/products.html*
- Omron Electronics LLC: *http://ociweb.omron.com/*
- Panasonic: *http://www.mew-europe.com/*
- Parker: *http://www.parker.com/ead/cm1.asp?cmid=307*
- Rockwell Automation: *http://www.automation.rockwell.com/*
- Schneider Electric: *http://www.us.telemecanique.com/*
- Siemens: *http://www2.automation.siemens.com/meta/index_76.htm*

Chapter 1 questions and problems begin on page 551.

CHAPTER

2

Input Devices and Output Actuators

2-1 GOALS AND OBJECTIVES

The primary goal of this chapter is to present the description and operation of typical mechanical and electrical input devices and output actuators that interface with a Programmable Logic Controller (PLC). A secondary goal is to introduce the student to input and output wiring techniques, which are important in connecting these devices to a PLC. Finally, troubleshooting tips are provided to give the student some insight into typical failure modes of input and output devices.

After completing this chapter you should be able to:

- Identify and describe various manually operated switches such as toggle switches, push button switches, and selector switches.
- Identify and describe various mechanically operated switches such as limit switches, flow switches, level switches, pressure switches, and temperature switches.
- Identify, describe, and make an application selection for proximity sensors and photoelectric sensors.
- Identify and describe the following output devices: solenoids, relays, contactors, motor starters, lamps, and alarms.
- Describe methods for powering and connecting input and output devices to a PLC.
- Develop control diagrams that use relay ladder logic which describes the operation of the sequential control system using various input and output devices.
- Troubleshoot input switches and sensors and output actuators and indicators.

2-2 MANUALLY OPERATED INDUSTRIAL SWITCHES

Manually operated switches that are connected to a PLC as input devices perform a simple on-off function, where the term *manually operated* implies that a person physically moves the switching mechanism. All on-off switches share similar contact configurations; however, their appearance, size, and mounting methods are often quite different.

The most common manually activated switches that interface to PLCs are the toggle switch, push button switch, selector switch, and push wheel, all of which are discussed in the following subsections. Before we discuss these manually operated switches, let's review the following terms that are used to describe switches and contact configurations.

- **Pole:** The term *pole* refers to an internal conductor in the switch that is moved by the switching mechanism. Switches can have any number of poles, but most switches used as PLC input devices have from one to three.
- **NC and NO:** The abbreviation *NC* stands for *normally closed* and refers to switch contacts that are in the closed position when the switch is off. Similarly, *NO* stands for *normally open* and refers to switch contacts that are not closed when the switch is off. The term *normally* implies that the switch operator has a nominal starting on or off position.
- **Throw:** There are usually just two terms, *single throw* and *double throw*, associated with a switch's throw specification. When a switch has both an NC and NO contact it has a double throw; if only an NO contact is present, then it has a single throw. Some double throw switches have a center off position and as a result both of the contacts are NO.
- **Operator:** The *operator* is the mechanical mechanism that is moved to cause a change in the switch contacts. Operators are manually moved or mechanically operated by linking to a movement in the production process or to a human operator.

2-2-1 Toggle Switches

Toggle switches are illustrated in Figure 2-1, and contact configurations in Figure 2-2. Note that the pole and throw specifications are combined to describe switch contact configurations; for example, single pole double throw (SPDT). The dotted lines connecting the poles indicate that they are all a part of a single switch, and all poles are changed by a single operation of the switch operator. In order to prevent inadvertent switch activation, some toggle switches requiring a two-step operation are used. First, the lever handle of the toggle switch must be pulled out, and then it must be moved to the desired position.

EXAMPLE 2-1

Select a switch configuration and draw a solution for the following control problem. A single switch should start a 110 VAC motor and control two 28 VDC indicator lamps. An illuminated red lamp indicates that the motor is not powered and an illuminated green lamp indicates that the motor has power applied.

SOLUTION

The solution requires a single switch to control a 110 VAC motor and two 28 VDC lamps so a two pole switch is needed (one pole for the lamps and one for the motor). Only one lamp should be on at a time, so a double throw action is necessary. The circuit in Figure 2-3 provides the necessary control. The switch has a double pole double throw contact configuration, and the dashed line between the poles indicates that the same operator changes both poles.

Other switch configurations are the *double break* contacts and the *make before break* contacts. Make before break contacts are found in some selector switches. In switches with double break contacts, both ends of the pole move when the switch is activated. Figure 2-4 illustrates the contact schematic for an SPST and SPDT double break selector switch. Note that *DB* is added to the abbreviation (SPSTDB) to indicate that a double break contact is present. Example 2-2 illustrates how double break contacts are used in a control circuit.

The SPDT switch operations that we have discussed have the NC contacts opening before the NO contacts can close. In this case, the pole is not in contact with either the NC or NO contacts for a few milliseconds. This operation is classified as

FIGURE 2-1: Electronic circuit toggle switches.

(a) Toggle switches

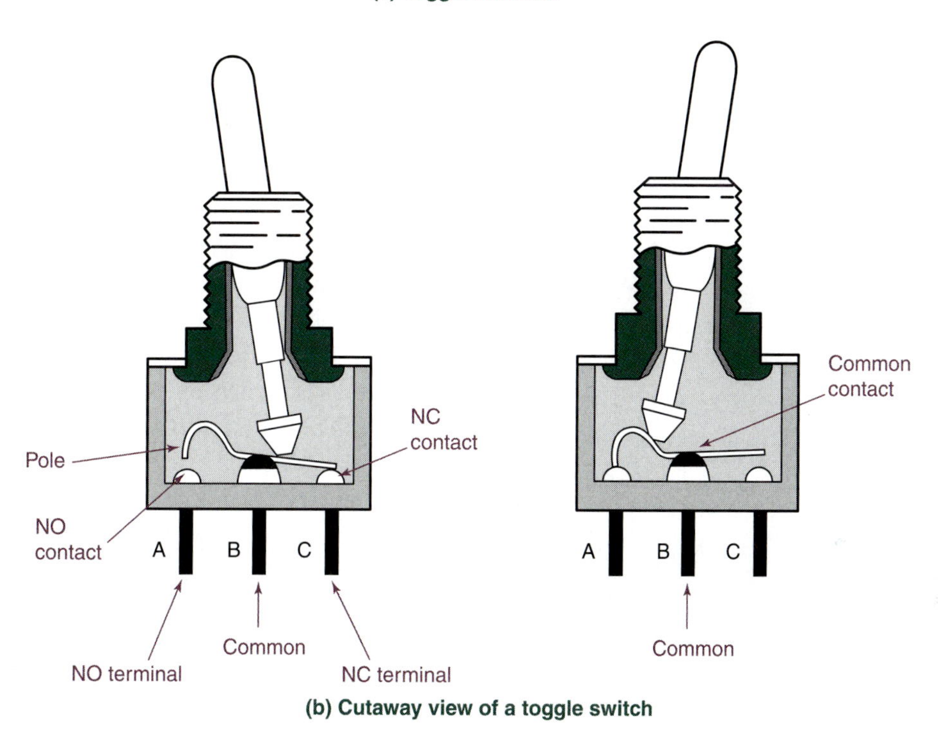

(b) Cutaway view of a toggle switch

Photo courtesy of NKK Switches—www.nkkswitches.com; and Rehg and Sartori, Industrial Electronics, *1st Edition, © 2006. Reprinted by permission of Pearson Education, Inc., Upper Saddle River, NJ.*

EXAMPLE 2-2

Draw the circuit for the following control problem using double break and non-double break contacts. Power to a 110 VAC motor and a 28 VDC lamp must be controlled so that the motor is powered when the switch is on and the red light is on when the switch is off.

SOLUTION

An SPDTDB switch contact is used for this application because the switched voltage for each device is different. The double break contacts permit different voltages to be switched by the NC and NO contacts. If a DB configuration, as shown in Figure 2-5, were not available, then a DPDT switch would have to be used with a circuit similar to Figure 2-3. Note that the SPDT switch shown in Figure 2-2 could not be used because the pole has a common terminal for the NC and NO contacts, and as a result the single pole could not be used to switch two different voltages.

FIGURE 2-2: Contact configurations.

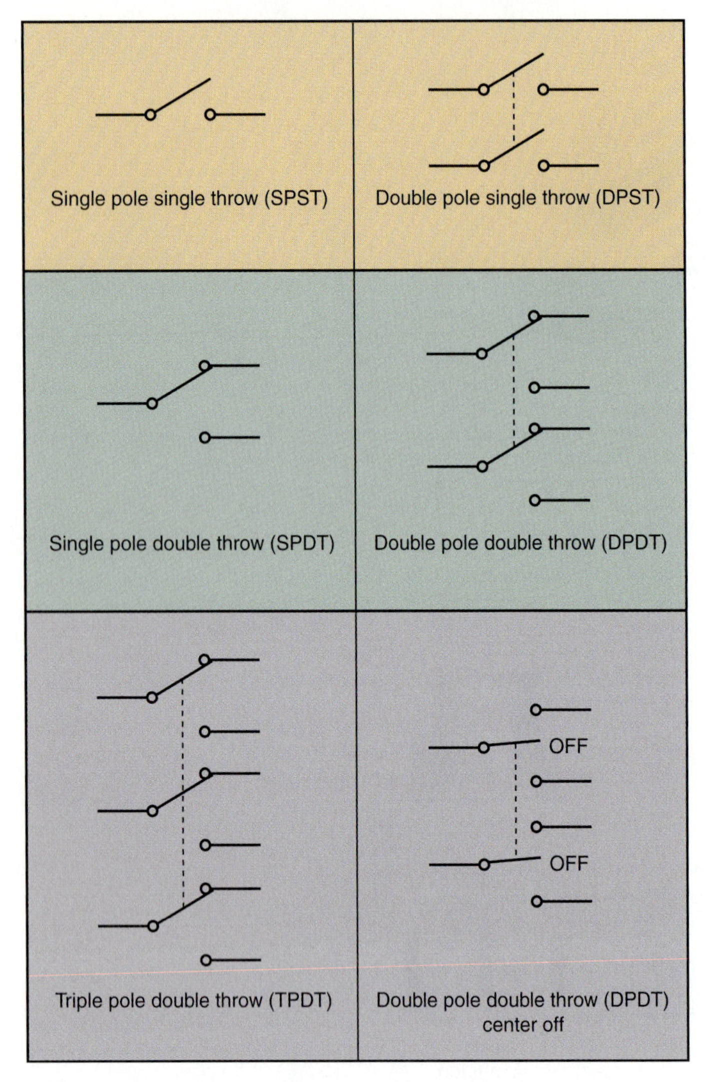

Rehg and Sartori, Industrial Electronics, *1st Edition, © 2006. Reprinted by permission of Pearson Education, Inc., Upper Saddle River, NJ.*

a *break before make*, which is the most common type of contact transfer. However, in some control circuits a problem can result if the switched device is not connected to either contact during the switching operation. The *make before break* contact transfer solves this problem. In this type of transfer, the NO contacts close before the NC contacts open. The make before break configuration is frequently used in rotary switches.

2-2-2 Push Button Switches

Push button switches, like toggle switches, are available in many different configurations, as Figure 2-6 illustrates. Most push buttons are of the momentary contact type, so that the on condition requires the button to be held in the down position. In a maintain contact push button switch, the button is pressed once to close the NO contacts, and then pressed a second time to return the NO contacts to their open condition.

Industrial push button switches are available in four styles: *no guard, full guard, extended guard,* and *mushroom button.* No guard or extended head switches have the button extending beyond the enclosing cylinder, as shown in Figure 2-6(b). Fully guarded or flush head switches have the button flush with the enclosing cylinder, as shown in Figure 2-6(a), and extended guard switches have the button below the enclosing cylinder, as shown in Figure 2-6(c). *Mushroom* switches are shown in Figure 2-6(d), (e), and (f), and are so named because the button has a large, circular, mushroom-shaped surface. Mushroom switches are most commonly used for applications where a device must be easily and quickly turned off, like an emergency stop control. The guarded switches are used where an application is turned on, so the guard prevents device activation if an object accidentally leans against the switch.

Momentary contact mushroom push buttons typically have the down position as the momentary contact, while the spring returns the button out to the normal position. Maintain contact push buttons have two configurations. One is a push-pull button that is shown in Figure 2-6(d). When the button is pressed it remains depressed until the button is pulled out to return the switch to the normal position. In some models, such as that shown in Figure 2-6(f), the mushroom button must be rotated before it will return to the normal position. Another maintain contact option requires the button to be pressed once to turn the switch on and pressed a second time to turn the switch off.

The style shown in Figure 2-6(g) has two push buttons interlocked so that one button turns the switch on and the second returns the switch to the normal position. Switch vendors offer both illuminated and non-illuminated buttons in a variety of terminal configurations. The abbreviation most often used to identify a push

FIGURE 2-3: Motor and lamp control circuit.

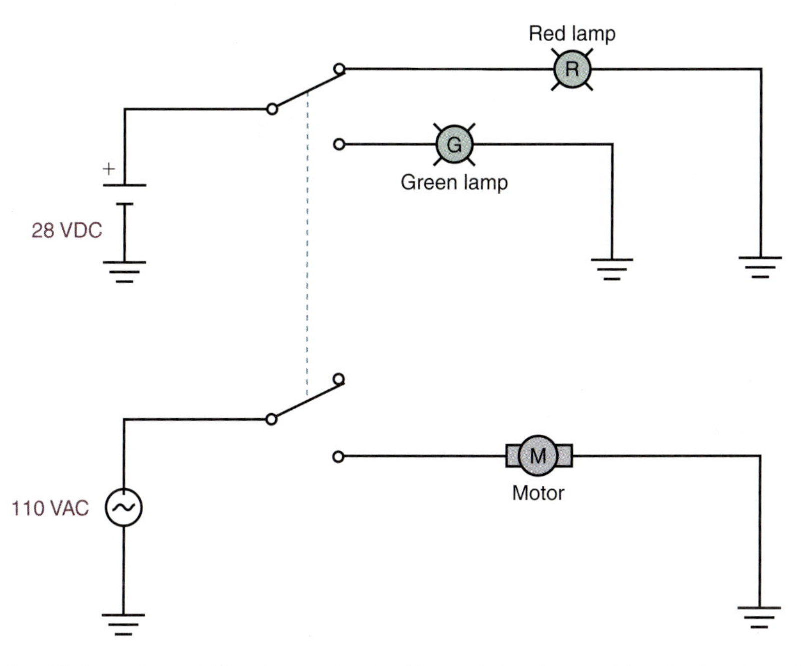

Rehg and Sartori, Industrial Electronics, *1st Edition, © 2006. Reprinted by permission of Pearson Education, Inc., Upper Saddle River, NJ.*

FIGURE 2-4: Double break switch contacts.

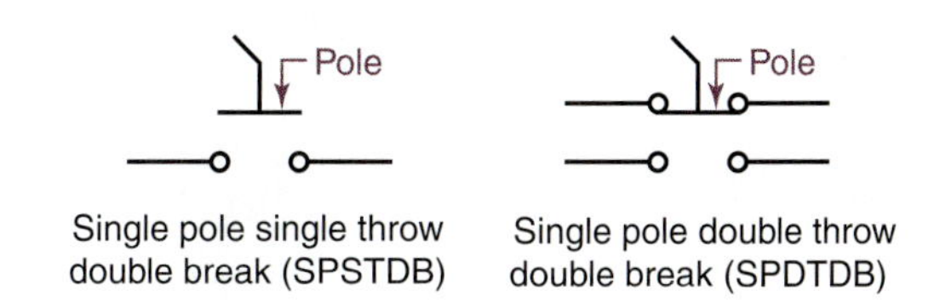

Rehg and Sartori, Industrial Electronics, *1st Edition, © 2006. Reprinted by permission of Pearson Education, Inc., Upper Saddle River, NJ.*

FIGURE 2-5: Example of DB control circuit.

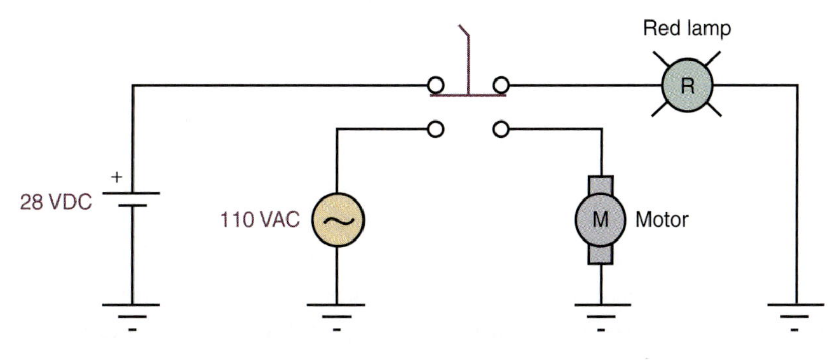

Rehg and Sartori, Industrial Electronics, *1st Edition, © 2006. Reprinted by permission of Pearson Education, Inc., Upper Saddle River, NJ.*

FIGURE 2-6: Push button switch operators.

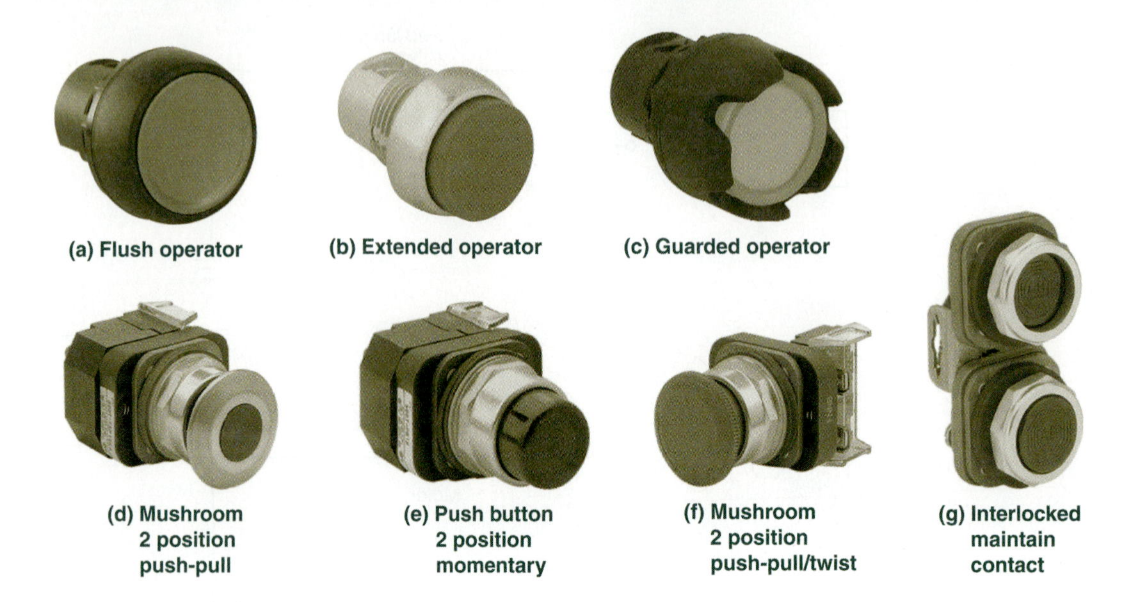

Courtesy of Rockwell Automation, Inc.

FIGURE 2-7: Selector switches.

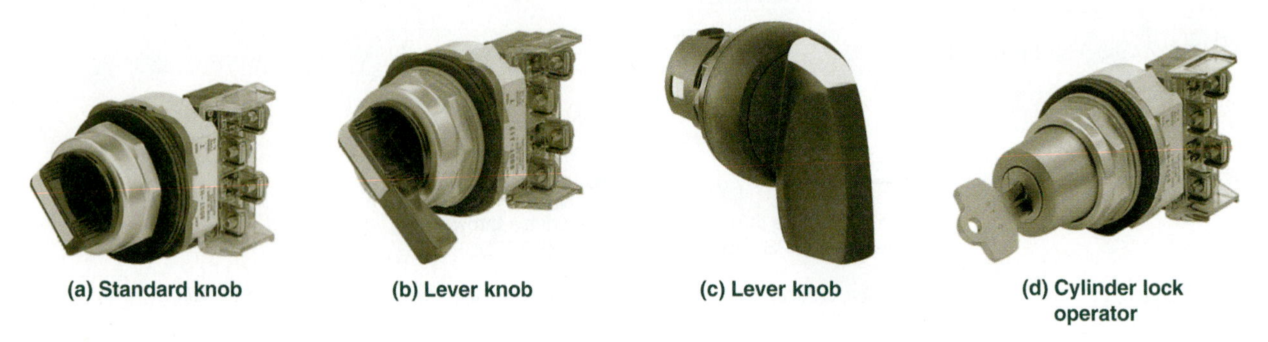

Courtesy of Rockwell Automation, Inc.

button switch is PB. If multiple push buttons are used they are identified as PB1, PB2, PB3, and so on.

Two other switch types that are included in the push button category are tactile switches and membrane switches. Tactile push button switches operate like a standard push button but with a much lower operating force. Membrane switches are created using layers of polycarbonate or polyester insulating material and conductive links. As a result, the switches can be integrated directly into a machine panel or instrument faceplate.

2-2-3 Selector Switches

Selector switches are typically two-, three-, and four-position switches with a number of different knob options (see Figure 2-7). The larger winged knob is designed for workers with gloves. The selector knob can be replaced with a keyed cylinder for key switch operation as shown in Figure 2-7(d). Selectors offer the maintain contact option at every switch position or momentary contact in many of the switch positions. The abbreviation most often used to identify a selector switch is SS. If multiple selector switches are used they are identified as SS1, SS2, SS3, and so on. Clockwise or

FIGURE 2-8: Switch symbols.

PUSH BUTTONS							
Momentary Contact					Maintained Contact		Illuminated
Single Circuit		Double Circuit	Mushroom Head	Wobble Stick	Two Single Ckt.	One Double Ckt.	
N.O.	N.C.	N.O. & N.C.					

(a) Push-button symbols

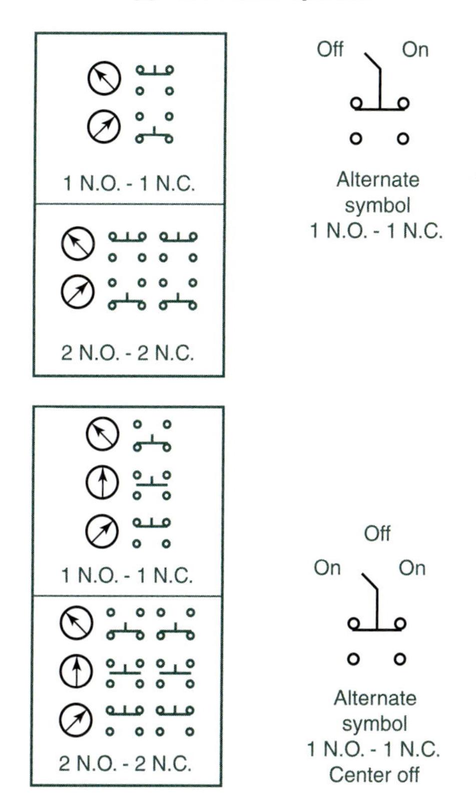

(b) Selector symbols

Rehg and Sartori, Industrial Electronics, *1st Edition, © 2006. Reprinted by permission of Pearson Education, Inc., Upper Saddle River, NJ.*

counterclockwise rotation is used to change the condition of the switch contacts. All of the standard contact configurations, such as SPST, SPDT, DPST, and DPDT, are available in both push button and selector switches.

Figure 2-8 illustrates the standard drawing symbols used for push button and selector switches. The push button symbols, Figure 2-8(a), are illustrated for momentary and maintained contact. The symbol for *wobble* lever (four-way toggle) represents a switch that has a handle. The handle triggers a change in contacts when the handle is moved up and down or left and right. The illuminated push button allows the switch face to be illuminated. The selector symbols are shown in Figure 2-8(b).

2-3 MECHANICALLY OPERATED INDUSTRIAL SWITCHES

Mechanically operated industrial switches are automatically opened or closed by a process parameter such as position, pressure, or temperature. The

FIGURE 2-9: Limit switches shown in five configurations.

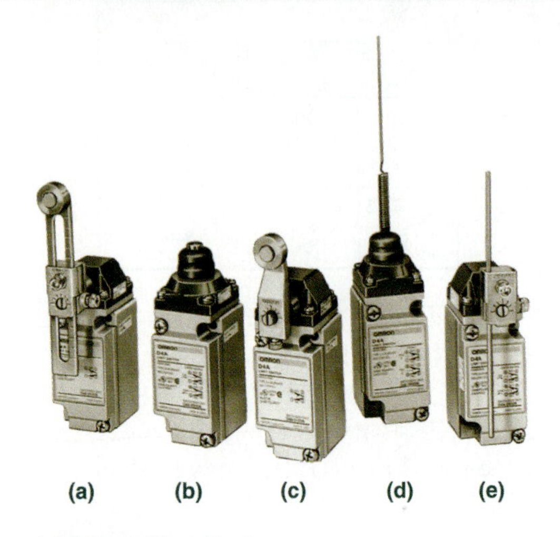

(a) Adjustable roller lever
(b) Plunger
(c) Standard roller lever
(d) Wobble lever
(e) Adjustable rod lever

Courtesy of Omron Electronics LLC.

switch contacts open or close when a predetermined process set point is reached. The following subsections discuss the mechanically operated switches typically interfaced with a PLC.

2-3-1 Limit Switches

The *limit switch* is the most commonly used mechanically operated industrial switch. Limit switch contacts are typically activated with a cam that rotates a lever, by depressing a plunger, or by tripping a wobble lever. Figure 2-9 shows five different limit switch configurations: (a) adjustable roller lever, (b) plunger, (c) standard roller lever, (d) wobble lever, and (e) adjustable rod lever.

Figure 2-10 illustrates the standard schematic symbols and control symbols for limit switches. The symbols in Figure 2-10(a) through (d) indicate that the switch is a momentary contact, and the symbol in Figure 2-10(e) indicates a maintained contact switch. The symbols in Figure 2-10(a) and (b) illustrate NO and NC contacts on the top and NO held closed and NC held open on the bottom. Machine schematics are normally drawn with the

FIGURE 2-10: Schematic symbols for limit switches—momentary and maintained contacts.

Limit				
Normally open	Normally closed	Neutral position		
		NP	Actuated	Maintained position
Held closed	Held open		NP	
(a)	(b)	(c)	(d)	(e)

(a) – (d) Momentary contact types
(e) Maintained contact type

Rehg and Sartori, Industrial Electronics, *1st Edition, © 2006. Reprinted by permission of Pearson Education, Inc., Upper Saddle River, NJ.*

machine in the off or starting condition, so the latter two symbols are used to show contacts that are activated when the machine has not yet started. The symbol in Figure 2-10(c) is a center neutral contact with the top contact held closed in Figure 2-10(d). The abbreviation most often used to identify a limit switch is LS. If multiple limit switches are used they are identified as LS1, LS2, LS3, and so on.

2-3-2 Flow Switches

Flow switches are used to detect a change in the flow of a liquid or a gas in a pipe or duct. The switch illustrated in Figure 2-11(a) is activated when the flow of liquid through the pipe moves the operator suspended in the flow. Switches are available for liquids with varying viscosities flowing in pipes and for detection of air and gas movement in ducts. The schematic symbol used for flow switches is illustrated in Figure 2-11(b). The abbreviation used for the flow switch is FL. If multiple flow switches are used, then they are identified by FL1, FL2, FL3, and so on.

2-3-3 Level Switches

Level switches or float switches are discrete switches used for control of liquid and granular material levels in tanks and bins. Level switches, like the three shown in Figure 2-12, use the vertical position of a float on a liquid surface to trigger a change in the switch contacts. The switch includes a snap action mechanism that provides quick-make and quick-break contact operation and adds hysteresis at the trigger point. *Hysteresis* is a separation between the activation point and the deactivation point of the switch. With hysteresis present, once the switch turns on, it remains on until the level is moved past the turn on point to the deactivation point. Hysteresis is needed to keep the level switch from cycling between on and off at a single activation point due to shock, vibration, or a varying level. Hysteresis is sometimes referred to as a dead band. Think of the thermostat in your home. If you set the temperature to 72 degrees, the heater typically turns on at 71 and off at 73, thus providing a dead band, which prevents the thermostat from rapidly cycling on and off.

Level switches are available in two configurations: *open tank*, Figure 2-12(a) and (c), and *closed tank*, Figure 2-12(b). Open tank models are used with tanks that are not sealed and are open to atmospheric pressure. Closed tank models are used in applications where the tank is sealed and could be pressurized. Figure 2-12(d) illustrates the NO and NC schematic symbols for the float switch.

FIGURE 2-11: Flow switch in liquid flowing through a pipe.

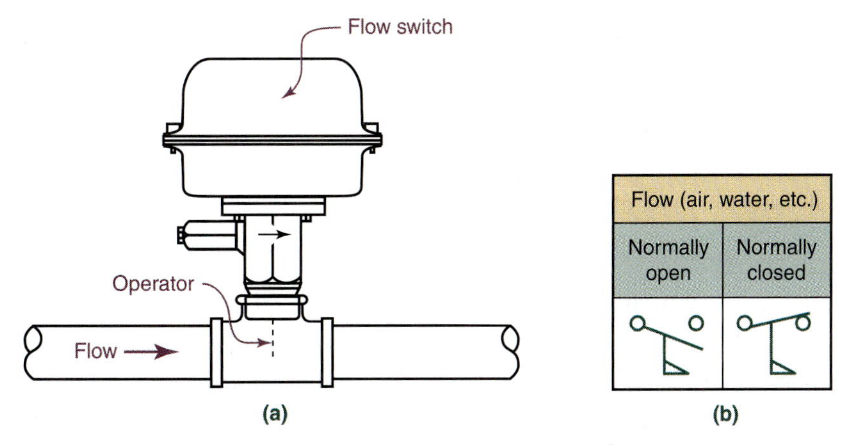

Rehg and Sartori, Industrial Electronics, *1st Edition, © 2006. Reprinted by permission of Pearson Education, Inc., Upper Saddle River, NJ.*

FIGURE 2-12: Level-activated lever switches.

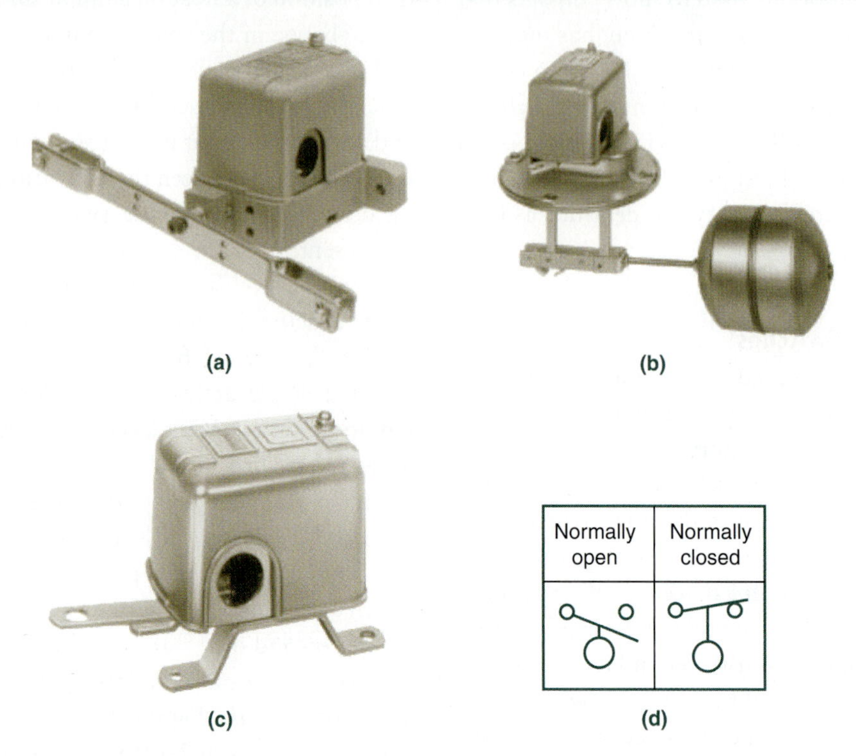

Courtesy of Square D/Schneider Electric.

2-3-4 Pressure Switches

Discrete pressure switches change the open or close condition of a contact based on the pressure applied to the device by water, air, or another fluid such as oil. Pressure switches are either absolute (trigger at a specific pressure value) or differential (trigger on the difference between two pressures). A common application for a pressure switch is to control the air pressure in a tank by turning on an air compressor motor whenever the tank pressure falls below a set value. Switches are available with fixed switching pressures or with an adjustable trigger level.

Pressure switches have two primary components: a *movable component* that is displaced by applied fluid pressure and a set of *snap-action contacts*. The fluid pressure is usually applied to the inside of a closed bellows or on one side of a bellows-shaped diaphragm, causing the ends of the bellows to expand or the diaphragm to displace. In other switches a piston is used to provide the displacement from the applied pressure. The electrical contacts are mechanically linked to the moving component so at a specific displacement the contacts change states (open to close or close to open). The contacts have a snap action so that when the trigger level is reached they rapidly move from the current state to the opposite state. The pressure switch in Figure 2-13(a) is a non-adjustable differential device. Differential switches trigger on the difference between two applied pressures. The schematic symbol for the pressure switch is illustrated in Figure 2-13(b), along with a typical contact configuration. Figure 2-15 shows additional contact symbols for pressure switches.

2-3-5 Temperature Switches

Discrete temperature switches, often called *thermostats,* cause electrical contacts to change state at a specific temperature. This contact change on a temperature switch is triggered by the expansion of a fluid inside a sealed chamber. Figure 2-14 illustrates a temperature switch where the sealed chamber includes the exterior capillary tubing and stainless steel cylinder along with a bellows inside the switch body. The fluid that fills this chamber has a high coefficient of expansion with

FIGURE 2-13: Differential pressure switch.

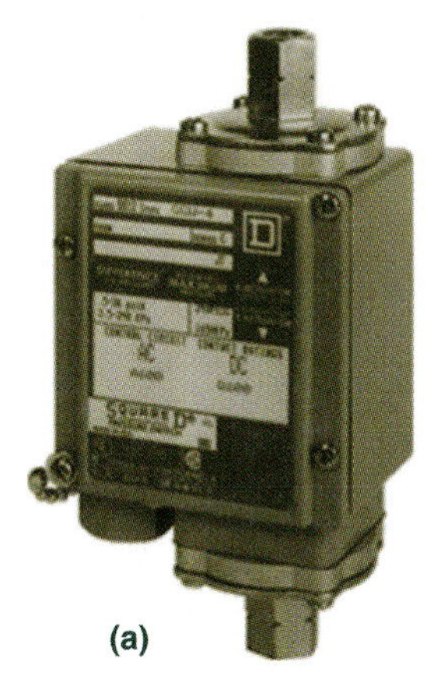

(a)

Contact Arrangement	Contact Symbol
1 N.O. - 1 N.C.	

(b)

Courtesy of Square D/Schneider Electric.

FIGURE 2-14: Temperature switch.

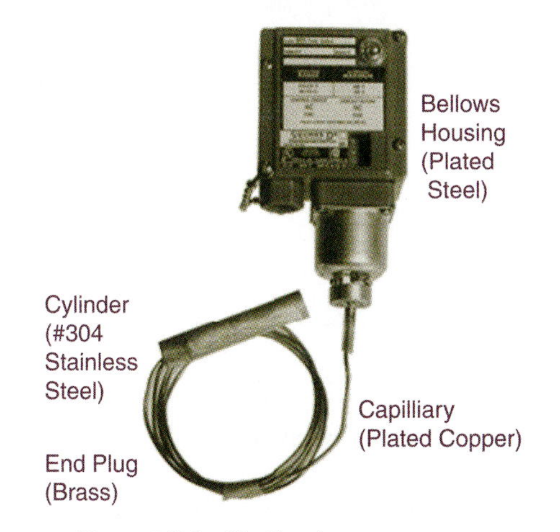

Courtesy of Square D/Schneider Electric.

increased temperature. As the stainless steel cylinder is heated the fluid expands, increasing the pressure against every surface in the sealed chamber. As a result, the bellows expands with the increased pressure and causes electrical contacts to change state.

Figure 2-15 illustrates the symbols and contact configurations for the pressure switches and temperature switches; a description of each of the symbols is also provided.

2-3-6 Control Diagrams

A control diagram is a type of symbolic language that describes the electrical/electronic operation of an industrial system. Control diagrams have numerous names such as *ladder logic, ladder*

EXAMPLE 2-3

Draw a ladder rung to turn on a red pilot lamp when three selector switches have the following conditions: switch S1 is off, switch S2 is on, and switch S3 is off. All switches are single pole double throw.

SOLUTION

The ladder rung solution is shown in Figure 2-16. Note that the switch contacts are considered inputs (left side of the rung) and the pilot lamp is considered an output (right side of the rung). The series connection produces the AND function. In other words, the input is true if you have S2 AND NOT S1 AND NOT S3. All switches have NO and NC contacts as they are double throw types. Because the output is active when S1 and S3 are off, the NC contacts on those switches must be used. Switch S2 is on for an active output, so the NO contact on that switch is used. Continuity through the input logic is achieved only when S1 and S3 are off (not true) and S2 is on (true), which results in turning on the lamp.

FIGURE 2-15: Pressure and temperature control symbols.

Symbol		Description
Pressure controls	Temperature controls	
		Automatic operation
		Single pole double throw — automatically opens or closes on rise or fall.
		Single pole double throw — slow-acting contact with no snap action. Contacts close on rise and close on fall with an open circuit between contact closures.
		Single pole single throw, normally closed — closes on rise.
		Single pole single throw, normally closed — opens on rise.
		Single pole single throw, normally open — closes on rise.
		Single pole single throw, normally closed — opens on rise.
		Two circuit, single pole single throw, normally open — a common terminal is connected to two separate contacts that close on rise.
		Two circuit, single pole single throw, normally closed — a common terminal is connected to two separate contacts that open on rise.
		Manual reset
		Single pole single throw, normally open — contacts open at a predetermined setting on fall and remain open until system is restored to normal run conditions, at which time contacts can be manually reset.
		Single pole single throw, normally closed — contacts open on rise and remain open until system is restored to normal run conditions, at which time contacts can be manually reset.
		Single pole double throw, one contact normally closed — contact opens on rise and remains open until system is restored to normal run condition, at which time contact can be manually reset. A second contact closes when the first contact opens.
		Single pole single throw, normally closed — contacts close on fall and remain closed until system is resorted to a higher predetermined setting.

FIGURE 2-16: Solution for Example 2-3.

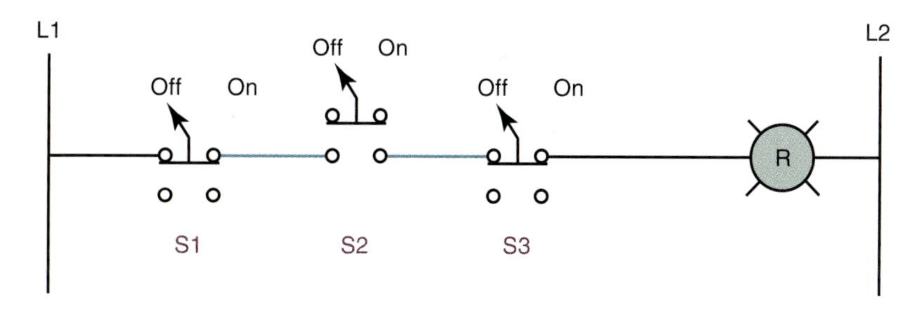

Rehg and Sartori, Industrial Electronics, *1st Edition, © 2006. Reprinted by permission of Pearson Education, Inc., Upper Saddle River, NJ.*

EXAMPLE 2-4

Draw a ladder rung to turn on a solenoid valve when the flow switch (FL1) is activated or when the selector switch (S1) is activated. All contacts are single pole double throw.

SOLUTION

The ladder rung solution is shown in Figure 2-17. Note that the switch contacts are considered inputs (left side of the rung) and the solenoid valve is considered an output (right side of the rung). The parallel connection is considered the OR function. In other words, the input is true if you have FL1 OR S1. All devices have NO and NC contacts as they are double throw types. The normal condition for the contacts is float set point is not reached (no flow) so the NO contacts are open, and the selector is in the off position with NO contacts open. The output should be active (on) when the flow is present OR the selector is turned on.

FIGURE 2-17: Solution for Example 2-4.

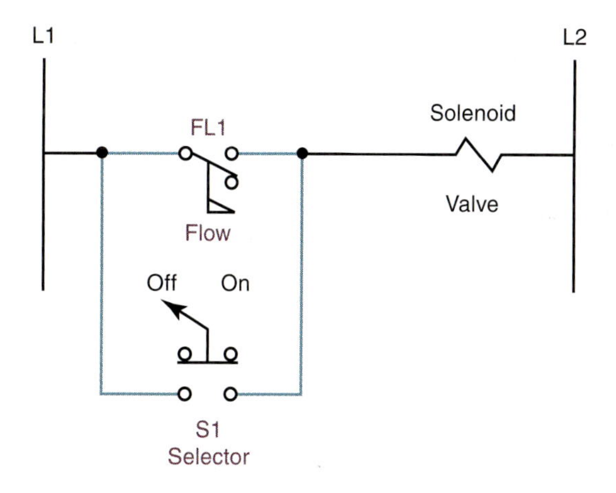

Rehg and Sartori, Industrial Electronics, *1st Edition, © 2006. Reprinted by permission of Pearson Education, Inc., Upper Saddle River, NJ.*

diagrams, relay ladder logic, motor control diagrams, two-wire diagrams, and *three-wire diagrams.* When each is used depends on the type of system that the drawing represents. For example, systems that include only relays are called relay ladder logic, but when they describe the control provided by a PLC they are called just ladder logic or ladder diagrams. Let's take a look at three examples of ladder diagrams using the switches that have been discussed (Examples 2-3, 2-4, and 2-5).

At the present time the use of the terms *true* and *active* (for the on condition) or *not true* and *not active* (for the off condition) for devices is not used because the operation of the electrical circuit is understood with just the *on* and *off* reference. However, PLC operation will be discussed where the switch condition is separated from the ladder logic representation. Therefore, starting to use these terms here prepares you for the next sections.

2-4 INDUSTRIAL SENSORS

Industrial sensors are the eyes, ears, and tactile senses of the PLC in an automated system. Strategically mounted sensors provide the automation system with the same data that an operator gathers using the five human senses. For example, in Figure 2-19 a sensor verifies that a part has the proper number of holes prior to insertion of a mating part into the holes. The sensor in the figure bounces a light beam off of the object back to an

EXAMPLE 2-5

Draw the ladder rung that turns on a motor contactor and a green pilot lamp for the following limit switch conditions: LS1 or LS2 are true and LS3 or LS4 are not true. Remember that the term *true* or *active* means a switch is activated and *not true* or *not active* means it is in the normal position. All switches are double pole double throw.

SOLUTION

The ladder rung solution is shown in Figure 2-18. Note that the switch contacts are considered inputs (left side of the rung) and the contactor and pilot lamp are considered outputs (right side of the rung). The parallel-series connection is the AND/OR function. In other words, the input is true if you have (LS1 OR LS2) AND (NOT LS3 OR NOT LS4). All switches have NO and NC contacts as they are double throw types. Since either switch LS1 or LS2 must be activated or switched to make the outputs true, the NO contacts are used in the ladder rung. Because either switch LS3 OR LS4 must be in its normal condition, the NC contacts are used in the ladder rung. Note that there are two parallel sets of limit switches and two outputs, a green pilot lamp and a motor contactor. When multiple outputs are present on a single rung, they are always in parallel.

FIGURE 2-18: Solution for Example 2-5.

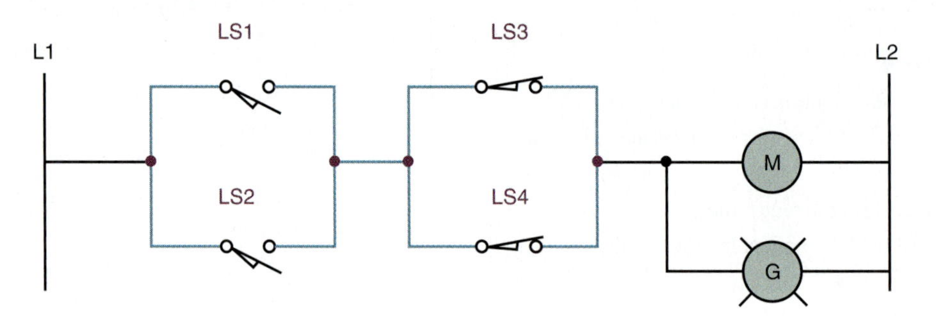

Rehg and Sartori, Industrial Electronics, *1st Edition, © 2006. Reprinted by permission of Pearson Education, Inc., Upper Saddle River, NJ.*

FIGURE 2-19: Hole inspection with sensor.

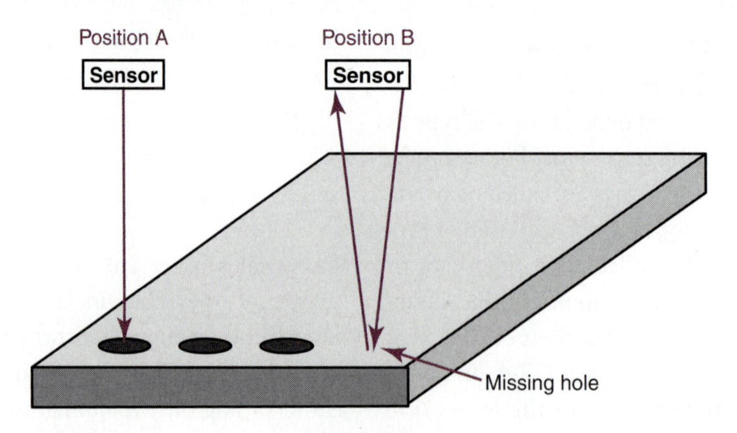

Rehg and Sartori, Industrial Electronics, *1st Edition, © 2006. Reprinted by permission of Pearson Education, Inc., Upper Saddle River, NJ.*

internal receiving device. If a hole is present, the beam goes though the hole (sensor position A) and does not return. If a hole is not present, the beam bounces back (sensor position B).

A vast number of sensors are available with unique combinations of sensing performance, output characteristics, and mounting options. Individual sensors are often used to sense and control a single machine function. In other applications, numerous sensors are used to control every operation of an automated system. Sensing devices fall into two categories:

- **Contact devices,** which physically touch the parameter being measured.
- **Non-contact devices,** which sense or measure the process parameter without physically touching it.

The non-contact sensor is a solid-state sensor with no moving parts, unlike the mechanical contact sensor, and it is faster and more reliable than the mechanical device. Contact and non-contact sensors have either a *discrete sensing capability*, which provides on and off states and is discussed in this section, or an *analog sensing capability*, which measures a range of input conditions and is discussed in Chapter 12.

Discrete sensors have a single trigger point and provide two states—*on* and *off*. The sensor depicted in Figure 2-19 is a discrete sensor. Another example of this type of switch is the thermostat that controls the heating systems in homes. A single temperature level (72° F) is set, and the heating system cycles on and off as the house temperature moves below and above the thermostat temperature setting. Discrete solid-state sensors use electronic circuits to perform the sensing function, and provide the on and off output states through a variety of output contacts and circuits.

Typical discrete contact sensors are those switches that were discussed in Section 2-3. Discrete non-contact sensors fall primarily into two categories: *proximity sensors* and *photoelectric sensors*.

2-4-1 Proximity Sensors

As the name implies, *proximity sensors* measure conditions without physically touching the part. Proximity sensors are used to automate discrete part production in manufacturing systems. Proximity sensors operate with an *inductive* or *capacitive* sensing capability, which generates a magnetic or electrostatic field and senses when the field has been breached. Others have an *ultrasonic* capability to make measurements.

Inductive proximity sensors. Inductive proximity sensors operate under the electrical principle of *inductance* and detect the presence of a ferrous or nonferrous metal part when it comes within the magnetic field generated by the sensor's coil. The only requirement for the part material is that it can conduct a current. Inductive proximity sensors are available in various package types with 70 percent of all sensors used falling into the cylindrical threaded barrel type. The shape of the part to be sensed and the sensing application dictate the type of package shape for best operation. Figure 2-20 shows two barrel inductive sensors.

Inductive proximity sensors are comprised of a coil, oscillator, and a detector, which generate an output as shown in the block diagram in Figure 2-21. The coil and the oscillator produce a magnetic field, and the shape of the field is determined by two factors: the core of the coil, which is made from highly permeable ferrite material, and the degree of shielding around the coil. The shield can be either an internal shield

FIGURE 2-20: Barrel inductive proximity sensors.

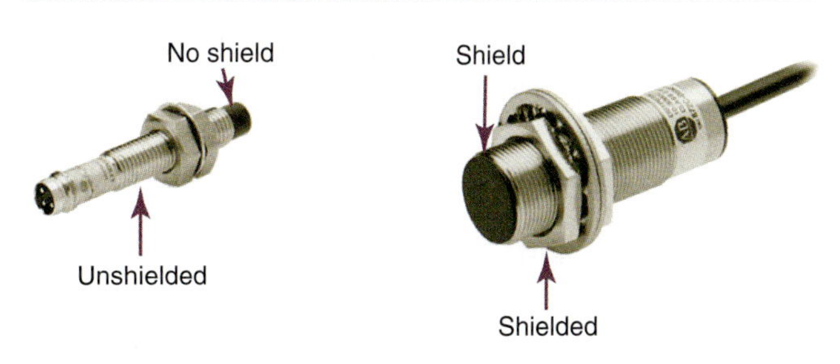

Courtesy of Rockwell Automation, Inc.

FIGURE 2-21: Inductive proximity sensor block diagram.

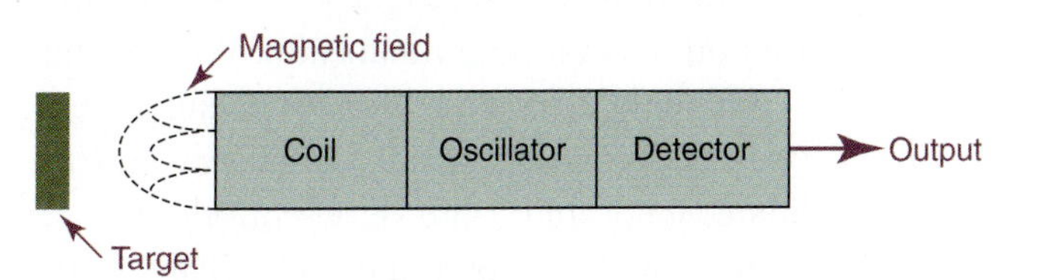

FIGURE 2-22: Target movement with respect to the inductive proximity sensor.

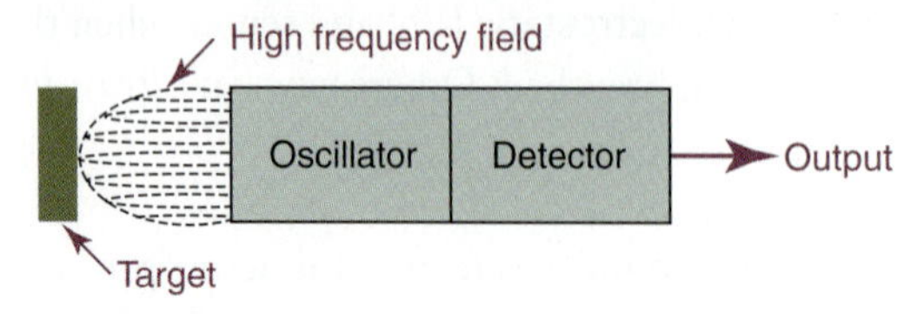

Rehg and Sartori, Industrial Electronics, *1st Edition, © 2006. Reprinted by permission of Pearson Education, Inc., Upper Saddle River, NJ.*

integrated into the sensor (see Figure 2-20) or a shield produced by the metal in which the sensor is mounted. In either case the shielding causes the field to be more focused; in other words, the shielded sensor has a smaller sensing range than the unshielded sensor.

The movement of a target with respect to an inductive proximity sensor is illustrated in Figure 2-22. If a target enters the high frequency magnetic field, eddy currents are induced on the target material. This transfer of energy to the target causes the oscillation amplitude to drop. The detector recognizes the decrease in amplitude and produces an output. Based on the operation of the inductive sensor circuit, the performance characteristics of the sensor include:

- Detection of all materials that are electrical conductors.
- Detection not limited to magnetic materials.
- Detection of stationary and moving objects.
- Preferred target is a flat, smooth object.
- Energy levels are low so that they do not create radio frequency interference or generate heat in the target.

Capacitive proximity sensors. Capacitive proximity sensors operate under the electrical principle of *capacitance* and detect the presence of a part when an object of any material type comes within the electric field established by the capacitor plate(s) in the sensor. Capacitive sensors sense both metallic and nonmetallic objects and are commonly used in the food packaging industry to check product inside containers and to validate fluid and solid levels inside tanks. However, the lower cost and high reliability of

FIGURE 2-23: Capacitive proximity sensor block diagram.

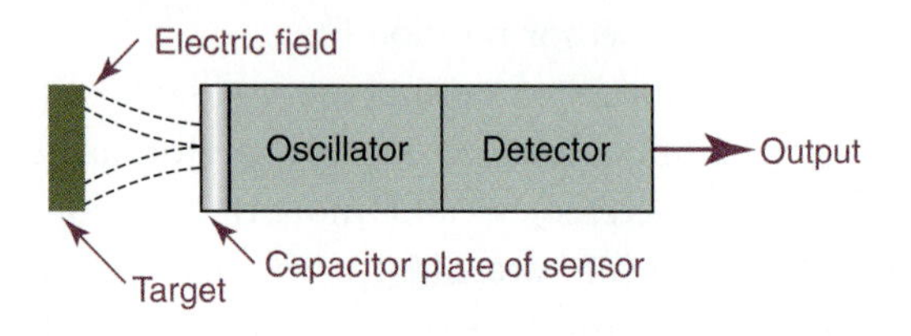

inductive sensors makes them the first choice for industrial automation over the capacitive types.

Capacitive proximity sensors use a changing dielectric to change the value of a capacitor in the sensor's oscillation circuit. Recall that the capacitance of a capacitor is determined by the size of the plates, plate separation distance, and the dielectric between the plates. Figure 2-23 provides a block diagram of a capacitive proximity sensor. Note that the target acts as the capacitor's second plate. As an approaching target interacts with the electrostatic field created by the sensor, the capacitance of a capacitor inside the sensor is changed. As the oscillation amplitude increases, the oscillator output voltage increases, causing the output of the detector to change.

Capacitive proximity sensors can sense conducting materials farther away than nonconducting, and the larger the target mass, the farther the sensing distance. Capacitive sensors are more sensitive to changes in temperature and humidity than the inductive sensor and have the following performance characteristics:

- Detected materials include conductors, insulators, plastics, glass, ceramics, oils and greases, water, and all materials with a high moisture content or dielectric constant greater than 1.2.
- Detection of stationary and moving objects.
- Preferred target is a flat, smooth object.
- Operation at low energy levels so that it does not create radio frequency interference or generate heat in the target.

Ultrasonic proximity sensors. Ultrasonic proximity sensors bounce sound waves off a target object and measure the time it takes for the sound waves to return, similar to sonar. The measured time is directly proportional to the distance or height of the target. Highest performance is achieved under the following conditions:

- Ideal target objects have a flat, smooth surface. Sensing distance is reduced when the object is rounded or uneven.
- Objects must be inside the ultrasonic pulse cone, which is 4 degrees or less from the center axis.
- Soft materials, such as foam or fabric, are difficult to detect because they don't reflect the sound waves adequately.
- Reflective surfaces must be positioned to reflect the ultrasonic waves back to the receiver.
- Object temperatures must be less than 100° C.

Any environmental conditions that deaden sound, such as disturbances in the air, can reduce the effectiveness of the sensor.

Proximity sensor applications. Inductive proximity sensor applications are shown in Figure 2-24. In Figure 2-24(a), (b), (c), and (e),

FIGURE 2-24: Inductive sensor applications.

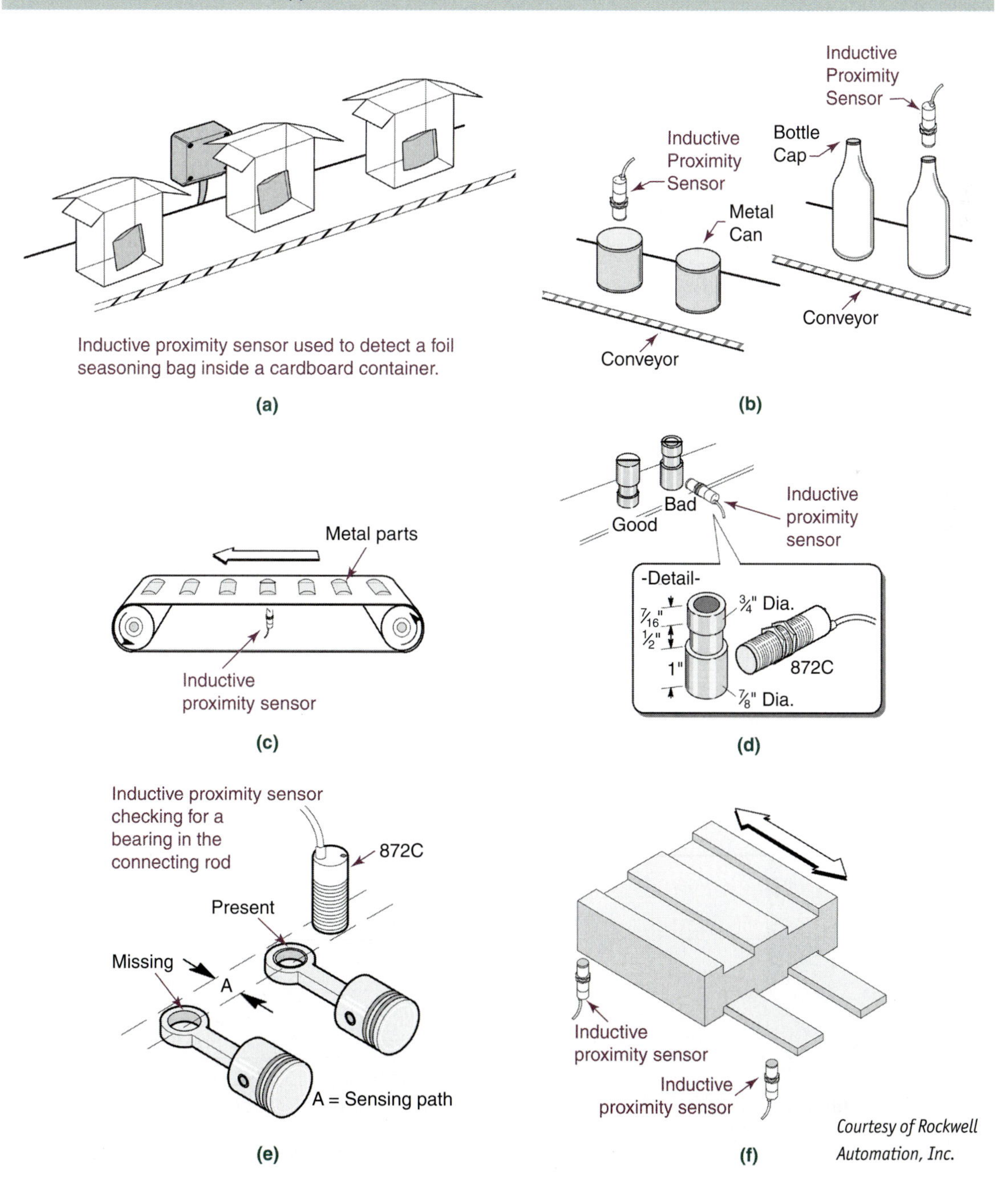

Courtesy of Rockwell Automation, Inc.

the sensors are used to check for the presence of a part in manufacturing. In Figure 2-24(d) the sensor detects parts that are not oriented properly, and in Figure 2-24(f) the limits of travel of the surface grinding plate are detected. Study the applications until you understand how these sensors are being used.

Capacitive and ultrasonic proximity sensor applications are illustrated in Figure 2-25. The first three applications, Figure 2-25(a) through (c), use a capacitive proximity sensor, while the last two, Figure 2-25(d) and (e), use an ultrasonic sensor. There are three level-detection applications illustrated in Figure 2-25(a), (b), and (d). In the first two, capacitive sensors are used with the sensor placed through the wall of the bin to sense the granular material and looking through the bin wall in the liquid application. To sense the liquid

FIGURE 2-25: Capacitive and ultrasonic proximity sensor applications.

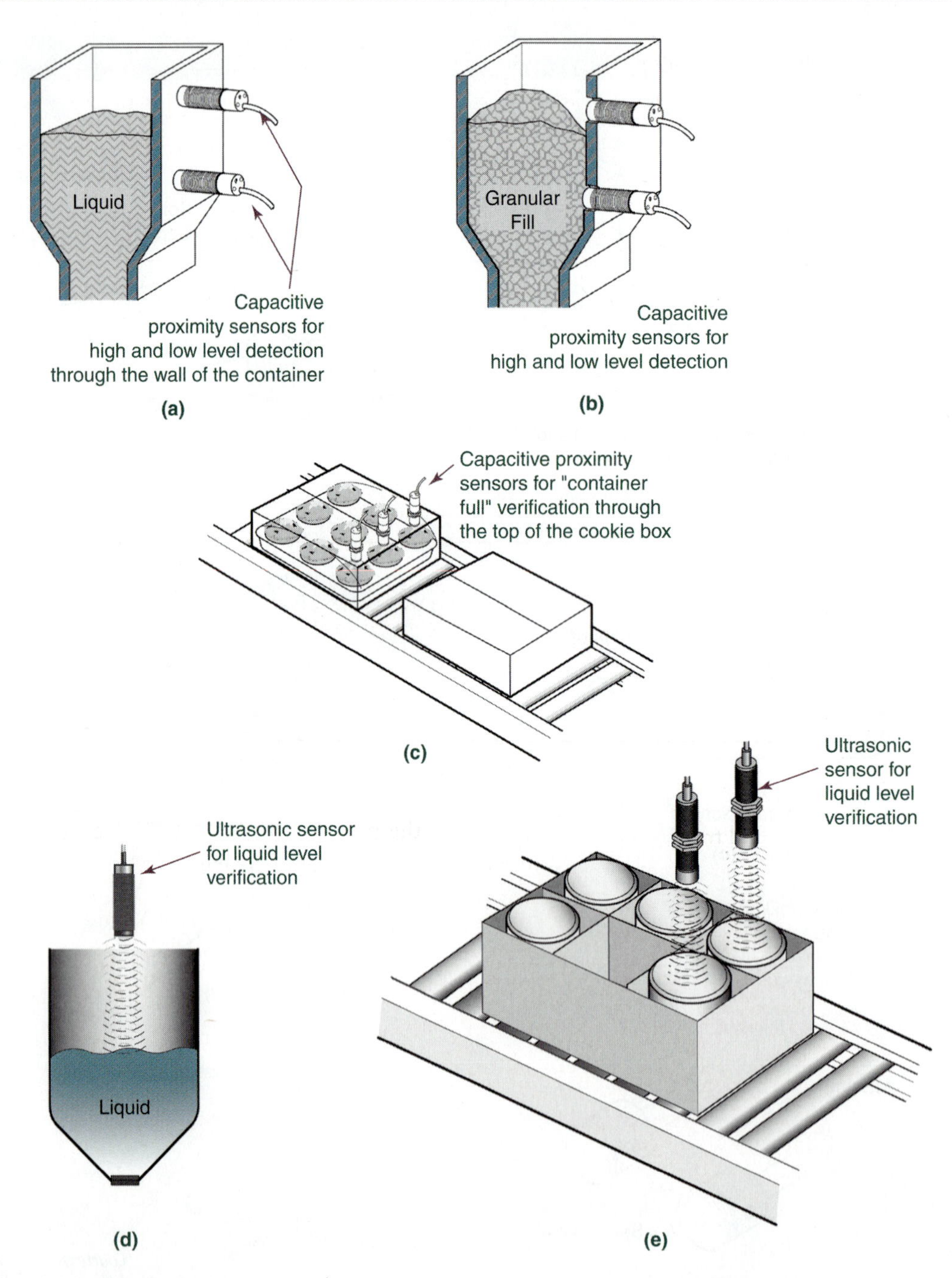

Courtesy of Rockwell Automation, Inc.

through the wall the liquid must have a significantly higher dielectric constant than the material used for the bin, and the bin wall must be thin. In the third level-detection application, an ultrasonic sensor is used to detect the level of the liquid in the tank. In the last two applications, Figure 2-25(c) and (e), sensors are use to verify the presence of products. In the first, three capacitive sensors verify that cookies are packaged in the box. Note that the sensors look through the box lid to check for the cookies. In the second application ultrasonic sensors, set for discrete mode operation, determine that the box is properly filled with cans. Study the applications until you understand how these sensors are being used.

2-4-2 Photoelectric Sensors

Photoelectric sensors are used in many applications and industries to provide accurate detection of objects without physical contact. Two components present in all photoelectric sensors are *a light source* and a *receiver* used to detect the presence of the light source. In its most basic form, a photoelectric sensor is just like a limit switch, where a beam of light replaces the limit switch's mechanical actuator. Photoelectric devices sense the presence of an object or part when the part either breaks a light beam or reflects a beam of light to a receiver. The change in light could be the result of the presence or absence of the object, or as the result in a change of the size, shape, reflectivity, or color of an object. Photoelectric sensors operate over distances from 5mm (0.2 in) to over 250 m (820 ft). Some laser devices can operate at ranges of 304.5 m (1000 ft).

Understanding the operation of photo sensors is the first step in using and troubleshooting them effectively. Photoelectric sensors are comprised of some or all of the following:

FIGURE 2-26: LED photoelectric sensors.

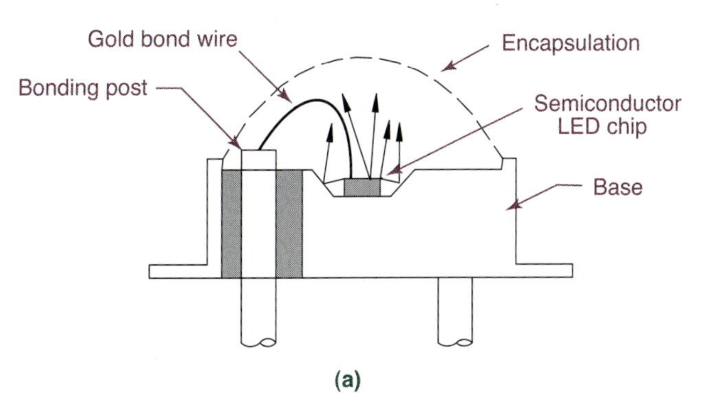

(a)

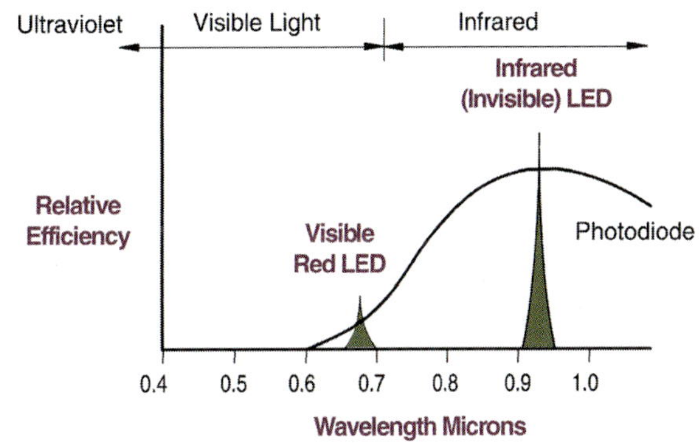

The invisible (infrared) LED is a spectral match for this silicon phototransistor and has much greater efficiency than a visible (red) LED.

(b)

Courtesy of Rockwell Automation, Inc.

- **Light source:** typically a light emitting diode (LED). The LED configuration used in photo sensors is illustrated in Figure 2-26, showing that infrared generates the most light and the least heat of any LED color. In many applications, a visible red beam of light is used, but other colors, such as visible red, blue, and yellow, are used in special applications where specific colors or color contrasts are important. When long distances between the source and receiver are necessary, sensors use lasers as a light source. In those applications distances of 304.5 m (1000 ft) are possible. The laser is usually a red color so that alignment between the light source and receiver is enhanced.
- **Light detector:** the component used to detect the presence of the light source. It produces a change in current directly proportional to the amount of light falling on the light detector. Photodiodes or phototransistors are the robust solid-state components that are most often used for the light detector. The wavelengths of the light for the source and receiver are often matched to improve sensing efficiency.
- **Lenses:** used with LED light sources and photo detectors to narrow or focus the light beam area (Figure 2-27) and increase the sensor's range. When the source and receiver beams or fields of view are narrow, alignment

FIGURE 2-27: Photoelectric sensor lens.

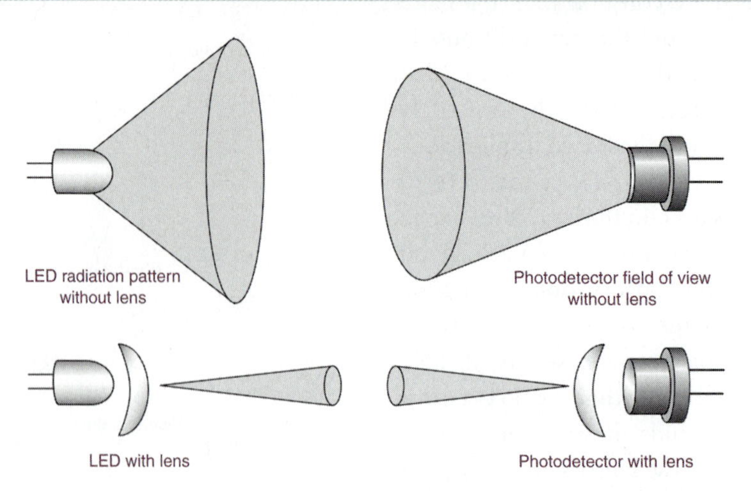

Courtesy of Rockwell Automation, Inc.

FIGURE 2-28: Comparison of photoelectric sensor output devices.

Output Type	Strengths	Weaknesses
Electromechanical Relay *AC or DC switching*	• Output is electrically isolated from supply power • Easy series and/or parallel connection of sensor outputs • High switching current	• No short circuit protection possible • Finite relay life
FET *AC or DC switching*	• Very low leakage current • Fast switching speed	• Low output current
Power MOSFET *AC or DC switching*	• Very low leakage current • Fast switching speed	• Moderately high output current
TRIAC *AC switching only*	• High output current	• Relatively high leakage current • Slow output switching
NPN or PNP Transistor *DC switching only*	• Very low leakage current • Fast switching speed	• No AC switching

Courtesy of Rockwell Automation, Inc.

can be difficult. However, when a wider field of view is required, then the sensor has a shorter overall range.

- **Logic circuit:** modulates the LED light source, amplifies the signal from the light detector, and determines if the output state should be changed.
- **Output device:** signals a PLC that a change has occurred. Figure 2-28 provides a comparison between the various output devices; study the table until you can name the options.

Photoelectric sensor operating modes. Photoelectric sensors operate in the following different sensing modes:

- **Through beam** or transmitted beam has the light source and receiver packaged in separate housings.
- **Retroreflective** or reflex has both the light source and receiver in one housing.
- **Polarized retroreflective** has polarizing filters in front of the light source and receiver that are 90 degrees out of phase with each other.

- **Diffused** or photo proximity has the light source and receiver in the same package like the retroreflective mode; however, it uses the detected part to reflect the light beam back to the receiver.

In the *through beam mode* the light source and receiver are packaged in separate housings. These two units are positioned opposite each other so that the light from the source (S) shines directly on the receiver (R) as shown in Figure 2-29. Targets must break or block the beam between the light source and the receiver. Through beam sensors provide the longest sensing distances and the highest level of operating margin. In photoelectric sensors the term *margin* is used to describe the light level reaching the receiver. With a margin of one, the receiver has the minimum level of light needed to detect that a beam is present. A graph of distance versus margin is included in the sensor figures. When an application dictates the use of photoelectric sensors, a margin greater than one is used.

In the *retroreflective mode* as shown in Figure 2-30 both the light source and receiver are packaged in a single housing. The light beam

FIGURE 2-29: Photoelectric sensor—through beam mode.

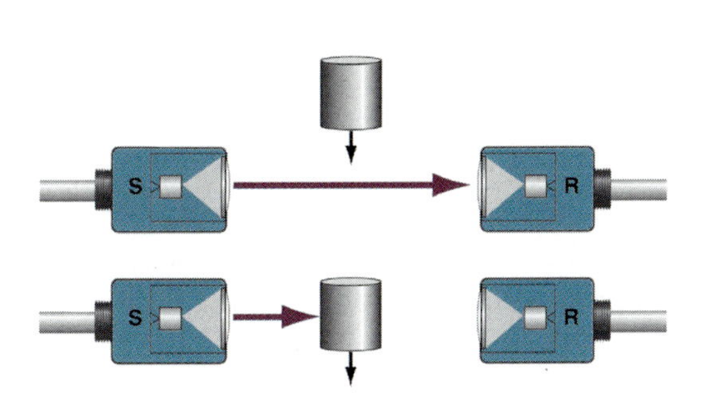

Courtesy of Rockwell Automation, Inc.

FIGURE 2-30: Photoelectric sensor—retroreflective mode.

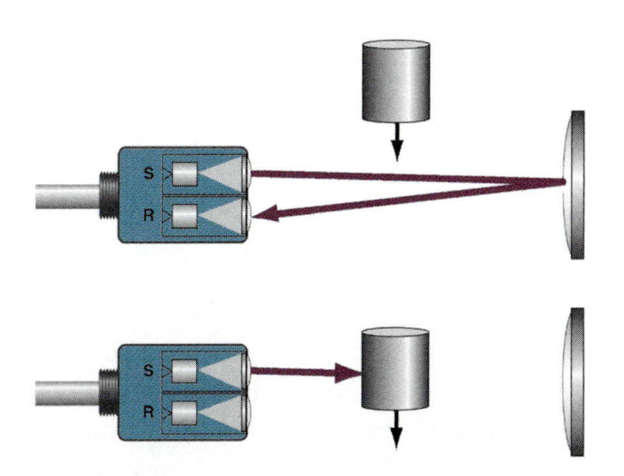

Courtesy of Rockwell Automation, Inc.

emitted by the light source is reflected by a special reflective device, called a *target*, back to the receiver where it triggers a change in the output device. The object is detected when it breaks this light beam and the output device changes states. Special reflectors or reflective tapes with special glass beads or corner-cube reflective surfaces are used for the retroreflective target; as a result, the target does not have to be aligned perfectly perpendicular to the sensor. Misalignment of a reflector or reflective tape of up to 15 degrees will not significantly reduce the operating margin of the sensing system.

In the *polarized retroreflective mode* the sensors have polarizing filters in front of the light source and receiver that are 90 degrees out of phase with each other. Study the illustration in Figure 2-31 until the filter orientation is clear, noting that the emitted light from the source has a horizontal polarization, and the only light that can reach the receiver *must* have a vertical polarization. When a highly polished or shiny object breaks the beam, the horizontally polarized light from the emitter is reflected back toward the receiver. The receiver's vertical filter blocks the light from reaching the receiver, so the receiver is not triggered. As a result, the sensor's output device is off. When no object is present, the emitted horizontally polarized light is reflected back by the special target in a nonpolarized condition. Some of the nonpolarized light passes through the vertical polarizing filter at the receiver and the receiver is activated. This causes the output device to turn on.

FIGURE 2-31: Photoelectric sensor—polarized retroreflective mode.

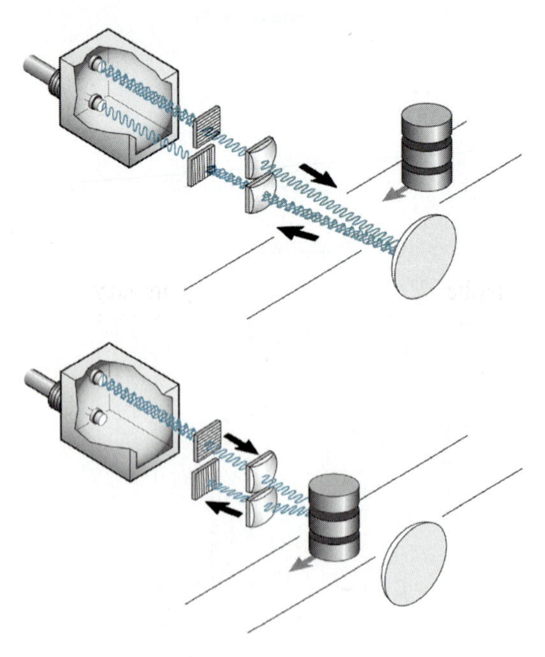

Courtesy of Rockwell Automation, Inc.

Access to both sides of the object to be sensed is required for transmitted beam and retroreflective sensors to establish a beam that can be interrupted by the passing object. It is often not possible to obtain access on both sides of an object. In these applications, a *diffused mode* is the only photoelectric sensor choice. Diffused mode sensors have the light source and receiver in the same package like the retroreflective mode; however, they do not use a special target to reflect the light beam back to the receiver. The object to be sensed becomes the target that reflects the light beam from the emitter back to the receiver. Light from the source striking the surface of the sensed object is scattered or reflected at all angles. The small portion reflected back to the sensor is detected by the receiver and used to trigger the output device. This is called a *standard diffused mode* and is illustrated in Figure 2-32.

Detecting targets positioned close to reflective backgrounds can be particularly challenging, and may require a special type of diffused mode sensor. There are five additional types of diffused mode sensors offered by vendors for automation control. Other types include *sharp cutoff diffused, fixed focus diffused, wide angle diffused, background suppression diffused,* and *distance measurement diffused.*

While many applications can be handled by any of these sensing modes, each offers specific strengths and weaknesses. The typical application plus advantages and cautions are summarized in Figure 2-33. Note that there are transmitted beam (through beam), two types of retroreflective, five types of diffused, and fiber optics sensors that operate in all modes. In some photoelectric sen-

FIGURE 2-32: Photoelectric sensor—standard diffused mode.

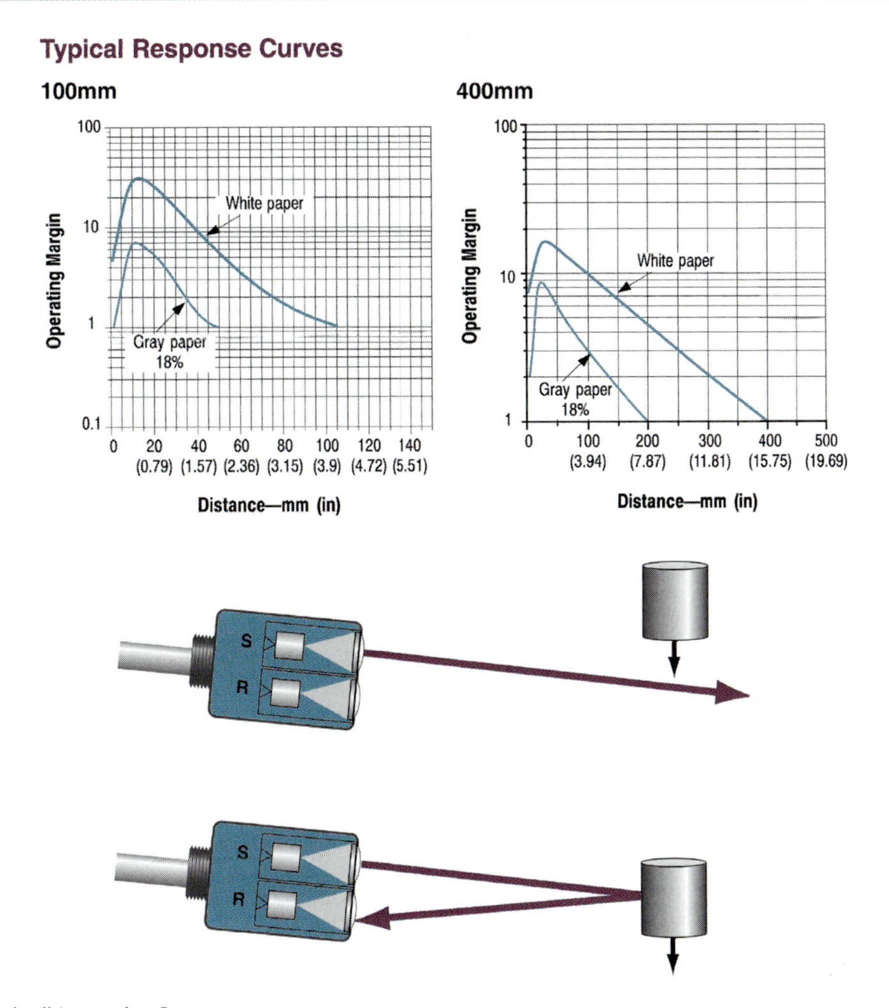

Courtesy of Rockwell Automation, Inc.

sor applications, specialized lighting is often required where imaging of an object is difficult.

Photoelectric sensor functions. All the photoelectric sensors function in the same basic manner, with a light source and the light detector; however, the electronics associated with the sensors have the following operational options:

- **Light ON–Dark ON:** Light ON occurs when the output becomes high or active when the source illuminates the receiver. Dark ON works in the opposite mode. These modes are selected using either a switch or through output terminal selection.
- **Output Control:** Sensors are available in a variety of output circuits which match the interfacing needs of the automation controllers. The modes of operation vary widely as well. The table in Figure 2-33 lists the photoelectric sensor modes, their typical application area, and their advantages and disadvantages.
- **On and Off Delay Timers:** Delay timers provide a variable delay to a change in the output after the input stimulus triggers an output change. The delay occurs as the sensor is changing from on to off so that the off condition is delayed by the selectable value.

FIGURE 2-33: Comparison of photoelectric sensor modes.

Photoelectric Sensing Modes Advantages and Cautions

Sensing Mode	Applications	Advantages	Cautions
Transmitted Beam	General purpose sensing Parts counting	• High margin for contaminated environments • Longest sensing distances • Not affected by second surface reflections • Probably most reliable when you have highly reflective objects	• More expensive because separate light source and receiver required, more costly wiring • Alignment important • Avoid detecting objects of clear material
Retroreflective	General purpose sensing	• Moderate sensing distances • Less expensive than transmitted beam because of simpler wiring • Ease of alignment	• Shorter sensing distance than transmitted beam • Less margin than transmitted beam • May detect reflections from shiny objects (use polarized instead)
Polarized Retroreflective	General purpose sensing of shiny objects	• Ignores first surface reflections • Uses visible red beam for ease of alignment	• Shorter sensing distance than standard retroreflective • May see second surface reflections
Standard Diffuse	Applications where both sides of the object cannot be accessed	• Access to both sides of the object not required • No reflector needed • Ease of alignment	• Can be difficult to apply if the background behind the object is sufficiently reflective and close to the object
Sharp Cutoff Diffuse	Short-range detection of objects with the need to ignore backgrounds that are close to the object	• Access to both sides of the object not required • Provides some protection against sensing of close backgrounds • Detects objects regardless of color within specified distance	• Only useful for very short distance sensing • Not used with backgrounds close to object
Background Suppression Diffuse	General purpose sensing Areas where you need to ignore backgrounds that are close to the object	• Access to both sides of the target not required • Ignores backgrounds beyond rated sensing distance regardless of reflectivity • Detects objects regardless of color at specified distance	• More expensive than other types of diffuse sensors • Limited maximum sensing distance
Fixed Focus Diffuse	Detection of small targets Detects objects at a specific distance from sensor Detection of color marks	• Accurate detection of small objects in a specific location	• Very short distance sensing • Not suitable for general purpose sensing • Object must be accurately positioned
Wide Angle Diffuse	Detection of objects not accurately positioned Detection of very fine threads over a broad area	• Good at ignoring background reflections • Detects objects that are not accurately positioned • No reflector needed	• Short distance sensing
Fiber Optics	Allows photoelectric sensing in areas where a sensor cannot be mounted because of size or environment considerations	• Glass fiber optic cables available for high ambient temperature applications • Shock and vibration resistant • Plastic fiber optic cables can be used in areas where continuous movement is required • Insert in limited space • Noise immunity • Corrosive areas placement	• More expensive than lensed sensors • Short distance sensing

Courtesy of Rockwell Automation, Inc.

Photoelectric sensor applications. The applications in Figure 2-34 illustrate many of the photoelectric sensor operating modes described in the table in Figure 2-33. The sensors used in Figure 2-34(a) and (f) are operating in the diffused mode. In Figure 2-34(a) *fiber optic units* extend the diffused sensor some distance from the amplifier out to the sensing point in the application. Here the presence of part of a product and an installed cork are checked as the bottle passes a point on the conveyor. In Figure 2-34(f) a self-contained sensor is used with two different types of transparent objects. The application in Figure 2-34(b) uses a retroreflective mode sensor to determine when the paper rolls have reached the desired diameter. The applications in Figure 2-34(c) and (d) both use the through beam mode with a light source in one leg of the sensor and the receiver in the other leg. The sensor in Figure 2-34(d) uses a sensor controller from the sensor manufacturer to provide process control based on the sensor response. The final application in Figure 2-34(e) uses two sensors to measure the difference in the height of two surfaces on a part. Note that a standard discrete diffused sensor is used to indicate when the part is in position and then the sensor controller records the two distance measurements from the sensors.

At this point the type of non-contact sensor (proximity or photoelectric) and the sensing mode (inductive, capacitive, through beam, retroreflective beam, or diffused beam) has been discussed. So let's take a look at selecting the sensor type and mode of operation based on some typical applications.

FIGURE 2-34: Photoelectric sensor applications.

Detecting presence of cork

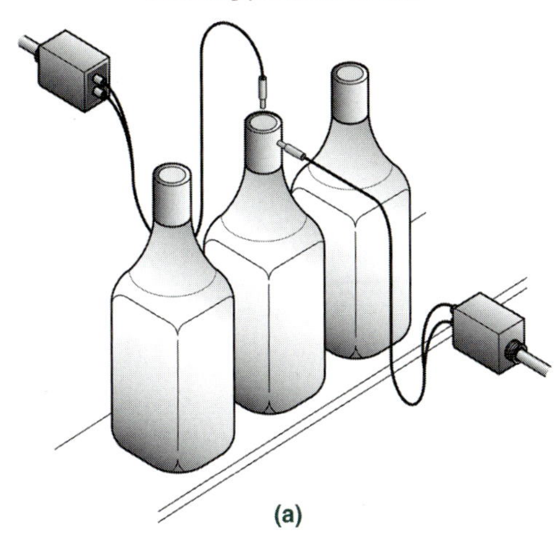

(a)

Detecting diameter of paper roll

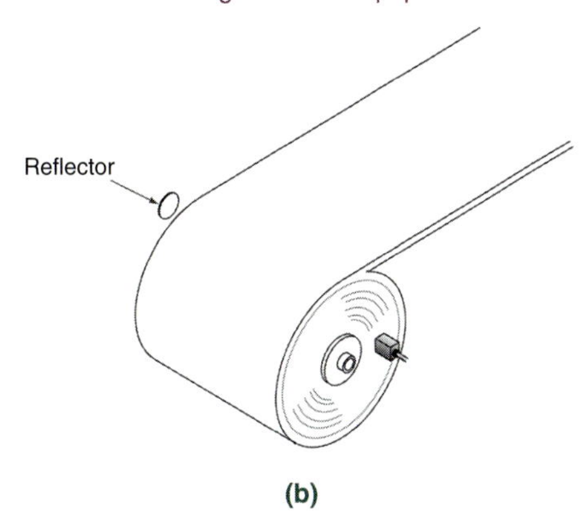

(b)

Detecting edge of material to activate cutter

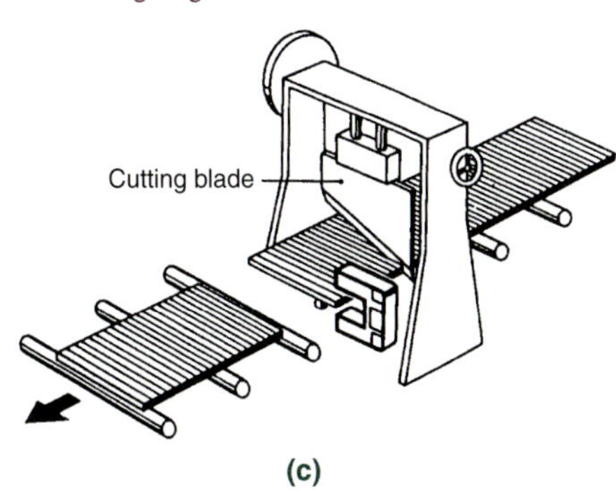
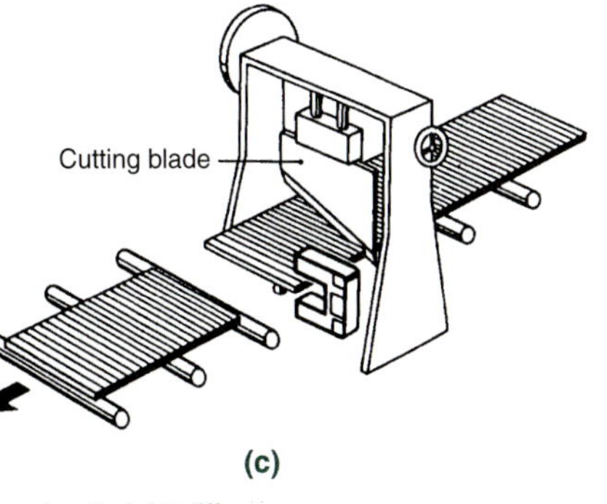

(c)

Detecting marks on packaging film to adjust roller and cutter speed

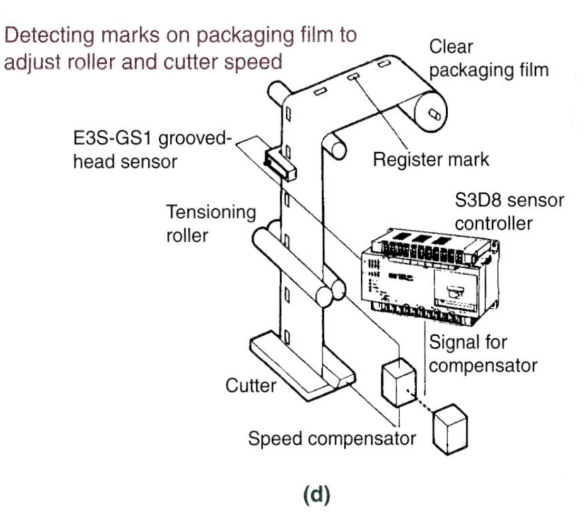

(d)

Measuring height difference

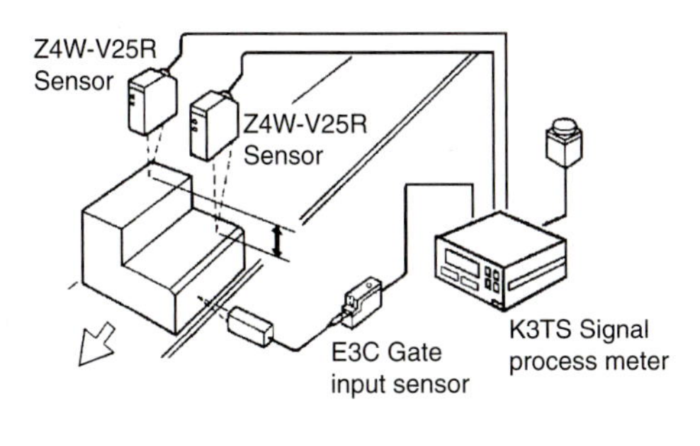

(e)

Sensing of transparent objects

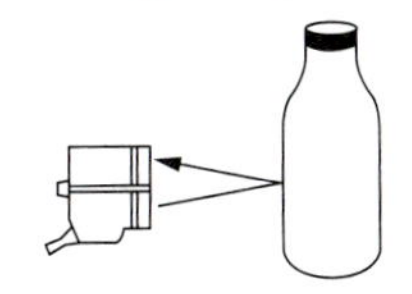

Typical examples
(1) Sensing of transparent or translucent objects.
(2) Sensing of transparent greases, film, or plastic plates.
(3) Sensing of the liquid level.

Sensing of objects through a transparent cover

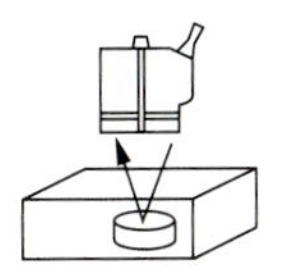

Typical examples
(1) Sensing of the contents in a transparent case.
(2) Sensing of the position of meter pointer.

(f)

Courtesy of Rockwell Automation, Inc.

EXAMPLE 2-6

Select a sensor type and sensing method or mode for each of the following application situations.

a. Count boxes (18 inches square) on a 24-inch conveyor belt with complete access to both sides of the conveyor.
b. Count shiny thermos bottles moving in a production machine. Access to both sides of the conveyor is permitted, but sensor must be 18 inches from parts.
c. Detect small metal screws coming from a bowl feeder on a pair of metal rails.
d. Detect a small black relay on a printed circuit board with a highly reflective surface. Sensor must be located 6 inches above the relay and board.
e. The level of shampoo must be verified as the clear plastic bottles move down a conveyor. The only sensor location is 0.5 inches from one side of the passing bottle.
f. Suggest an alternate sensor for the application in Figure 2-34(a).
g. Detect the leading edge of a plate moving down a conveyor. Access is available only from the top.
h. Detect shiny round plastic cans at the input to a labeling machine. The sensor must be mounted 1.5 inches from the can, and a reflector could be mounted on the other side of the object.
i. Detect the presence of a metal slug at the input to a forging machine. There is a single side access restriction, but the sensor can be mounted as close to the slug as necessary.

SOLUTION

a. Use a through or transmitted beam since it is the first choice when access to both sides of the object is available.
b. The transmitted beam is not used since reflections could reduce reliability. Use a polarized retroreflective for reliable detection.
c. The metal rails prevent the use of an inductive proximity. Use a fixed focus diffused with the focus placed on the screws. Wide angle diffused cannot be used because of all the other hardware that would be in the field of view.
d. The main problem is the highly reflective background, nonmetallic object, single-side access, and the 6-inch sensing distance. Distance is too long for capacitive, so background suppression diffused must be used.
e. The clear plastic bottle would permit either a capacitive or diffused sensor to be used. However, the close sensing distance eliminates the diffused because of the blind zone in a diffused sensor. A second fixed focus diffused would be used to verify a bottle was present by detecting the cap, and then the level could be verified.
f. Two capacitive proximity sensors could be used in this application as an alternate solution.
g. Transmitted beam and fixed focus diffused are the sensors of choice for edge detection. With access from only one side the fixed focus diffused must be used.
h. The application could support a retroreflective sensor, except that the close mounting distance would place the cans in the blind zone. Therefore a wide angle diffused is the best sensor for this application.
i. An inductive proximity is the best choice for this application.

2-5 INTERFACING INPUT FIELD DEVICES

Interfacing input field devices to the PLC involves the physical connection of the device, the powering of the device, the sizing of the wiring to ensure adequate current carrying capability, and the routing of the wiring to minimize electrical interference and safety. The term *field* is used to designate that the device is not part of the PLC. The PLC inputs are generally configured into two types—inputs share a common return or input pairs are totally isolated. Both types are typically otpo-isolated from the input field devices.

2-5-1 Powering Input Field Devices

Sensors typically have the following power voltage ratings: 10–30 VDC, 20–130 VAC, 90–250 VAC, and 20–250 VAC/DC. These values bracket the supply voltages available on the factory floor. AC sensors power the load and sensor from the same power source, but most DC sensors require a separate DC supply that isolates the sensor electronics from the AC power line.

Two-wire and three-wire sensor outputs. Sensors outputs are commonly divided into two categories: two-wire and three-wire. Two-wire devices have an output switch that is placed in series with the PLC and the power source. Figure 2-35(a) illustrates the two-wire switch. Note that when the switch is closed, the current flows into the PLC input, through the PLC electronics, and returns to the power source via the PLC common. Figure 2-35(b) illustrates the three-wire switch. The three-wire switch operates as a two-wire switch but has a third wire that is connected to the PLC common. Generally, the three-wire switch is only for electronic sensors that have an output wire plus two power wires.

FIGURE 2-35: Two-wire and three-wire sensors.

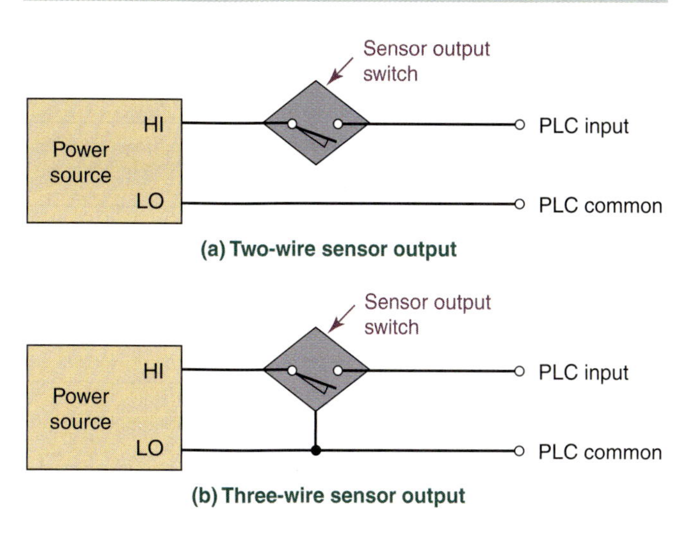

2-5-2 Input Wiring

The PLC input is both a *physical interface* for the connection of wires and an *electrical/data interface* to determine the on/off state or level of the signal from the attached field device. PLC input circuitry (covered in Chapter 1) varies with the manufacturers of the equipment, but generally the input circuitry signal conditions the input voltage before it's fed into an opto-isolator integrated circuit. The purpose of the opto-isolator is to isolate the incoming voltage and grounds from the rest of the PLC circuitry. Figure 2-36 depicts PLC input opto-isolators for a DC input and for an AC input. The series resistor limits the current into the PLC input from the switch or sensor connected to the input. For the DC unit, the input polarity must be observed for the opto-isolator to turn on. For the AC unit, the opto-isolator will turn on with either polarity. This AC unit is typically designated as an AC/DC-type of PLC input module and is used for both AC and DC inputs because the input polarity does not matter. However, caution should be taken when dealing with the hot and neutral lines of an AC input. In general a PLC input module has either all inputs isolated from each other with no common input connection or groups of inputs share a common connection. For example, inputs are grouped in fours so every four inputs has a common ground. Figure 2-37 illustrates both these input module configurations. Figure 2-37(a) shows PLC input circuits connected to a common terminal, and Figure 2-37(b) shows PLC totally isolated input circuits.

FIGURE 2-36: PLC opto-isolator inputs.

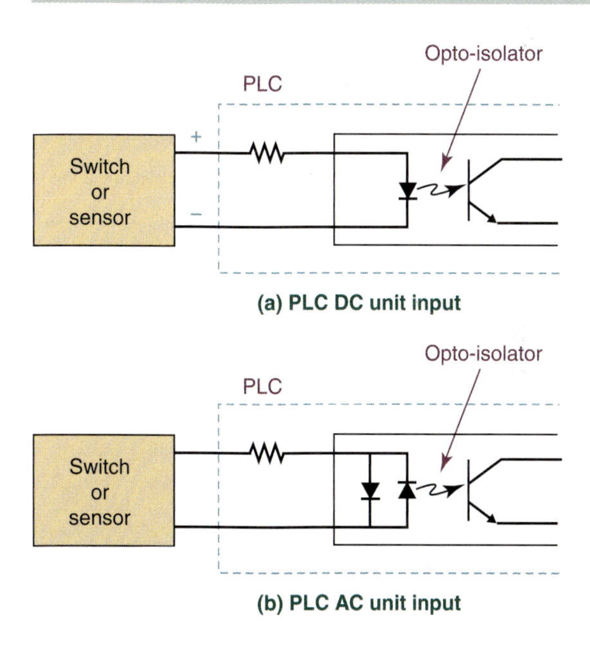

The input wiring for a sensor and PLC input is illustrated in Figure 2-38. In this interface, current flows out of the PLC input (a sourcing input), and into the sensor output terminal (a sinking output). Note the direction of current flow as specified by the arrow. The PLC input and the

FIGURE 2-37: PLC inputs with common terminal configurations.

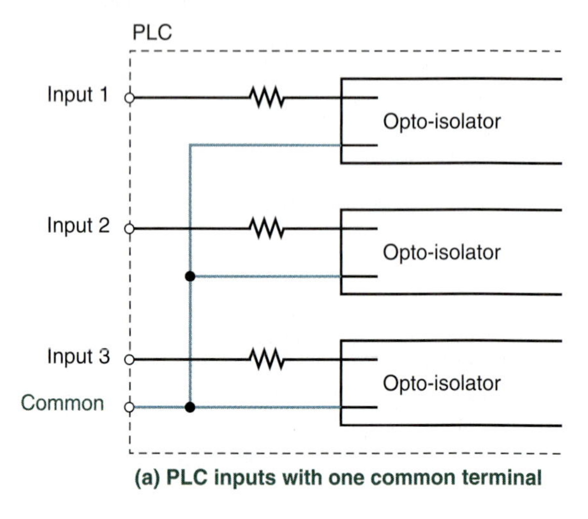

(a) PLC inputs with one common terminal

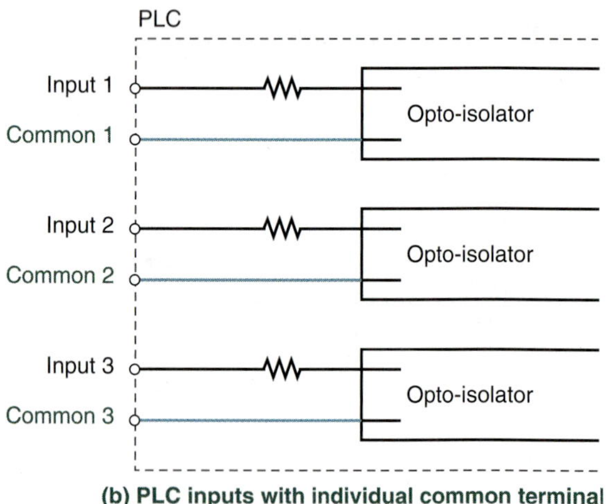

(b) PLC inputs with individual common terminal

sensor output can be either current sourcing (current flows out from from the terminal when active) or current sinking (current flows into the terminal when active). Let's examine this important concept of current sinking and current sourcing.

2-5-3 Current Sinking and Current Sourcing Devices

When wiring a sensor to the PLC input, these rules must be followed:

- Current sinking sensors must be matched to current sourcing PLC inputs.
- Current sourcing sensors must be matched with current sinking PLC inputs.

Reread the rules until you have them memorized. If you interface a sensor to a PLC and do not follow the rule, the interface will not function, and damage may result for the sensor and/or the PLC input circuitry.

When applied to DC type sensors, an NPN is a sinking transistor, which is shown in Figure 2-38, and a PNP is used for sourcing. Referring to Figure 2-38, the NPN (sinking) sensor output has the emitter connected to the return of a power supply and the collector to the PLC input; thus current flows out of the PLC and into the sensor.

FIGURE 2-38: Sensor and PLC input interface.

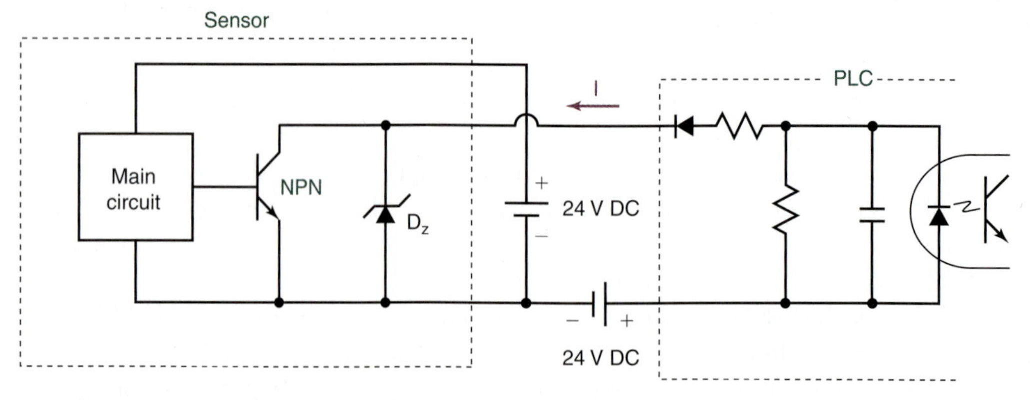

Rehg and Sartori, Industrial Electronics, *1st Edition, © 2006. Reprinted by permission of Pearson Education, Inc., Upper Saddle River, NJ.*

When a PNP (sourcing) sensor output is used, the emitter is connected to a positive voltage of a power supply and the collector to the PLC input; thus current flows out of the sensor and into the PLC. In both cases, when the sensor is off, no current flows through the PLC. Read Appendix B for additional PLC module interface circuitry information.

2-6 ELECTROMAGNETIC OUTPUT ACTUATORS

An *electromagnetic actuator* is any device that contains a magnetic winding or coil that converts electrical energy into mechanical movement. The common types of these actuators driven by a PLC are solenoids, relays, contactors, and motor starters. Discussions of these types of actuators and their electrical symbols are found in the following subsections.

2-6-1 Solenoid-Controlled Devices

Before we discuss devices let's first look at solenoids. *Solenoids* convert electrical energy directly into linear mechanical motion. The basic DC solenoid illustrated in Figure 2-39 has two components: a *coil of wire* and an *iron core plunger*. In the de-energized case, Figure 2-39(a), the only force acting on the iron core is the spring pushing it out of the coil. When the switch is closed, Figure 2-39(b), current through the coil creates an electromagnet with a magnetic flux that flows from the top of the coil into the bottom of the coil. This magnetic field pulls the iron core into the coil because the magnetic force is greater than the spring force. The stroke is the difference between the de-energized and energized positions of the core.

AC-powered solenoids are more common in industry and include a third component called a *frame*. Figure 2-40(a) illustrates a cutaway view for each of three time intervals of solenoid operation; the solenoid schematic symbol is shown in Figure 2-40(b). Study the cutaway of the AC solenoid until you recognize all the parts and the three time intervals: (1) just after the coil is energized, (2) as the plunger, called an *armature*, is closing, and (3) fully energized when the armature is pulled in.

A very common solenoid-controlled device is the valve. A *solenoid valve* is an electromechanical device that is used to control the flow of air or fluids such as water, inert gas, light oil, and refrigerants. Solenoid valves are the simplest output device and are used extensively in industrial

FIGURE 2-39: Basic solenoid.

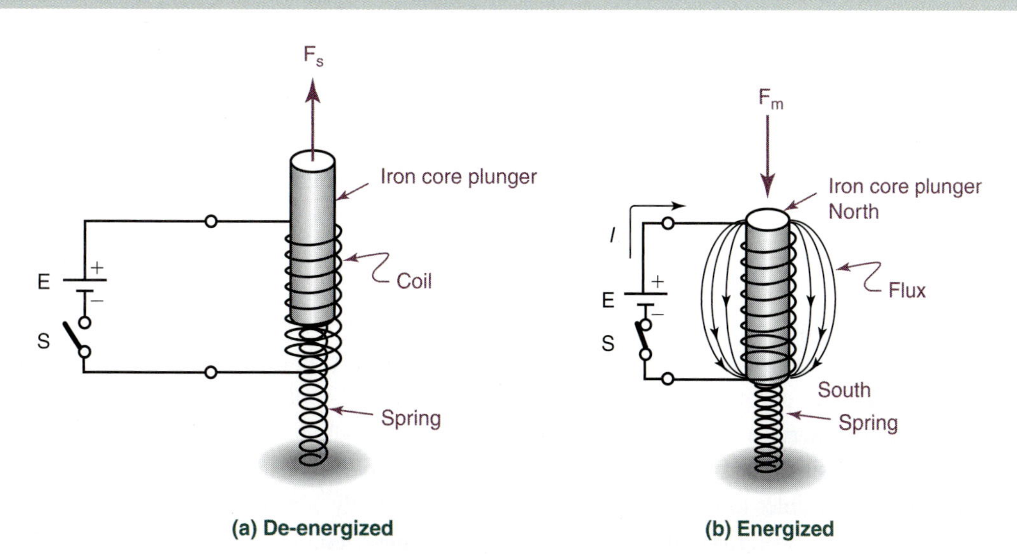

Rehg and Sartori, Industrial Electronics, *1st Edition, © 2006. Reprinted by permission of Pearson Education, Inc., Upper Saddle River, NJ.*

FIGURE 2-40: AC solenoid.

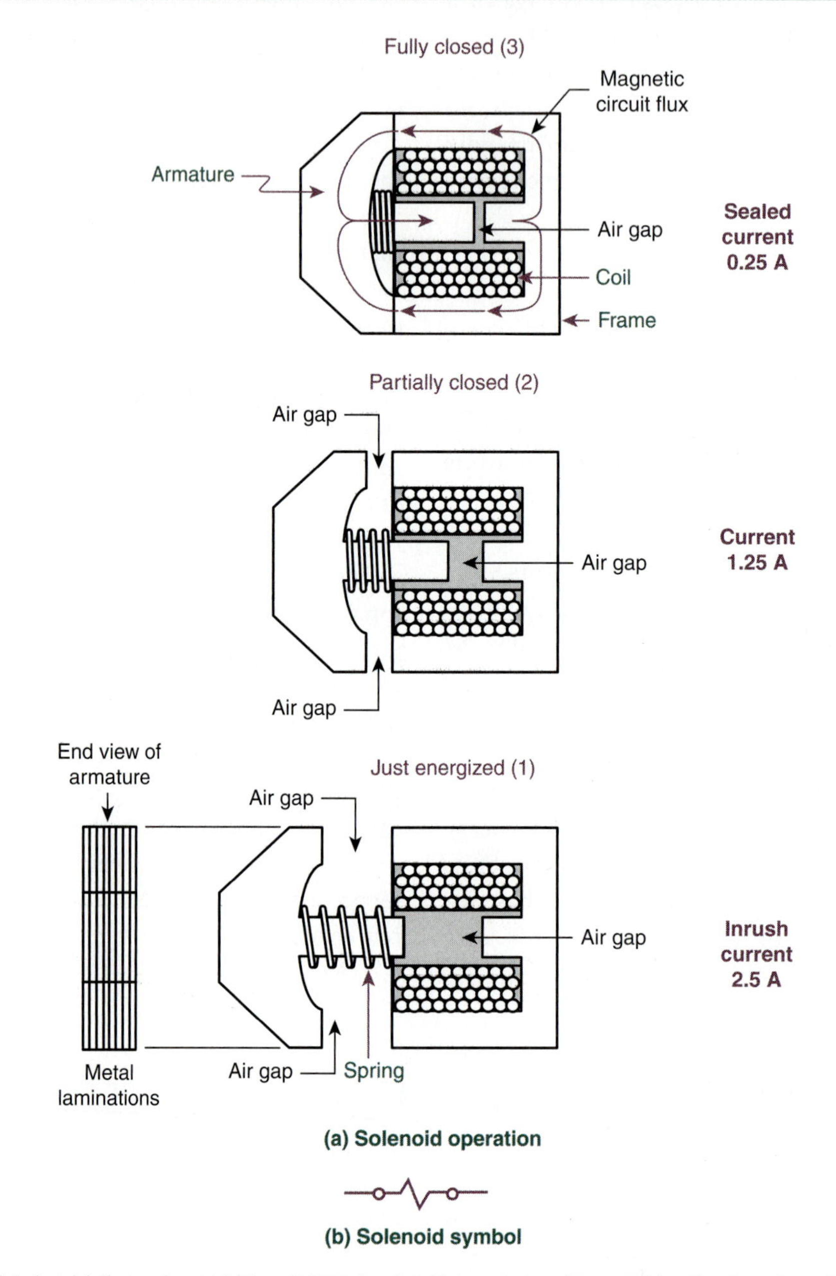

(a) Solenoid operation

(b) Solenoid symbol

Rehg and Sartori, Industrial Electronics, *1st Edition, © 2006. Reprinted by permission of Pearson Education, Inc., Upper Saddle River, NJ.*

control systems. Figure 2-41 illustrates the basic construction of a normally closed, two-position solenoid valve. A cutaway view and an exploded view are depicted. The spring that is wrapped around the plunger exerts a force on the pilot valve, holding it in position and allowing no flow through the valve body, inlet to outlet. This is the de-energized condition. When the coil is energized, a magnetic field is generated, moving the plunger, pilot valve, and piston assembly, allowing flow through the valve body. The magnetic field overcomes the spring force, which is pushing in the opposite direction of the magnetic field. The most common fluid controlled by this valve type is oil in a hydraulic cylinder or air in a pneumatic cylinder. Also available in the marketplace are solenoid valves in the normally open position, which close when the coil is energized.

Solenoid valves are generally limited to small-diameter pipes and are not applicable for direct

FIGURE 2-41: Solenoid valve.

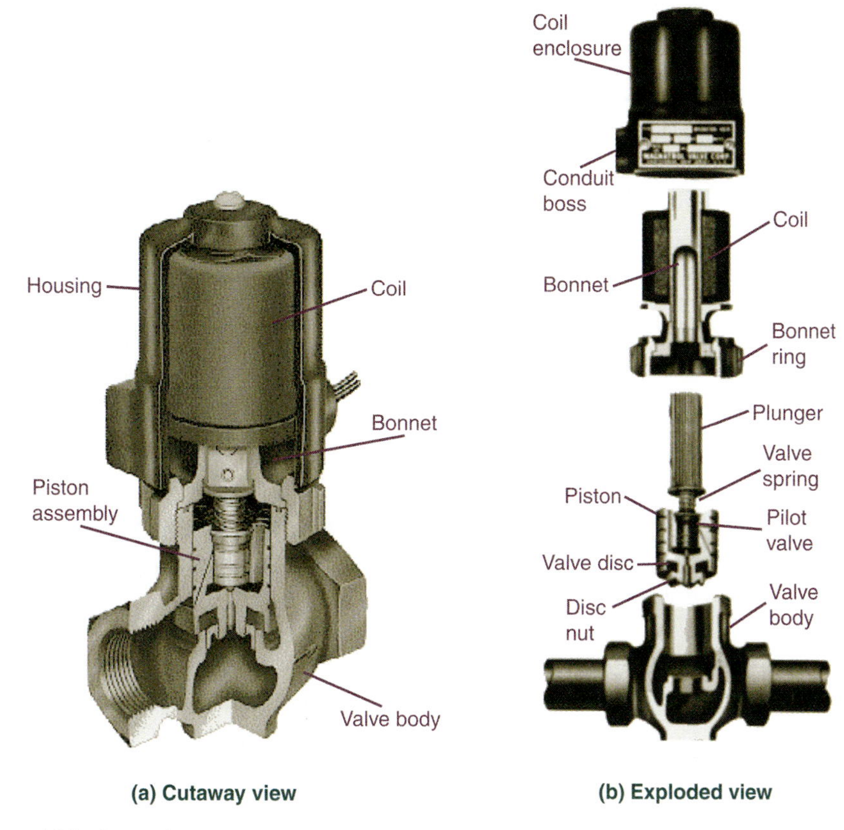

Courtesy of Magnatrol Valve Corporation.

control of fluid through large-diameter pipes. However, the solenoid valve is used as a pilot or lead valve to control the flow to a pneumatic-actuated valve, which controls the flow through the large-diameter pipe. Pneumatic- or air-assisted control valves are opened and closed by an electrical signal that is converted to air pressure (3–15 psi), allowing the valve opening to be anywhere between full open and full closed. These valves are mainly used to control the flow of liquid through large-diameter pipes and in explosive atmosphere applications such as chemical processing and spray painting. Figure 2-42 illustrates a pneumatic-assisted control valve—a three-way, two-position smaller solenoid valve controls a larger pneumatic-actuated valve. Note the solenoid valve symbol pictures the de-energized state as indicated by the arrows, indicating that the exhaust is connected to the piping system, and by a T symbol indicating the inlet is closed. When the solenoid is energized, the exhaust is closed and the inlet is connected to the piping system, and air flows to the diaphragm housing. The diaphragm compresses the spring, opening the main valve and allowing air to flow through the pipe. The air pressure in the diaphragm is let out through the exhaust port when the solenoid is de-energized, and the spring inside the diaphragm closes the main valve. This is a discrete application of a smaller valve switching 3 to 15 psi on and off to a larger control valve.

2-6-2 Control Relays

Control relays are remotely operated switches that are one of the oldest control devices still in common use. They are the combination of electromagnets and solenoids with switch contact configurations. There are numerous relay models to satisfy every control requirement. Industrial control relays are shown in Figure 2-43.

Relays have two primary functions: (1) control of a large current and/or voltage with a small electrical signal and (2) isolation of the power used to control the action from the power that must be

FIGURE 2-42: Pneumatic-assisted control valve.

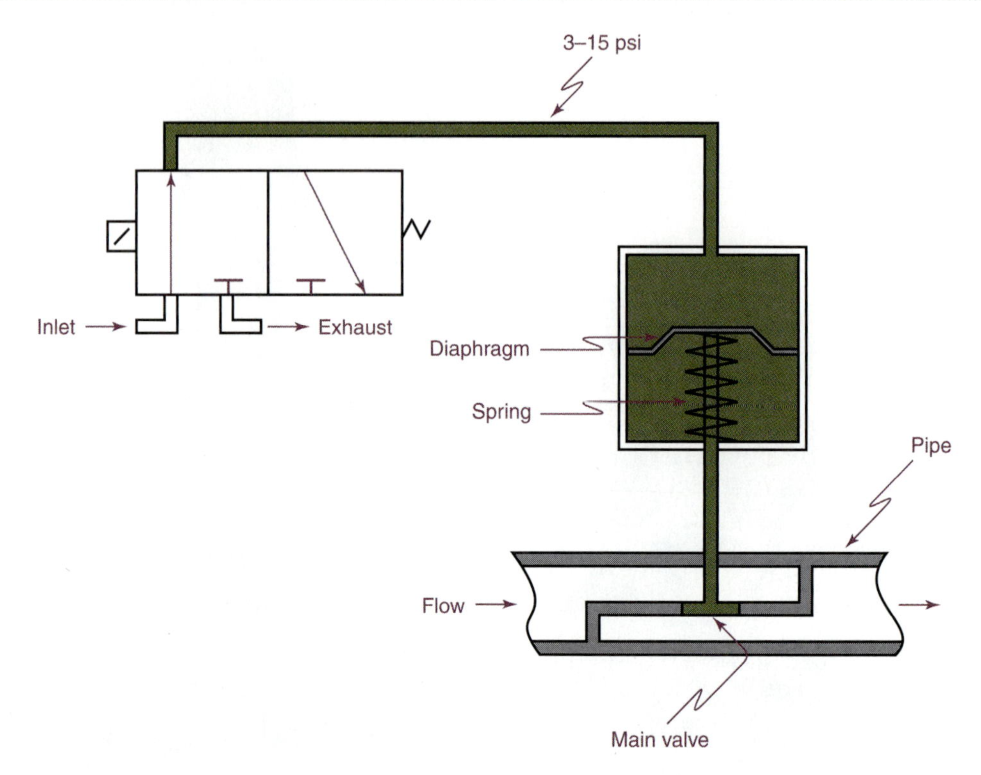

Rehg and Sartori, Industrial Electronics, *1st Edition, © 2006. Reprinted by permission of Pearson Education, Inc., Upper Saddle River, NJ.*

FIGURE 2-43: Industrial control relays.

Rehg and Sartori, Industrial Electronics, *1st Edition, © 2006. Reprinted by permission of Pearson Education, Inc., Upper Saddle River, NJ.*

switched to cause some action. Usually relay coil currents are typical well below 1 amp, whereas contact ratings for industrial relays are at least 10 amps. In effect, a relay acts as a binary (on or off) amplifier. This binary amplifier effect is illustrated by the control circuit example in Figure 2-44. In the example, a relatively small electronic switch located in a PLC at some distance from the relay controls the switching of a 24 VDC source across the coil of the relay located at the site of the load. When the relay coil is energized the contacts switch the large voltage and current to the load. Eliminating the relay and using the electronic switch in the PLC to accomplish this is not possible because of the high AC voltage and current that would have to be routed to and switched by the PLC. The second function is illustrated in the example as well. The control voltage is DC, and controlled voltage is AC with both of the sources and grounds isolated from each other. Study the circuit until the operation and advantage of the relay is clear.

Relay Contact Ratings

The most important consideration when selecting relays, or relay outputs on a PLC, is the *rated voltage and current*.

- **Rated voltage:** the suggested operation voltage for the coil. If the voltage is too low the relay may fail to operate and voltages too high can shorten the relay's life. PLC output modules have a wide voltage range to work with most relay voltage requirements.

- **Rated current:** the maximum current before contact damage, such as welding or melting, occurs.

The contact ratings for a typical industrial control relay are illustrated in Figure 2-45. Note the large variation between allowable DC and AC load currents for approximately the same voltage.

Referring to Figure 2-45, the 120 VAC has the break contact current maximum at 6 amps, whereas at about the same DC voltage the maximum break current is 0.55 amps. The primary reason for this wide variation is the small air gap between open contacts, which is not adequate to break a sustained DC arc from an inductive load. However, this lower limit is adequate for the control logic applications where this type of relay is generally used. Occasionally the required load on relay contacts may be slightly higher than their nominal rating. Current capacity is increased by connecting contacts in parallel, and arc suppression is improved by connecting multiple contacts in series. The parallel contacts must be on the same relay so that the parallel contacts close and open at the same time.

Relay contact configurations. The contact configuration and terminology covered in the first part of the chapter for switches applies equally well to relay contacts. The two primary specifications that transfer to relays are number of poles and contact condition—normally open (NO) or normally closed (NC). A relay armature may actuate more than one set of contacts, so as in switches the relay could be a multiple pole device. Relays typically have one to eight poles. Contacts are normally open, normally closed, or a combination of the two. As with switches, the "normal" state of a relay's contacts is the condition present when

FIGURE 2-44: Relay control circuit.

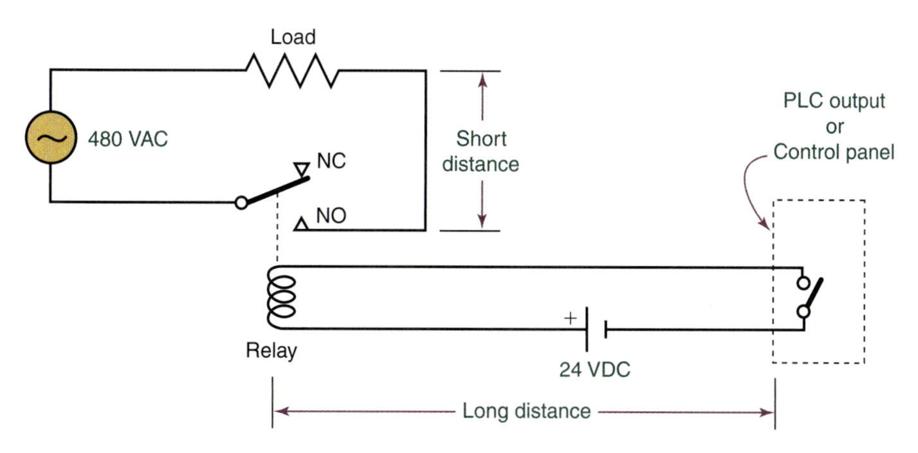

Rehg and Sartori, Industrial Electronics, *1st Edition, © 2006. Reprinted by permission of Pearson Education, Inc., Upper Saddle River, NJ.*

FIGURE 2-45: Control relay contact ratings.

Contact Ratings

AC Ratings								DC Ratings			
Volts	Inductive 35% Power Factor						Resistive 75% Power Factor	Volts	Inductive		
	UL Rating	Make		Break		Cont.	Make, Break &		UL Rating	Make & ▲	Cont.
		Amps	VA	Amps	VA	Amps	Cont. Amps			Break Amps	Amps
120	A600	60	7200	6	720	10	10	125	Q600	0.55	2.5
240		30	7200	3	720	10	10	250		0.27	2.5
480		15	7200	1.5	720	10	10	600		0.10	2.5
600		12	7200	1.2	720	10	10				

▲ 69 VA maximum up to 300 volts.

Courtesy of Square D/Schneider Electric.

FIGURE 2-46: Relay schematic symbols.

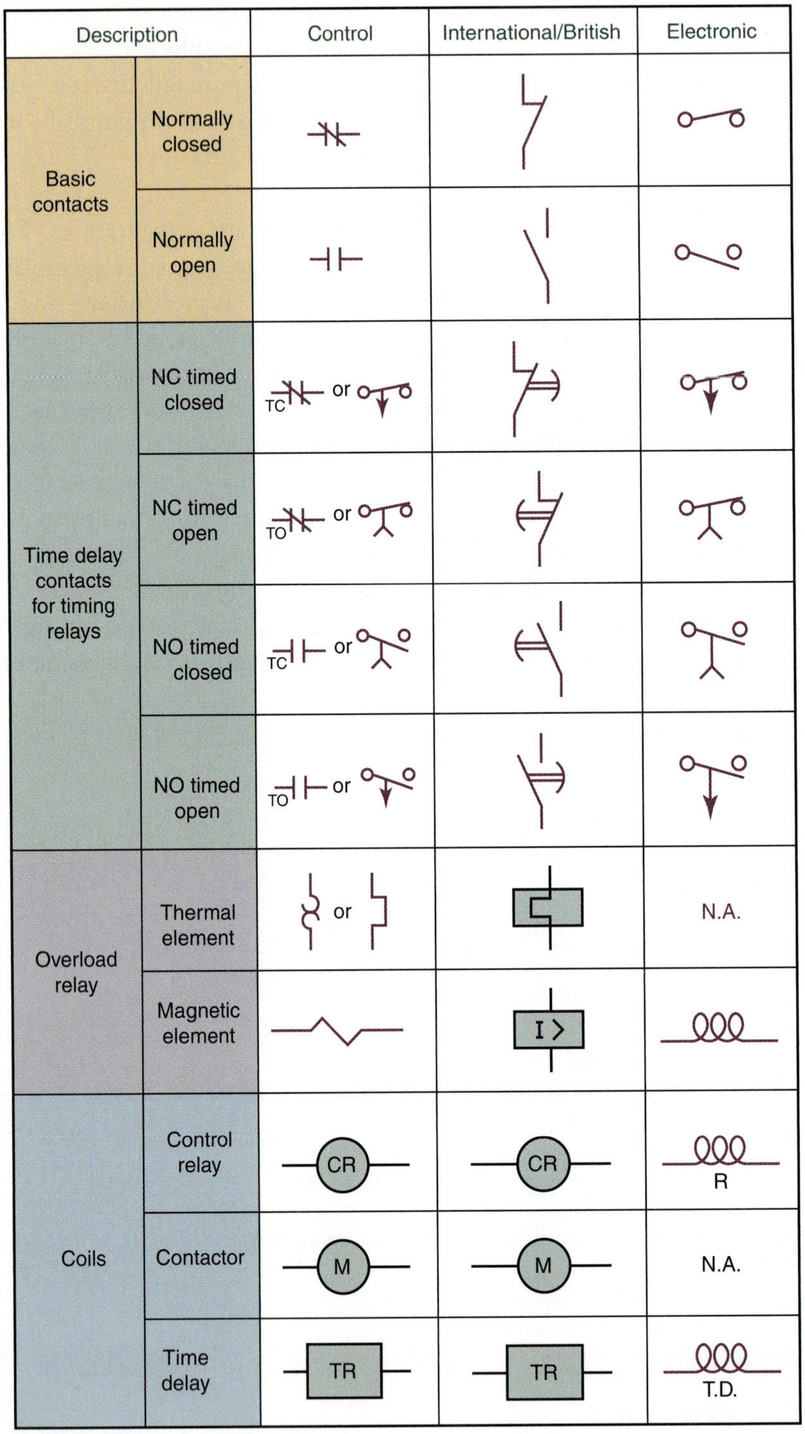

Description		Control	International/British	Electronic
Basic contacts	Normally closed			
	Normally open			
Time delay contacts for timing relays	NC timed closed	TC or		
	NC timed open	TO or		
	NO timed closed	TC or		
	NO timed open	TO or		
Overload relay	Thermal element	or		N.A.
	Magnetic element		I >	
Coils	Control relay	CR	CR	R
	Contactor	M	M	N.A.
	Time delay	TR	TR	T.D.

Rehg and Sartori, Industrial Electronics, *1st Edition, © 2006. Reprinted by permission of Pearson Education, Inc., Upper Saddle River, NJ.*

the coil is de-energized, just as it comes in the box from the manufacturer. In Figure 2-44, the relay is shown in the de-energized condition and the states of the contacts are labeled. The symbols used in schematics to identify relay coils and contacts are different for an electronic schematic versus a control wiring drawing. In addition, the international standard has another representation. The table in Figure 2-46 lists the symbols used for each type of circuit schematic.

Holding contacts. *Holding* or *seal-in contacts* provide a method of maintaining current flow after a momentary switch has been pressed and released. The holding contacts carry the full power of the rung output. In general, a holding contact is connected in parallel with the momentary switch as shown in Figure 2-47. When the NO start push button PB1 is pressed, current flows through the NC stop push button PB2 to the motor starter M. The auxiliary contact M1 of the starter is connected in parallel with PB1, keeping the starter coil energized after PB1 is released.

2-6-3 Latching Relays

The *latching relay* is a relay whose contacts remain open and/or closed even after power has been removed from the coil. Thus, if a relay position must be maintained in an automated process after power is removed, the latching relay provides that function. The latching relay has a latch coil and an unlatch coil. Figure 2-48 depicts a latching relay with two coils labeled coil A and coil B. Energizing coil A produces a magnetic field, which opposes the magnetic field of the permanent magnet in circuit A, and is great enough to break the armature free and snap into a closed position. The armature remains in that position upon removal of power from coil A, but it will snap back upon energizing coil B. Because the operation depends upon cancellation of a magnetic field, the polarity indicated in the figure must be applied to the coils.

Figure 2-49 illustrates the control diagram for a latching relay in the unlatched state. In this state the solenoid SOL1 is de-energized. When the push button PB1 is momentarily pressed, the latch coil L is energized, setting the relay in the latched position. The relay contact is closed, thus completing the circuit and energizing SOL1. When the push button PB2 is momentarily pressed, the unlatch coil U is energized, setting the relay in the unlatched position. The relay contact is open, thus opening the circuit and de-energizing SOL1.

2-6-4 Contactors

Contactors, shown in Figure 2-50, are relays designed to switch large currents from large voltage sources. Contactors have multiple contacts so that both lines of a single-phase source and all three lines of a three-phase source can be switched. In addition to the contacts used to switch the primary voltage, there are usually one or more contacts, called *auxiliary contacts,* for use in the contactor control circuit. These auxiliary contacts are often limited to 120 VAC and may be either normally open or normally closed.

FIGURE 2-47: Seal-in contact.

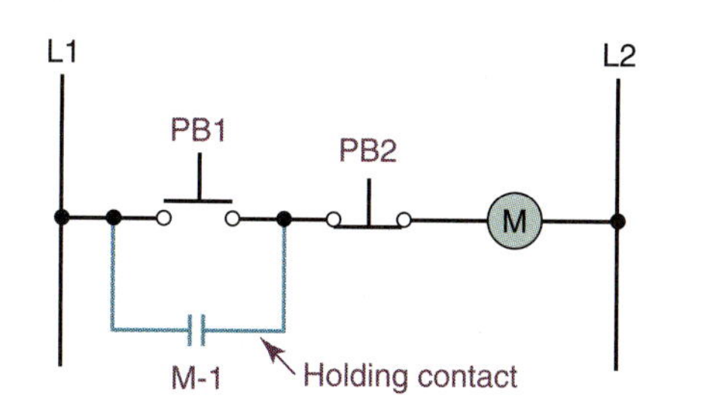

FIGURE 2-48: Latching relay operation.

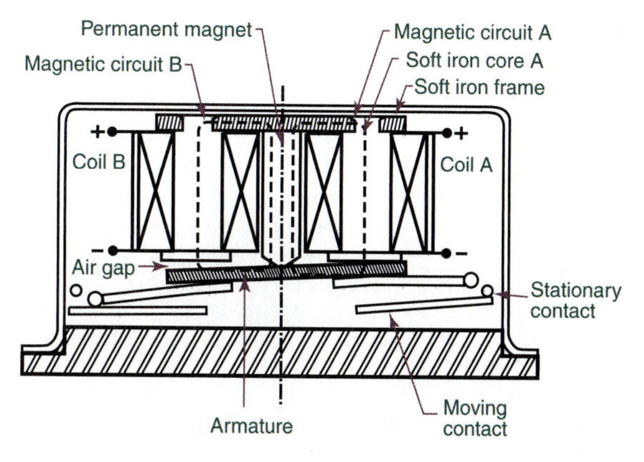

Courtesy of Teledyne Relays.

FIGURE 2-49: Latching relay control diagram.

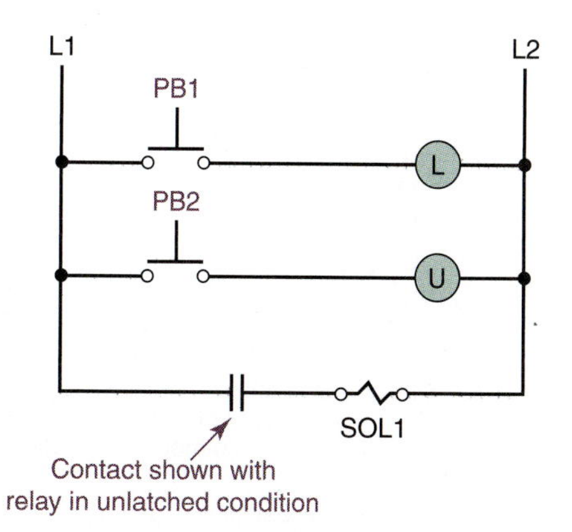

FIGURE 2-50: Contactors.

Courtesy of Square D/Schneider Electric.

Contactors also have an *arc-quenching system* to suppress the arc formed when the contacts carrying inductive current open. Contactors for AC and DC loads are quite different in design because of the need to prevent arcing when contacts open. DC contactors are designed to handle DC current, as well as the greater difficulty of breaking a DC arc.

2-6-5 Motor Starters

A typical motor starter includes contacts, which were discussed in the previous subsection, an *overload block* or thermal unit, which provides overcurrent protection for the motor with a thermal *overload contact*, which is opened when an overload is detected. The thermal unit resets to its normal state when the overload is removed. Most thermal units are manufactured to operate in one of the following three basic overload methods.

- **Eutectic Alloy Overload Method** uses a eutectic alloy, which is similar to solder but with a lower melting point, to control a ratchet-pivot assembly that controls the overload contacts.
- **Bimetallic Overload Method** uses two dissimilar metals, bonded together, to provide the movement that opens the overload contacts.
- **Phase-loss Sensitivity Overload Method** is similar to the bimetallic overload method except that the difference in current in any of the phases is used to trip the trip-bar assembly, activating the overload and cutting off current flow to the motor.

Figure 2-51 shows a motor starter and a cutaway showing one set of contacts and the ther-

FIGURE 2-51: Starter operation.

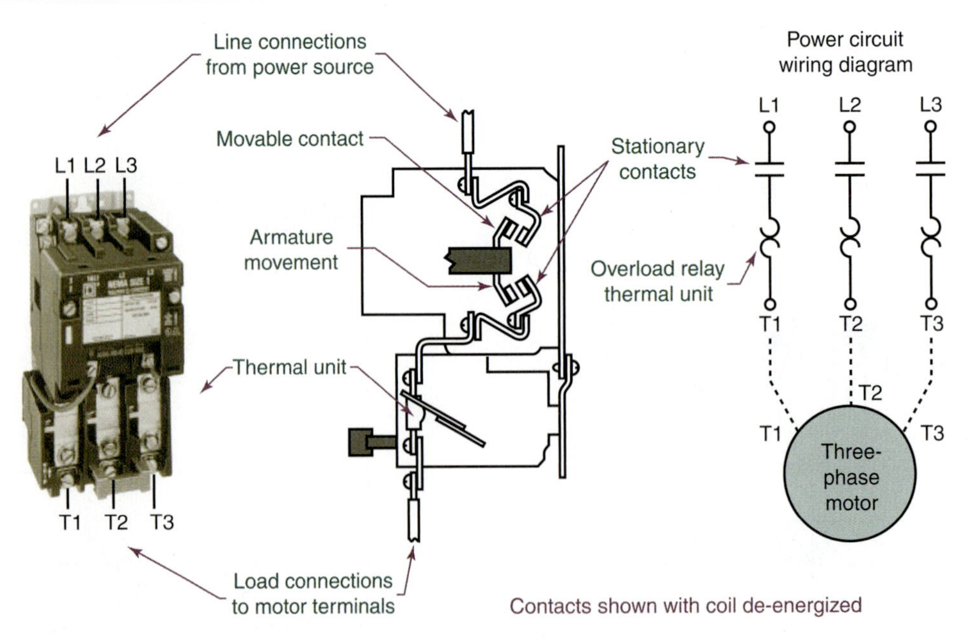

Courtesy of Square D/Schneider Electric.

EXAMPLE 2-7

Draw the control diagram with relay ladder rungs for controlling the movement of a container on a conveyor where

- The depression of a start button starts a motor, which starts a conveyor moving.
- The motor stops when the container reaches a specific position or when the stop button is depressed.
- A red lamp indicates that the conveyor is stopped, and a green lamp indicates that the conveyor is moving.

SOLUTION

Refer to Figure 2-52, the control diagram, as you read the following description.

1. The depression of the start button PB1 energizes the control relay CR if the normally closed limit switch LS is not activated and the stop button PB2 is not depressed.
2. The CR-1 contact closes, sealing in CR even if PB1 is released.
   - The CR-2 contact opens, turning off the red lamp.
   - The CR-3 contact closes, turning on the green lamp.
   - The CR-4 contact closes, energizing the motor contactor M, which starts the motor, moving the container as the conveyor moves.
3. The (NC) limit switch LS is activated (contacts open) when the container reaches the specified position, de-energizing CR, which opens contact CR-1.
   - The CR-2 contact closes, turning on the red lamp.
   - The CR-3 contact opens, turning off the green lamp.
   - The CR-4 contact opens, de-energizing the motor contactor, which stops the motor, the conveyor, and the container.

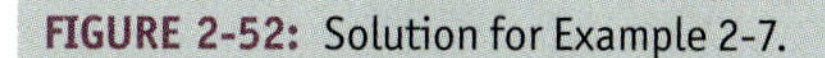

FIGURE 2-52: Solution for Example 2-7.

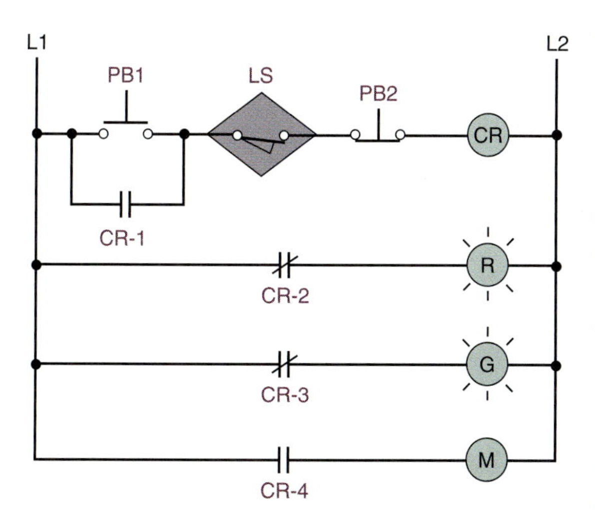

mal unit. The incoming three-phase AC power, typically 480 volts or higher for motors 1 horsepower or greater, is connected to L1, L2, and L3 in the figure. The motor attaches to the other end at T1, T2, and T3. The motor starter contacts and a thermal overload unit separate the power lines from the motor. The coil and armature inside the motor starter close all three contacts to start the motor. The contacts are normally open so that the power is broken whenever the motor starter's coil is not energized.

2-7 VISUAL AND AUDIO OUTPUT DEVICES

In addition to driving output actuators, the PLC turns on various visual and audio devices to indicate a specific condition and/or warning. For example, if a process parameter has been exceeded a lamp could be illuminated, or if an unplanned event occurs a flashing light and an audible alarm could be initiated. These devices have been briefly mentioned previously, but they are discussed in detail in the following subsections.

2-7-1 Pilot Lamps

Pilot lamps are industrial grade lamps used in control panels and machine front panels to indicate events and conditions in the system. The variety of lamp models is as varied as the applications that use them. Figure 2-53 shows two types and the schematic symbols for each. Both models permit testing of the lamp. The top lamp

FIGURE 2-53: Pilot lamp types.

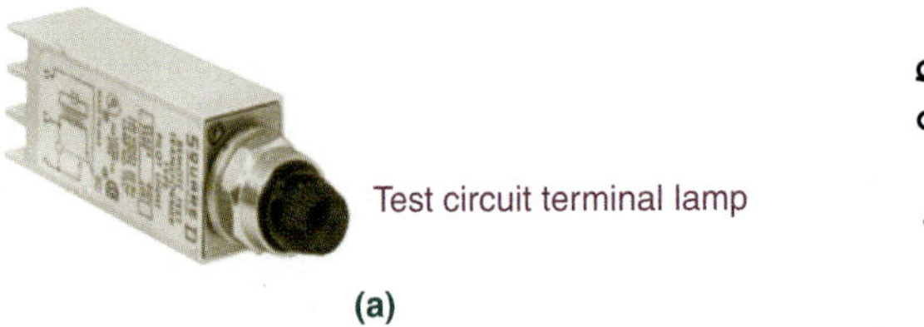

Courtesy of Square D/Schneider Electric.

has a push to test switch that is activated by applying pressure to the lens face. The second has test terminals at the rear that can be wired to a test switch to activate the lamp. Lamp voltages include 6.3, 28, and 120 volts, and the lens cap color options include clear, red, green, amber, and yellow. The letter (first letter of the color, for example, R for red) inside the schematic symbol indicates the lens color.

2-7-2 Horns and Alarms

Industrial horns are available in several different sizes that offer a large range of sound output levels—from localized to broad-scale—to alert operators of machine conditions almost anywhere on the plant floor. Figure 2-54 illustrates a typical horn and the schematic symbol for the horn. The horn in the figure is available in sizes from 3-inch to 7-inch cubes, where the size is a function of the output sound level capability—the larger the horn the higher the sound level capability. Input voltages typically fall within the ranges of 10 to 24 VDC or 24 to 240 VAC. Horn features include multi-tone capability, tone selection, and volume control, allowing users to assign different tones to various alarm situations. Using selectable tone options, users can select a tone and frequency that contrasts with ambient noise in the external environment, helping to ensure that the alarm signal can be clearly distinguished. For example, users can select a lower decibel sound with a high contrasting frequency for maximum distinctness and minimal noise. The multi-tone/circuit feature also allows a single horn to produce up to three different tones for three different conditions, essentially acting as three horns in a single housing. For applications requiring greater visibility, the light and horn combination has a strobe light option that can be combined for effective visual and audio signaling.

FIGURE 2-54: Industrial horns.

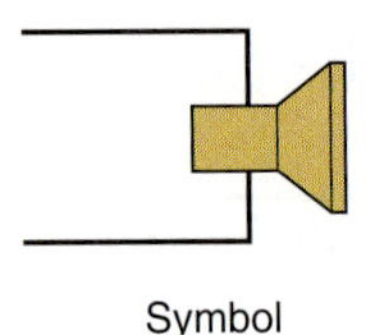

Symbol

Courtesy of Rockwell Automation, Inc.

FIGURE 2-55: Industrial panel alarms.

Courtesy of Rockwell Automation, Inc.

FIGURE 2-56: PLC relay outputs with common terminal.

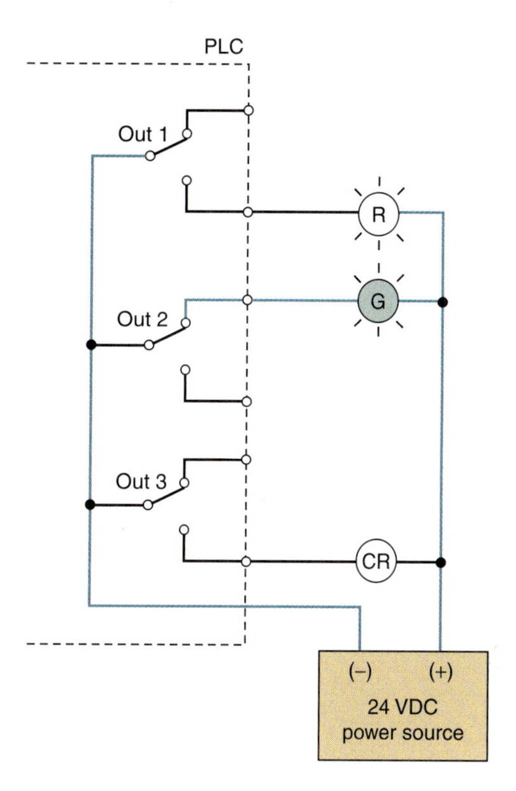

Figure 2-55 illustrates a variety of panel-mounted *industrial alarms*, and the schematic symbol for an alarm. These alarms can provide sound outputs in the range of 80 to 105 dB and are available in 12 VAC/DC, 24 VAC/DC, 120 VAC, and 240 VAC. Alarm features include continuous or pulsing sound outputs, a flashing or steady LED lamp, and a strobe light version. Horns and alarms are ideal for applications where operators monitor and maintain multiple pieces of equipment.

2-8 INTERFACING OUTPUT FIELD DEVICES

Interfacing output devices to a PLC generally involves a PLC output module with either a relay or a solid-state output to which the field device is attached. Relay outputs are used for high-current requirements in the one- to two-ampere range or for power isolation, whereas solid-state outputs are used to control low-power DC circuitry with transistors or low-power AC circuitry with triacs. As with the interface to PLC inputs, PLC outputs are available with a common terminal or totally isolated.

2-8-1 Powering Output Field Devices

Typically, the power for output field devices is not supplied by the PLC but provided separately, allowing the wide range of requirements to be satisfied. The power includes solenoid and relay coil voltage, motor voltage (single-phase and three-phase), and the power for lamps, horns, and alarms. Vendor device specifications detail the power requirements for the field devices. The user must ensure that the current requirements of the devices are compatible with the PLC output capability.

2-8-2 Output Wiring

Using the *PLC relay outputs* involves wiring to relays that share a common terminal or connecting to relays that are totally isolated. Relay outputs that share a common terminal are shown in Figure 2-56. Note that the common terminal is connected to the wiper of all the relays, and each relay has an NC and an NO contact, which are available for wiring to a field device. Note the 24 VDC power source is connected to all

FIGURE 2-57: PLC relay outputs with isolated common terminals.

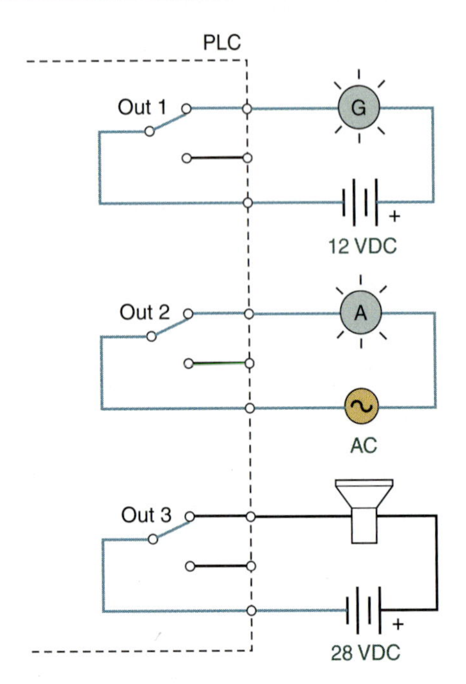

Note: Out1, Out2, and Out3 are relay contacts

FIGURE 2-58: PLC current sinking and current sourcing outputs.

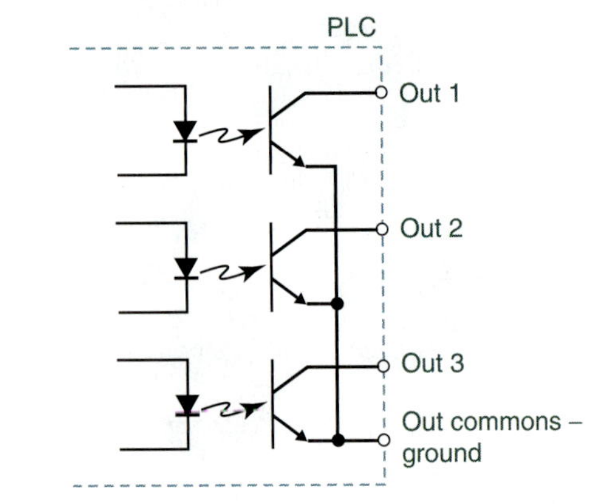

(a) PLC solid state outputs-current sinking type

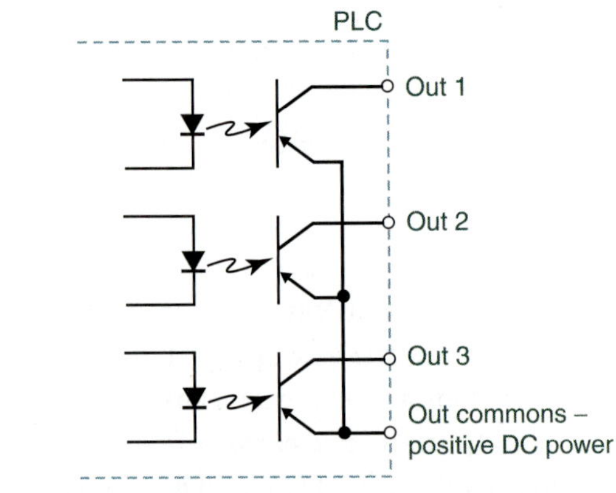

(b) PLC solid state outputs-current sourcing type

Note: Outputs are driven by transistors - NPN for the sinking and PNP for the sourcing

the field devices plus the PLC common terminal. Observe that the PLC relays are in the de-energized state and the red lamp and the relay CR are connected to NO contacts and are off, but the green lamp is connected to an NC contact and is on.

PLC output relays that are totally isolated from each other are illustrated in Figure 2-57 and show the field devices that they are driving. Note each relay output controls a different device, each with its own power source. Output 1 is connected to a DC voltage-powered lamp, output 2 is connected to an AC voltage-powered lamp, and output 3 is connected to a DC voltage-powered alarm. Observe that when the PLC relays are de-energized, both lamps are on and the horn is off.

The PLC solid-state outputs are for the most part driven by transistors or logic elements and generally share a common terminal, although triac outputs to switch AC loads are available in the totally isolated configuration. As with the transistor PLC input, the transistor PLC output is either current sinking or current sourcing, as shown in Figure 2-58. Note that the transistors are driven by an opto-isolator and that they share a common terminal. Let's examine Figure 2-59,

FIGURE 2-59: Field devices connected to PLC outputs.

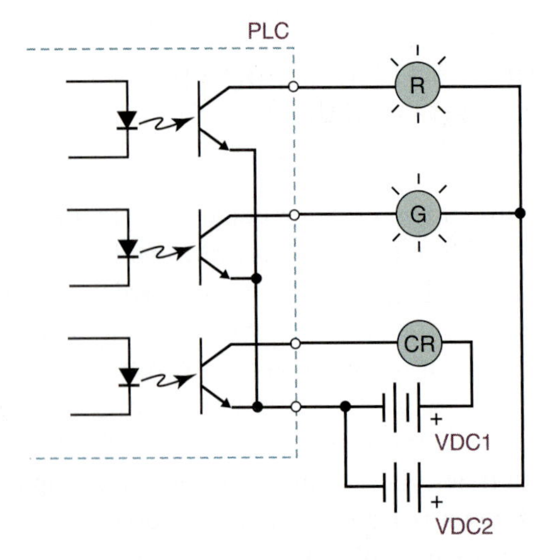

Note: Outputs are driven by NPN transistors

which depicts field devices connected to the PLC solid-state outputs. Note that the two lamps are powered by the same power source, whereas the relay coil is powered from a different source. However, the commons of both power supplies are connected to the common of the transistor outputs. Each of field devices remains off until the transistor driver turns on.

2-8-3 Current Sinking and Current Sourcing Devices

When wiring a field device to a DC voltage PLC output, these rules must be followed:

- **Sinking outputs:** current must flow through the field device and into the current sinking PLC output to ground.
- **Sourcing outputs:** current must flow out of the current sourcing PLC output and through the field device to ground.

For the current sinking output (an NPN transistor), the negative side of the power source must be connected to the PLC output common terminal. For the current sourcing output (a PNP transistor), the positive side of the power source must be connected to the PLC output common terminal. Reread these rules until you have them committed to memory. Additional PLC module interface circuitry is discussed in Appendix B.

2-9 TROUBLESHOOTING INPUT AND OUTPUT DEVICES

Every minute a production system is not producing due to a problem, the company is losing money. That is why competent system troubleshooters are worth their weight in gold. The best way to make your mark in a career is to demonstrate that you can solve problems, especially troubleshooting machine and system failures.

The primary message is that you must have a troubleshooting game plan because random and unrelated actions to solve the problem are seldom successful. In general, to troubleshoot a problem, you work from the outside of the machine or system and work your way inside. This implies a series of steps. It is tempting to jump quickly to a conclusion as to the problem, and occasionally you may be correct. The methodical process may appear to take longer, but it leads to a correct solution more often. Also, the more experience you have following a plan, the more often you are correct when you decide to make an intuitive guess. Here is a suggested plan.

1. Outside analysis comes first. Take an overall look at the system and note all the symptoms. Get symptom reports from more than one person if you cannot observe the symptoms in person. Determine if the problem is in the machine or system or in the users of the hardware. Most PLCs have indicator lights on the input/output modules to aid in troubleshooting.
2. Analyze the interfaces. The places where mechanical and electrical systems come together are often the places where they fail. Inputs, outputs, control devices, and power supplies are often interfaces where problems occur. Check those that could relate to the symptoms recorded.
3. Start to move into the system following an electrical path or signal flow from an interface using techniques to isolate problems to individual sections, circuits, or components using a combination of system outputs, test equipment, and an experienced engineering sense (a high level of common sense).
4. Identify the most likely component for the problem. When you have a high degree of confidence that you have identified the malfunctioning component or interface, then it is often faster to just replace it to verify your suspicion.
5. Identify the cause of the failure. Finding the problem is only half of the task for a competent troubleshooter. Identifying why the failure occurred is the second part. It may just be component life expectancy, but it is important to determine if anything from a user or system standpoint shortened the device's life.

6. Verify that the fix is permanent. The system should be taken to the limits of its operation over a test period to verify that the problem is truly fixed. The length of time and limits used depends on the problem that was removed.

2-9-1 Troubleshooting Switches

All of the switches covered in this chapter have a common troubleshooting approach. Switch problems can be grouped into two categories: *operator problems* and *contact problems.*

Operators such as a switch handle or push button are the mechanical elements that force the contacts from their normal position to the opposite state. Often the operator will change positions from on to off but the contacts may not move. When operators fail, the contacts may remain in either the normal position or the opposite one. Operator problems usually affect all the contacts in the multipole multi-throw configuration.

Contact problems on switches include always open, always closed, or excessive resistance for closed contacts. If a contact is forced to carry excessive current or has arced, the contacts fuse together and never open. The same switching conditions can cause the contacts to be burned, and the resistance when closed becomes much higher than normal.

When a switch is suspected for the system problem the following procedures can be used.

- If a contact should be open, then the voltage measured across the contacts should be the voltage being switched.
- If a contact should be closed, then the voltage measured across the contacts should be near zero. A higher reading indicates excessive contact resistance or open.
- If excessive contact resistance is suspected, then one of the switch wires should be removed and resistance measured with an ohmmeter.
- If a switch is not closing a contact, then the jumper across the contacts will test to see if the problem is fixed.
- If a switch is not opening, then removal of one of the switched wires will verify that problem.

2-9-2 Troubleshooting Relays

Relay problems can be grouped into the two categories just described for switches plus a third category, *relay coil hum*. The operator for a relay is the electromagnetic coil or solenoid plus the mechanical linkage to the contact poles. All of the problems and troubleshooting procedures discussed for switch operators and contacts apply to relays as well.

Because relay contacts are switched by the action of a solenoid or electromagnet, improper drive currents to the coil can cause some problems. A specified minimum amount of current through the coil, called the *pull-in current*, is required to positively "pull in" the armature to actuate the contact(s). After the armature is pulled in it takes less magnetic field flux and less coil current to hold it there. As a result, a value of coil current significantly lower than the pull-in value must be reached before the armature "drops out" to its normal state. This current level is called the *dropout* current. If the improper drive currents are present relays may not pull in completely, causing higher contact resistance and overheating of the coils. Armatures that do not fully close AC relays cause coil overheating due to the lower coil reactance values, resulting in higher sustained coil currents. Also, relays may fail to drop out if the coil current is not falling into the drop-out range.

Devices using AC solenoids will have one or more single turn coils, called shading rings, embedded in the face of their magnetic armature assembly. Without this coil the armature would tend to drop out whenever the AC voltage drops toward zero, and then be pulled in as the voltage and magnetic field reverses. This produces a humming noise in AC relays. The shading ring, which minimizes this noise, produces an induced magnetic field that is out of phase with that of the applied power. This holds the armature in between power reversals. Over time, shading rings tend to crack from the pounding of the armature faces. When this happens, the solenoid will become very noisy, coil current will increase, and premature failure will result.

2-9-3 Troubleshooting Proximity Sensors

The following tips may be helpful if the problem appears to be in the sensor or sensor amplifier. Sensors from different vendors have different operating characteristics, so the first requirement is to know how the sensors in the system operate. In fact, some sensors have adjustment screws to set switching points with details in vendor literature. The following order does not indicate a preferred sequence. Review all the tips and consider those that apply to the current sensor problem.

- Verify that the sensor has power in the specified range. Checking other operating equipment connected to the same power bus is a method for testing for power.
- Verify that all the amplifier settings are correct. Many sensor amplifiers have a sensitivity adjustment, so make sure that the protective seal is still in place.
- Verify that all switch settings are correct.
- Use the operation indicator on the sensor or sensor amplifier to determine if the sensor electronics recognize that a part is present. An on condition for the operation indicator usually indicates that the output transistor or relay is operating. Some devices with the output set to the normally open (NO) operation will have the operation indicator on when an object is sensed. The opposite is true for an NC setting. A testing method is to move the part toward the sensor along the same path used in the process and determine how close the required sensing distance is to the maximum value.
- Verify that a foreign object is not creating a problem on one of the sensing heads.
- Verify that the velocity of the parts past the sensor does not exceed the frequency response of the unit.
- Verify that the sensing distance was not reduced due to a change in the ambient temperature or supply voltage.

2-9-4 Troubleshooting Photoelectric Sensors

The following tips may be helpful if the problem appears to be in the sensor or sensor amplifier. The first rule is to know the operation of the sensor in the system.

- Verify that the sensor has power in the specified range. Verifying power to other operating equipment connected to the same power bus is one method for testing for power.
- Verify that all the amplifier settings are correct. Many sensor amplifiers have a sensitivity adjustment, so make sure that the protective seal is still in place. Verify that all switch settings are correct.
- Use the operation indicator on the sensor or sensor amplifier to determine if the sensor electronics recognize that a part is present. An on condition for the operation indicator usually indicates that the output transistor or relay is operating. Some devices that have the output set to light ON mode have the operation indicator on when light is striking the sensor. The opposite is true for a dark ON mode setting.
- Verify that the lenses are clean and free of foreign objects.
- Verify that the velocity of the parts past the sensor does not exceed the rise and fall time for the unit.
- Verify that the sensing distance was not reduced due to a change in the ambient temperature or supply voltage.

Chapter 2 questions and problems begin on page 553.

CHAPTER

3

Introduction to PLC Programming

3-1 GOALS AND OBJECTIVES

Chapter 3, like Chapter 1, covers numerous PLC topics that are necessary to know before specific PLC instructions are introduced. One goal of this chapter is to cover topics that focus on how the PLC operates, such as number systems, bit/word addressing, memory allocations, input and output addressing, and scanning cycle. A second goal is to cover topics that focus on PLC programming. The final goal is to introduce the three Allen-Bradley processors—PLC 5, SLC 500, and ControlLogix—by grouping the PLC 5 and SLC 500 topic together where they are similar but keeping the ControlLogix separate because it is quite different from the other two. Although the topics may appear to be unrelated, they are all operational concepts needed in Chapters 4 through 16 when programs are built with the many PLC instructions.

After completing this chapter you should be able to:

- Convert to and from hexadecimal, decimal, octal, and binary numbers.
- Describe the input, output, internal relay, and status addressing format for rack/slot-based PLC 5 and SLC 500 and tag-based Logix PLC processors from Allen-Bradley.
- Describe PLC scan time and the use of examine if closed and examine if open instructions.
- Write PLC programs and interpret PLC programs with the following logic combinations: AND, OR, AND/OR, and OR/AND.
- Describe the function of sealing instructions.
- Describe the operation of standard and latched PLC outputs.
- Describe PLC memory organization, data types, and use of internal binary bits, bytes, and words.
- Develop ladder logic programs with input instructions and output coil combinations.
- Describe and use program design techniques to develop a PLC program.

3-2 NUMBER SYSTEMS

Everyday tasks, like buying gasoline or groceries, employ the decimal or base 10 number system. This system is also used in manufacturing automation to input parameter values or to display the value of system variables. In addition to decimal, four other number systems are used in automation: *binary, octal, binary coded decimal (BCD),* and *hexadecimal*. It is important to understand all the number systems if you work with a process controlled by a PLC; however, hexadecimal is used most often. Number systems are introduced in the text at the beginning of the chapter in which they are used for a PLC instruction. As a result, binary, octal, and decimal are at the start of this chapter, binary arithmetic is in Chapter 6, and BCD and hexadecimal are in Chapter 7.

3-2-1 Number System Basics

All number systems have a *base* or *radix*. In the decimal system the base is 10, which means that 10 symbols (0 through 9) are used to represent decimal numbers. In addition to the base, the quantity specified by a number is a function of the place or position of the digits in the number. For example, 1 represents a quantity of one and 100 represents a quantity of one hundred. In the first number, the 1 occupies the zero position, and it has a value of 1×10^0 or 1×1 or just 1. Note that any non-zero number raised to the zero power is equal to 1. In the second number, the 1 is in the two's position, and it has a value of 1×10^2 or 1×100 or 100. The other two positions that hold zeros do not change the final value because they will contribute 0×10^0 or 0 and 0×10^1, which is also 0. Therefore, the number value is 100. The value of a decimal number depends on the digits that make up the number and the place or position value of each digit. The *position values* are illustrated by the number 4562.6 in Figure 3-1. Locate the position values, the base, and the digit values in the figure.

Note that the position values increase from right to left, and the zero position is to the left of the decimal point. Figure 3-1 also illustrates how the value of a decimal number is determined by adding the product of the *digits* and their *position weight values*. Note in the figure that the sum of the products is 4562.6^{10}, which is read as "four thousand five hundred and sixty-two point six base 10." The 10 in the subscript indicates the base or radix for the number. This discussion of decimal notation is used to generalize the conversion of a number from any base to base 10. The formula is:

$$\text{Number}_{10} = \text{the summation of all positions digits} \times \text{Base}^{\text{Position value of the digit}}$$

The formula is the summation of all position digits times the given number's base raised to the power of the position value of the digit. Using Figure 3-1, verify that the demonstrated conversion process employs this general formula. Although engineers use the decimal system, the computer uses the binary number system for all of its internal operation.

3-2-2 Binary System

Computers and PLCs make logical decisions and perform mathematical calculations using electronic circuits. The easiest and least expensive electronic circuits are *discrete designs* that have two states: on or off. The number system used in these electronic systems has two digits, 0 and 1, has a base of 2, and is called a *binary number system*. The binary system must be able to use just two digits to represent every numerical value required by a control system.

FIGURE 3-1: Weight values and position values.

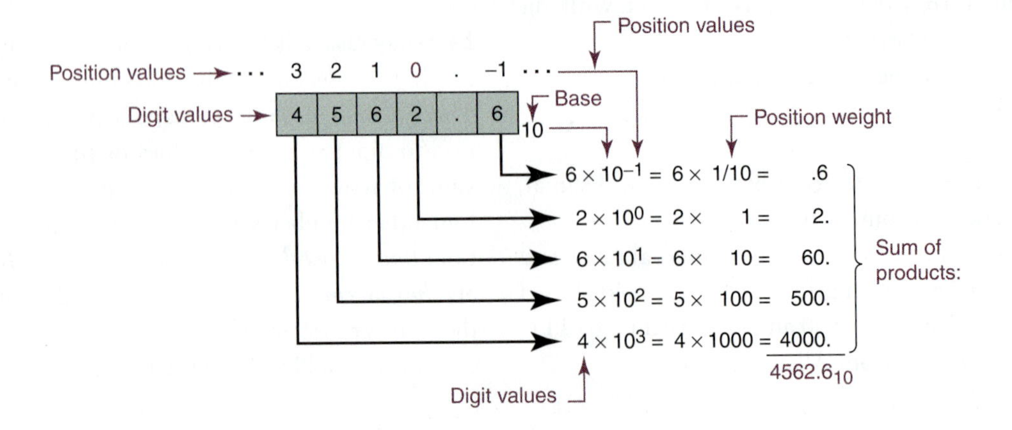

Figure 3-2 compares the binary system (base 2) to the decimal system (base 10), demonstrating how the base 2 system counts from 0 to 18_{10}. The base of a number is indicated by placing the base as a subscript on the least significant digit. For example, the 18_{10} in the previous sentence means 18 in the base 10 number system.

The two systems are identical for the decimal numbers, 0 and 1. However, after 1, the binary system runs out of digits. Therefore, a 2 in base 10 is a one zero in the base 2 system. Another distinction between the systems is that 10 in the binary system is pronounced "one zero base two" and not "ten." The binary for 18_{10} is 010010_2 and is pronounced "one zero zero one zero base two" and not "ten thousand ten." Note that leading zeros are normally not identified when reading the value. Also, the "base two" at the end is usually dropped.

Binary and decimal conversion. The PLC works in binary and the world works in decimal, so knowing how to convert numbers between the two systems is critical. Figure 3-3 illustrates the conversion process from binary to decimal. Find the location of the position values, digit values, and position weight for the binary number in the figure.

The digit value in position 0 in the figure is a 1. Each digit value produces a number that is part of the sum of products in Figure 3-3. The digit value of 1 in position 0 contributes a 1 to the sum of products. The digit value of 1 in position 1 produces a 2 in the sum of products. Observe the digit values for the rest of the positions and note the value produced in the sum of products.

The conversion process in Figure 3-3 indicates that the values in the sum of products column results from multiplying the digit value by the digit weight. For example, for position 0 the sum of products value is $1 \times 1 = 1$, and for position 3 it is $0 \times 8 = 0$. So if the digit value is a 0 the sum of products value is always a 0, but if the digit value is a 1 then the position weight value determines the sum of products value. The position weight is the base 2 raised to the power of the position value. Study the figure for the conversion of each digit value so you see that 10110111_2 is equal to 183_{10}. In summary, each digit value is multiplied by its position weight and then added to get the decimal equivalent.

On the other hand, conversion from decimal to binary, illustrated in Figure 3-4, uses a series of divisions by 2, the binary base value. Study

FIGURE 3-2: Decimal, binary, and octal number systems.

	$Decimal_{10}$		$Binary_2$						$Octal_8$ ← Base	
Position values →	1	0	5	4	3	2	1	0	1	0
	0	0	0	0	0	0	0	0	0	0
	0	1	0	0	0	0	0	1	0	1
	0	2	0	0	0	0	1	0	0	2
	0	3	0	0	0	0	1	1	0	3
	0	4	0	0	0	1	0	0	0	4
	0	5	0	0	0	1	0	1	0	5
	0	6	0	0	0	1	1	0	0	6
	0	7	0	0	0	1	1	1	0	7
	0	8	0	0	1	0	0	0	1	0
	0	9	0	0	1	0	0	1	1	1
	1	0	0	0	1	0	1	0	1	2
	1	1	0	0	1	0	1	1	1	3
	1	2	0	0	1	1	0	0	1	4
	1	3	0	0	1	1	0	1	1	5
	1	4	0	0	1	1	1	0	1	6
	1	5	0	0	1	1	1	1	1	7
	1	6	0	1	0	0	0	0	2	0
	1	7	0	1	0	0	0	1	2	1
	1	8	0	1	0	0	1	0	2	2

FIGURE 3-3: Conversion from binary to decimal.

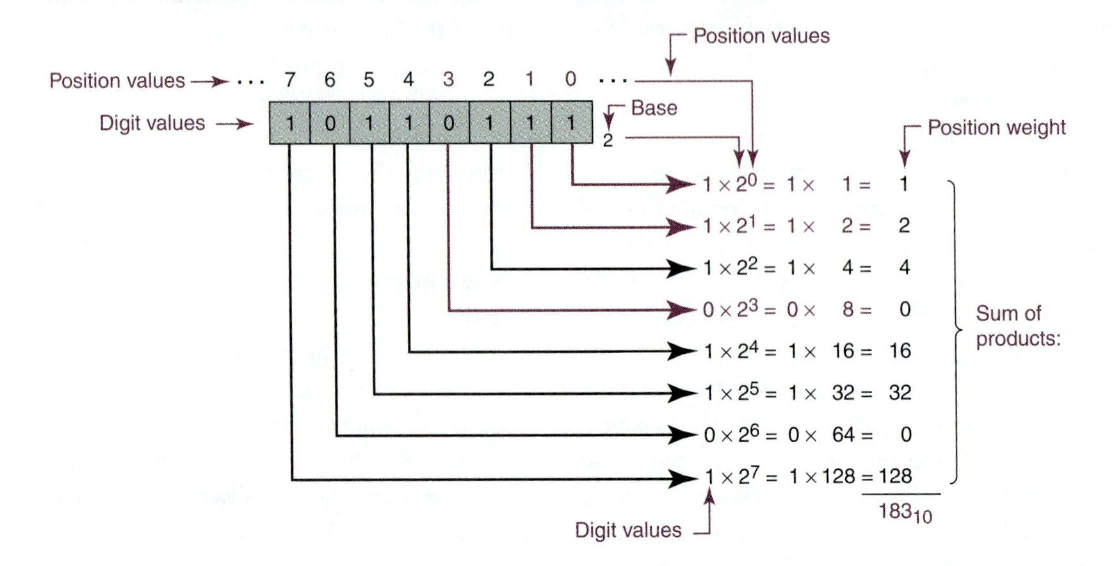

FIGURE 3-4: Conversion from decimal to binary.

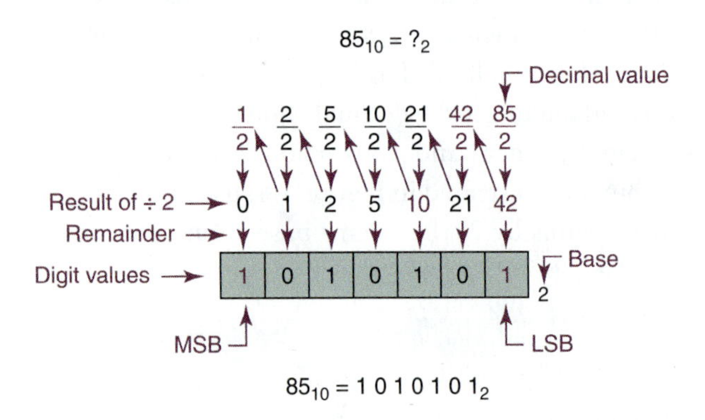

the process in the figure until you can see where the result of the division and the remainder are placed.

The remainder of the first division (1 in the figure) is the least significant bit (LSB) value of the binary conversion, and the remainder of the last division (1 in the figure) is the most significant bit (MSB) value. The result of each division (i.e., 42 in the first division) is used for the next division (i.e., 42/2 in the figure). A review of the Figure 3-4 indicates that 85_{10} is equal to 1010101_2.

3-2-3 Octal Number System

The octal number system, shown in Figure 3-2, has a base or radix of 8, which means that eight symbols, 0 through 7, are used to represent octal numbers. As a result of not using digits 8 or 9, the digits in the zero position value increases from 0 to 7, then repeat after 7. Octal numbers, like their binary counterpart, are pronounced in a special manner. For example, the octal number 7261 is pronounced "seven two six one base 8" and not "seven thousand two hundred sixty-one." The octal number system, popular in the early days of computers and PLCs, easily converts from binary to octal and back to binary.

Octal conversions. Octal numbers can easily be converted to binary using the method illustrated in Figure 3-5(a). Each octal digit, 0 through 7, in the number 2713_8 is replaced with its three-bit binary equivalent. Review the binary values for octal digits 0 through 7 in Figure 3-2. The three-bit binary groups are appended together to form the binary equivalent, 10111001011_2, of the octal value. Conversion from binary back to octal, also illustrated in Figure 3-5(a), is equally straightforward. One can see that the binary number is partitioned into groups of three bits, starting at the least significant bit end of the binary value. Leading zeros are added if necessary to obtain the final group of three. Then each partition of three-bit binary numbers is converted to its equivalent 0 through 7 octal value, using the chart in Figure 3-2. Study the conversion for the values in Figure 3-5(a) until conversion between octal and binary numbers is

FIGURE 3-5: Octal conversion.

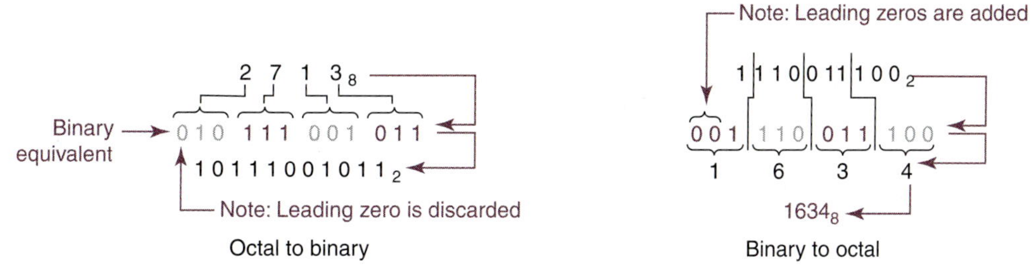

(a) Octal to binary and binary to octal conversions

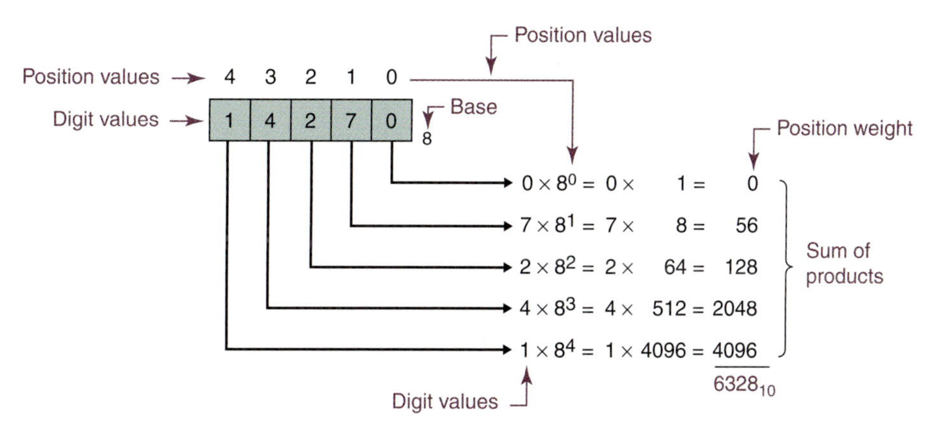

(b) Octal to decimal conversion

understood and the binary values for 0 through 7 are committed to memory.

Conversion from octal to decimal, illustrated in Figure 3-5(b), uses the formula introduced earlier for binary numbers.

$$\text{Number}_{10} = \Sigma \text{ position digits} \times 8^{\text{Position value of the digit}}$$

This formula evaluates as the summation of all position digits times the radix (8) raised to the power of the position value. The conversion of 15_8 is performed as follows:

$$\text{Number}_{10} = 1 \times 8^1 + 5 \times 8^0 = 1 \times 8 + 5 \times 1 = 8 + 5 = 13_{10}$$

This result is verified in Figure 3-2 where $13_{10} = 001101_2 = 15_8$. Conversion from base 10 to base 8 can be performed by first converting from decimal to binary and then converting the binary value into octal.

A knowledge of number systems is necessary to understand the operation of some PLC instructions, and binary notation is used throughout the text when referring to Boolean (on and off) signals.

3-3 HEXADECIMAL NUMBER SYSTEM

3-3-1 Introduction

The binary number system requires many digits to represent a large value, for example, to represent the decimal value 202 requires eight binary digits. Note that the decimal number 202 is only three digits and, thus, represents numbers much more compactly than does the binary numbering system. However, decimal numbers are not usable in computer systems.

When dealing with large values, binary numbers quickly become too unwieldy. A shorthand method of expressing large values is the hexadecimal (hex) numbering system, which is a popular numbering system in the PLC. Hexadecimal numbers offer the following two features:

- Hex numbers are very compact
- It is easy to convert from hex to binary and binary to hex.

As in the binary system, which uses the base 2, and the octal system, which uses the base 8, and the decimal system, which uses the base 10, the hexadecimal system uses the base 16. These 16 digits are represented by the numbers 0 through 9 and the letters A, B, C, D, E, and F.

Hexadecimal numbers are used extensively in PLC comparison instructions and instructions that incorporate masks. Therefore, a good grasp of the conversions described in this Section are necessary to evaluate the comparison actions and the instruction results.

3-3-2 Comparison of Numbering Systems

The following table lists the decimal numbers 0 through 16 and the binary, octal and hexadecimal equivalents. Note that a lower case "b" is appended to the binary number, a lower case "q" is appended to the octal number and a lower case "h" is appended to the hexadecimal number.

Decimal	Binary	Octal	Hex
00	0000b	00q	00h
01	0001b	01q	01h
02	0010b	02q	02h
03	0011b	03q	03h
04	0100b	04q	04h
05	0101b	05q	05h
06	0110b	06q	06h
07	0111b	07q	07h
08	1000b	10q	08h
09	1001b	11q	09h
10	1010b	12q	0Ah
11	1011b	13q	0Bh
12	1100b	14q	0Ch
13	1101b	15q	0Dh
14	1110b	16q	0Eh
15	1111b	17q	0Fh
16	10000b	20q	10h

This table provides all the information that is needed to convert from one number base into any other number base for the decimal values from 0 to 16.

3-3-3 Binary to Hex Conversion

Conversion from a binary number to hex number is accomplished by the following two steps:

1. Break the binary number into 4-bit sections from the LSB to the MSB.
2. Convert each 4-bit binary number to its hex equivalent.

For example, the binary value 1010 1111 1011 0010 is broken into 4-bit sections starting from the LSB and continuing toward the MSB, and then it is converted into the hex number AFB2.

- 1010111110110010 (Binary number to be converted)
- 1010 1111 1011 0010 (Binary number in 4-digit sections)
- A F B 2 (Hex number)

Let's look at another example, where the binary value 10010011001011 is converted into the hex number 24CD.

- 10 0100 1100 1101
- 2 4 C D

Note that there are only 2 digits in the most significant section, and it is not unusual to have less than 4 digits in the most significant section. That is why it is important that when breaking the binary number into 4-bit sections, you start from the LSB.

3-3-4 Hex to Binary Conversion

It is also easy to convert from a hex number to a binary number and is accomplished by the following two steps:

1. Convert the hex number to its 4-bit binary equivalent.
2. Combine the 4-bit sections by removing the spaces.

For example, the hex value AFB2 will be written:

- A F B 2 (Hex number to be converted)
- 1010 1111 1011 0010 (Binary equivalent)
- 1010111110110010 (Binary digits combined)

3-3-5 Hex to Decimal Conversion

The conversion from a hex number to a decimal number is accomplished by the following four steps:

1. Multiply each hex digit by its weighted value
2. Convert the hex digits to their decimal equivalent
3. Convert the hex weighted values to decimals and complete the multiplication
4. Add all the values.

Let's use the hex number AFB2h from a previous example and convert it to its decimal equivalent of 44,978.

- $A \times 16^3 + F \times 16^2 + B \times 16^1 + 2 \times 16^0$ (Weighted values)
- $10 \times 16^3 + 15 \times 16^2 + 11 \times 16^1 + 2 \times 16^0$ (Decimal equivalents)
- $10 \times 4096 + 15 \times 256 + 11 \times 16 + 2 \times 1$ (Conversion)
- $40{,}960 + 3{,}840 + 176 + 2 = 44{,}978$ (Multiplications & result)

3-3-6 Decimal to Hex Conversion

The conversion of a decimal number to a hex number is slightly more difficult. The typical method to convert from decimal to hex is repeated division by 16. While repeated subtraction by the weighted position value is another method, it is more difficult for large decimal numbers.

3-3-7 Repeated Division Method

For this method, divide the decimal number by 16, and convert the remainder to a hex number, which is the least significant digit of the final hex number. This process is continued by dividing the quotient by 16 and converting its remainder until the quotient is 0. When performing the division, begin as the least significant digit (right) and each new digit is the next more significant digit (the left) of the previous digit.

This method can be clarified by reviewing the following table, which provides an example of the repeated division by 16 method. Let's use the number 44,978, and verify that the equivalent hex number is AFB2h.

Division	Quotient	Remainder	Hex Digit
44978/16	2811	2	2
2811/16	175	11	B
175/16	10	15	F
10/16	0	10	A

Arranging the hex digits from right to left yields the number AFB2h. Review this example until you understand the method. Having introduced the number system, we now address the storage of numbers in the PLC.

3-4 BITS, BYTES, WORDS, AND MEMORY

Binary numbers have special notations, as illustrated in Figure 3-6. A single binary digit is a *bit*, and eight bits are a *byte*, pronounced like the word *bite*. Two bytes or 16-bits is a *word,* and two words or 32 bits represents a *double word*.

PLC memory is organized using either bytes, single words, or double words. For example, most older PLCs use 8-bit or 16-bit memory words, and newer systems, like ControlLogix from Allen-Bradley, use double word (32 bits) as the default. The memory block, introduced in Figure 1-2, indicates that the memory has three binary interfaces: *data, control,* and *address*. A general description of each interface states that the data bus carries parameter values and PLC instructions, the control bus provides the logical control for movement of instruction and data, and the address bus carries the binary address number for all the binary values stored in memory.

FIGURE 3-6: Bits, bytes, and words.

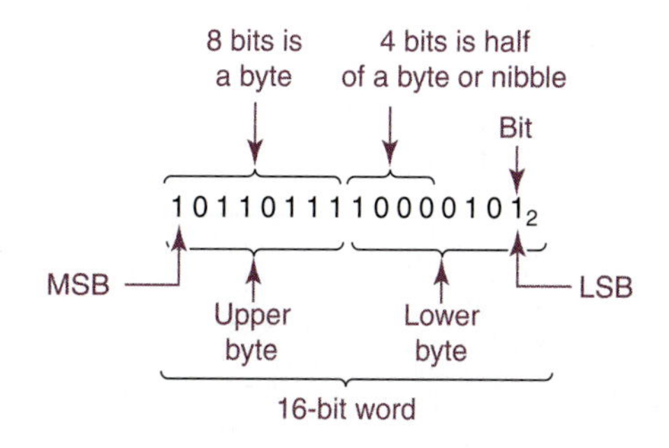

Memory is organized into blocks of consecutive bytes or words. Figure 3-7 shows a 16-bit 1K memory layout, where 1K memory represents 1024 locations. Each memory location has a number like the street address number of a house, and points to a specific word in memory. The address numbers for the 1K memory layout in the figure are 0 to 1023_{10} or 0000000000_2 to 1111111111_2. Each of the 1024 16-bit words has a unique 10-bit address. When the address bits are increased to 11 the addressable memory increases to 2048 16-bit words. This relationship between the memory size and the required number of address bits is present as the memory grows to the size used in current PLCs. The following are some examples of PLC memory limits.

- Automation Direct D4-450 processors: 60-kilobyte models
- Allen-Bradley ControlLogix processors: 2-, 4-, and 8-megabyte models
- Modicon Quantum processors: up to 8 megabytes available
- Siemens Samitac S7 processors: up to 20 megabytes available

Many PLCs are organized using memory words in a structure called a register. An understanding of the register structure is useful when PLC instructions are used in a program.

3-5 PLC MEMORY AND REGISTER STRUCTURE

Although knowledge of the details of the PLC memory technology is not necessary to program and implement PLCs, an understanding of program and data storage and the location of processor status bits is important. For example, when a program is created, one must know where the PLC places input and output data and where the CPU saves operational status information. This text presents the memory organization for the Allen-Bradley PLC 5, SLC 500, and ControlLogix processors models.

FIGURE 3-7: PLC 1K memory block.

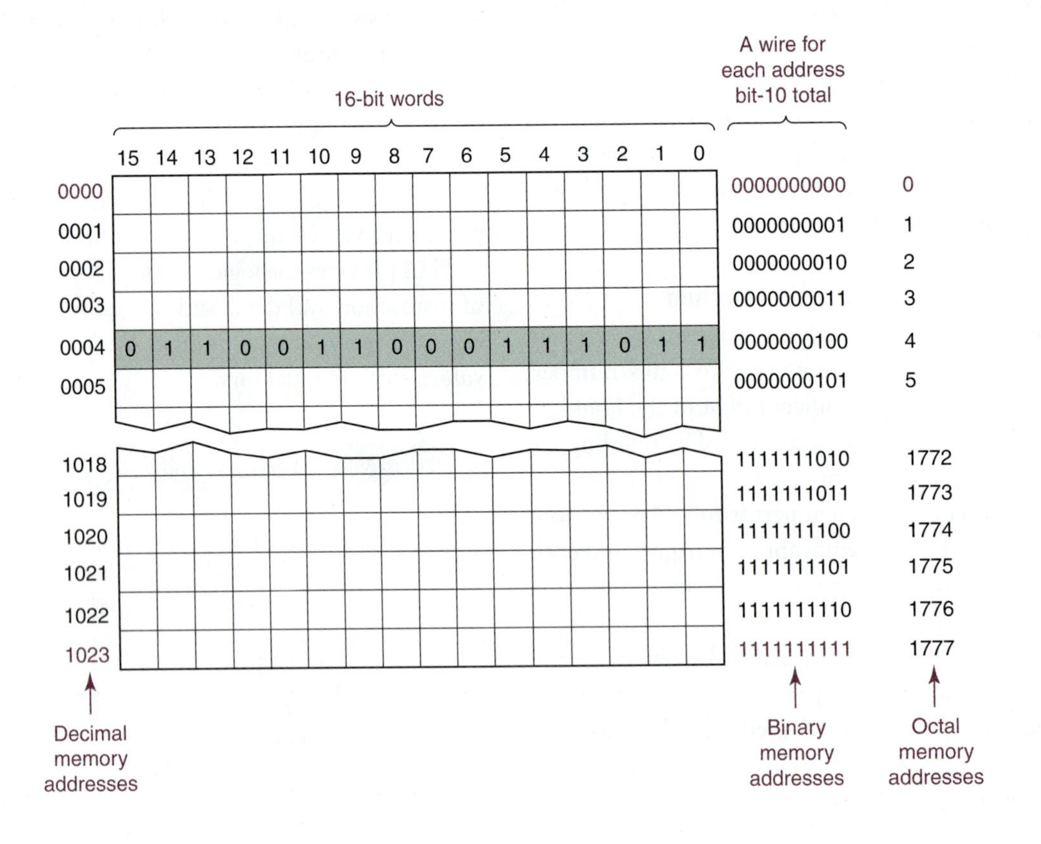

Since it is necessary to learn one vendor's PLC system well, the Allen-Bradley (AB) PLCs, which represent the largest share of U.S. installations, are the primary focus of this text. With a good knowledge of the three AB systems, it is not difficult to learn a second manufacturer's PLC system.

3-5-1 Allen-Bradley Memory Organization

Allen-Bradley PLCs have two distinctly different memory structures identified by the terms *rack-based systems* and *tag-based systems*. The rack address-based system, introduced in Chapter 1, has it roots in the early PLC systems.

Rack-based memory. Figure 3-8 represents the common memory structure for most rack/slot address–based PLCs, like the PLC 5 and SLC 500, as a two-drawer filing cabinet. An examination of the figure shows that one drawer is for *program* files and the second is for *data* files.

Each of the two file groups (data and program) in the figure is subdivided into instruction-specific files and file types.

Data files. The data files for the SLC 500, Figure 3-8, are sub-divider into 10 *designated* file types (0 to 9) and a user-defined file type with 246 files (10 through 255). Each file holds a number of instructions or memory locations based on the instruction stored in the file. The instructions are stored in the files using 8-, 16-, or 32-bit memory words. The number of bytes, words, or double words present depends on the type of folder examined. For example,

FIGURE 3-8: Rack/slot-based memory.

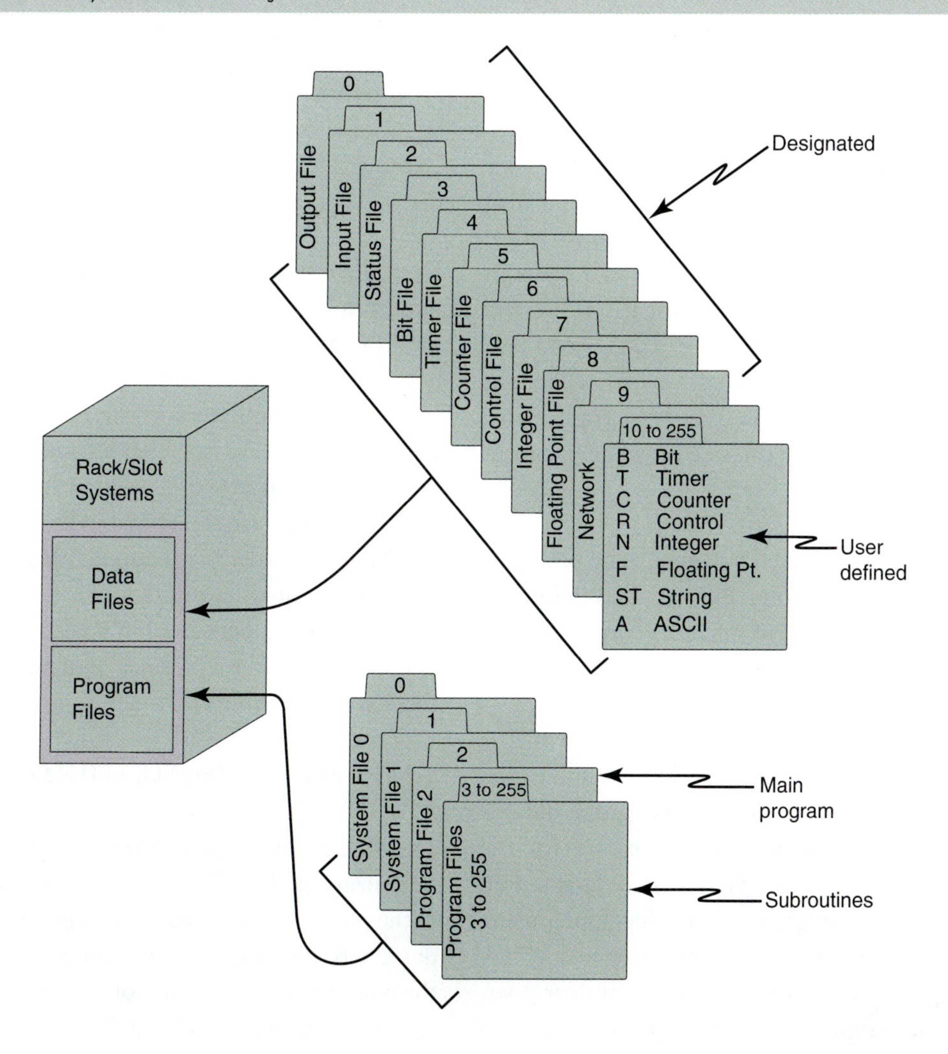

folder 0 (Output File) may have 31 words used to store output data status. Folder 7 (Integer File) may have 1000 words used to store data in an integer format. The memory map depicted in Figure 3-8 has blocks of 16-bit words assigned to files 0 through 8. With this memory allocation technique, output data is assigned a specific part of memory, input is assigned another, and all the files have specific locations where their values are stored. Therefore, in order to know the status of a field device connected to an input module, one must know the specific memory location in which to look for that data.

A separate block of memory is designated as *user-defined* space. In this location, the programmer has the option to designate what type of data (i.e., timers, counters, integer values, or any of the designated file types) will be stored. User-defined space is important when the available memory assigned to a file has been consumed.

Program files. In Figure 3-8 the program files, a part of processor memory, store the ladder logic programs. Two program file types, *system* and *program,* exist. They are subdivided into *system functions* (file 0 and 1), *main program* (file 2), and *subroutine programs* (files 3 through 255). File 2 is the default location for the main program, and files 3 through 255 are available for subroutines called from the main program. The size of data and program memory is determined by the size of the memory in the PLC.

Tag- or variable-based systems. The new generation of Allen-Bradley PLCs, called the Logix series, uses a different memory organization, called a *tag-based* memory system. In this system all data (i.e., output status, input status, integer values, or any of the designated file or data types) are assigned a variable name called a tag. In Logix systems the program data is stored in a tag, so it is not necessary for the programmer to know where that data is stored in memory. A program can be developed using only tag names, and the processor tracks where all tag values are stored. The programmer does, however, assign input and output terminals to input and output tags before the program is executed.

With the two types (rack-based or tag-based) of PLC memory systems defined, the memory organization for each of the Allen-Bradley processors can be introduced. The PLC 5 is addressed first, followed by the SLC 500 and then the ControlLogix.

3-5-2 Allen-Bradley PLC 5 Memory Organization

The PLC 5 divides processor information into the two file groups, program and data, as described in the previous section. The data files area stores processor information obtained from input modules, results sent to output modules, and other system data. The PLC 5 is a rack/group-based memory allocation system with specific memory allocated for I/O data based on the rack and group assigned to the I/O module. In PLC 5 notation the place where modules are loaded is called a *chassis* (the unit into which modules are placed) instead of a rack, because a rack may not be limited to one chassis. The structure of the data files 0 through 999 are illustrated in Table 3-1. Study the table carefully and then continue.

Data file content. Table 3-1 shows the 16 data files (numbers 0 through 8 and 3 to 999) used in the PLC 5, as well as the user-defined file numbers 9 to 999. Each file type has an identifier indicated by the letters in the File Type Indentifier column. The associated file number is listed plus the number of words used to store the data. Some file types use structures, which are groups of memory words. For example, there are 1000 timer structures permitted. This means that 1000 timers are permitted in a ladder logic program. Each timer or timer structure requires three words to hold the timer data.

3-5-3 Allen-Bradley SLC 500 Memory Organization

The SLC 500 is similar to the PLC 5 in memory organization. The SLC also divides information in the processor into two file groups, program and data, with the same type of information stored in the data files. The structure of data files 0 through 255 is illustrated in Table 3-2.

TABLE 3-1 PLC memory organization.

File Type	File Type Identifier	File Number	Maximum Number of 16 bit Words or Structures
Output image	O	0	32 words[1]
Input image	I	1	32 words[1]
Status	S	2	12 words[1]
Bit (binary)	B	3	1000 words
Timer	T	4	1000 structures (3000 words)[2]
Counter	C	5	1000 structures (3000 words)[2]
Control	R	6	1000 structures (3000 words)[2]
Integer	N	7	1000 words
Floating point	F	8	1000 structures (2000 words)[2]
ASCII	A	3 to 999	1000 words
BCD	D	3 to 999	1000 words
Block transfer	BT	3 to 999	1000 structures (6000 words)[2]
Message	MG	3 to 999	585 structures (32760 words)[2]
PID	PD	3 to 999	399 structures (32718 words)[2]
SFC status	SC	3 to 999	1000 structures (3000 words)[2]
ASCII string	ST	3 to 999	780 structures (32769 words)[2]
User defined	–	9 to 999	6

1. This is the number for a PLC 5/11 and 5/20. The number increases for PLC 5/30 through 5/80.
2. The structure value indicates the number of these file types that are permitted. Structures require multiple words to store the files data.

The table shows 10 ten data files (numbers 0 through 9) used in the SLC series and the user-defined file numbers 10 through 255. Compare Tables 3-1 and 3-2 to understand the differences between the PLC 5 and SLC 500 systems.

Data file content. Data files are organized by the type of data they contain using the SLC 500 instruction and processor notation. The files commonly used on all SLC PLCs include the following:

- **Output (file 0):** This file stores the on or off condition at the output terminals for the output module associated with this memory register.
- **Input (file 1):** This file stores the on or off condition at the input terminals for the input module associated with this memory register.
- **Status (file 2):** This file stores controller information used for troubleshooting controller and program problems.
- **Bit (file 3):** A bit consists of one **b**inary dig**it** and is often referred to as a Boolean type of data element. Bit files are used most often for bit (relay logic)-type program development.
- **Timer (file 4):** This file stores the data for each timer used in a program. The data includes timer accumulator and preset values, plus all status and output bits.
- **Counter (file 5):** This file stores the data for each counter used in a program. The data includes counter accumulator and preset values, plus all status and output bits.
- **Control (file 6):** This file stores the length, pointer position, and status bits for specific PLC instructions.

TABLE 3-2 SLC memory organization.

File Type	File Type Identifier	File Number	Maximum Number of 16 bit Words or Structures
Output image	O	0	31 words[1]
Input image	I	1	31 words[1]
Status	S	2	164 words[1]
Bit (binary)	B	3	256 words
Timer	T	4	256 structures (768 words)[1]
Counter	C	5	256 structures (768 words)[1]
Control	R	6	256 structures (768 words)[1]
Integer	N	7	1000 words
Floating point	F	8	256 structures (512 words)[1]
Network	x[2]	9	256 words
User Defined	x[3]	10 to 255	246 words

1. The structure value indicates the number of these file types that are permitted. Structures require three words to store the file data.
2. If non SLC 500 devices exist on the Allen Bradley DH-485 network link, use this area for network transfer. You can use either binary (B) or integer (N) file types by specifying the appropriate letter for x. Otherwise, you can use file 9 for user-defined files.
3. Use this area when you need more binary, timer, counter, control, integer, floating-point, or network files that will fit in the reserved files. You can use binary (B), timer (T), counter (C), control (R), integer (N), floating-point (F), or transfers (B and/or N) file types by specifying the appropriate letter for x. You cannot use this area for output image, input image, and/or status files.

- **Integer (file 7):** This file typically includes 256 16-bit words for the storage of unsigned or signed integer values. The storage elements can be addressed at the word and bit level. The range of stored signed integer values is –32,768 to +32,767, and the range for unsigned values is 0 to 65,635.
- **Floating point (file 8):** This file stores single precision non-extended 32-bit numbers that include their whole and decimal components. The range of values stored is $+/- 1.1754944 \times 10^{-38}$ to $+/- 3.4028238 \times 10^{+38}$. Some PLCs reserve 64-bit memory locations for even larger scientific notation values.
- **Network (file 9):** Network data
- **User-defined (files 10–255):** These files can be used to create any file type from 3 through 9. They are used to expand the number of data files available to the programmer.

This chapter describes the output (0), input (1), and bit (3) file types in detail and indicates the type of processor data stored in the status file (file 2). The remaining file types are covered in Chapters 4 through 11.

Program file content. Program files contain controller information, the main ladder program, interrupt subroutines, and all subroutine programs. These files in the SLC system are:

- **System Program (file 0):** This file contains various system-related information and user-programmed information, such as processor type, I/O configuration, processor file name, and password.
- **Reserved (file 1):** This file is reserved.
- **Main Ladder Program (file 2):** This file contains the main ladder logic program.

- **Subroutine Ladder Program (files 3 to 255):** Subroutines are ladder logic programs called from the main ladder logic program or other subroutine program. All subroutine programs are placed in this program area.

The memory organization for the ControlLogix processor is the final processor covered.

3-5-4 Allen-Bradley Logix System Memory Organization

Variable- or tag-based systems are used in all of today's new PLC models, such as the Allen-Bradley's Logix family of processors. In this type of addressing system, field device inputs and outputs, internal relays, and data values are assigned variable names, like the variables used in programming languages such as BASIC or C. Instead of the term *variable,* Allen-Bradley uses the term *tag* and defines it as follows.

A tag is a text-based name for an area of the controller's memory where data is stored.

In the ControlLogix controller, tags are a mechanism for allocating memory, referencing data in programs, and monitoring data. The minimum memory allocation for a tag is 4 bytes (16 bits).

At some point in the design, tags that represent input and output field devices are assigned to an I/O module and a specific module terminal number. When the variables are defined, a wide assortment of data types are available.

Variable data types. The most frequently used ControlLogix data types are divided into five groups: *Boolean, integer, real, strings,* and *user defined.* The Boolean group has only one data type defined, but other groups, such as integer, have eight definitions of integer data that are available for selection by the programmer. In the user-defined category, the system gives the automation designer and PLC programmer the option of creating a unique data type specific to the process being controlled. The tag data can be *restricted* to the local program or can be *global* so it is available for all programs and tasks within the controller. Input and output data types are all global.

There are two predefined data types: *basic data types* and *structured data types.* Table 3-3 lists and specifies the more commonly used basic data types, including discrete, integer, and real. Note the description, size, and range for each of the commonly used types.

Structured data types are created for predefined functions and instructions used in ladder logic. As seen in the list of the structured data types in Table 3-4, commonly used instructions such as timers and counters are a specific data type. Note the names of each group and the data stored.

ControlLogix program organization. In the rack/group- or rack/slot-based systems, the programmer assigns the data from an input field device to an input instruction on a ladder rung by specifying the memory location for the data.

TABLE 3-3 Data types for IEC language variables.

Data Type	Bits			
	31 16	15 8	7 1	0
Bool	not used			0 or 1
Sint	not used		−128 to +127	
Int	not used	−32,768 to +32,767		
Dint		−2,147,483,648 to +2,147,483,647		
Real	-3.40282347E^{38} to −1.17549435E^{-38} (negative values) 0 1.17549435E^{-38} to 3.40282347E^{38} (positive values)			

TABLE 3-4 Tag-based system-structured data types.

Structure Data Type	Type of Data Stored
Counter	Control structure for the counter instructions
Timer	Control structure for the timer instructions
Control	Control structure for the array instructions
Motion Instructions	Control structure for the motion instructions
Motion Group	Control structure for the motion group
PID	Control structure for the PID instructions
Axis	Control structure for an axis
Message	Control structure for the message instructions

Therefore, the programmer must know the input module and terminal number for the field device because the address is based on that information.

In the tag-based system, the allocation of variable names for program values is not tied to specific memory locations in the memory structure. For example, in the ControlLogix system, the programmer has three options in creating tags:

1. Leave the ? in the tag location in the ladder and then define a tag at a later time.
2. Define all of the tags for the programming project before the program is developed.
3. Define the tags as the program is entered.

After tags are defined, those that represent input or output field devices are assigned the input or output terminal where the field device is attached. In addition, the programmer must be familiar with the memory structure used in the Logix system. The memory structure for the ControlLogix PLC is illustrated in Figure 3-9.

This memory structure uses the following terms:

Project: A *project* (the large white project box in Figure 3-9) is a collection of all of the program's elements. When a project is created the new controller dialog box requests the following information:

- Controller model and software revision number
- Project name and description
- Rack size and slot location of the processor
- File folder name into which the program is stored

Task: A *task,* associated with a program, has two functions. First, it holds the information necessary to schedule the program's execution. Second, it sets the execution priority for one or more programs. ControlLogix supports up to 32 tasks. Two types of tasks are used: *continuous* and *periodic*.

- Continuous tasks execute non-stop. At the creation of a new project, a continuous task (white task box in Figure 3-9) is created.
- Periodic tasks interrupt the continuous task and execute for a fixed length of time at specific time intervals. Whenever the time period for the task expires, the task executes one last time. The periodic rate can be from 1 ms to 2000 s with a default of 10 ms. For example, a periodic task, like the one in Figure 3-10, is used to store production information at regular fixed intervals.

Program: Although each task requires at least one *program* (white program box inside white task box in Figure 3-9), a task can have as many as 32 separate programs. Only one program in a task can execute at any one time.

Routine: *Routines* provide the executable code for the project, using a specific programming language such as ladder logic.

Main routine: When a program executes, its main routine executes first. The main routine is used to call (execute) other routines (subroutines).

Subroutine: A subroutine is any routine other than the main routine.

FIGURE 3-9: Tag-based program structure for ControlLogix.

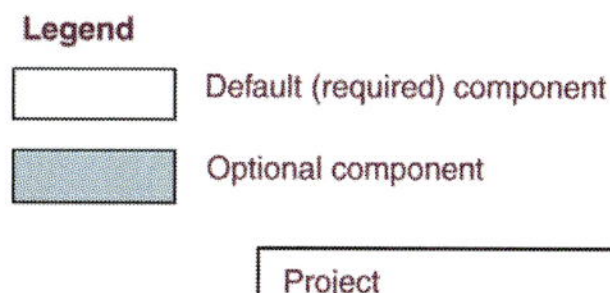

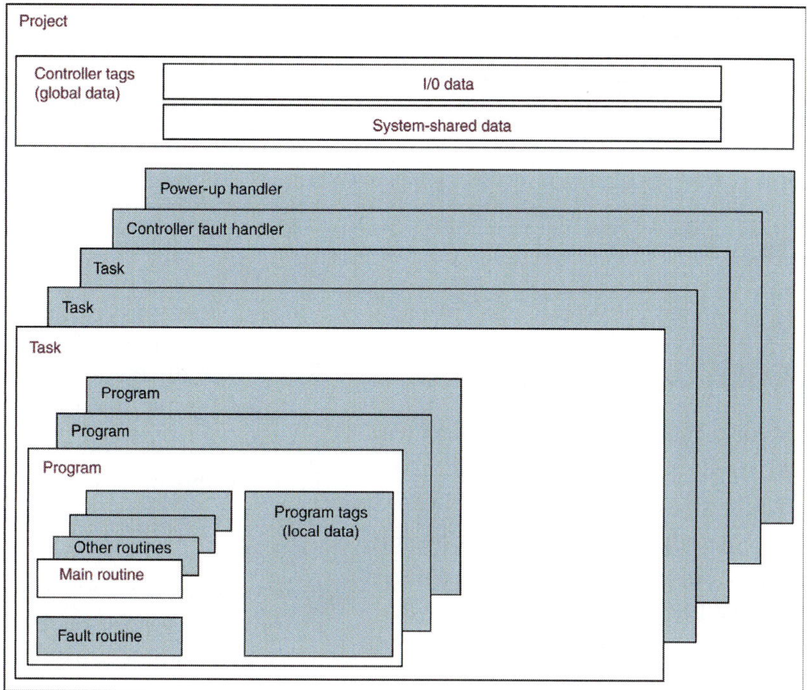

FIGURE 3-10: Periodic and continuous tasks.

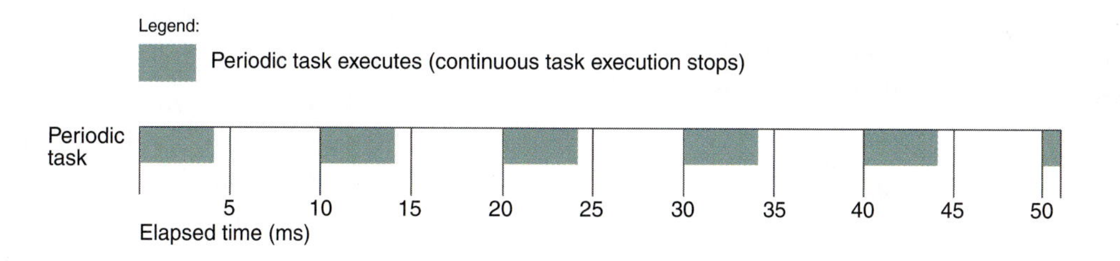

Program tags: The *program tags* (blue box in Figure 3-9) area is a memory location in which to save variable information such as the local tag names and data types.

Controller tags: The *controller tags* (white box in project box in Figure 3-9) area is a memory location in which to save global variable information.

The program structure illustrated in Figure 3-9 is represented in the RSLogix 5000 software as shown in Figure 3-11. The first folder, Controller, appended with the project name (Controller_1 in the figure), includes global tags and system data. The next folder, Tasks, has all of the program code including tasks, programs, and routines. In addition, the program tags are saved in this area.

This section of the text described the memory and register structures used in the three Allen-Bradley processors. The PLC 5 and SLC 500 had similar structures but the ControlLogix used tags in place of specific memory locations. The next step is to introduce the addressing format for the three PLC processors based on the memory and register structure.

FIGURE 3-11: Program structure and RSLogix 5000 left panel.

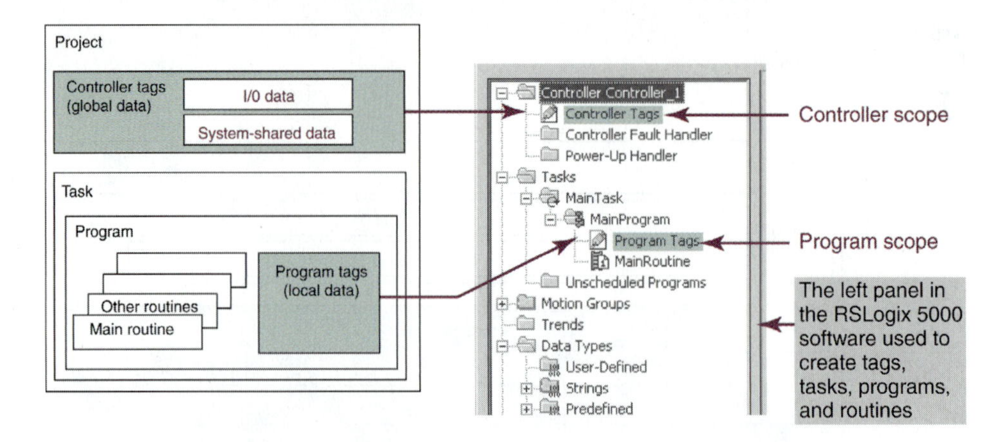

3-6 INPUT AND OUTPUT ADDRESSING

An engineer or technician must be familiar with both rack-based and tag-based addressing schemes because all three Allen Bradley PLC processors are frequently found in automation control schemes. As a result, all three are covered starting with the PLC 5.

3-6-1 PLC 5 Rack/Group-Based Addressing

Input and output (I/O) addressing connects the physical location of a field device at a terminal on an I/O module to a bit location in the processor memory. Therefore, before creating a PLC 5 program using RSLogix 5, the system is configured by performing the following pre-program tasks:

1. Select input and output modules.
2. Determine field device terminations to the modules.
3. Define the memory address of field devices and data.

The PLC 5 uses a rack and group number, called *rack/group addressing*.

Defining terminals, groups, and racks. The main purpose of a PLC is to control output field devices, like valves, using inputs from switches and sensors. To use field device data in a program, the condition of input and output devices must be saved in memory. The part of processor memory that saves input and output addresses is the *input image table,* called file I, and *the output image table,* called file O (see Table 3-1). Each bit of the input and output image tables is associated with a terminal on an input or output module that can be wired to a field device. Learning this process starts with defining *terminals, groups,* and *racks* for the PLC 5.

Terminals: A *terminal* is an attachment point for field devices on PLC input and output modules and is associated with specific space in processor memory. Terminals are sometimes called I/O points.

Groups: A *group* includes 16 input terminals and 16 output terminals. Therefore, a group consists of one input word (16 bits) and one output word (16 bits).

Logical rack: A *logical rack* is a set of 8 groups (numbered 0 to 7). Since a group is one input word and one output word, a logical rack has 8 input words and 8 output words, or 128 input terminals (8 words, 16 bits each) and 128 output terminals (8 words, 16 bits each). PLC 5 processors can support up to 24 racks with the zero rack in the PLC 5 system reserved for the processor.

In summary, the terminal numbers increase from zero *counting in octal* to the maximum number for the module. Every input and output terminal has a corresponding input and output bit in the input word and output word. A group is composed of an input word and an output word. A rack represents 8 groups (numbers 0 to 7) and may or may not be the same as the physical chassis in which the modules are inserted.

PLC 5 Modes and memory allocation. Usually group 0 is associated with the slot 0 in the chassis, and group 7 is a slot toward the right end. The PLC 5 is configured using three addressing modes: *half-slot, one- or single-slot,* and *two-slot* addressing. These addressing modes are defined as:

Half-slot: In *half-slot addressing* the PLC is configured with two groups to a slot, so that the slot can accept a module that has two input words (32 terminals) and two output words (32 terminals) for field device attachment. One group (16 inputs and 16 outputs) fills one-half of the slot.

One- or single-slot: In *one-slot addressing* the PLC is configured with one group to a slot, so that the slot can accept a module that has one input word (16 terminals) and one output word (16 terminals) for field device attachment. Since one group (16 inputs and 16 outputs) fills the slot, the slot number and the group number are the same.

Two-slot: In *two-slot addressing* the PLC is configured with one group distributed over two adjacent slots, so that the adjacent slots can accept a module that has one input word (16 terminals) and one output word (16 terminals) for field device attachment. One group (16 inputs and 16 outputs) fills two slots.

As Figure 3-12 indicates, the chassis is configured by setting switches. In the PLC 5, the relationship between the slot locations and the group number is determined by two factors: (1) the number of input and output terminals on the module placed into the slot, and (2) the selected addressing mode for the chassis. Single-slot addressing is the only one described and used in this text because when 16-terminal or point input or output modules are used in this mode, the group and slot numbers are the same. This characteristic makes addressing easier to understand. Figure 3-13 illustrates an example of 16-terminal input and output modules using single-slot addressing. Note that the slot numbers at the top match the group number at the bottom.

Chassis slots 6 and 7: In Figure 3-13, slot 6 has a 16-point input module and slot 7 has a 16-point output module. With single-slot addressing, group 6 (slot 6) and group 7 (slot 7) identify the bits for these I/O points. As a result, word 6 in the input image table and word 7 in the

FIGURE 3-12: PLC 5 slot addressing mode switch setting.

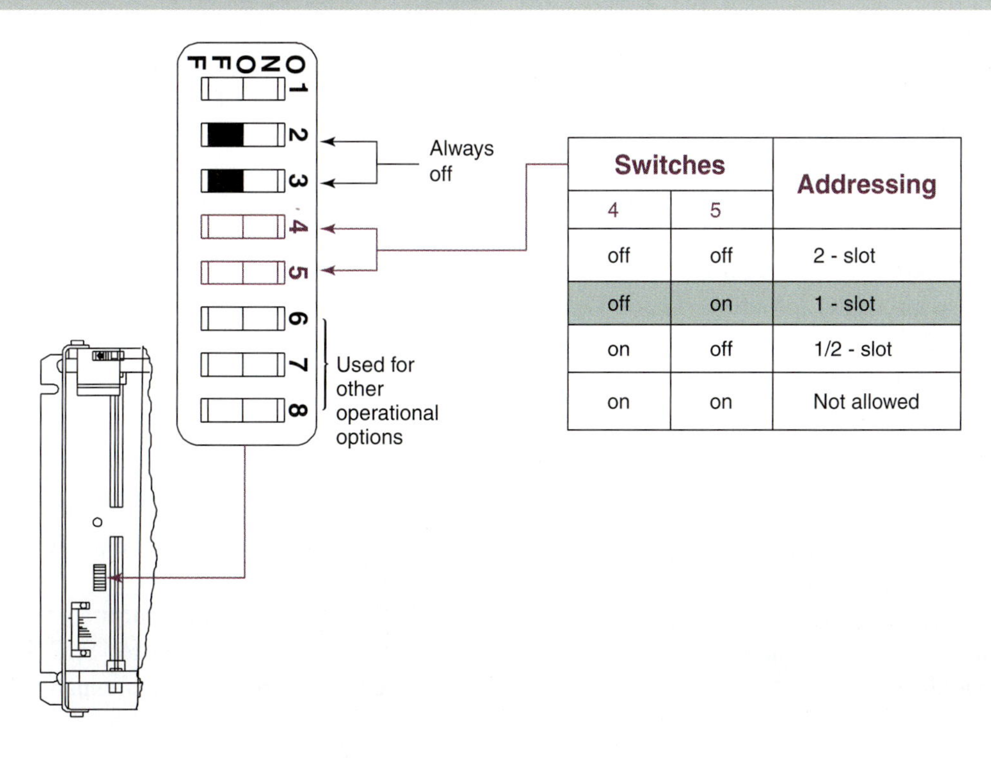

Switches		Addressing
4	5	
off	off	2 - slot
off	on	1 - slot
on	off	1/2 - slot
on	on	Not allowed

FIGURE 3-13: PLC 5 single-slot addressing mode.

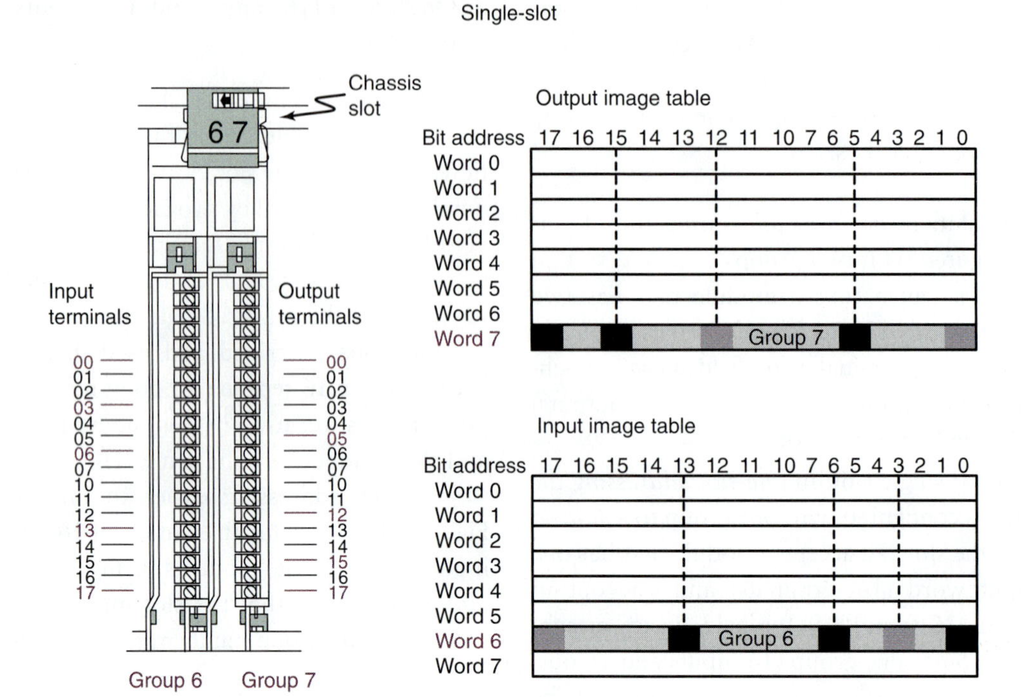

output image table are allocated to group 6 and group 7, respectively. Each group can have both an input word and an output word, but these modules are just an input and output type. As a result, word 7 in the input image table and word 6 in the output image table are not used.

With the input and output memory organization covered for the PLC 5 system, the addressing format for input and output bits is discussed next.

PLC 5 Rack/Group I/O addressing. The PLC 5 system with single-slot addressing, illustrated in Figure 3-14, has the following format for addressing I/O.

> *The structure starts with either an I for inputs or an O for outputs, and uses a colon (:) as a delimiter or separator. The next two digits are the rack number, which is followed by the I/O group number. The forward slash (/) separates the final two digits, which are the terminal numbers or input or output image table bit numbers.*

The illustration in Figure 3-14 shows a single-slot addressing scheme with 16-point input and output modules located in adjacent slots 4 and 5 in rack 01. Because single-slot addressing is used, the slot number is also the group and word number used in the address. The addresses on the ladder rung below the figure use the format previously described for PLC 5 systems. Note that the RSLogix 5 ladder logic programming software places the bit number below the rung and the rest of the address above the line. All numbers in the address are in the *octal* number system.

3-6-2 SLC 500 Rack/Slot-Based Addressing

The addressing of I/O in the SLC model PLC is similar to the PLC 5 single-slot addressing previously described. However, the SLC uses slots and PLC 5 uses groups. The programming software used for the SLC system is RSLogix 500.

SLC 500 series input and output addressing. Input and output addressing for SLC 500 was introduced in Chapter 1 and is expanded upon

FIGURE 3-14: PLC 5 address syntax for inputs and outputs.

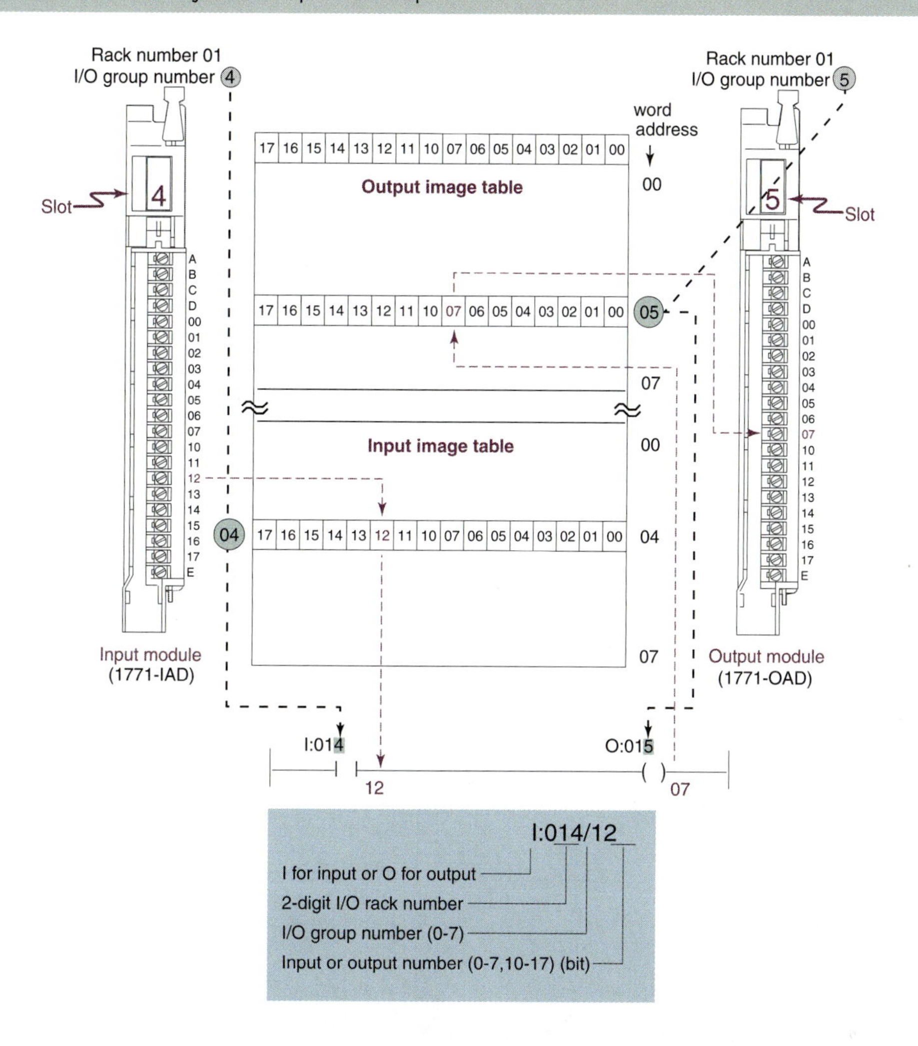

EXAMPLE 3-1

Determine the input and output addresses for the bits with light blue blocks in Figure 3-13. Assume the modules are in rack 01.

SOLUTION

a. Figure 3-13 output: **0:017/0**—rack 01, group 7, and bit 0
b. Figure 3-13 output:**0:017/12**—rack 01, group 7, and bit 12
c. Figure 3-13 input: **I:016/3**—rack 01, group6, and bit 3
d. Figure 3-13 input: **I:016/17**—rack 01, group 6, and bit 17

Note that the octal number system is used for the terminal numbers on the PLC 5 modules.

here. The illustrations in Figure 3-15(a) through (d) provide the required information for learning the SLC rack/slot addressing process. Note that the PLC rack in the figure has four slots, with the processor in slot 0 and an input module in slot 1. The module has 16 inputs (IN0 to IN15) and three field devices connected to input terminals 2 (momentary push button switch), 6 (NO limit switch held closed), and 12 (a three-wire proximity sensor). In order to reference the

FIGURE 3-15: Rack/slot addressing for the SLC 500 input and output modules.

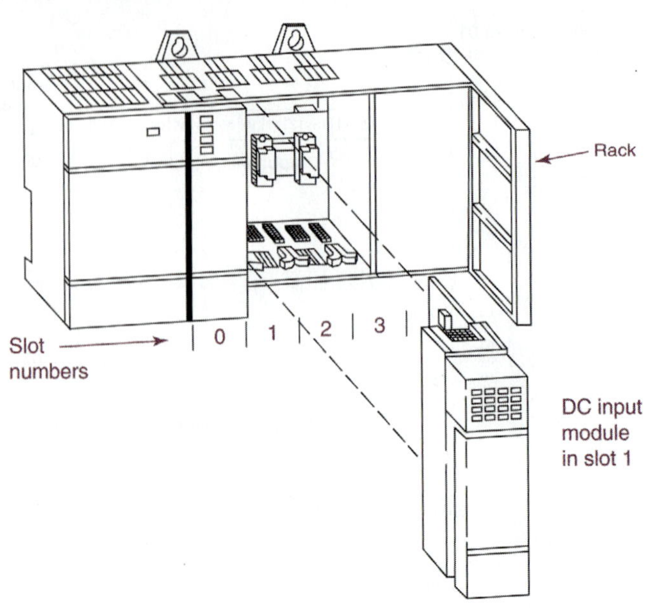

(a) SLC 500 four slot rack

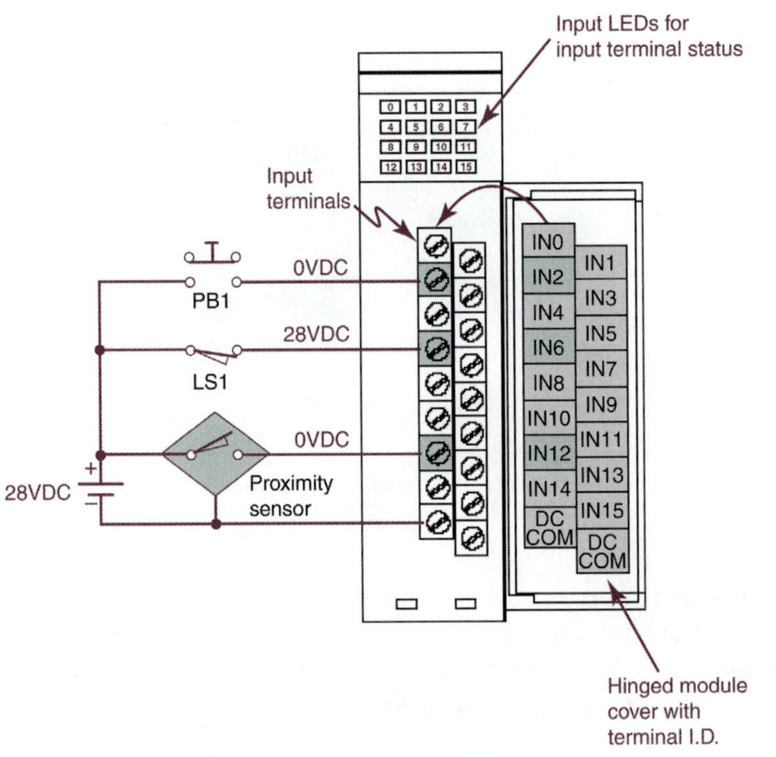

(b) SLC 500 input module

FIGURE 3-15: (Continued).

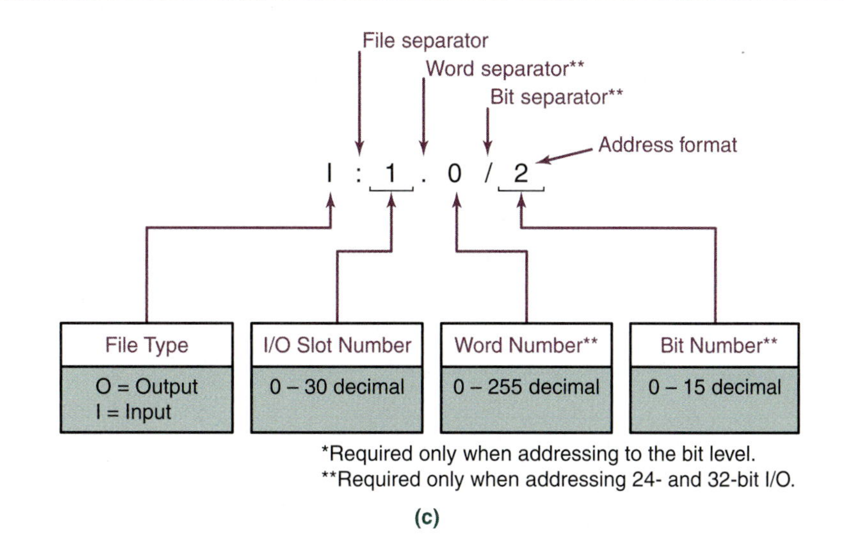

(c)

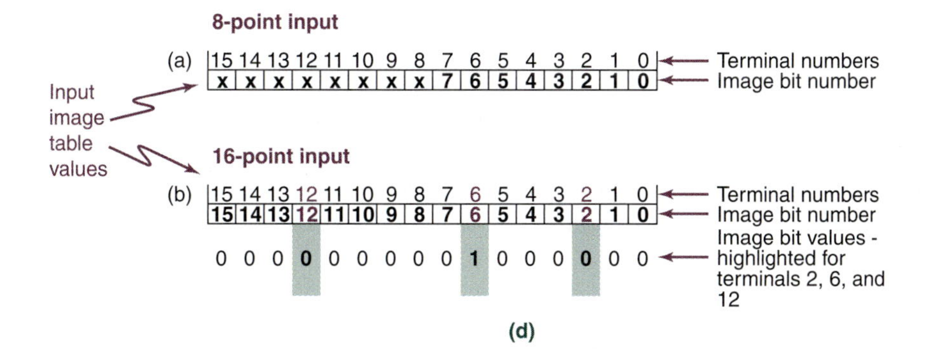

(d)

three field devices in a ladder program, their addresses must be known. The following address format description is illustrated in Figure 3-15(c).

The structure begins with either an I for inputs or an O for outputs and uses a colon (:) as a delimiter or separator. The next digit, the slot number, is followed by the period (.) delimiter. The next digit, the word number (only needed if the addressed word is not 0), is followed by the forward slash (/) delimiter, which separates the word number from the final two digits for the terminal number.

The values used for the address in Figure 3-15(c) are for the momentary pushbutton switch connected to terminal IN 2 in Figure 3-15(b).

Two input register or image table examples are illustrated in Figure 3-15(d): an 8-point input module and a 16-point input module. If the number of inputs exceeds 16, the address must include the word number (0 for inputs 0 to 15 and 1 for inputs 16 and higher). The module in Figure 3-15(b) is a 16-point input module. Note that the input image table values (0s and 1s) for the field devices in Figure 3-15(b) are displayed in Figure 3-15(d). The open push button and proximity sensor have 0 volts at the terminal, so the register value is 0. The limit switch has 28 volts present, so its register value is 1.

The illustrations in Figure 3-15(a) through (d) are for an input module in slot 1, but a similar approach is used when an output module is added to the PLC. For example, if an output module is placed in slot 2, the I in the address in Figure 3-15(c) changes to an O, and the slot changes from 1 to 2. The output terminal to which the field device is connected determines the bit number. Again, a word number is necessary only if the number of outputs exceeds 16. Examples 3-2, 3-3, and 3-4 illustrate this addressing concept.

EXAMPLE 3-2

An Allen-Bradley SLC system uses a rack like the one illustrated in Figure 3-15(a), with the 16-point input DC module in slot 3. Determine the address for a discrete input signal attached to terminal 5 of the module.

SOLUTION

The address for the discrete DC input would be

I:3/5

The letter I indicates that it is an input, the 3 indicates that the DC input module is in slot 3, and the 5 indicates that the discrete field device input signal wire is connected to terminal 5.

EXAMPLE 3-3

An Allen-Bradley SLC system uses a rack like the one illustrated in Figure 3-15(a), with the 24-point input DC module in slot 2. Determine the address for a discrete input signal attached to terminal 19 of the module.

SOLUTION

The address for the discrete DC input would be

I:2.1/3

The letter I indicates that it is an input, the 2 indicates that the DC input module is in slot 2, and the 1 indicates that it is the second word. Two words are needed because there are more than 16 terminals. Because terminal 16 is bit 0 in word 1, terminal 19 is bit 3 in the second word. Figure 3-16 illustrates the relationship between terminals and register bits for input and output modules in the SLC 500 system. Find bit 3 in word 1 and verify that it is terminal 19 on the module.

EXAMPLE 3-4

An Allen-Bradley SLC system uses a rack like the one illustrated in Figure 3-15(a), with an 8-terminal DC output module in slot 1. Determine the address for the field device attached to terminal 7 of the module.

SOLUTION

The address for the discrete DC output would be

O:1/7

The letter O indicates that it is an output, the 1 indicates that the module is in slot 1, and the 7 indicates that the field device output signal wire is connected to terminal 7.

FIGURE 3-16: Addressing 32-point input and output modules for the SLC 500 PLC.

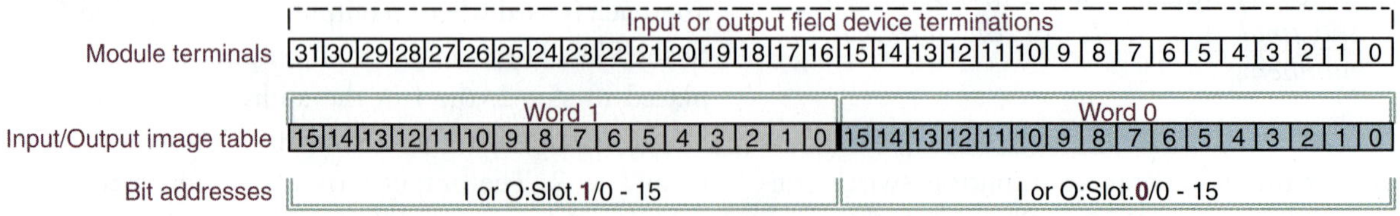

Input or output field device terminations

Module terminals: 31 30 29 28 27 26 25 24 23 22 21 20 19 18 17 16 15 14 13 12 11 10 9 8 7 6 5 4 3 2 1 0

Input/Output image table: Word 1: 15 14 13 12 11 10 9 8 7 6 5 4 3 2 1 0 | Word 0: 15 14 13 12 11 10 9 8 7 6 5 4 3 2 1 0

Bit addresses: I or O:Slot.**1**/0 - 15 | I or O:Slot.**0**/0 - 15

Allen Bradley SLC 500 addressing for a 32-point input or output module

I/O Addressing. The display of the input and output registers for an SLC 500 controller is illustrated in Figure 3-17. Note that output slot 5 and input slot 1 have 32-point modules installed. The output words are O:5.0 and O:5.1, and the inputs are I:1.0 and I:1.1. The addressing options when single and multiple words are present are illustrated in the output register data file. The address O:8/14 indicates that the field device is attached to terminal 14, and the module is in slot 8. No word (O:8.0/14) element is present since the 0 word is assumed if none is entered.

If two words are present, then the elimination of the word element is still an option. For example, the bit addressed O:5.1/4 indicates that the bit is in locaton 4 of word 1 for the module in slot 5. On the module's word 1 location 4 is terminal number 20, since numbering of terminals starts at 0 and goes to 31. As a result, a second addressing option would be O:5/20 with the word element eliminated and the bits numbered 0 to 31, identical to the terminals. Bits in word 0 are numbered 0 to 15 and in word 1 they are numbered 16 to 31. If additional words are present they would continue to count in this fashion. It is clear from this discussion that using memory locations for program files offers a number of disadvantages in data addressing. For example, inputs and outputs cannot be given a variable name associated with their field device, and all I/O wiring address assignments must be completed before programming for the PLC can begin.

3-6-3 Other Vendors' Rack/Slot PLC Addressing

Most companies standardize automation control to a single PLC vendor to make programming and troubleshooting of the system controller easier.

FIGURE 3-17: SLC 500 input and output registers.

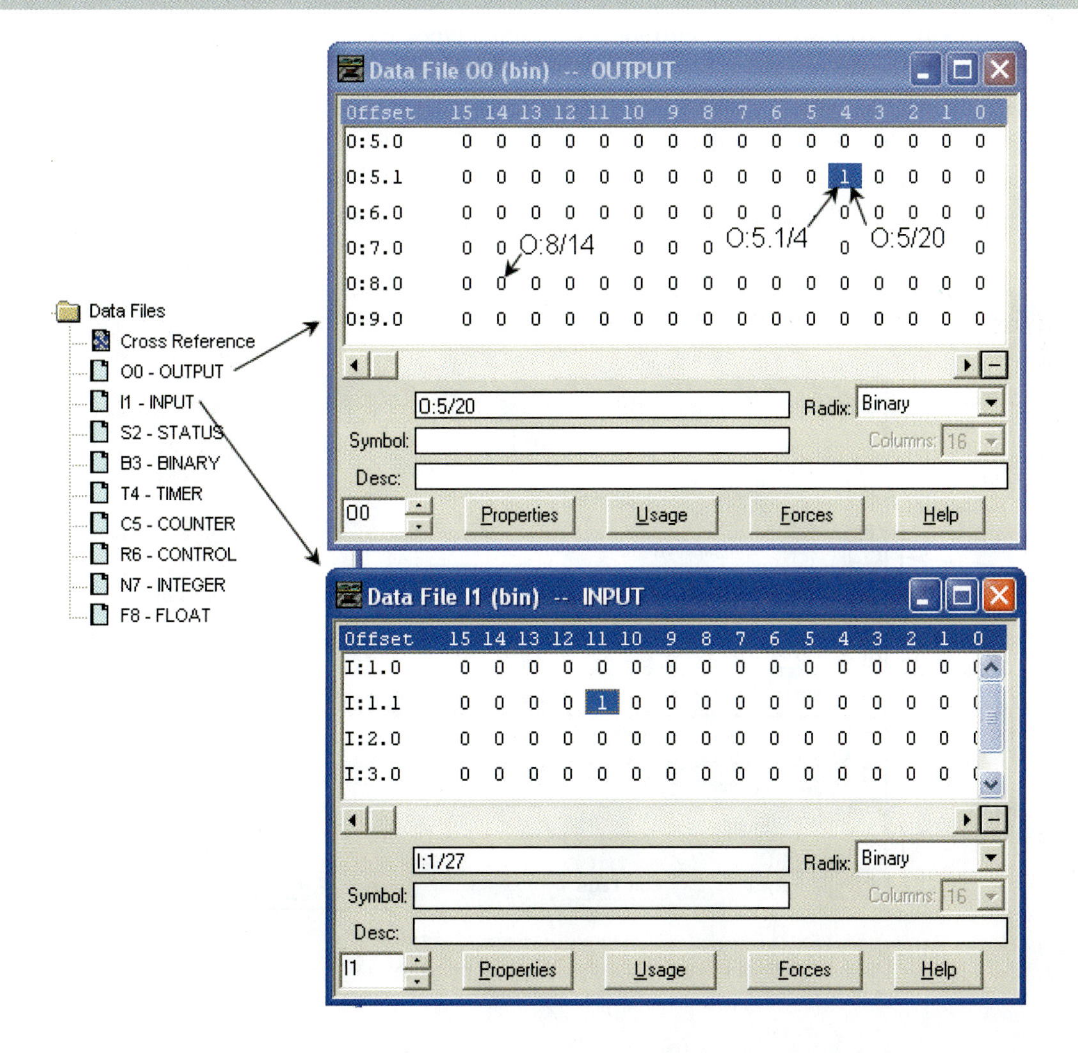

Unfortunately, this does not mean that only one type of PLC is used. Many automation systems purchased from vendors use a PLC as the imbedded machine controller. The PLC used is often one selected by the system builder and cannot be changed. One must be familiar with the operation and programming language of that PLC in order to troubleshoot machine problems. Therefore, someone working in manufacturing automation will need to know several PLC systems. No two PLC manufacturers have identical programming formats, but PLC procedures and instructions are similar to the Allen-Bradley languages covered in this text. When one is learned well, the others can be learned quickly.

3-6-4 Tag-Based Addressing

In the previous sections the older rack/group and rack/slot formats for addressing inputs and outputs was described. In each case, the programmer was obligated to use an address that was mapped to the specific memory location where that data was stored by the processor. In the current generation of PLCs, like AB's ControlLogix, the programmer assigns a *tag name* in the RS Logix 5000 software for field devices, process data, and most other ladder logic instructions. The type of tag needed—for example, Boolean (on or off), integer, real, timer, counter, or control—is specified by assigning a *data type* to the tag name. Review some of the examples of basic and structured data types described in Section 3-5-4 and listed in Tables 3-3 and 3-4.

With this approach, the storage location of the Boolean input or output tag no longer is required; instead, the programmer creates a tag to represent the data. All program development proceeds with just the tag names and the data types assigned. Later, input and output variables are matched with the pin number on the respective modules where the field devices are connected.

Creating tags in the tags window. The tags window is used to create program tags as the ladder logic is developed. Figure 3-18 provides an example of a tags window. Compare the data element, tag name, and type of data listed under those column headings.

Note that the tag names identify by name or describe the data that the tag represents, and the data types are like those listed earlier in the chapter in Tables 3-3 and 3-4. The tags have the same form for all types of data, which is a major benefit offered by this new generation of PLCs.

As shown in Figure 3-19(a), the tag window is opened from either the Controller tags or

FIGURE 3-18: Program tags window for creating tags.

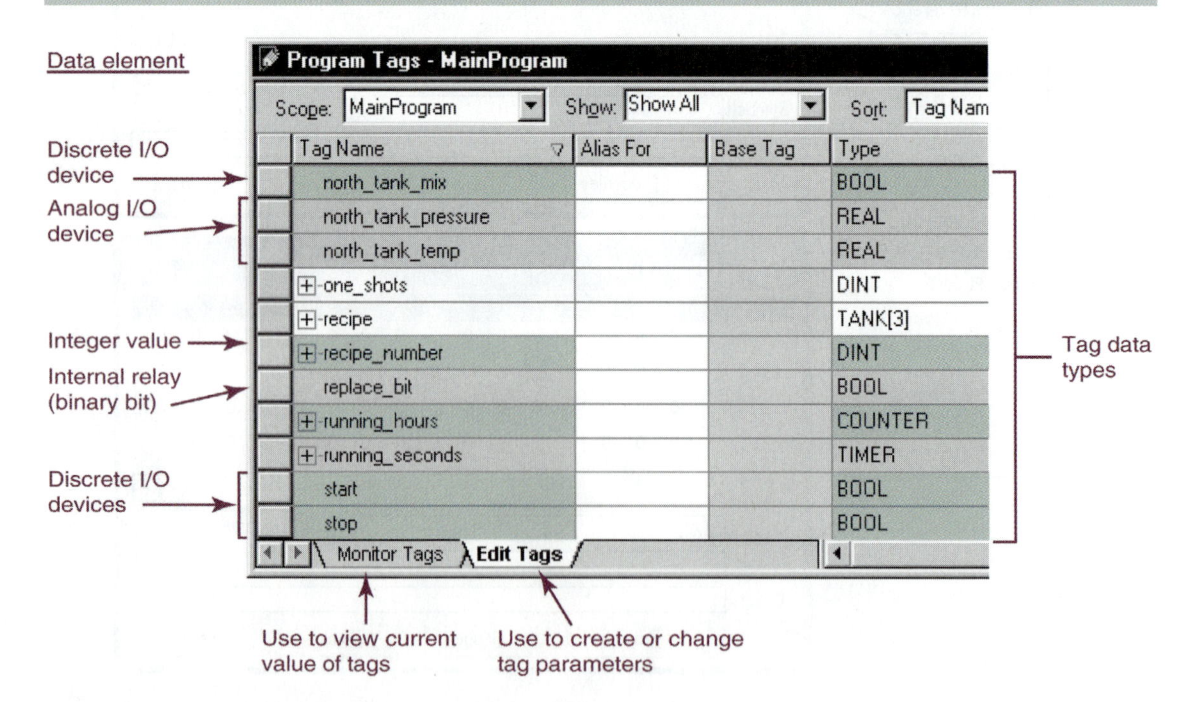

FIGURE 3-19: Creating tags and setting the alias.

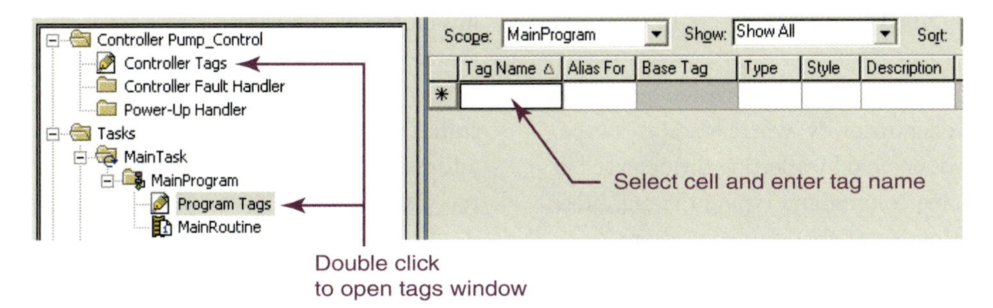

Double click
to open tags window

(a) Opening the tag window

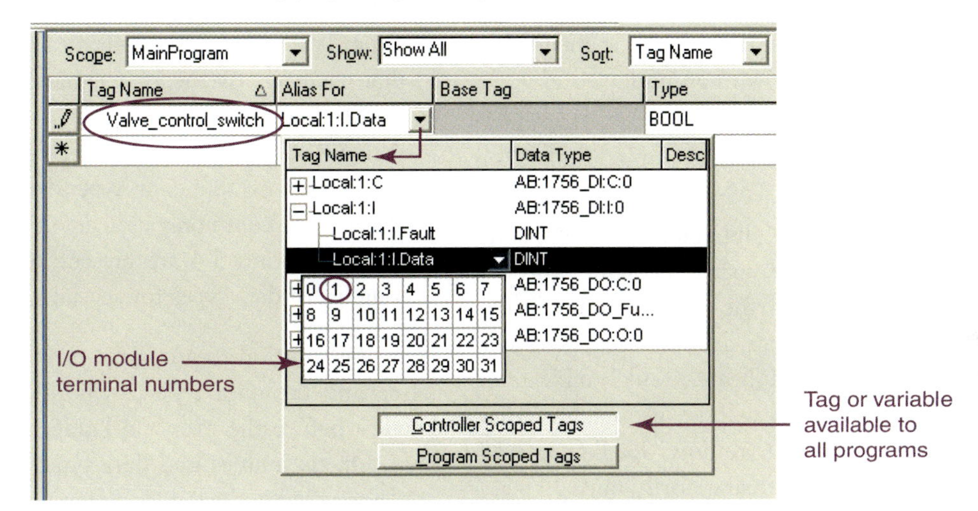

(b) Assigning I/O terminal numbers and address

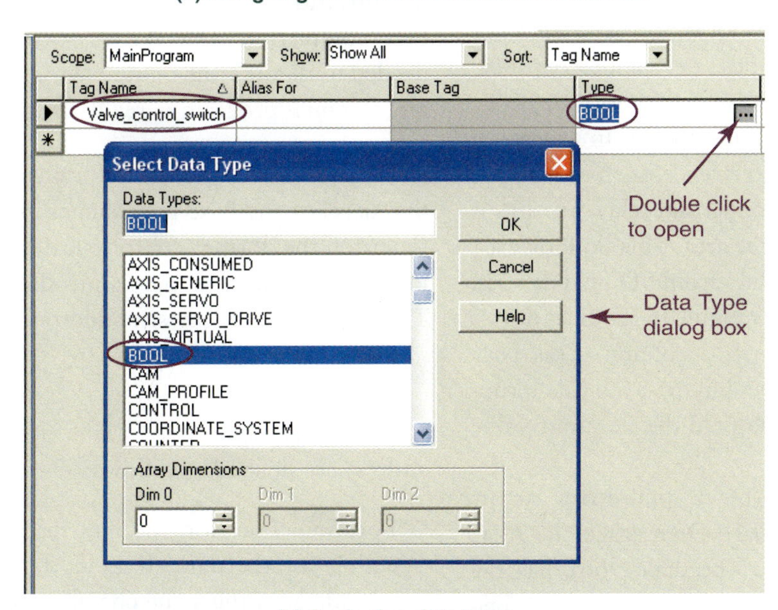

(c) Assigning data types

Program Tags file listing. Each of the four visible columns—Tag Name, Alias For, Base Tag, and Type—is described next.

Tag Name: A *tag name* is assigned to an area of the controller memory where data is stored. The name is entered by selecting the open cell in the tag name column and typing the tag. Note the following requirements and good practices:

- Tags can only include numbers, letters (not case sensitive), and single underscores (_) with a maximum length of 40 characters.
- Tags must begin with a letter or underscore but cannot end with an underscore.
- When entering tags, mixed case is used for ease of reading, for example, Tank_1 or Tank1 and not TANK_1 or TANK1.
- Tags are sorted alphabetically in the software; therefore, using the same word to start tags from the same manufacturing area keeps them grouped together. For example, Tank1_heater, Tank1_mixer, and Tank1_ sensor will be grouped in that order in the tag window. See Figure 3-18 for an example using north_tank.

Alias For: *Alias for* is a tag that references memory defined by another tag. The Alias For column links the data's internal memory location to the descriptive name for the value listed in the Name Tag column. In general, if a tag is an internal reference, such as a binary bit for an internal or virtual relay, an alias is not necessary, and the area is blank. However, if tags represent values from I/O field devices, a path through the module to the field device must be created in this column. This path gives the processor a link to a module terminal point where the field device input value is present.

In Figure 3-19(b), a pull-down arrow appears after *Local:1:I.Data* when the Alias For cell is selected. The dialog box has the address locations for all I/O modules. When *Local:1:I* is expanded two files appear: *Local:1:I.Fault* and *Local:1:I.Data*. When *Local:1:I.Data* is selected, a second drop-down arrow appears. This arrow displays a dialog box [Figure 3-19(b)] for all of the terminal numbers on the input module located at the address *Local:1:I.Data*. Next, the tag name (Value_control_switch) used in the program is linked to the input terminal number 1 where the field device represented by the tag name is connected. This step makes the logical address, called the tag name, an alias for the physical address located in *Local:1:I.Data.1*.

Base Tag: A *base tag* is a tag that represents the location where the data is stored in memory. In most cases, the address entry in the Base Tag column is the same as the entry in the Alias For column.

Type: The *Type* column indicates the data type that has been assigned to the tag name. When the heading Type is selected, Figure 3-19(c), the dialog box link illustrated in the figure is active and produces the data type dialog box as shown. The commonly used data types listed in Tables 3-3 and 3-4 are present, along with a long list of data types for special applications.

In the solution to a control problem, the tag name and tag data type can be identified and entered before the program ladder logic is created. Or, tag names and data types can be created as needed in the ladder design. In either case, identification of the physical addresses for the tags (completion of the Alias For column) is not needed until all of the module wiring is finished, and the final program is ready for field testing.

Alias and base tag address format. A study of the Alias For and Base Tag columns in Figure 3-21 indicates the format for physical addresses of input and outputs in the ControlLogix system. The format for the physical address is:

Location : Slot : Type . Member . Submember . Bit

The terms used in the format for the physical address are defined as follows:

Location: Location specifies the network location for the data. If LOCAL is used, the module is in the same chassis or DIN rail as the controller. If ADAPTER_NAME is used, it identifies the remote communication adapter, such as a DeviceNet remote I/O block or bridge module.

Slot: Slot specifies the slot number of the I/O module in its chassis or DIN rail.

EXAMPLE 3-5

Create tag names, data types, and aliases in a tag window for the tank control problem described in Chapter 1 [Figure 1–7(a) and Figure 1–17] using the PLC configuration and field device wiring illustrated in Figure 3-20.

SOLUTION

Study the solution in Figure 3-21. Note how tag names were chosen to represent the input and output field devices and how the choice of names caused similar data to be grouped together. Also, one tag, Valve_control_bit, is an internal memory bit used like the control relay in the original solution in Chapter 1.

Two additional columns, Style and Description, are displayed. In the Style column, the programmer indicates the desired number system radix for display of variable values. In the optional Description column each bit is described. This description appears above the instruction in the ladder logic if that display option is selected.

FIGURE 3-20: PLC configuration for Example 3-5.

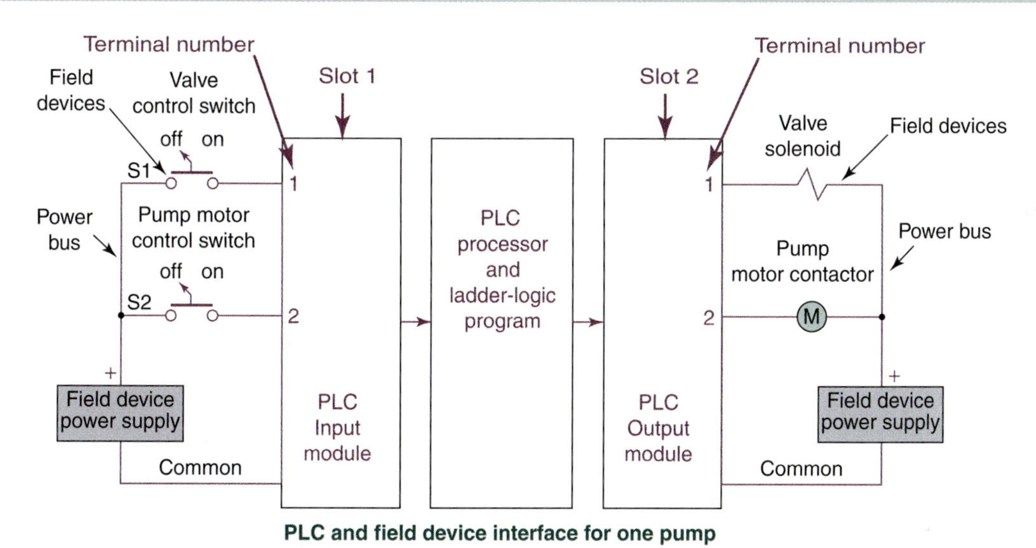

FIGURE 3-21: Tag window for pump control—Examples 3-5 and 3-6.

Pump control tag data

Scope: MainProgram Show: Show All Sort: Tag Name

Tag Name	Alias For	Base Tag	Type	Style	Description
Pump_motor_control_switch	Local:1:I.Data.2(C)	Local:1:I.Data.2(C)	BOOL	Decimal	Start switch for pump motor
Pump_motor_starter	Local:2:O.Data.2(C)	Local:2:O.Data.2(C)	BOOL	Decimal	PLC output bit for pump motor...
Valve_control_bit			BOOL	Decimal	Internal relay for valve control
Valve_control_switch	Local:1:I.Data.1(C)	Local:1:I.Data.1(C)	BOOL	Decimal	Start switch for valve solenoid
Valve_solenoid	Local:2:O.Data.1(C)	Local:2:O.Data.1(C)	BOOL	Decimal	PLC output bit for valve conta...

Type: Type specifies four types of data: *I* for input, *O* for output, *C* for configuration, and *S* for status.

Member: Member specifies the type of data that the module can store. For digital modules a DATA member usually stores the input or output bit values, and for an analog module a Channel member (CH#) usually stores the analog channel values.

Submember (optional): Submember is specific data related to a member.

Bit (optional): Bit specifies the specific terminal on a digital I/O module.

EXAMPLE 3-6

Define each term for the base tag associated with the Valve_control_switch and the Valve_solenoid in Figure 3-21.

SOLUTION

a. The base tag or physical address for the tag Valve_control_switch is Local:1:I.Data.2(C). Local indicates that the module is in the same rack as the processor, 1 indicates that the module is in slot 1 in the rack, I indicates that the module is an input, Data indicates that it is a digital input, 2 indicates that the selector switch is attached to terminal 2 on the module, and C indicates that it is a controller tag with global access.

b. The base tag or physical address for the tag Valve_solenoid is Local:2:O.Data.2(C). Local indicates that the module is in the same rack as the processor, 2 indicates that the module is in slot 2 in the rack, O indicates that it is an output module, Data indicates that it is a digital output, 2 indicates that the valve solenoid is attached to terminal 2 on the module, and C indicates that it is a controller tag with global access.

Two types of delimiters, colons (:) and periods (.), are used to separate the information presented in the physical memory call-out. Note that bits are not delimited by a slash (/) as in the PLC 5 and SLC 500. If the address is a control-type tag, a (C) is placed at the end of the address to indicate a controller tag with global scope.

After comparing all of the base tags in Figure 3-21 with the physical memory address format, one should be able to identify all parts of the physical address.

3-7 INTERNAL CONTROL RELAY BIT ADDRESSING

Relay ladder logic uses control relays to represent the binary values in the logical solution of control problems. Review the operation of the relay ladder logic in Figure 1-7 in order to see how control relays are used in the tank control problem. The number of poles on the mechanical relays is determined by the number of contacts needed in other rungs. PLC Ladder logic uses internal memory bits as internal relays, which are sometimes called *virtual control relays*. Since virtual relays are just internal memory bits, the number of virtual relays used and therefore the number of instructions with the virtual relay addresses in other rungs is limited only by the PLC memory size. Addressing these internal memory bits in a rack/slot-based system will use a format similar to that required for field devices. Internal relays in a tag-based system require only the identification of the tag name and the setting of the data type to Boolean. The following sections cover each type.

3-7-1 PLC 5 and SLC 500 Binary Bit Addressing

Review the memory layout for the Allen-Bradley PLC 5 and SLC 500 family of processors introduced earlier in Tables 3-1 and 3-2, respectively. Note that file 3, the bit file, is designated by the letter *B* in the address. The format for addressing bit files in both of these PLC systems is illustrated in Figure 3-22(a).

The binary bit address starts with a *file letter*, *B*, to indicate that a binary bit or word is referenced. The value after the file letter is the *file number*, which indicates the block of memory designated for use in storing binary bit data. File number 3 is the default binary bit file in both the PLC 5 and SLC 500 PLC. The only difference is the number of addressable bits available in file 3 for each processor. The SLC 500 has 256 (numbers 0 to 255) 16-bit words and the PLC 5 has 1000 (0 to 999) 16-bit words in file 3. Additional storage is available for binary bits when the default locations are full.

The colon is a file delimiter or separator between the file number and element or word number. The forward slash separates the word number from the bit address or the *bit number*.

FIGURE 3-22: Binary bit addressing in SLC 500 PLCs.

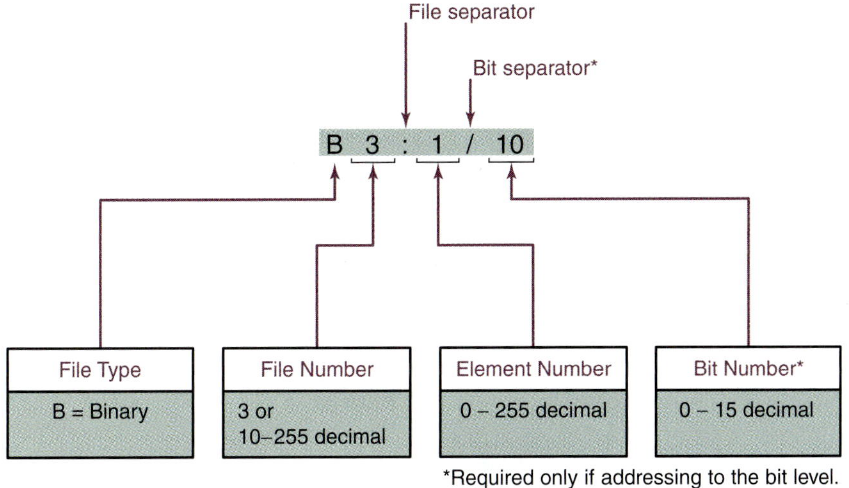

(a) Binary bit addressing syntax

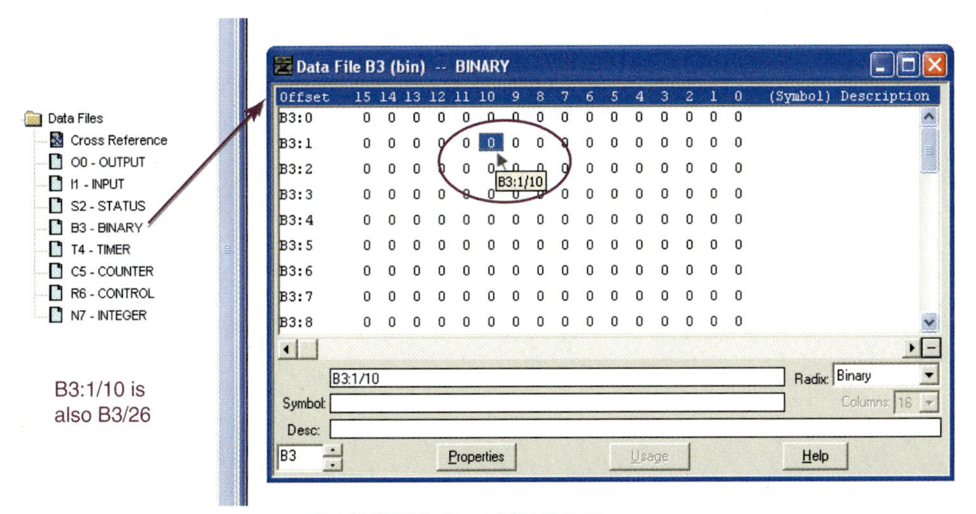

(b) SLC 500 view of file 3 data

EXAMPLE 3-7

Convert the PLC ladder logic solution for the tank problem in Figure 1-17(b) to a solution using an Allen-Bradley SLC system. Obtain module slot and terminal numbers from Figure 3-20. Let bit zero of word zero in file 3 be used for the internal relay.

SOLUTION

Study the solution in Figure 3-23(a). PLC input and output instructions replace the field devices in the relay ladder logic solution. The internal relay replaces the control relay, and the binary address for the internal relay is used to reference the output coil in rung 0 and the input instructions in rungs 1 and 2. Figure 3-23(b) describes the address notation for input and output instructions and for internal bits. Note how the ladder has been documented with a ladder description, rung descriptions, and instruction descriptions. All PLC programs should have instruction descriptions, rung comments, and ladder file titles. Some of the solutions in this book do not have this full documentation in order to reduce figure size.

Review input, output, and binary bit addressing for an understanding of how the generic addresses and the ladder from Chapter 1 were converted to the SLC 500 addressing format using slot and terminal numbers.

FIGURE 3-23: Solution for Example 3-7.

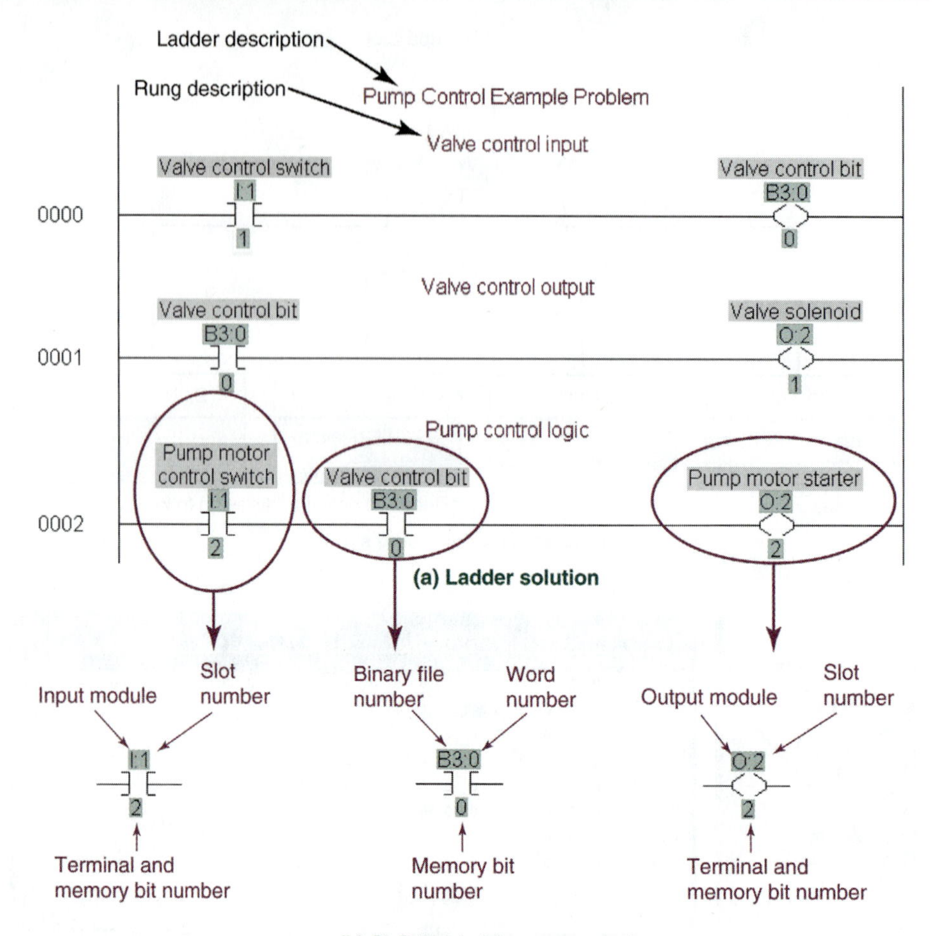

The bit number, 0 to 15, references the specific bit in the 16-bit word. If the forward slash and this last bit field is omitted, the address references the entire word.

For another addressing option, omit the element or word number so that the address format would be

B3/ [bit number]

Because the word or element number (0 to 255) is omitted, the bits of the 256 words are numbered consecutively from 0 to 4095. This format is used less frequently than the format described in Figure 3-22(a).

Figure 3-22(b) illustrates a binary bit file for the SLC 500 PLC. The first nine words (B3:0 to B3:8) are displayed with their value, 0 or 1, for each bit. Each row begins with the first three fields of the internal bit address (B3:0, for example); to complete the address one adds a forward slash and the bit location in the word. An example, B3:1/10 (bit 10 in word 1), is highlighted. The alternate address for this example is B3/26 (binary bit 26 out of the possible 4096 bits).

A *radix* drop-down box in the lower-right corner is used to change the display's radix value. The radix selected for display is usually decimal or binary, depending upon how the data is used in the PLC program.

Examples of PLC 5 and SLC 500 binary bit addressing are provided in Figure 3-24. Note that an example with the element or word number

FIGURE 3-24: PLC 5 and SLC 500 internal relay address examples.

Allen Bradley – PLC 5 and SLC 500 System	
B3:5/14	Bit 14, word or element 5 – File 3 (0 to 255 words) is reserved for binary bit data.
B3:240/7	Bit 7, word or element 240 – File 3 (0 to 255 words) is reserved for binary bit data.
B15:0/14	File 15 (user defined), Bit 14 and word or element 0 – User defined file 15 word 0 is defined as a binary bit file by the B prefix.
B3:9	Bits 0 to 15, word or element 9 – Since the bit field is omitted, the address references all the bits in word or element 9.
B3/157	Bit 13, word or element 10 – Since the word field is omitted, the bit address is the bit location when counted consecutively from bit zero in word zero. The bit number for the number 15 bit in any consecutive word is equal to (the word number × 16) –1. For this example, the 15th bit in word 10 is (10 × 16) –1 = 159, so the 13th bit is 157.

omitted is provided for bit addressing. A formula and process for finding the value of the bit for this format are also provided.

3-7-2 ControlLogix Binary Bit Addressing

An internal control relay is created in the ControlLogix system by creating a tag (either program or controller type or scope) and assigning a Boolean type to the tag. The Valve_control _bit tag, created for the pump control problem and illustrated in Figure 3-21, is an example of an internal memory bit or control relay in the Logix system.

EXAMPLE 3-8

Convert the PLC ladder logic solution for the tank problem in Figure 1-17(b) to a solution, using an Allen-Bradley ControlLogix system. Obtain tag names and base tag addresses from Figure 3-21.

SOLUTION

Figure 3-25 presents another solution for the tank system in Figures 1-7 and 3-20. Note that the binary address for the internal relay is used to reference both the output coil in rung 0 and the input instructions in rungs 1 and 2. Compare this solution with the SLC 500 ladders described earlier and verify that it operates the same. Note also how much more descriptive the solution becomes when tag names are used in place of the memory addresses in the SLC solution.

The virtual relays used in the last two examples used non-retentive memory bits. A second type, called retentive memory, is also available and is discussed next.

3-7-3 Retentive and Non-retentive Memory

Retentive memory retains the memory state (0 or 1) when the PLC is cycled from on to off and back to on. Non-retentive memory locations are reset to 0 whenever the PLC is turned on. All PLCs offer both types of memory for the binary bits. Allen-Bradley PLC software provides both retentive [-(L)- and –(U)-] and non-retentive [-( )-] coil instructions as a programming option. Non-retentive memory permits an output bit to become true when the rung is true and then return to a false state when the rung is false or the PLC is turned off. The retentive bit is set when the rung is true and remains set when the rung is false or when the PLC power is removed. Retentive memory is used when a *warm restart* must be a part of the system operation. In a warm restart situation the automation must maintain the value of some variables if the PLC or automation system is shut down by a power failure or system fault. Retentive memory permits the system to be restarted with memory locations holding the values that were present when the execution was halted.

Retentive memory for control of process machinery components, such as conveyors, pneumatic and hydraulic actuators, and motors, may present safety hazards. For example, when a latched output bit turns on a conveyor motor, that motor is in the on condition until the bit is unlatched. If conveyor power is lost, the conveyor stops; however, when power is restored the conveyor motor immediately turns on and the conveyor starts moving. Operators or maintenance technicians who are unaware that the conveyor motor is controlled by a retentive memory bit could be injured if they are working on

FIGURE 3-25: Solution for Example 3-8.

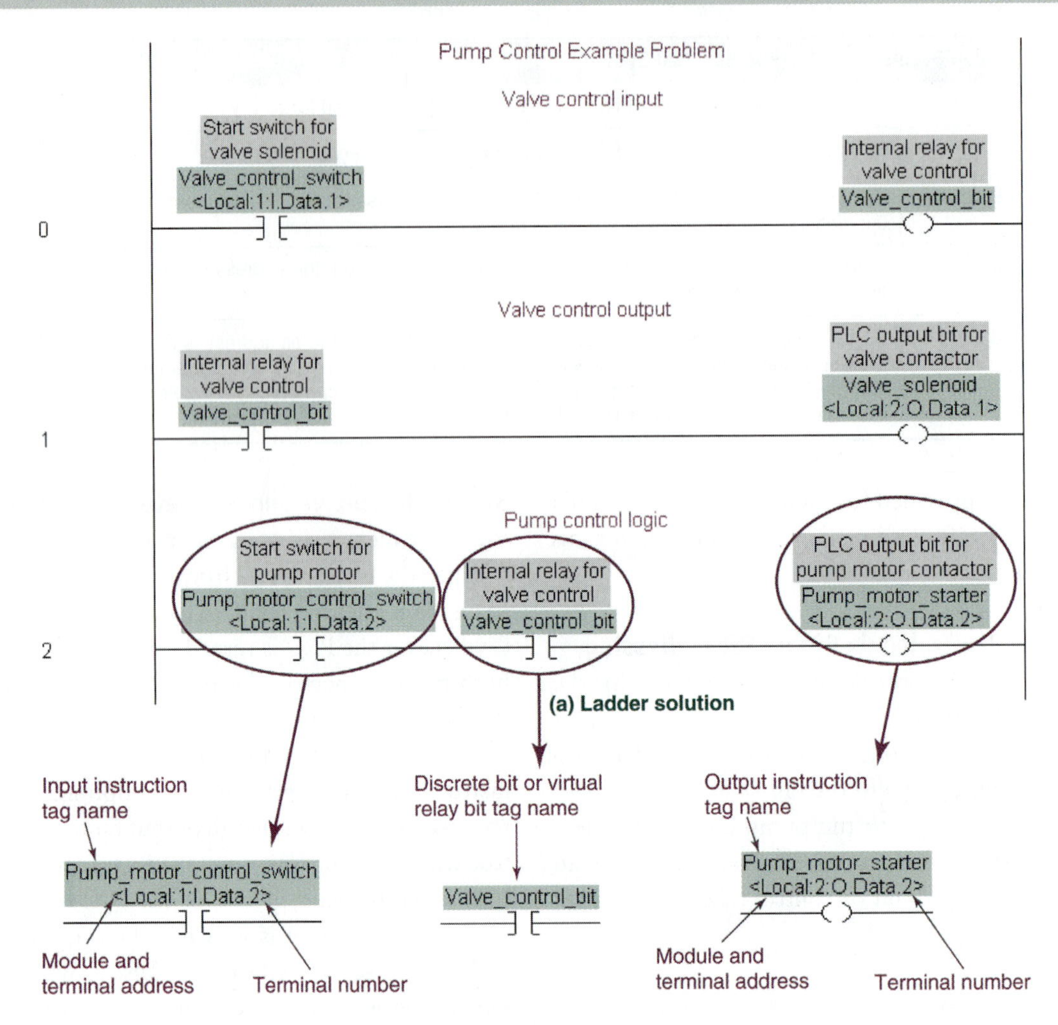

the motor and conveyor drive or if they are near the conveyor belt when the system starts to move at the restoration of power.

PLCs have a third type of stored memory information called *status data*, which is discussed next.

3-8 STATUS DATA ADDRESSING

Most PLCs have a *status memory* allocation that stores important data related to the operation of the processor. In the PLC 5 and SLC 500 models, instructions can set or reset processor status bits; in other cases, a program error during execution will set error bits in the status memory area. In some cases, the programmer may access status data and use that data to modify the program execution. Generally, the programmer does not use the status bits or words in the course of programming most applications. It is important, however, to be aware that this processor status data is available and accessible when specific control situations demand its use. The PLC 5 and SLC 500 have status memory allocations and the ability to address some status data.

3-8-1 PLC 5 and SLC 500 Status Data Addressing

The SLC 500 PLC saves status data in 163 words of memory. In some cases a single word will have 16 bits of different status data. In others, blocks of words are used to store data from a single event, such as the *Global Status File* of network data stored in words 100 to 163. Figure 3-26

FIGURE 3-26: SLC 500 status memory addressing of first pass bit.

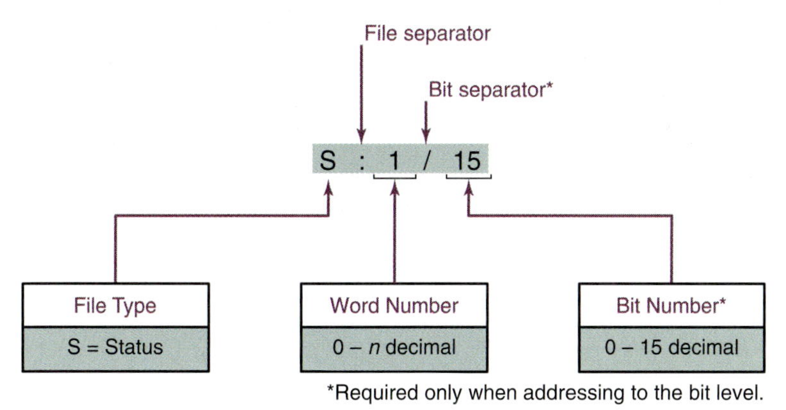

(a) Status addressing

First-Pass Bit S:1/15

Use the bit to initialize your program as the application requires. When this bit is set by the processor, it indicates that the first scan of the user program is in progress (following power up in the RUN mode or entry into a REM Run or REM Test mode). The processor clears this bit following the first scan.

When this bit is cleared, it indicates that the program is not in the first scan of a REM Test or REM Run mode.

(b) First-pass bit definition

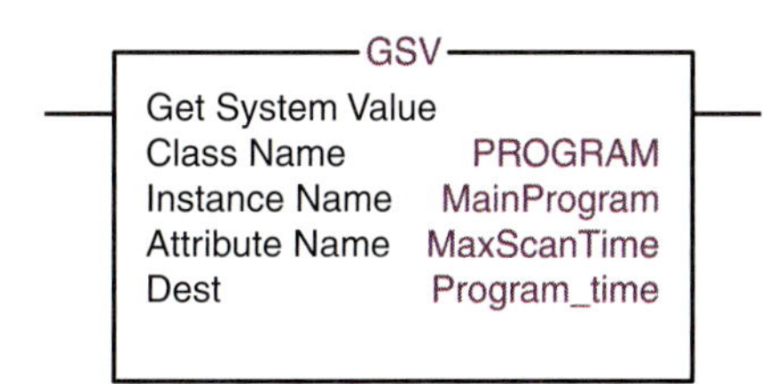

(c) ControlLogix command to get status information

illustrates the format for addressing data in the PLC 5 and SLC 500 status file.

One should recognize the format used for file addressing in the both PLC systems. Figure 3-8 indicates that file 2 is the status file, and the delimiters in Figure 3-26(a) continue to be a colon between file type and word, with a forward slash to separate the word from the bit. Figure 3-26(b) illustrates a typical status file that is often used in programs. The status bit, addressed by S:1/15, is called the *first pass bit*. Its definition and operation can be studied in Figure 3-26(b).

The information presented thus far covers the first four address files (0 – outputs, 1 – inputs, 2 – status, and 3 – binary bits) in the PLC 5 and SLC 500 system. Later chapters describe the remaining files (files 4 through 8).

3-8-2 Logix System Status

Unlike the PLC 5 and SLC 500 models, the Logix models have no status file. Instead they use two instructions, *GSV (Get System Value)* and *SSV (Store System Value)*, to get and set controller status data that is stored in objects. The class names used in the instructions include *AXIS, CONTROLLER, CONTROLLERDEVICE, CST, DF1, FAULTLOG, MESSAGE, MODULE, MOTIONGROUP, PROGRAM, ROUTINE, SERIALPORT, TASK*, and *WALLCLOCKTIME*.

When the GSV instruction is used in a program, it retrieves the specified status information stored as an attribute under the object name. For example, Figure 3-26(c) shows a GSV instruction that will retrieve the longest program execution time from a variable named *PROGRAM*.

MainProgram.MaxScanTime. Note that the object has three parts listed in the figure: Class Name, Instance Name, and Attribute Name. The GSV instruction moves the longest program execution time to a tag called Program_time specified as the destination (Dest).

3-9 ALLEN-BRADLEY INPUT INSTRUCTIONS AND OUTPUT COILS

In the PLC ladder logic solutions presented thus far, only the *examine if closed* (-| |-) input XIC instruction and the output energized (-( )-) output OTE instruction have been used. A second instruction (XIO) and symbol (-|/|-) called *examine if open*, is often required for the input field devices in ladder logic. Also, two retentive output instructions, OTL and OTU, *output latch* (-(L)-) and *output unlatch* (-(U)-), respectively, are required for the ladder outputs. Figure 3-27 presents information on the two inputs, and Figure 3-30 covers the three outputs. This information applies equally to PLC 5, SLC 500, and ControlLogix processors and programming.

3-9-1 Examine if Closed and Examine if Open Instructions

The names *examine if closed* and *examine if open* plus the shorthand notation XIC and XIO, used for Allen-Bradley (AB) PLCs, are used in this text. Figure 3-27 lists a number of other terms that are widely used to describe these two logic symbols and input instructions. The figure also indicates the address format used most often for all three AB PLCs. Read the descriptions in the figure again and review the ladder rung in Figure 3-28.

The rung in Figure 3-28 has an XIC (-| |-) input and an output energize symbol (-( )-) shown. For the output to be *true (activated)*, the input XIC instruction must provide *continuity* or *power flow*. In relay ladder logic, the input contacts apply a voltage and pass a current to a relay coil at the output. Since PLC ladder logic is only a logical representation, no real current flows. However, the concept of power flow or continuity through the XIC instructions is still used. Row 5 in Figure 3-27 focuses on the conditions that produce continuity and power flow through the XIC and XIO instructions in a PLC ladder rung, and row 6 describes the conditions that interrupt continuity and power flow through the same instructions.

The XIC's input image table or register bit must be a 1 to produce continuity, but for continuity in XIO the bit must be a 0. The bit is a 1 if the input field device connected to the input terminals has a closed contact; it is a 0 if the contact of the field device is open. Thus a true XIC instruction needs continuity in the field device contact, and a true XIO instruction needs no continuity or an open in the field device contact. The XIO and XIC can also have an address from another internal reference such as an output bit address or an internal relay bit address. Again, if the XIC's address bit is a 1 then continuity is present, and the XIO's reference bit must be a 0 for continuity. Review row 5 until the conditions that produce continuity are clear.

Row 6 describes the conditions necessary to *interrupt* continuity. These conditions are just the opposite of what is required to produce continuity. If the input field device contact is open, then the XIC has no continuity or no logical power flow. Similarly, if the input field device for the XIO instruction has a closed contact, then no continuity is present and the instruction has no logical power flow. Again, review row 6 until the conditions that interrupt continuity are clear.

You may wonder how current sinking and sourcing input modules affect the operation of the XIC and XIO instructions just discussed. The type of input has no effect since a closed field device contact makes both a current sinking and current sourcing input active (a 1 is placed in the input image table). Likewise, an open field device contact makes both types of module inputs not active (a 0 is placed in the input image table).

Field device contacts versus PLC instructions. An important distinction must be made between the normally open (NO) and normally closed (NC) physical contacts used in input and output field devices, such as the contacts of a push button switch and the ladder input logic symbols XIC (-| |-) and XIO (-|/|-) Input field devices, like switches, have physical contacts that present a voltage or a ground to the input module of the PLC. The symbols used for both the NO and NC physical switch contacts and the XIC and XIO ladder logic symbols are identical.

FIGURE 3-27: Ladder contacts—Examine if closed and examine if open.

	Symbol	—| |—	—| / |—
1	Allen Bradley names	Examine If Closed	Examine If Open
2	Abbreviation	XIC	XIO
3	Other names used	Examine On Examine If On Open Contact Symbol Normally Open Contact	Examine Off Examine If Off Closed Contact Symbol Normally Closed Contact
4	Possible addresses	**PLC 5 and SLC 500** Input image table bit address, i.e., I:2/3. Output image table bit address, i.e.,, O:3/12. Any internal relay bit address, i.e., B3:5/8. Any bit level address. **ControlLogix** Any tag with a Boolean data type. Any tag associated with an input module terminal.	**PLC 5 and SLC 500** Input image table bit address, i.e., I:2/3. Output image table bit address, i.e., O:3/12. Any internal relay address, i.e., B3:5/8. Any bit level address. **ControlLogix** Any tag with a Boolean data type. Any tag associated with an input module terminal.
5	Continuity, power flow, true, active, closed contacts, or highlighted if	Input field device contacts are closed. The input image table bit is a 1. The reference address (i.e., B3:5/8 or O:3/12) is true (1).	Input field device contacts are open. The input image table bit is a 0. The reference address (i.e., B3:5/8 or O:3/12) is false (0).
6	No continuity, no power flow, false, not active, open contacts, or not highlighted if	Input field device contacts are open. The input image table bit is a 0. The reference address (i.e., B3:5/8 or O:3/12) is false (0).	Input field device contacts are closed. The input image table bit is a 1. The reference address (i.e., B3:5/8 or O:3/12) is true (1).
7	Operational comments	• The symbol looks like a normally open relay contact, but no current flows since the ladder rung is just a logical representation of the input logic necessary to activate the output. • To evaluate the condition of this logic symbol we ask the processor to EXAMINE IF (the field device is) CLOSED. If it is, then this symbol has continuity. If the field device is OPEN, then no continuity is present. • The address or memory bit location associated with this symbol applies the condition of that address or bit to the symbol. If the address is true, or 1, then this symbol has continuity. If it is false, or 0, there is no continuity.	• The symbol looks like a normally closed relay contact, but no current flows since the ladder rung is just a logical representation of the input logic necessary to activate the output. • To evaluate the condition of this logic symbol we ask the processor to EXAMINE IF (the field device is) OPEN. If it is, then this symbol has continuity. If the field device is CLOSED, then no continuity is present. • The address or memory bit location associated with this symbol applies the condition of that address or bit to the symbol. If the address is false, or 0, then this symbol has continuity. If it is true, or 1, there is no continuity.

FIGURE 3-28: Continuity and power flow in a ladder logic rung.

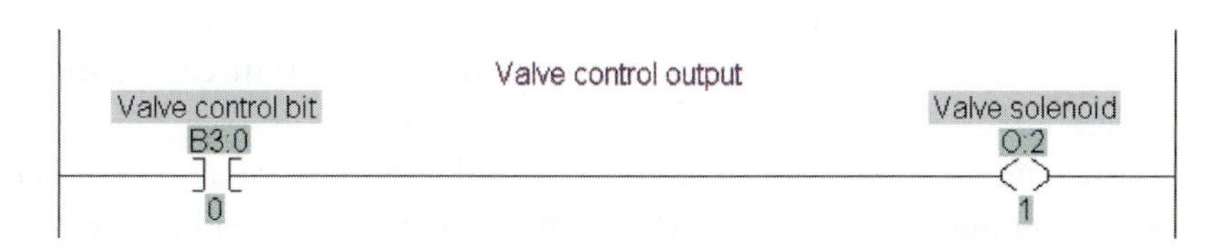

However, the input logic symbols in PLC ladders are virtual contacts or logic symbols because they are created in software and represent Boolean memory values. Therefore, they are called *instructions* and not *contacts.* Frequently, PLC literature refers to XIC (-| |-) and XIO (-|/|-) instructions as normally open or normally closed contacts or just contacts in general. To avoid confusion, this text uses the term *contact* for field devices and the term *instruction* for any input XIC or XIO symbol addressed by a PLC memory bit.

Linking field devices with XIC and XIO instructions. Figure 3-29 shows the interface between an input field device (limit switch) and XIC and XIO instructions in a ladder logic rung. A study of the figure indicates that the limit switch has normally open (NO) contacts when not actived and closed contacts when activated by the automation process.

An *input image table*, also called an input memory register, has a bit associated with each terminal of the input module. The image table or register bit is a 0 when when the limit switch is in the normally open position, and it is a 1 when the field device contacts are held closed. Figure 3-29(a) depicts this convention in the XIC example.

Two choices for the instruction placed in the ladder logic rungs are examine if closed (XIC) and examine if open (XIO). The XIC instruction symbol (-| |-) looks like a normally open set of contacts, and from a virtual power flow analogy it works accordingly. When its memory address bit or image table bit is a 0, continuity is *not* present. However, when these bits are a 1, the instruction produces continuity. Figure 3-29(a) illustrates these two conditions. When an instruction produces continuity the RSLogix software grays the symbol to indicate an active state with continuity.

The XIO ladder logic instruction symbols (-|/|-) look like a normally closed set of contacts. Note that continuity is present when the input image table bit is a 0 and the field device contacts are open. However, when the field device contact closes and the image table bit is a 1, the XIO instruction is not true and no continuity is present. Figure 3-29(b) illustrates both of these conditions. Note that the symbol is grayed out, indicating continuity and power flow when the input image table bit is a 0.

A ladder input instruction and the input field device are linked in the instruction address by the input module terminal where the field device is attached. Since the ladder instruction in Figure 3-29 has an address of 1:3/10, it is the input module in slot 3 and terminal 10 (SLC 500 notation). The figure shows that the condition of the ladder instruction, either active or not active, is determined by the bit value in the input module register for that terminal, either 1 or 0, respectively. In Figure 3-29, shading is used to indicate the instructions producing continuity. The use of the terms *examine if closed* and *examine if open* is confusing but necessary because these instructions represent PLC memory bits, not switch contacts. Review Figure 3-29 again to be sure you understand how field devices and input instruction bits interact.

Ladder logic terminology. This text uses the terms *true, active or continuity* and *false, not active or no continuity* to describe the conditions of the XIC and XIO instructions. In other reference material additional terms for continuity include *power flow, contacts closed,* and *highlighted.* For no continuity terms include *no power flow, contacts open,* and *not highlighted.* These terms merely indicate that the input instruction condition is either allowing logical power flow or interrupting it.

3-9-2 Output Energize, Output Latch, and Output Unlatch Instructions

Figure 3-30 provides information on three of the most commonly used output instructions: output energize (OTE), output latch (OTL), and output unlatch (OTU). The OTE output is similar to a relay coil in relay ladder logic, and the OTL and OTU instructions are like a latching relay (described in Chapter 2) with a latch and unlatch coil.

Allen-Bradley output energize—OTE instruction. Column 1 in Figure 3-30 lists the OTE basics. The address of an output is usually a reference to an output field device connected to an

FIGURE 3-29: Linking input field devices with XIC and XIO ladder logic commands.

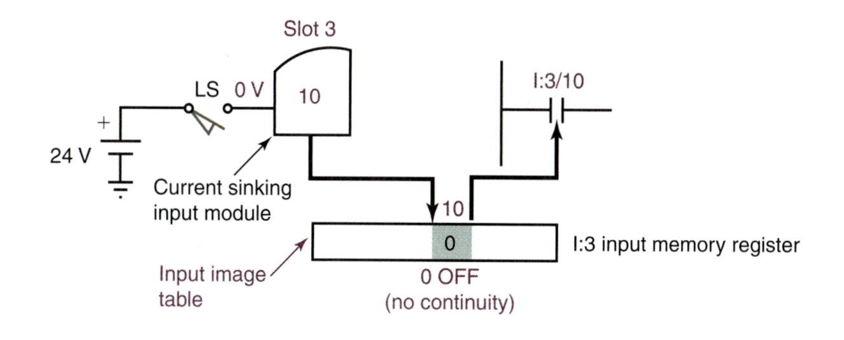

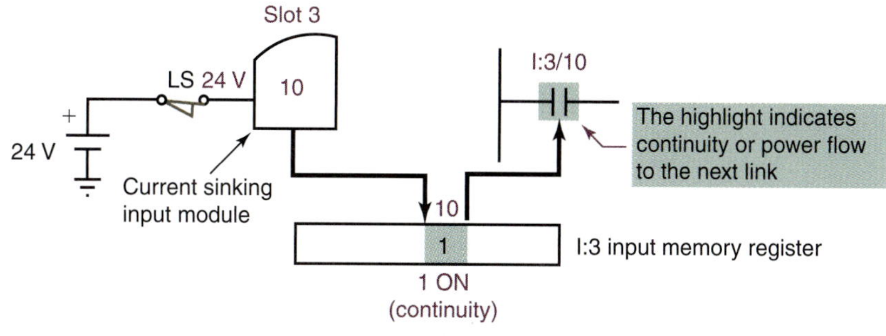

(a) Examine if closed example

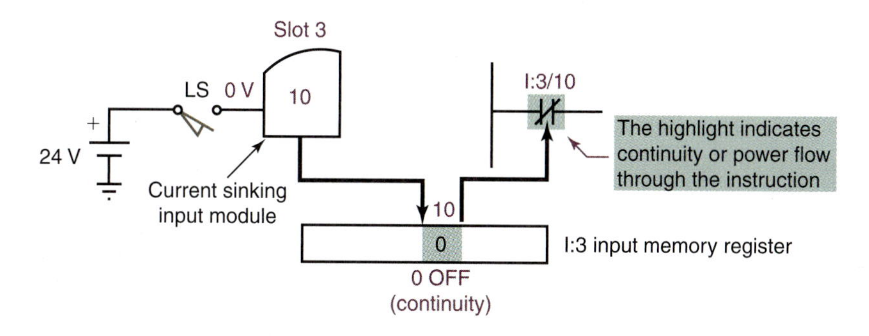

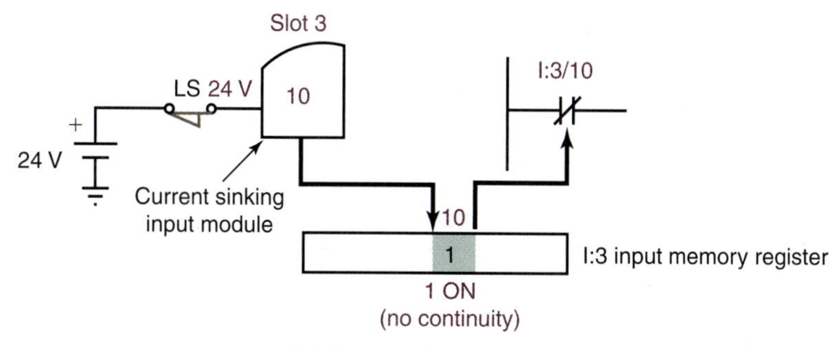

(b) Examine if open example

Source: Rehg and Sartori, Industrial Electronics, © 2006, p. 583. Reprinted by permission of Prentice Hall, Upper Saddle River, NJ.

FIGURE 3-30: Ladder coils—Output energize, output latch, and output unlatch.

	1	2	3
Symbol	—( )—	—(L)—	—(U)—
Allen Bradley names	**Output Energize**	**Output Latch**	**Output Unlatch**
Abbreviation	**OTE**	**OTL**	**OTU**
Other names used	Output Coil Output Instruction	Retentive Coil Retentive Memory	Retentive Coil Retentive Memory
Possible addresses	**PLC 5 and SLC 500** Output image table bit address, i.e., O:3/12. An internal relay bit address, i.e., B3:5/8. **ControlLogix** Any tag with a Boolean data type. Any tag associated with an output module terminal.	**PLC 5 and SLC 500** Output image table bit address, i.e., O:3/12. An internal relay bit address, i.e., B3:5/8. **ControlLogix** Any tag with a Boolean data type. Any tag associated with an output module terminal.	**PLC 5 and SLC 500** Output image table bit address, i.e., O:3/12. An internal relay bit address, i.e., B3:5/8. **ControlLogix** Any tag with a Boolean data type. Any tag associated with an output module terminal.
The output symbol is true, active, or highlighted and the bit in the output image table is true or a 1 if:	The input logic is true. The rung segment to the left of the symbol is true.	The input logic sets the retentive bit for this Output Latch to true or 1.	The input logic is true. When the Output Unlatch is true the retentive bit for this address is reset to false or 0.
Operational comments	• The OUTPUT ENERGIZE symbol is analogous to a relay coil in relay ladder logic. However, here it represents the output field device connected to the output module. • The output image table bit for this energized OUTPUT instruction is set to 1 when the ladder rung is TRUE. • The output image table bit for this energized OUTPUT instruction is set to 0 when the ladder rung is FALSE.	• The OUTPUT LATCH is analogous to a latching coil for a latching relay in relay ladder logic. However, here it is a command that sets a retentive memory bit to 1. • This retentive output is set when the rung input logic is true and remains set until a rung with an OUTPUT UNLATCH instruction with the same address becomes true. • The precautions described in Section 3-7-3 on retentive memory apply to this instruction as well.	• The OUTPUT UNLATCH is analogous to an unlatching coil for a latching relay in relay ladder logic. However, here it is a command that clears a retentive memory bit to 0. • The OUTPUT UNLATCH command has the same output image table address as the OUTPUT LATCH command that set the retentive bit. • Retentive memory instructions should be used cautiously because of the safety issues described in Section 3-7-3.

output module. Therefore, it could be any valid output address used in the PLC 5, SLC 500, or ControlLogix systems. Read the information and review the sample addresses for each of the outputs in the figure.

The output instructions are true or active if the input logic instructions provide power flow or continuity to the output. Therefore a path must exist from the left power rail (vertical bar) through rung links that are true and input instructions that are grayed out. Figure 3-31 provides a visual picture of how power flow occurs and how it forces the output instruction to a true state.

Note that link 1 connected to the left power rail is always true. The remaining links (2, 3, and 4) are true if the instructions to the left of them are true. Ultimately, the output is true if the link to

FIGURE 3-31: Ladder rung power flow.

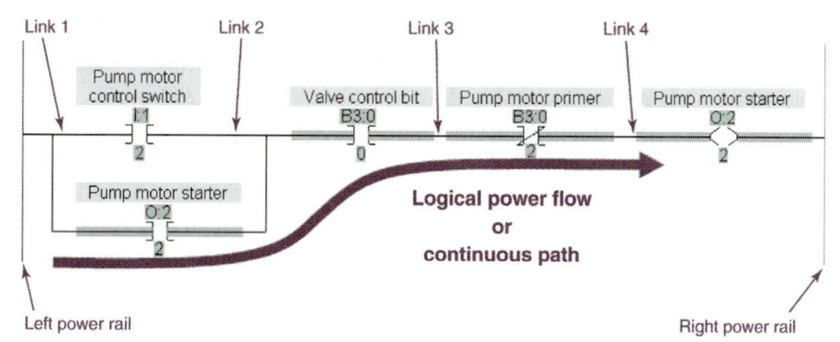

(a) Rung power flow or continuity example

Instruction address	Instruction type	Status	Register bit value	Condition	Instruction highlight
I:1/2	XIC	False	0	No power flow	Clear
O:2/2	XIC	True	1	Power flow	Gray
B3:0/0	XIC	True	1	Power flow	Gray
B3:0/2	XIO	True	0	Power flow	Gray

(b) Contact information

FIGURE 3-32: OTL and OTU instructions.

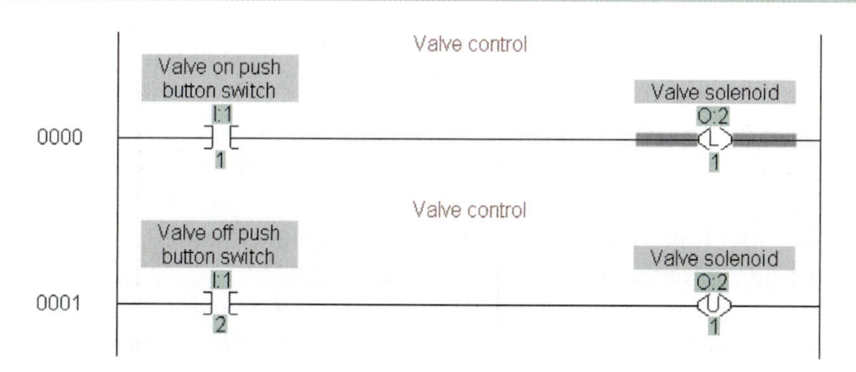

its left (link 4 in the figure) is true. When an XIC instruction is true, the bit value in its input register or image table is a 1, power flow is present, and the instruction is highlighted. In contrast, if the bit value in the XIO's register is a 0, the instruction has power flow and is highlighted. The OTE instruction is implemented with the same format used in all of the Allen-Bradley processors.

Allen-Bradley output latch and unlatch—OTL and OTU instruction. The OTL and OTU instruction information is listed in the last two columns in Figure 3-30. These two instructions perform like the latching relay described in Chapter 2.

The output latch turns on the output image table bit for this output when the OTL instructions are true. That bit remains true (1) until it is cleared (set to a 0) by the action of another rung, using an OTU instruction. These instructions have *retentive memory registers* that remain in their current state (0 or 1) until changed by another instruction.

As illustrated in Figure 3-32, the OTL and OTU instruction sequence is used when the automation system requires an output to remain on after a momentary field device input switch has cycled from close to open or remain on after a power loss is restored.

Note in the figure that the solenoid valve output is true (latched), but the input instruction

(I:1/1) has no power flow (not highlighted). If input I:1/2 in rung 2 becomes true, then the valve is turned off by the OTU instruction. The addresses for the OTL and OTU instructions must be the same, such as O:2/1 in this example. All of the Allen-Bradley processors implement the OTL and OTU instructions with the same format.

Output latch versus sealing instructions. The latch and unlatch output instructions are retentive, which means that the state of those outputs is unchanged even when power is removed from the processor. The safety issues associated with retentive memory in PLCs was addressed earlier in this chapter. An alternate approach to the use of latching outputs is the use of *sealing instructions*. Figure 3-33 illustrates a sealing instruction implementation for an SLC 500 program. Note that the figure uses the address of the output for an input instruction that is across the start instruction. Many references use the term *sealing contact* instead of *sealing instruction*.

The ladders in Figures 3-32 and 3-33 perform the same control. A valve is turned on with one depression of a push button and turned off with the depression of a second push button. The rung in Figure 3-33 shows the condition after the valve on push button switch is pressed and released. At the start, all instructions, except I:1/2, are false (no power flow). When I:1/1 is pressed the OTE output is true (power flow through I:1/1 and I:1/2). If the OTE is true (1), then the instruction O:2/1 (valve solenoid sealing instruction) is true because it uses the address of the OTE output. This instruction *seals around* the momentary push button, so that I:1/1 can return to a false state (no power flow), and the output is held in the true condition.

The output returns to the false state (0) when the push button I:1/2 is pressed because the instruction I:1/2 becomes false (no power flow). Therefore, the OTE output bit is false (0), and the sealing instruction returns to the no power flow state. This rung has no inherent safety issues because it is built without retentive memory bits. Accordingly, when power is lost in the processor all of the input and output bits clear to a false (0) state. When power is restored the valve solenoid is off and must be turned on with the push button.

3-10 INPUTS, OUTPUTS, AND SCAN TIME

The operation of input instructions and output coil instructions was addressed in the last section. The elapsed time from a change in an input field device to the resulting change in the output field device is dictated by the PLC *scan* and *scan time*. Therefore, it is important to understand these concepts.

3-10-1 Scan Time

PLC ladder logic programs have multiple rungs; therefore, a process must be used to coordinate the changes in input instructions on all of the rungs and on the associated changes in the outputs. The process used is a *PLC scan,* which is illustrated in Figure 3-34 and described next.

1. *Read the status of all input field devices* by transferring a 0 or a 1 to the input registers, based on the open or closed condition of the field device and voltage level present at the input module terminals. This concept was introduced in Figure 3-29.

FIGURE 3-33: Sealing instruction operation.

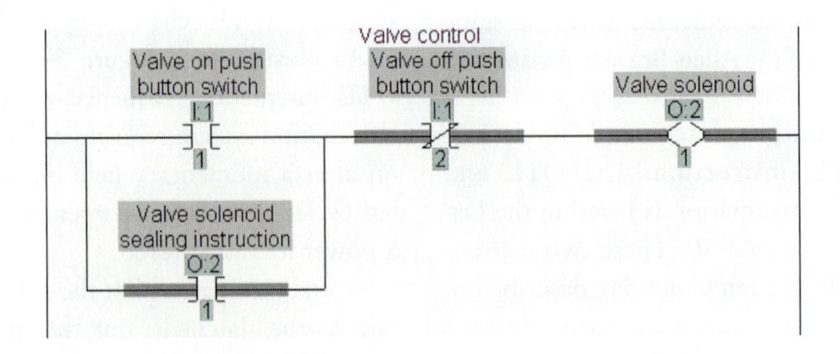

FIGURE 3-34: PLC scan cycle.

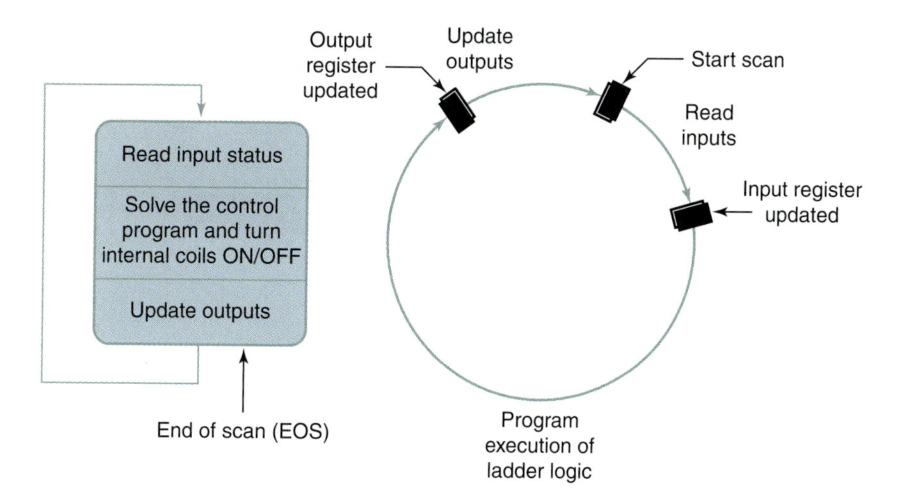

2. Perform the following four steps:
   a. Change the highlight condition of all input instruction symbols to reflect the condition of their bits in the input register.
   b. Evaluate the Boolean logic present on each rung starting with rung 0.
   c. Determine the condition of the ladder link (true or false) connected to the output.
   d. Set the output instruction true (1) if the link is true (input logic on the rung is true) or set it false (0) if the link is not true (input logic on the rung is false).
3. Update the output bit addressed by the ladder output instruction in the output register. This bit either controls an output field device connected to an output module terminal or controls an internal relay.

If the output coil is true, then its output register bit is a 1, and the output module turns on the output field device attached to that terminal. If the output coil is false, then the associated register bit is a 0, and the output module turns off the output field device attached to that terminal.

The time required for a scan, called *scan time*, is not constant, but it is affected by the number of rungs in the program, or the amount of memory used to store the program. A good rule of thumb is 1 millisecond per 1000 bytes of program memory used. Scan time is also affected by the varying time required to execute some of the program instructions. Scan times can vary from 0.2 $\bar{\mu}$s to 50 ms.

3-10-2 Linking Inputs and Outputs

Figure 3-35 illustrates the complete operation of a PLC from input field device to output field device. This figure also illustrates the scan steps described in Figure 3-34, displaying four views of field device input data moving through the PLC during a scan to the ouput field device.

The figure's four parts indicate the flow of information during a full PLC scan.

Start scan and read inputs: Figure 3-35(a) indicates the condition of the system prior to the start of a scan. Note that the normally open (NO) limit switch (LS) wired to terminal 10 is held closed by the process, but the input register is still false (0) and does not reflect that change. All input terminals are scanned.

Input register update: In Figure 3-35(b), the 16-bit input register is updated. In this example, bit 10 changes from a 0 to a 1.

Program execution of ladder logic: In Figure 3-35(c), the ladder logic program is executed by first updating all instructions on the input side of the rungs. Here the input instruction logic is evaluated to determine if power flow from the left rail to the output is present. All true rungs will have the output instruction changed from false to true. In this example,

FIGURE 3-35: Scan cycle—sequence of operation.

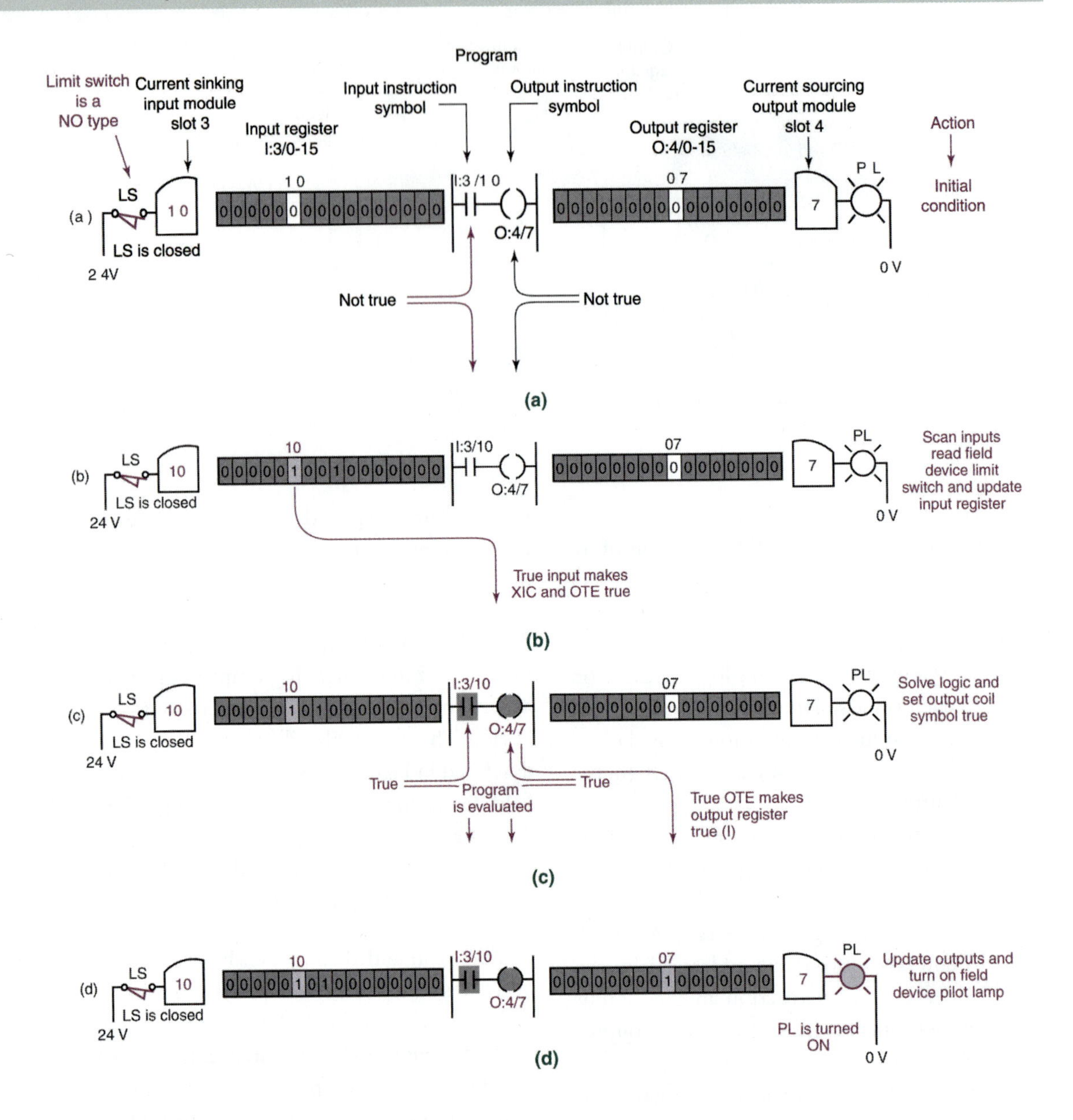

the XIC instruction, I:3/10, becomes true (continuity is present, and power flow occurs). Because the link connected to the output instruction coil, O:4/7, is true, the output instruction becomes true (highlighted).

Update output register and outputs: Figure 3-35(d) illustrates how the output register is updated, and bit 7 changes from a 0 to a 1. This change in the output register causes the output module in slot 4 to turn on the pilot lamp connected to output terminal 7.

Although Figure 3-35 has the register notation for an Allen-Bradley SLC 500, the scan cycle described applies to all Allen-Bradley PLCs. It is assumed that the field devices and the input and output modules have been matched using the current sinking and current sourcing techniques outlined in Chapters 1 and 2. At this point, you should be able to determine the condition (true or false) of an output field device given a ladder logic program and the condition and type of the input field device.

FIGURE 3-36: Process tank and PLC interface and program for Example 3-9.

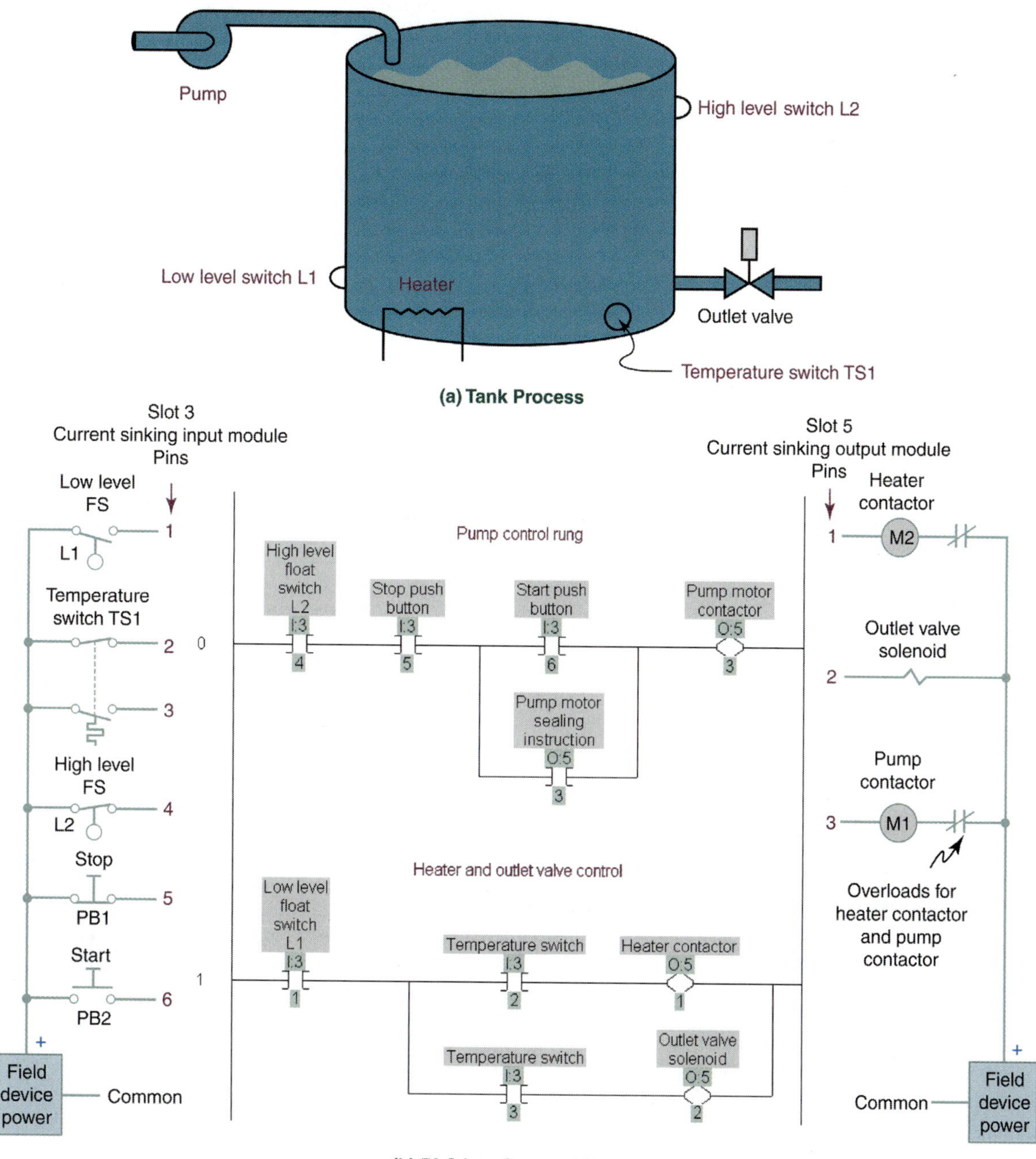

3-10-3 Process Tank Application

The system in Figure 3-36, a process tank with an input pump, heater, and drain valve, is used for liquid heating applications commonly found in the chemical and food processing industries.

The input field devices include:

- Two float switches: a low level (L1) that is normally open (NO) and a high level (L2) that is normally closed.
- A temperature switch: one switch (TS1) with a trigger value of 185° F that has one normally closed (NC) contact and one NO contact.
- Two momentary push button switches: an NC stop (PB1) and an NO start (PB2).

The output field devices include:

- Two contactors with overloads: one for the heater and one for the pump.

- A solenoid operated outlet valve.
- The following description of the operation refers to Figure 3-36:

The mixing tank shown in the figure is used to heat a liquid. Note the location of the field devices just described. A start push button, PB2, is used to initiate the processes, and a stop push button, PB1, is used to interrupt it. The three outputs are turned on based on the following logic:

Fill pump = PB2 AND NOT L2

The start PB is pressed, and the liquid is not above the high level float switch. PB2 has a sealing instruction because it has momentary contacts.

Heater contactor = L1 AND NOT TS1

Since the low level float switch is active (liquid above the switch level) and the temperature is not above 185° F, TS1 is not active. The NC contact of the temperature switch is used.

Outlet valve = L1 AND TS1

Since the low level float switch is active (liquid above the switch level) and the temperature is above 185° F, TS1 is active. The NO contact of the temperature switch is used.

The PLC ladder logic in Figure 3-36 implements these three logic equations in rungs 0 and 1. Note that rung 1 has two outputs.

EXAMPLE 3-9

The ladder logic in Figure 3-36 provides the control for the heater tank. Given the following condition of the input field devices before a scan, determine the condition of the three outputs.

- Liquid is above L2.
- The temperature is at 200° F.

Use the ladder logic program in Figure 3-36 to predict what output field devices will be on at the end of the scan.

Solution

First, translate the given process conditions to the status of the field devices. For example, with liquid above L2 both level swithes are active, and with the temperature above 185° F the temperature switch is active. The best way to see how these actions change the ladder logic is to build a table that starts with the condition of field devices and progresses down to the condition of the rung link before the output instruction. The following table displays the results.

A study of the system status and the table indicates that the two normally closed field device contacts on L2 and TS1 are forced open when the devices are active. The open condition of the field device contacts makes the XIC instruction with that address false. Therefore, rung 0 is false, and the rung link after I:3/2 is false. This result forces the pump motor and heater contactors to the off condition.

Conversely, I:3/1 and I:3/3 XIC instructions are both true since their controlling field contact is closed. As a result, there is continuity from the left rail of the ladder logic to the outlet valve solenoid. Therefore, the outlet valve is open, and the tank is draining.

System Parameters	Input Field Devices and Logic Rung Status			
Field devices	L1 (NO)	L2 (NC)	TS1 (NO)	TS1 (NC)
Input module pin number	1	4	3	2
Field device condition	Active	Active	Active	Active
Field device contacts	Closed	Open	Closed	Open
Input device address	I:3/1	I:3/4	I:3/3	I:3/2
Rung link status	True	False	True	False

With an understanding of the scan cycle and of the technique for linking inputs and outputs in a PLC, one can easily evaluate ladder logic programs. Understanding the PLC scan process is also useful for troubleshooting problems in PLC-controlled systems. The design of a ladder logic program for a specific system, however, is more difficult. The next section describes the selection of the correct ladder logic instructions (XIC or XIO), given input and output field device configurations and the required system operation.

3-11 PLC PROGRAM DESIGN AND RELAY LADDER LOGIC CONVERSION

The need for design skills in ladder logic programming has two origins:

- *Many relay ladder logic control systems still exist in industry and need to be converted to a PLC solution.* One needs to connect the field devices to PLC input and output modules and then design a PLC ladder logic program that replicates the existing control in the relay ladder logic.
- The PLC has become the de facto standard device for the sequential control of production machines and for an increasing number of process control applications. As a result, one needs to know how to design PLC ladder logic programs for new applications.

A process change must produce a specific response from the process control system. In the Figure 3-36 example, when the liquid in the tank reaches the high-level mark, the fill pump must turn off. The designer must work with three elements to achieve control:

1. The input field device
2. The PLC program
3. The output field device

The input field device can be either a switch or a sensor and can have either normally open or normally closed contacts. The type of contacts used affects the design. The PLC program can use either XIC or XIO input instructions and OTE or latch-type output instructions. The output field device can be either *direct acting* (if the OTE is true, then the field device is on) or *inverse acting* (if the OTE is true, then the field device is off). In many cases, the type of input field device, the contact configuration, and the action of the output field device are fixed by the process equipment. Therefore, the designer concentrates on selecting the correct PLC input instructions in order to achieve the proper operation.

3-11-1 Examine if Closed and Examine if Open Selection

Students learning PLCs are often confused when selecting the correct XIC or XIO input instructions for the PLC ladder rung. An input field device with an NC contact configuration does not necessarily have an examine if open (-|/|-) instruction type representing it in the PLC rung. PLC input rung instructions must be selected using a process that looks at the input field device contact type (NO or NC) and its operational condition (activated or not activated). In addition, the input rung instruction selected, XIC or XIO, is affected by the requirement for the output field device to be either on or off. Therefore, learning a process that will *always* work is critical.

The selection table in Figure 3-37 illustrates a selection strategy. Design statements usually begin with a statement describing how system outputs should respond when input switches and sensors change states. For example, in Figure 3-36 the liquid continues to heat after the liquid reaches the maximum level, and the pump is turned off. When the liquid temperature is higher than the set point, the temperature switch is activated (the NO contact of the TS1 temperature switch is held closed), and the outlet valve is turned on or opened. When the design is performed, the temperature switch has an instruction on a ladder rung using either an XIC or XIO instruction. The instruction type selected, XIC or XIO, depends on how the control statement relates the action of the temperature switch to the desired action of the drain valve. In addition, the type of contacts present on the input field device affect the input instruction. The chart leads students through the selection process based on these factors. The process of using the chart is the same for all Allen-Bradley processors and is enumerated as follows:

FIGURE 3-37: XIC and XIO selection guide.

0	1	2	3	4	5	6
Row number	Input field device type	Input field device actuation	Field device contact condition	Output field device condition	Ladder logic OTE output instruction condition	Ladder logic input contact
1	NO	Not actuated	Open contacts	Off	Not true	Examine if closed XIC (-| |-)
2	NO	Not actuated	Open contacts	On	True	Examine if open XIO (-|/|-)
3	NO	Actuated	Closed contacts	Off	Not true	Examine if open XIO (-|/|-)
4	NO	Actuated	Closed contacts	On	True	Examine if closed XIC (-| |-)
5	NC	Not actuated	Closed contacts	Off	Not true	Examine if open XIO (-|/|-)
6	NC	Not actuated	Closed contacts	On	True	Examine if closed XIC (-| |-)
7	NC	Actuated	Open contacts	Off	Not true	Examine if closed XIC (-| |-)
8	NC	Actuated	Open contacts	On	True	Examine if open XIO (-|/|-)

1. Identify the input field device and determine the contact type (NO or NC) used. In column 1 of the chart, find the four corresponding rows, NO or NC, for that contact type.

   Tank example: The NO contacts of the TS1 temperature switch are used for control of the drain valve so rows 1 to 4 in the chart are selected.

2. Determine if the input field device is actuated or not actuated for the desired change in the output field device. From the four rows chosen in step 1, find the two rows in column 2 of the chart that correspond to this condition.

   Tank example: The temperature switch must be activated (NO contacts held closed) for the outlet valve to be on or opened. Rows 3 and 4 are selected because they represent NO contacts with an activated field device. Note in column 3 that an active NO switch has closed contacts, which sets the input register bits to 1.

3. Determine the desired condition, on or off, for the output field device in column 4. Locate the corresponding row from the two rows selected in step 2.

   Tank example: Since the outlet valve must be turned on (column 4) or opened, the parameters across row 4 describe the operation of the input and output field devices for this problem. For example, an active NO input field device is turning on the ladder output OTE instruction, so that the output field device is on. Therefore, the 24 volts appled to the input pin must make the rung true.

4. From column 6 select the ladder rung input instruction type that satisfies this application.

   Tank example: The rung input instruction must be examine if closed or XIC (-| |-) for this application. The XIC instruction is true when the input terminal is high (+24 volts) and the input register bit is a 1. This true XIC instruction allows power flow to the output coil, which energizes the outlet valve.

Although the process was demonstrated using input and output field devices, it works equally well with internal input bits and internal output bits. After working through several examples, the process for picking XIC and XIO instructions becomes clear and dependence on the chart in Figure 3-37 diminishes.

3-11-2 Multiple Inputs

Most PLC programs, like the example ladder logic in Figures 3-31 and 3-36, have more than one power flow path in the rung and have more than one input per rung. The logic in these multiple input rungs falls into these six categories: *AND, OR, AND/OR, OR/AND, INVERSION, and FEEDBACK logic*. Figure 3-38 shows the relay ladder equivalent circuit, ladder logic equivalent program, gate logic equivalent, and Boolean equation for the six types. A brief description of each follows.

AND logic: All inputs must be true for the rung to be true [Figure 3-38(a)].

OR logic: If any input is true, then the rung is true [Figure 3-38(b)].

OR/AND logic: If all inputs in at least one AND group are true, then the rung is true [Figure 3-38(c)].

AND/OR logic: If at least one input in all OR groups is true, then the rung is true [Figure 3-38(d)].

INVERSION logic: The state of a rung with an XIC instruction is the inverse of the state of the address of the XIC instruction [Figure 3-38(e)].

FEEDBACK logic: The feedback from the output in the form of a sealing instruction holds the output in the true state [Figure 3-38(f)].

Multiple input instruction rungs are not difficult to understand. For example, the logic equation for the AND logic in Figure 3-38(a) is $X = AB$ and is read as X is true if A AND B are true. One can see from the relay schematic that both switches must be on for the output to be on. The same logic is represented in the ladder logic. Equations with AND logic are usually written in two forms, which are $X = AB$ or $X = A \cdot B$. The AND function is assumed if two letters are placed next to each other or if a dot (•) is used between the terms.

A second example, Figure 3-38(b), shows an OR logic equation, $X = A + B$. This equation is read as X is true if either A OR B are true. Note that the plus sign (+) is used to indicate that the OR function is used between the terms. The relay schematic shows this relationship electrically. The light is on if either LS1 OR LS2 is closed or active.

EXAMPLE 3-10

An application must use a normally closed (NC) sensor to turn off an output field device (output image table bit is a 0) when the sensor is not activated (NC contact in normal position). Determine the type of ladder logic instruction symbol required for this application.

SOLUTION

The chart in Figure 3-37 is used to find the logic instruction by selecting rows in the following order (Note: the shaded area is for the previous example):

1. Rows 5 to 8 for the NC contacts on the field device (column 1).
2. Rows 5 and 6 for the not activated field device (NC contacts – closed) requirement (column 2).
3. Row 5 because the output device must be off (column 4).
4. Therefore, the ladder rung instruction symbol (row 5 and column 6) is examine if open XIO (-|/|-)

Verify the operation. The not activated NC field device contact puts a 1 in the input image table and causes the XIO instruction to be false. A false XIO does not have continuity or power flow, and the output is false. A false output turns off the associated output field device. This results in the desired operation.

When a person just learning PLC programming reads this more complex problem statement, it is not immediately clear that an XIO instruction would be used. The selection chart leads the programmer through the selection process. After a short time using the chart and selecting input instrutions for the eight combinations, the chart can be discarded as no longer necessary.

FIGURE 3-38: Logic configurations.

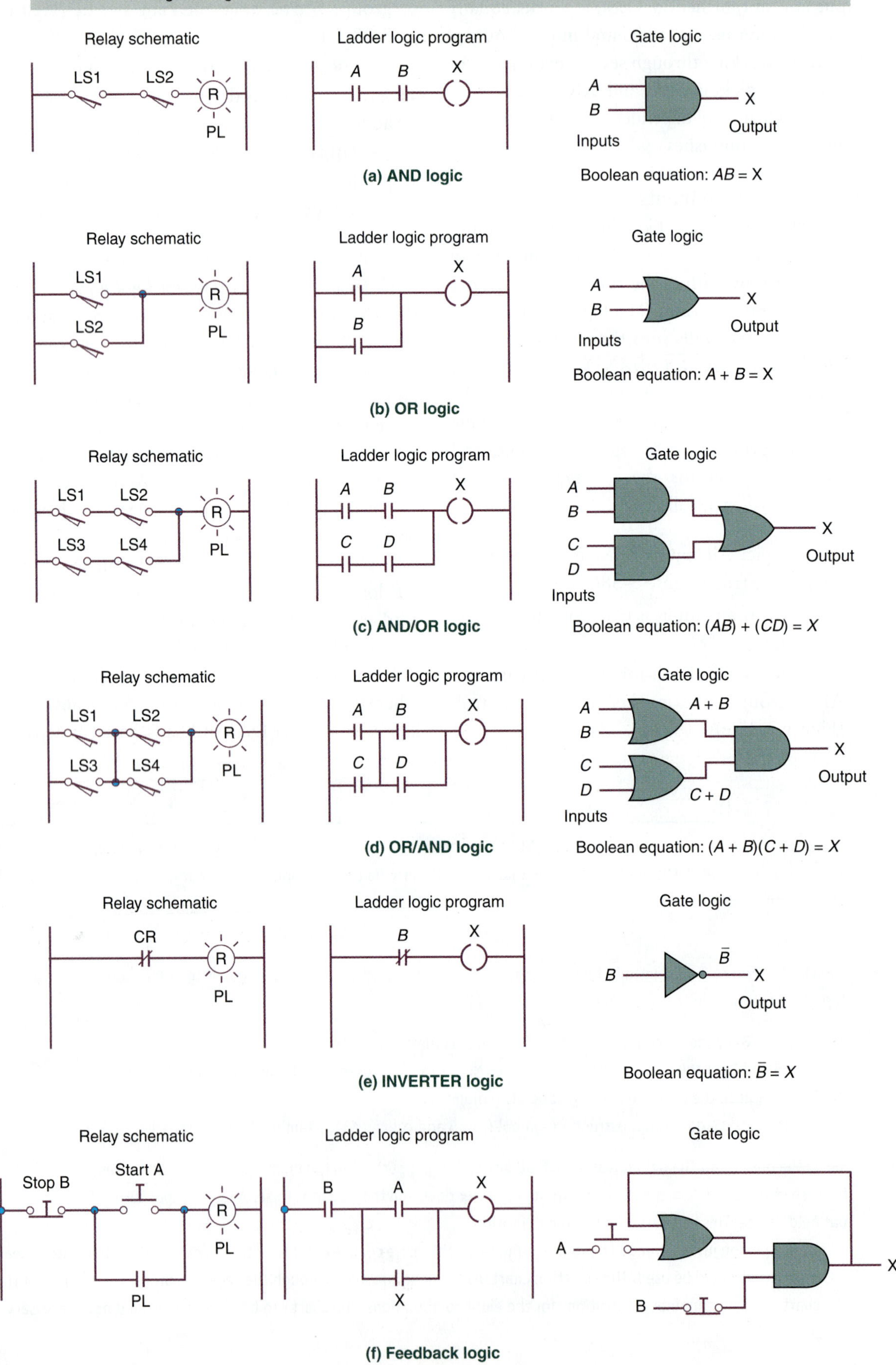

EXAMPLE 3-11

Write the Boolean equation for the Y output rung in Figure 3-39(a).

SOLUTION

Instructions A, B, C, D, and E are an OR/AND configuration, F is an AND, and G and H are an OR type. The equation is

$$Y = [(\text{not } A \cdot B \cdot C) + (D \cdot \text{not } E)] \cdot F \cdot (\text{not } G + H)$$

EXAMPLE 3-12

Determine the condition of the input switches (activated or not activated) for power flow in Figure 3-39(a). The sensor contact configurations are as follows: A is NO, B is NC, C is NO, D is NO, E is NC, F is NO, G is NC, and H is NO.

SOLUTION

Use the table in Figure 3-37 to solve this problem, beginning with input instruction A (NO contacts) in the ladder logic in Figure 3-39(a).

1. Find the four rows in column 1 that agree with the input contacts. Instruction A is NO, so rows 1 to 4 are selected.
2. Find the two rows from rows 1 to 4 that agree with the ladder rung instruction type in column 6. Instruction A is an XIO instruction, so the two rows are rows 2 and 3.
3. Find the one row from rows 2 and 3 that has the output true in column 5. Instruction A is highlighted, so power flow is present and the output is true. Thus row 2 describes the operation of this instruction and can be used to find the condition of the NO field device. In this case the field device is *not active* (column 2).

Use this process to step through the solution for the remaining instructions. Verify results with the following answers.

The input condition for the remaining switches is as follows: B field device is not activated, C field device is activated, D field device is activated, E field device is not activated, F field device is activated, G field device is activated, and H field device is activated.

Take each of the remaining examples in Figure 3-38, beginning with the relay schematic, and determine what limit switches must be closed (active) for the light to be illuminated. Then see if that agrees with the logic equation given in the last column. The input logic combinations described in this section apply to the PLC 5, SLC 500, and ControlLogix processors.

Figure 3-39 illustrates the many variations on these standard configurations. Four power flow combinations are presented because of different combinations of active input instructions. The power flow path follows highlighted instructions. An unbroken path of these instructions from the left rail to the output instruction causes the output to be true.

FIGURE 3-39: Multiple input power flow example (Note: Bold contacts have continuity.).

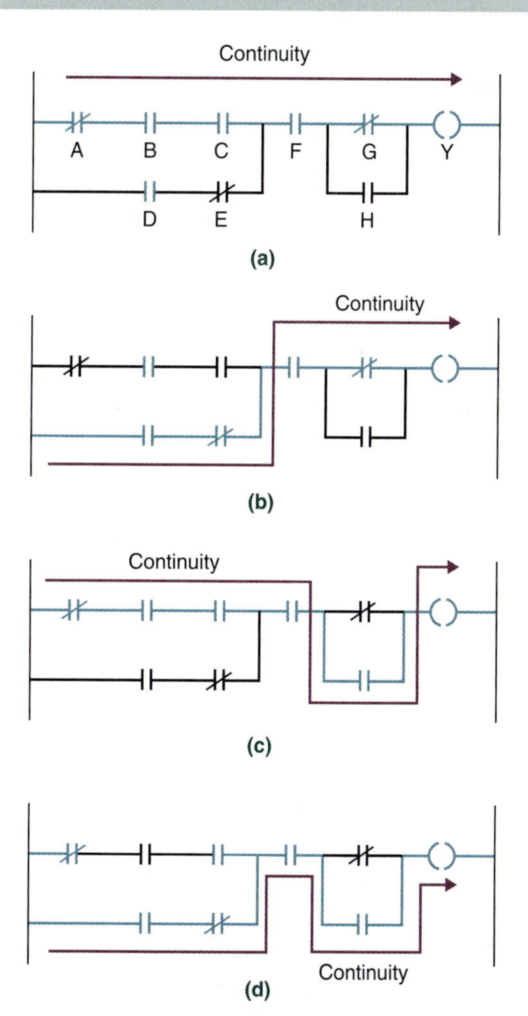

The combination of NC or NO field device switch contacts, XIC or XIO rung instructions, and a rung power flow requirement for a true output makes the selection of sensor activation conditions difficult to determine. The selection table (Figure 3-37) helps to provides correct answers. However, it also helps you see how all of the variables interact, so that dependance on the chart is eventually eliminated.

3-11-3 Multiple Outputs

PLCs also can have branches at the output side of the rung. Figure 3-40 illustrates several multiple output configurations. Note that the illustrations in Figure 3-40(a) and (b) use SLC 500 addressing, but the format could be used by any Allen-Bradley PLC processor. The ControlLogix processor permits both the older format and the addition of multiple outputs linked on the same ladder rung, as shown in Figure 3-40(c). In this figure all outputs are OTE, but any PLC output instruction could be used in place of the OTE instructions. Review the ladders in Figures 3-38, 3-39, and 3-40 for the many different instruction configurations.

3-11-4 Empirical Program Design

Programmable logic controller programming is accomplished using two techniques: *empirical* and *structured*. The empirical programming approach is frequently used with ladder logic programs; the structured technique is used with higher-level PLC programming languages such as Structured Text and Sequential Function Charts. The empirical approach is described here and structured approaches are presented in the second part of the text.

The empirical approach, the solution technique used most often for PLC ladder logic program design, develops the ladder solution one rung at a time. The frustration for designers is that this design approach yields numerous solutions that will all work. It is not easy to determine if the solution generated is the optimal solution available. Any design procedure is just a guideline, and the designer has many options for the implementation. The options selected are dictated by company preferences and designer experience. The design of the ladder logic for the process problem in Figure 3-36 is used to illustrate how the empirical process works, using the following seven steps:

1. As completely as possible, write a description of the problem and process control sequence. PLC ladder logic is best suited for process problems in which a

FIGURE 3-40: Multiple ladder outputs.

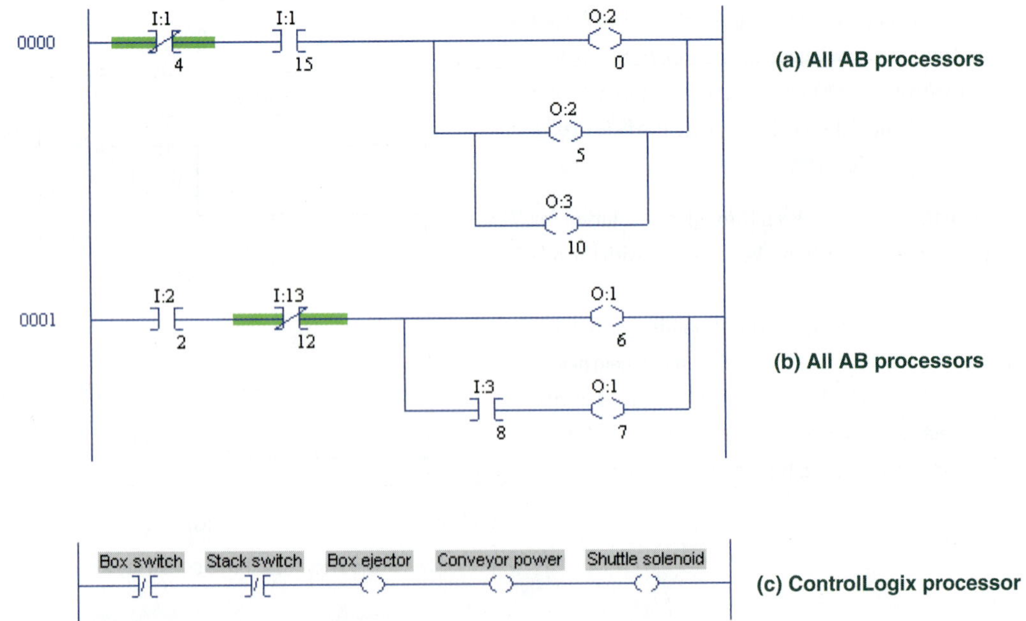

sequence of events must be controlled. The following problem statement describes the process in Figure 3-36:

Example problem statement: A mixing tank like the one shown in Figure 3-36(a) is used to heat a liquid. Note the location of all devices. A start push button is used to initiate the processes, and a stop push button is used to stop the pump. The inlet pipe's pump delivers fluid to the tank. Two level switches, L1 [low level float switch (FS)] and L2 [high level float switch (FS)], control the fill and empty cycles in the process. When the start switch is pressed the fill pump is activated. Liquid rises in the tank until the L2 FS is activated and stops the fill cycle. The heater is activated in the fill cycle when the liquid activates the L1 FS. The heater turns off when the process liquid exceeds the set point temperature (185° F) and makes temperature sensor TS1 active. Due to the temperature of the input liquid, the set point temperature is not reached until after the tank has finished filling. When the tank is full and the liquid is at the desired temperature, TS1 is active, the output valve is turned on (opened), and the tank empty cycle starts. The output valve is closed at the end of the empty cycle when the liquid level is below L1 FS.

2. Make a list of all output devices and all input devices with the type of contact used (manual input switches, process switches, and sensors).

   Example output and input device list: The process has three output devices: pump contactor, heater contactor, and outlet solenoid valve. The process has five input devices: stop momentary push button PB1 (NC), start momentary push button PB2 (NO), float switch Ll (NO), float switch L2 (NC), and temperature switch TS1 (NO contacts close when the temperature is above 185° F) . Because the switches are all single pole double throw, one set of NC and NO contacts is available. The contact type listed is connected to the input module as illustrated in Figure 3-36. *Note that only the NO contact from the TS1 temperature switch is used in this design.*

3. Use the process description to write a logic equation that makes each output device active (on). Indicate the input conditions that make the output active.

   Example output logic equations: The logic equations for the three outputs are:

   Pump motor contractor = NOT PB1 AND PB2 (sealed with output instruction) AND NOT L2

   The pump motor contactor is true (pump is running) when the start switch is pressed (NO contact is held closed) and sealed by the pump output OTE instruction, *AND* when the stop switch is not active (NC contact is closed), *AND* when the high level float switch L2 is not active (fluid is below high level switch, and so NC contact is closed).

   Heater contractor = LI AND NOT TSI

   The heater contactor is true (liquid is heating) when the low level float switch L1 is active (fluid is above low switch level, so the NO contact is held closed), *AND* the temperature switch TS1 is not active (fluid temperature is below 185° F, so the NO contact is open).

   Outlet solenoid value = LI AND TSI

   The outlet valve solenoid is true (valve is open) when the low level float switch L1 is active (fluid is above the low level switch, so the NO contact is held closed) *AND* when the temperature switch TS1 is active (temperature is above 185° F, so the NO contact is held closed).

4. For a PLC 5 or SLC 500 design, one must complete a wiring diagram for input and output devices, illustrating their connection to the input and output modules. Ladder logic input instruction addresses must include module terminal numbers for these PLC modules. If a CompactLogix system is used, then the tag names for all field device components and contacts

must be identified. Assigning input and output alias addresses for module terminal numbers can be performed at the same time, but the ladder logic programming is not dependent upon it.

Example input and output wiring diagram solution: Figure 3-41 illustrates the wiring for the input and output field devices. Wire labeling and numbering that adheres to company policy can be added at this time to the diagram.

5. Enter the SLC 500 ladder logic diagram by creating the rung(s) for each logic equation. Use the ladder instruction selection guide in Figure 3-37 to select an XIC or an XIO instruction for each input field device contact. Because Allen-Bradley ladders execute left to right and top to bottom, the rungs are created and listed in the execution order used by the sequential process. Also, this arrangment makes the program easier to read and understand. Finally, the finished ladder logic is documented and annotated.

   Example ladder logic solution: The ladder logic in Figure 3-41 produces the control required for the problem statement. Study this solution to the problem and compare it to the ladder logic solution for the same problem in Figure 3-36. Both work equally well, but the solution in Figure 3-36 uses both the NO and NC contacts on the temperature switch. As a result, an XIO instruction is needed for this solution in rung 1 for the heater contactor. The sealing instruction, O:5/3, opens after pump power is removed by liquid above the high level float switch so the pump does not restart during the drain cycle.

6. Test the solution to see if it meets the control requirements specified in step 1. If changes are needed, modify or add rungs to fix any problems.

   Example ladder modification: A serious problem not discussed in the original Figure 3-36 solution requires additional ladder modifications. When the NO contacts of the liquid temperature switch (TS1) are closed by a liquid temperature greater than 185° F the heat output (O:2.0) is false because the XIO instruction (I:1/1) in the heater rung is false (no power flow). A temperature greater than 185° F also opens the drain valve because TS1 causes the I:1/1 XIC instruction in the valve rung to be true. However, the liquid temperature likely will drop below 185° F as the liquid drains, causing TS1 NO contacts to open, so that the heater turns back on and the outlet valve closes. When the liquid temperature is again at the proper level, the drain cycle continues. This cycling will occur until L1 (low level float switch) is open and the tank is empty. This cycling of the heater is not a problem because it holds the liquid at the proper temperature; however, the closing of the outlet valve is a problem. The outlet valve rung must be modified by placing a sealing instruction with the outlet valve address around the XIC instruction in the valve rung. Now when the temperature drops below 185° F the valve is sealed on, and the TS1 instruction no longer controls the valve rung. Figure 3-42 illustrates this modified rung.

7. Document the solution based on company standards. Add rung comments and instruction descriptions that would give an engineer or technician who is not familiar with the design sufficient information to troubleshoot any production or control system problems.

The empirical approach described by the seven-step process works well for small ladder logic applications. However, as the control problem gets more complex, the designer's ability to predict all of the interaction amoung the rungs is limited, and numerous fixes and patches are necessary during design testing. As a result, everyone except the original designer finds the program more difficult to read and understand. Compare the fix added to the solution in Figure 3-42 with the original solution in Figure 3-41. Also, compare this design solution with the original solution in Figure 3-36 and with the relay ladder logic solution in Chapter 1.

Another problem created by this approach is the nonsequential nature of the program structure. Although the PLC scans the program from the first rung down to the last, turning on out-

FIGURE 3-41: Empirical design for tank control example.

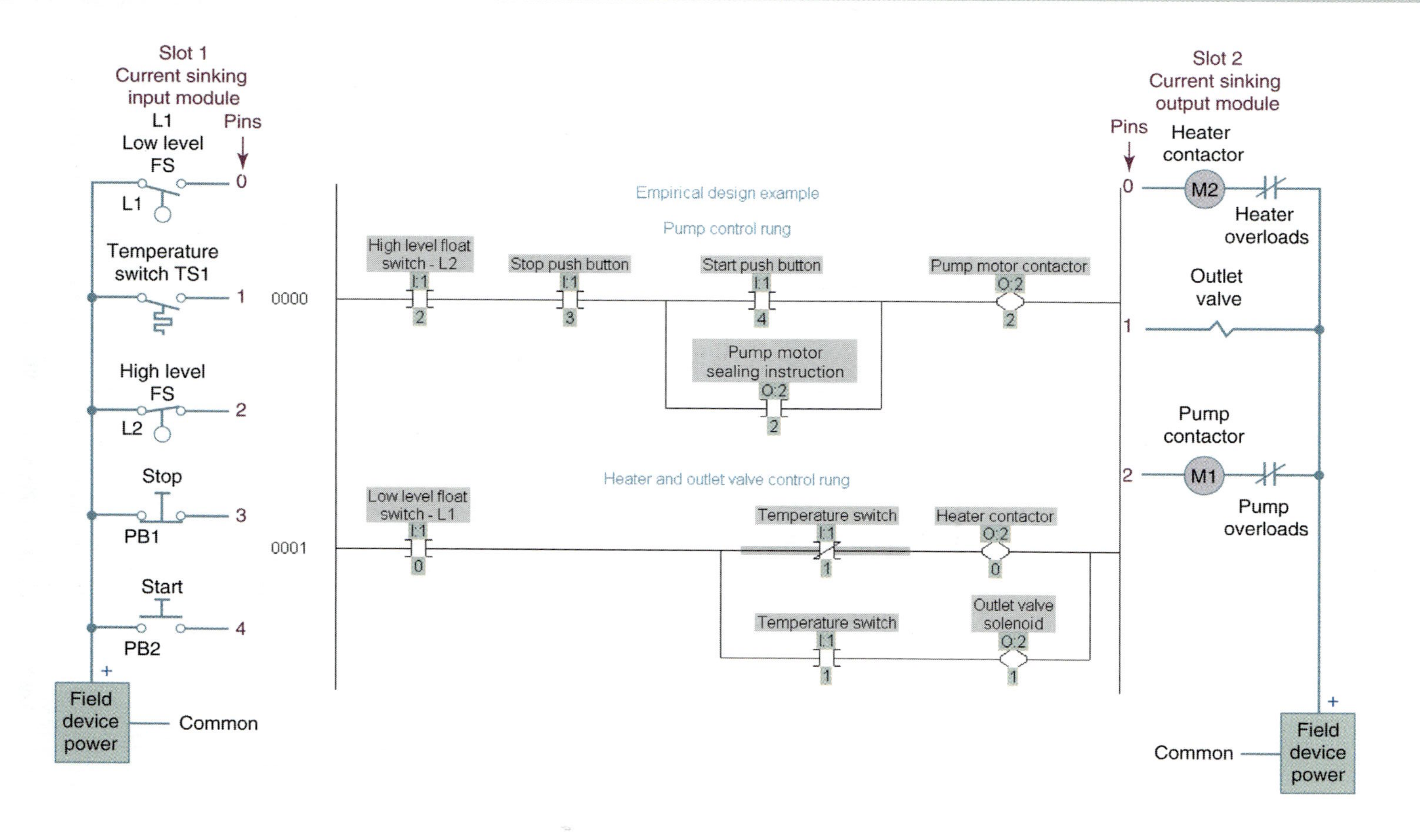

FIGURE 3-42: Fix to tank control design.

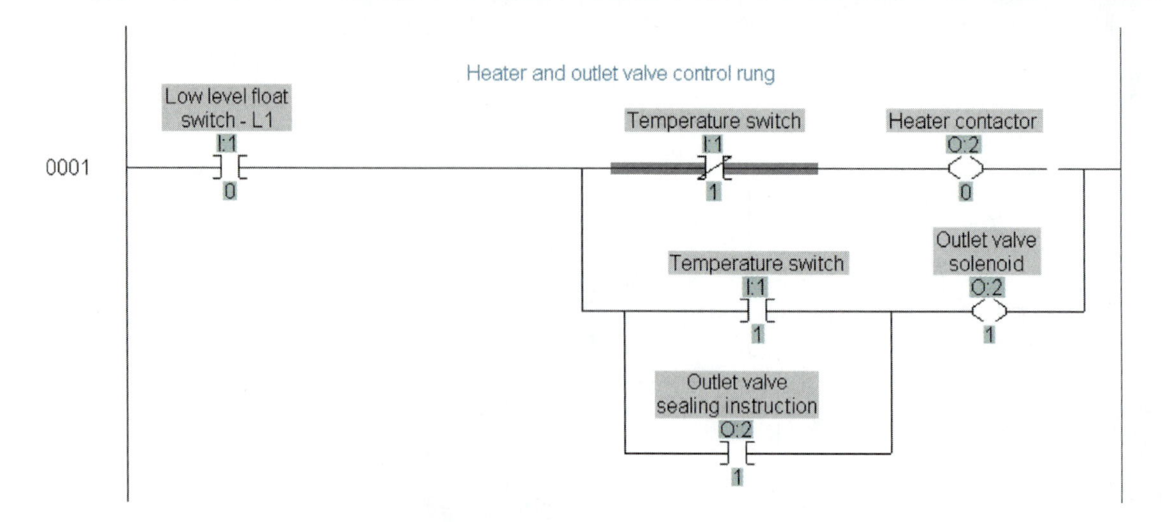

puts based on the input logic, the sequential change in the output states does not necessarily follow in that order. For example, the output on the last rung may turn on first and be followed by another output somewhere else in the ladder. It is not uncommon for a large system control problem to have a ladder program with several thousand rungs present. Determining the output sequence and machine operation in such a large system is more difficult if the program was developed using this empirical technique.

The previous example, as well as the next two, refer to the SLC 500 processor. If the PLC 5 were used, addresses would be adjusted for the single slot addressing mode and for rack number. If the ControlLogix were used, tag names with appropriate data types would be used for addressing.

EXAMPLE 3-13

Design a ladder logic system to provide two-handed control for a production machine, so that the operator uses both hands to initiate the start cycle of the machine.

Solution

Figure 3-43 illustrates the circuit that satisfies this control requirement. The necessary logic equation is Machine on = AND PB2. Note that a simple AND logic statisfies the need for two-handed control.

EXAMPLE 3-14

Design a ladder logic system to drive two pneumatic actuators through the part-loading cycle illustrated in Figure 3-44(a). The two pneumatic actuators and the gripper have sensors (NO contacts) to indicate the end positions of the actuators. A cycle is started when the start selector switch is on and when the pickup sensor indicates that a part is in the pickup location. Use the limit sensors to control actuator and gripper sequences.

Solution

Figure 3-44(b) illustrates the circuit satisfying the control requirement. The logic equations necessary for control are:

Axis X down = SEL1 AND PTS AND YIN AND GOP

Gripper close = XDN (sealed by GRV) AND NOT YOUT

Axis Y out =YUP AND GCL

Where YIN and YOUT are Y axis in and out sensors, XUP and XDN are X axis up and down sensors, GOP and GCL are gripper open and close sensors, and GRV is the gripper valve. Verify that the Boolean equation for each output is implemented in Figure 3-44(b) ladder logic. Move the robot through its cycle starting at the home position, letting the end-of-axis sensors change and verifying that the ladder logic provides a good solution. Note that the two-axis pneumatic robot is driven by the end-of-axis sensors so that optimum speed is attained.

Note that not GCL is used for GOP and Not GOP is used for GSL in the final design. The Not GCL makes sure that the gripper is fully closed before the X axis starts back up. The Not GOP makes sure that the gripper is fully opened before returning to the home position.

FIGURE 3-43: Ladder logic for two-handed control.

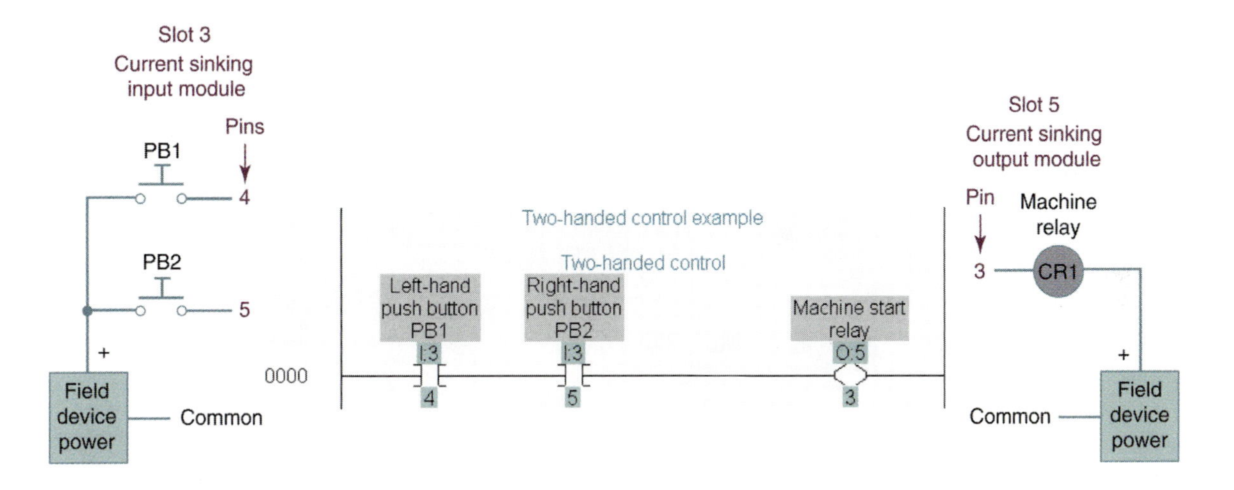

FIGURE 3-44: Two-axis robot part loader.

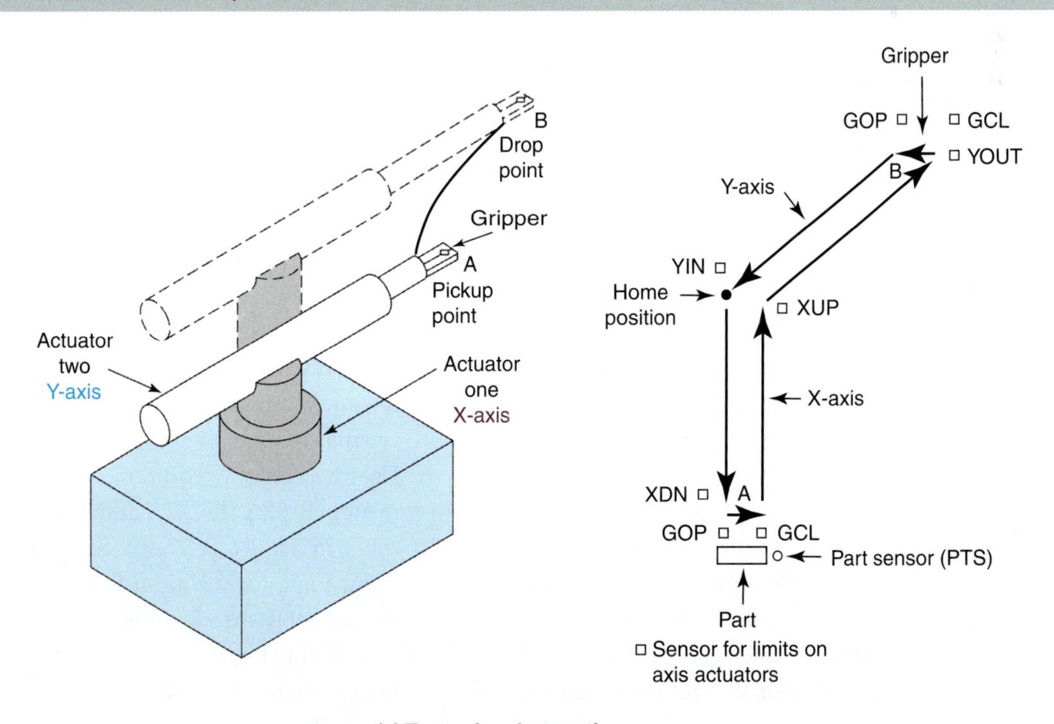

(a) Two-axis robot motion

FIGURE 3-44: (Continued).

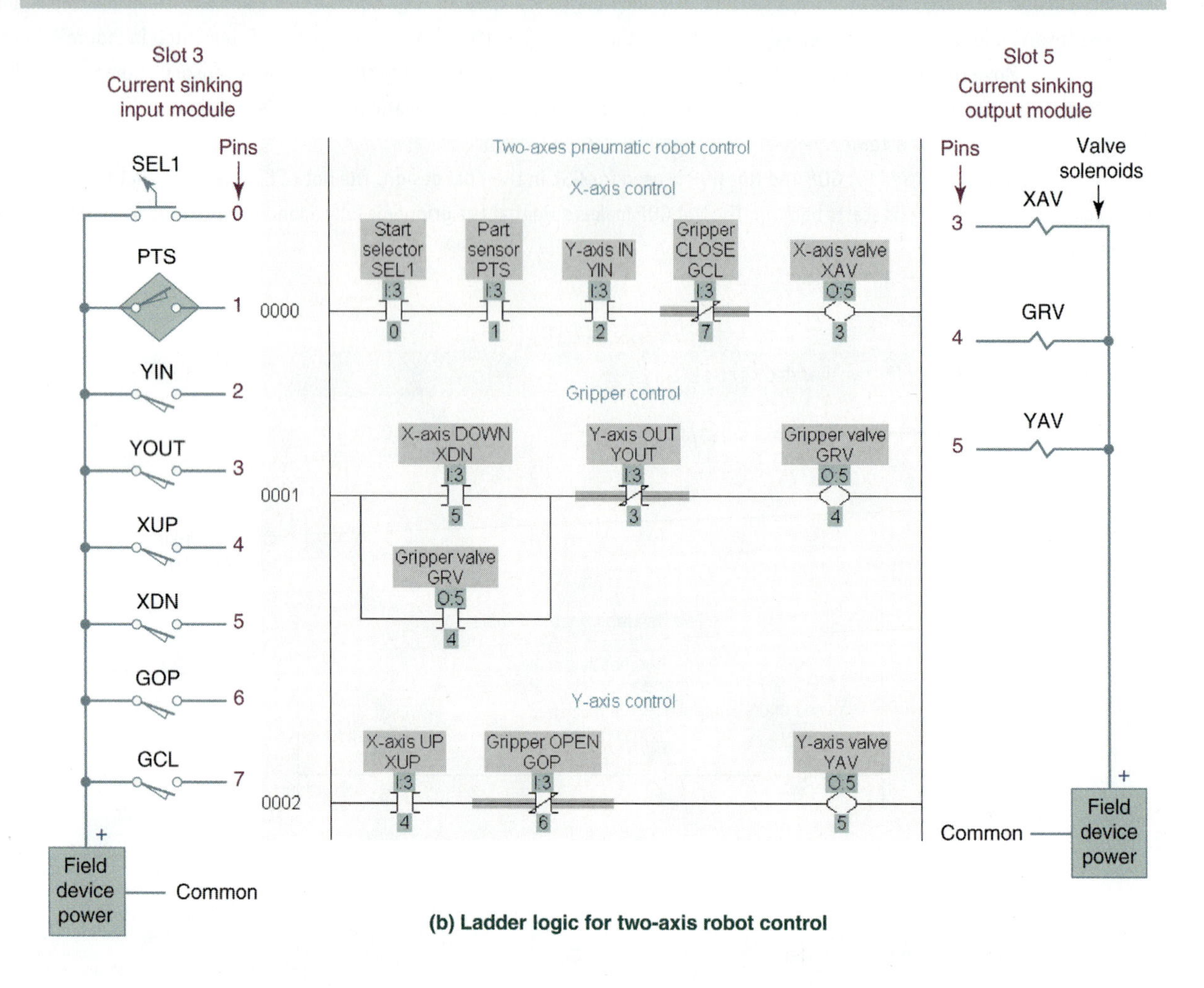

(b) Ladder logic for two-axis robot control

3-11-5 Converting Relay Logic to PLC Solutions

In addition to developing new PLC design, engineers and technicians are often asked to convert existing relay ladder logic to a PLC solution. The following technique is used:

1. Study the relay ladder logic for an understanding of how the control system functions. Write logic equations for each rung, in order to define the input logic for an active output condition.
2. Circle all of the elements on the relay ladder logic that are physical input field devices or output field devices. Note whenever multiple contacts on an input field device were used. Also, record the type of input contact (NO or NC) and condition (active or not active) when the output is on.
3. Select input and output modules whose voltages and current ratings are consistent with the input and output field devices.
4. Draw the input and output interface for all input and output field devices. Connect only one contact for each input field device to the input module.
5. Create a ladder logic rung for every relay logic rung. Replace every output device with an OTE instruction. If the output was a control relay, replace it with a binary bit (virtual relay).
6. Use the control requirements established in step 1, plus the XIC and XIO selection table in Figure 3-37, to create the rung input logic for all input field devices and virtual relay instructions.
7. Check the logic to verify that the PLC ladder logic performs exactly like the original relay ladder logic.

EXAMPLE 3-15

Convert the relay ladder logic in Figure 3-45(a) to a PLC implementation.

SOLUTION

1. Study the relay ladder logic for an understanding of how the control system functions. Write logic equations for each rung to define the input logic for an active output condition. Note that only the NO contact of the LS8 limit switch is used.

Outputs CR1 and PL3 = (PB14 OR CR1) AND LS7

Outputs SOL3 UP and CR2 = (PS7 AND CR1) AND SOL

Outputs SOL4 FWD = LS8 AND NOT LS9 AND CR2

Outputs SOL5 DWN, CR3, and PL4 = [PB1 OR (NOT LS8 AND CR3)] AND NOT PB2 AND NOT CR2

2. Circle all the elements on the relay ladder logic that are physical input field devices or output field devices [See Figure 3-45(b)]. Note whenever multiple contacts on an input field device were used. Also, record the type of input contact (NO or NC) and condition (active or not active) when the output is active or on. Multiple contacts were used on LS8, but the solution uses only a NO contact from LS8.

PB1	NO	active
PB2	NC	not active
PB14	NO	active
LS7	NO	active
LS8	NO	active for one rung and not active for another rung
LS9	NC	not active
PS7	NO	active
SOL	NO	active

FIGURE 3-45: Relay ladder logic for converstion to ladder logic.

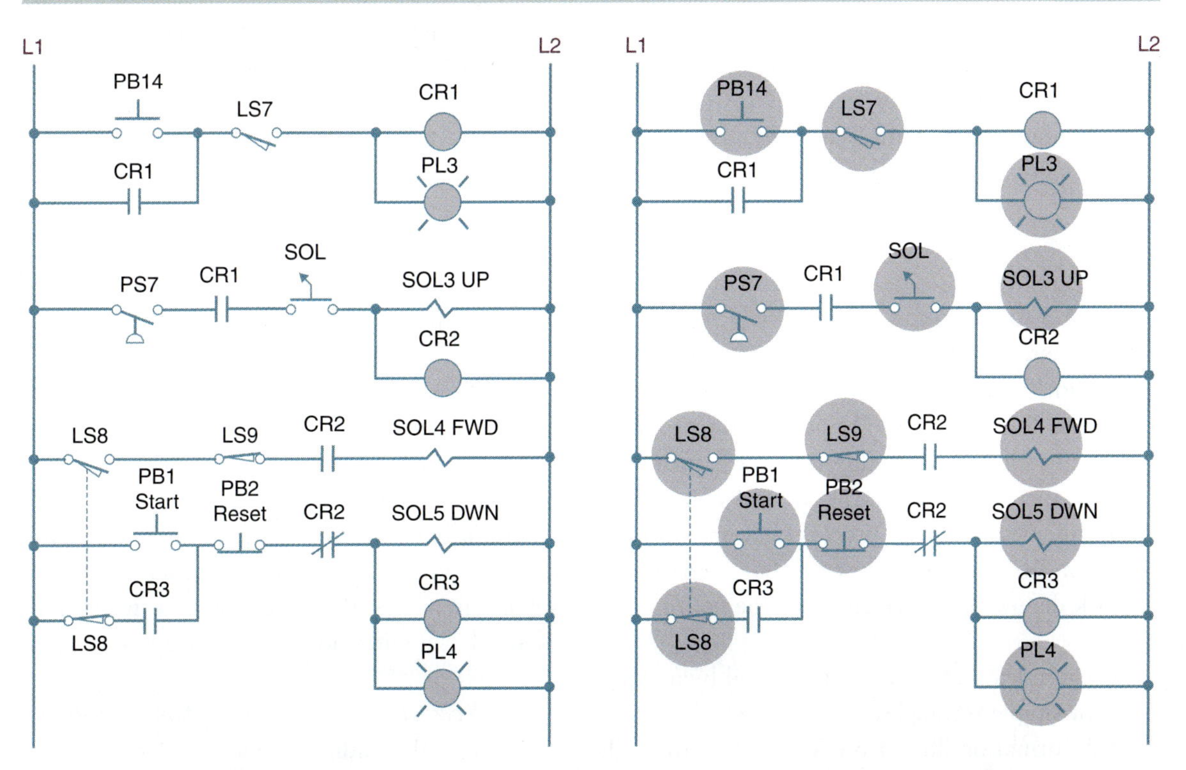

3. Select input and output modules with voltages and current ratings consistent with the input and output field devices. Input and output modules are 120 VAC.
4. Draw the input and output interface for all input and output field devices. Connect only one contact for each input field device to the input module. See Figure 3-46 for the interface solution.
5. Create a ladder logic rung for every relay logic rung. Replace every output device with an OTE instruction. If the output was a control relay, replace it with a binary bit. Using both the control requirements established in step 1 and the XIC and XIO selection table in Figure 3-37, create the rung input logic for all input field devices and internal relay instructions. See Figure 3-46 for the ladder logic solution.
6. A check of the logic verifies that the PLC ladder logic performs exactly like the original relay ladder logic.

3-12 TROUBLESHOOTING LADDER LOGIC CONTROL SYSTEMS

The PLC's reliability record is unmatched when compared with other microprocessor-controlled devices. For example, the statistically calculated mean time between failure rate is close to 5 years for some Allen-Bradley PLC products. Even with this great record, failures in the total automation system do occur and must be corrected. The system includes all of the input and output field devices, the PLC processor and modules, the ladder logic or other type of automation programs, and other devices and actuators present in the system. To locate the fault, an organized approach must be implemented.

3-12-1 System Troubleshooting Tools

Such an organized approach requires tools. The three important ones are *system block diagrams*, *bracketing*, and *signal flow analysis*.

Block diagrams. A block diagram is a set of rectangles used to describe all the parts of a system. Figure 3-47 illustrates a block diagram for the tank control problem. Note that each field device is a block, input and output modules are blocks, and the PLC is a block. The parts of the program that impact the signal flow also are included. Study the system in the figure and compare it with the description of process tank operation in Section 3-11-4. The following characteristics of block diagrams are evident from the figure.

- Complex systems are represented by a series of simple rectangles.
- Information flows from left to right through the rectangles.
- Systems, subsystems, and program structures can be represented.

Because the system block diagrams are not provided by the equipment vendor, in most cases the engineer must create one for troubleshooting requirements. A block diagram is easily generated as follows:

1. Make a list of all of the system components that would be replaced in the advent of a failure.
2. Arrange the list of system components with inputs at the top and outputs at the bottom. The remaining items that are between inputs and outputs should be arranged in the order that signals or information flows through them.
3. Put all of the block diagram components into rectangles and connect the blocks with signal flow lines, based on the system operation.

A troubleshooting tool, called bracketing, is used with the block diagram to isolate a fault.

Bracketing. Bracketing is a technique that uses markers to identify the portion of the system block diagram in which the fault exists. The brackets are used initially on the system block diagram, but they can be moved to a schematic diagram when the fault is narrowed down to a single circuit. A three-step process is used to establish the initial location for the brackets.

1. *Record and study all system symptoms.* List all conditions that vary from normal operation.

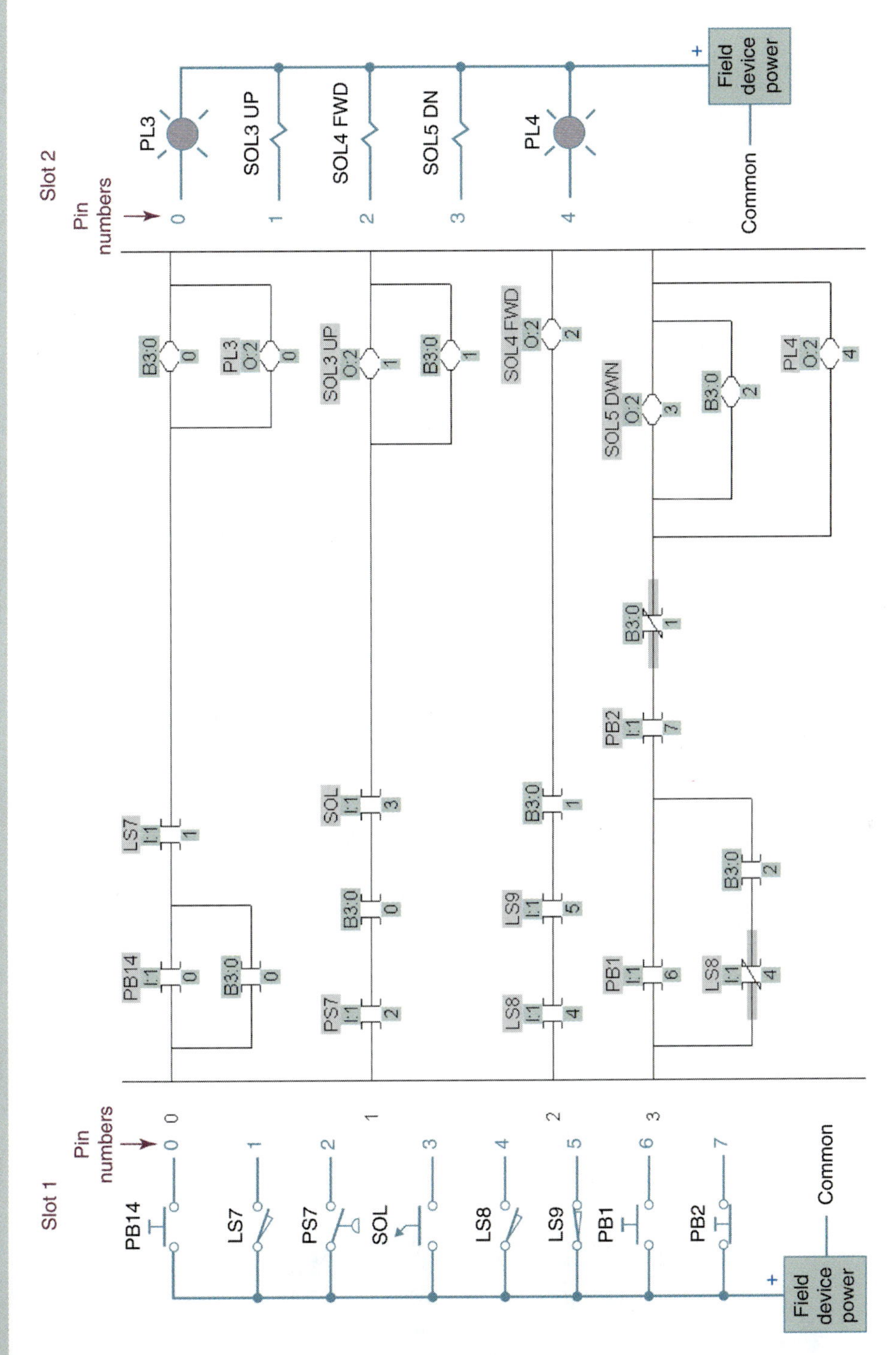

FIGURE 3-46: PLC ladder logic for relay ladder logic.

FIGURE 3-47: Tank control block diagram and signal and power flow.

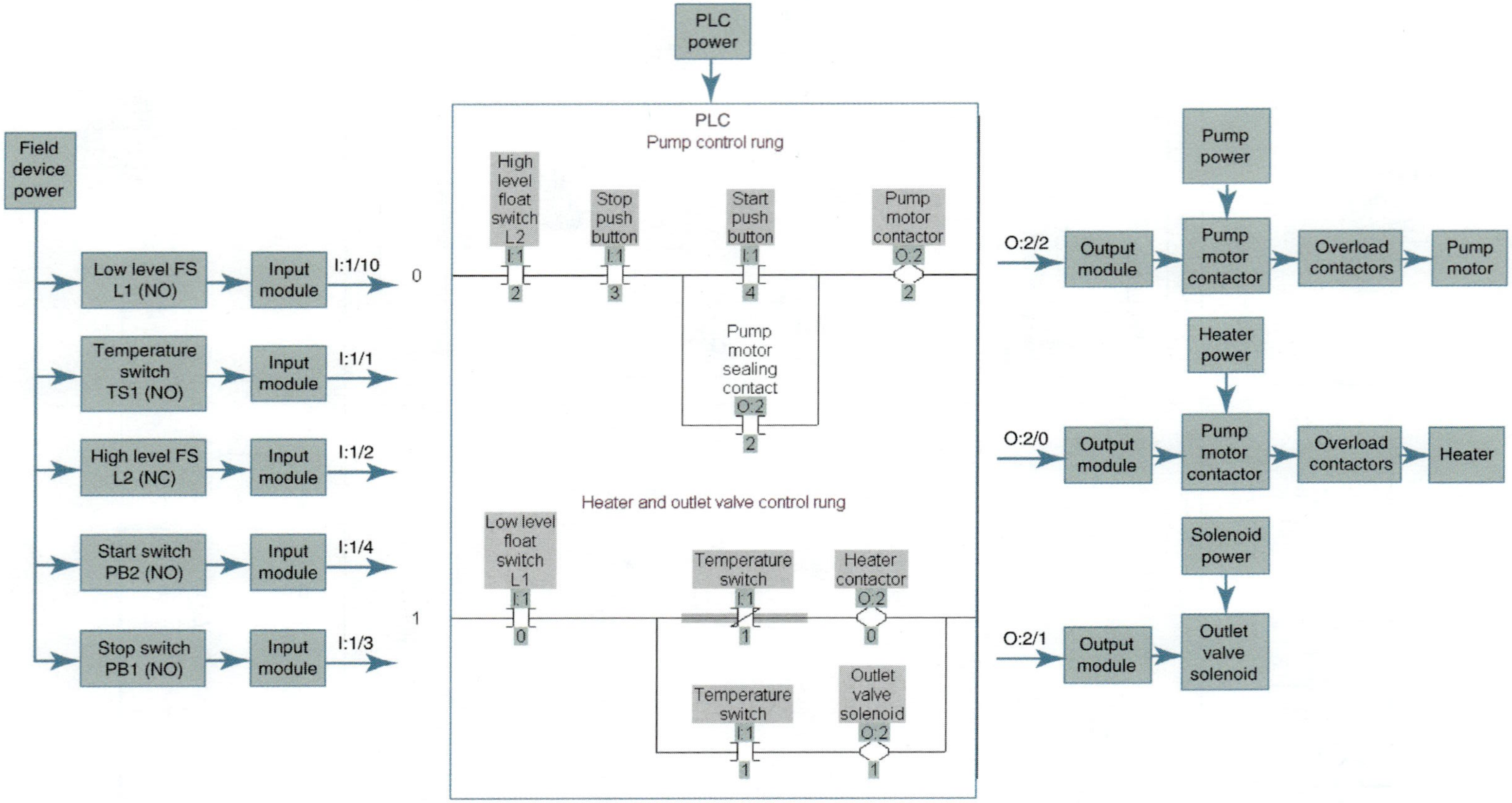

2. *Locate points on the system block diagram where abnormal operation is occurring.* Place a right bracket ( ]) after each abnormal block. Note that the signal flow on the block diagrams is from left (inputs) to right (outputs). In general, the right brackets are placed on the right side of output blocks that are not operating.
3. *Along the signal flow path, move to the left from each bad bracket until normal operation is observed.* Place a left bracket ([ ) to the right of the block where a normal output was detected. In general, the left bracket initially is placed with a minimum of detailed testing or only with easily performed tests.

The application of the brackets to the system block diagram is often just a mental process for experienced troubleshooters. However, the beginner should physically mark them on the block until all the troubleshooting techniques are fully internalized. The brackets tell us that the fault is somewhere in the system between the left and right brackets. Before proceeding with additional checks on the faulty system, it is important to learn the concepts associated with signal flow in the block diagrams.

Signal flow. Signal flow is generally divided into two groups, called *power* and *information.* Power flow illustrates how power is delivered to all the components in the system, and information flow indicates how information or data flows from source(s) to destination(s).

The five distinct types of signal flow topologies are *linear, divergent, convergent, feedback,* and *switched.* The user must recognize each type of path because a unique troubleshooting approach is associated with each of the topologies. A description of each type follows.

Linear: The linear signal path is a series connection of blocks, like the input and output blocks in Figure 3-47.

Divergent: A divergent signal path is present when a single block feeds two or more blocks or when a path on the left divides into two or more paths going to the right. In Figure 3-46 the power flow options in rung 1 of the ladder logic program illustrate the divergent type of signal and power flow.

Convergent: A convergent signal path is present when signals from two or more blocks feed into a single block or when two or more paths on the left merge into one path going to the right.

Feedback: A feedback signal flow is created when part of the output signal is diverted back to the system input and added to the input signal. All of the sealing instructions are examples of this type of signal or power flow.

Switched paths: Switched signal flow paths include linear, divergent, and convergent paths, with switches present to change the flow of the signal.

Every system has a signal flow block diagram that is a combination of these five topologies. Topology analysis speeds the troubleshooting process.

Signal flow analysis. The five topologies (linear, divergent, convergent, feedback, and switched) introduced in the previous section have rules that can be used to speed the search for the faulty system component. The topology rules that cause bracket movement follow.

Linear rule: When brackets enclose a linear set of system blocks to be tested, the first test point should be at or just before the midpoint of the bracketed area. If the signal is faulty, then the right bracket ( ]) moves to that point because the fault is to its left. If the signal is valid, however, then the left bracket ([ ) moves to that point because the fault is in the blocks to the right of this point. Application of this *divide-and-conquer rule* eliminates half of the components with a single check. In the PLC signal flow the ladder logic is included in the flow path. Although tests in the middle of the ladder are not possible, the concept can still be applied.

Divergent rule: When brackets enclose system blocks with a divergent path, the stage before the divergence is fault-free if any of the divergent paths are normal. For example, assume that brackets ([ ]) enclose the power supply that provides power to three other blocks in the diagram. If one can verify that power is being delivered to one of the system blocks, then the power supply is not the problem. As a result, the left bracket ([ ) is moved to the output side of the power supply.

Convergent rule: When brackets enclose system blocks with a convergent path, the following two rules must be applied:

- Rule 1: If all convergent inputs are required to produce a valid output, then a valid output indicates that all input paths are fault-free.
- Rule 2: If only one convergent input is required to produce a valid output, then each input must be checked to verify that the input paths are fault-free.

Feedback rule: When brackets enclose system blocks with a feedback path, a change or modification to the feedback path is used to indicate normal operation of the closed-loop system.

Switched rule: If brackets contain linear, divergent, or convergent topologies that are changed by a switch setting, it is necessary to observe the system when the switch is moved to another position. If the trouble disappears, then the problem is in the signal flow path of the previous switch position. Switched paths are some of the easiest to troubleshoot.

3-12-2 Troubleshooting Sequence

A general set of steps can be developed for use on system faults. The sequence is as follows:

1. *Define the problem*: Gather all of the symptom data for the failure, as well as information on the state of the system when the fault occurred.
2. *Decide what needs to be tested*: Put the brackets on the system block diagram (physically or mentally) in order to narrow the search for the fault to that part of the system where faulty operation has been verified.
3. *Decide what type of test to perform*: Initially, apply tests that do not require measurements inside system boxes. Start with changes in the signal flow path in order to perform broad system tests that will eliminate some of the suspect units between the brackets. As the brackets narrow the type of test becomes more precise with increased use of test equipment. When tests fail to provide any new information on the location of the fault, move to the next level of more precise tests.
4. *Correct the problem*: After the faulty unit or component is identified, the problem is corrected by adjustment or replacement of the unit or component.
5. *Verify correct operation*: After a problem is corrected, thoroughly test the system in order to verify that the applied fix corrected all of the system problems.
6. *Determine the cause of the failure*: getting the system back into operation is only half of the solution.

The first three steps in the sequence are repeated as often as necessary until the fault is located. Symptoms are studied, possible faulty units are bracketed, and the easiest possible changes in the signal flow are applied to eliminate some suspected units. As the brackets move closer together, the tests on suspected units introduce more precise test equipment. Throughout the process, symptoms continue to be analyzed. The analysis includes time to think about the problem in order to reflect on results of previous tests and possible future tests. If possible, determine what caused the unit or component to fail and recommend system changes that would prevent a similar failure in the future.

3-12-3 Troubleshooting Input and Output Modules

The general troubleshooting approach introduced in Section 3-12 allows the technician and engineer to view the system as a series of process elements or boxes. Figure 3-48(a) depicts the linear signal flow from the high level float switch L2 to the pump motor. The flow is repeated in

FIGURE 3-48: Signal flow path.

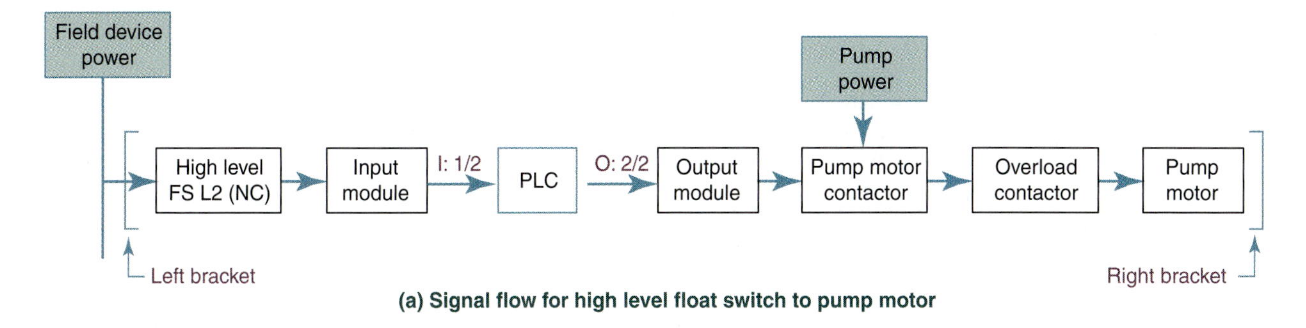

(a) Signal flow for high level float switch to pump motor

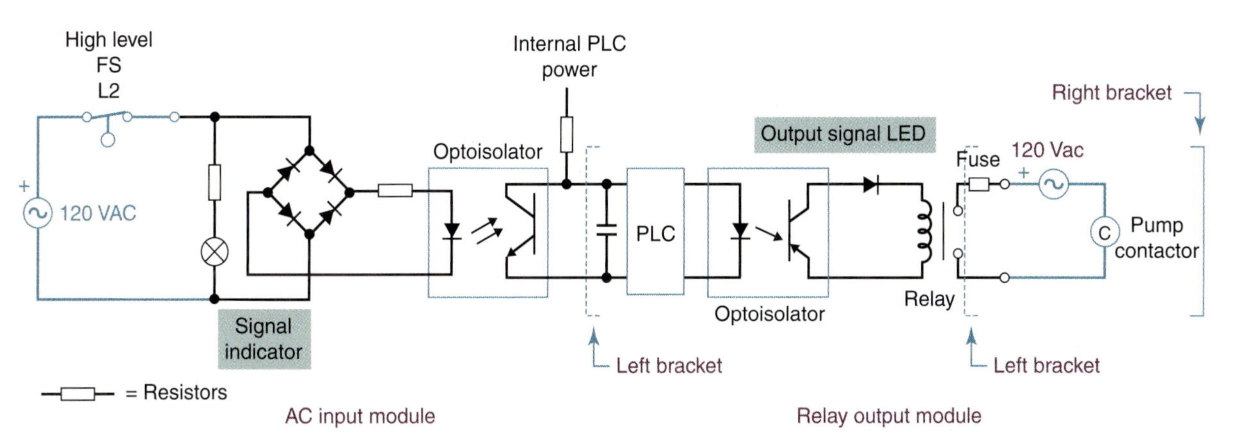

(b) Input and output module schematics

Figure 3-48(b), with typical internal circuits for an input module and an output module inserted into the boxes. Note PLC input and output modules have circuit indicators that are on if a signal is present. Each I/O point displays an indicator on the module's face. Figure 3-49 shows the front of a module with the 16 indicators grouped near the top. In the illustration only the numbers for ports 0, 3, 12, and 15 are present, but all are present on actual modules. These indicators are ideal for troubleshooting PLC system problems. The following troubleshooting sequence for the tank control solution in Figures 3-41 and 3-47 illustrates this point:

- The fill pump does not respond when the start push button is pressed, and the tank is empty. It was verified that the start switch is operating properly and that the source power to all devices is present. Therefore, the problem is located in a module between the left and right brackets on the signal flow in Figure 3-48(a).
- Because the indicator for input 2 on the input module is on, the NC float switch is closed and a voltage is present at terminal 2 of the module. The left bracket moves to the output of the input module. (See new dashed bracket.)
- Because the indicator for output 2 on the output module is on, the PLC logic made O:2/2 active. The left bracket moves to the right side of the output relay contacts in Figure 3-48(b). (See new dashed bracket.)

At this point, more than half of the control circuit has been removed from the troubleshooting problem by using signal indicators on the PLC modules. Some output modules that use fuses to protect the output circuits may have a blown fuse indicator [Figure 3-49(b)]. If the module's fuse is blown, then the indicator is on. If individual ports are fused then the module must be removed to check for a bad output point fuse. Note that the NO output relay contacts in Figure 3-48(b) are fused.

FIGURE 3-49: LED indicators on I/O modules.

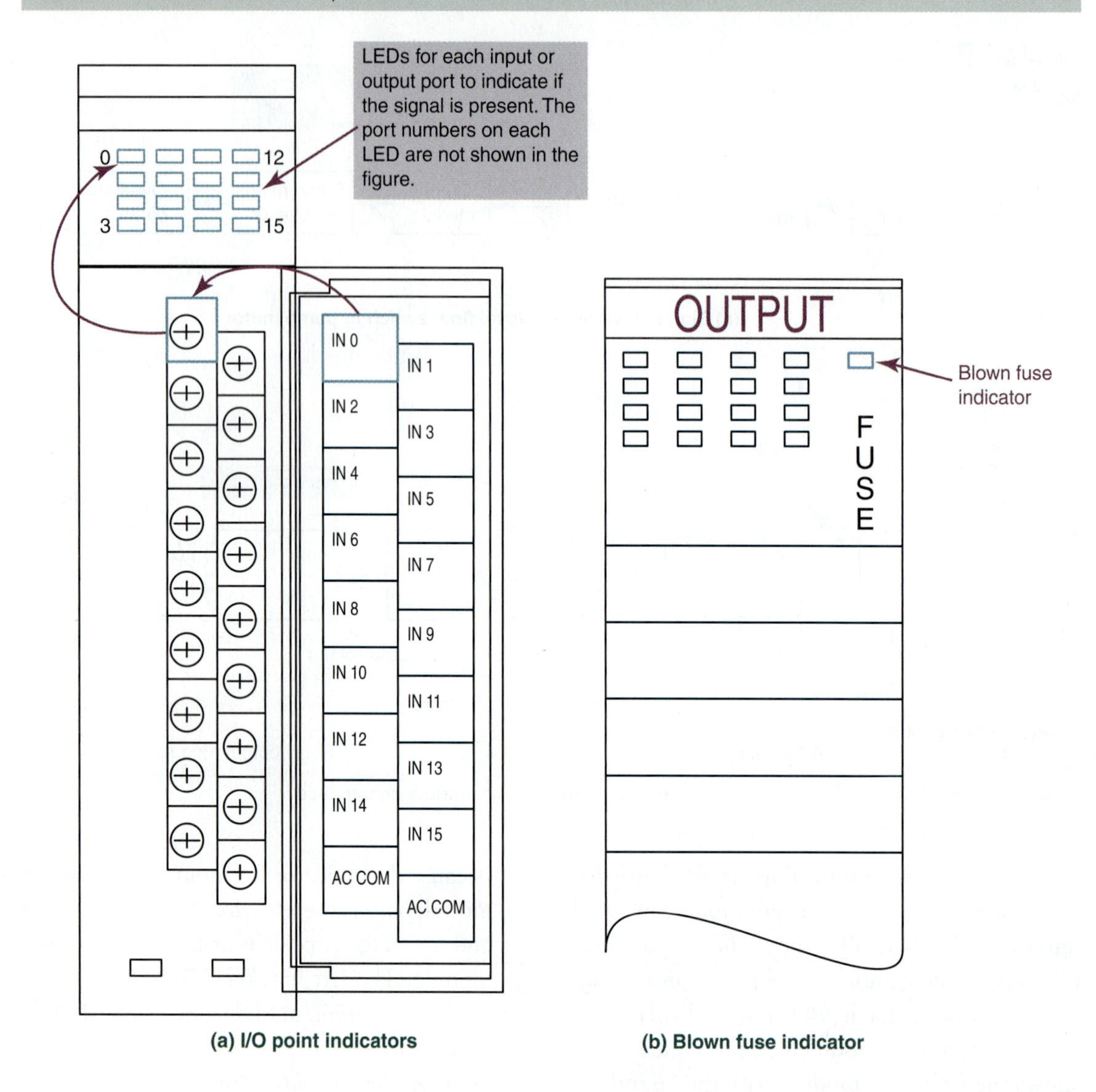

(a) I/O point indicators

(b) Blown fuse indicator

- After the output module is removed, it is determined that the fuse for output 2 is bad. After a good fuse is installed the system operates normally.
- Other techniques to troubleshoot I/O modules efficiently include:
  - If the module appears to be the problem, swap a similar module from another slot in order to see if the problem is solved or merely moves to a new location in the PLC rack.
  - If an individual I/O point is suspect, change the field device termination to another I/O point. Change the program and test to see if the problem has been corrected.
  - Use the forcing function. Turn on or off an input or rung from the programming console in order to see how the system responds. Inputs can be forced regardless of the state of the input field device.

Chapter 3 questions and problems begin on page 556.

CHAPTER

4

Programming Timers

4-1 GOALS AND OBJECTIVES

There are two principal goals of this chapter. The first goal is to provide the student with information on the operation and functions of hardware timers—both mechanical and electronic. The second goal is to show how the programmable timer instructions are applied in Allen-Bradley PLCs that are used in industrial automation systems.

After completing this chapter you should be able to:

- Describe the operation of a mechanical timing relay.
- Explain the differences between timed contacts and instantaneous contacts of a timing relay.
- Describe the difference between mechanical and electronic timing relays.
- Compare and contrast retentive timers to non-retentive timers.
- Describe the operation of TON, TOF, and RTO timer instructions.
- Describe the operation of cascading timers.
- Develop ladder logic solutions using timer instructions for the Allen-Bradley PLC 5, SLC 500, and ControlLogix series of PLCs.
- Convert relay ladder logic timer rungs to their PLC equivalent.
- Troubleshoot system problems associated with I/O modules and ladder rungs with timer instructions.

4-2 MECHANICAL TIMING RELAYS

Mechanical timing or time delay relays, as the name implies, have *fixed* or *variable* delay incorporated into their design that suspends the movement of the contacts when the coil is energized, de-energized, or both. Timers are a critical part of industrial automation and are necessary in sequential processes where a machine follows a set operational sequence with some steps assigned a specific time span. In relay ladder logic the timers are called *timing relays* because a contact closure is associated with the timing function. Knowledge of timing relays is important because relay ladder logic implementations continue to be used in small control applications and where higher current levels must be switched. In addition, relay ladder logic program with mechanical timers must be converted to a PLC implementation; as a result, an understanding of mechanical timer operation is necessary for a successful conversion. The schematic symbols for the four basic types of timing relays are illustrated in Figure 4-1.

FIGURE 4-1: Schematic symbols for timing relays.

Description	Control	International/British	Electronic
Normally open timed closed NOTC (a)	TC or		
Normally closed timed open NCTO (b)	TO or		
Normally open timed open NOTO (c)	TO or		
Normally closed timed closed NCTC (d)	TC or		

Rehg and Sartori, Industrial Electronics, 1st Edition, © 2006, Reprinted by permission of Pearson Education, Inc., Upper Saddle River, NJ

Mechanical timing relays use pneumatics to develop the time delay by the controlled release of air through an orifice during the expansion or compression of a bellows. The time delay period is set by positioning a needle valve to vary the amount of orifice restriction. The pneumatic timing relay provides *on-delay* and *off-delay* timing options with a range of 0.05 to 180 seconds and an accuracy of plus or minus 10 percent of the set time. However, pneumatic timers tend to drift over time, thus requiring periodic adjustment. Both AC and DC switching types are available with a typical switching current range of 6 to 12 amps and a voltage range of 120 to 600 volts. The continuous current is typically 10 amps.

4-2-1 Timed Contacts

Timed contacts have a *fixed* or *adjustable* delay action set by the pneumatic timing process. Time delay relay contacts are specified as either normally open (NO) or normally closed (NC), with the additional requirement that the delay operates in the direction of closing or in the direction of opening. The four basic types of time delay relay contacts fall into two groups: on delay and off delay.

On-delay timing relays. The normally open and normally closed timed contacts for *on-delay* timing relays have special names. The normally open are called *normally open, timed close* (NOTC) contacts, and the normally closed are called *normally closed, timed open* (NCTO) contacts. The two types of contacts operate as follows:

Normally open, timed closed (NOTC): The control and electronic symbols and the timing diagram for *normally open, timed closed* on-delay timing relays are shown in Figures 4-1(a) and 4-2(a), respectively. After the relay coil is energized, the timed normally open (NO) contacts remain open until after the time delay value. After the time delay (5 seconds in the figure), the timed contacts change state (NO contacts close) and remain in that new state as long as the coil is energized. When the coil is de-energized, the timed contacts immediately return to their initial state (NO contacts open).

Normally closed, timed open (NCTO): The symbols and timing diagram for *normally closed, timed open* on-delay time delays are shown in Figures 4-1(b) and 4-2(b), respectively. After the relay coil is energized, the timed normally closed (NC) contacts remain closed until after the time delay value. After the time delay (5 seconds in the figure), the timed contacts change state (NC contacts open) and remain in that new state as long as the coil is energized. When the coil is de-energized, the timed contacts immediately return to their initial state (NC contacts closed).

The action of the NOTC and NCTO contacts could also be described as the action of an NO and an NC contact on an on-delay time relay.

Off-delay timing relays. The normally open and normally closed timed contacts for *off-delay* timing relays also have special names. The normally open are called *normally open, timed open* (NOTO) contacts, and the normally closed are called *normally closed, timed closed* (NCTC) contacts. The two types of contacts operate as follows:

Normally open, timed open (NOTO): The control and electronic symbols and the timing diagram for *normally open, timed open* off-delay timing relays are shown in Figures 4-1(c) and 4-2(c), respectively. After the relay coil is energized, the timed NO contacts immediately close and remain in that new state as long as the coil is energized. When the coil is de-energized, the timed contacts remain in the changed state (the NO contacts close) until the set time delay value is reached. At the end of the time delay (5 seconds in the figure), the timed contacts return to their initial state (NO contacts open). Note that the delay starts after power is removed from the coil.

Normally closed, timed close (NCTC): The symbols and timing diagram for *normally closed, timed closed* off-delay timing relays are shown in Figures 4-1(d) and 4-2(d), respectively. After the relay coil is energized, the timed NC contacts immediately open and remain in that new state as long as the coil is energized. When the coil is de-energized, the timed contacts remain in the changed state (NC contacts open) until the set time delay value is reached. At the end of the time delay (5 seconds in the figure), timed contacts return to their initial state (NC contacts closed). Note that the delay starts after power is removed from the coil.

The action of the NOTO and NCTC contacts could also be described as the action of an NO and an NC contact on an off-delay time relay. In addition to the timed contacts on timing relays, *instantaneous* contacts are also present.

4-2-2 Instantaneous Contacts

Instantaneous contacts operate independently from the timing process, like standard control relay contacts. When the coil is energized the contacts change states; when the coil is de-energized they return to their normal states. An illustration of the instantaneous contact on each type of delay is provided in Figure 4-2; the schematic symbols are the same as a basic relay contact. Note that the contact state change coincides with the waveform of the coil voltage.

EXAMPLE 4-1

Draw the relay ladder diagram for an application where a motor is started 10 seconds after a start momentary push button is depressed and is stopped when a stop momentary push button is depressed.

SOLUTION

Figure 4-3 illustrates the solution, where TMR1 is the NOTC time delay coil, contact TMR1-1 is an instantaneous contact, and contact TMR1-2 is a timed contact. The instantaneous contact seals in the momentary start push button after it's released, and the normally open, timed closed contact activates the motor after the 10-second delay. Both TMR1 contacts are associated with one timer.

FIGURE 4-2: Timing relay timing diagrams.

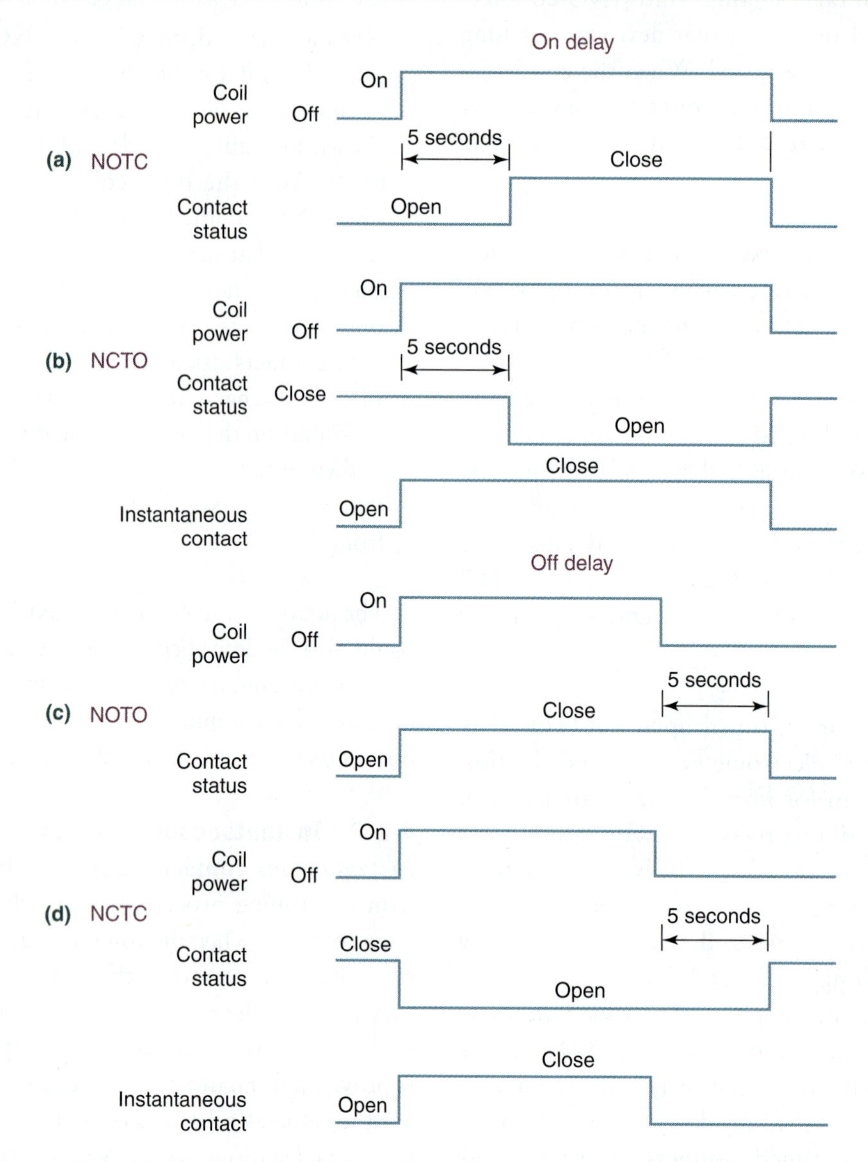

Rehg and Sartori, Industrial Electronics, 1st Edition, © 2006, Reprinted by permission of Pearson Education, Inc., Upper Saddle River, NJ

FIGURE 4-3: Relay ladder diagram for Example 4-1.

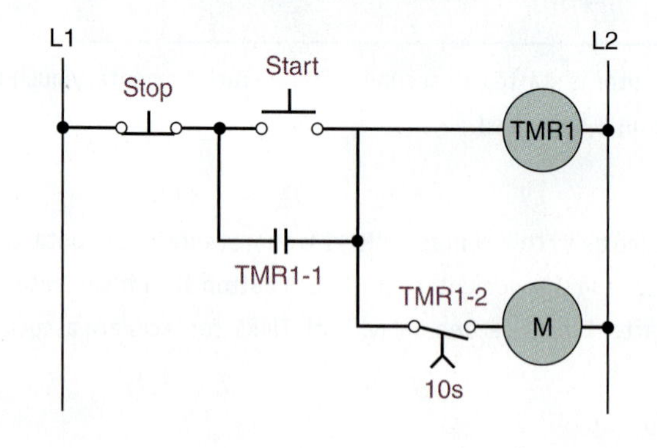

4-2-3 Timing Relay Operation

The operation is based on the pneumatic control illustrated in Figure 4-4. Study the drawing until all the components are familiar, and refer to the figure as you read the following description.

- When the solenoid plunger (10) is retracted from the push rod (11), it allows the spring (3) located inside the synthetic rubber bellows (1) to push the timing mechanism plunger (4) upward.
- As the plunger rises, it causes the over-center toggle mechanism (5) to move the snap-action

FIGURE 4-4: Pneumatic timing mechanism.

(a) Pneumatic timer

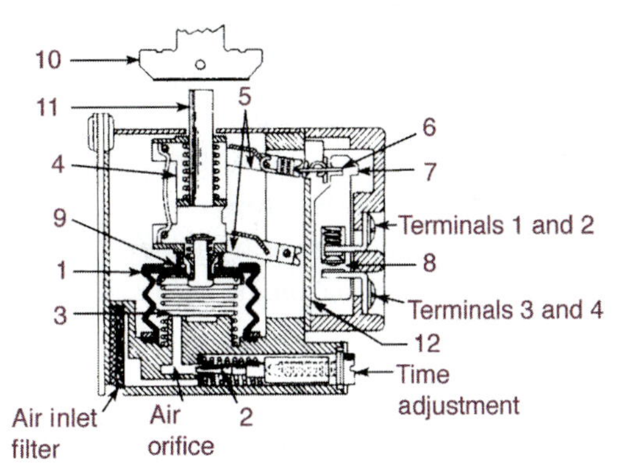

(b) Cutaway of pneumatic timer

Source: (a) Courtesy of Square D/Schneider Electric and (b) Courtesy of Rockwell Automation, Inc.

toggle blade (6) upward, which in turn picks up the push plate (7), which carries the movable contacts (8).

- The speed with which the bellows can expand is determined by the setting of the needle valve (2). If this needle valve is nearly closed, a maximum length of time will be required for air to pass it and permit the bellows to expand. The needle valve setting determines the time interval that must elapse between the release of the solenoid actuator and expansion of the bellows to switch the contact.
- When the push rod (11) is again depressed by the solenoid plunger (10), it forces the timing mechanism plunger (4) to the lower position, exhausting the air through the release valve (9) and resetting the timer almost instantaneously.

4-2-4 Selecting Timing Relays

Timing relays are selected based on the following operational characteristics:

- Length of time delay required
- Range of timing values required for the machine or process
- Timing options required for the process
- Repeatability and accuracy of the timed delay required for the process
- Current rating, configuration, and quantity of timed contacts and/or instantaneous contacts required for the control

4-3 ELECTRONIC TIMING RELAYS

Electronic timing relays are more accurate and repeatable than pneumatic timing relays, plus they provide an economical solution for applications requiring basic timing functions. Figure 4-5(a) depicts a typical electronic timing relay, and Figure 4-5(b) depicts a multifunctional timing relay. The typical timing relays provide the timing functions as described in the previous section and operate with a supply voltage in the 24 to 48 VDC range or the 24 to 240 VAC range. The solid-state electronics provide timing settings from 0.05 seconds to 60 hours with a timing accuracy of plus or minus 5 percent of the set time and an excellent repeatability of plus or minus 0.2 percent.

The multifunction electronic timing relay is typically microprocessor controlled and provides 10 or more timing functions, which are variations of the on-delay and off-delay timed outputs plus

FIGURE 4-5: Timing relays.

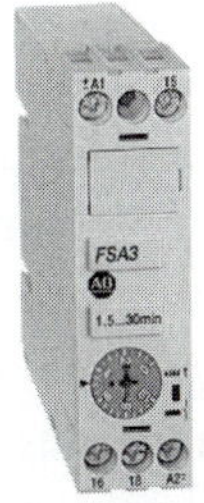

(a) Electronic timing relay

(b) Multifunction timing relay

Courtesy of Rockwell Automation, Inc.

some pulsed output options. With the variety of timing functions and ranges available, the multifunction relay eliminates the need for additional auxiliary relays in complex applications, saving installation time and reducing parts and labor costs. Go to the Allen-Bradley Web site at http://www.ab.com and select timers to see the numerous output combinations that are available for electronic timers. These electronic timing relays are stand-alone devices and not a part of a PLC.

4-4 PLC TIMER INSTRUCTIONS

Timer instructions are important in PLC applications where the time for a machine's cycle times is critical, or when some time delay is needed between process sequences. PLC timers are *output instructions* that provide the same functions as on-delay and off-delay mechanical timing relays and electronic time delay relays.

PLC timers offer numerous advantages over their mechanical and electronic counterparts. PLC time settings can be easily changed and the quantity of timers can be changed through programming without wiring modifications. In addition, the PLC timer is highly accurate and repeatable because its time delays are generated in the PLC processor. The accuracy of the timed event may be affected, however, if the program has a large number of rungs and therefore a long scan time.

4-5 ALLEN-BRADLEY TIMER INSTRUCTIONS

The timer instructions for the Allen-Bradley (AB) PLC 5, SLC 500, and Logix processors operate in nearly identical fashion. Therefore, most of the example solutions in this chapter use the SLC 500 instructions; however, PLC 5 and ControlLogix instructions are used in a few examples to illustrate the differences in the three systems.

AB has three timer instructions discussed in this chapter: timer on-delay (TON), timer off-delay (TOF), and retentive timer on-delay (RTO). The next sections prepare for the discussion of these instructions by introducing the timer ladder logic symbol, timer parameters, and the function of the timer Boolean bits and integer registers.

FIGURE 4-6: PLC timer block.

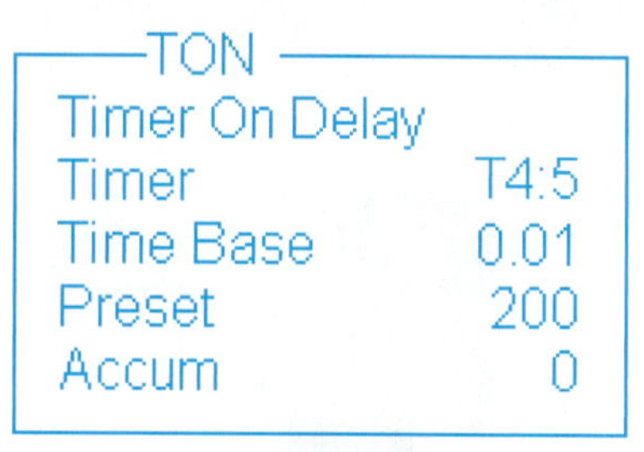

(a) Allen Bradley

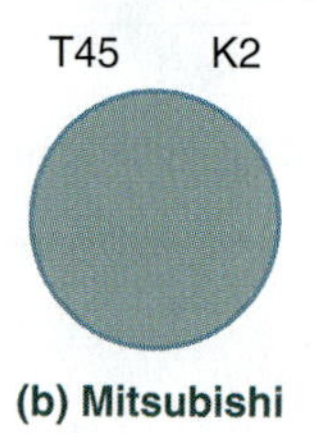

(b) Mitsubishi

4-5-1 Allen-Bradley Timer Symbol and Parameters

All three AB timer instructions are represented as blocks in the ladder logic with three (ControlLogix) or four (PLC 5 and SLC 500) data parameters. Figure 4-6(a) shows the TON timer instructional block for the PLC 5 and SLC 500. In other PLC brands the timer uses a symbol like that in Figure 4-6(b), and in some cases they use the symbol of a timing relay discussed in Section 4-2. Each PLC manufacturer represents the data inside the block slightly differently, but the parameters generally include the same information. The timer blocks for the PLC 5 and SLC 500 are illustrated in Figure 4-7 and the block for the ControlLogix processor is illustrated in Figure 4-8. The four parameters required for a timer include *timer number, time base, preset value,* and *accumulator value.* Refer to Figures 4-7 and 4-8 as you read the following descriptions.

- **Timer Number and Tag Name:** The Allen-Bradley PLC 5 and SLC 500 timer instructions, Figure 4-7(a) and (b), use a timer file, T4, for all timers and attach a unique number to identify the specific timer. For example, T4:0, T4:1, and T4:2 are three timers numbered 0, 1, and 2. The colon (:) is a delimiter used to separate the file number and the timer number. The number of timers allowed in file T4 is 256 (numbers 0 to 255); if more timers

FIGURE 4-7: PLC 5 and SLC 500 TON timer instructions.

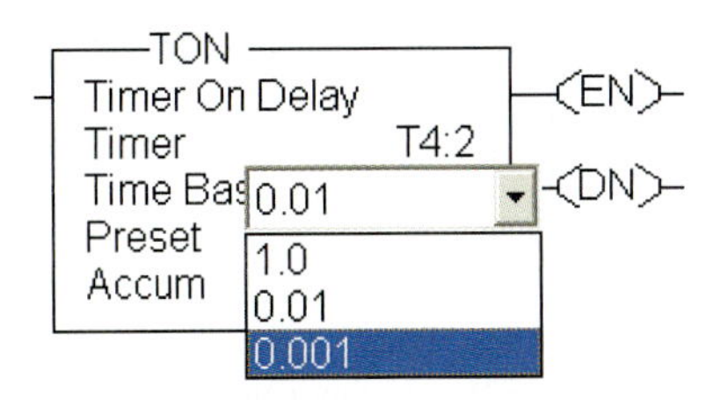

(a) SLC 500 timer instruction

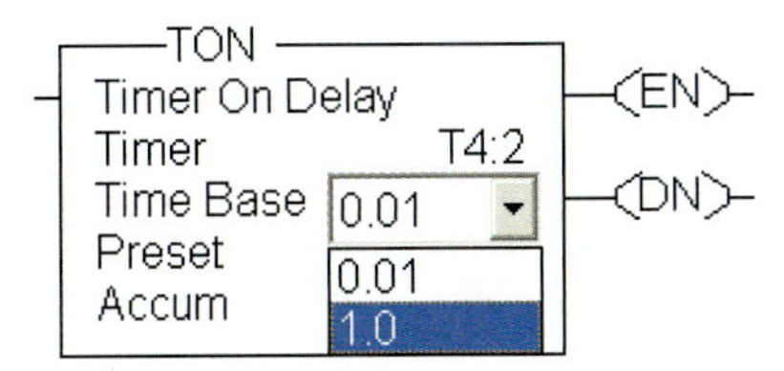

(b) PLC 5 timer instruction

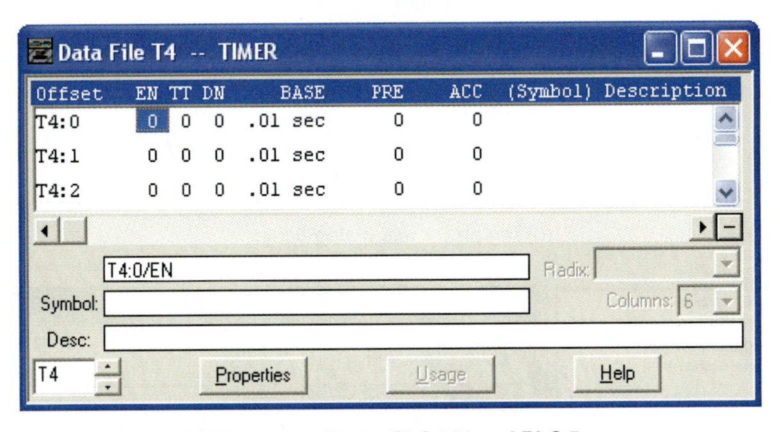

(c) Timer date file for SLC 500 and PLC 5

are needed files T10 through T255 can be used, with each holding 256 timers. The timer database file is shown in Figure 4-7(c) with the current value of all parameters displayed for each timer. Timer parameters can be entered directly into the timer instructions or into this database file dialog box.

The Logix processors, Figure 4-8(a), use a tag name for the timers, such as Pump_timer. The descriptive tag name makes it easier to know what function the timer serves in the control system. Any valid tag name (see Chapter 3 for tag name rules) can be used, but the name must be declared using the programming software tag properties dialog boxes illustrated in Figure 4-8(b). The tag name typed into the timer instruction appears at the top of the dialog box when the tag is validated. The description (optional), tag type, and data type are added to complete the validation. The description can be any text desired, and the tag type used most often is *Base*. The data type, *TIMER*, must be selected or typed. A pop-up Select Data Type dialog box appears when the selection box button at the right of the data type line is double-clicked.

The timer tag database is shown in Figure 4-8(c). The database is accessed by double-clicking the Program Tags file in the file menu. This dialog box offers two views of the timer database: monitor tags or edit tags. To view tag values the monitor tags tab is selected at the lower left of the dialog box and the display in Figure 4-8(c) is displayed. The values for all timer variables are displayed. Note that the Logix system has some additional variables compared to the PLC 5 and SLC 500

FIGURE 4-8: ControlLogix timer instruction.

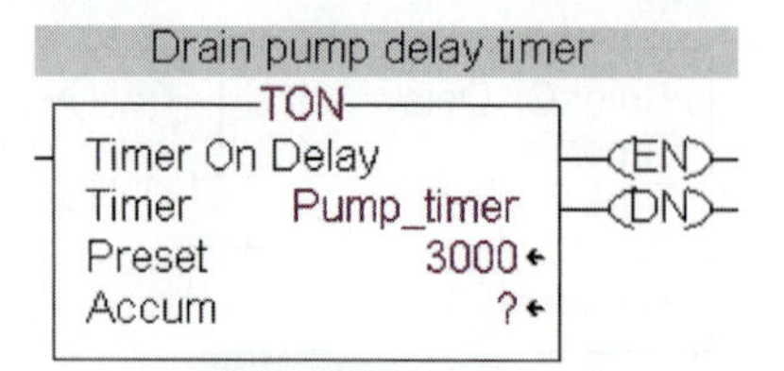

(a) ControlLogix timer instruction

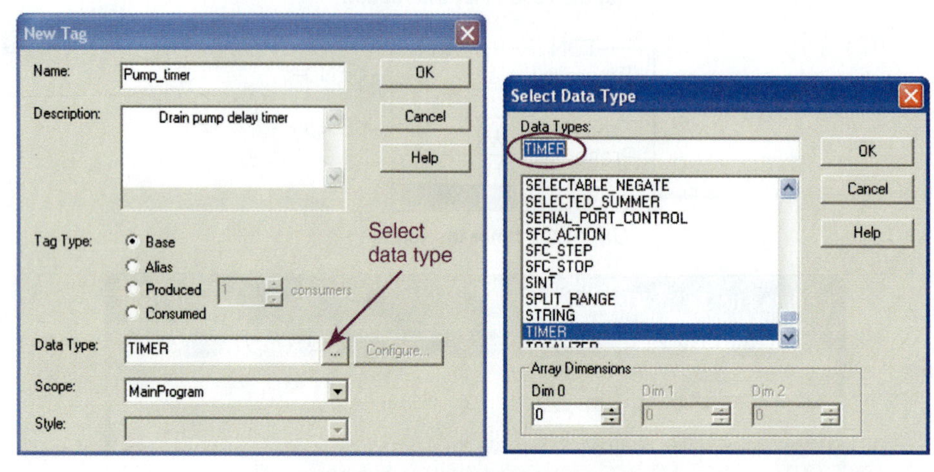

(b) Tag properties dialog boxes

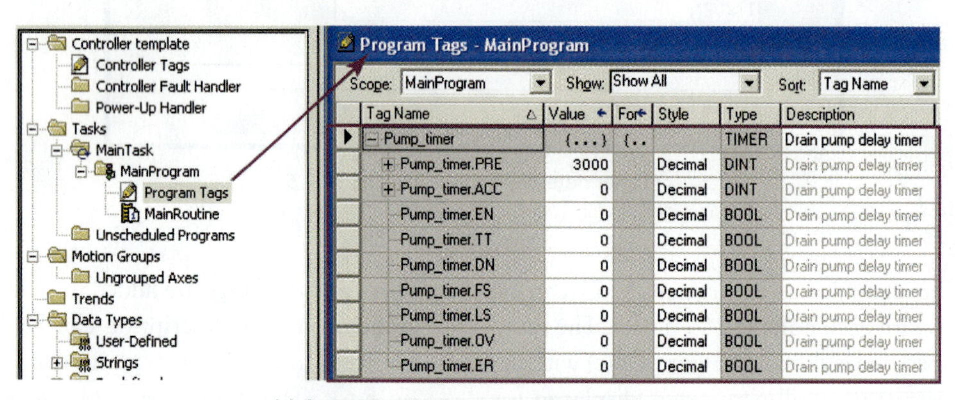

(c) ControlLogix timer tag database file

systems. If the other tab, edit tags, is selected then changes to the timer database are entered.

- **Time Base:** PLC timers increment from 0 to a preset value in time segments of 1, 0.1, 0.01, and 0.001 seconds. The *time base* indicates the incremental change in the accumulator value when the timer instruction is active. For example, if the preset holds 1000 and the time base is 0.01, then the time delay is configured for 10 seconds (1000×0.01). Figure 4-7(a) illustrates the time base options for the SLC 500 and Figure 4-7(b) shows the options for the PLC 5. Note that the SLC 500 has a time base value of 0.001 seconds listed, but it is not supported and cannot be used.

The ControlLogix timer, Figure 4-8(a), has two variations from the PLC 5 and SLC 500 models. First, the *time base* selection field is absent since it has a *fixed time base* of 0.001 seconds, and second, the timer number is replaced by a tag name.

- **Preset Value (PRE):** This integer value is the number of time increments that the timer must accumulate to reach the desired time delay. For example, if the time base is 0.01 and the preset value is 200, then the time delay is 2 seconds (0.01 × 200). The range of preset value for the PLC 5 and SLC 500 timers is from 0 to +32,767. If a timer preset value is a negative number, a runtime error occurs.

The preset value for the ControlLogix timer in Figure 4-8(a) and (c) is 3000. The timer time is 3 seconds (3000 × 0.001) since the time base is fixed at 0.001 seconds for each increment of the accumulator. The range for the ControlLogix preset value is −32,768 to +32,767 for integers, but it is in the +/− 2 million range for double integers.

- **Accumulator Value (ACC):** The accumulator value indicates the number of increments that the timer has accumulated while the timer rung and instruction are active. The ranges of values permitted for the accumulator are the same as those given for the preset value. The accumulator value is reset to zero when the timer is reset, and the non-retentive timers are reset when the rung and instruction are false.

Configuring a timer includes: selecting the timer number or tag name, selecting a time base (SLC 500 and PLC 5 only), and entering a preset value for the time delay required. In rare situations an accumulator value other than zero is entered. The three timer bits used in timer ladder logic control are described next.

4-5-2 Allen-Bradley Timer Bits

The three AB timer models (PLC 5, SLC 500, and Logix) and all three Allen Bradley timer types have the same three Boolean bits for ladder logic control. Their names and descriptions follow.

- **Timer Enable Bit (EN):** The enable bit is true when the rung input logic is true, and the enable bit is false when the rung input logic is false. When the EN bit is true the timer accumulator is incrementing at the rate set by the timer time base.
- **Timer Timing Bit (TT):** The timer timing bit is true only when the accumulator is incrementing. TT remains true until the accumulator reaches the preset value. When the accumulator value is equal to or greater than the preset value, the timer timing bit is returned to a false condition. In other words, the TT bit indicates when timing action is occurring and can be used to control timed events in automation applications.
- **Timer Done Bit (DN):** The DN bit signals the end of the timing process by changing states from false to true or from true to false depending on the type of timer instruction used.

4-5-3 Allen-Bradley TON, TOF, and RTO Instructions

The three types of Allen-Bradley timer instructions include: *on-delay timer* (TON), *off-delay timer* (TOF), and *retentive timer* (RTO). The truth tables in Table 4-1 describe the conditions that cause a true or false state on the timer output bits (EN, TT, and DN) for each timer type. This truth table applies to timers from all three Allen-Bradley processors. Read the truth table before continuing.

The action of the timer enable bit is the same for all three types; namely, it is true if the timer instruction rung logic is true and false if the logic is false. However, the timer action created by the enable varies with the three different timer types. Review Table 4-1 to verify this operation.

The timer timing (TT) bits of TON and RTO are true when the accumulator (ACC) is less than the preset value AND the timer is enabled. The TOF has the same operation except that the enable bit is false. All three of the timers have a different logic requirement for the TT to be false. Also, the done bit on each timer has unique true and false conditions. Review TT and DN bit operation in the table.

The most frequently used timer instruction, TON, has an active DN bit if the ACC is equal to or greater than the preset (PRE) value AND the timer enable bit remains true. Compare this with the logic for the other two.

The timing diagrams of the TON, TOF, and RTO timers are illustrated in Figure 4-9. Study each timer in the figure and note the condition of the TT and DN bits as the EN bit transitions from false to true and back to false. Compare the operation illustrated in the timing diagram with the description of the output bit operation in Table 4-1. Note the operation of the TT and DN bits if the EN

TABLE 4-1 Timer output bit truth table.

On-delay timer output bits	are TRUE if	are FALSE if
Timer enable	Timer rung is true.	Timer rung is false.
Timer timing	Timer rung is true AND the accumulator value is less than the preset value.	Timer rung is false OR the accumulator value is equal to or greater than the preset value OR the timer done bit is true.
Timer done	Timer rung is true AND the accumulator value is equal to or greater than the preset value.	Timer rung is false OR the timer rung is less than the preset value.

(a) Truth table for the on-delay timer output bits (TON)

Off-delay timer output bits	are TRUE if	are FALSE if
Timer enable	Timer rung is true.	Timer rung is false.
Timer timing	Timer rung is false AND the accumulator value is less than the preset value.	Timer rung is true OR the accumulator is equal to or greater than the preset value OR the done bit is false.
Timer done	Timer rung is true OR the timer timing bit is true.	Timer rung is false AND the accumulator value is equal to or greater than the preset value.

(b) Truth table for the off-delay timer output bits (TOF)

Retentive timer output bits	are TRUE if	are FALSE if
Timer enable	Timer rung is true.	Timer rung is false.
Timer timing	Timer rung is true AND the accumulator value is less than the preset value.	Timer rung is false OR the accumulator value is equal to or greater than the preset value.
Timer done	The accumulator value is equal to or greater than the preset value.	Reset instruction is initiated OR the timer rung is true but the accumulator is less than the preset value.

(c) Truth table for the retentive timer output bits (RTO)

bit changes before the preset value is reached by the accumulator.

A summary of the general operation of a TON, TOF, and RTO timer with a 15-second preset value follows. The description applies to timers from all three Allen-Bradley processors. Refer to Table 4-1 and Figure 4-9 as you read each timer's description.

On-delay Timer. The *on-delay timer* (TON) in Figure 4-9(a) starts timing (15 second delay) when the timer's ladder rung becomes true. The true rung forces the enable (EN) bit to true, causes the accumulator to start incrementing by the values set in the time base, and makes the timer timing (TT) bit true. The done

FIGURE 4-9: Timing diagrams.

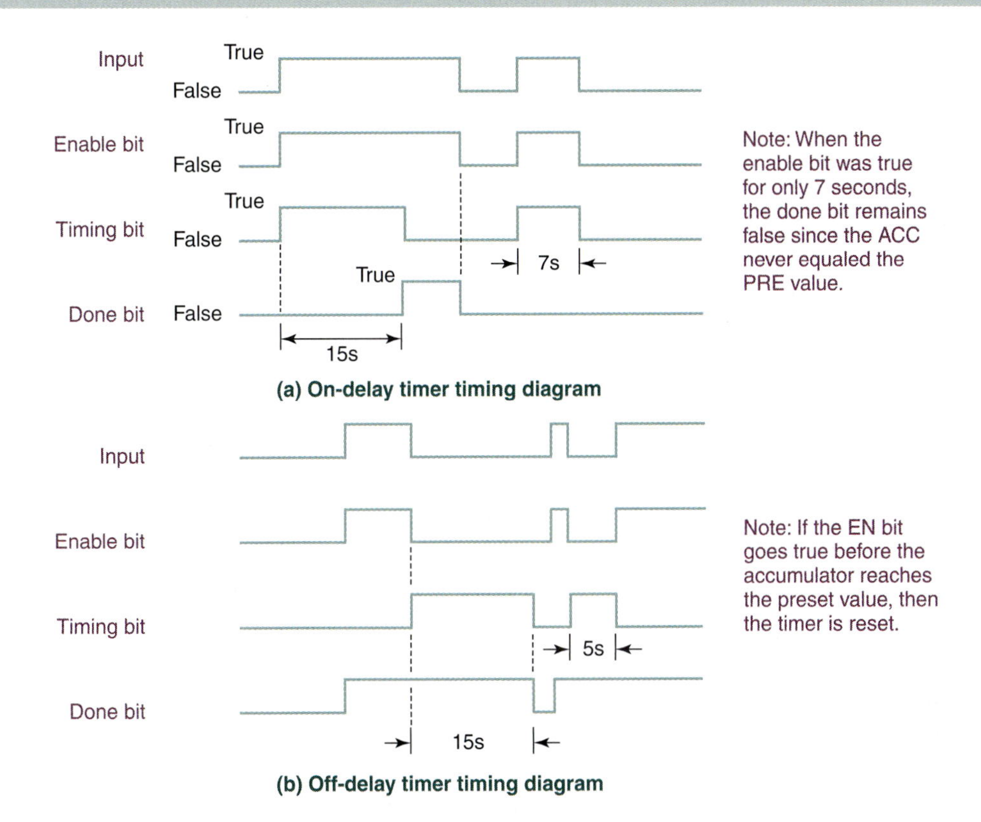

(DN) bit becomes true when the accumulator value reaches the preset time value. The DN bit remains true until the timer's rung returns to the false state, making the EN bit false. This reset action also returns the accumulator to a zero value. The condition of the timer instruction is determined by the input logic on the rung; therefore, timer operation is controlled by the associated input field device(s). If the EN bit returns to a false condition before the accumulator reaches the preset value [the 7 second pulse in Figure 4-9(a)], then the timer is reset and the DN bit remains false (no change).

A photocopier is an example of an on-delay timing function. When the print button is pressed, the operation of the photocopier is not started for some time period (an on-delay) to permit the copier to heat up before starting to make copies.

Off-delay Timer. For the off-delay timer (TOF) in Figure 4-9(b), the done bit is true and the accumulator is set to zero when the ladder rung and enable bit are true. No changes in the timer bits occur until the ladder rung and enable bit return to the false state in Figure 4-9(b). At this point the accumulator starts incrementing toward the 15 second preset value with the increment set by the time base. When the accumulator value equals the preset value, the timer done bit goes from true to false. If the EN bit returns to a true condition, [the 5 second pulse in Figure 4-9(b)], before the accumulator reaches the preset value, then the timer is reset and the DN bit remains true (no change).

As an example of an off-delay timing function, think about the light in an automatic garage door opener. When the garage door opener is activated, the light comes on when the door starts to open. The door motor turns off when it is open, but the light remains on (an off-delay) a preset period of time before it is extinguished.

Retentive Timer. The retentive timer, RTO, accumulates time whenever it is active, which means that the timer retains the accumulated

FIGURE 4-9: (Continued).

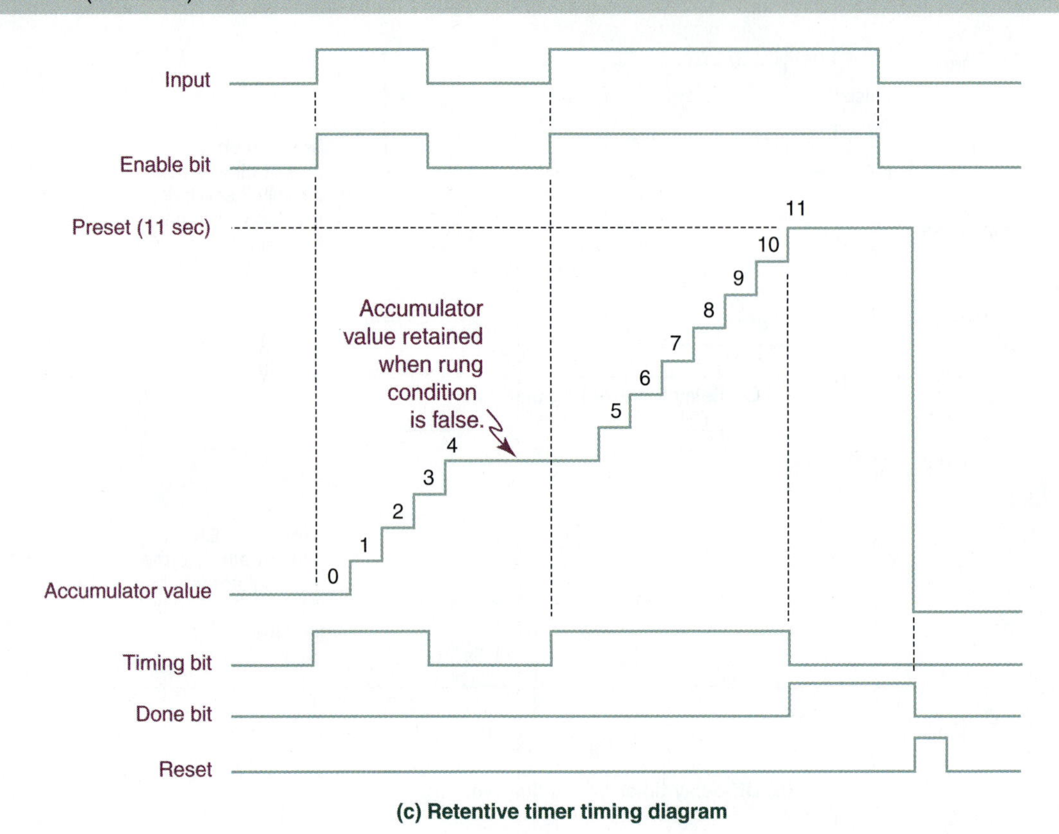

(c) Retentive timer timing diagram

time even if the rung is not active or power to the PLC is lost. As a result, the accumulator retains the current time value and starts incrementing from that value when the ladder rung and enable bit once again go true. The time base sets the time increment for the accumulator change and the preset value indicates the desired time delay. The done bit goes true and the timer timing bit goes false when the accumulator value is equal to or greater than the preset value. The retentive timer retains its current time when power is lost or when the timer rung is false. The only method of resetting a retentive timer is by a *reset instruction* that has the address as the timer.

Figure 4-9(c) illustrates the operation of a retentive timer with a preset value of 11 seconds. When the timer rung becomes active, the enable bit is true and the timer accumulator (ACC) begins to increment. When the rung is false, the ACC holds the current value, which is 4 in the figure. When the input returns to a true state, the ACC begins incrementing from 4 until it reaches the preset value of 11. At the preset value, the ACC stops incrementing and the retentive timer done bit (DN) is true. The figure shows how a reset (RES) instruction is used to reset the ACC to zero and return the timer done bit to a false state.

The accumulator of the retentive timer operates like the trip mileage indicator on the instrument panel in your car. As you drive, the indicator displays your accumulated miles. When you stop for gas the display holds the number. It then continues accumulating as you start driving again. When you finish the trip, you manually reset the display to zero.

With the operation of the TON, TOF, and RTO timer instruction covered, the next section describes how each timer parameter and bit is addressed.

4-6 ALLEN-BRADLEY TIMER PARAMETER AND BIT ADDRESSING

The timer parameters and control bits described in the last two sections are stored in the processor memory. The format for storing the PLC 5 and SLC 500 parameters is the same, but is quite different for the Logix family of processors.

4-6-1 PLC 5 and SLC 500 Timer Memory Map

The PLC 5 and SLC 500 processors use three words in memory to store control bit values and operational parameters. Figure 4-10 illustrates how the timer memory for these processors is organized. Each block of words is identified with the timer number; for example, a three-word block would be addressed as T4:5. This three-word block holds the data for timer 5.

Word 0 is the control word with the control or timer output bits (EN, TT, and DN) stored in the three most significant bits. These output bits are Boolean data types, so their values in the timer memory map are either 0 or 1. The preset value is stored in word 1 and the accumulator value is stored in word 2. Figure 4-10 illustrates the layout of the three bits (EN, TT, and DN) and two words (PRE and ACC) that can be addressed for system control.

The address structure for timers in the PLC 5 and SLC 500 processors uses the following format:

Tf:e.s/b

Each element in the timer address format is defined in the following table.

Element	Description	
T	The T indicates that the address is a timer file.	
f	The default value for f is 4. File 4 supports 256 timer instructions (T4:0 to T4:255). If more than 256 timers are needed in a program, then additional files (10 to 255) are available. Each of these files supports 256 timers (T9:0 to T9:255).	
:	Element delimiter	
e	Element number, e, is the number of the timer.	For file 4, e has a range of 0 to 255 timers. The same range is used for e if files 10 to 255 are used.
.	Word delimiter	
s	Word number, S, indicates one of the three timer words.	The value of S ranges from 0 to 2 because each timer has three addressable words.
/	Bit delimiter	
b	Bit number, b, is the bit location in the timer words.	The range is 0 to 15 for all three timer words, but bits 13, 14, and 15 are the only ones used for word 0.

Example timer data addresses are listed in the following table. Study the timer address structure (Tf:e.s/b) and the description of each address element in the previous table, and then verify that you understand what timer data is being addressed by each of the following examples.

Addressing Examples	Description
T4:0/15 or T4:0/EN	Enable bit of timer number 0
T4:2/14 or T4:2/TT	Timer timing bit of timer number 2
T4:15/13 or T4:15/DN	Done bit of timer number 15
T4:5.1 or T4:5.PRE	Preset value of timer number 5
T4:10.2 or T4:10.ACC	Accumulator value of timer number 10
T4:20.1/0 or T4:20.PRE/0	Bit 0 of the preset value of timer number 20
T4:3.2/11 or T4:3.ACC/11	Bit 11 of the accumulator value of timer number 3
T4:25/DN	The done bit for timer 25 in timer file 4
T4:255/TT	The timer timing bit for the last timer (255) in timer file 4
T9:0.ACC	The accumulator word for timer 0 in timer file 9
T9:255.PRE	The preset word for the last timer (255) in timer file 9
T255:255/EN	The enable bit for last available timer in the system

FIGURE 4-10: Timer output bit image map.

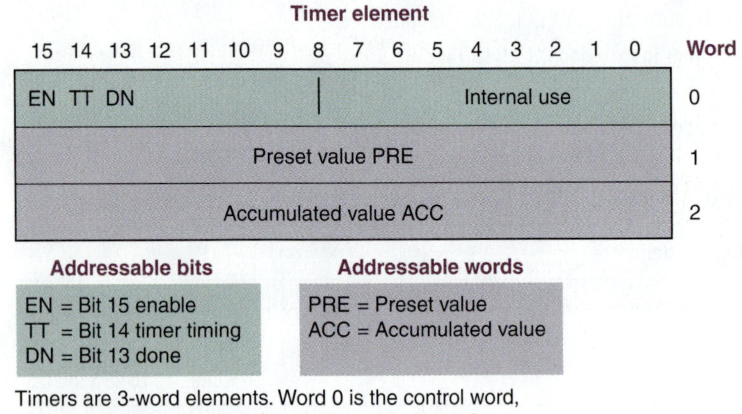

These addressing examples indicate all of the possible addressing modes that are available for PLC 5 and SLC 500 timers. Note that parameters are addressed based on the bit or word number or with the mnemonic for that bit or word. For example, in the first example bit 15 is also the enable (EN) bit. Also, in the fourth example the preset is addressed as a .1 for word 1 or as .PRE for preset word. These bit and word addresses are used in any other PLC instruction where a timer bit or word address is permitted. For example, in Chapter 6, move instructions will use the corresponding word addresses which are used with counters to transfer preset values to counters.

4-6-2 ControlLogix Timer Addressing

The format for addressing ControlLogix timers is simplified with the use of tag names for each timer. Figure 4-11 illustrates how the data for timer, *running_seconds*, is displayed in the Program Tags dialog box. Note that the Edit Tags tab is selected at the bottom of the dialog box, so this box could be used to enter parameter data. The Monitor Tags tab could be selected to examine the value of timer bits and words.

The ControlLogix timers have the EN, TT, and DN output bits and PRE and ACC parameter words found in the PLC 5 and SLC 500 PLCs plus the four other data values displayed in Figure 4-11.

FIGURE 4-11: ControlLogix output bit and parameter addressing.

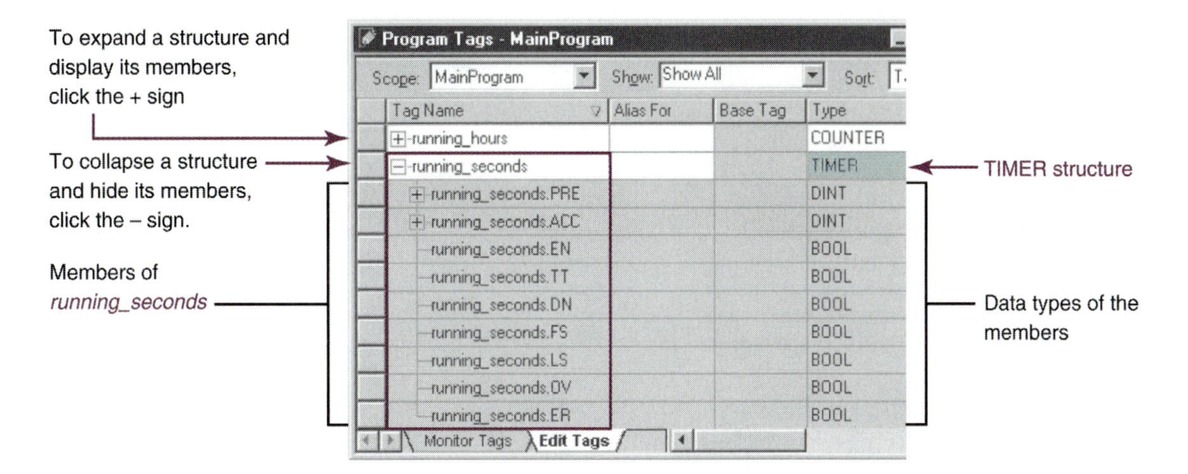

4-6-3 Timer Contacts versus PLC Instructions

An important distinction was made in Chapter 3 between the normally open and normally closed physical contacts on input field devices, and the XIC and XIO instructions used for ladder input logic. This distinction is also carried over to the timers. Mechanical timing relays have physical contacts and electronic timers have either physical contacts or solid-state switches to control output devices. However, the virtual timers in PLCs are created in software; as a result, they have memory bit outputs and the PLC timer is called an instruction.

There is, however, a relationship between the mechanical and electronic timers and their PLC counterparts. The PLC TON timer is the same as an on-delay timing relay, and the TOF timer is the same as the off-delay timing relay. The TON done bit is like a normally open, timer closed timing relay contact, and the TOF done bit is like the normally open, timed opened relay contact. The instantaneous contacts on the timing relays are equivalent to the enable bit on both types of PLC timers.

4-7 PROGRAMMING ALLEN-BRADLEY TON AND TOF TIMER LADDER LOGIC

Sections 4-5 and 4-6 presented an overview of timer instructions, including setting timing parameters and addressing timer data. This section uses that information to develop timer ladder logic for machine and system control. The first section looks at six ladder configurations that are used to build most timer ladder logic. Learning this standard timer ladder logic is important because most timer ladder solutions are just variations from these six standard ladder configurations.

The second half of this section covers a number of example problems that demonstrate how the three Allen-Bradley timer instructions are used for automation control.

4-7-1 Standard Ladder Logic for Allen-Bradley TON Timers

Automation programs that include timers use a standard set of timer ladder logic configurations. Learning these common timer ladder solutions is a great way to start the study of timer applications. The standard TON solutions for common control problems are listed in Figure 4-12.

The first set of standard rungs, Figure 4-12(a) through (d), illustrates timer ladder configurations triggered by field device switches with momentary or continuous types of contacts. This set also covers the different output options for TON timers. Read the description of the timer operation in the figure as you study the standard timer ladder logic. It is clear that the simplest timer applications require two rungs, one for the timer instruction and one for the output device being controlled. In addition, a maintain contact input for the timer is the simplest to implement. The momentary contact inputs

FIGURE 4-12: Standard ladder logic rungs for TON timers.

Application	Standard Ladder Logic Rungs for TON Timers
Turn on an output device for a set time period when a maintain contact input device changes states. The input field device is a NO selector switch. If the switch is closed, then the timer rung, EN bit, and TT bit are true because instruction I:1/1 is true. Therefore, the T4:1/TT instruction has continuity and the O:3/0 output is true. The TT bit and output are on for the preset time value.	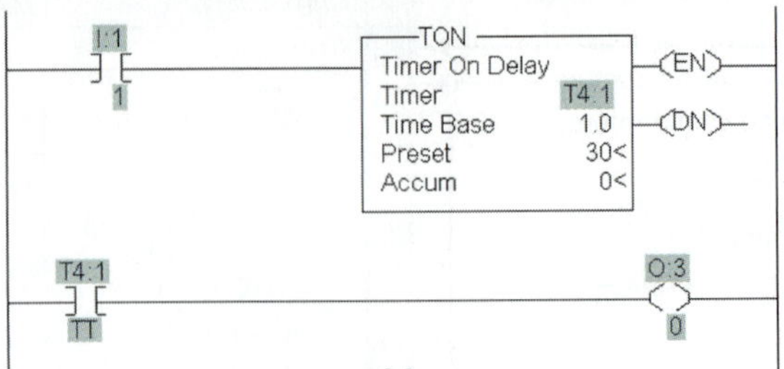 I:1 1 — TON Timer On Delay, Timer T4:1, Time Base 1.0, Preset 30<, Accum 0< — (EN) (DN) T4:1 TT — O:3 0 (a)
Turn on an output device after a set time period when a maintain contact input device changes states. The input field device is a NO selector switch. If the switch is closed, then the timer rung, EN bit, and TT bit are true because instruction I:1/1 is true. After the preset time value the done bit is true. Therefore, the T4:1/DN instruction has continuity and the O:3/0 output is true. The output is on until the timer input rung goes false.	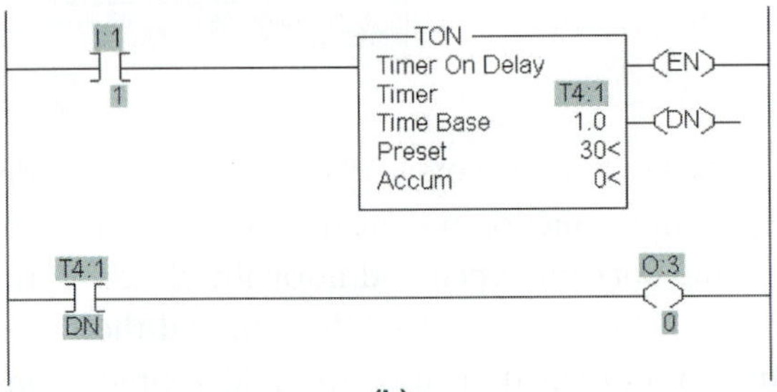 I:1 1 — TON Timer On Delay, Timer T4:1, Time Base 1.0, Preset 30<, Accum 0< — (EN) (DN) T4:1 DN — O:3 0 (b)
Turn on an output device for a set time period when a momentary contact input device changes states. The input field device is a NO push button switch. If the switch is closed, then the timer rung, EN bit, and TT bit are true because instruction I:1/1 is true. The active T4:1/TT bit seals the input instruction and makes the O:3/0 output true. The TT bit and output are on for the preset time value.	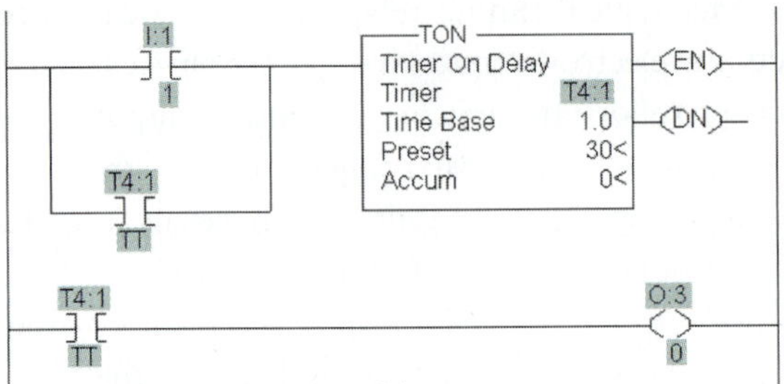 I:1 1 (parallel T4:1 TT) — TON Timer On Delay, Timer T4:1, Time Base 1.0, Preset 30<, Accum 0< — (EN) (DN) T4:1 TT — O:3 0 (c)
Turn on an output device for one PLC scan after a set time period when a momentary contact input device changes states. The input field device is a NO push button switch. When the switch is closed, then the timer rung, EN bit, and TT bit are true because instruction I:1/1 is true. The active T4:1/TT bit seals the input instruction for the momentary switch. After the preset time value the done bit is true. Therefore, the T4:1/DN instruction has continuity and the O:3/0 output is true. One scan later, T4:1/TT goes false and the timer is reset. This causes the output, O:3/0, to be true for one scan time.	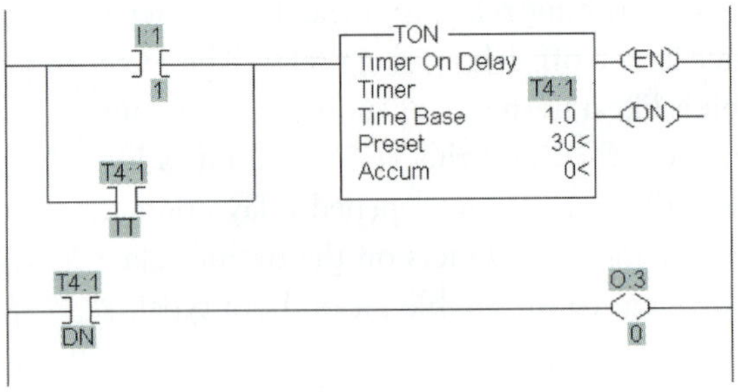 I:1 1 (parallel T4:1 TT) — TON Timer On Delay, Timer T4:1, Time Base 1.0, Preset 30<, Accum 0< — (EN) (DN) T4:1 DN — O:3 0 (d)

FIGURE 4-12: (Continued).

Application	Standard Ladder Logic Rungs for TON Timers Pulse Generators
Turn on an output device with a pulse sequence where one half of the duty cycle has a variable time and the other has a time equal to the scan time of the ladder. The input field device is a NO selector switch. If the switch is closed, then the timer rung, EN bit,and TT bit are true because instruction I:1/1 is true and the pulse timer done bit is false. As a result, the timer is incrementing the ACC at a 1 second rate. When the ACC equals the preset (5 seconds), the timer DN bit is true. This makes the timer rung false because the XIO instruction has no power flow since the timer DN is true. Therefore, the timer resets, which makes the DN bit is true for only one scan, and the timer starts the timing process again. The pulse output timing diagrams for the two output options are illustrated below the ladder logic. The done bit is used to generate one waveform, and the timer timing bit is used to generate the other. Note that the outputs are virtual relays (memory bits) since most field devices could not respond fast enough for this narrow pulse. The pulse is used, however, for logic control in the program.	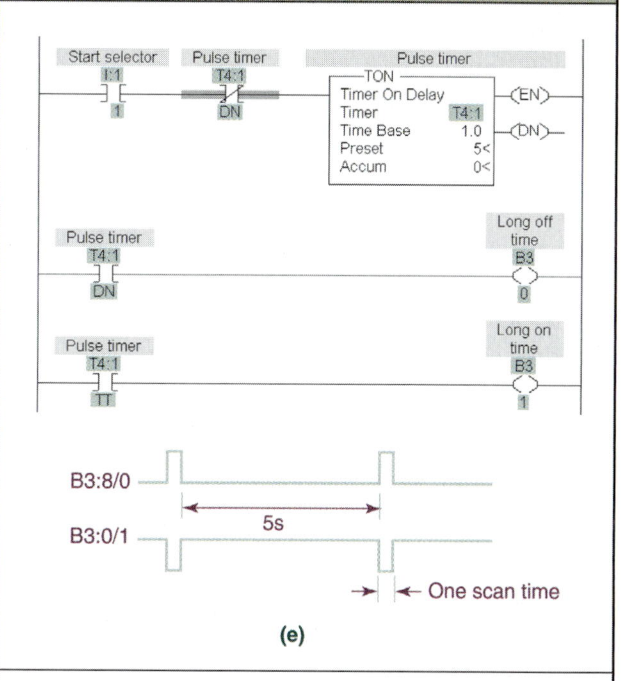 (e)
Turn on an output device with a pulse sequence where both halves of the duty cycle have a variable time. The input field device is a NO selector switch. If the switch is closed, then the T4:2 timer rung and the timers EN bit and TT bit are true because instruction I:1/1 is true. As a result, the timer is incrementing the ACC at a 1 second rate. When theT4:2 ACC equals the preset (4 seconds), the timer DN bit is true. This makes the T4:3 timer rung true, and that timer starts incrementing its ACC toward the preset value of 2 seconds. While the T4:3 ACC is incrementing, the output O:3/2 is true because the T4:2 done bit remains true. When the T4:3 timers ACC equals 2, the done bit of T4:3 becomes true. This resets both the T4:2 timer (XIO instruction in timer T4:2 rung is true) and the T4:3 timer held on by the T4:2 done bit. The output also returns to the false state when T4:3 ACC reaches 2 seconds. Note that the up time for the pulse is determined by the T4:3 since that timer determines how long the output is true. The pulse output timing diagram is illustrated below the ladder logic.	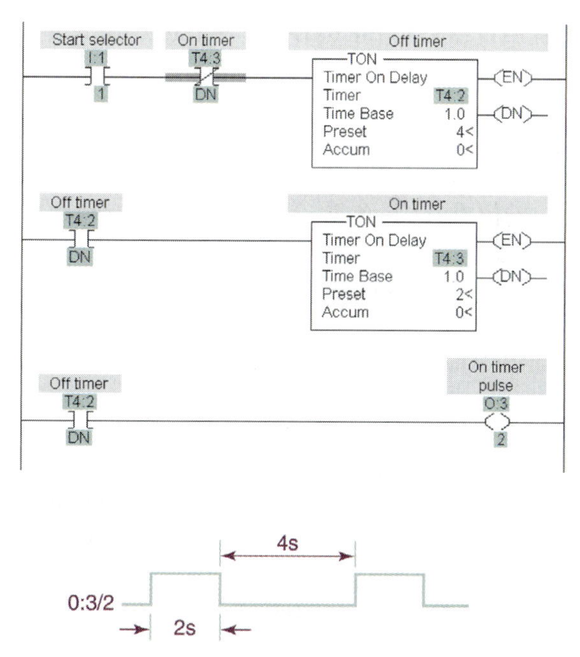 (f)

requires a sealing instruction because the TON timer instruction must be true until the timer reaches the preset value. The timer's TT bit is used to perform the required sealing of the input. The standard circuits in Figure 4-12(c and d) also demonstrate how a timer is used to turn on an output for a set time period or turn on one after a set time period.

The second set, Figure 4-12(e) and (f), features two configurations for using timers to create pulse generators. The pulse generator in Figure 4-12(e) is called a regenerative clock because it uses an output of the timer reset itself. The XIO instruction in the timer rung is addressed with the timer done to reset the timer whenever the done bit changes from false to true. The TON instruction is reset whenever the timer rung is false; as a result, the false done bit resets this regenerative circuit so the timer instruction done bit is true for only one scan time. This output pulse (one scan time wide) is too narrow to drive most field devices, so this ladder configuration is used most often for logic control within the PLC program. This configuration is used in the next two chapters for control of other PLC instructions.

The other pulse generator, Figure 4-12(f), provides a variable duty cycle that is a function of the preset values for the two timers. The done bit from the first timer, T4:2, is used to control the output, O:3/2, and to make the second timer, T4:3, active. T4:3 determines how long the output is on before it resets the system and timing sequence is restarted. Therefore, the second timer controls the on time for the pulse and the first timer controls the off time. This is a regenerative timer using two timer instructions. Note that other program rungs can be inserted between the timer rungs and the output rung. Read the description of the pulse generator operation in the figure as you study the standard pulse generator ladder logic.

The Allen-Bradley SLC 500 timer is used to create the example solutions in all six TON ladders. However, PLC 5 and ControlLogix timers could be used and the operation of the ladder logic would be unchanged. The only difference would be the addresses used for the input logic for the PLC 5 and the use of tags for the ControlLogix.

4-7-2 Standard Ladder Logic for Allen-Bradley TOF Timers

The standard TOF solutions used in control problems are listed in Figure 4-13. The standard rungs illustrate how the TOF timer performs with a selector switch for the input field device. The selector contacts, used in Figure 4-13(a), remain in the open position long enough for the timer to time out or complete the timing process. The field device in Figure 4-13(b) could be either a selector with maintain contacts or a push button with momentary contacts. The time duration for a true done bit is the sum of the time the input is held closed and the preset time value. The choice is dictated by the system control requirements. The timing diagrams are included because the operation of TOF timers is often more difficult to understand than their TON counterparts. Note that other program rungs can be inserted between the timer rung and the output rung. Read the description of the timer operation in the figure as you study the standard timer ladder logic.

The Allen-Bradley SLC 500 timer is used to create both TOF example solutions; however, PLC 5 and ControlLogix TOF timers could be used and the operation of the ladder logic would be unchanged. The only difference would be the addresses used for the input logic in the PLC 5 and the use of tags for the ControlLogix.

4-7-3 Allen-Bradley TON and TOF Timer Applications

This section includes a number of timer applications and example problems that demonstrate how TON and TOF timers for the three Allen-Bradley processors are used. The applications describe the use of a timer in a control requirement, and the examples show you how a control problem is stated and illustrate one workable solution.

In addition, the standard time ladder logic, which is the basis for the solution, is indicated. What becomes clear is that just a few timer configurations are used to solve most of the timer problems in automation control.

Pump delay control logic application. In some large pump applications power to the pump motor is delayed while auxiliary circuits open valves or initiate priming operations. The ladder logic in Figures 4-14(a and b) illustrates a pump

FIGURE 4-13: Standard ladder logic rungs for TOF timers.

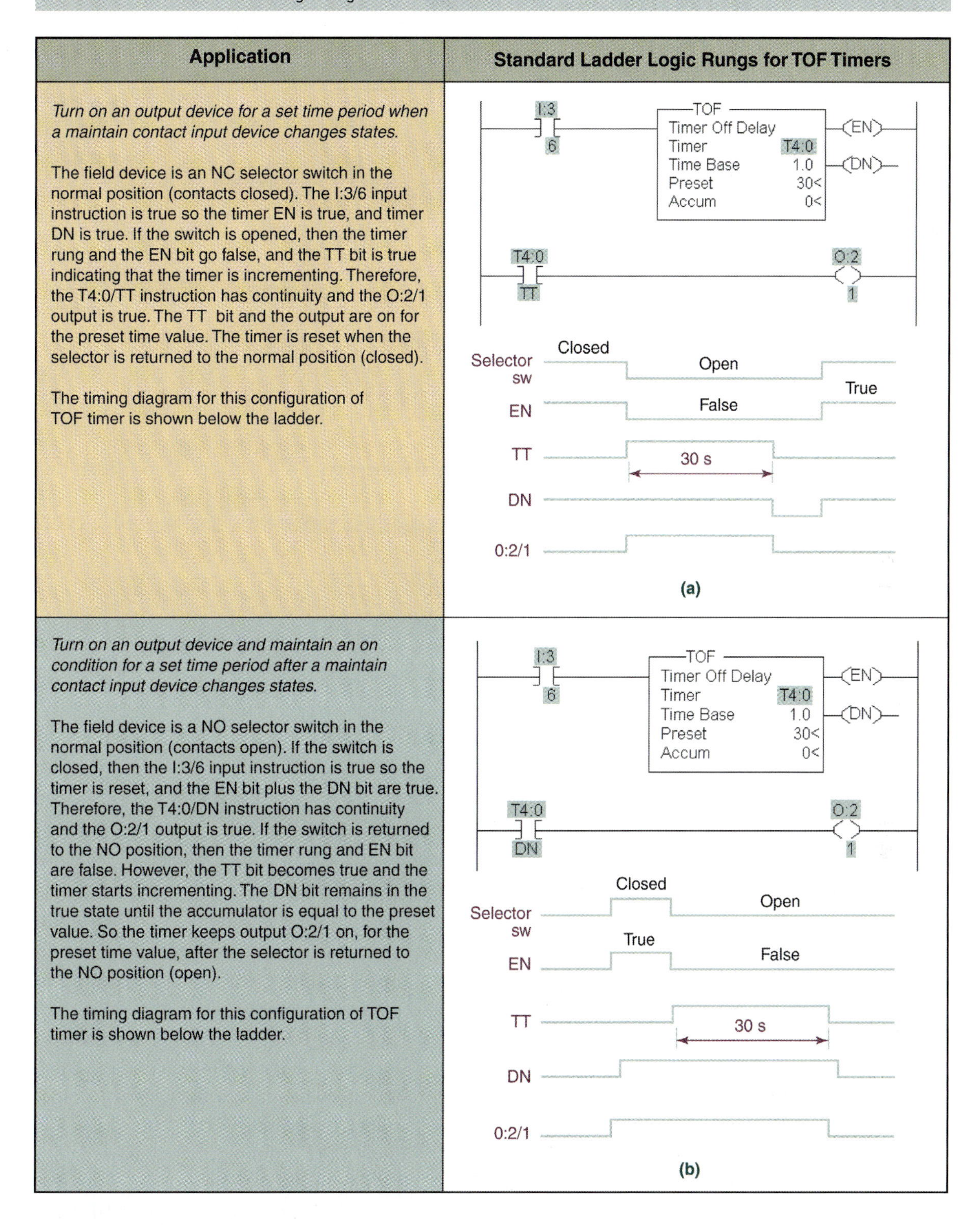

Application	Standard Ladder Logic Rungs for TOF Timers
Turn on an output device for a set time period when a maintain contact input device changes states. The field device is an NC selector switch in the normal position (contacts closed). The I:3/6 input instruction is true so the timer EN is true, and timer DN is true. If the switch is opened, then the timer rung and the EN bit go false, and the TT bit is true indicating that the timer is incrementing. Therefore, the T4:0/TT instruction has continuity and the O:2/1 output is true. The TT bit and the output are on for the preset time value. The timer is reset when the selector is returned to the normal position (closed). The timing diagram for this configuration of TOF timer is shown below the ladder.	(a)
Turn on an output device and maintain an on condition for a set time period after a maintain contact input device changes states. The field device is a NO selector switch in the normal position (contacts open). If the switch is closed, then the I:3/6 input instruction is true so the timer is reset, and the EN bit plus the DN bit are true. Therefore, the T4:0/DN instruction has continuity and the O:2/1 output is true. If the switch is returned to the NO position, then the timer rung and EN bit are false. However, the TT bit becomes true and the timer starts incrementing. The DN bit remains in the true state until the accumulator is equal to the preset value. So the timer keeps output O:2/1 on, for the preset time value, after the selector is returned to the NO position (open). The timing diagram for this configuration of TOF timer is shown below the ladder.	(b)

start delay timer using a TON PLC timer instruction and the associated timing diagram. When selector A, an input field device switch, is active the timer begins incrementing toward the preset value in 0.01 second increments. Two seconds later, when the preset value is reached ($0.01 \times 200 = 2$ seconds), the timer done bit becomes active and the pump contactor is turned on. When SS1 is opened, the timer resets and the pump turns off. As in all TON timers, the accumulated value

FIGURE 4-14: On-delay timer diagrams.

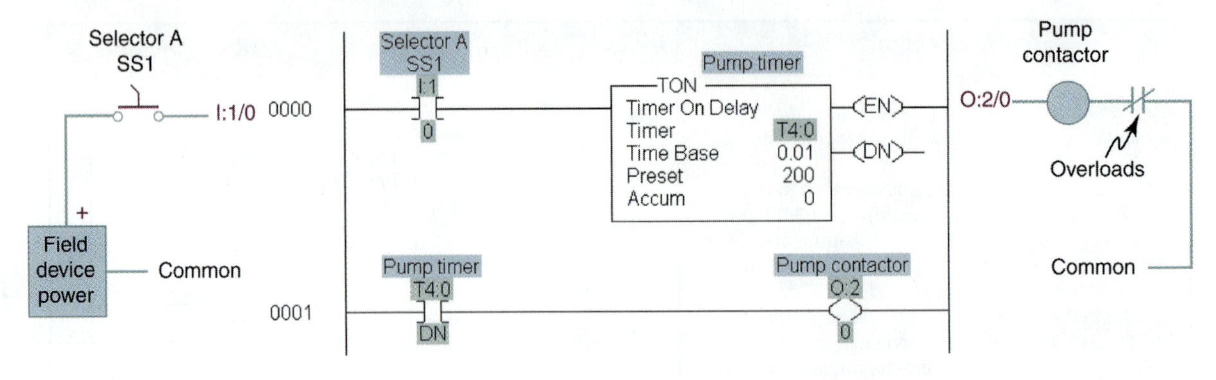

(a) Ladder diagram

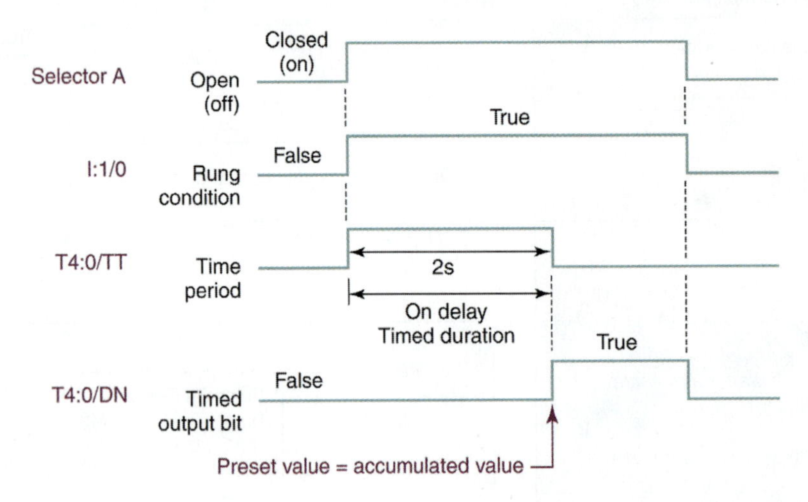

(b) Timing diagram

is automatically reset to zero when the enable bit goes from a 1 to a 0. Note that this timer example is the standard configuration shown in Figure 4-12(b). All of the timer applications are either one of the standard timer ladder logic configurations or some combination of those configurations.

Traffic Light Control Application. Figure 4-15 depicts a ladder diagram where the active and done bits from three on-delay timers are used to turn on and off traffic lights—the red light is on for 32 seconds, the green light is on for 27 seconds, and the amber light is on for 5 seconds. Note that the timers are numbered T4:0, T4:1, and T4:2; their preset times are 32, 27, and 5 seconds, respectively. Refer to Figure 4-15 as you read the following operation of the timers.

1. Before power is applied, all timer EN, TT, and DN bits are false, all examine if closed (XIC) instructions (-| |-) are not active (no continuity), and all examine if open (XIO) instructions (-|/|-) are not active (continuity).
2. At power on the T4:2/DN bit is false, so the examine if open instruction in rung 0 is true, which makes the rung true. Since rung 0 is true, the T4:0/EN bit is true, and the T4:0/TT timing cycle is started. The two input instructions in rung 3 are true (T4:0/EN bit is true and the T4:0/DN false), so output O:2/0 is true and the red light is turned on.
3. After 32 seconds, the T4:0/DN done bit is a 1, making rung 3 and output O:2/0 false. When the address on an examine if open instruction is a 1, the instruction is false

FIGURE 4-15: Ladder diagram for traffic light control.

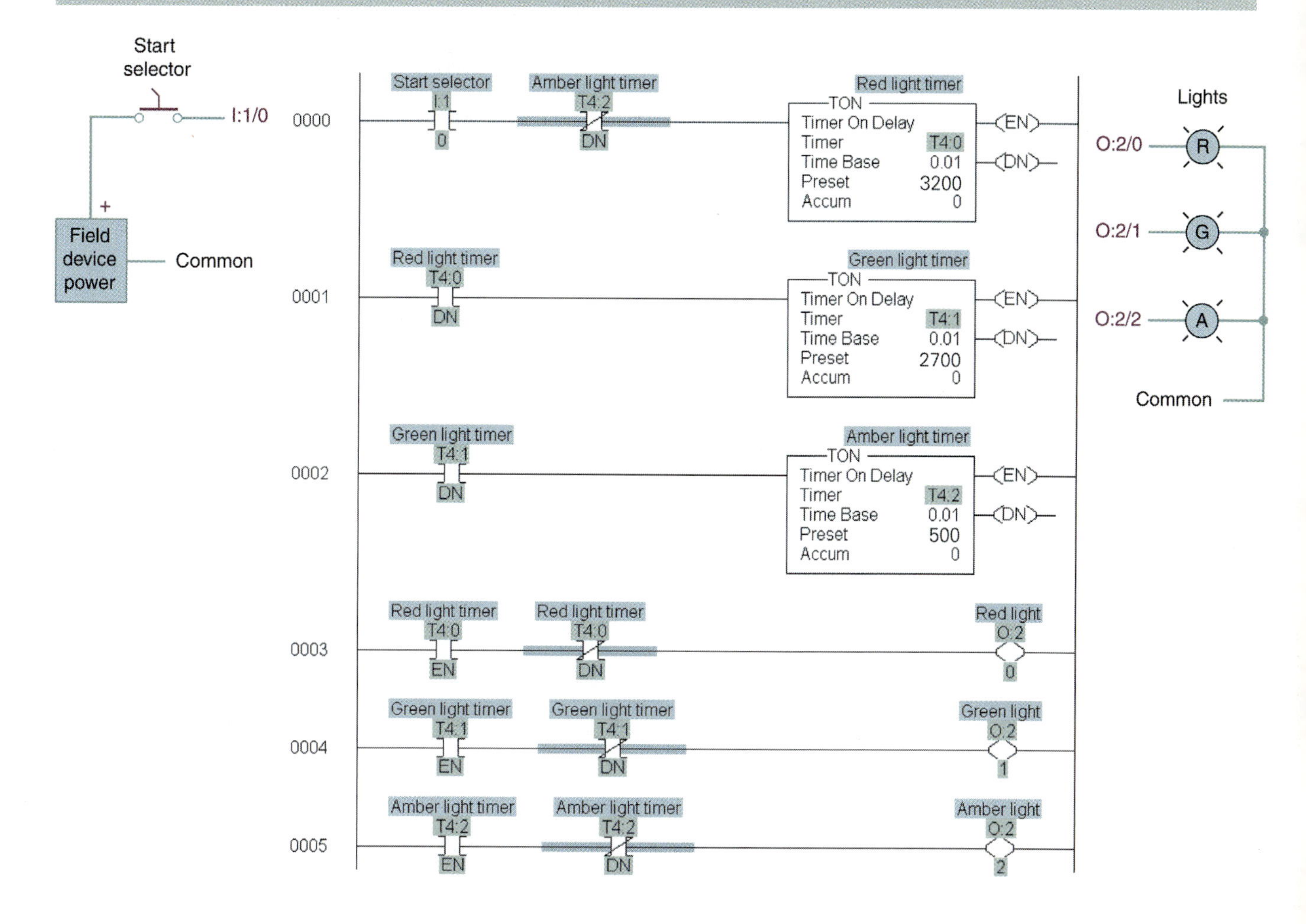

and continuity is removed, which turns off the red light. In addition, the T4:0/DN addressed XIC instruction in rung 1 is true, which starts the green light timer. Rung 4 is also true because the T4:1 timer enable bit is true and the T4:1 done bit is false. This makes output O:2/1 active and turns on the green light.

4. After an additional 27 seconds, the T4:1/DN bit goes true, making rung 4 false and extinguishing the green light. In addition, rung 2 is true, which starts the amber timer. As a result, rung 5 and the O:2/2 output are true, thus illuminating the amber light.
5. After an additional 5 seconds, the T4:2 done bit goes true, making rung 5 false and extinguishing the amber light. Also, the true T4:2/DN bit makes rung 0 and the T4:0 timer false. This causes the T4:0 done bit to go false, which makes rung 1 false, causing the T4:1 timer and T4:1/DN bit to go false. The change in T4:1/DN makes rung 2 and T4:2 false. With this change in T4:2 and an active start selector, rung 0 returns to the active state.
6. With rung 0 true again, the previous timing sequence is repeated.

All of the timers in this example are modifications of the standard timer ladder logic in Figure 4-12(b).

Machine guard lock and indicator application. Production machines often lock out the operator while the machine is processing parts plus a fixed time for the machine to come to a stop. The ladder logic in Figure 4-16(a) uses the selector switch, which starts the process and triggers a TOF timer. The timer controls an output that

FIGURE 4-16: Off-delay timer diagrams.

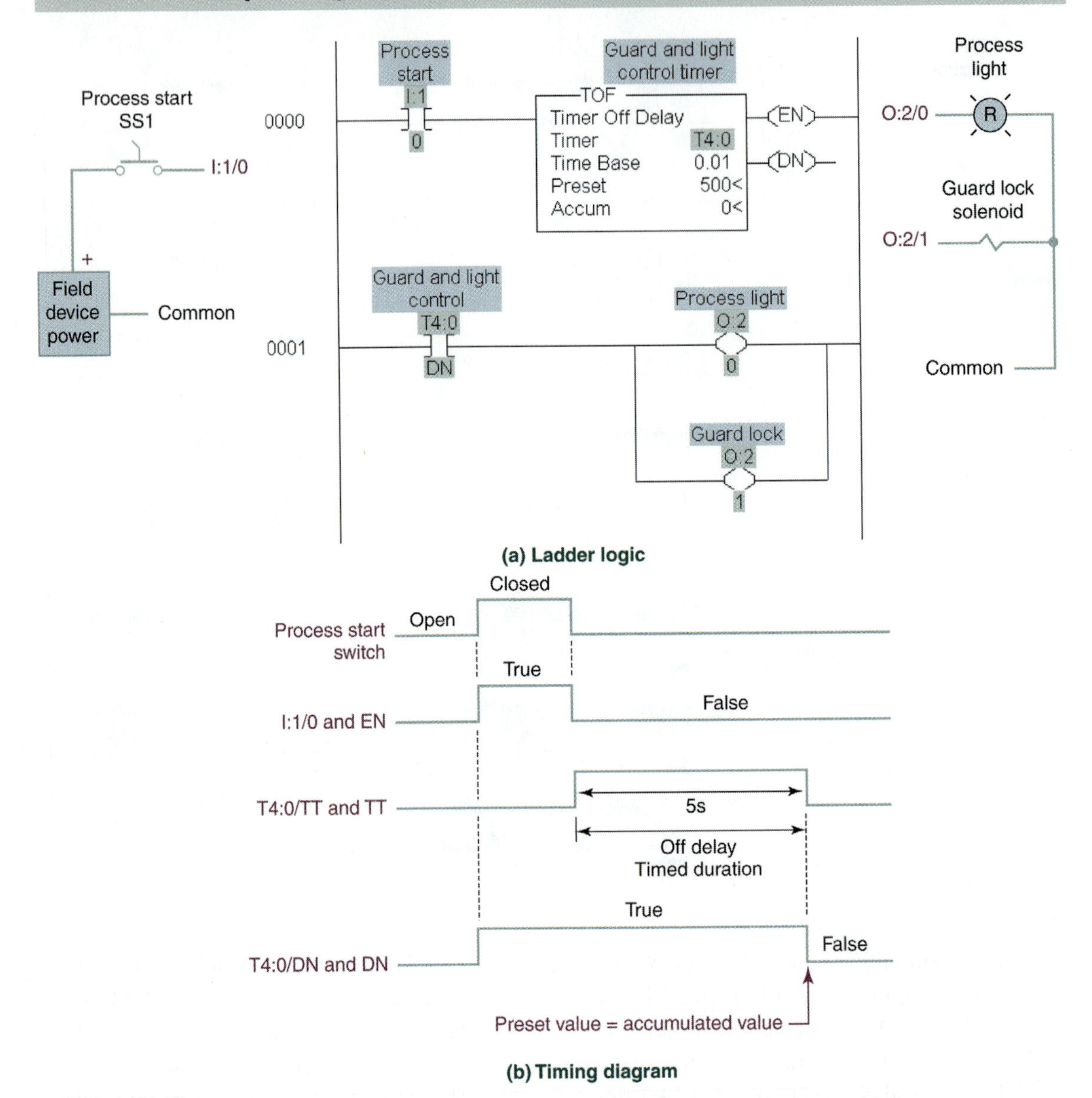

locks the machine doors at the start of the process and lights a doors locked indicator. The operator turns off the machine when the part is finished and the TOF timer keeps the doors locked for an additional 5 seconds for the motor to come to a stop. Figure 4-16(b) displays the timing diagram for this operation. When the NO selector switch is true, the *machine start* input instruction is true, the T4:0/EN and T4:0/DN bits are true, the accumulator value is reset to zero, and rung 0 becomes active. As a result, rung 1 is true because T4:0/DN is true and the machine door is locked and the process light is illuminated. When the machine is turned off, the switch contacts return to the NO state, rung 0 returns to the false state, and the timer accumulator begins incrementing toward the preset value while T4:0/DN remains true. When the preset value is reached ($500 \times 0.01 = 5$ seconds), the timer output (T4:0/DN) becomes false and the door is unlocked and the process light is extinguished.

Note that this timer example is the standard configuration in Figure 4-13(b), but it has a maintain contact switch. Compare the ladder logic and the timing diagrams for the TOF timer in Figures 4-13(b) and 4-16 to see how the standard ladder logic was adapted to this application.

EXAMPLE 4-2

Draw a ladder diagram for a pumping system where the pump requires a 5-second delay before pumping; when the pump is shut off, it requires a 15-second delay before it can be restarted. Start and stop switches are NO momentary contact push buttons.

SOLUTION

Refer to Figure 4-17(a), the ladder solution, and 4-17(b), the timing diagram, as you read the following description.

The pump control is implemented with T4:0, an on-delay timer, and T4:1, an off-delay timer. The activation of the momentary start switch makes rung 0 true, which initiates the on-delay timer (T4:0/EN and T4:0/TT are true). Since the start switch is

FIGURE 4-17: Pumping system ladder diagram for Example 4-2.

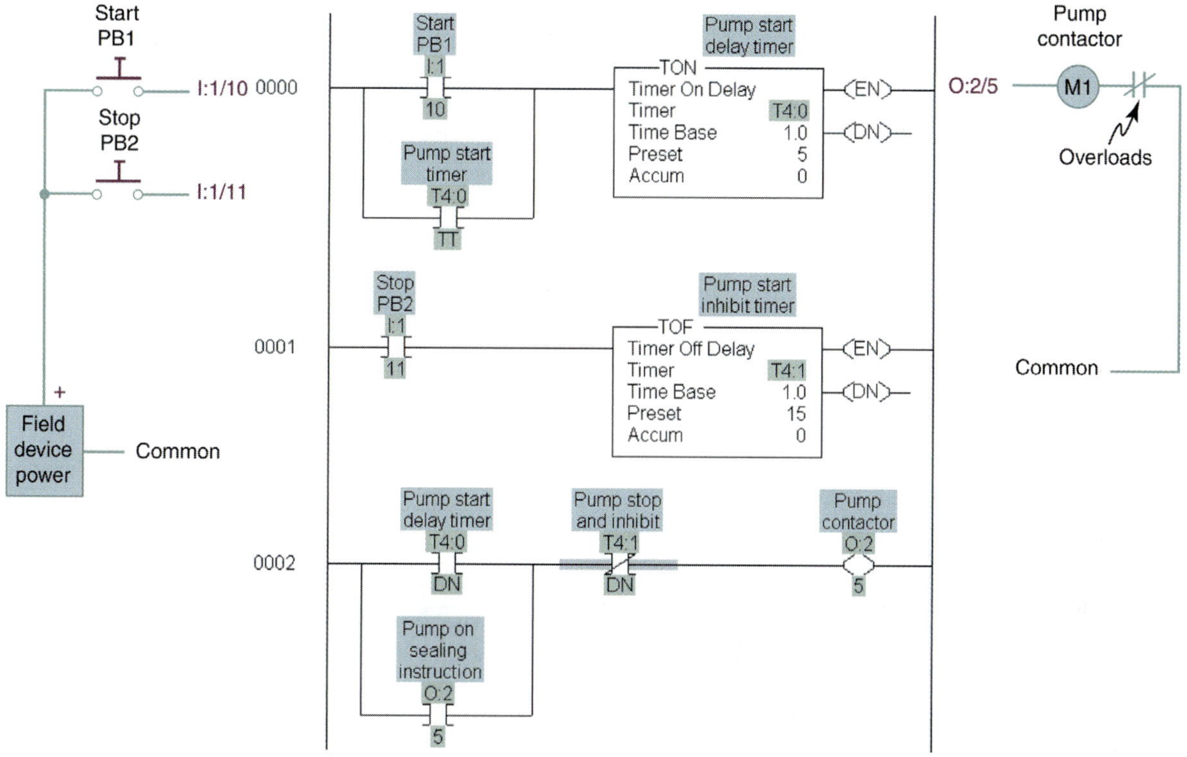

(a) Ladder logic

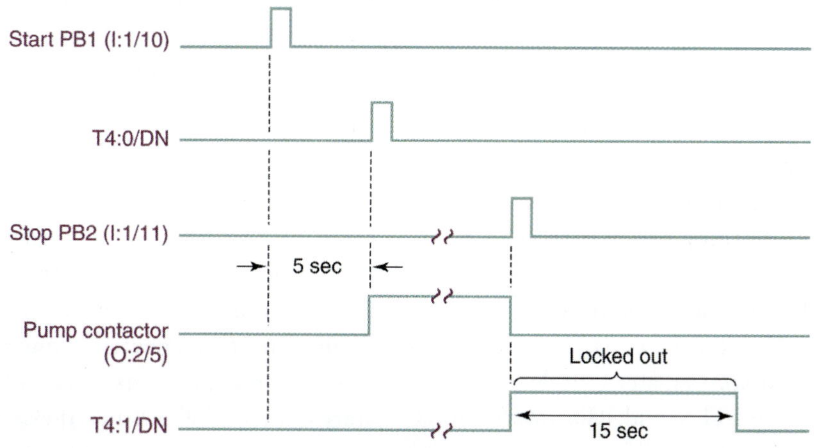

(b) Timing diagram

a momentary contact type, the I:1/10 instruction must be sealed with the T4:0 timer timing bit (T4:0/TT) to keep the rung active while the timer is incrementing the accumulator for the on-delay time of 5 seconds. Upon completion of the on-delay time of 5 seconds, the T4:0 timer done bit (T4:0/DN) is active, which makes the XIC instruction in rung 2 true. The XIO instruction in rung 2 is also true (continuity) because the T4:1 timer done bit (T4:0/DN) addressing the XIO instruction is false or 0. As a result, the pump output, O:2/5, is true, so the pump starts 5 seconds after the start switch is pressed. The sealing instruction (O:2/5 around T4:0/DN in rung 2) is necessary because the T4:0/DN bit starts the pump after 5 seconds but is a 1 or true for only one scan. The sealing instruction in rung 2 keeps the pump on after the delayed start.

T4:0/DN is true for only one scan because the timer resets immediately after the preset time is reached. This occurs because T4:0/TT bit is used to seal the XIC start instruction in rung 0, which is a momentary start push button. At 5 seconds the timer timing bit becomes false, which makes the T4:0 timer rung false and the timer resets. As a result, one scan after the done bit is true it returns to the false or 0 state. Review the operation of the standard ladder logic in Figure 4-12(c), which is used for this pump delay timer.

Now when the stop PB2 is pressed, the I:1/11 instruction in rung 1 is true, and the T4:1/DN bit of the TOF timer in rung 1 is true. When the stop switch is released, the T4:1 TOF timer starts timing and keeps the T4:1/DN bit true for 15 seconds. Thus the initiation of the stop push button makes the XIO instruction in rung 2 false because the T4:1/DN output is true. This action stops the pump because output O:2/5 false. Rung 2 is held in this false condition by the XIO instruction for the duration of the T4:1 time, so the start push button cannot restart the pump for 15 seconds. Upon completion of the off-delay time of 15 seconds, the T4:1/DN output becomes false, the XIO instructions returns to true state, and the pump can be restarted. Note that this example uses standard timer ladder logic from Figures 4-12(c) and 4-13(b).

4-8 ALLEN-BRADLEY RETENTIVE TIMERS

Review the operation of the retentive timer in Table 4-1(c) and Figure 4-9(c). The retentive timer (RTO) operates the same as a TON timer, except the accumulator (ACC) is not reset when the timer enable returns to the false state. The accumulator will continue to increment from the previous value whenever the EN bit goes from false to true. When the ACC equals the PRE value the timer timing bit goes false and the done bit becomes true. The done bit remains in that state until a reset (RES) instruction for the timer is executed. The reset instruction is covered in the next section. Compare and study Table 4-1(c) and Figure 4-9(c) until you understand the logical operation of an RTO timer instruction.

The RTO instruction operates the same for all three Allen-Bradley processors. The RTO ladder logic symbol for the PLC 5, SLC 500, and ControlLogix systems is the same as their TON symbol. After the reset instruction is introduced in the next section, an example is used to illustrate how the RTO and RES instructions operate.

4-8-1 Reset Instruction for RTO Timer and Other Allen-Bradley Instructions

Since the retentive timer does not automatically reset itself, a reset instruction is used to return the timer accumulator to zero and turn off the done bit. The reset (RES) instruction must have the same program address as the timer you want to reset. The reset instruction can reset the timer at any time during its operation and is independent of the input conditions. The reset instruction is also used for the TON and TOF timers and with other Allen-Bradley instructions covered in later chapters. The operation of the reset instruction is the same for all three types of Allen-Bradley PLC processors. Example 4-3 illustrates the operation of an RTO and RES instruction in an automation system.

Heater sequential control application. In large furnaces the electric heaters are often turned on or off in a sequence to control the heating and cooling of the product. In this application three heaters come on at the same time and remain on as long as the momentary start switch is held. When the switch is released the heaters turn off in sequence at 30-second

EXAMPLE 4-3

The pumping system in Figure 4-17 from Example 4-2 has an additional requirement to shut the pump down and illuminate a red pilot lamp after four hours of operation. The illuminated pilot lamp indicates that it's time to check the pump since it moves very abrasive material. A NO momentary push button reset switch is used to reset the system when the maintenance is completed. Draw the new ladder diagram for the pumping system with these additional requirements.

SOLUTION

Refer to Figure 4-18 as you read the following description.

The operation of rungs 0, 1, and 2 are similar to the ladder logic in Example 4-2; the last three rungs are new. Every time the pump is running, output O:2/5 (rung 2) is active. This makes the XIC instruction (O:2/5) and the retentive timer (T4:2) in rung 3 active, which increments the accumulated time in T4:2. Note that the retentive timer preset value is 14,400 seconds, which is 4 hours. When the accumulated number equals 14,400 seconds, the retentive timer done bit (T4:2/DN) is true, which causes a true

FIGURE 4-18: Pumping system ladder diagram for Example 4-3.

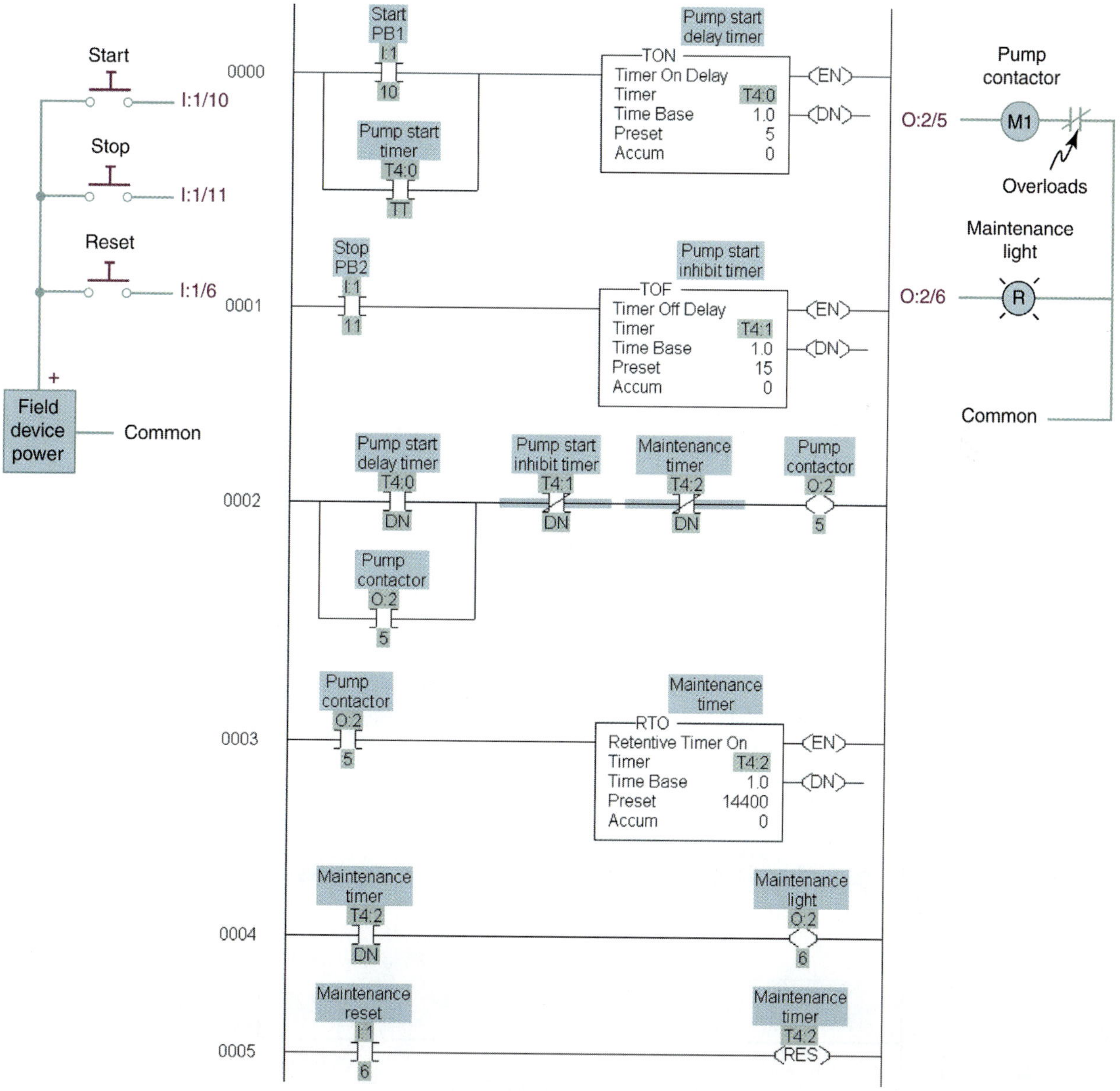

condition on output 0:2/6 in rung 4, and the maintenance light to illuminate. A second T4:2/DN bit, added in rung 2, is assigned to the examine if open instruction (-|/|-) called maintenance timer. This XIO instruction prevents the pump from being restarted until after the retentive timer is reset and the done bit is a 0. Finally, when the reset push button switch is pressed the I:1/6 instruction in rung 5 is true, which makes RES true. This resets the accumulator in the retentive timer to zero and turns off the T4:2/DN. Note that the address on the reset instruction is the address of the RTO instruction to reset.

intervals. Figure 4-19(a) depicts a ladder diagram using the ControlLogix syntax where the done bits from three off-delay timers (TOF) are used to turn on three heaters and then sequentially turn them off at the 30-second interval rate. (Note that the arrowed lines in the ladder

FIGURE 4-19: Control of heaters with off-delay timers.

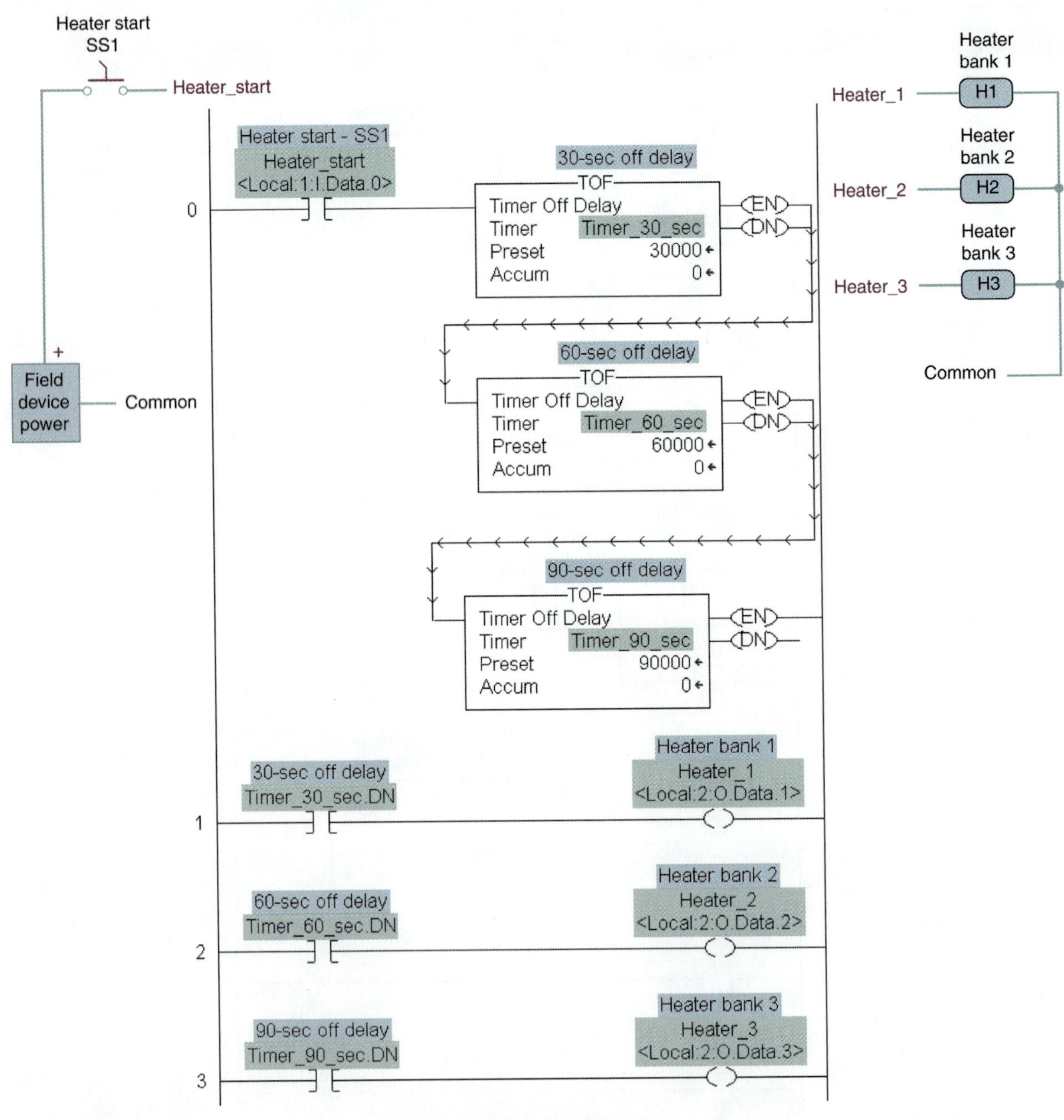

(a) Ladder diagram (ControlLogix)

FIGURE 4-19: (Continued).

Program Tags - MainProgram

Scope: MainProgram Show: Show All Sort: Tag Name

Tag Name	Alias For	Base Tag	Type	Style	Description
Heater_start	Local:1:I.Data.0(C)	Local:1:I.Data.0(C)	BOOL	Decimal	Heater start - SS1
+-Pump_timer			TIMER		Drain pump delay timer
+-Timer_30_sec			TIMER		30 sec off delay
+-Timer_60_sec			TIMER		60 sec off delay
+-Timer_90_sec			TIMER		90 sec off delay
Heater_1	Local:2:O.Data.1(C)	Local:2:O.Data.1(C)	BOOL	Decimal	Heater bank 1
Heater_2	Local:2:O.Data.2(C)	Local:2:O.Data.2(C)	BOOL	Decimal	Heater bank 2
Heater_3	Local:2:O.Data.3(C)	Local:2:O.Data.3(C)	BOOL	Decimal	Heater bank 3
*					

(b) Tag database file (ControlLogix)

logic are part of the ControlLogix display format to indicate that the rung was too large to be displayed in a single horizontal line.) The timers are identified by their tag names, Timer_30_sec, Timer_60_sec, and Timer_90_sec. When the heater start switch, tag Heater_start, is held closed, all three TOF timer EN and DN bits transition to true, so the outputs (tags Heater_1, Heater_2, and Heater_3) in rungs 1, 2, and 3 are true. As a result, heater banks 1, 2, and 3 are turned on. After the start [normally open (NO) momentary selector] switch is released, all three timers start timing. Heater 1 turns off after 30 seconds, heater 2 turns off after 60 seconds, and heater 3 turns off after 90 seconds. This application uses the standard ladder logic described in Figure 4-13(b) with a momentary contact for the trigger.

This example illustrates the use of a TOF timer with an NO momentary contact field device. Since the TOF timing operation is triggered with a true to false transition of the timer rung, the momentary selector switch in this example makes it an ideal trigger for the TOF timer. The TOF timer done bit becomes true when the rung is true and remains true until the accumulator reaches the preset value. As a result, the combination of an NO momentary switch and the done bit of a TOF timer is ideal for this timed off control of an output. In comparison, if a TON timer is used, then the rung must remain true until the preset value is reached. This requires a maintain contact switch or a sealing instruction if a momentary switch is used [see Figure 4-12(c) and (d)].

There are several changes in the ladder logic since the ControlLogix processor is used. The timer outputs can be placed in series on a single rung instead of having three parallel outputs. Also, tag names are used in place of the file number addresses used with the PLC 5 and SLC 500 systems. Finally, the aliases for the tag names are included to identify the input and output module racks, the slot numbers, and the terminal numbers. Review the solution's tag names and the information in the tag name database displayed in Figure 4-19(b). Each of the timer cells can be expanded (click on the + in front of the tag name) to display all the timer data. You may want to review this addressing format in Chapter 3, and then read this solution again.

4-9 CASCADED TIMERS

When one timer's output triggers another timer's input, those timers are referred to as *cascaded*. Cascaded timers are used when there is a need of a time delay that exceeds the maximum time delay capability of a single timer. Figure 4-20 illustrates cascaded timers used to achieve an extended time delay of 43,200 seconds, or 12 hours. The

FIGURE 4-20: Cascaded timers.

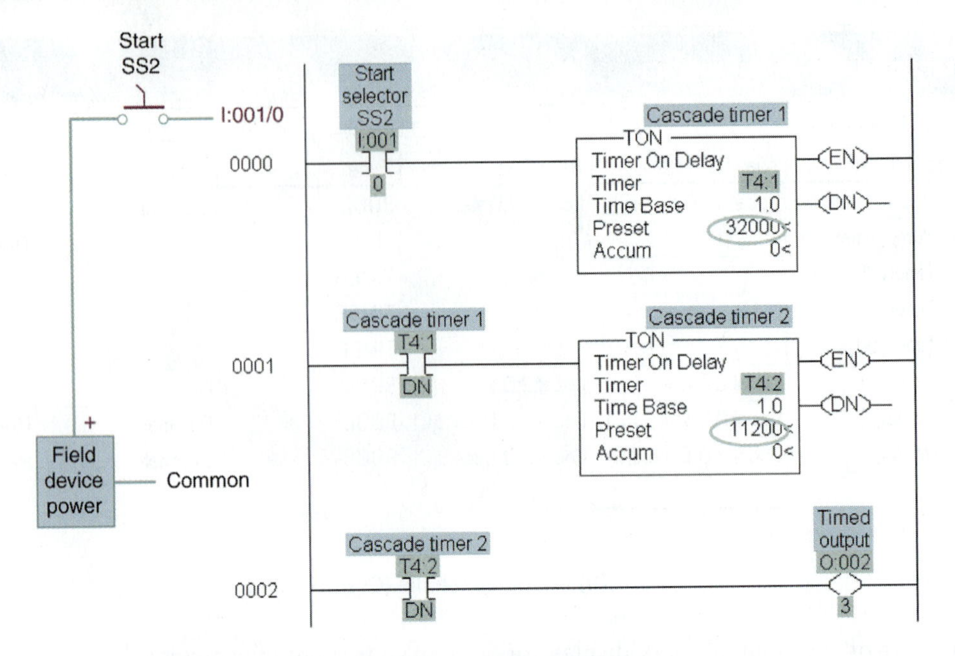

(a) Ladder diagram (PLC-5)

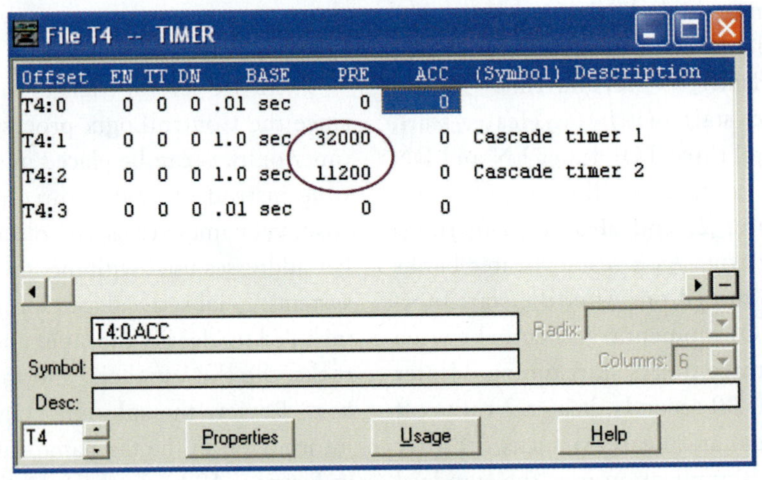
File T4 -- TIMER

Offset	EN	TT	DN	BASE	PRE	ACC	(Symbol) Description
T4:0	0	0	0	.01 sec	0	0	
T4:1	0	0	0	1.0 sec	32000	0	Cascade timer 1
T4:2	0	0	0	1.0 sec	11200	0	Cascade timer 2
T4:3	0	0	0	.01 sec	0	0	

(b) Timer data file (PLC-5)

addressing format for the PLC 5 processor is used in this example solution.

Note that in Figure 4-20 the timer done bit (T4:1/DN) of timer 1 is used to make rung 1 active and start the second timer. The preset value of T4:1 is 32,000 (the maximum of this timer) and the preset value of T4:2 is 11,200. When the start switch is closed, I:001/0 is true, rung 0 becomes true, and T4:1 begins to increment the accumulator. After the T4:1 accumulator reaches 32,000, its done bit becomes true, causing the T4:2 timer accumulator to begin to increment. When the T4:2 timer accumulator reaches 11,200, its done bit turns on O:002/3 and signals a time delay equal to 43,200 seconds. Figure 4-20 indicates that one timer's done bit is the input to another timer, hence these timers are cascaded.

Except for the input and output instruction addressing, the PLC 5 ladder logic is the same as the SLC 500 displayed in the earlier figures. The timer data file for the PLC 5 system is shown in Figure 4-20(b). Although four timers are shown, only the two used in the program have the data illustrated. Timer data values can be changed using this pop-up file display. You may want to review the PLC 5 addressing format in Chapter 3, and then read this solution again.

4-10 EMPIRICAL DESIGN PROCESS WITH PLC TIMERS

The empirical design process, introduced in Section 3-11-4, is an organized approach to the design of PLC ladder logic programs. However, the term *empirical* implies that some rework of the design after it is finished is often necessary. If you tried some of the designs at the end of Chapter 3, then you may understand that the process often does not lead directly to a complete design. This troublesome aspect of the empirical process should become more obvious when timers are added into the design process.

4-10-1 Adding Timers to the Process

The first step in using timers in PLC ladder designs is to know the operation of the three types of timers summarized in Table 4-1 and Figure 4-9 and all the standard timer circuits illustrated in Figures 4-12 and 4-13. Stop now and review them if necessary.

When a timer is added to a ladder it affects two rungs: one rung that makes the timer instruction (TON, TOF, RTN) active, and a second rung that uses a timer output (EN, TT, or DN) to control a system parameter. The complete empirical process is listed in Section 3-11-4; modifications for timers follow and are listed in the solution to Example 4-5.

Step 1: (Write the process description): Include a complete description of time delay(s) required in the process. Note especially the trigger for the delay(s) that is required, the outputs that are delayed, and if it is an on-delay or an off-delay.

Step 2: (Write Boolean equations for all field devices): One of the Boolean expressions should indicate the logic necessary to enable the timer(s). Also, timer output bits should be added into the other Boolean expressions where timers are controlling process outputs.

Examples 4-4 and 4-5 demonstrate how timer ladder design is added into the design process.

EXAMPLE 4-4

Design a ladder logic system to provide two-handed control for a production machine. Two-handed control requires that the operator use both hands to initiate the start cycle of the machine. However, operators tape down one of the hand controls with duct tape so that they can load the machine with one hand and start it with the other. The safety demands a two-handed control circuit with anti-tie down capability. The left and right start push buttons must be operated within a half-second window or the machine will not start. A simpler solution without anti-tie down was developed in Example 3-13; review that before continuing.

SOLUTION

As long as the second hand switch closes within 0.5 second after the first switch closure, then the machine would be allowed to start. If the two switch closures fall outside this 0.5-second window, the machine is off. A look at the standard timer ladder logic in Figure 4-12 indicates that circuit (b) could be used. However, rung 2 is changed as follows:

- The two push button start switches are added to rung 2.
- The XIC instruction addressed by the timer DN bit is changed to an XIO instruction.

The Boolean logic to start the machine is:

Machine on = LH_PB AND RH_PB AND NOT START_INHIBIT

The first push button contact that closes starts the 1 second timer. If the second push button's contact closes outside the 1-second window, then the timer opens the start circuit in rung 0 before the second push button contact can close and start the machine. The machine is inhibited because the timer done bit, *Start_inhibit.DN,* becomes true and the XIO instruction addressed by this bit in rung 0 is then false. The two hand PBs must be in parallel in order that either can enable the timer so the logic equation is:

Timer enable = LH_PB OR RH_PB

The circuit for a ControlLogix processor satisfying the control requirement is illustrated in Figure 4-21(a). Review the ladder logic operation to verify that it satisfies the control description.

The tag dialog boxes are displayed in Figure 4-21(b) and (c). The TON timers in ControlLogix have a fixed 0.001 second time base so a preset of 1000 produces a 1-second delay (1000×0.001 s $= 1$ s). The tag data base in Figure 4-21(b) is in the monitor mode and shows the current value of all tags; as a result, the 1000 appears as the preset value. When the system is running all parameter can be monitored. Figure 4-21(c) illustrates the edit tag mode for the tag data base. Interface data is presented and the method used for entering instruction descriptions using a drop down text box is illustrated.

EXAMPLE 4-5

It is common in automation systems to use timers to set the extension and retraction time for a pneumatic actuator when the cylinder does not have end-of-travel sensors. The pneumatic robot in Figure 4-22(a) is used for material handling and the axes and gripper cylinders do not have end-of-travel sensors. Use 2 seconds for actuator extension and retraction and to open and close the gripper. A cycle is started when the start selector switch is on and when the pickup sensor indicates that a part is in the pickup location. Use the timing diagram in Figure 4-22(b) and the interface information in Figure 4-22(c).

Solution

When empirical programming is used, there are numerous valid solutions to control problems of this type. One may be more efficient (less ladder rungs) than others, but all work equally well.

The cycle time for the robot is 12 seconds with the following sequence starting when the start selector and part sensor are true: X-axis down (4 seconds), gripper closes (6 seconds), X-axis up (2 seconds), Y-axis extends (4 seconds), Y-axis retracts (2 seconds). A study of the timing diagram reveals that some of the actions of the actuators overlap. For example, the X-axis is down for 4 seconds and the gripper closes during the last 2-seconds that the X-axis is down. Verify this overlap on the waveform. Also, there are three waveforms (X-axis timer, gripper timer, and Y-axis timer) that specify the motion of the actuators. However, there are two waveforms (gripper on delay and Y-axis on delay) that are just used to delay the start of those axes motions. For example, the gripper solenoid must be turned on (closed) 2 seconds after the start of the cycle, so a 2-second timer is used to achieve this delay and also to trigger the start of the gripper timer. The cycle is synchronized (cycle timer) with a pulse that occurs every 12 seconds. The axes and gripper waveforms are listed in Figure 4-22(b). Study these waveforms and the ladder logic in Figure 4-22(c) for the X-axis control, Figure 4-22(d) for the gripper control, and Figure 4-22(e) for the Y-axis control as you proceed.

The following steps are added to the empirical design process in Chapter 3 when timers are present.

1. *Draw a timing diagram for all outputs.* The first step in the discrete control of a sequential machine is to generate a timing diagram that shows the on/off sequence for each of the actuators and other field devices. If the timing of input switches and sensors is important, then they are included as well.

   The timing diagram for the robot in this example is displayed in Figure 4-22(b), and the ladder solution is shown in Figure 4-22(c), (d), and (e).

2. Use the timer operation descriptions in Table 4-1 and Figure 4-9 plus the standard timer ladder logics in Figures 4-12 and 4-13 to identify the type of timer ladder(s) to consider for each waveform. Each waveform requires the timer instruction in one rung and the timer output in a second rung.

   All the waveforms in this example could be produced with TON timers. The cycle timer ladder (rung 0) is found in the standard ladders logic, Figure 4-12(e). The standard ladder in Figure 4-12(d) is used for the output timers (rungs 1 and 2, 4 and 5, and 7 and 8) and the standard ladder in Figure 4-12(c) is used for the output on delay timers (rungs 3 and 4, and 6 and 7).

FIGURE 4-21: Two-handed machine control with anti-tie down.

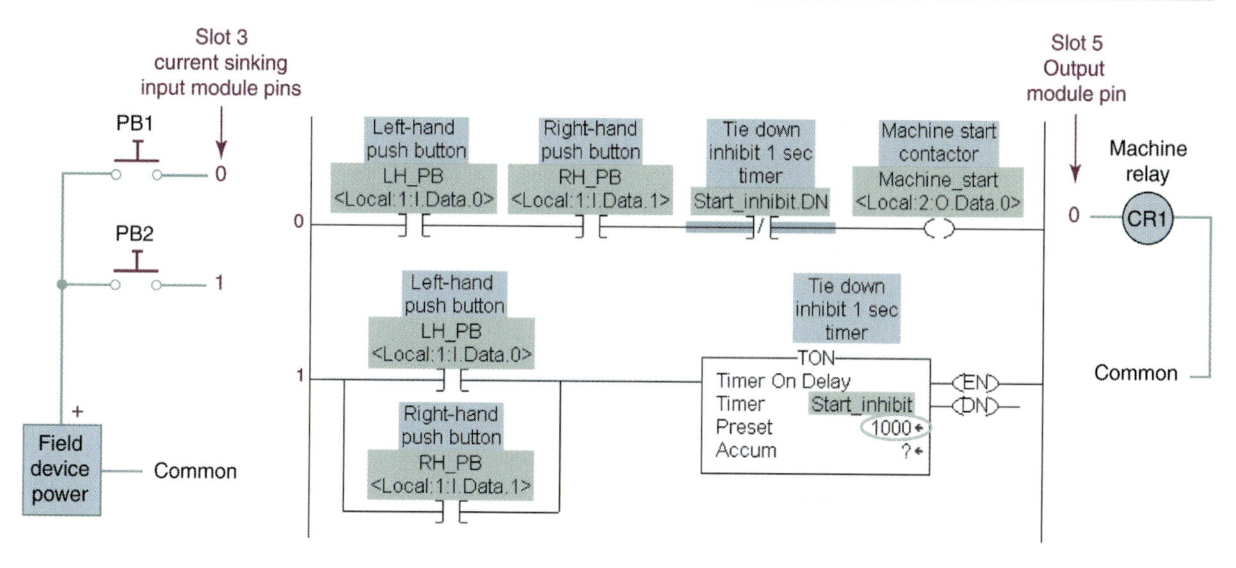

(a) Anti-tie down ladder logic

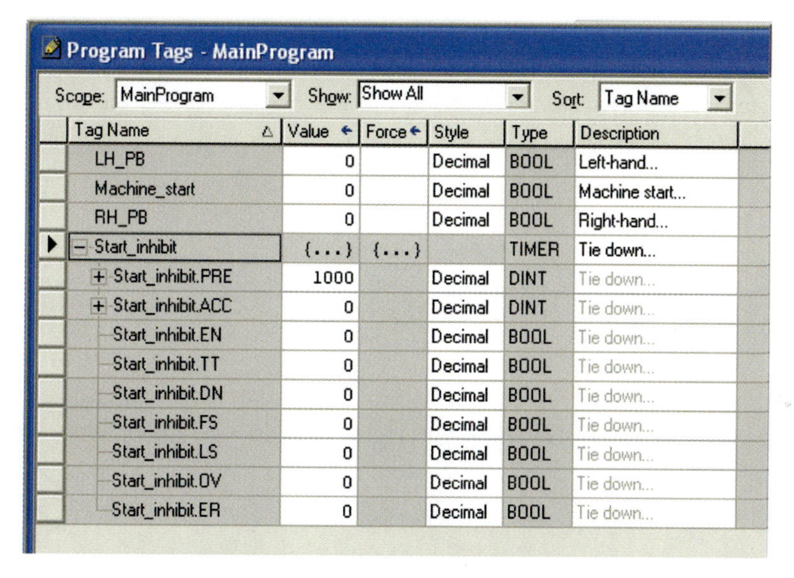

Program Tags - MainProgram

Scope: MainProgram Show: Show All Sort: Tag Name

Tag Name	Value	Force	Style	Type	Description
LH_PB	0		Decimal	BOOL	Left-hand...
Machine_start	0		Decimal	BOOL	Machine start...
RH_PB	0		Decimal	BOOL	Right-hand...
− Start_inhibit	{...}	{...}		TIMER	Tie down...
+ Start_inhibit.PRE	1000		Decimal	DINT	Tie down...
+ Start_inhibit.ACC	0		Decimal	DINT	Tie down...
Start_inhibit.EN	0		Decimal	BOOL	Tie down...
Start_inhibit.TT	0		Decimal	BOOL	Tie down...
Start_inhibit.DN	0		Decimal	BOOL	Tie down...
Start_inhibit.FS	0		Decimal	BOOL	Tie down...
Start_inhibit.LS	0		Decimal	BOOL	Tie down...
Start_inhibit.OV	0		Decimal	BOOL	Tie down...
Start_inhibit.ER	0		Decimal	BOOL	Tie down...

(b) Tag value dialog box

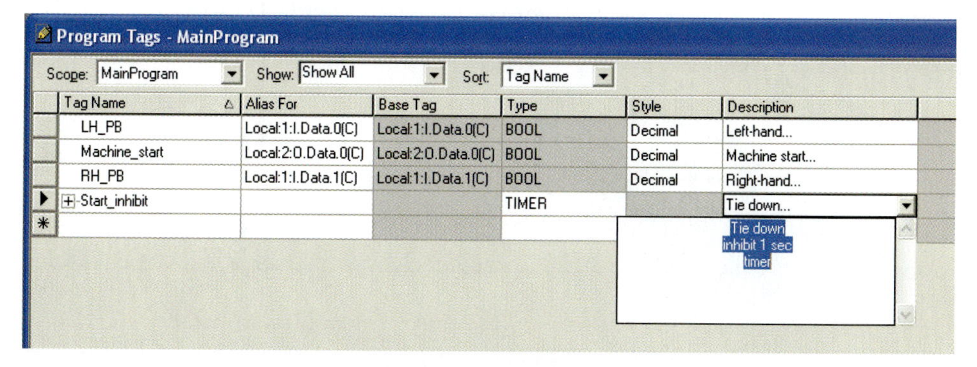

Program Tags - MainProgram

Scope: MainProgram Show: Show All Sort: Tag Name

Tag Name	Alias For	Base Tag	Type	Style	Description
LH_PB	Local:1:I.Data.0(C)	Local:1:I.Data.0(C)	BOOL	Decimal	Left-hand...
Machine_start	Local:2:O.Data.0(C)	Local:2:O.Data.0(C)	BOOL	Decimal	Machine start...
RH_PB	Local:1:I.Data.1(C)	Local:1:I.Data.1(C)	BOOL	Decimal	Right-hand...
+ Start_inhibit			TIMER		Tie down...
*					

Tie down inhibit 1 sec timer

(c) Edit tag dialog box

3. Start with the waveform for the initial sequential machine action and work through each step or stage in the machine operation. Sequential machines operate in steps and often the previous step triggers the following step.

   In this example the process cycle timer is addressed first, then the motion of the X-axis, then the gripper, and finally the Y-axis. The completion of one timed operation triggers the next timer process.

4. Write the input Boolean logic equation to control the timer instruction and the output actuator. This is often a trial-and-error technique where you try a solution and then modify it.

   In rung 1, for example, X-AXIS AND PTS AND (CYCLE TIMER DN sealed by X-AXIS TIMER TT). The sealing instruction is needed because the cycle timer done bit is only true for one scan.

5. Link the standard timer ladders together and verify that the solution satisfies the problem requirements.

   For this example, see the robot control ladder in Figure 4-22(c) through (e).

   The following comments summarize the operation of the robot control ladder solution.

- Rung 0 is a pulse generator (preset value establishes the time between cycle start pulses), and placing the done bit on an XIO instruction in the timer's input logic makes the pulse width equal to one scan.
- The Start selector switch, SEL1, is placed in the logic rung for each timer (rung 0, 1, 3, and 6) so that the system can be reset with that instruction.
- The instructions used to make the timer instruction active are all done bits (rungs 1, 3, 4, 6, and 7) that are only active for one scan. As a result, these timer activation bits are sealed with the timer timing bit to keep the instruction active until the accumulator is equal to the preset values.

FIGURE 4-22: Two-axis pneumatic robot control.

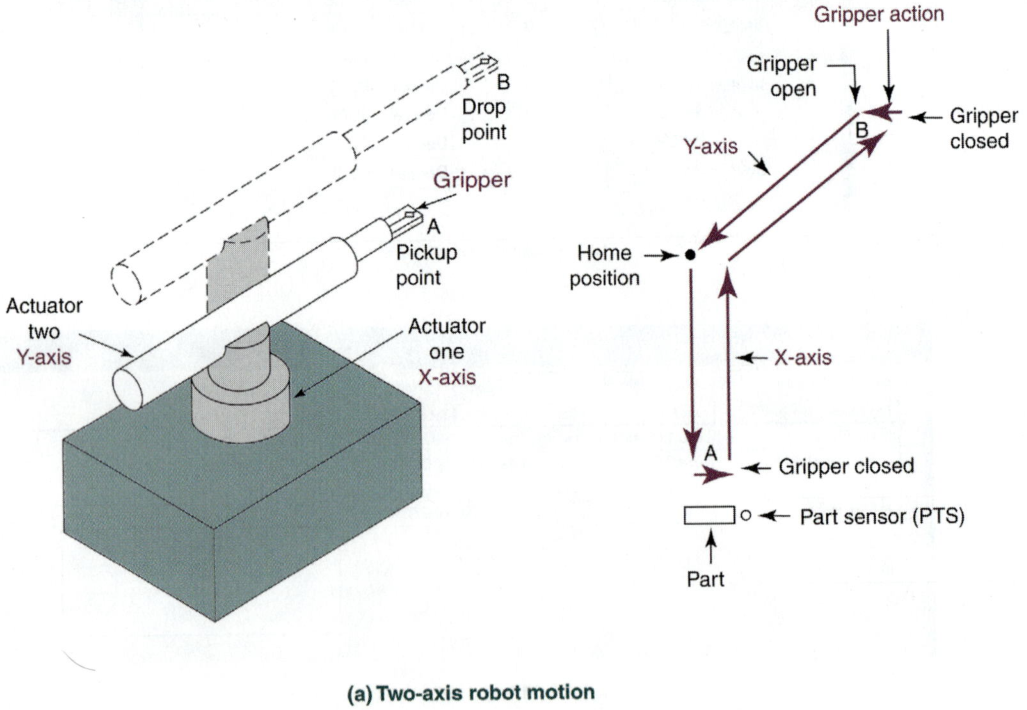

(a) Two-axis robot motion

FIGURE 4-22: (Continued).

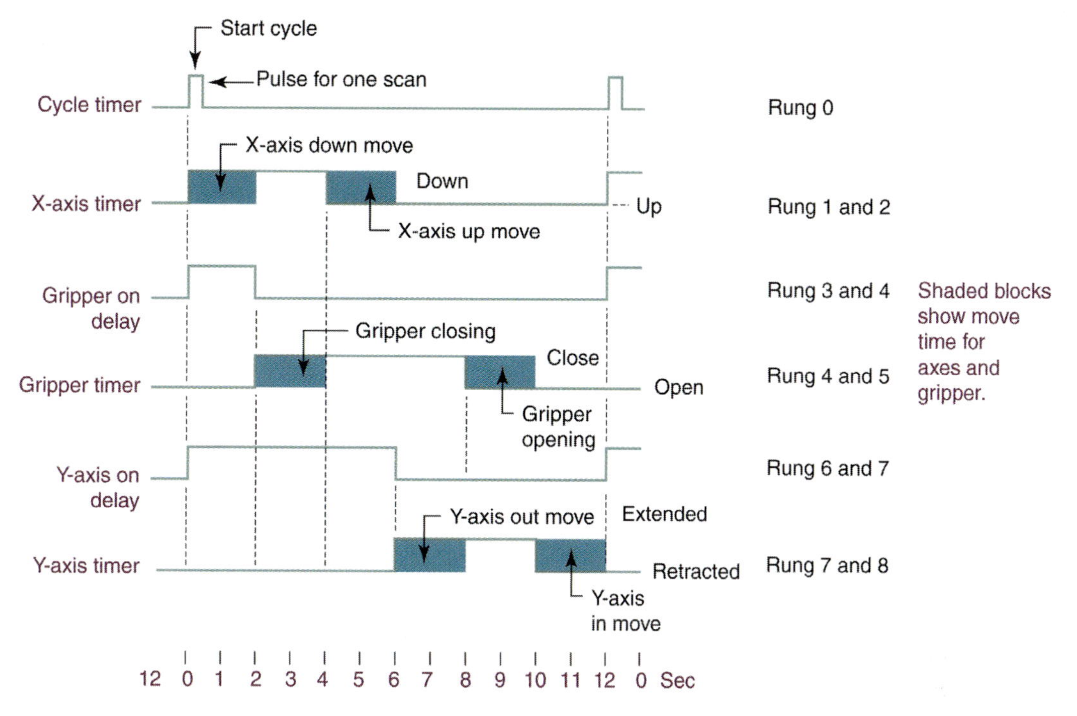

(b) Two-axis timing diagram

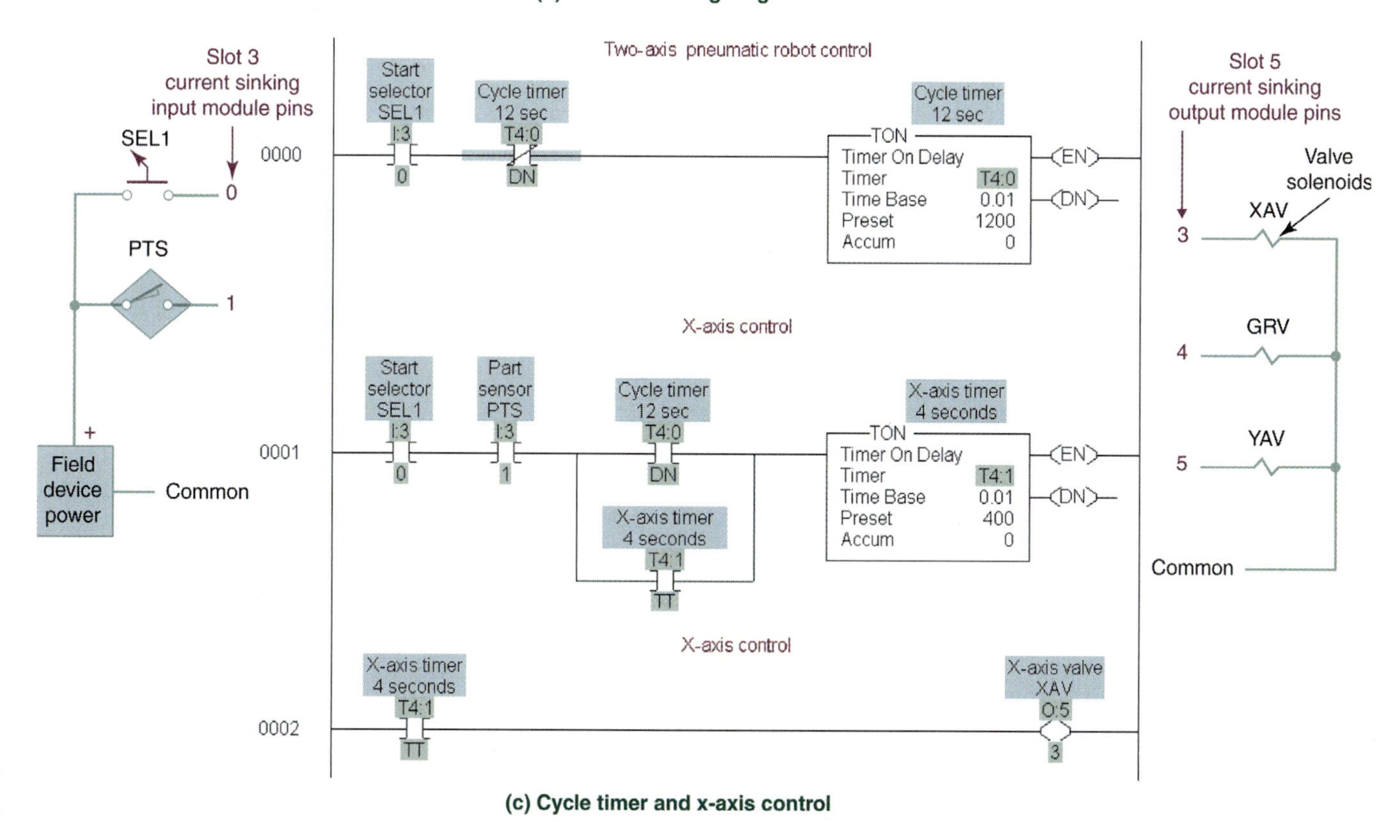

(c) Cycle timer and x-axis control

FIGURE 4-22: (Continued).

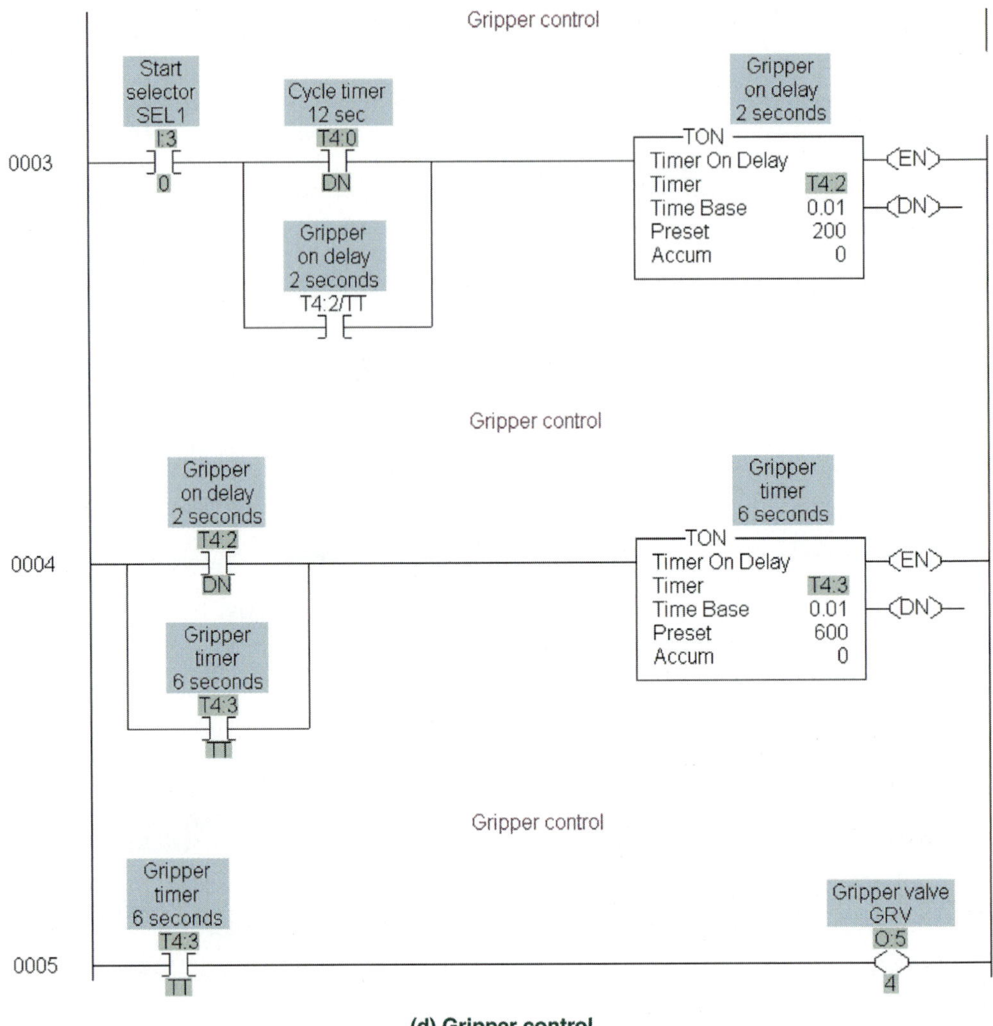

(d) Gripper control

4-11 CONVERSION OF RELAY LOGIC TIMER LADDERS TO PLC LOGIC

Conversion of relay ladder logic to an equivalent PLC ladder solution was introduced in Section 3-11-5 with relay ladders containing only input instructions and output coils. When mechanical or electronic timers are present in the relay ladders, they must be converted as well. The following conversion rules for timers are appended to the initial rule set in Section 3-11-5.

1. Replace the on-delay relay timer (NO timed closed type) with a TON PLC timer.
2. Replace the off-delay relay timer (NO timed open type) with a TOF PLC timer.
3. Select the time base so that the timing resolution meets the requirements of the application.
4. Set the preset value so that the product of the time base and the preset value equal the delay time value.
5. In an on-delay conversion, replace the NO timed contact with an XIC instruction addressed with the done bit of the PLC timer. If the NC relay timer contact is used, then replace it with an XIO instruction addressed with the done bit of the PLC timer.
6. In an off-delay conversion, replace the NO timed contact with an XIC instruction

FIGURE 4-22: (Continued).

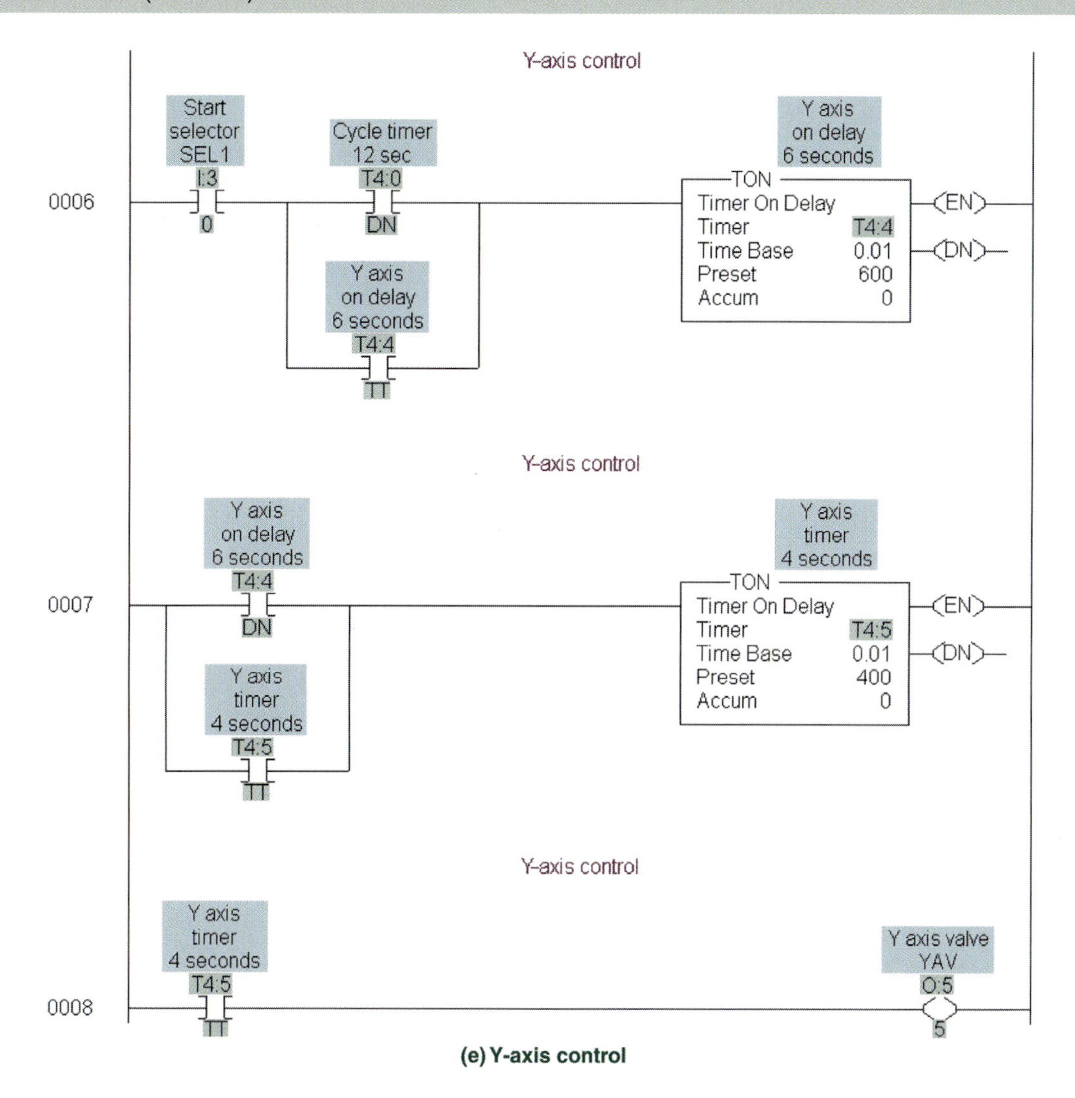

(e) Y-axis control

addressed with the done bit of the PLC timer. If the NC relay timer contact is used, then replace it with an XIO instruction addressed with the done bit of the PLC timer.

7. If an instantaneous NO contact on the mechanical or electronic time delay relay is used, then use an XIC instruction address with the enable bit from the PLC timer.
8. If an instantaneous NC contact on the mechanical or electronic time delay relay is used, then use an XIO instruction address with the enable bit from the PLC timer.

The conversion of input field devices—switches and sensors—plus output field devices—actuators, and contactors—follows the guidelines specified in Section 3-11-5.

4-12 TROUBLESHOOTING LADDER RUNGS WITH TIMERS

Some guidelines and a systematic procedure for troubleshooting PLC systems were presented in Section 3-12. In this chapter the troubleshooting of the timer instruction is addressed.

4-12-1 Troubleshooting Timer Ladder Logic

The most difficult ladder timer programs to verify are those with multiple cascaded timers with small preset time values. Execution is often too

EXAMPLE 4-6

Convert the relay ladder timer circuit in Figure 4-3 to a PLC solution using a PLC 5 system from Allen-Bradley.

SOLUTION

Refer to Figures 4-3 and 4-23(a) as you read the solution. The TMR1 mechanical timer is an on-delay timer with an NO timed close contact (TMR1-2) and an NO instantaneous contact (TMR1-1). The TMR1 relay is replaced by a TON timer (T4:0). The TMR1-1 instantaneous contact is replaced by an XIC PLC instruction with the address T4:0/EN. The TMR1-2 NO timed closed contact is replaced by an XIC PLC instruction with the address T4:0/DN. The input contacts and output coil are replaced with PLC logic symbols as illustrated in Figure 4-23(a). Now the ladder is examined to determine if simplification is possible.

The solution in Figure 4-23(a) functions exactly like the wired relay ladder logic; however, the ladder logic can be simplified. The PLC ladder in Figure 4-23(b) works equally well but does not have the stop push button contact in rung 1. The stop instruction is not necessary because the timer is reset when the stop instruction in rung 0 is active. A timer reset makes the done bit 0, which turns off the motor.

The I/O Configuration dialog box for the PLC 5 is illustrated in Figure 4-23(c). Note that all the data necessary for addressing an instruction are provided. Double-clicking on the Chassis Type data opens the list of the modules present.

FIGURE 4-23: Relay ladder logic conversion.

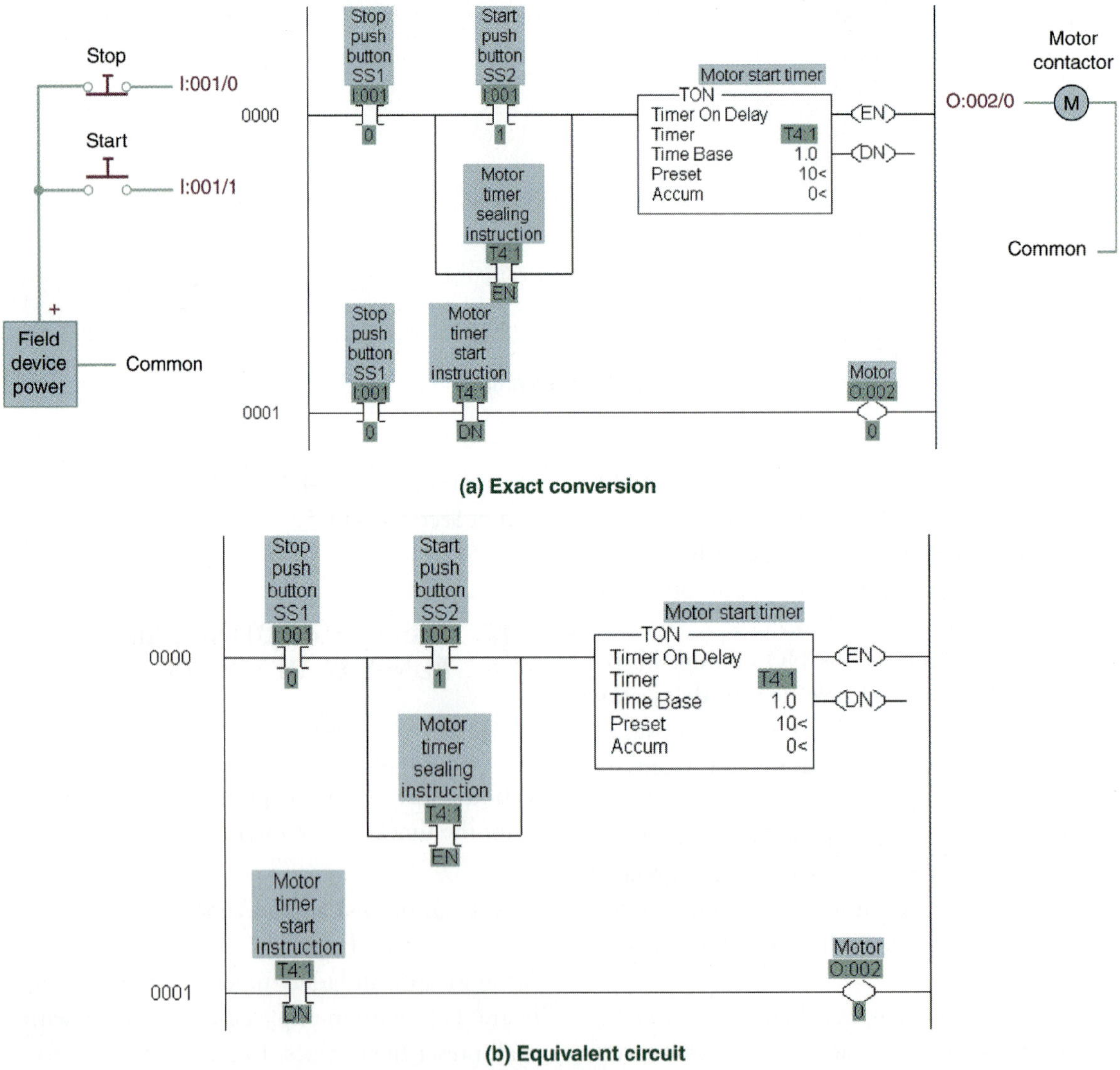

Stop
I:001/0
Start
I:001/1
+
Field device power
Common
0000
Stop push button SS1 I:001 0
Start push button SS2 I:001 1
Motor timer sealing instruction T4:1 EN
Motor start timer
TON
Timer On Delay
Timer T4:1
Time Base 1.0
Preset 10<
Accum 0<
EN
DN
Motor contactor
O:002/0 M
Common
0001
Stop push button SS1 I:001 0
Motor timer start instruction T4:1 DN
Motor O:002 0

(a) Exact conversion

0000
Stop push button SS1 I:001 0
Start push button SS2 I:001 1
Motor timer sealing instruction T4:1 EN
Motor start timer
TON
Timer On Delay
Timer T4:1
Time Base 1.0
Preset 10<
Accum 0<
EN
DN
0001
Motor timer start instruction T4:1 DN
Motor O:002 0

(b) Equivalent circuit

FIGURE 4-23: Relay ladder logic conversion.

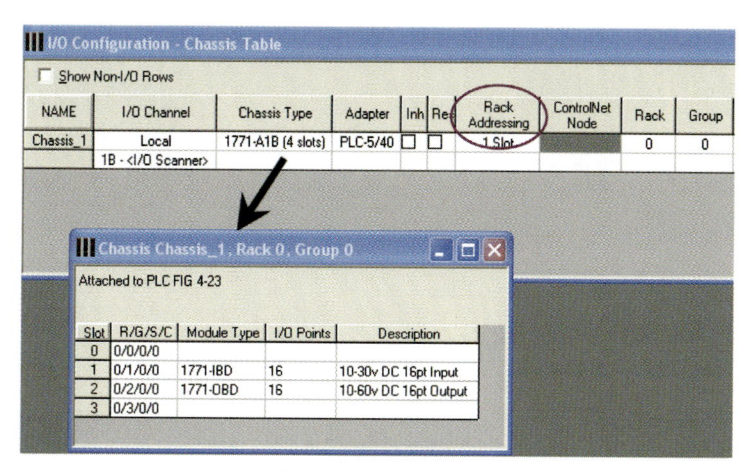

(c) PLC 5 I/O configuration

fast to determine if the syntax is correct. One or more of the following suggestions may help.

- Test the timers starting with the first in the sequence, and then add one timer at a time until the total sequence is operational.
- If the preset times are very small, increase all times proportionately to initially verify the correct sequential operation.
- Use the timer dialog boxes like those illustrated in Figure 4-20(b) for PLC 5 and SLC 500 and in Figure 4-21(b) for the ControlLogix timer instructions to track timer data as the program executes.

Often PLC programs cannot be tested on the manufacturing process, so the use of PLC simulators from Allen-Bradley for the three processors is necessary for program verification and troubleshooting. The simulator permits full execution of the ladder program in the off-line mode to verify that proper operation of the system was achieved in the ladder program.

4-12-2 Temporary End Instruction

The temporary end (TND) instruction is useful for troubleshooting any PLC program, but it is especially helpful for timers. The TND instruction is an output instruction and is shown in Figure 4-24 in rung 2. TND is used to progressively debug a program, or conditionally omit the balance of your current program file. It is placed as an output on the rung with input instruction logic. When the logic preceding this output instruction is true, TND stops the processor from scanning the rest of the program file, updates the I/O, and resumes scanning at rung 0 of the main program. If the TND instruction's rung is false, the processor continues the scan until the next TND instruction or the END statement.

The robot program in Figure 4-22(c) is modified by placing the TND instruction after rung 1. The modified program is shown in Figure 4-24. The done bit on the timer makes the TND rung true and program terminates after rung 1. This is a way to verify that the X-axis timer ladder logic is operating properly. The instruction could be moved through the program and an axis test added with each move down the ladder logic.

4-13 LOCATION OF THE INSTRUCTIONS

The location of instructions from this chapter in the Allen-Bradley programming software is indicated in Figure 4-25.

FIGURE 4-24: Temporary end instruction.

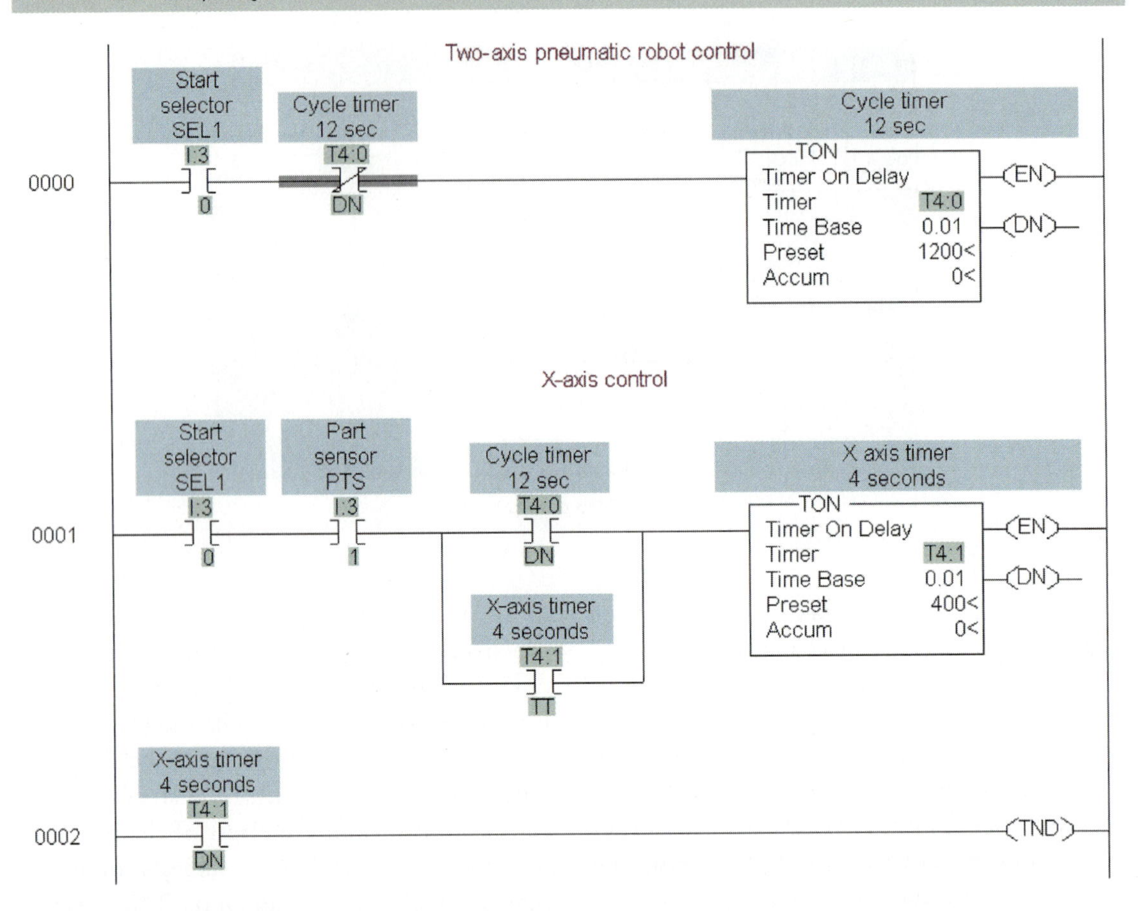

FIGURE 4-25: Location of instructions described in this chapter.

Systems	Instructions	Location
PLC 5, SLC 500	TON, TOF, RTO	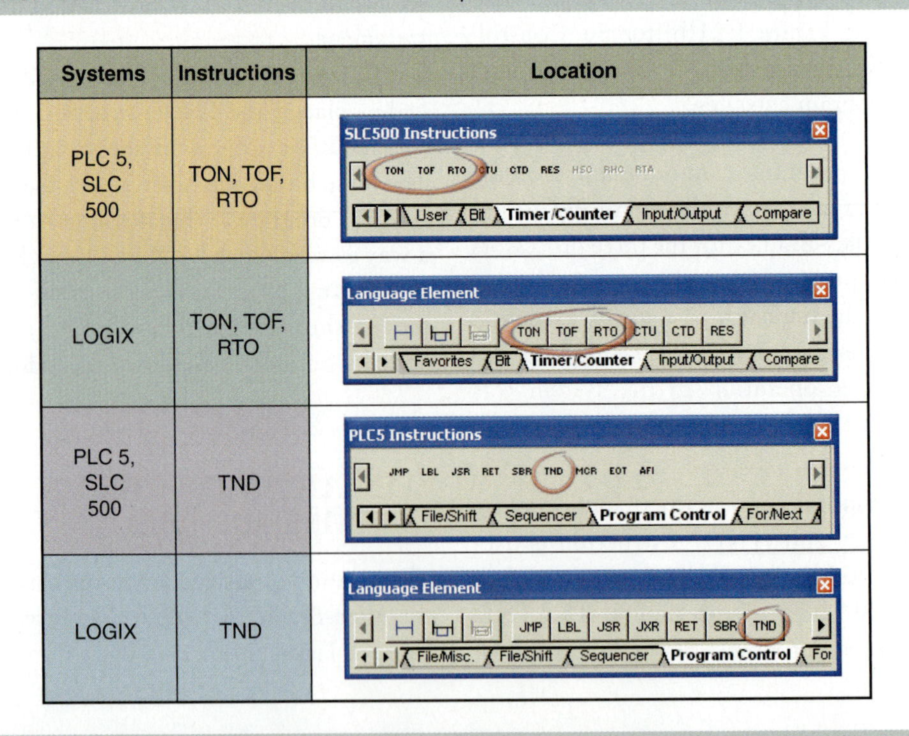
LOGIX	TON, TOF, RTO	
PLC 5, SLC 500	TND	
LOGIX	TND	

Chapter 4 questions and problems begin on page 560.

CHAPTER

5

Programming Counters

5-1 GOALS AND OBJECTIVES

There are two principal goals of this chapter. The first goal is to provide the student with information on the operation and functions of hardware counters—both mechanical and electronic. The second goal is to show how the programmable counter instructions are applied to the solution of automation problems using Allen-Bradley PLCs.

After completing this chapter you should be able to:

- Describe the operation of mechanical and electrical counters.
- Describe the program data used to define counter operation.
- Generate and analyze ladder diagrams for the up counter, down counter, and up/down counter in industrial applications.
- Implement cascade counters to achieve high counting requirements.
- Use the done bit, enable bit, and overflow/underflow bits to control automated systems.
- Develop ladder logic solutions using counter instructions for Allen-Bradley PLC 5, SLC 500, and Logix systems.
- Convert relay ladder logic counter rungs to their PLC equivalent.
- Describe troubleshooting techniques for counter ladder logic.

5-2 MECHANICAL AND ELECTRONIC COUNTERS

Mechanical counters, illustrated in Figure 5-1, use shaft rotations to increment or decrement numerical wheels, thus displaying an accumulated count. Counters are available with a push button reset, a lever reset, or no reset, and actuating the reset causes the counter display to indicate all zeros. The counter without a reset is typically used as an elapsed time meter.

Electronic counters, illustrated in Figure 5-2, have an LCD numerical readout and can count up, down, or up and down depending on the application. After selecting the operating mode and function, input pulses drive the counter's electronics, which in turn drive the LCD readout. Front panel reset, remote reset, or no reset styles are available.

A typical application of electronic counters is shown in Figure 5-3, where a photoelectric

FIGURE 5-1: Mechanical counters.

Courtesy of Redington Counters, Inc.

FIGURE 5-2: Electronic counters.

Courtesy of Redington Counters, Inc.

sensor in the through beam mode is used to count soft drink bottles moving along a conveyor. The output of the receiver is the input to an electronic counter.

5-3 INTRODUCTION TO ALLEN-BRADLEY COUNTERS

PLC counter instructions are an important industrial automation application. Allen-Bradley and other vendors have *up* and *down* counter instructions, with some vendors offering an *up/down* counter instruction as well. Allen-Bradley permits up/down counting through programming with individual up and down instructions. *Counters* are similar to timers, except counters accumulate the changes in an external trigger signal whereas timers increment using an internal clock. PLC counters are generally triggered by a change in an input field device that causes a false to true transition of the counter ladder rung. PLC counters are output instructions that serve the same function in control systems as mechanical and electronic counters. Specifically, counters turn on or turn off an output field device after the counter accumulator has reached a preset value.

PLC count settings can be easily modified and counters can be added to an application through the PLC software without wiring modifications. Since the PLC counters are virtual devices existing only in the PLC software, the number of counters available is large and is only limited by file number allocation. In addition, PLC counters have extremely high accuracy and repeatability

FIGURE 5-3: Application of electronic counter.

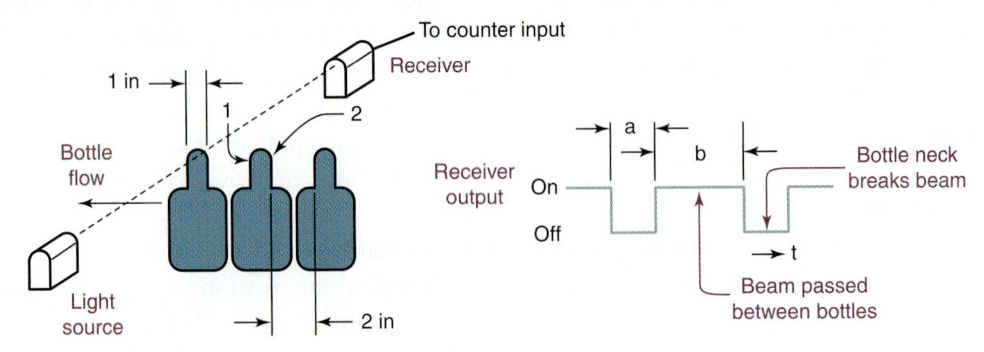

Source: Rehg and Sartori, Industrial Electronics, *1st Edition, © 2006, Reprinted by permission of Pearson Education, Inc., Upper Saddle River, NJ.*

because the counting function is implemented in software using microprocessor technology.

Allen-Bradley counters, illustrated in Figure 5-4, are schematically represented as a block with counter parameters displayed inside. Counter parameters include:

- **Counter Number:** This number identifies the counter file and data. Allen-Bradley PLC 5 and SLC 500 processors use a counter file, identified by the letter C, that has a default file number of 5. A numeric value is appended to the C5 file to indicate the counter number. For example, C5:0, C5:1, and C5:250 are three counters numbered 0, 1, and 250. The SLC 500 supports up to 256 counters in file C5, and the PLC 5 supports 1000 counters. If additional counters are required, user-defined files (10 to 255 for the SLC and 9 to 999 for the PLC 5) are used. The colon (:) is a delimiter used to separate the file number and the counter number.

  The Logix processors use a tag name for the counters, such as Product_count. This tag is then assigned a data type named *counter*. The counter tag or number must be unique and only one up or down counter can have that specific number or name symbol in a ladder solution.
- **Preset Value (PRE):** This is the count value or set point that the counter must accumulate before the counter output is active or true. In Allen-Bradley PLC 5 and SLC 500 systems, the PRE value has a range of −32,768 to +32,767. The values are stored in binary form with the negative number stored as a 2's complement. Signed numbers and 2's complement arithmetic are covered at the beginning of the next chapter. The ControlLogix system permits preset values between +2,147,483,647 or −2,147,483,648, which is the range of values for a double integer.
- **Accumulated Value (ACC):** This register or tag stores the accumulated number of counts or false to true transitions of the counter rung. The ACC is generally set to zero at the beginning of the count. In PLC 5 and SLC 500 systems the ACC values can range from −32,768 to +32,767. Any value in that range can be loaded into the accumulator, but the accumulator value will always be zero when the counter is reset. The ControlLogix system permits the count value to reach +2,147,483,647 or −2,147,483,648, before an overflow indication bit is generated.

FIGURE 5-4: PLC counter block.

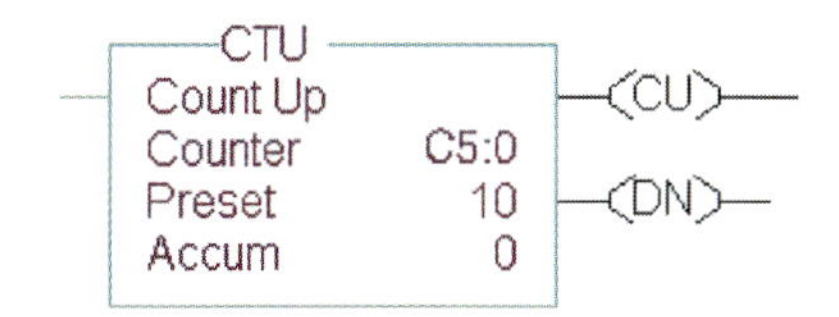

The PRE and ACC values are 16 bit integers stored in memory registers for the PLC 5 and SLC 500 systems. When they are referenced, the address must specify the integer word value for that parameter. For example, if the accumulator is addressed as *C5:6.ACC*, then that references the 16 bit integer register with the accumulator value for counter 6. An address such as *C5:4.PRE* would address the integer word with the preset value for the number 4 counter.

Tags are used in the ControlLogix processor to identify the counters. So an address such as *Counter_1.ACC* references the 32 bit accumulator value for the counter named *Counter_1*. A similar address for the preset value would be *Counter_1.PRE*.

As a result, the PRE and ACC values for all Allen-Bradley PLCs can be set or used by other PLC instructions by just referencing the address of these parameters. In later chapters, the ACC values are frequently referenced in other PLC input instructions for program control.

5-3-1 Counter Output Bits

The bits used to control program flow and to turn on and turn off output field devices in all Allen-Bradley processors include: *count up (CU) enable bit*, *count down (CD) enable bit*, *count up overflow (OV) bit, count down underflow (UN) bit*, and *done (DN) bit*. A truth table describing

TABLE 5-1 Counter truth table.

Up-counter bits	are TRUE if	are FALSE if
Counter enable	Counter rung is true.	Counter rung is false or reset instruction is initiated.
Counter overflow	Accumulated value wraps around from positive maximum value to negative maximum value.	Accumulated value is equal to or less than the positive maximum value.
Counter done	The accumulated value is equal to or greater than the preset value.	The accumulated value is less than the preset value.

(a) Truth table for the up-counter bits

Down-counter bits	are TRUE if	are FALSE if
Counter enable	Counter rung is true.	Counter rung is false or reset instruction is initiated.
Counter underflow	Accumulated value wraps around from negative maximum value to positive maximum value.	Accumulated value is equal to or greater than the maximum negative value.
Counter done	The accumulated value is equal to or greater than the preset value.	The accumulated value is less than the preset value.

(b) Truth table for the down-counter bits

counter output bit operation is illustrated in Table 5-1. Review that table and then read the following bit descriptions.

- **Count Up (CU) Enable Bit:** CU is active or true when the input logic on the counter rung makes the up counter rung true or active. The CU enable bit is off when the up counter rung is false or inactive.
- **Count Down (CD) Enable Bit:** CD is active or true when the input logic on the counter rung makes the down counter rung true or active. The CD enable bit is off when the down counter rung is false or inactive.
- **Count Up Overflow (OV) Bit:** OV is associated with an up counter and is active or true when the counter increments above the maximum positive value or +32,767 for PLC 5 and SLC 500 processors. On the next up count the counter will wrap around to the maximum negative number, or −32,768 for these PLCs. For example, if the ACC is at +32,767, then the ACC will have a −32,768 value after the next false to true transition of the input logic. For all additional false to true transitions of the input the ACC will increment toward 0. The range for the counter accumulator is illustrated in Figure 5-5.
- **Count Down Underflow (UN) Bit:** UN is associated with a down counter and is active or true when the counter decrements below the maximum negative value, or −32,768 for PLC 5 and SLC 500 processors. On the next down count the counter will wrap around to the maximum positive number or 32,767 for

FIGURE 5-5: PLC 5 and SLC 500 accumulator range.

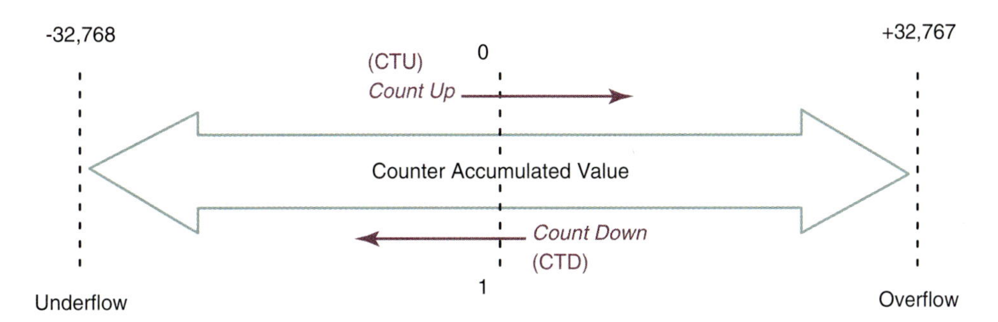

those Allen-Bradley PLCs. For example, if the ACC is at −32,768, then the ACC will have a +32,767 value after the next false to true transition of the input logic. For all additional false to true transitions of the input the ACC will decrement toward 0. Review Figure 5-5 for the range of the accumulator and values for the UN bit.

- **Done (DN) Bit:** DN is true or a 1 for all counters when the ACC value is equal to or greater than the PRE value. It is false or a 0 when the ACC is less than the PRE value.

Three PLC counter functions are introduced in the following sections of this chapter: the *up counter* (CTU), the *down counter* (CTD), and the *up/down counter,* which is a combination of the CTU and CTD. PLC counters are retentive, that is, whatever number is in the counter accumulator at power shutdown remains unchanged upon power-up. Also, the reset (RES) instruction sets the accumulator count to zero for the Allen-Bradley up and down counters.

5-4 ALLEN-BRADLEY COUNTER AND RESET INSTRUCTIONS

The counter instructions for the Allen-Bradley (AB) PLC 5, SLC 500, and Logix processors operate in nearly identical fashion. Therefore, most of the example solutions in this chapter use the SLC 500 instructions; however, PLC 5 and ControlLogix instructions are used in a few examples to illustrate the differences in the three systems.

AB PLCs have two counter instructions, count up (CTU) and count down (CTD). In addition, the reset (RES) instruction is used to initialize both types of counters. The SLC 500 and MicroLogix PLCs also have a high-speed counter (HSC) instruction. Counter addressing for each of the Allen-Bradley processors is addressed in the next several sections.

5-4-1 PLC 5 and SLC 500 Counter and Reset Addressing

The counter instructions for the PLC 5 and SLC 500 processors use the same default address file, C5, and have the same instruction structure. The user-defined file numbers for the SLC 500 and PLC 5 start with 10 and 9 respectively. The counter registers and control bits are located in three words, just as in the AB timer instructions. Review the bit and word layout illustrated in Figure 5-6. Note that word 0 contains the output bit data, word 1 is the preset value, and word 2 is the accumulator value. The addresses of all bits and registers can be used by other instructions.

Addressing counter registers and outputs. The address structure for counters in the PLC 5 and SLC 500 are similiar and use the following format:

Cf:e.s/b

Figure 5-7 describes the address format. The file number, f, is 5, but file numbers 10 through 255 for the SLC 500 are also available if the 256 counters in file 5 are not sufficient. The element number, e, is the counter identification number. Each file (5 and 10 through 255) has 256 additional counters available. The counter word number, s, identifies which of the three

FIGURE 5-6: PLC 5 and SLC 500 three-word counter files.

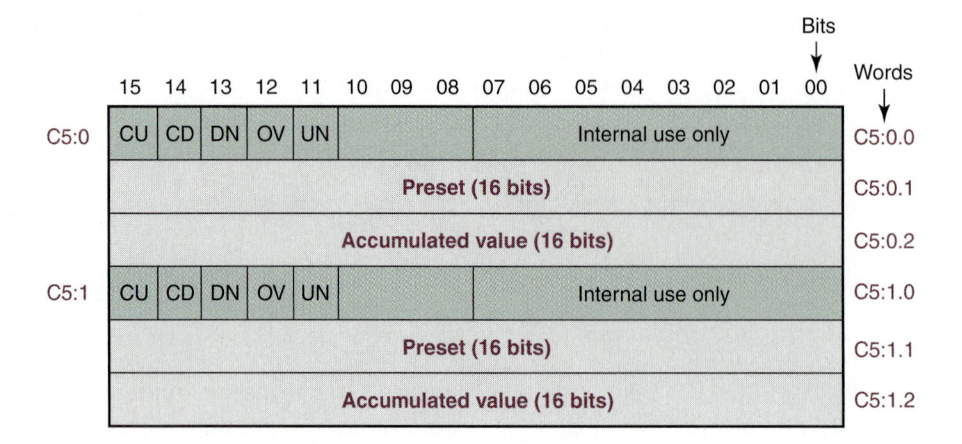

words (0, 1, or 2) in each counter file is being addressed. The first words for the two counters in Figure 5-6 are C5:0.0 and C5:1.0. The final entry in a counter address is the bit number, b, within the counter word. Each output bit is addressed using either the mnemonic letters, such as DN for the done bit, or the bit location in the register, for example, C5:0/13 for the done bit (Figure 5-6). Example counter data addresses are listed in Table 5-2. Study the counter address structure (Cf:e.s/b) and the description of each address element in Figure 5-7, and then verify that you understand what counter data is being addressed by each example in Table 5-2.

FIGURE 5-7: Counter address form for PLC 5 and SLC 500.

Element description

C	The C indicates that the address is a Counter.	
f	File number. For SLC 500 processors the default is 5. File 5 supports 256 counters (C5:0 to C5:255). If more than 256 counters are needed, then file numbers 10 to 255 are available. Each of these supports 256 counters (C10:0 to C9:255).	
:	Element delimiter	
e	Element number, e, is the number of the counter.	For the SLC 500 file 5, e has a range of 0 to 255 counters. The same range is for e if files 10 to 255 are used. These are 3-word elements. The range is 0 to 255.
.	Word delimiter	
s	Word number, s, indicates one of the three counter words.	The value of s ranges from 0 to 2, because each counter has three addressable words.
/	Bit delimiter	
b	Bit number, b, is the bit location in the timer word.	The range is 0 to 15 for all 3 counter words, but bits 10, 11, 12, 13, 14, and 15 are the only ones used for word 0.

TABLE 5-2 Counter address format for PLC 5 and SLC 500.

Address Level	Address Format	Description
Bit	C5:0/15 or C5:0/CU	Count up enable bit – true if the counter rung is active
Bit	C5:0/14 or C5:0/CD	Count down enable bit – true if the counter rung is active
Bit	C5:0/13 or C5:0/DN	Done bit – true if the accumulator is equal to or greater than the preset value
Bit	C5:0/12 or C5:0/OV	Overflow bit – true if the accumulator is greater than +32,767
Bit	C5:0/11 or C5:0/UN	Underflow bit – true if the accumulator is less than –32,768
Word	C5:0.1 or C5:0.PRE	Preset value of the counter
Word	C5:0.2 or C5:0.ACC	Accumulated value of the counter
Bit	C5:0.1/0 or C5:0.PRE/0	Bit 0 of the preset value
Bit	C5:0.2/0 or C5:0.ACC/0	Bit 0 of the accumulated value

Figure 5-8(a) illustrates a PLC 5 and SLC 500 count up instruction, while Figure 5-8(b) illustrates the same parameter data in the counter data file dialog box. Compare the parameter data for counter 4 (C5:4) in both formats. The preset value is 25 for the count up instruction (CTU) and the instruction description is *Machine cycle counter*. When the PLC program is running, the current accumulator value is displayed in the instruction block and in the data file.

FIGURE 5-8: PLC 5 and SLC 500 counter instruction.

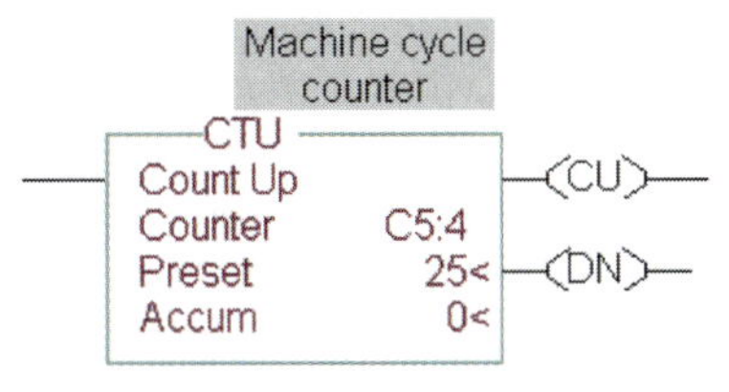

(a) PLC 5 and SLC 500 counter instruction

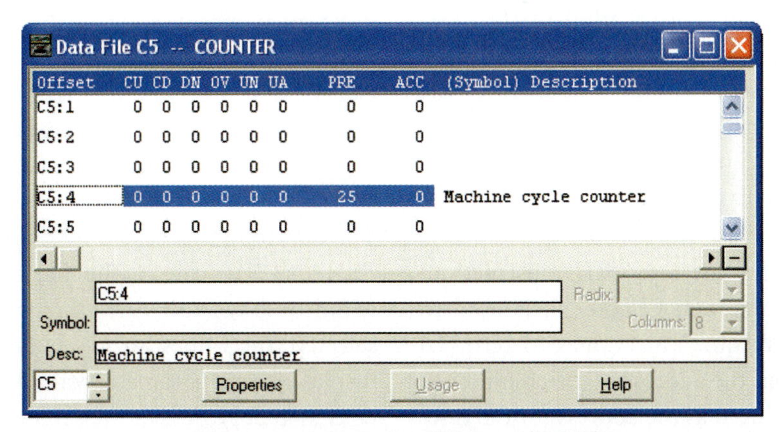

(b) Counter data file for PLC 5 and SLC 500

The count down instruction would have a similar display.

High-speed counter. The SLC 500 and MicroLogix processors have a high-speed counter (HSC) instruction built into the hardware. The HSC instruction counts high-speed pulses at a maximum pulse rate of 8k Hz. The HSC, a variation of the CTU counter, counts false to true transitions at input terminal I:0/0. The HSC is a hardware counter operating asynchronously or independent of the ladder program scan. As a result, the HSC does not count rung transitions. This means that the maximum pulse rate for the counter is not limited to the scan rate of the PLC. In contrast, the CTU counter instruction is a software instruction counting transitions (false to true) of the rung logic, so the maximum pulse rate is limited by the scan rate of the ladder program. The HSC's status or output bits and accumulator values are non-retentive, whereas the CTU instruction has retentive output bits and ACC values.

When the rung containing the HSC output instruction is enabled, the HSC instruction counts false to true transitions at input terminal I:0/0. A false input value interrupts the counting. Only fixed I/O controllers, like the MicroLogic system, that have 24 VDC inputs can use the HSC instruction, and only one HSC instruction is allowed per controller. To use the HSC with a fixed controller, a jumper in the controller must be clipped. The status bits, registers, and addressing for the HSC are the same as that used for the standard CTU counter. The HSC is always C5:0 and reads inputs on I:0/0. The HSC adds one additional status bit, called the update accumulator (UA), which is located in bit 10 of word C5:0.0 of the HSC counter data register. When the UA bit is true the accumulator register, C5:0.2, is updated to the count value in the HSC hardware counter. In other SLC 500 and PLC 5 models, a high-speed counter module is used to capture high-speed pulses.

Reset instruction. The reset (RES) instruction, introduced for timers in Chapter 4, is used to return counter accumulator values to zero. To perform this reset, the address of the counter, for example C5:3, is used as the address for the reset instruction. The overflow and underflow bits are also reset to zero when the accumulator value is reset to a value within the normal operating range. The done bit may also be reset depending on how the preset value compares with a zero accumulator value. It is important to remember that the ACC register and all output bits are held in the zero state as long as the RES instruction rung is true. The RES instruction releases the counter instruction to start counting when the reset instruction rung is false. Thus the counter is reset and disabled while the input logic of the reset rung remains true.

5-4-2 Logix Counter Instructions

The counter instruction for the ControlLogix PLC is the same as that illustrated for the PLC 5 and SLC 500 in Figure 5-8(a), except for the address format. For example, the SLC 500 counter number, such as C5:3, is replaced with a tag name such as Machine_cycle_counter. The tag name makes it easier to know the counter function in the control system. Figure 5-9(a) shows an RSLogix 5000 software ladder rung with two counters. Note that the multiple output counters are shown in series; this is another improvement offered by the 5000 software. When the input instruction Can_switch becomes active, the rung goes from false to true and both counters in the rung increment. Counters are named with any valid tag name (see Chapter 3 for tag name rules), then the Tag Properties dialog box illustrated in Figure 5-9(b) is used to assign a *COUNTER* data type to the tag. The tag name entered, for example *Total_count*, appears at the top, and the description, tag type, and data type must be added. The description can be any text desired and the default *base* type is used most often. The data type, *COUNTER*, must be selected or typed. A pop-up selection dialog box [see the dialog box on the right in Figure 5-9(b)] appears when the selection box button at the right of the Data Type window is clicked.

The counter database is shown in Figure 5-9(c). The database is accessed by double-clicking the *Program Tags* file in the file menu. With the Monitor Tags tab (lower left of illustration) selected, the values for all counter variables are displayed and parameter values can be entered or changed. Note that the preset values from the

FIGURE 5-9: Logix counter instruction.

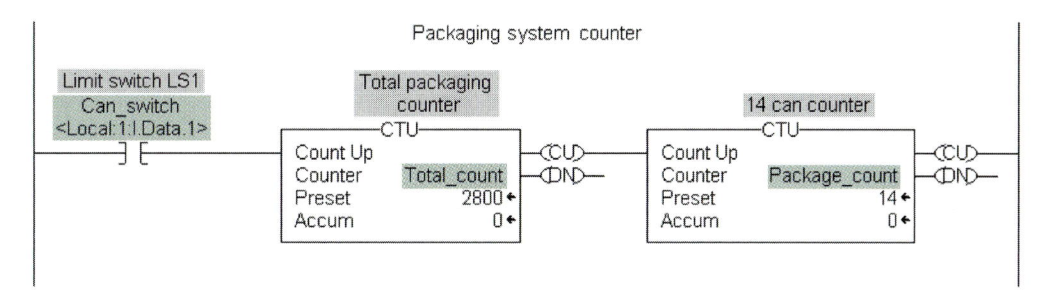

(a) RSLogix 5000 ladder with counters

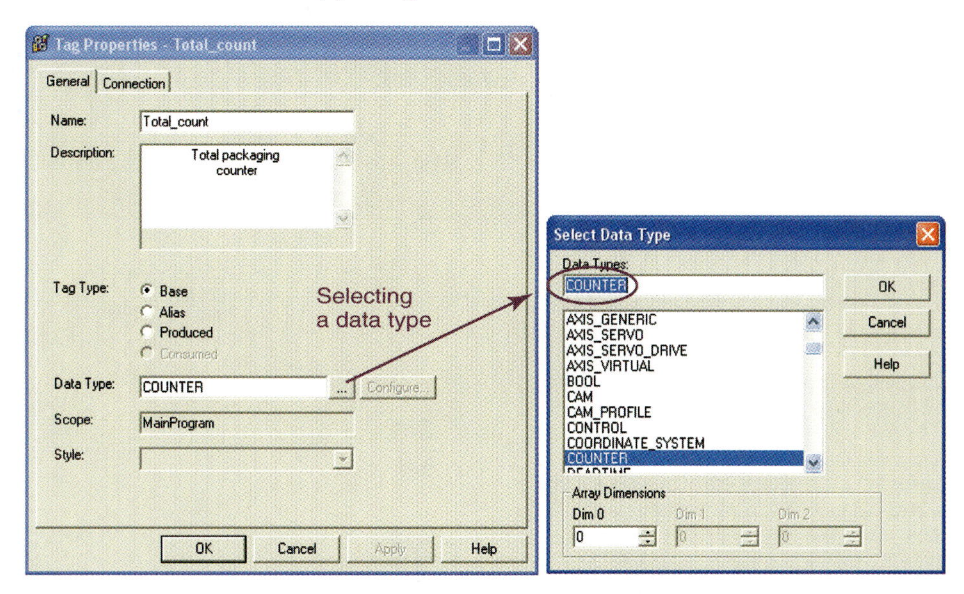

(b) Tag properties dialog box and data type list

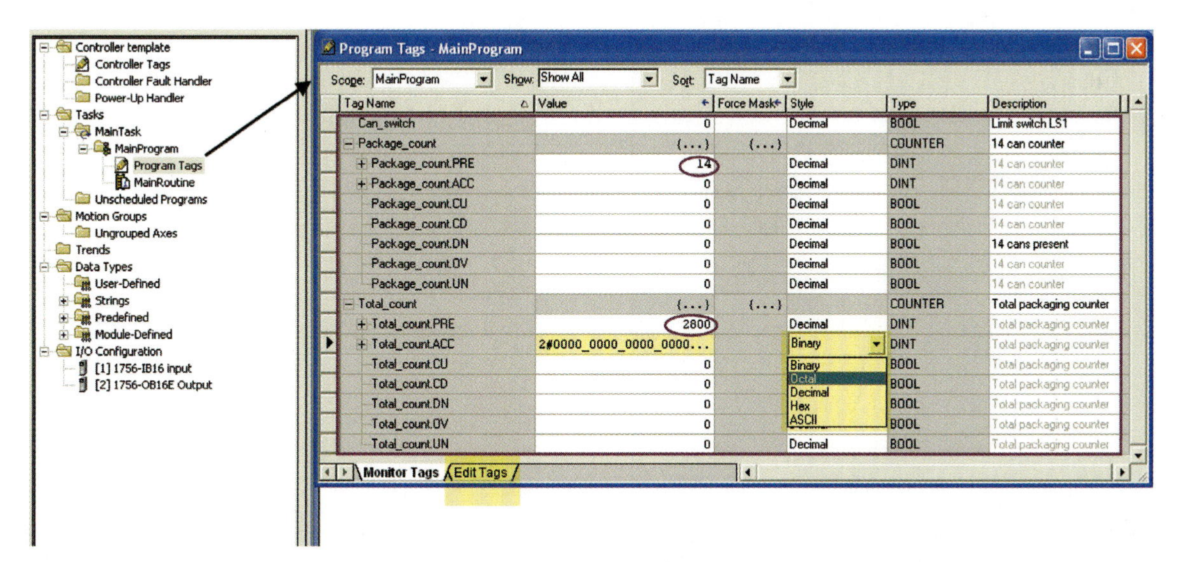

(c) Program tags dialog box

ladder display are visible. Also, the radix or base of the value is changed with the drop-down selection box (see blue highlight) that appears when you click the tag cell in the style column. The ACC display was changed from decimal to binary in the figure. Note that five radix values are offered for display of the register data. The ACC is for a double integer decimal value (two 16-bit words), so the binary representation must be 32-bit words, but the cell is not expanded. Thus only the upper 16 bits of the 32-bit word are visible.

A second input and output database is displayed when the Edit Tags tab at the bottom left of Figure 5-9(c) is selected. This representation is used to assign input and output pins for field devices, but it is not as important for internal instructions such as counters.

5-4-3 Standard Ladder Logic for Counters

Standard counter ladder logic configurations are used when counter instructions are necessary in a control program. Figure 5-10 illustrates four

FIGURE 5-10: Standard ladder logic rungs for CTU and CTD counters.

Application	Standard Ladder Logic Rungs for CTU and CTD Counters
Turn on an output field device after a preset number of false to true transitions of the input logic or turn off an output field device after a preset number of counter inputs. The input field devices are NO sensors. Every time the sensor becomes active, the counter rung and the CU bit transition from false to true and the counter accumulator increments up one count. When the ACC is equal to 10, the DN bit is set true. This causes output O:3/1 to become true and O:3/2 to become false. The counter ACC is reset to 0 when the RES instruction is true or active. The address of the RES instruction is the same as the counter. A CTU counter is shown but a CTD counter could be used for a count down application. Also, a SLC 500 format is presented, but the PLC 5 or ControlLogix could be used as well.	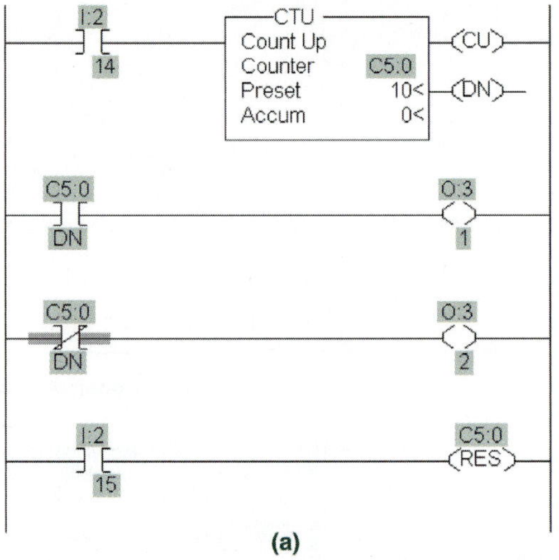 I:2 / 14 — CTU Count Up, Counter C5:0, Preset 10<, Accum 0< — (CU) (DN) C5:0 DN — O:3 / 1 C5:0 DN (NC) — O:3 / 2 I:2 / 15 — C5:0 (RES) **(a)**
Turn on an output field device after a preset number of false to true transitions of the input logic or turn off an output field device after a preset number of counter inputs. Hold the ACC value at the PRE value when the ACC value equals the PRE value. The input field devices are NO momentary push button switches. Every time I:2/14 becomes active, the counter rung and the CU bit transition from false to true and the counter accumulator increments up one count. When the ACC is equal to 9, the DN bit is set true. This opens the counter input rung and freezes the ACC at 9. This also causes O:3/1 to become true and output O:3/2 to become false. The counter ACC is reset to 0 when the RES instruction is true or active. Note that the address of the RES instruction is the same as the counter. A CTU counter is shown but a CTD counter could be used for a count down application. Also, a SLC 500 format is presented, but the PLC 5 or ControlLogix could be used as well.	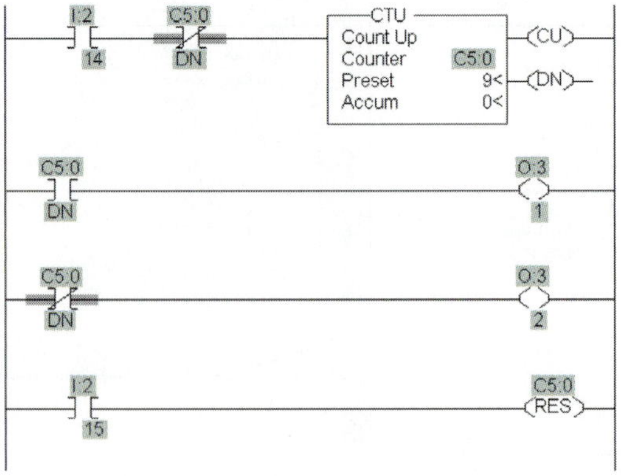 I:2 / 14, C5:0 DN (NC) — CTU Count Up, Counter C5:0, Preset 9<, Accum 0< — (CU) (DN) C5:0 DN — O:3 / 1 C5:0 DN (NC) — O:3 / 2 I:2 / 15 — C5:0 (RES) **(b)**

configurations with features that can be mixed and matched to solve most ladder logic requirements for counters. Read the description of the standard counter ladder configurations in the figure before continuing.

CTU/CTD Standard Ladder Logic: The application addressed in Figure 5-10(a) is for the discrete control of output field devices based on measuring the number of events that occur in a process. The counter in the figure records the number of false to true transitions of input instruction I:2/14 based on changes of an input field device. When the correct count is reached (ACC is equal to PRE), the counter done bit (DN) turns on or off field devices, depending

FIGURE 5-10: (Continued).

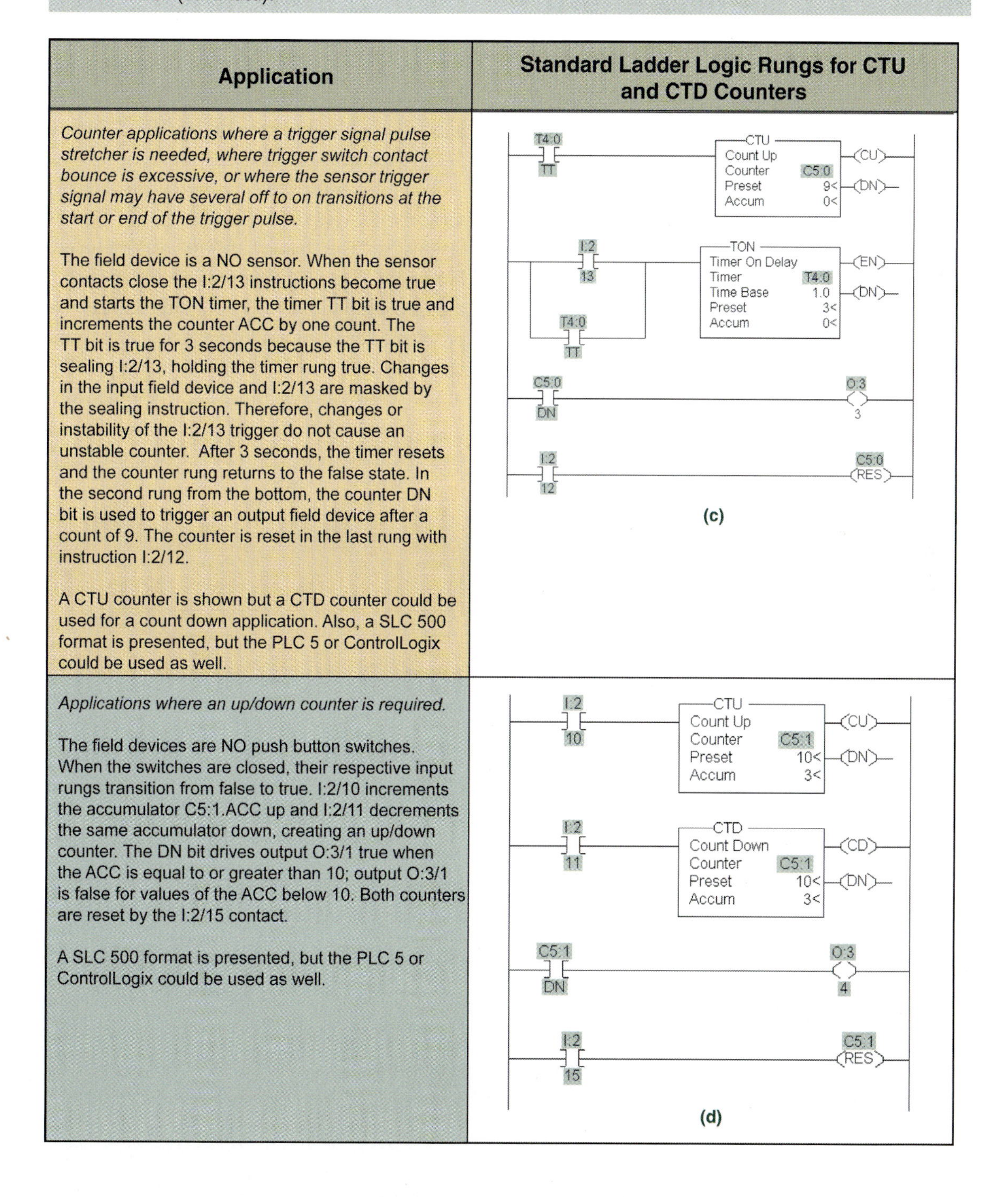

Application	Standard Ladder Logic Rungs for CTU and CTD Counters
Counter applications where a trigger signal pulse stretcher is needed, where trigger switch contact bounce is excessive, or where the sensor trigger signal may have several off to on transitions at the start or end of the trigger pulse. The field device is a NO sensor. When the sensor contacts close the I:2/13 instructions become true and starts the TON timer, the timer TT bit is true and increments the counter ACC by one count. The TT bit is true for 3 seconds because the TT bit is sealing I:2/13, holding the timer rung true. Changes in the input field device and I:2/13 are masked by the sealing instruction. Therefore, changes or instability of the I:2/13 trigger do not cause an unstable counter. After 3 seconds, the timer resets and the counter rung returns to the false state. In the second rung from the bottom, the counter DN bit is used to trigger an output field device after a count of 9. The counter is reset in the last rung with instruction I:2/12. A CTU counter is shown but a CTD counter could be used for a count down application. Also, a SLC 500 format is presented, but the PLC 5 or ControlLogix could be used as well.	(c)
Applications where an up/down counter is required. The field devices are NO push button switches. When the switches are closed, their respective input rungs transition from false to true. I:2/10 increments the accumulator C5:1.ACC up and I:2/11 decrements the same accumulator down, creating an up/down counter. The DN bit drives output O:3/1 true when the ACC is equal to or greater than 10; output O:3/1 is false for values of the ACC below 10. Both counters are reset by the I:2/15 contact. A SLC 500 format is presented, but the PLC 5 or ControlLogix could be used as well.	(d)

on how the output instruction rung is configured. In Figure 5-10(a) an XIC instruction is used to turn on output O:3/1, and an XIO instruction is used to turn off output O:3/2 when the preset count is reached. In this application the counter will continue to record changes in the field device after the PRE value is reached. In this and all the remaining examples, the RES instruction is used to reset the counter ACC to zero. The RES instruction has the same address as the counter that is being reset. An up counter is shown, but this ladder rung would work with a down counter as well.

The configuration in Figure 5-10(b) is like Figure 5-10(a) except that an XIO instruction, addressed with the counter done bit, is added to the input logic for counter. This XIO instruction permits changes of the input field device to transition the counter until the ACC is equal to the PRE value. At this point, the active DN bit makes the XIO instruction false and the rung can no longer cause the counter to increment. The ACC is reset to zero and counting is restored with the RES instruction. An up counter is shown, but this ladder rung would work with a down counter as well.

CTU/CTD Stretched Trigger Standard Ladder Logic: The ladder in Figure 5-10(c) solves an interesting process problem. Study the bottle counting application in Figure 5-3 where a sensor is used to detect the presence of a bottle. In applications like these, you often get an output from the sensor as illustrated in Figure 5-11(c) and (d). The multiple outputs at the start and end of the detection are a result of the bottle wobbling as the trigger starts and ends. The sensor beam is broken, then established, and then broken again as the bottle neck breaks the beam. The same scenario occurs when the bottle neck leaves the beam. As a result, you get three counting pulses [Figure 5-11(e)] where only one should have been present. Study Figure 5-11 to see how this is occurring.

The problem is solved in two ways. First, a sensor with hysteresis is used so that the oscillations at the trigger point are reduced. However, if the bottle oscillations are large, then sensor hysteresis will not fix the false count problem. The second solution is the ladder configuration in Figure 5-10(c). The first transition of the sensor starts a timer and the timer timing bit is used to trigger the counter instruction. Any oscillations of the bottle are ignored and

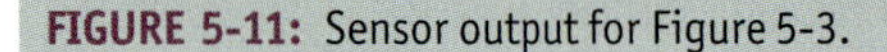

FIGURE 5-11: Sensor output for Figure 5-3.

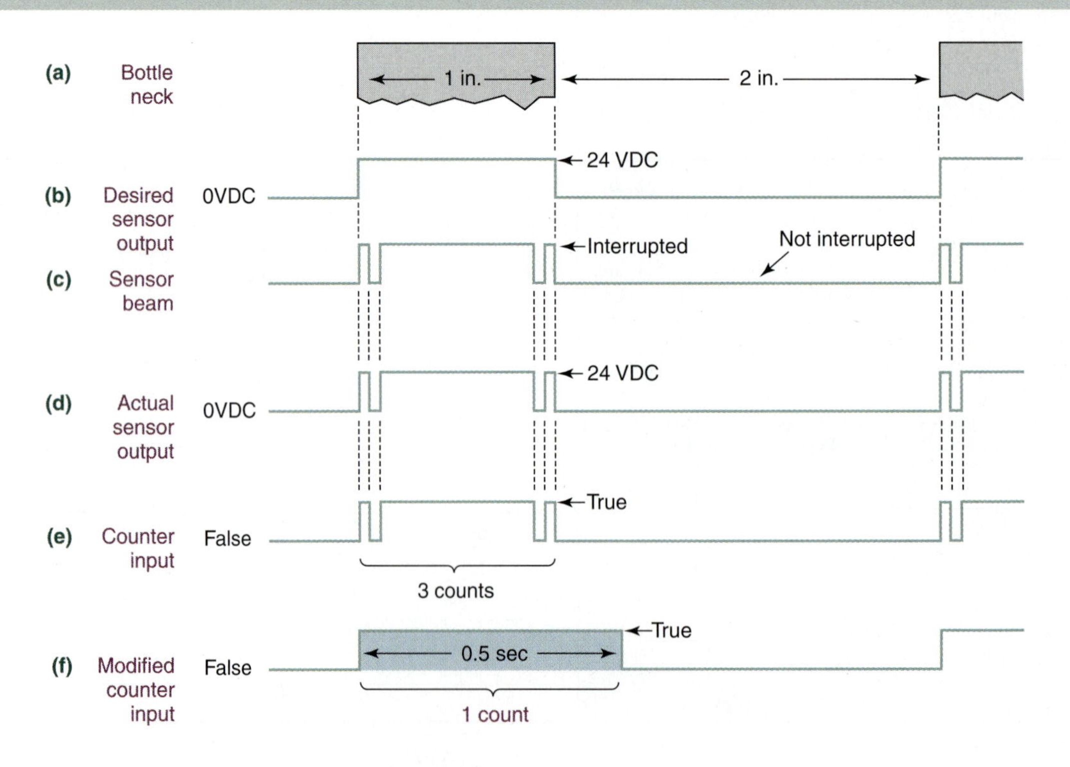

the correct count is made. The timer preset must be large enough to cover the original sensor pulse, but less than the period for the sensor output, as shown in Figure 5-11(f). The ACC is reset to zero with the RES instruction. An up counter is shown, but this ladder rung would work with a down counter as well. Example 5-1 illustrates this point.

Study the solution in Figure 5-10(c) until it is clear how this standard counter ladder logic solves the problem of the bottle counter in this example.

CTU/CTD Up/Down Standard Ladder Logic: The standard counter in Figure 5-10(d) is an up/down counter configured from an up counter and a down counter. Each counter is triggered from a different bit in the input image table (I:2/10 and I:2/11 in the figure), but the up counter and down counter have the same address, C5:1. Since each counter changes the same counter accumulator register (C5:1.ACC), the accumulator increments with every false to true transition of I:2/10 and decrements with every false to true transition of I:2/11. Some PLC systems have an up/down counter instruction. In that case, the counter symbol has two input lines: one to increment, or count up, and one to decrement, or count down. Also, some PLCs have the reset action built into the counter instruction. In Figure 5-10(d), the counter is reset with a RES instruction in the fourth ladder rung.

Study the four standard counter configurations in Figure 5-10. Every counter application you encounter can be explained using some combination of the techniques present in these four basic counter circuits. Note that the ACC in the counter is reset to zero and that counting is restored with the RES instruction.

5-4-4 Allen-Bradley Up Counters

The AB up counter (CTU) is an output instruction that increases the accumulator value by one for every false to true transition of the counter's rung. The done bit is true when the accumulator reaches the value stored as the preset. The CTU operation is the same for all three Allen-Bradley PLCs.

The up counter application in Figure 5-12 turns a green light on and an amber light off after the accumulated value reaches 4. Review the ladder logic in Figure 5-12(a) and the timing diagram in Figure 5-12(b). When the push button switch PB1 closes, the I:1/0 bit of the input image table changes from a 0 to a 1. This makes the counter rung true and increments the accumulator in C5:0 CTU instruction. When the ACC is less than the PRE value, the amber light is on since the done bit address on the XIO instruction is false. When the ACC is equal to or greater than the PRE value (four or more transitions of PB1), the green light is turned on and the amber is extinguished. When the push button switch PB2 is pressed, the XIC instruction, I:1/1, activates the reset instruction, which forces the accumulated value to zero. Counting resumes when PB2 is released and PB1 is closed. Study the ladder logic and timing diagram until the operation of the up counter is understood.

The CTU applications in Examples 5-2 and 5-3 permit a comparison between the SLC 500 and ControlLogix versions of the count up instruction. Review the standard ladder logic for counters in Figure 5-10 since they are the basis for all counter ladder logic designs.

EXAMPLE 5-1

The bottles are moving at a rate of 3 inches per second in Figure 5-3. The ladder in Figure 5-10(c) is used to drive the bottle counter. Determine the delay for the pulse stretched timer in this application.

SOLUTION

Study the timing diagram for the problem shown in Figure 5-11. The bottles have a 3-inch separation from the front edge of one bottle to the front edge of the next bottle. If the bottles are traveling at a fixed rate of 3 inches per second, then it takes 1 second to cover those 3 inches of travel. Setting the PRE for the timer at 0.5 seconds (half of the total travel time) provides a single trigger pulse for the counter as each bottle passes.

FIGURE 5-12: Up counter application.

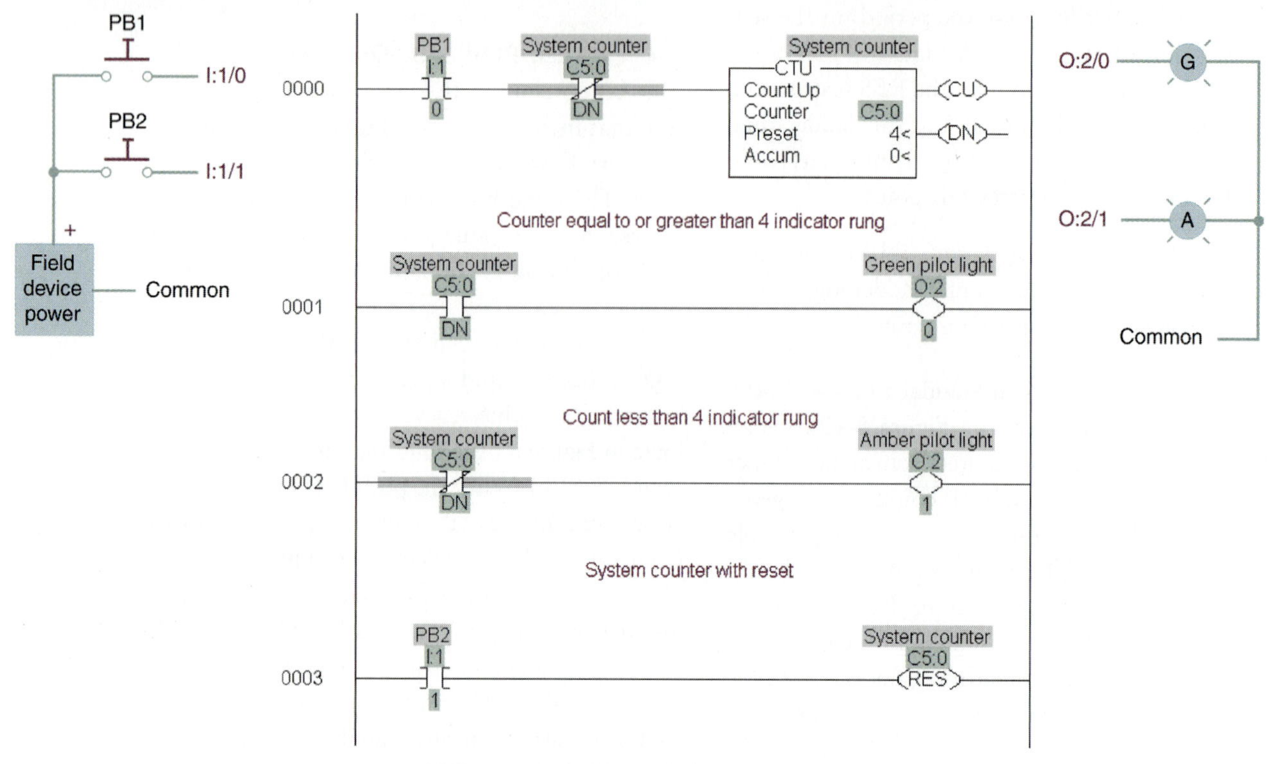

(a) Up counter ladder logic for a PLC 5 and SLC 500

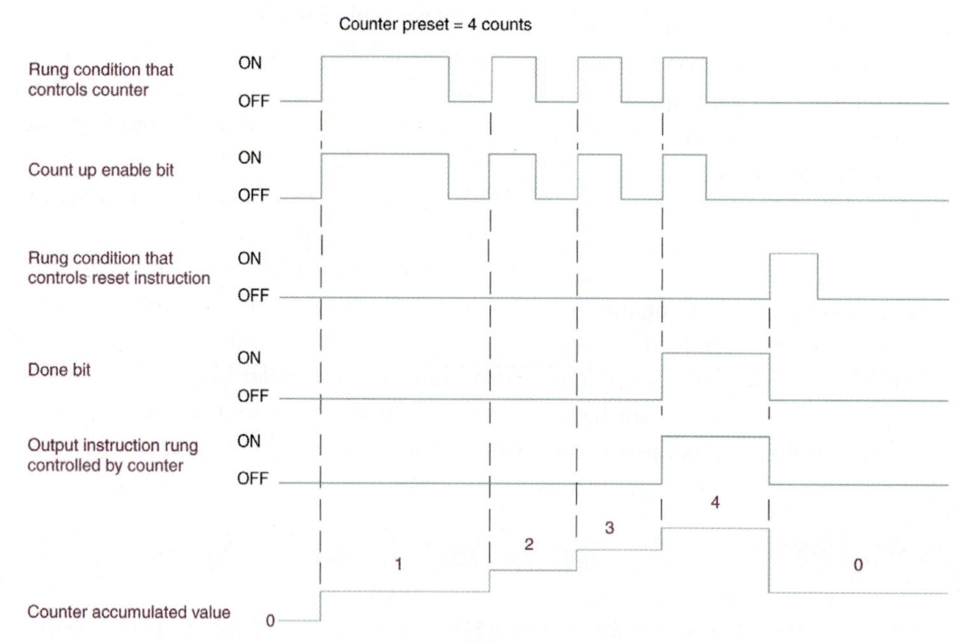

(b) Up counter timing diagram for ladder in (a)

EXAMPLE 5-2

Design the ladder logic for an industrial application that packages canned vegetables supplied by a conveyor. When 12 cans are detected by a current sourcing proximity sensor, a packaging operation is initiated. The production line must package 200 boxes of 12 cans per shift. When 200 packages have been completed, a red stack light is illuminated. While the system is packaging cans, a green stack light is illuminated. A total count of cans packaged per shift should also be recorded.

SOLUTION

Refer to the ladder logic in Figure 5-13 as you read the following solution.

When a can activates sensor S1, the XIC input instruction I:1/10 is true, which sends a false to true transition to counters C5:0 and C5:1. Note that the preset values of the counters are:

- C5:0 has a preset value of 3000, and records the maximum amount of cans on the conveyor per shift.
- C5:1 has a preset value of 12, which is the amount of cans per package.
- C5:2 has a preset value of 200, which is the maximum amount of packages that can be completed per shift.

The *unit total counter* done bit, C5:1/DN, drives three operations: (1) it increments C5:2, which is the package counter, (2) it turns on the *packaging operation* for the 12 cans just counted, and (3) it resets the C5:1 accumulator to zero, which prepares this CTU instruction to count the next set of 12 cans. CTU instruction C5:2 uses the false to true transition of the C5:1 done bit to count the number of packages completed per shift. When the C5:2/DN bit is false, the green stack light is on, indicating that 200 packages have not been completed. When C5:2 reaches 200, its done bit turns off the green light and turns on the red pilot light. Finally, push button switch PB1 initiates reset instructions for counters C5:0, C5:1; and C5:2 so that the packaging control is ready for the next shift.

ControlLogix solution for packaging automation. The SLC 500 solution for Example 5-2 is converted to the equivalent solution for the ControlLogix PLC and displayed in Figure 5-14. Compare the two solutions and notice that the Logix solution permits more than one output per rung, so the ladder logic appears less cluttered. In addition, the use of tag names for symbols and data values makes it easier to read and interpret the ladder logic solution. The operation of the two ladder solutions is identical. If the input logic of a rung makes the rung true, all outputs on the rung are active. Note only the red light is implemented and tag data is displayed in Figure 5-14(b).

Example 5-3

Modify the ladder diagram for the packaging control solution in Figure 5-13 for Example 5-2 to detect an interruption of the can flow on the conveyor and then to:

- Turn off the can conveyor motor and green stack light.
- Turn on an amber stack light.

SOLUTION

Refer to Figure 5-15 as you read the following description.

The four additional rungs in Figure 5-15 are added to the automation application solution in Figure 5-13. Detection of a jam on the can conveyor is achieved by verifying that counter C5:1 increments every 3 seconds.

A standard start/stop rung is used with the addition of an internal relay bit, B3:0/0, addressing an XIO instruction in rung 0. The start/stop field devices are momentary push buttons, with the start an NO contact configuration

FIGURE 5-13: Ladder diagram for the packaging system—Example 5-2.

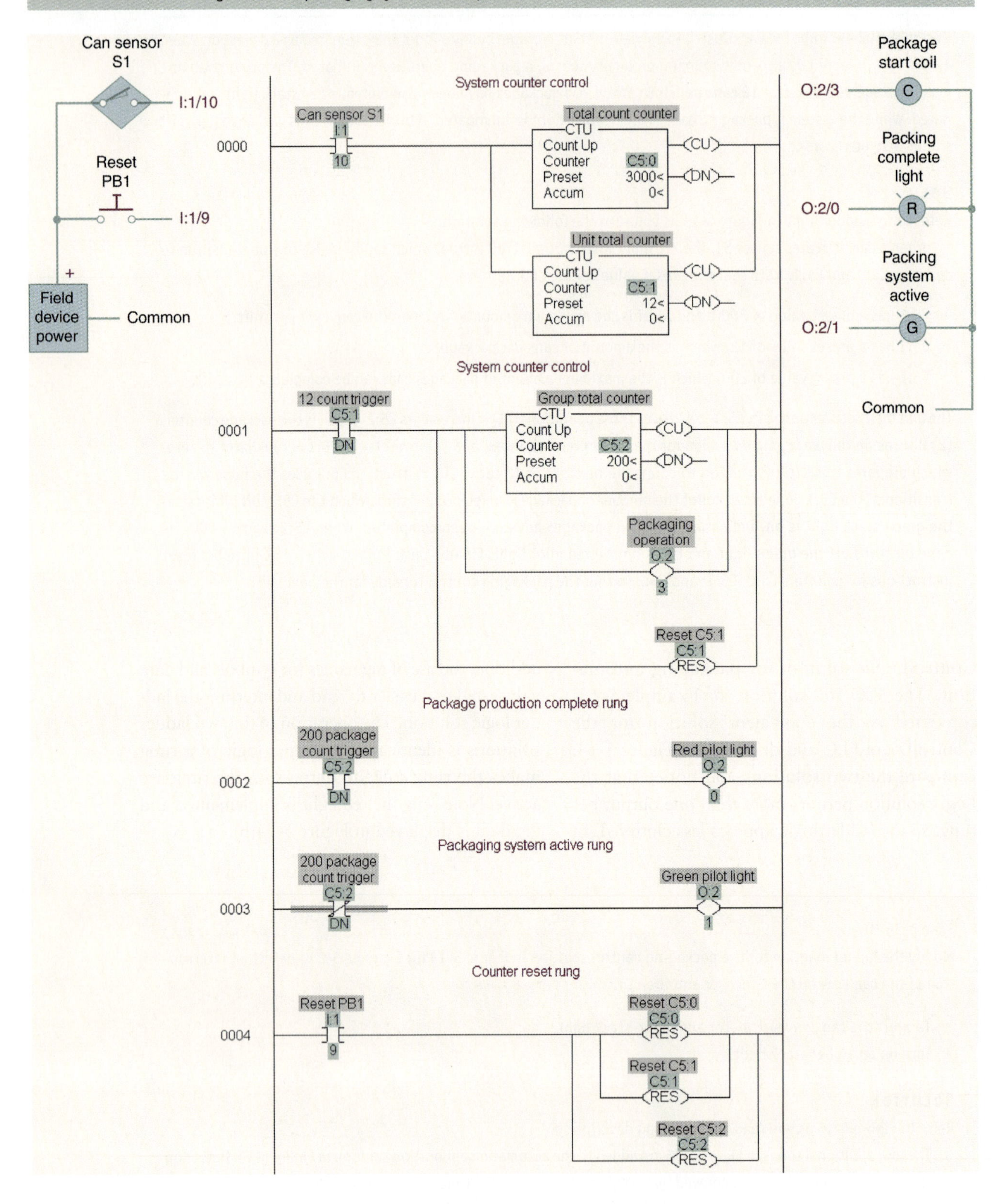

and the stop an NC contact. The virtual relay XIO instruction, B3:0/0, gives another rung the option of turning off the conveyor motor contactor. Note that for safety:

- The motor overloads are hardwired in series with the motor contactor coil.
- An auxiliary contact from the motor contactor is used as an input to provide a sealing instruction for the start push button.

Usually the start switch is sealed by an instruction with the address from the motor contactor OTE instruction (0:2/10 in Figure 5-15). However, the following description illustrates why this would produce an unsafe operating condition.

FIGURE 5-14: ControlLogix ladder for the can packaging system.

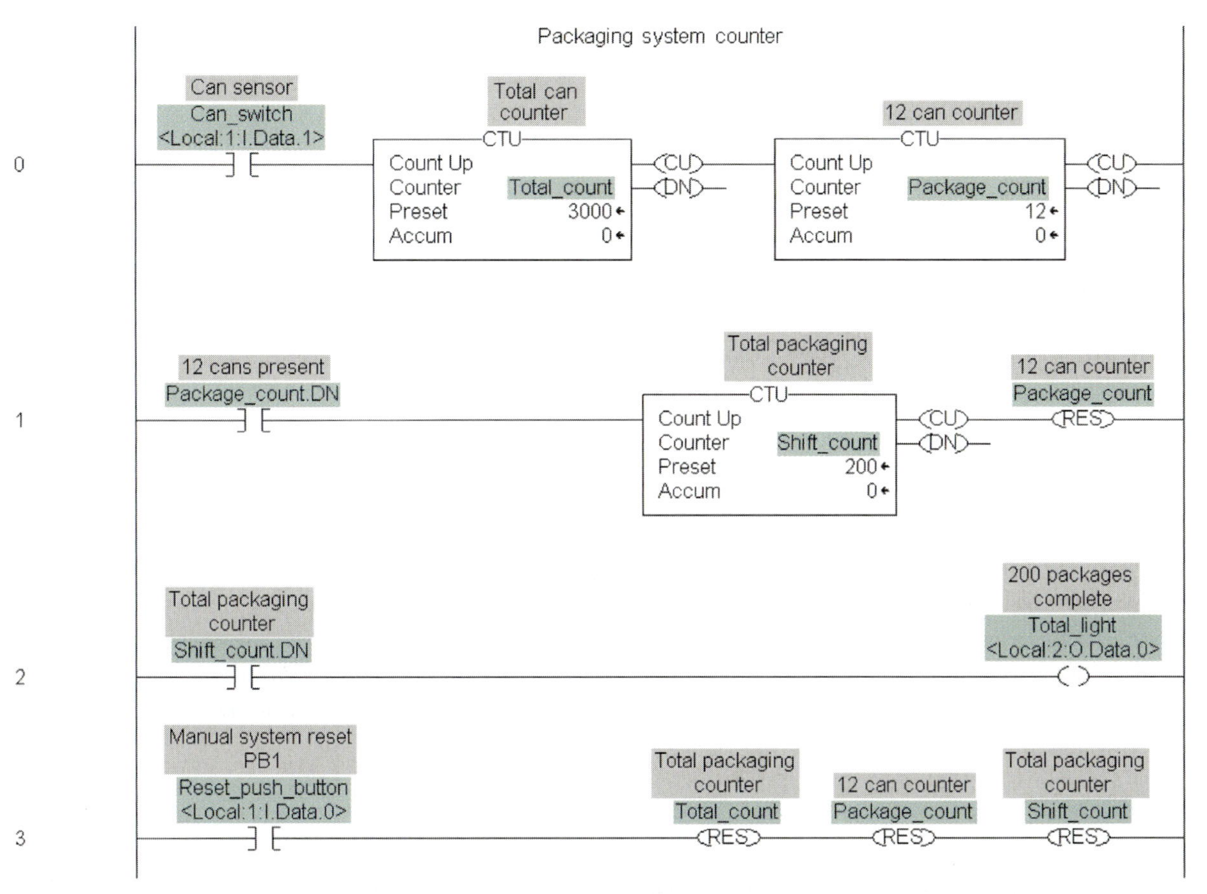

(a) ControlLogix solution for can packaging

Program Tags - MainProgram

Scope: MainProgram Show: Show All Sort: Tag Name

Tag Name	Alias For	Base Tag	Type	Style	Description
Can_switch	Local:1:I.Data.1(C)	Local:1:I.Data.1(C)	BOOL	Decimal	Limit switch LS1
+ Package_count			COUNTER		14 can counter
Reset_push_button	Local:1:I.Data.0(C)	Local:1:I.Data.0(C)	BOOL	Decimal	Manual system reset PB1
+ Shift_count			COUNTER		Total packaging counter
+ Total_count			COUNTER		Total packaging counter
Total_light	Local:2:O.Data.0(C)	Local:2:O.Data.0(C)	BOOL	Decimal	200 packages complete

(b) ControlLogix program tag dialog box data

FIGURE 5-15: Packaging system jam detection ladder diagram—Example 5-3.

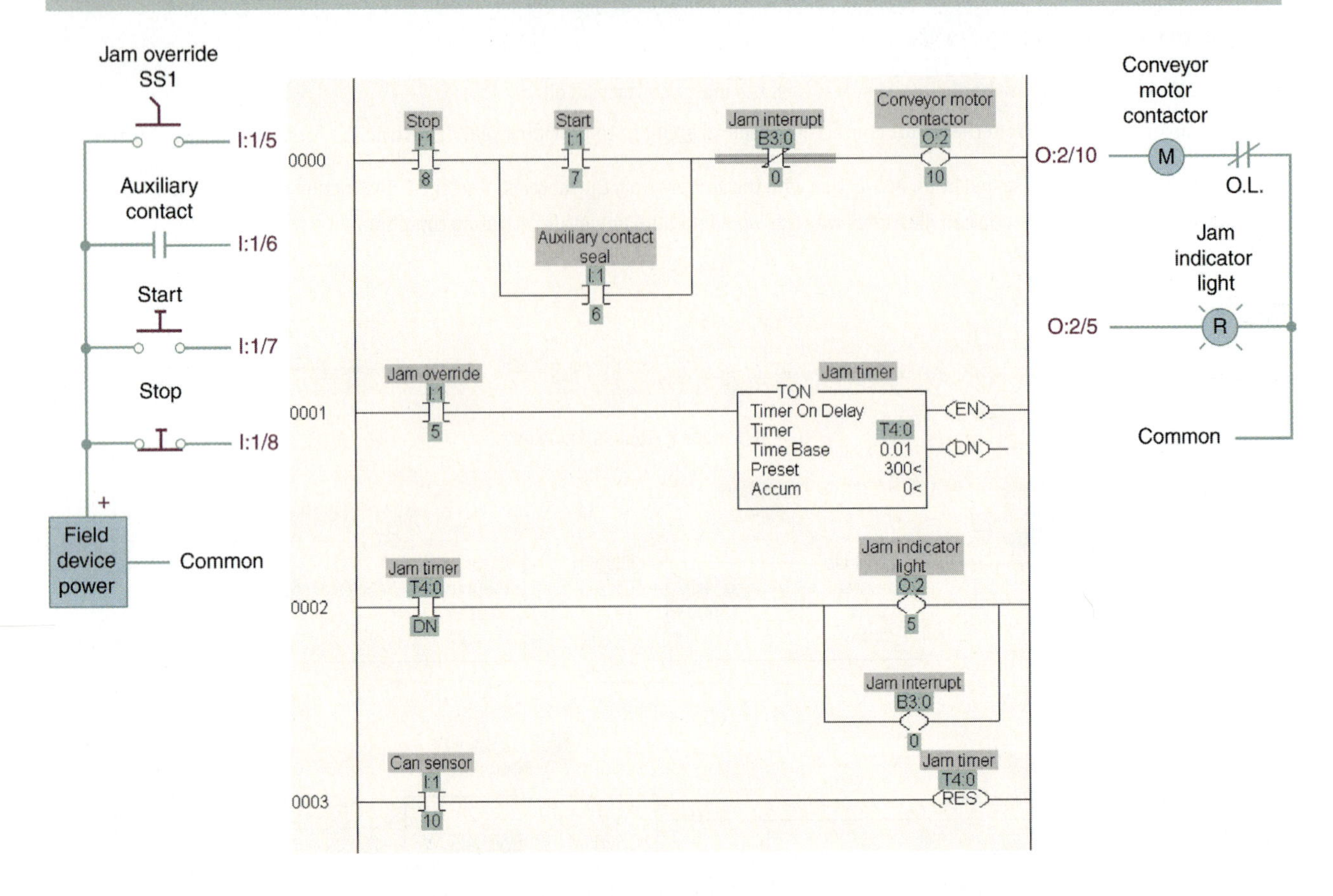

Assume that the sealing contact for the start I:1/7 is addressed to the O:2/10 output, and the conveyor is started by pressing the start PB. A jam on the conveyor causes the conveyor motor overload contacts to open the motor contactor field circuit, and the conveyor motor turns off. If the start switch is sealed by the O:2/10 address, then the O:2/10 output remains true. The field device contactor coil continues to have power applied by the ladder rung, but it is open in the field wiring. When the jam is removed and the overloads reset, the conveyor will start moving immediately and that could present a danger of injury to maintenance or machine operators.

The unsafe condition is removed by using an NO auxiliary contact from the motor contactor as an input field device, and assigning this address to the start PB sealing instruction. If the contactor overload contacts trip, the motor contactor turns off and the auxiliary contacts open. This forces output O:2/10 to a false state. Thus when the jam is removed and the overloads are reset, the conveyor does not restart until the start PB is again depressed.

Four rungs are used to detect a conveyor jam, as shown in Figure 5-15. Rung 0 is a start/stop rung for the conveyor motor. This was always present, but it was not shown in the solution in Figure 5-13 because it did not play a part in that solution. Here it is used to control the conveyor contactor when the system is started with the start push button and also when the conveyor is halted by a jam. Rung 1 has a TON timer that is enabled by an NO start selector switch used to activate this jam detection logic. The timer has a 3-second preset value and is enabled by input I:1/5. If the timer is enabled for more than 3 seconds, then the done bit is true and rung 2 lights the jam indicator and turns off the conveyor motor. However, without a jam present, the can sensor, I:1/10, is true every 1.25 seconds and the can sensor instruction in rung 3 resets the timer at that time interval. This resets the timer accumulator to zero, and the timing process restarts.

If a jam is detected, the stack light output and the virtual relay B3:0/0 are true. The XIO instruction for the B3:0/0 bit makes the motor contactor output false and drops out the start circuit. The system will not restart until the start PB is pressed after the jam is cleared. Review the ladder logic and this description until you understand how the timer in this automation system acts as a *watch dog* timer.

5-4-5 Allen-Bradley Down Counters

The function of the *down counter* is to decrement its accumulated value on the false to true transitions of the counter rung. The operation of the down counter is identical to that of the up counter except that the value in the accumulator is decremented until the preset value is reached and the output becomes true. A down counter could be used to count the number of cans dispensed from a soft drink machine; when the machine is empty the counter could turn on a sold out light. However, an up counter can be used in this application as well. The CTU would count every can sold with the preset number set to the maximum number of cans that the machine can hold. When the accumulator reaches the preset value the sold out light is turned on. In fact, a down counter is seldom used by itself; it's generally used with an up counter to form an up/down counter. The conditions to set and reset the output bits of a down counter are shown as a truth table in Table 5-1(b), and the timing diagram for the down counter is illustrated in Figure 5-16. Before you continue, review the table and timing diagram until you are familiar with the operation of the down counter output bits.

5-4-6 Allen-Bradley Up/Down Counters

Figure 5-17 illustrates the up/down counter timing diagram. The up/down counter in the figure provides an output as long as the count is equal to or greater than 4. The preset value is 4, and

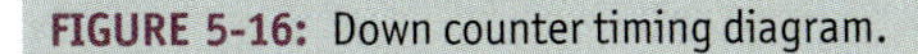

FIGURE 5-16: Down counter timing diagram.

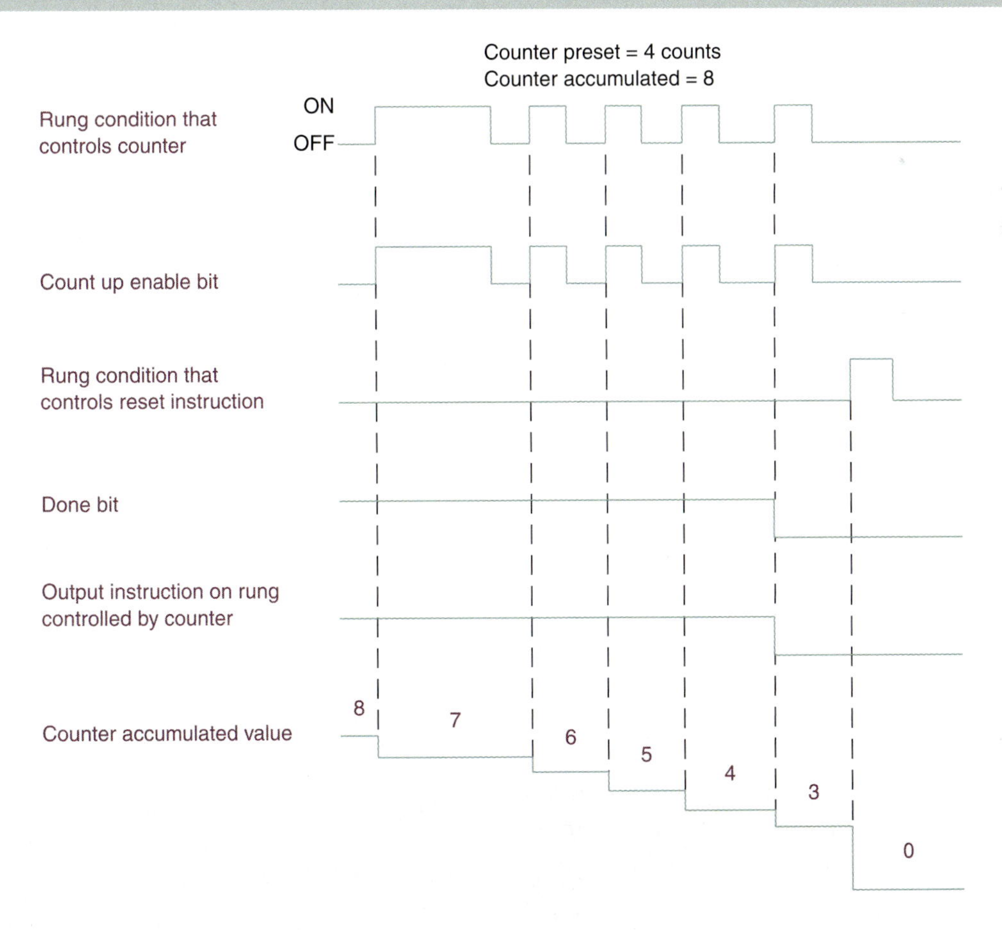

FIGURE 5-17: Up/down counter timing diagram.

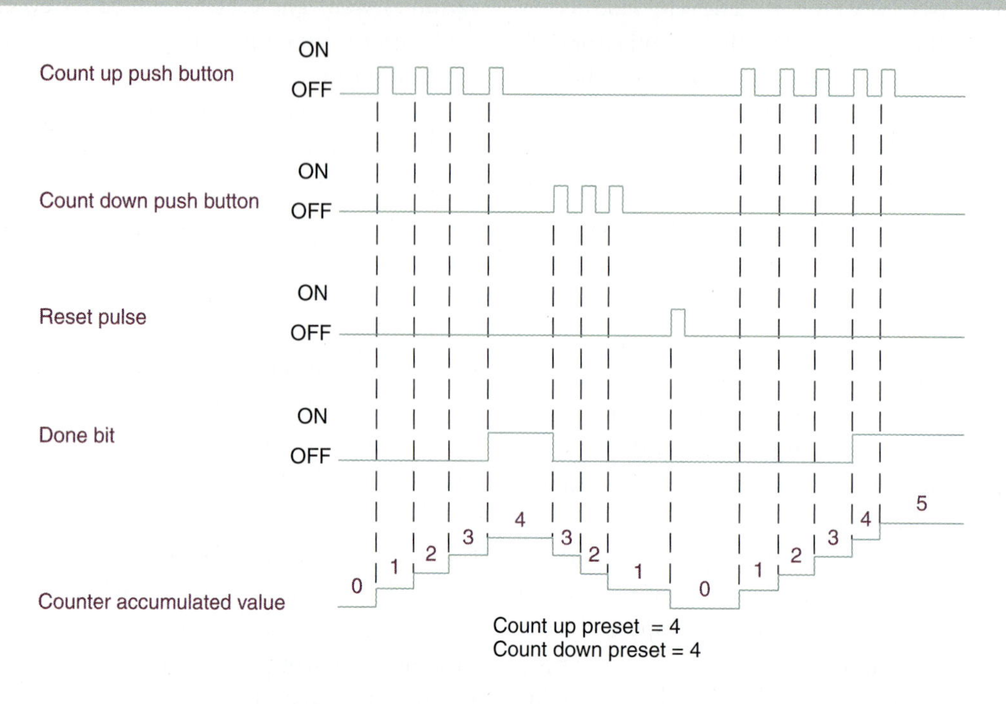

EXAMPLE 5-4

Draw the ladder diagram using counters to count the number of cars entering and leaving an airport parking garage. The parking garage can hold 308 cars; when it is full a red light is illuminated.

Solution

Refer to Figure 5-18 as you read the following description.

When a car enters the garage, the limit switch on the entrance gate, LS1, is activated so rung 0 is true. The up counter increments when the rung cycles from false to true. In similar fashion, when a car exits the garage, the exit gate limit switch, LS2, is activated, and a similar false to true transition of rung 1 decrements the down counter by 1. Because the up counter and the down counter are the same counter number (C5:0), they are at the same program address and the accumulated value is the same for both. When the accumulated value reaches 308, the done bit C5:0/DN turns on the red full lot light. No reset rung is provided because the lot always has some cars present, so the ACC value will not be zero. Once a month management does an inventory late at night and uses the counter data dialog box, Figure 5-8(b), to update the ACC counter.

the accumulated value starts at 0. The count up pulses increment the counter by 1, the count down pulses decrement the counter by 1, and the reset returns the counter to 0. Note that the output is true when the count is greater than or equals 4 and false when the count is 3 or less.

5-4-7 Allen-Bradley One-Shot Instructions

One-shot instructions are used with numerous PLC instructions. They are used with counter reset instructions when the counter must start a new count before the reset input contacts return to their normal state. One-shot instructions are available on most PLC systems, but can be built using standard ladder logic on all PLCs. One-shot operation is defined as follows:

> *One-shot instructions make the rung output active for one scan even if the input instruction triggering it remains active for multiple scans.*

FIGURE 5-18: Ladder diagram for parking lot counter—Example 5-4.

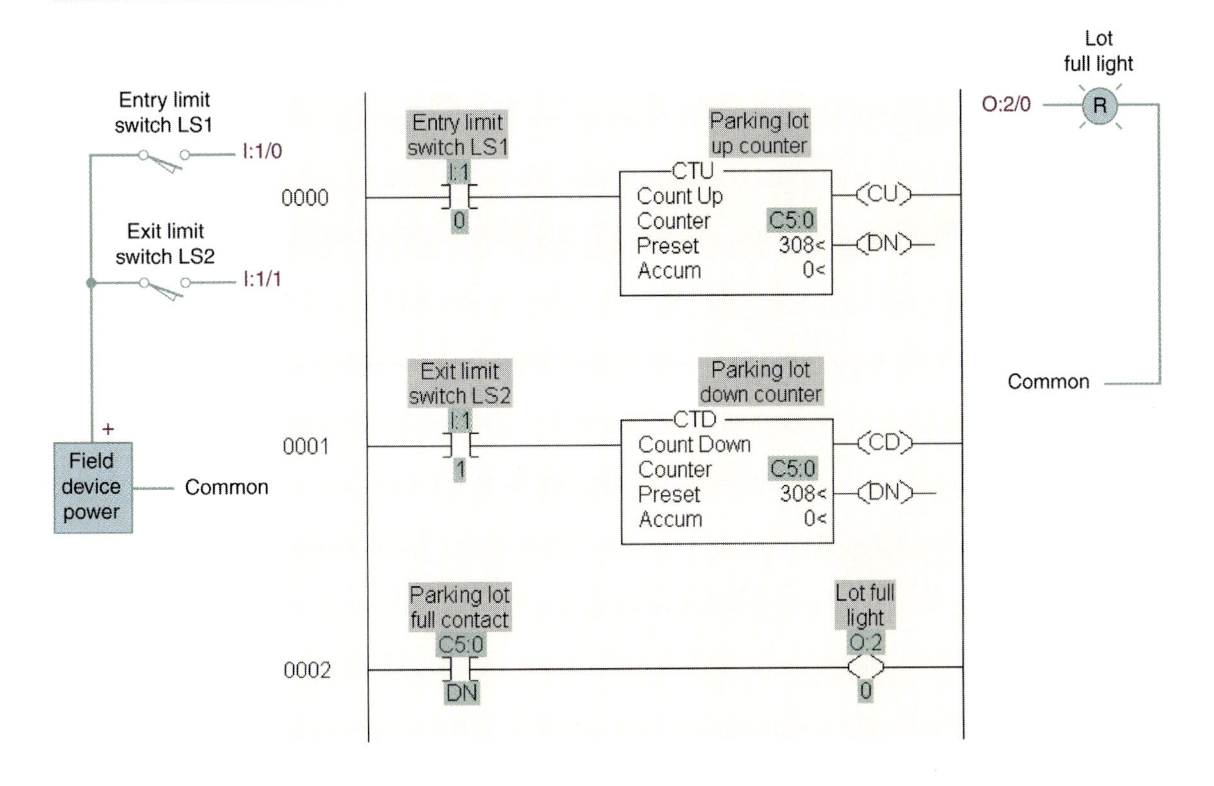

The three processors, PLC 5, SLC 500, and ControlLogix, use similar but not identical one-shot instructions. The standard ladder logic rungs for all Allen-Bradley's one-shot instructions are illustrated in Figures 5-19 (PLC 5), 5-21 (SLC 500), and 5-22 (ControlLogix). The timing diagrams in Figure 5-20 illustrate the pulse action for one-shot instructions. Read the descriptions for these example ladder rungs and study the timing diagrams before continuing.

PLC 5 ONS One-Shot instructions. Allen-Bradley has three one-shot instructions for the PLC 5 processor that are illustrated in Figure 5-19. The ONS one-shot instruction is illustrated in Figure 5-19(a), and this instruction is available on all PLC 5 processor models. The timing diagram for this one-shot instruction is shown in Figure 5-20(a). Note that the instruction format starts with standard input logic instruction(s), which is followed by the ONS instruction symbol and then the output instruction. The input can be any valid input instruction for the PLC 5, and the output instruction can be any valid output instruction for the processor. The one-shot instruction is triggered by a false to true change in the input logic, so it is an input rising instruction. Remember it is a rising edge from the rung logic that triggers the ONS instruction regardless of the type of change that occurred in the field device. The address for the ONS symbol is either a binary or integer bit address, for example, B3:0/3 or N7:10/8. This address stores the previous value of the ONS instruction and cannot be used in any other ladder rung. The ONS instruction, which must be located on the rung directly next to the output instruction, makes the output true for only one scan regardless of how long the input logic is true.

PLC 5 OSR and OSF one-shot instructions. The OSR is a one-shot rising and the OSF is a one-shot falling instruction. Both are output instructions placed on the right end of the rung and operate in similar fashion. The OSR is triggered by a false to true change (rising or leading edge) in the input logic and the OSF by

FIGURE 5-19: Standard ladder logic rungs for PLC 5 one-shot instructions.

Application	Standard Ladder Logic Rungs for PLC 5 One Shot Instructions
Turn on an output instruction or a bit for one scan with either a rising or falling edge of an input instruction. **ONS** The ONS, one shot instruction, is used to make an output instruction (RES) true for one scan when the input logic on the rung goes from false to true. The ONS must be assigned a Boolean bit (B3:0/39) that is not used anywhere else in the program. Also, the ONS instruction is the final instruction for the input logic and cannot be in an OR branch. **OSR** The OSR, one shot rising instruction, is an output instruction (O:001) used to make an output bit true for one scan. The OSR is triggered by a rising edge on the ladder rung input logic (I:001/1) as illustrated in Figure 5-20(a). The OSR must be assigned a Boolean bit (word 17 and bit 7) that is not used anywhere else in the program. **OSF** The OSF, one shot falling instruction, is an output instruction used to make an output bit (O:002) true for one scan. The OSF is triggered by a falling edge on the ladder rung input logic (I:001/2) as illustrated in Figure 5-20(b). The OSF must be assigned a Boolean bit (word 18 and bit 4) that is not used anywhere else in the program.	Part sensor I:001 0 — B3 [ONS] 39 — Counter reset C5:0 (RES) **(a)** Limit switch I:001 1 — OSR One Shot Rising; Bit Address B3/17; Source Bit 7; Dest O:001 — (OB) (SB) **(b)** Heat switch I:001 2 — OSF One Shot Falling; Bit Address B3/18; Source Bit 4; Dest O:002 — (OB) (SB) **(c)**

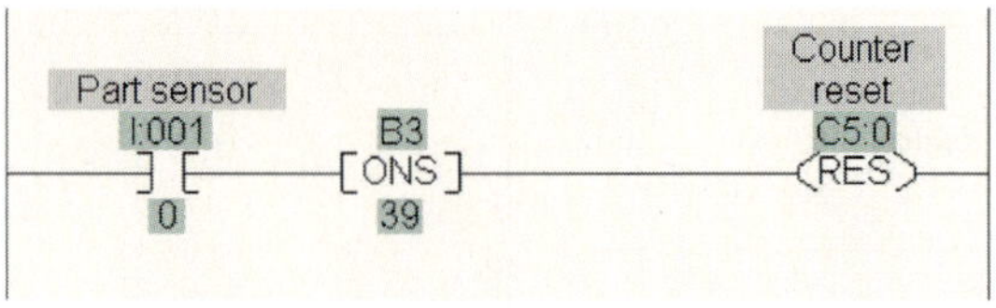

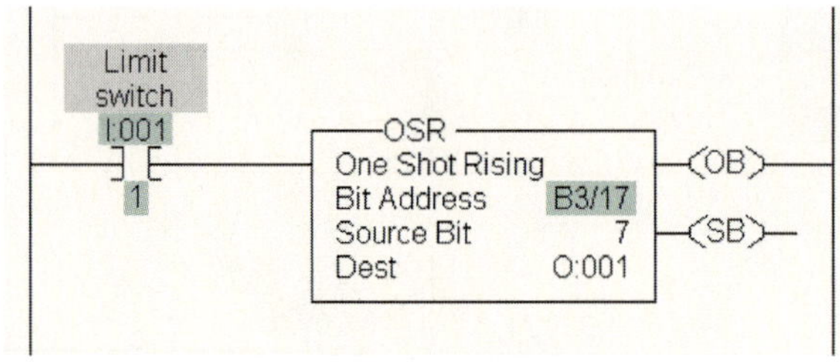

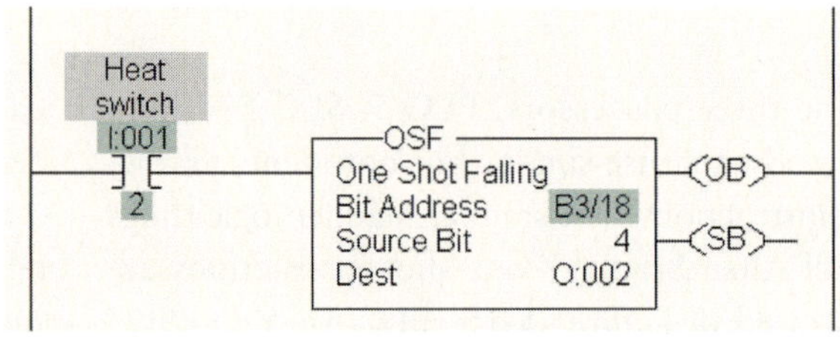

FIGURE 5-20: One-shot timing diagrams.

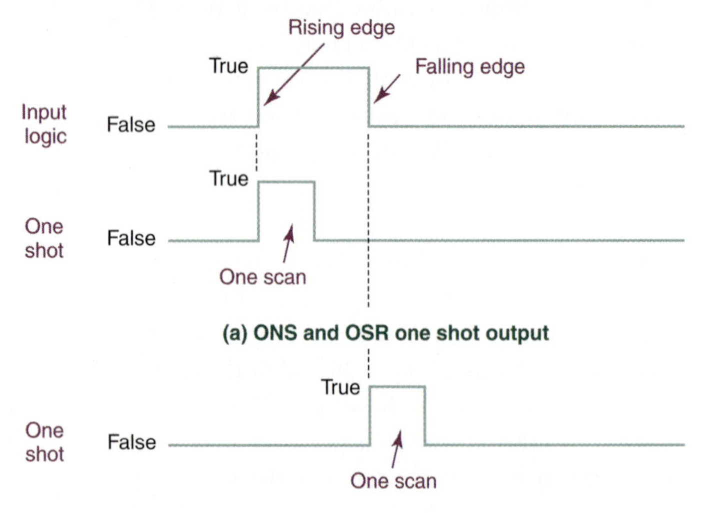

a true to false change (falling or trailing edge). These commands are available only on the enhanced model of the PLC 5. The standard ladder implementation of this instruction is illustrated in Figure 5-19(b) and (c). Note that the three parameters entered in the instruction are:

Bit Address: The bit address used to store the status of the one-shot instruction, such as B3/17.

Source Bit: The bit address of the output word address that will be true for one scan, such as bit7 in word O:001.

Dest: The destination or output word, such as O:001, that has the bit held true for one scan.

FIGURE 5-21: Standard ladder logic rungs for SLC 500 one-shot instructions.

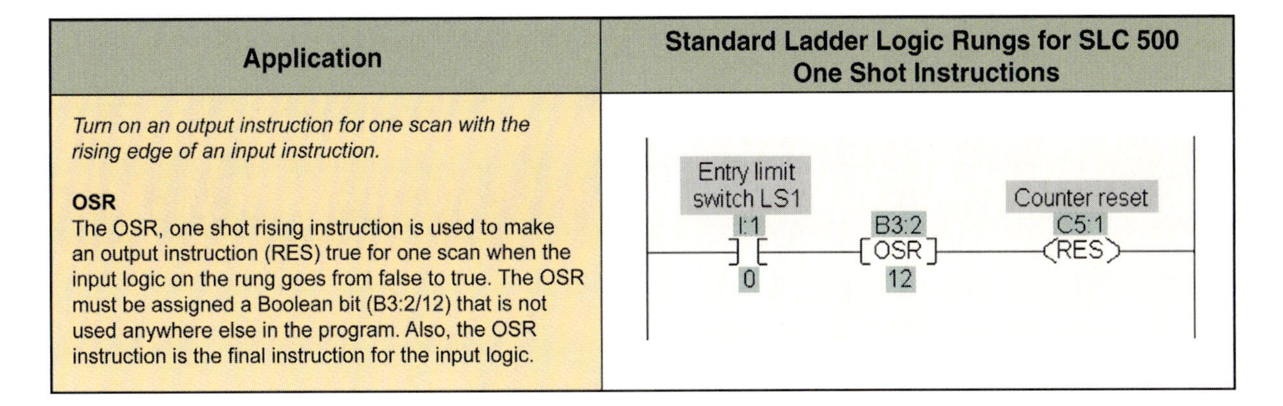

Application	Standard Ladder Logic Rungs for SLC 500 One Shot Instructions
Turn on an output instruction for one scan with the rising edge of an input instruction. **OSR** The OSR, one shot rising instruction is used to make an output instruction (RES) true for one scan when the input logic on the rung goes from false to true. The OSR must be assigned a Boolean bit (B3:2/12) that is not used anywhere else in the program. Also, the OSR instruction is the final instruction for the input logic.	Entry limit switch LS1 I:1 0 — B3:2 OSR 12 — Counter reset C5:1 RES

The timing diagrams for the OSR and OSF instructions are shown in Figure 5-20(a) and (b), respectively. The reset instruction used with timers, counters, and control instructions is illustrated in Figure 5-19(a).

SLC 500 OSR one-shot instructions. The OSR instruction is illustrated in Figure 5-21, and the timing diagram is displayed in Figure 5-20(a). This instruction is available on all SLC processor models, and the instruction format is identical to the ONS for the PLC 5. The input can be any valid input instruction for the SLC PLC, and the output instruction can be any valid output instruction for the processor. The one-shot instruction is triggered by a false to true change in the input logic, so it is an input rising instruction. The address for the OSR symbol is either a binary or integer bit; for example, B3:2/3 or N7:0/8. This bit stores the previous value of the OSR instruction and cannot be used in any other ladder rung. The OSR makes the output instruction true for one scan independent of the number of scans that the input is true.

ControlLogix ONS, OSR, and OSF one-shot instructions. The ControlLogix standard circuits are illustrated in Figure 5-22 and the timing diagrams are like those shown in Figure 5-20. The one-shot instructions in ControlLogix have the same ladder format as those in the PLC 5 with the exception of the addressing. The one-shot instructions in ControlLogix use tag names for all instruction and parameter addressing. In addition, the output bit uses a tag name in place of the output bit number and word number used in the PLC 5.

One-Shot applications with counters. One-shot instructions are often required in counter applications. An example describes this best, so review the counter ladder example in Figure 5-23. Assume that the counter is counting parts entering a production queue using a part detection sensor. When the queue has four parts present, the counter done bit, C5:1/DN, stops the part flow into the queue and triggers a machine loader that puts the four parts into a production machine. The machine then makes I:1/1 active to reset the counter for the next four-count sequence. This parts loaded signal from the machine keeps the reset instruction true for 5 seconds, and during that time parts continue to flow into the queue. However, the counter accumulator is held at 0 because the reset instruction is active for 5 seconds. If a new part enters the queue and activates the part detection sensor, I:1/0, while the machine reset instruction is true, then the part is not counted.

The problem is solved with the modified reset rung in Figure 5-24. Now the RES instruction is active for only one scan, which resets the counter in milliseconds. Even though I:1/1 is true for 5 seconds, the C5:1 can start counting one scan after the reset logic is active. Study this example of a one-shot application until it is clear how the ONS instruction is used.

FIGURE 5-22: Standard ladder logic rungs for ControlLogix one-shot instructions.

Application	Standard Ladder Logic Rungs for ControlLogix One Shot Instructions
Turn on an output instruction or a bit for one scan with either a rising or falling edge of an input instruction. **ONS** The ONS, one shot instruction, is used to make an output instruction (Shift_count) true for one scan when the input logic on the rung goes from false to true. The ONS must be assigned a tag (One_shot) of the Boolean data type that is not used anywhere else in the program. Also, the ONS instruction is the final instruction for the input logic. **OSR** The OSR, one shot rising instruction, is an output instruction used to make an output tag (Output_1) with a Boolean data type true for one scan. The OSR is triggered by a rising edge on the ladder rung input logic as illustrated in Figure 5-20(a). The OSR must be assigned a tag (Storage_bit_1) of the Boolean data type that is not used anywhere else in the program. **OSF** The OSF, one shot falling instruction, is an output instruction used to make an output tag (Output_2) with a Boolean data type true for one scan. The OSF is triggered by a falling edge on the ladder rung input logic as illustrated in Figure 5-20(b). The OSF must be assigned a tag (Storage_bit_2) of the Boolean data type that is not used anywhere else in the program.	Can sensor Can_switch <Local:1:I.Data.1> One_shot ONS Total packaging counter reset Shift_count RES **(a)** Can sensor Can_switch <Local:1:I.Data.1> OSR One Shot Rising Storage Bit Storage_bit_1 Output Bit Output_1 OB SB **(b)** Can sensor Can_switch <Local:1:I.Data.1> OSF One Shot Falling Storage Bit Storage_bit_2 Output Bit Output_2 OB SB **(c)**

FIGURE 5-23: Machine queue parts counter.

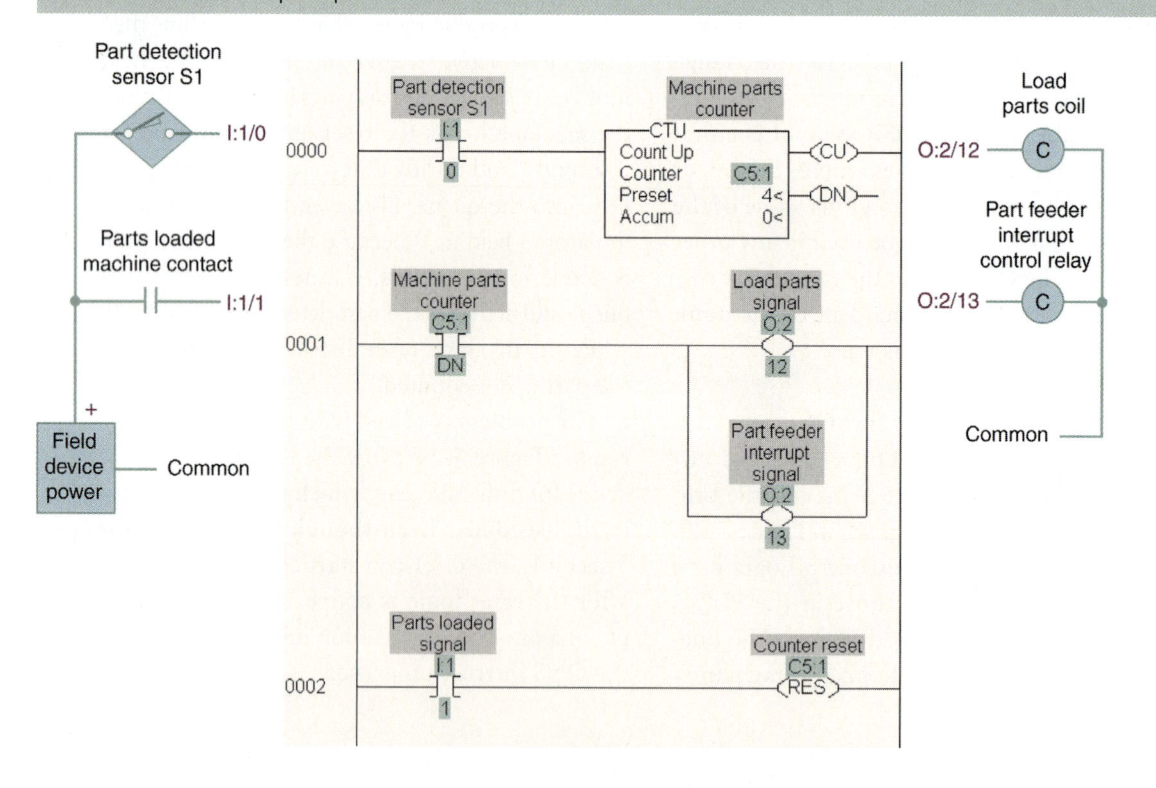

5-5 CASCADED COUNTERS

Some applications require the counting of events where the total number of events exceeds the maximum number allowable per counter. SLC 500 up counters, for example, have a maximum count of 32,767. *Cascaded counters* are used to extend the count to a value high than the individual counter maximum. As in cascaded timers, the output of one counter is the input to another. Figure 5-25 depicts a ladder diagram that illustrates cascaded counters, where a red light is turned on after 60,000 counts.

The first rung has C5:1 counting the input pulses generated by I:1/0. After C5:1 counts to 30,000, the C5:1 done bit goes true and performs two functions: it makes the rung for the C5:1 counter false so C5:1.ACC remains at 30,000, and it connects the second counter, C5:2, to the input instruction I:1/0. After C5:2 has counted to 30,000, the C5:2 done bit goes true. When the C5:1 and the C5:2 done bits are both true, the red light is on. The C5:2 ACC remains at 30,000 because C5:2/DN on the XIO instruction in the counter rung makes the counter instruction false.

5-6 EMPIRICAL DESIGN PROCESS WITH PLC COUNTERS

The empirical design process, introduced in Section 3-11-4, is an organized approach to the design of PLC ladder logic programs. However, the term *empirical* implies that some degree of trial and error is present. This troublesome aspect of the

FIGURE 5-24: Modified ladder rung to fix timing problem on the machine queue parts counter.

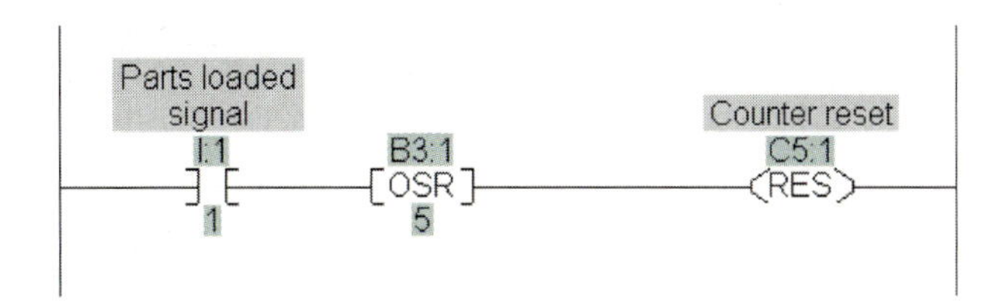

empirical process should become more obvious when counters are added into the design process.

5-6-1 Adding Counters to the Process

The first step in using counters in PLC ladder design is to know the addressing and operation of the up, down, and up/down counters and all the standard counter circuits illustrated in Figure 5-10. In addition, the operation of the reset instruction, RES, must be clear and the operation of the one-shot instructions for the various processes must be learned. Stop now and review all of these if necessary.

When a counter is added to a ladder it affects three rungs: one rung to make the counter command (CTU or CTD) active, a second rung that uses a counter output (CU, CD, OV, UN, or DN) to control a system parameter (in most counter applications the DN bit is the only output bit used), and a third rung to reset the counter accumulator. The complete empirical process is listed in Section 3-11-4; the modifications for counters are as follows:

FIGURE 5-25: Cascade counters.

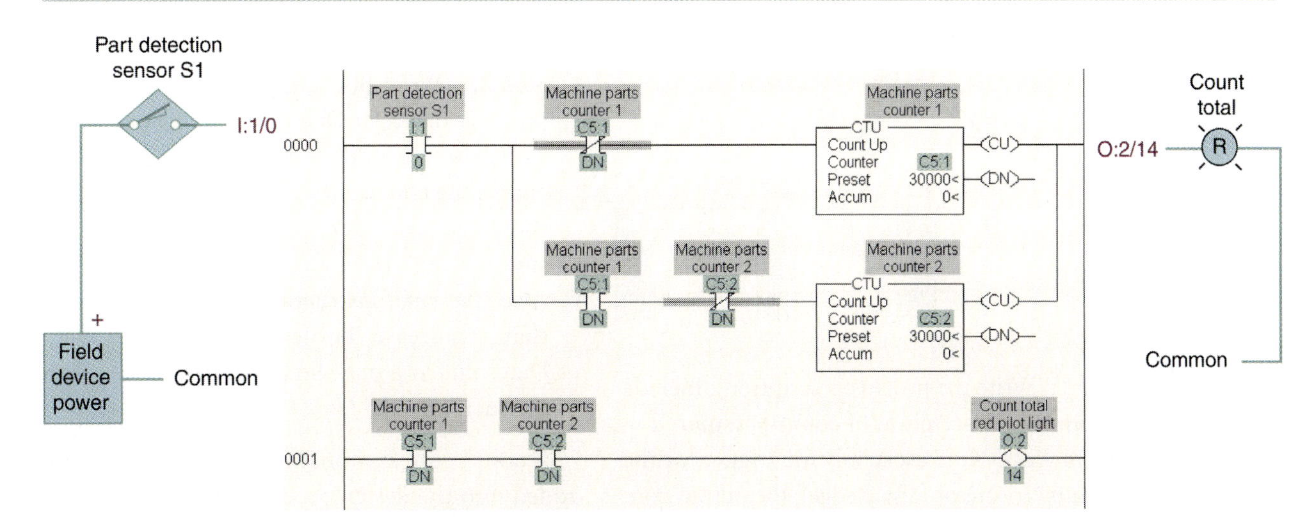

EXAMPLE 5-5

A conveyor system, illustrated in Figure 5-26, sorts boxes so that each chute receives 10 boxes. The operation is as follows:

- Gates 1 and 2 are up, sensor S1 counts 10 boxes for chute 1, and then gate 1 drops.
- Sensor S2 counts 10 boxes for chute 2 and then gate 2 drops.
- Sensor S3 counts 10 boxes for chute 3 and then gates 1 and 2 are raised and the process starts over.

With average conveyor speed, it takes 4 seconds for the boxes to enter the chute after the sensor detects them. Draw the PLC 5 ladder logic necessary for this control problem.

SOLUTION

Refer to Figure 5-26 as you read the empirical design process used. The empirical design is as follows:

- The operation of the system is described in the problem statement.
- The system inputs are sensors 1, 2, and 3 and the outputs are C5:0 (chute 1), C5:1 (chute 2), C5:2 (chute 3), gate 1 actuator, and gate 2 actuator.

Before the equations are written, study the standard ladder logic options for the PLC 5 system counters shown in Figure 5-10 The solution in Figure 5-10(b) is selected because the counter done bit remains active and can be used to control the gates.

The logic equations for the outputs are:

- C5:0 count input = sensor 1 AND NOT C5:0/DN
- C5:1 count input = sensor 2 AND C5:0/DN AND NOT C5:1/DN
- C5:2 count input = sensor 3 AND C5:1/DN AND NOT C5:2/DN
- T4:0 timer input = C5:2/DN AND NOT T4:0/DN
- RES = T4:0/DN (timer 0 has a preset value of 6 seconds). Note that the reset instruction resets all counters. The timer is reset by its own done bit.
- Gate 1 = C5:0/DN (counter 0 has a preset value of 10)
- Gate 2 = C5:1/DN (counter 1 has a preset value of 10)

The sensor and actuator wiring plus ladder logic solution is illustrated in Figure 5-27.

The timer is used to permit the tenth box to reach chute 3 before the system is reset and all gates are raised and the process is restarted.

FIGURE 5-26: Conveyor distribution system.

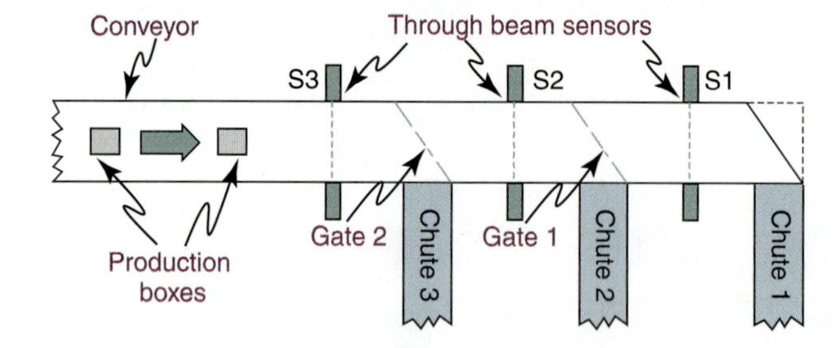

Step 1: (Write the process description): Include a complete description of count(s) required in the process. Note especially the trigger for the count(s) (rising or falling edge), the output controlled by the counter DN bit, and how the counter is reset.

Step 2: (Write Boolean equations for all output field devices): One of the Boolean expressions should indicate the logic necessary to enable the counter(s). The counter output bits should be added into the other Boolean expressions as required by the process description. Also, the reset instruction, RES, is an output that requires a Boolean equation as well. Determine if a one-shot is needed for the RES instruction.

Example 5-5 demonstrates how counters are added into the design process.

FIGURE 5-27: Box sorter ladder logic—Example 5-5.

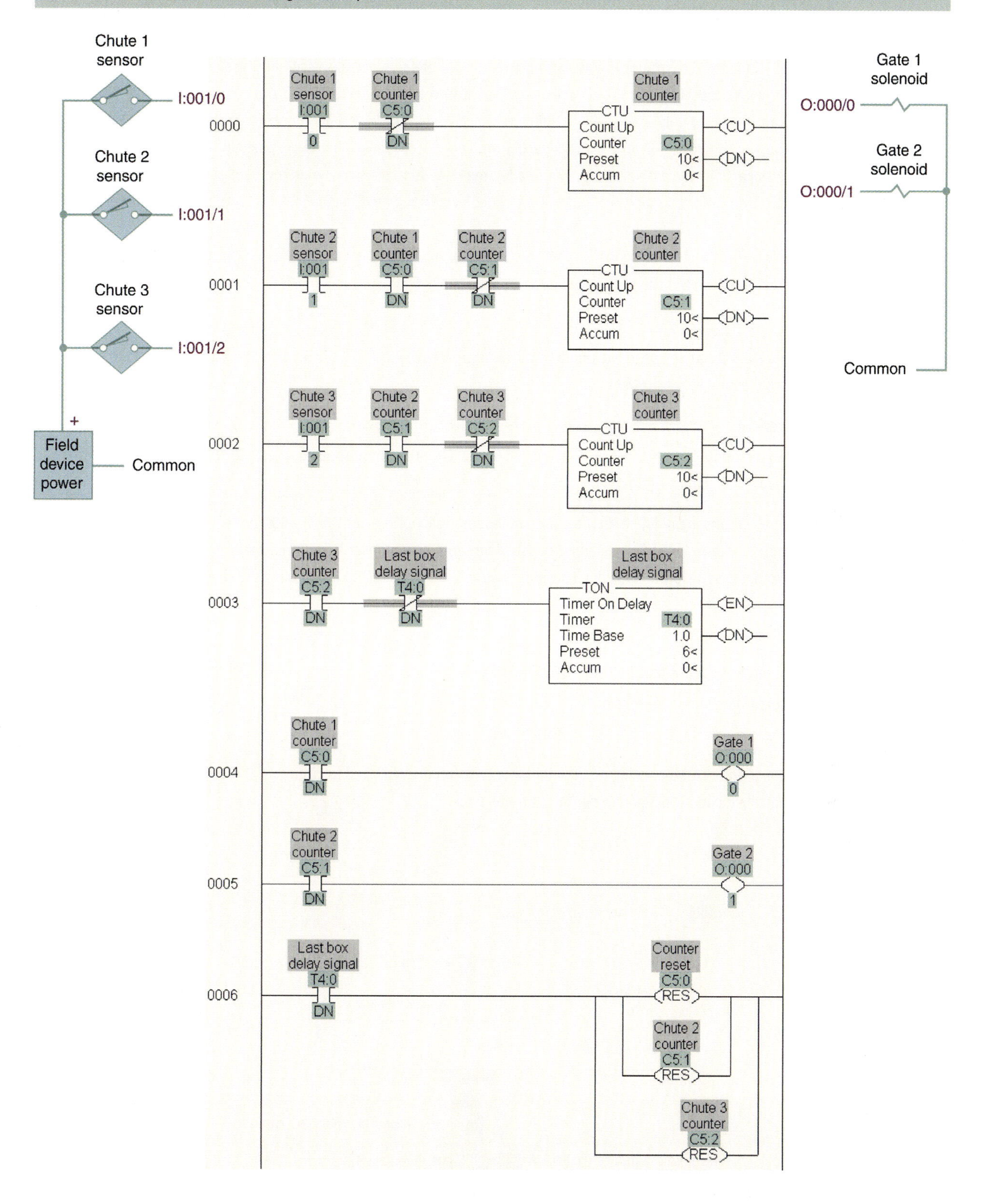

EXAMPLE 5-6

The assembly system in Figure 5-28 is typical of many used in automation systems. This system will be used for program design in several chapters. The design for two control elements is addressed in this example.

The pneumatic ejector pushes parts off the conveyor into a four-part queue. If the queue is full, then parts are passed to another assembly station down the conveyor. Sensor 1 detects a part for this assembly machine and triggers the ejection process. Sensor 2 detects parts in the hold area, sensor 3 detects parts in the stop area, and sensor 4 detects parts in the assembly machine ready area. The stop actuator holds the next part to be released, and the hold actuator holds the top three parts in the queue when the next part is released. The assembly system takes the part from the ready area and adds it to the product.

Design a control system using SLC 500 instructions that manages the queue by ejecting a part if the queue is less than 4 and passing parts if the queue is equal to 4. Draw the SLC 500 ladder logic necessary for this control problem.

SOLUTION

Refer to Figure 5-29 as you read the empirical design process used. The empirical design for managing the ejector and assembly queue is as follows:

- The operation of the system is described in the problem statement.
- The system inputs are sensors 1 and 4, and the outputs are C5:0 (count up counter), C5:0 (count down counter), and pneumatic actuator (ejector).

Before the equations are read, study the standard ladder logic options for the SLC 500 system counters, Figure 5-10. The solution in Figure 5-10(d) is selected because an up/down counter is required for this application. Also, the timing diagram for this problem is illustrated in Figure 5-17. Note that the done bit is active went the counter ACC is at 4.

The logic equations for the outputs are:

- C5:0 count up input = sensor 1 AND NOT C5:0/DN
- C5:0 count down input = sensor 4
- Pneumatic ejector = sensor 1 AND NOT C5:0/DN

The sensor and actuator wiring and ladder logic solution is illustrated in Figure 5-29. Review the last two examples so that the design process for counters is understood.

FIGURE 5-28: Assembly system part queue control.

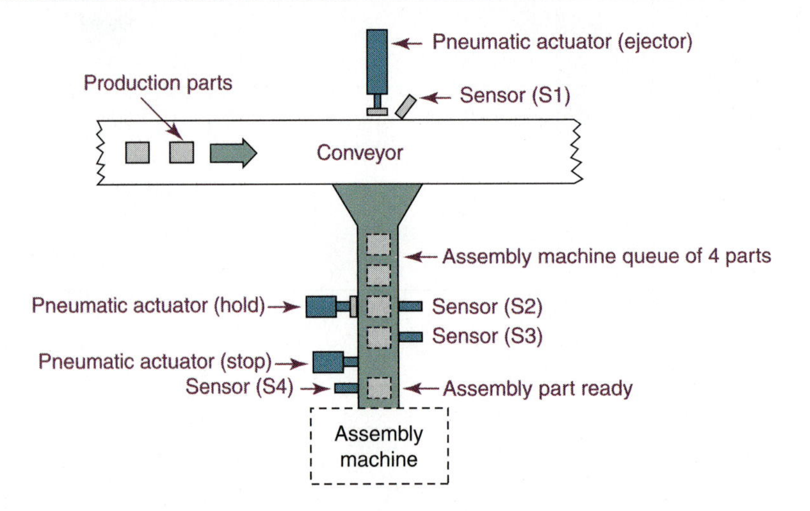

FIGURE 5-29: Queue control ladder logic—Example 5-6.

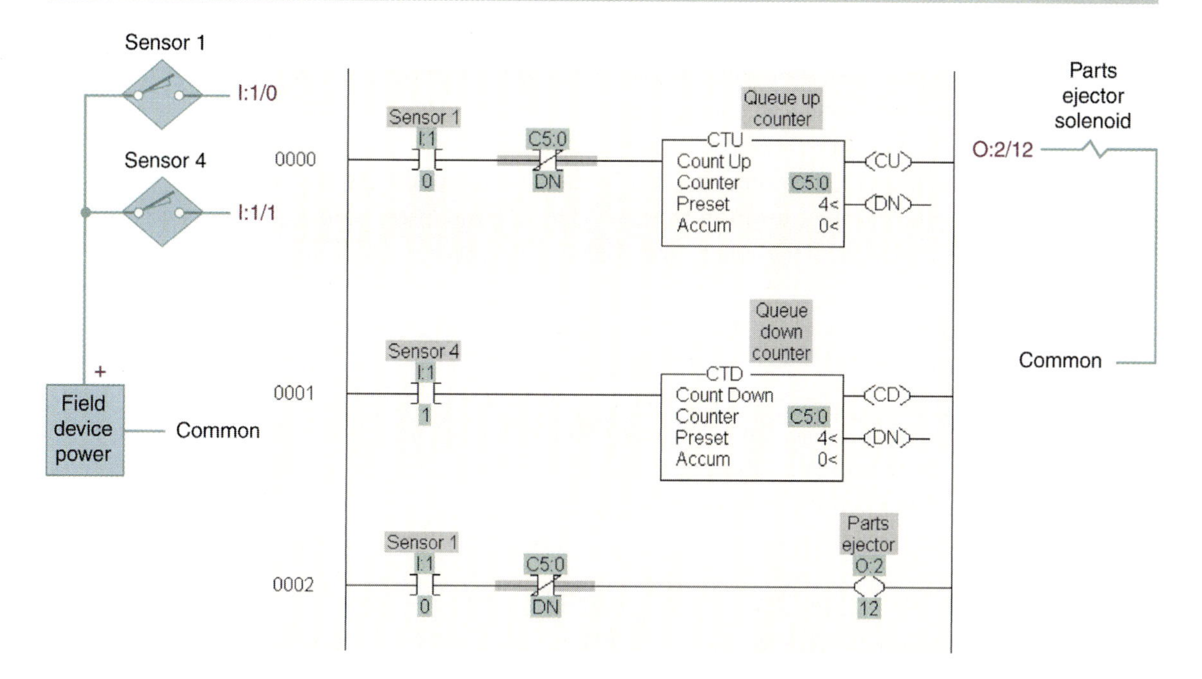

5-7 CONVERSION OF RELAY LOGIC COUNTER LADDERS TO PLC LOGIC

Conversion of relay ladder logic with contacts and timers to an equivalent PLC ladder solution was introduced in Chapters 3 and 4. When mechanical or electronic counters are present in the relay ladders, they must be converted as well. The following conversion rules for counters are appended to the initial rule set in Section 3-11-5.

1. Replace the mechanical or electronic counter operated as an up counter with the PLC up counter.
2. Replace the mechanical or electronic counter operated as a down counter with the PLC down counter.
3. Set the preset value so that the PLC instruction turns on the done bit at the same count value used in the mechanical or electronic relay ladder device.
4. Replace the mechanical counter contacts and electronic counter solid-state output with the PLC equivalent. For example, NO contacts in relay logic are replaced with an XIC instruction in the PLC output rung. If NC contacts are present use an XIO instruction. The controlled device should be the same, but care must be taken to match the current sinking versus sourcing needs of the field device to the opposite specifications on the output module.

The conversion of input field devices—switches and sensors—plus output field devices—actuators and contactors—follows the guidelines specified in Section 3-11-5.

5-8 TROUBLESHOOTING COUNTER LADDER LOGIC

The most difficult counter programs to verify are those with multiple cascaded counters and one or more reset instructions. Use the following suggestions for troubleshooting counters.

- Test the counters starting with the first in the sequence, and then add one counter at a time until the total sequence is operational. Use the temporary end instruction, TND, described in Chapter 4.
- If the reset instructions are present, determine if all necessary process executions driven by

counter bits are performed before the counter is reset.

- Use the suspend instruction, SUS, to verify status values for all registers and bits at critical points in the ladder. The SUS instruction is described in the next subsection.
- If the count is inconsistent, then verify that the period (from a false to true transition to the next false to true transition) of the counter logic transition is not shorter than the scan time.
- Be aware of situations where counter values are used to update internal PLC memory bits and inhibit process actions, since the scan time and internal update time is much faster than most process events. The counter problem described in Section 5-8-2 illustrates this type of programming situation.
- Use the counter dialog boxes like those illustrated in Figure 5-8(b) for PLC 5 and SLC 500 and in Figure 5-9(c) for the ControlLogix counter to track changes in counter parameters and tags.

5-8-1 Suspend Instruction

The suspend instruction, SUS, is used to trap and identify specific conditions during system troubleshooting and program debugging. A program can have multiple suspend instructions, each controlled by a different input instruction address. The SUS is added to the cascade counter example in Figure 5-25 and redrawn in Figure 5-30. When true, this output instruction places the controller in the suspend or idle mode. The suspend ID, number 100 in the figure, must be

FIGURE 5-30: Ladder logic with SUS instruction.

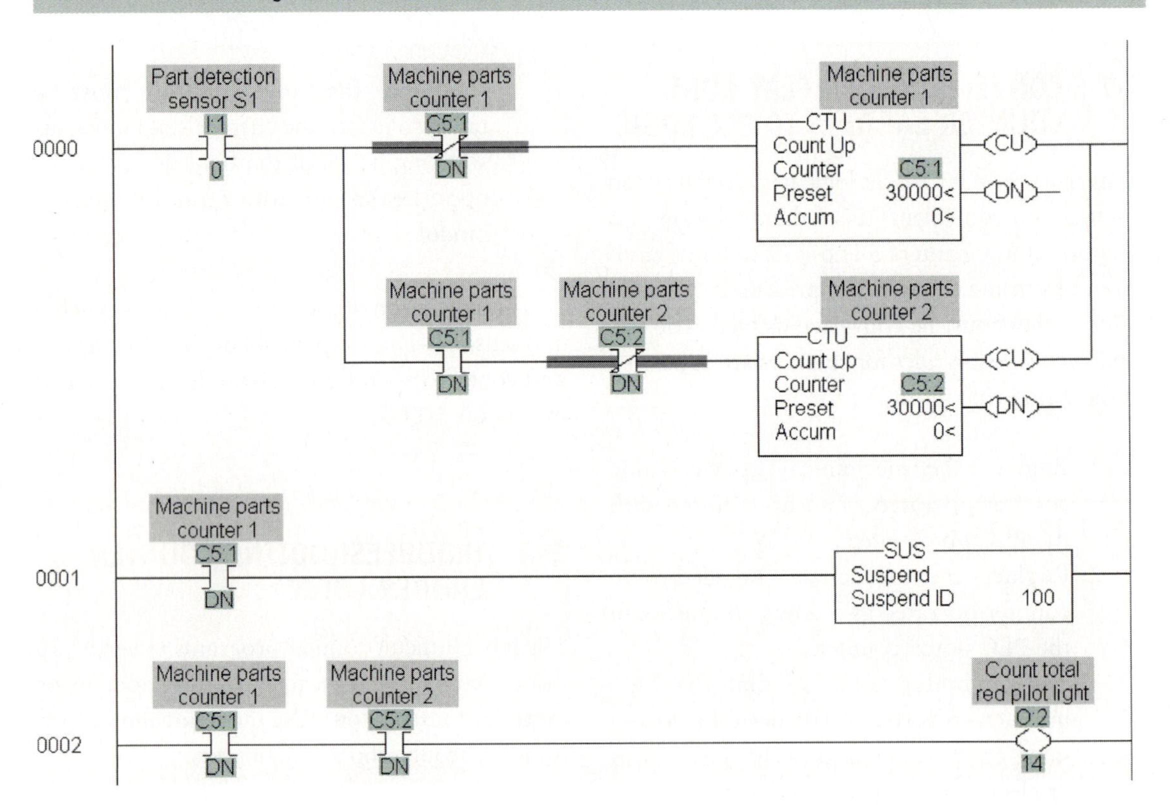

selected by the programmer and entered in the instruction. When the SUS instruction executes, the ID number is written in word 7 (S:7) of the status file. If multiple suspend instructions are present, then this will indicate which SUS instruction was active. The suspend file (program or subroutine number identifying where the executed SUS instruction resides) is placed in word 8 (S:8) of the status file. All ladder logic outputs are de-energized, but other status files have the data present when the suspend instruction is executed.

5-8-2 Process Speed versus Scan Time

Problems often occur where counter values are used to inhibit process actions because of the relative speed of the scan is usually much faster than the process event. The ladder logic from Example 5-6 in Figure 5-29 illustrates this situation. The following problem was observed as the program in Figure 5-29 was executing.

> *When the queue is at three parts and a fourth part arrives at ejector sensor 1, the count in the queue increases to 4 but the fourth part is not ejected into the queue.*

Study the following two equations for the control and see if you can identify the problem.

C5:0 count up AND NOT C5:0/DN
Pneumatic AND NOT C5:0/DN

Note that both of those outputs (counter C5:0 and ejector) are triggered by the same logic and that a feedback is present since one of the input conditions, NOT C5:0/DN, is generated by counter C5:0, the output for that input logic. Also, the time required for incrementing C5:0 is the scan rate, which is relatively fast (10 to 100 msec), and the time required for the ejector to extend and return is relatively long (>1500 msec) because it is a pneumatic system.

In the ladder solution the up counter is incremented from 3 to 4 when the fourth part makes sensor 1 active. Thus the queue indicates that four parts have arrived, and the C5:0/DN is now true. The pneumatic ejector output, O:2/12, is also true when the fourth part arrives at sensor 1. However, within one scan (10 to 100 seconds) the C5:0/DN bit is true, making the XIC instruction with the C5:0/DN address false so power flow to the pneumatic ejector output is present for only one scan. As a result, the ejector never moves because its response time is much larger. The solution is to add a timer using the standard ladder configuration in Figure 4-12(c) that turns on an output for a set period of time to drive the ejector. The following changes would be necessary: 1) change the parts ejector output in rung 2 to a binary bit B3:0/0; 2) Add the two rungs from the standard to the solution; 3) Change the I:1/1 XIC instruction in the standard to the binary bit B3:0/0; 4) Change the OTE output in the standard from O:3/0 to O:2/12 parts ejector. Now the ejector output is on for a set time for every part ejection.

5-9 LOCATION OF THE INSTRUCTIONS

The location of instructions from this chapter in the Allen-Bradley programming software is indicated in Figure 5-31.

FIGURE 5-31: Location of instructions described in this chapter.

Systems	Instructions	Location
PLC 5, SLC 500, LOGIX	CTU, CTD, RES	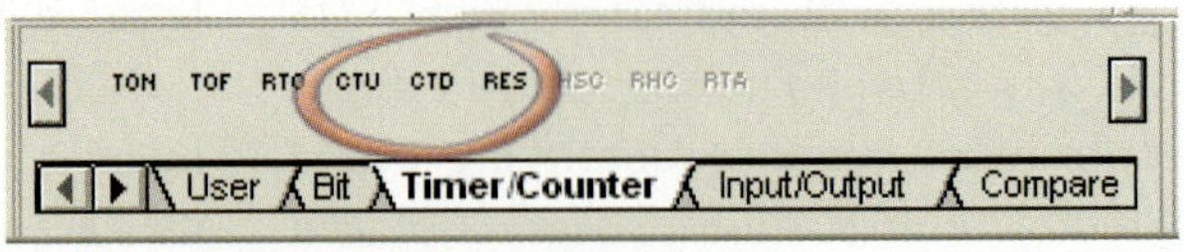 TON TOF RTO CTU CTD RES HSC RHC RTA User Bit Timer/Counter Input/Output Compare
PLC 5	ONS, OSR, OSF	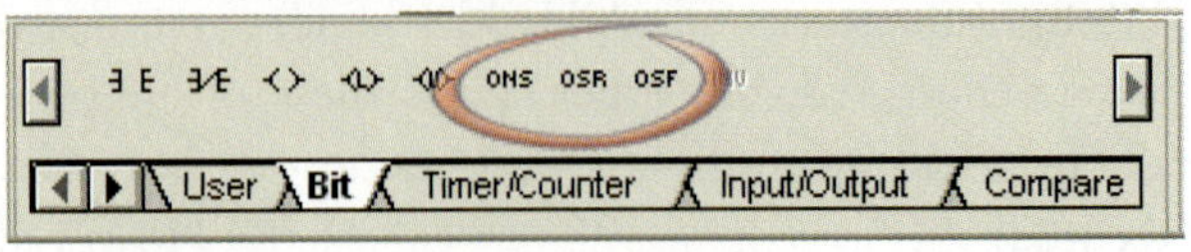 ONS OSR OSF User Bit Timer/Counter Input/Output Compare
SLC 500	OSR	 ONS OSR OSF DDT FBC User Bit Timer/Counter Input/Output Compare
LOGIX	ONS, OSR, OSF	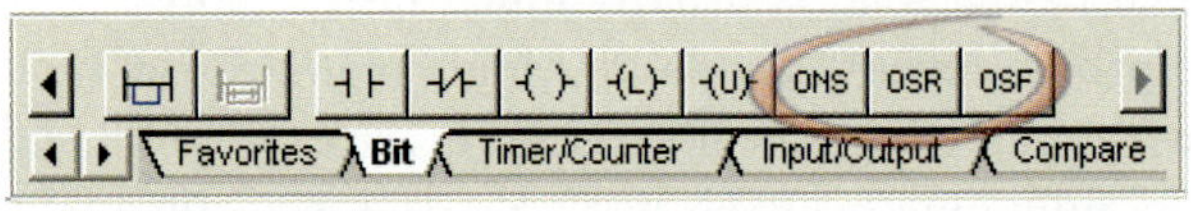 ONS OSR OSF Favorites Bit Timer/Counter Input/Output Compare
PLC 5	TND	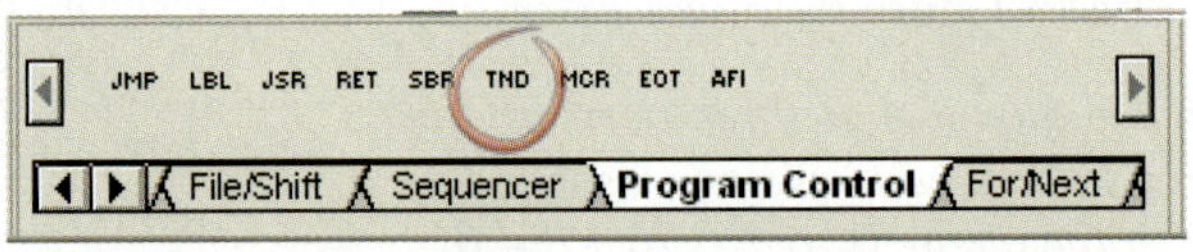 JMP LBL JSR RET SBR TND MCR EOT AFI File/Shift Sequencer Program Control For/Next
SLC 500	TND, SUS	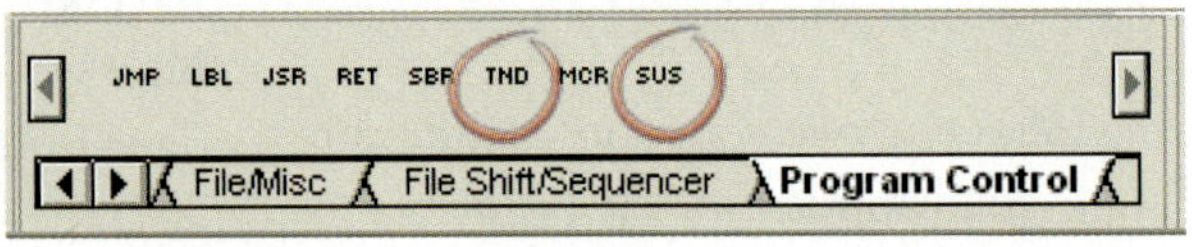 JMP LBL JSR RET SBR TND MCR SUS File/Misc File Shift/Sequencer Program Control
LOGIX	TND	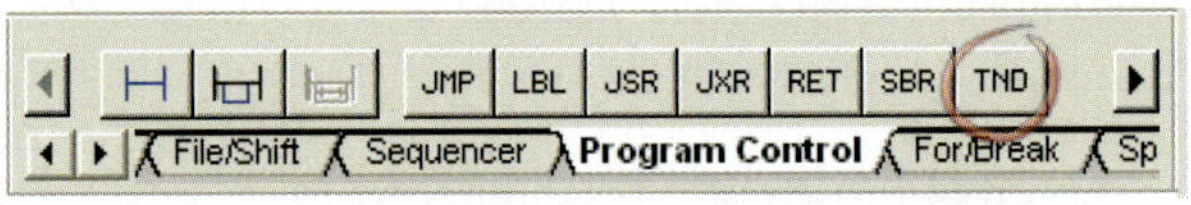 JMP LBL JSR JXR RET SBR TND File/Shift Sequencer Program Control For/Break Sp

Chapter 5 questions and problems begin on page 563.

CHAPTER
6

Arithmetic and Move Instructions

6-1 GOALS AND OBJECTIVES

There are three principal goals of this chapter. The first goal is to provide the student with information on binary arithmetic—adding, subtracting, multiplying, and dividing binary numbers. The second goal is to introduce the arithmetic and move instructions for the Allen-Bradley PLC 5, SLC 500, and Logix systems. The third goal is to show how the arithmetic and move instructions are applied to specific PLCs that are used in industrial automation systems.

After completing this chapter you should be able to

- Explain the concept of binary arithmetic.
- Describe one's and two's complement binary notation.
- Describe the arithmetic instructions (addition, subtraction, multiplication, division, square root, and clear) for the Allen-Bradley PLC 5, SLC 500, and Logix systems.
- Describe the move instructions (move, move with mask, and negate) for the Allen-Bradley PLC 5, SLC 500, and Logix systems.
- Develop ladder logic solutions using arithmetic and move instructions for the Allen-Bradley PLC 5, SLC 500, and Logix systems.
- Include arithmetic and move instructions in the empirical design process.
- Describe troubleshooting techniques for ladder rungs with arithmetic and move instructions.

6-2 BINARY ARITHMETIC

Before we discuss arithmetic PLC instructions, let's briefly examine binary arithmetic operations. These arithmetic operations include addition, subtraction, multiplication, and division.

The addition of binary numbers includes the following four sum combinations:

$$\begin{array}{r} 0 \\ \underline{+0} \\ 0 \end{array} \qquad \begin{array}{r} 0 \\ \underline{+1} \\ 1 \end{array} \qquad \begin{array}{r} 1 \\ \underline{+0} \\ 1 \end{array} \qquad \begin{array}{r} 1 \\ \underline{+1} \\ 0 \end{array} \text{ carry 1}$$

Note that the results of the first three additions are obvious, but the fourth result needs some explanation. In the decimal system, $1 + 1 = 2$, but in the binary system $1 + 1 = 0$ with a carry of 1 to the next most significant place value because only two digits are available—1 and 0. Example 6-1 illustrates addition operations in the binary system.

EXAMPLE 6-1

Add the following decimal numbers using the binary system:

a. 10 plus 2
b. 28 plus 11

SOLUTION

Study the carrying concept in the solutions until you are familiar with it.

a. 1 carry
1010
<u>0010</u>
1100, which is 12

b. 11 carry
11100
<u>01011</u>
100111, which is 39

The subtraction of binary numbers includes the following difference combinations:

$$\begin{array}{cccc} 1 & 1 & 0 & 0 \\ \underline{-0} & \underline{-1} & \underline{-0} & \underline{-1} \\ 1 & 0 & 0 & 1 \text{ borrow } 1 \end{array}$$

Note that the results of the first three subtractions are obvious, but in the fourth subtraction condition, borrowing a 1 from the next most significant place value is required. The borrowing concept in subtraction operations in the binary system is illustrated in Example 6-2.

The multiplication of binary numbers includes the following four product combinations:

$0 \times 0 = 0 \quad 0 \times 1 = 0 \quad 1 \times 0 = 0 \quad 1 \times 1 = 1$

EXAMPLE 6-2

Subtract the following decimal numbers using the binary system:

a. 12 minus 2
b. 28 minus 7

SOLUTION

Study the borrowing concept in the solutions until you are familiar with it.

a. 1 borrow
1100
<u>0010</u>
1010, which is 10

b. 11 borrow
11100
<u>00111</u>
10101, which is 21

EXAMPLE 6-3

Multiply 4 times 5 using the binary system

SOLUTION

```
   100
   101
   100
  000
 100
 10100, which equals 20
```

EXAMPLE 6-4

Divide 28 by 4 using the binary system

SOLUTION

```
        111, which is 7
   100)11100
        100
        110
        100
         100
         100
         000
```

Binary numbers are multiplied in the same method as decimal numbers—form partial products and add them together. Example 6-3 illustrates this method. Likewise, dividing binary numbers is also accomplished in the same manner as decimal numbers. Division is illustrated in Example 6-4.

The binary arithmetic operations that were demonstrated involved only positive numbers. How negative numbers are represented in the binary system and used in arithmetic operations is discussed in the next section.

6-3 SIGNED BINARY NUMBERS

In binary systems, the plus sign, indicating a positive number, and the minus sign, indicating a negative number, cannot be handled in arithmetic operations. One method of representing a binary number as positive or negative is to assign a *sign bit* to the number as the most significant bit. If the sign bit is a 0, then the number is positive; if the sign bit is a 1, then the number is negative. There are two other methods available for representing the sign of a binary number—the one's complement and the two's complement. To *complement* a binary number means to change it to a negative number. The one's complement is obtained by changing 1s to 0s and 0s to 1s. The two's complement is obtained by adding 1 to the one's complement. Table 6-1 shows a decimal number and its equivalent binary number with a sign bit as a one's complement and as a two's complement. Two's complement is typically the preferred method because you can perform subtraction using addition.

Two's complement is the way computers represent integers. To get the two's complement negative notation of an integer, you write out the number in binary. You then invert the digits, and add 1 to the result. Suppose we're working with 8-bit quantities and want to find how −28 would be expressed in two's complement notation. Use the following steps.

1. First write 28 in binary form, which is 00011100.
2. Set the 0s to 1s and the 1s to 0s, which is 11100011.
3. Add 1, which yields 11100100, which is –28 in two's complement.

TABLE 6-1 Binary representation of decimal numbers.

Decimal Number	Magnitude with Sign Bit	One's Complement	Two's Complement
+7	0111	0111	0111
+6	0110	0110	0110
+5	0101	0101	0101
+4	0100	0100	0100
+3	0011	0011	0011
+2	0010	0010	0010
+1	0001	0001	0001
0	0000	0000	0000
−1	1001	1110	1111
−2	1010	1101	1110
−3	1011	1100	1101
−4	1100	1011	1100
−5	1101	1010	1011
−6	1110	1001	1010
−7	1111	1000	1001

EXAMPLE 6-5

Subtract 12 from 69 in 8-bit binary using the two's complement.

Solution

The solution is $69 - 12 = 57$ or $69 + (-12) = 57$. When the negative 2's complement of 12 is added to 69, the result is 57.

01000101 (69)
11110100 (–12 in two's complement)
00111001 (57)

EXAMPLE 6-6

Subtract 69 from 12 in 8-bit binary using the two's complement.

Solution

The solution is $12 - 69 = -57$ or $12 + (-69) = -57$. Note that the answer starts with a 1 in the sign bit to indicate that the answer is negative.

00001100 (12)
10111011 (–69 in two's complement)
11000111 (–57)

6-4 ALLEN-BRADLEY ARITHMETIC INSTRUCTIONS

Arithmetic instructions, or *math instructions* as they're called in some literature, allow the PLC to perform arithmetic and trigonometric operations on the contents stored in memory or register locations. The mnemonic, name, and description of the numerous instructions available in the Allen-Bradley PLC systems are displayed in Figure 6-1(a). The description of each instruction indicates what values are entered along with the location where the result of the math or trig operation is stored. All of the instructions in the figure are output instructions that are active when the rung is true.

6-4-1 Structure for Arithmetic Instructions

The *addition (ADD), subtraction (SUB), division (DIV), multiplication (MUL), square root (SQR),* and *clear (CLR)* instructions are used most often. The rest of the arithmetic instructions are used less often and typically in special applications. The first five instructions listed in Figure 6-1(a)

FIGURE 6-1: Allen-Bradley math instruction groups and symbols.

Instruction		Descriptions
Mnemonic	**Name**	
ADD	Add	Adds source A to source B and stores the result in the destination.
SUB	Subtract	Subtracts source B from source A and stores the result in the destination.
MUL	Multiply	Multiplies source A by source B and stores the result in the destination.
DIV	Divide	Divides source A by source B and stores the result in the destination and the math register.
DDV	Double Divide	Divides the contents of the math register by the source and stores the result in the destination and the math register.
CLR	Clear	Sets all bits of a word to zero.
SQR	Square Root	Calculates the square root of the absolute value of the source and places the integer result in the destination.
SCP	Scale with Parameters	Produces a scaled output value that has a linear relationship between the input and scaled values.
SCL	Scale Data	Multiplies the source by a specified rate, adds to an offset value, and stores the result in the destination.
ABS	Absolute	Calculates the absolute value of the source and places the result in the destination.
CPT	Compute	Evaluates an expression and stores the result in the destination.
SWP	Swap	Swaps the low and high bytes of a specified number of words in a bit, integer, ASCII, or string file.
ASN	Arc Sine	Takes the arc sine of a number and stores the result (in radians) in the destination.
ACS	Arc Cosine	Takes the arc cosine of a number and stores the result (in radians) in the destination.
ATN	Arc Tangent	Takes the arc tangent of a number and stores the result (in radians) in the destination.
COS	Cosine	Takes the cosine of a number and stores the result in the destination.
LN	Natural Log	Takes the natural log of the value in the source and stores it in the destination.
LOG	Log to the Base 10	Takes the log base 10 of the value in the source and stores the result in the destination.
SIN	Sine	Takes the sine of a number and stores the result in the destination.
TAN	Tangent	Takes the tangent of a number and stores the result in the destination.
XPY	X to the Power of Y	Raise a value to a power and stores the result in the destination.

(a)

FIGURE 6-1: (Continued).

ADD
Add
Source A ?
?
Source B ?
?
Dest ?
?

(b) Math instruction symbol with three parameters

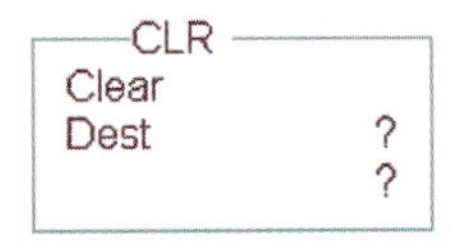

(c) Math instruction symbol with one parameter

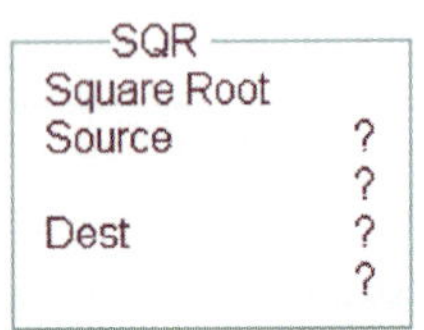

(d) Math instruction symbol with two parameters

use the ADD instruction format with three parameters illustrated in Figure 6-1(b). These arithmetic operations are performed using the two operands, *Source A* and *Source B*, and the result of the operation is stored in the destination (*Dest*).

The structure for the clear instruction, Figure 6-1(c), has a single destination register to indicate the register value that goes to zero when the instruction's rung is true. The remaining instructions in Figure 6-1(a) use the SQR instruction format with two parameters illustrated in Figure 6-1(d).

Data types for arithmetic operations. Most arithmetic operations in a PLC require only *single-precision arithmetic*, meaning the value of the operands and the result can be stored in one 16-bit register (one word). If the operation involves larger numbers, *double-precision arithmetic* is required. Double precision means that the PLC uses double the number of locations (32-bit registers) for the operation—two words for each operand and two words for the result. Table 6-2 indicates the register size and the numerical limits placed on data values for a number of different data types used with the math and trig instructions in the Logix processors. The PLC 5 and SLC systems have the same values except that they do not support double integers. Real data registers typically are double integers but can use three words to hold the value.

6-5 OPERATION OF ALLEN-BRADLEY ARITHMETIC AND MOVE INSTRUCTIONS

In the following subsections, the most commonly used arithmetic instructions and the move instructions are discussed. The discussion starts with the parameter requirements for the instruction registers and the updates to the status bits and math register that result from an instruction execution.

TABLE 6-2 Data types and binary value range for ControlLogix system.

Data Type	Bits			
	31 16	15 8	7 1	0
Bool	not used			0 or 1
Sint	not used		−128 to +127	
Int	not used	−32,768 to +32,767		
Dint		−2,147,483,648 to +2,147,483,647		
Real	−3.40282347E38 to −1.17549435E-38 (negative values)			
	0			
	1.17549435E-38 to 3.40282347E38 (positive values)			

Entering parameters. The values entered into the math instructions should conform to the following rules.

- **Source:** The *source* is the address(es) of the value(s) on which the mathematical or move operation is to be performed. This can be a word address or program constant value. An instruction that has two source operands does not accept program constants in both operands.
- **Destination:** The *destination* is the address where the result of the operation is placed. Signed integers in both the source and destination locations are stored in two's complement form.

Updates to arithmetic status bits. The arithmetic status bits are found in word 0, bits 0 to 3 in the controller status file, and the overflow trap bit is in word 5, bit 0. After an arithmetic instruction is executed, the arithmetic status bits in the status file are updated for the conditions described in Table 6-3.

The status bits are used in ladder programs to turn on warning lights when process calculations produce out of range results that could indicate a problem in the process. They also indicate when the result from an arithmetic operation exceeds the size of the destination register listed in the instruction.

Updates to the math register. Status word S:13 contains the *least* significant word of the 32-bit value of the MUL instruction. It contains the remainder for DIV instruction. Status word S:14 contains the *most* significant word of the 32-bit value of the MUL instruction. It contains the unrounded quotient for DIV instruction. S:13 and S:14 are not used when the math parameters are in the floating point or real data format.

Addressing floating point data file. The floating point file, F8 in the Allen-Bradley PLC 5 and SLC 500 systems, is used whenever fractional numerical data or numerical data with values greater than +32,767 or less than −32,768 are needed. Floating point data has two parts: an *integer* and an *exponent*. Two words are used to store floating point files, one for the integer and the second for the exponent. The Allen-Bradley SLC 500 systems support for floating point data registers in the math instructions starts with a model 503.

Floating point data is used most often with process systems where the data is *analog* and not *discrete*, and where field devices are analog process sensors and proportional control devices. In that setting analog input and output modules are required to pass the analog data into the PLC and back out for process control. The addressing used for floating point registers in the PLC 5 and SLC 500 systems is described in Table 6-4.

Floating point data is defined in the Logix processors by declaring the tag name a *Real data type* when the tag is defined.

6-5-1 Addition Instruction

The *addition instruction* (ADD) performs the addition of two values or operands that are placed in memory locations or registers and stores

TABLE 6-3 Status bits for arithmetic instruction.

Status Bit	Name	Description
S:0/0	Carry (C)	Bit is set if a carry is generated; otherwise it is cleared.
S:0/1	Overflow (V)	Bit is set if the result or value of the math instruction does not fit into the designated destination.
S:0/2	Zero (Z)	Bit is set if the result or value after a math, move, or logic instruction is a zero.
S:0/3	Sign (S)	Bit is set if the result or value after a math, move, or logic instruction is a negative (less than zero) value.
S:5/0	Overflow trap	The overflow trap bit (S:5/0) is set upon detection of a mathematical overflow or division by zero.

TABLE 6-4 Floating point register addressing for SLC 500.

Format		Explanation
Ff:e	**F**	Floating point file
	f	File number. Number 8 is the default file number (i.e., F8). A user-defined file number from 10 to 255 can be used if additional storage is required (e.g., F10 or F25).
	:	Element delimiter
	e	Element number. The element values range from 0 to 255 with each element using two words. As a result they are non-extended 32-bit numbers.
Examples	**F8:4** **F9:42**	Element 4, floating point file 8, Element 42, floating point file 10 (file 10 is a user-defined floating point file with 256 elements).

FIGURE 6-2: Allen-Bradley PLC 5 and SLC 500 systems addition instruction with PLC 5 input logic addresses.

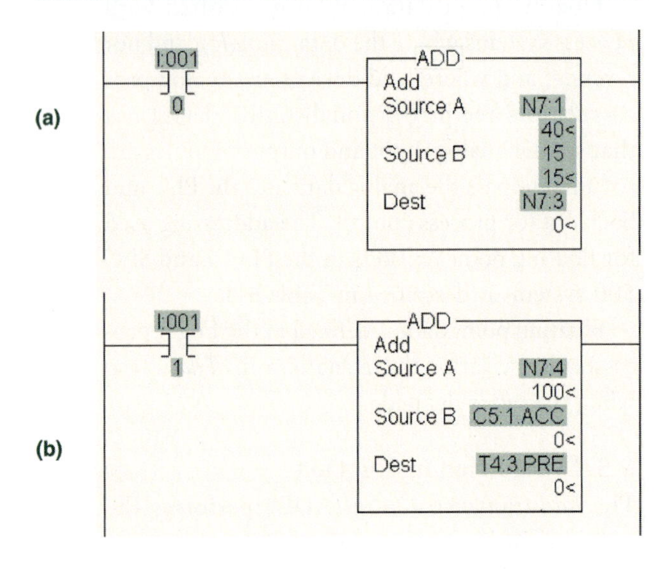

FIGURE 6-3: Allen-Bradley Logix system addition instruction.

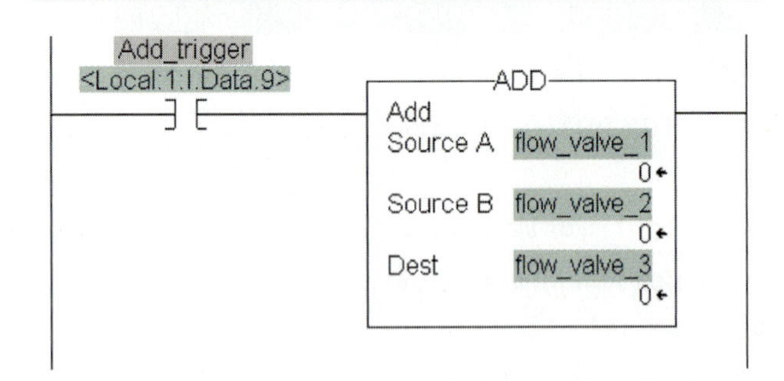

the result in another memory location or register. Figure 6-2 depicts the ADD instruction for the PLC 5 and SLC 500 systems. Figure 6-2(a) illustrates an ADD instruction where Source A points to an integer constant stored in N7:1, and Source B is set to a fixed value of 15.

Figure 6-2(b) illustrates a similar ADD instruction with two variables: an integer value in N7:4 and the value of the accumulator of counter C5:1. Note that the ADD instruction syntax is the same for PLC 5 and SLC 500, but the input address uses a PLC 5 format.

Note that in Figure 6-2(a) when I:001/0 is true, the value of Source A, N7:1 (40), is added to the fixed value (15) of Source B. The result of the addition is 55 and is stored in the destination, N7:3. In Figure 6-2(b), when I:001/1 is true, the value of Source A, N7:4 (100), is added to the value of Source B, C5:1.ACC, which is the accumulated value of counter number 1 in counter file 5. The result of the addition is stored in T4:3.PRE, which is the preset value of timer 3 in timer file 4.

In the ControlLogix system the operands in the ADD instruction are tag names as shown in Figure 6-3. Note that Source A is flow_valve_1, Source B is flow_valve_2, and the destination is flow_valve_3. The ControlLogix addition operation is the same as in the PLC 5 and SLC 500 systems.

6-5-2 Subtraction Instruction

The *subtraction instruction* (SUB) is shown in Figure 6-4. The value in Source B is subtracted from the value in Source A, and the result is stored in the destination. Note that when I:001/0 is true, the value of Source B, N7:2 (25), is subtracted from the value of Source A, N7:1 (45). The result of the subtraction is 20, and it is stored in the destination, N7:3. As in the ADD instruction, the operands can be constants or variables for the PLC 5 and SLC 500 systems and are tag names for the Logix systems. On all arithmetic instructions, one constant can be entered directly into the arithmetic instruction without using an integer register to hold it.

FIGURE 6-4: Allen-Bradley PLC 5 and SLC 500 systems subtraction instruction with PLC 5 input logic addresses.

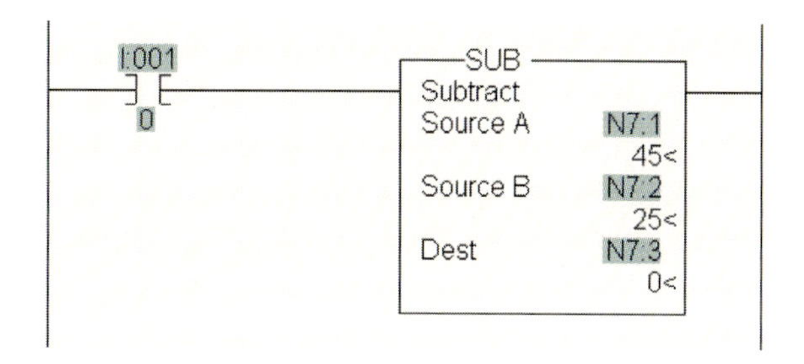

6-5-3 Multiplication Instruction

The *multiplication instruction* (MUL) is shown in Figure 6-5. The value in Source A is multiplied by the value in Source B, and the result is stored in the destination. Note that when I:001/0 is true, the value of Source A, N7:1 (20), is multiplied by the value of Source B, N7:2 (3). The result of the multiplication is 60, and it is stored in the destination, N7:3. The operands can be constants or variables for the PLC 5 and SLC 500 systems and are tag names for the Logix system.

FIGURE 6-5: Allen-Bradley PLC 5 and SLC 500 systems multiplication instruction with PLC 5 input logic addresses.

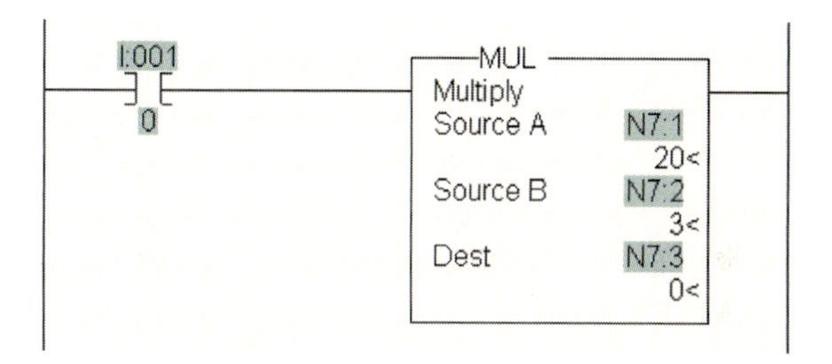

6-5-4 Division Instruction

The *division instruction* (DIV) is shown in Figure 6-6. The value in Source A is divided by the value in Source B, and the result is stored in the destination. Note that when I:001/0 is true, the value of Source A, N7:1 (40), is divided by the value of Source B, N7:2 (5). The result of the division is 8, and it is stored in the destination, N7:3. The operands can be constants or variables for the PLC 5 and SLC 500 systems and are tag names for the Logix system.

FIGURE 6-6: Allen-Bradley PLC 5 and SLC 500 systems division instruction with PLC 5 input logic addresses.

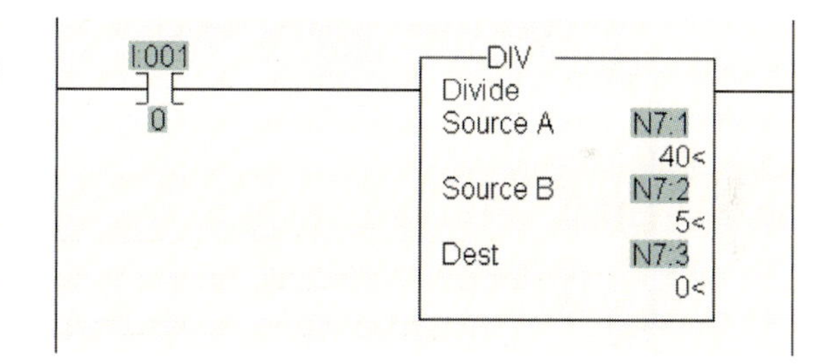

EXAMPLE 6-7

Design a ladder program with arithmetic blocks that converts degrees Fahrenheit to degrees Celsius.

SOLUTION

The formula that is implemented by the ladder program is

$$C = 5 \times (F - 32)/9$$

The solution is illustrated in Figure 6-7(a) using SLC PLC logic. The input and output values are shown in the data file view of integer registers N7:0 to N7:9 in Figure 6-7(b). The subtraction instruction subtracts Source B, N7:2 (32), from Source A, N7:1, which contains the value of the temperature 212 degrees Fahrenheit. The result of the subtraction operation (180) is stored in the destination, N7:3, which is also Source A of the multiply instruction. Source A is multiplied by Source B, N7:4 (5). The result of the multiplication operation (900) is stored in the destination, N7:5, which is also Source A of the division instruction. Source A is divided by Source B, N7:6 (9). The result, which is 100 degrees Celsius, is located in the destination, N7:7.

FIGURE 6-7: Solution for Example 6-7.

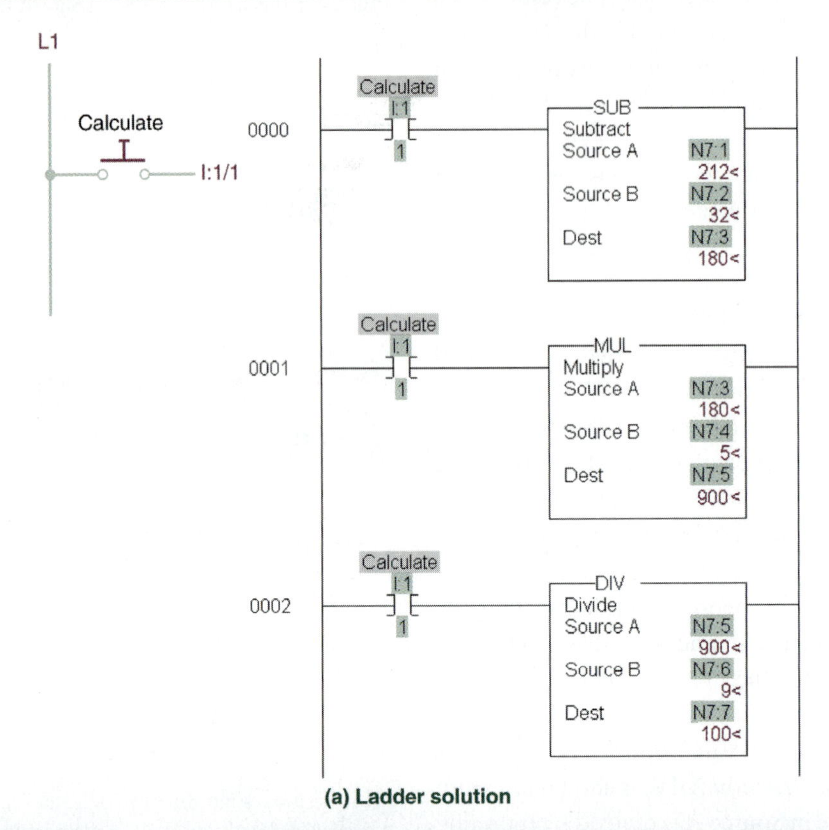

(a) Ladder solution

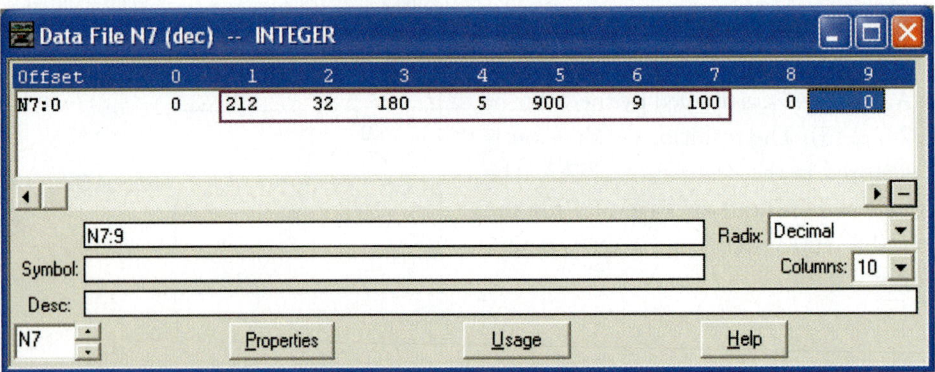

(b) Integer data register values

Field Device Power Rails. In previous chapters and problems, the input and output field device wiring was shown with all power sources and ground terminations illustrated. This was done so that the wiring of field devices to input and output modules could be learned. However, industry ladder documentation shows input contacts connected to a vertical power rail called L1 and all output field devices connected to a vertical power rail called L2. This simplified wiring convention will be used for the remainder of the ladder examples in the text. These power rails, L1 and L2, could be any positive DC voltage, an AC voltage, or a ground. It is assumed that the PLC input and output modules have been matched to the field device requirements for DC sinking/sourcing currents or an AC voltage level. A complete field wiring drawing is produced that indicates all power wiring and grounds. For input field devices, the only two things that must be known to analyze the ladder logic are the contact type (NO or NC) and the condition of the field device (activated or not activated). All output field devices are even simpler—if the OTE instruction is true the output field device is on, and if the OTE is false the field device is off.

EXAMPLE 6-8

Design a ladder program with arithmetic blocks that sets the upper and lower limits of a laboratory temperature chamber to plus and minus one-half percent of a manually set temperature.

SOLUTION

The solution is illustrated in Figure 6-8 using SLC 500 logic. The input and output values are shown in the data file view of floating point registers F8:0 to F8:9 in Figure 6-8(b). The multiplication instruction multiplies Source A, F8:1, which is the set temperature, by Source B, which is a constant value of 0.005. The result is stored in the destination, F8:3, which is also Source B of the subtraction and addition instructions. The results in F8:3 are subtracted from the set temperature, F8:1, and stored in the destination, F8:4, which is the lower temperature limit. Similarly, the results in F8:3 are added to the set temperature and stored in the destination, F8:5, which is the upper temperature limit.

FIGURE 6-8: Solution for Example 6-8.

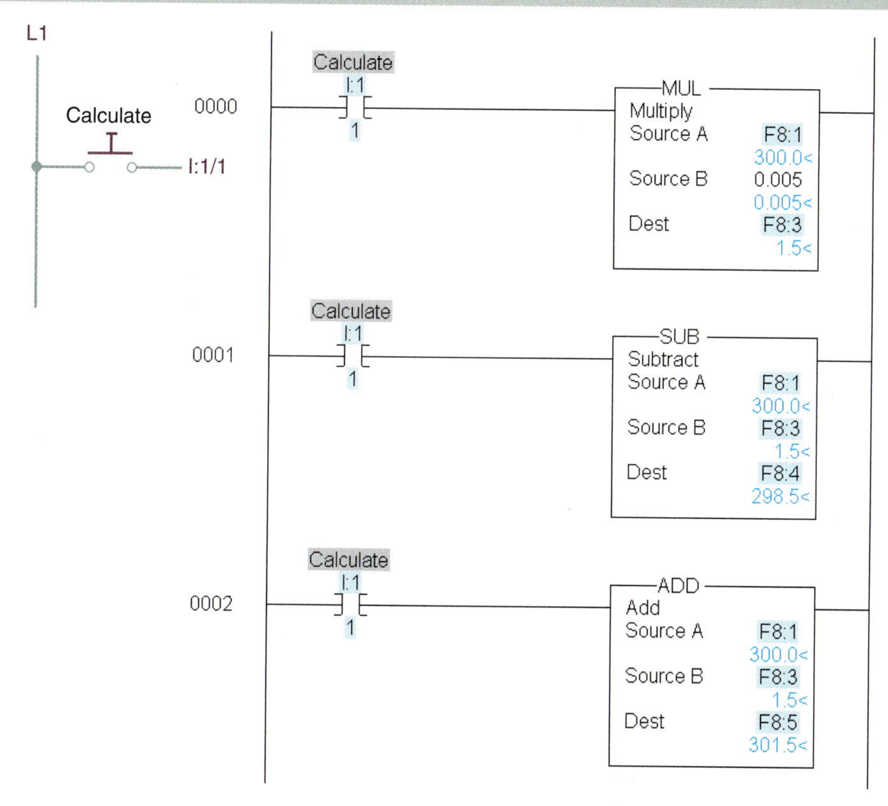

(a) Ladder logic

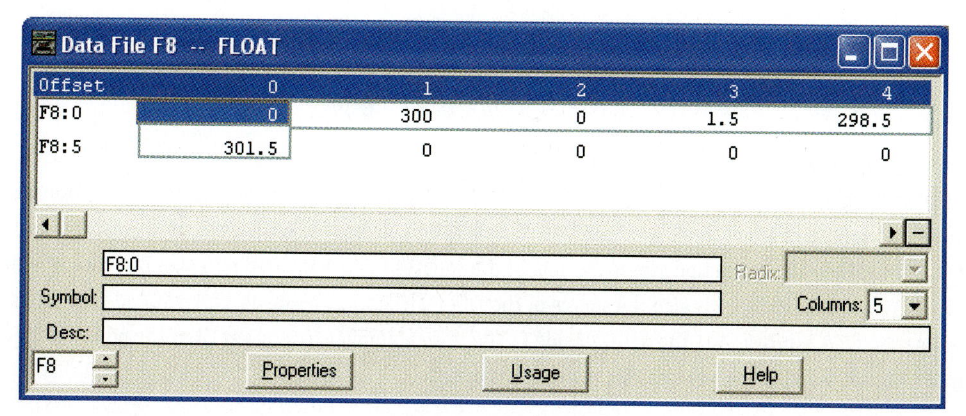

(b) Floating point data file

Changing register values. The integer and floating point registers shown in Figures 6-7(b) and 6-8(b) are the dialog boxes in the RSLogix 5 and RSLogix 500 computer-based software for the PLC 5 and the SLC 500, respectively. These boxes allow the user to view data results and to enter data into registers. Integer data is entered into an N7 register by first selecting the radix (lower-right drop-down menu) to be used for the value. Next, the data word in the register is selected by clicking on the current value. The new value is entered in the selected radix and then the Enter key is pressed. The new value appears in the register location and in the instruction field where that register is displayed. The changes can be made with the processor in program or run mode. If the radix is changed after a number is entered, then the display switches to the new radix value. This procedure works with the binary (B3) and floating point (F8) data files as well, except that the F8 registers can only be entered and displayed in the decimal radix. The Logix systems permit the same type of data value viewing in the program tags dialog box with the monitor tags tab selected. Data is entered and changed in the same dialog box but with the edit tags tab selected.

6-5-5 Square Root Instruction

The *square root* (SQR) instruction performs the mathematical operation of taking the square root of a constant or variable. As illustrated in Figure 6-9, the number whose square root is to be determined is in the source, and the result is placed in the destination. If the value of the source is negative, then the number in the destination is the square root of the absolute value of the source. Note that when I:001/0 is true, the square root of the value of the source, N7:1 (49), is taken. The result of the operation is 7 and is stored in the destination, N7:3. The operands can be constants or variables for the PLC 5 and SLC 500 systems and are tag names for the Logix system.

FIGURE 6-9: Allen-Bradley PLC 5 and SLC 500 systems square root instruction with PLC 5 input logic addresses.

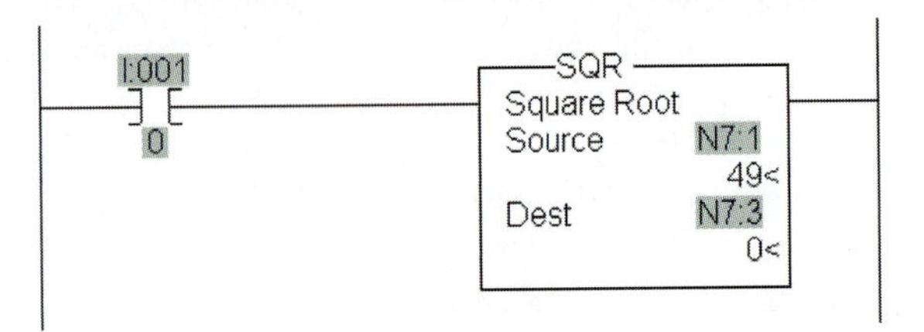

6-5-6 Move Instructions

The *move instructions* copy the contents of one memory location or register to another memory location or register. In this section, the following three move instructions are discussed:

- The *move* (MOV) instruction, which moves data from one location to another.
- The *move with a mask* (MVM) instruction, which moves only designated bits from one location to another.

EXAMPLE 6-9

Design a ladder program to calculate the hypotenuse of a right triangle whose side A is 6, side B is 8, and side C is the hypotenuse.

SOLUTION

The formula, which is the called the Pythagorean Theorem, is

$$C^2 = A^2 + B^2, \text{ or } C = (A^2 + B^2)^{1/2}$$

The solution is illustrated in Figure 6-10. The first MUL instruction multiplies Source A by Source B, which both are F8:1 (6), thus yielding side A squared that is stored in destination F8:3 (36). The second MUL instruction multiplies Source A by Source B, both of which are F8:2 (8). This operation yields side B squared, which is stored in destination F8:4 (64). The ADD instruction stores the sum of the squares (side A squared plus side B squared) in destination F8:5 (100). Note that F8:5 is also the source of the SQR instruction. The result of the SQR instruction is stored in destination F8:6, which is 10—the value of side C. The raise a number to a power (X to the power of Y) instruction, XPY, is another option where the MUL instructions are used.

FIGURE 6-10: Solution for Example 6-9.

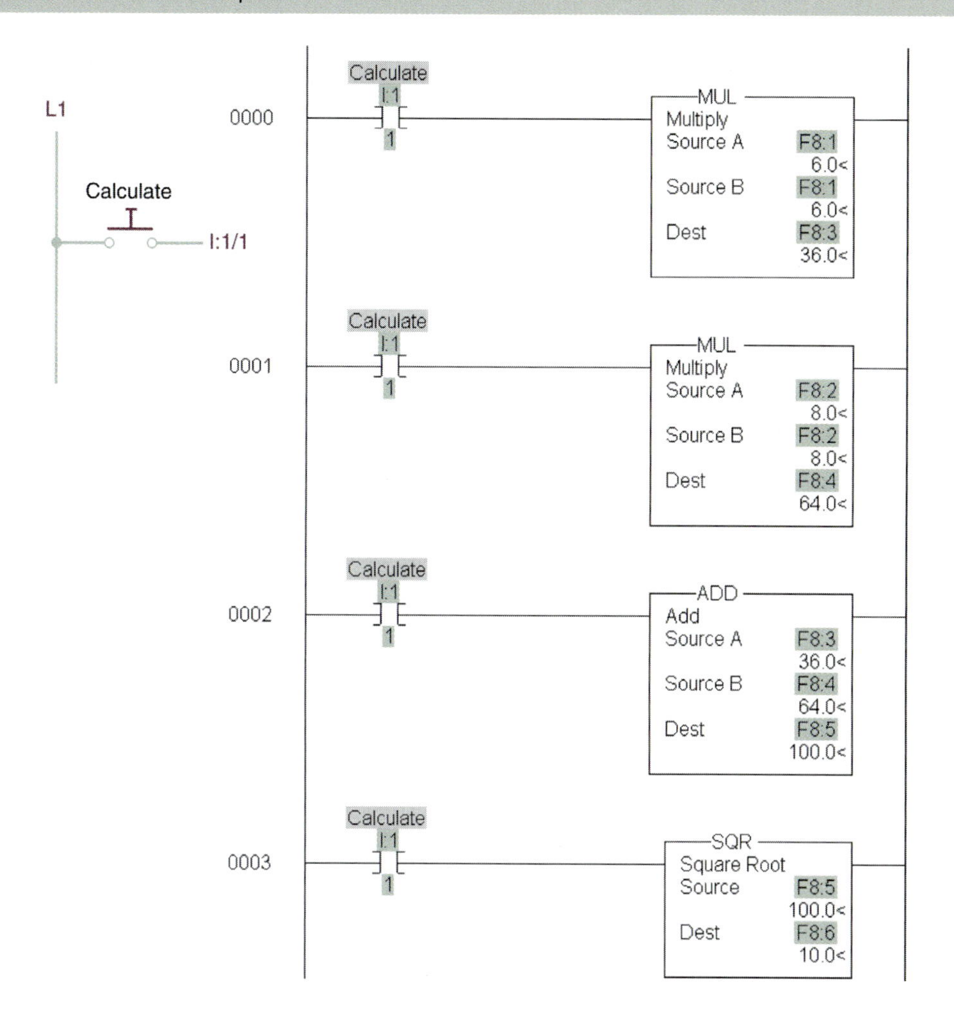

- The *negate* (NEG) instruction, which moves the negative representation of the data from one location to another.

Figure 6-11 illustrates all three move instructions for the SLC 500. Figure 6-11(a) illustrates the MOV instruction. When input contact I:1/1 is true, the contents of the source, N7:0, are moved to the location designated as the destination, N7:6. Figure 6-11(b) illustrates the MVM instruction. Now when I:1/1 is true, only the portion of the source, N7:1 (18998), that is aligned with the 1s in the mask, N7:2 (F0FFh), is moved to the destination, N7:3. Study the 1–0 bit pattern in Figure 6-11(b) until you are familiar with the move with a mask operation. The MVM instruction is effectively an AND operation between the source data and the mask data. When a 1 in the mask is ANDed with either a 0 or 1 in the source, the result in the destination is the value of the source bit. When a 0 in the mask is ANDed with the source bit, the destination bit is always a 0. Finally, Figure 6-11(c) depicts the NEG instruction. The negative representation of the contents of the source, N7:4, is moved to the contents of the destination, N7:5. Positive numbers are stored in binary format; negative numbers in two's complement.

6-6 STANDARD LADDER LOGIC FOR ALLEN-BRADLEY MATH AND MOVE INSTRUCTIONS

Several standard ladder logic configurations are used in the design of rungs that include math and move instructions. Figure 6-12 illustrates four configurations with features that can be mixed and matched to solve most ladder logic requirements for math and move instructions. Note that the configurations show an addition (ADD) instruction as the output, but any of the other math (SUB, DIV,

FIGURE 6-11: Allen-Bradley PLC 5 and SLC 500 systems move instructions with SLC 500 input logic addresses.

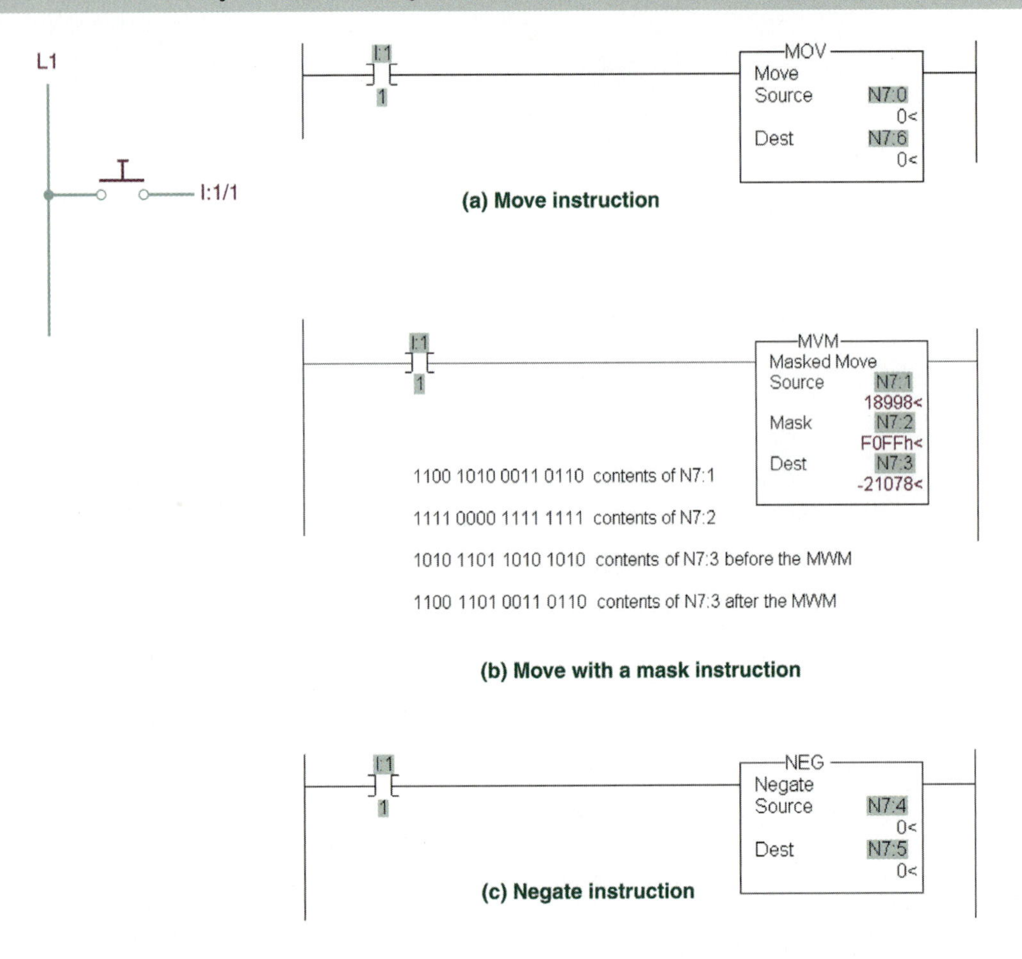

MUL, or SQR) instructions or the move (MOV) instruction could be substituted.

Math/Move standard A. The application addressed in Figure 6-12(a) features the use of an *unconditional rung* to activate an output instruction. In many applications the process requires that a mathematical manipulation or the transfer of data occur as often as possible. When no input logic is present the output instruction is true every scan. The sampling rate is equal to the scan rate because the output is true every scan. This configuration is used in applications that require continuous calculations or moves on real-time data to update the output register. For example, if a MOV output instruction is used and the source is the input address of an analog data channel from the output of a color sensor, then every scan of the color value is moved to a destination register.

Math/Move standard B. The configuration in Figure 6-12(b) uses one or more input logic instructions to determine when the math or move instruction is active. This is the more traditional implementation of the instructions. Output action occurs when the source register(s) is loaded and the input conditions make the rung true.

Math/Move standard C. The configuration in Figure 6-12(c) uses one or more input logic instructions to determine when the math or move instruction is active. In addition, a one-shot rising (OSR) instruction is placed just before the output instruction. This makes the math or move instruction active for only one scan. This configuration could be used, for example, to move data from a color sensor measuring the color of a part passing beneath it on

FIGURE 6-12: Standard ladder logic for all math and move instructions.

Application	Standard Ladder Logic Rungs for All Math and Move Instructions
Turn on an output math or move instruction as often as possible in the ladder logic. No input logic is present so the rung is always active. The result of this unconditional input condition is an addition instruction at the output that is executed with every scan. The sample rate for this active output is equal to the scan rate. While the addition instruction is used in the configuration example, any of the other math instructions (SUB, DIV, MUL, or SQR) or the move (MOV) instruction could be substituted. Also, this standard would be the same for the PLC 5 except for addresses and the same for ControlLogix except for the use of tags.	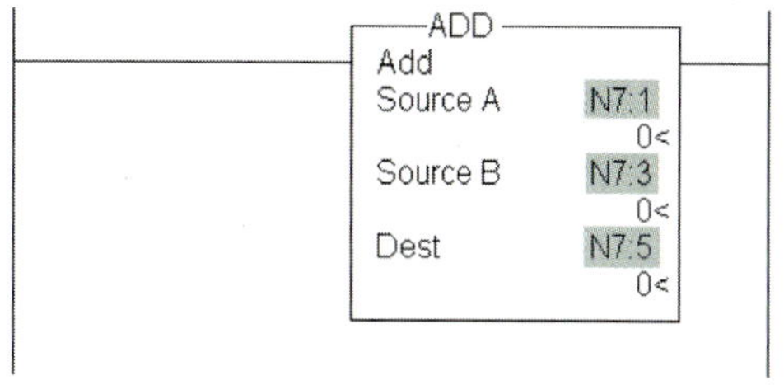 ADD Add Source A N7:1 0< Source B N7:3 0< Dest N7:5 0< (a)
Schedule the activation of an output math or move instruction by the action of an input field device(s) and the true condition of the input ladder logic. The input field device(s) and the input ladder logic in the addition instruction rung determine when the output is active and the addition is performed. As a result, the addition operation is executed on a schedule and at a rate determined by the field devices. While the addition instruction is used in the configuration example, any of the other math instructions (SUB, DIV, MUL, or SQR) or the move (MOV) instruction could be substituted. Also, this standard would be the same for the PLC 5 except for addresses and the same for ControlLogix except for the use of tags.	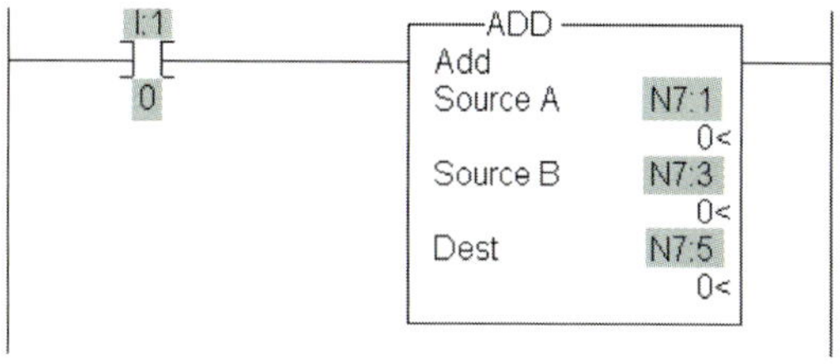 I:1 0 ADD Add Source A N7:1 0< Source B N7:3 0< Dest N7:5 0< (b)
Turn on an output math or move instruction for only one scan after the action of an input field device(s) and the true condition of the input ladder logic. The input field device(s) and the input ladder logic in the addition instruction rung determine when the output is active and the one shot assures that the addition is performed for only one scan. While the addition instruction is used in the configuration example, any of the other math instructions (SUB, DIV, MUL, or SQR) or the move (MOV) instruction could be substituted. Also, this standard would be the same for the PLC 5 except for input addresses and the OSR instruction would change to ONS. It is the same for ControlLogix except for the use of tags and the change to the ONS instruction.	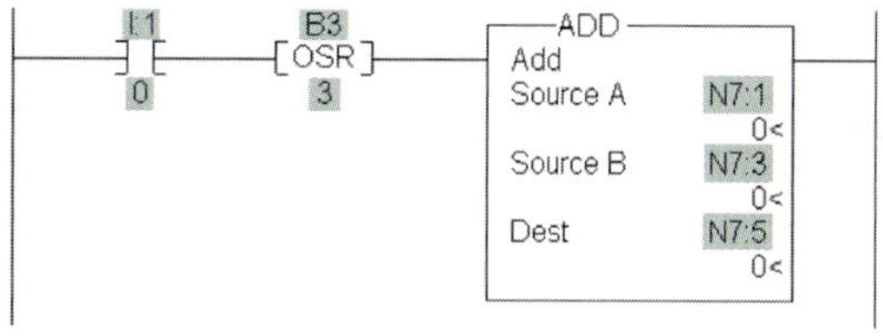 I:1 0 B3 OSR 3 ADD Add Source A N7:1 0< Source B N7:3 0< Dest N7:5 0< (c)

EXAMPLE 6-10

Design the ladder logic that multiplies the accumulator of counter C5:3 with the first four bits of the input word I:2.0. Place the results in N7:4.

SOLUTION

Rung 0 in the solution, Figure 6-13, uses standard ladder logic A to continuously move the counter accumulator values into N7:3 from counter C5:3. Rung 1 (standard ladder logic B) moves values from the input image table, register I:2.0, to N7:2 through a mask (000F) that filters out all but the first four bits. The move occurs when input I:1/0 is true (NO push button pressed), and the multiply occurs when rung 2 is true, which requires that both push buttons be pressed. The results are placed in N7:4.

FIGURE 6-12: (Continued).

Application	Standard Ladder Logic Rungs for All Math and Move Instructions
Use an ADD or SUB instruction to build an incremental adder or subtractor in an X = X + A or Y = Y - B implementation. When an ADD instruction is used in this format, the value in an integer or floating point register (X) is added to a process variable or tag value (B) and the result or answer is placed back into the original register (X). The one shot assures that the instruction is executed one time for every false to true change in the input logic. The same format can be used with the subtraction instruction. Also, this standard would be the same for the PLC 5 except for input addresses and the OSR instruction would change to ONS. It is the same for ControlLogix except for the use of tags and the change to the ONS instruction.	I:1 0 — B3 OSR 3 — ADD Add Source A N7:0 0< Source B N7:1 0< Dest N7:0 0< **(d)**

a conveyor. Figure 6-14 illustrates the timing diagram for the color sensor application. Note the following:

- The color values from the color sensor are only valid for a fixed time when the part is passing under the sensor. Values generated directly before and after the valid time are not usable.
- The trigger pulse produced by the sensor has a rising edge within the valid data region, but the falling edge of the pulse is outside the valid data region. The pulse width is not a valid trigger since part of the pulse falls outside the valid data window.
- The narrow one-shot pulse produces one scan that falls within the valid data value.

The move instruction is active for only one scan time due to the one-shot instruction, even though the trigger signal is true much longer. The one-shot instruction moves the data in the narrow window of time when the data is valid. If the one-shot instruction was not present, then the trigger pulse from the sensor would produce multiple scans that included invalid color data. In that case, the last

EXAMPLE 6-11

Design a ladder program to move the part color value from a color sensor similar to that illustrated by the timing diagram in Figure 6-14. Two values should be moved to N7:0 and N7:1, and the average value should be calculated and placed in N7:2. The color data comes from I:2.0, and the color sensor trigger is connected to terminal 0 on the input module in slot 4 of the SLC 500. The color value is valid for 500 milliseconds; the trigger pulse starts 5 milliseconds after the color value is valid and is true for 700 milliseconds.

SOLUTION

The solution in Figure 6-15 uses the standard ladder logic from Figure 6-12(c) in rung 0 and Figure 6-12(b) in rung 2 for moving a color value at two different points in the sensor output into registers N7:0 and N7:1, respectively. The first sample requires OSR instruction in the rung because the trigger pulse is true for 700 milliseconds, but the second sample just uses an XIC instruction with the timer done bit since T4:0/DN is true for only one scan. This timer is used to delay the second sample and move until 400 milliseconds after the color value is valid. Therefore, one sample is taken at the beginning of the color reading and one is taken near the end. The final two rungs use math instructions to create an average value from the two readings.

FIGURE 6-13: Solution for Example 6-10.

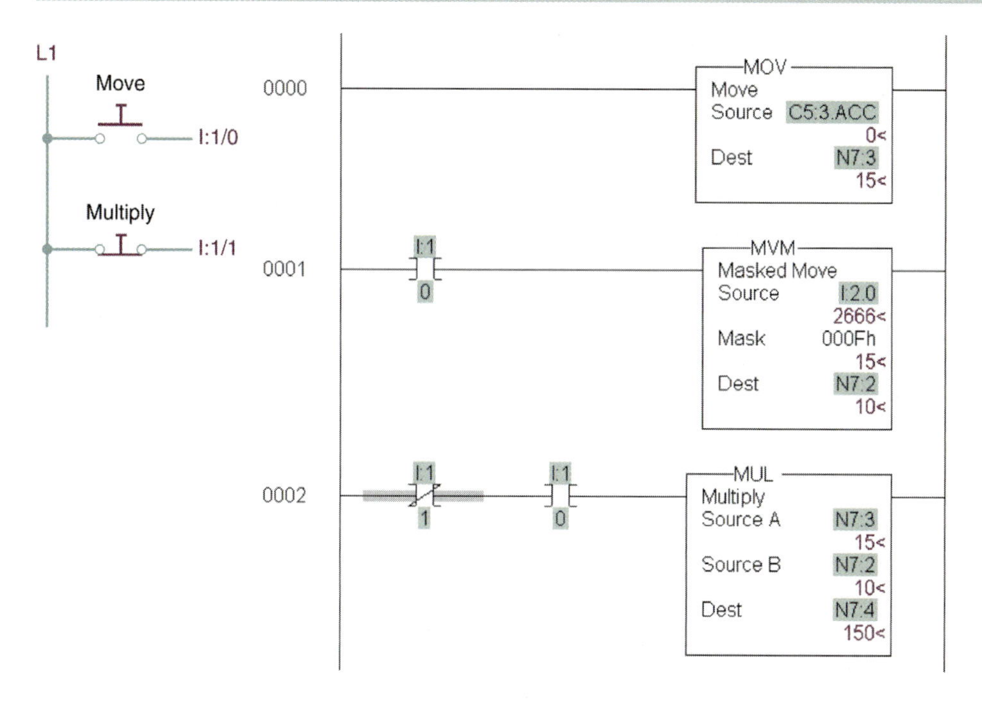

(a) Ladder logic

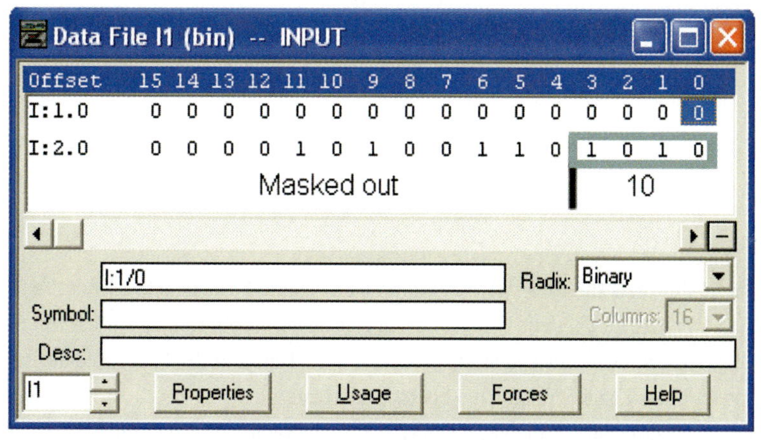

(b) Input image table and data file with masked out bits and passed bits shown. Note passed bits have a value of 10

value moved before the trigger pulse returns from true to false would be the value stored in the destination register, and that value would be during invalid color data. This configuration is the opposite of the ladder in standard ladder logic A, which activated the output on every scan.

Math/Move standard D. The configuration in Figure 6-12(d) uses one or more input logic instructions and the OSR instruction to determine when this math instruction is active. This implementation is used with addition (ADD) and subtraction (SUB) instructions to perform cumulative addition or subtraction. The syntax for the ADD instruction in the rung in Figure 6-12(d) is N7:0 = N7:0 + N7:1, where (destination) = N7:0 (Source A) + N7:1 (Source B) or N7:0 = N7:0 + N7:1, where Source A can be any integer or floating point register (in SLC 503 and higher) and Source B is an integer or floating point

FIGURE 6-14: Color sensing timing diagram.

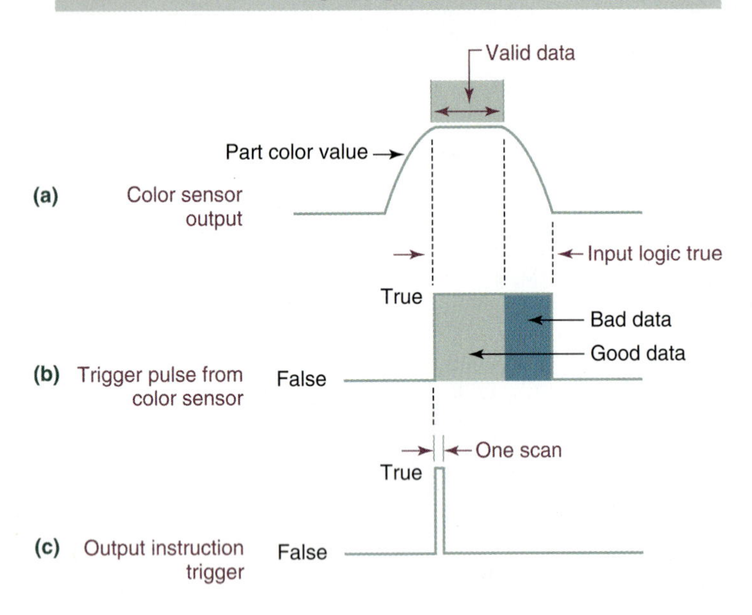

register that has changing values that must be added or subtracted. The value in Source A is added to the value in Source B and the sum is placed back into Source A as the destination register. This configuration is used when it is necessary to have a running total for the values in a register that is changing at regular intervals. If Source B is a fixed value then the register in Source A is incremented by that constant value with every execution.

When a math instruction rung is true, it will execute the instruction on every scan. Therefore, with this configuration it is important to have only one scan for each false to true transition of the input logic. The one-shot instruction takes care of this single scan requirement. This output action occurs when the Source B register is loaded and the input conditions make the rung true.

EXAMPLE 6-12

A vision camera scans the pallet bar codes as they pass, the number of parts on the pallet (not a constant value) is extracted from the bar code data, and that number is transferred to a PLC and placed in register N7:5. The B3:0/3 bit in the PLC is turned on by the vision system when the quantity data has transferred. The total number of pallets is stored in N7:4. Design a ladder program that keeps a running total for pallet parts in N7:6 and the number of pallets left in the batch in N7:4. Include logic for a normally open push button (PB) switch (I:1/4) that is used to clear register N7:6 at the end of a production run in preparation for the next batch.

Solution

Rung 0 in Figure 6-16 totals the pallet parts, and rung 1 in the figure subtracts 1 from the total number of pallets in the batch. The standard ladder logic in Figure 6-12(d) is used for both rungs. The last rung uses the clear instruction to reset the parts counter at the end of the production run. This same PB input could be used with a move instruction to load the number of pallets for the next batch into N7:4.

This example illustrates how many PLC programs have numerous solution options. For example, rung 0 could be replaced by two rungs. The first rung would add N7:5 to the total in N7:6 and place the results in a third register, N7:8. Then the second rung would move the new total in N7:8 to the total register N7:6. Also, rung 1 could be built with a count down counter that keeps the total number of remaining pallets to be processed stored in the accumulator.

A second lesson learned is the value of instruction descriptions. Compare the solution in Figure 6-16 with the solution in Figure 6-15. Notice how the instruction descriptions in Figure 6-16 make it easier to understand and learn the operation of the ladder logic compared to an undocumented ladder.

EXAMPLE 6-13

A limit switch operates a counter, which counts rubber balls coming off a conveyer for loading into a shipping carton. Four different size balls use the same conveyer and the same size carton. The carton can hold 96 balls

FIGURE 6-15: Solution for Example 6-11.

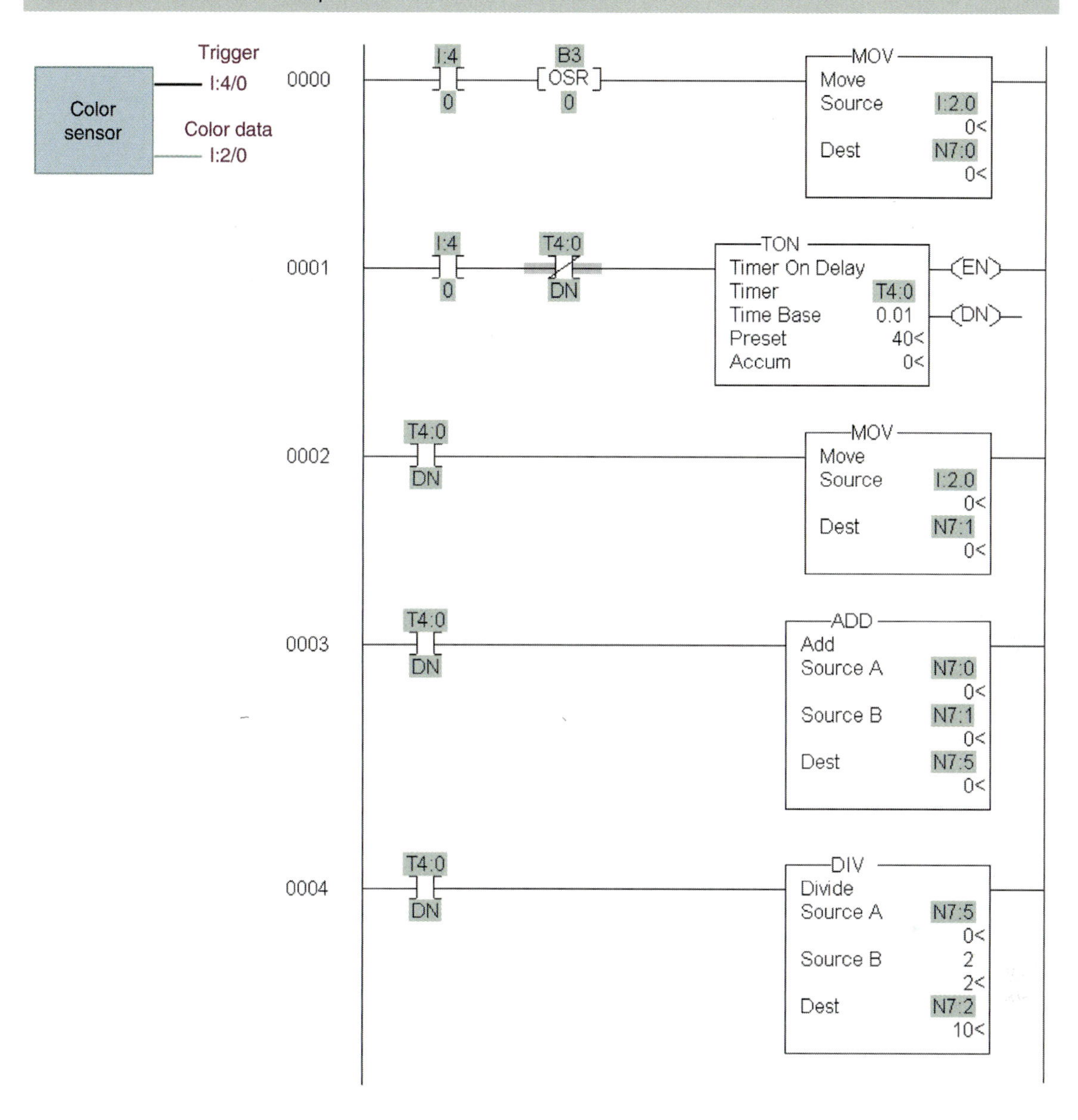

of type A, 48 balls of type B, 24 balls of type C, and 12 balls of type D. A pilot light is illuminated when the carton is full. Select the necessary input field devices and design a ladder logic program to satisfy this automation task.

Solution

The solution is illustrated in Figure 6-17. The field devices selected for the design include a NO reset PB switch, the NO contacts on the limit switch ball counter, and a ball selector pushbutton to identify the types of balls on the conveyor. The destination for each move instruction is the counter's preset register, and the source for each counter is an integer register that holds the desired box ball count for each size of ball. When the ball selector pushswitch activates input I:1/2, source N7:1 (96) is moved to C5:1.PRE, which is the counter's preset. When the pushswitch activates input I:1/3, source N7:3 (48) is moved to the counter's preset. When the pushswitch activates input I:1/4, source N7:4 (24) is moved to the counter's preset. When the pushswitch activates input I:1/5, source N7:5 (12) is moved to the counter's preset. The done bit of C5:1 turns on the full box pilot light, and the push button PB1 resets the counter accumulator to 0.

FIGURE 6-16: Solution for Example 6-12.

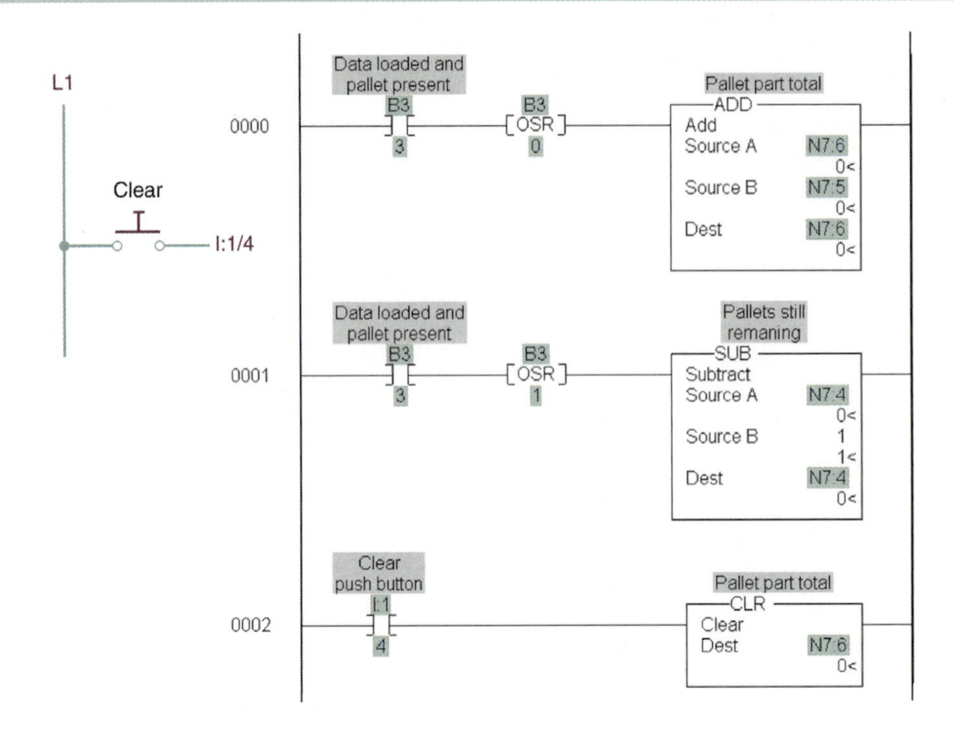

6-7 EMPIRICAL DESIGN PROCESS WITH MATH AND MOVE INSTRUCTIONS

The empirical design process, introduced in Section 3-11-4, is an organized approach to the design of PLC ladder logic programs. However, the term *empirical* implies that some degree of trial and error is present. As more instructions are added to the design, the process becomes more complex. However, the process used for new instructions is quite similar to that learned in the previous chapters.

6-7-1 Adding Math and Move Instructions to the Process

The first step in using mathematical and move instructions in PLC ladder designs is to know what parameter values are entered in the source and destination registers and all the standard math/move circuits illustrated in Figure 6-12. In addition, it is important to know when these output instructions should be true for only one scan, which dictates when a one-shot instruction is needed to trigger the outputs. Stop now and review these if necessary.

Math instructions. The use of math instructions implies a mathematical equation. Review Examples 6-7 and 6-9 and notice that their problem statements listed an equation that required the math instructions to solve. The number of rungs required depends on the complexity of the equation. Every mathematical operation (ADD, SUB, MUL, DIV, and SQR) requires a separate output, which is usually on a separate ladder rung. The complete empirical process is listed in Section 3-11-4; the modifications for math instructions are as follows:

Step 1: (Write the process description): Include a complete description of the numerical solution necessary for the stated problem. Note especially the type of numeric data present, integer versus real numbers, since it dictates the data type for the tag names in ControlLogix or register types needed for PLC 5 and SLC 500. Also, identify clearly the trigger used for the calculations. The trigger, which includes input field devices and internal logic, falls into two categories: conditional using input logic and unconditional with a trigger initiated every scan. The required standard ladder logic configurations are often determined at this point as well.

Step 2: (Write Boolean equations for all math instructions): A Boolean expression is needed

FIGURE 6-17: Solution for Example 6-13.

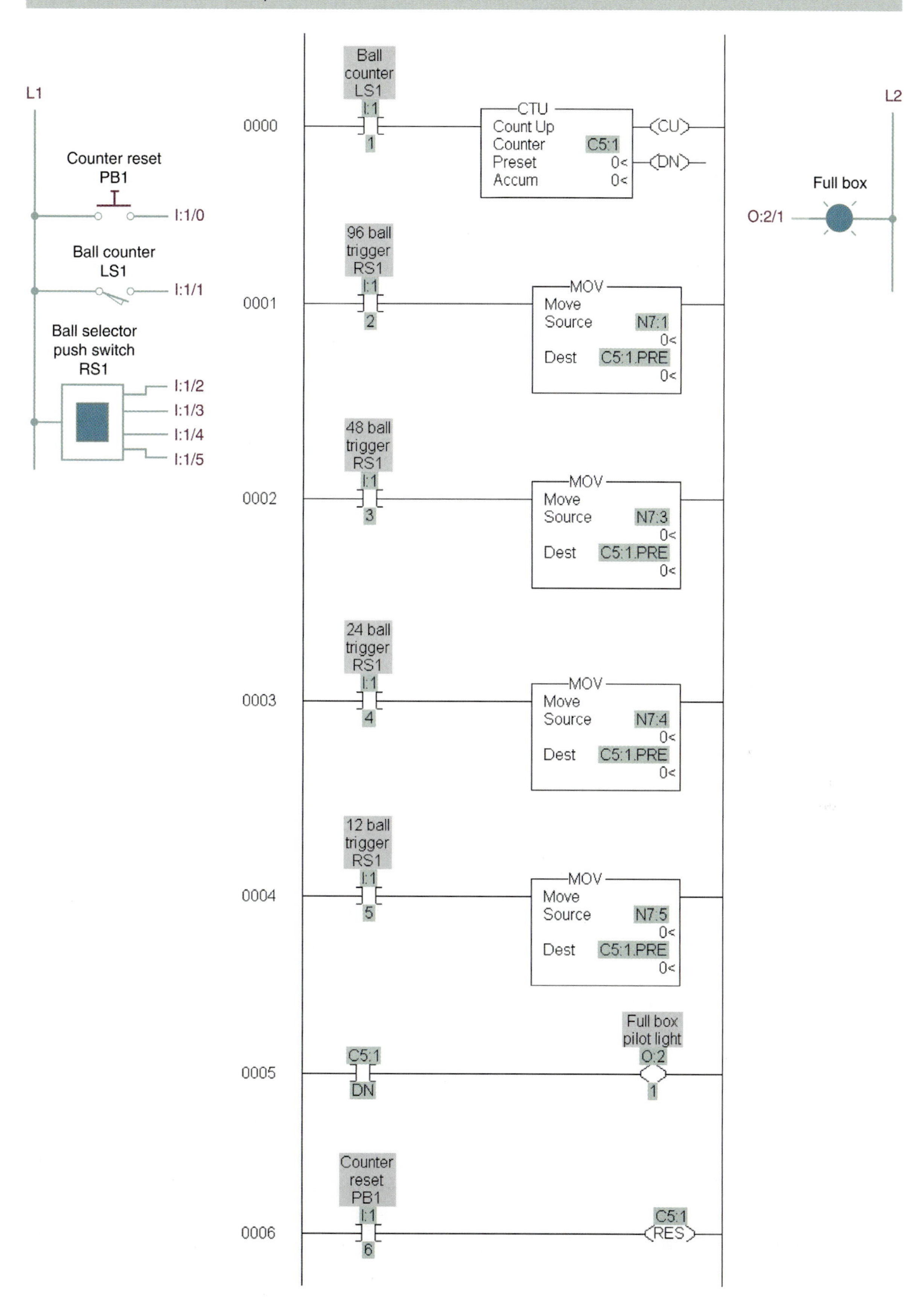

for each math instruction, but often the same Boolean expression is used to trigger all calculations.

Step 3: (Determine the number of math instructions necessary to solve the stated equation): Often the equation as stated in the process description can be simplified to reduce the number of math instructions required. For example, the equation $Y = A \times B + A \times C$ would require three instructions: two multiplications for the two products and an addition to add the products together. However, if the equation is written $Y = A \times (B + C)$, then only two instructions are required: an addition for the sum in the parentheses and a multiplication for the final product. Therefore, Step 3 is to simplify the expression if possible and then determine the types of math instructions needed for the solution.

Step 4: (Determine the order or sequence of operation of the math instructions for the simplified equation): The examples in Step 3 illustrate this point. Note in the first equation $(A \times B + A \times C)$ that the sequence would be two rungs with MUL instructions followed by a rung with an ADD instruction. In the second case $[A \times (B + C)]$, the sequence would be an ADD instruction followed by an MUL instruction. A general rule is to make calculations in the following order: first, computations inside parentheses; next, multiplication and division; and finally, addition and subtraction.

Move instruction. The move instruction is often used to support the loading of parameters into other PLC instructions. For example, the preset values of counters and timers are often changed by a move instruction. In other cases the move instruction is used to shift data from one register to another. The complete empirical process is listed in Section 3-11-4; the modifications for move instructions follow.

Step 1: (Write the process description): Include a complete description of movement of data between registers and between registers and instruction parameter locations. Also, identify clearly the trigger used for the moves. The trigger, which includes input field devices and internal logic, falls into two categories: conditional using input logic and unconditional with a trigger initiated in every scan.

Step 2: (Write Boolean equations for all move instruction triggers): A Boolean expression is needed to describe the trigger for each move instruction.

Examples 6-14 and 6-15 demonstrate how math and move instructions are added into the design process.

EXAMPLE 6-14

Example 5-2 had the following problem statement.

Design the ladder logic for an industrial application that packages canned vegetables supplied by a conveyor. When 12 cans are detected by a current sourcing proximity sensor, a packaging operation is initiated. The production line must package 200 boxes of 12 cans per shift. When 200 packages have been completed, a red stack light is illuminated. While the system is packaging cans, a green stack light is illuminated. A total count of cans packaged per shift should also be recorded.

Modify the conveyor ladder solution in Figure 5-14 for Example 5-2 as follows:

- A label-checking sensor verifies that all cans have labels attached. All cans without labels are ejected before packaging station.
- The number of ejected cans is counted and the total number of cans currently on the conveyor is determined.
- The number of ejected cans and the total number of cans on the conveyor are transferred to integer registers every 60 minutes.
- Select registers as needed.

Review the previous design and then design ControlLogix ladder rungs for the required modifications.

SOLUTION

The empirical design information includes:

- All data values present are integer data.
- The trigger for counting the cans without labels (bad_can_counter) can be either the missing label sensor input or the can ejector output.
- The trigger for the MOV and SUB instructions must be executed every 60 seconds, so an update timer with a preset value of 60,000 (60,000 units $\times$ 0.001 seconds per unit = 60 seconds) is required.

The standard ladder logic in Figure 6-12(b) is used since the SUB instruction must be triggered at regular intervals. The move instruction uses the same ladder configuration for the same reason. The following Boolean logic is used for the ladder rungs:

Ejector = bad_can_sensor

Bad_can_counter = ejector

Move instruction = transfer_timer/DN

SUB instruction = transfer_timer/DN

Based on this information the ladder solution in Figure 6-18 was developed for the Allen-Bradley ControlLogix PLC. Rung 0 uses a proximity sensor to detect cans that enter the conveyor system and to increment the accumulator of the counter Total_can_count. The can sensor is a diffused type photoelectric sensor used to reflect light from the bright metal can when the label is missing. The sensor does not detect cans with the paper label attached. Unlabeled cans are counted and ejected by rungs 1 and 2, respectively. Rung 3 is a regenerative timer used to generate a continuous stream of output pulses every 60 seconds. The done bit, used as a trigger for the move and subtraction instructions, is true for one scan. Rung 4 moves the bad_can_count to an integer register (Bad_can_total), and rung 5 calculates the total_cans_packaged from the total that entered the conveyor less the number of bad cans. Results are placed in tag Total_cans_packaged. Study the ladder logic solution and verify that the ladder satisfies the problem description in the example.

Notice how the Logix system's use of tag names for all variables and registers makes it easier to understand the ladder logic solution for the problem, even when instruction descriptions are not present.

EXAMPLE 6-15

Example 5-5 had the following problem statement.

A conveyor system, illustrated in Figure 5-26, sorts boxes so that each chute receives 10 boxes. The operation is as follows:

- *Gates 1 and 2 are up, sensor S1 counts 10 boxes for chute 1, and then gate 1 drops.*
- *Sensor S2 counts 10 boxes for chute 2 and then gate 2 drops.*
- *Sensor S3 counts 10 boxes for chute 3 and then gates 1 and 2 are raised and the process starts over.*
- *With average conveyor speed, it takes 4 seconds for the boxes to enter the chute after the sensor detects them.*

Modify the conveyor ladder solution in Figure 5-27 (Example 5-5) for the following operations.

- Determine the total box count for each chute.
- Determine the total box count for all three chutes.
- Move the counts to integer registers at the completion of each distribution cycle.
- Select registers as needed.

Review the previous design and then design SLC 500 ladder rungs for the required modifications. Note that the counter accumulator values generated in Figure 5-27 are used in the solution for this problem.

FIGURE 6-18: Solution for Example 6-14.

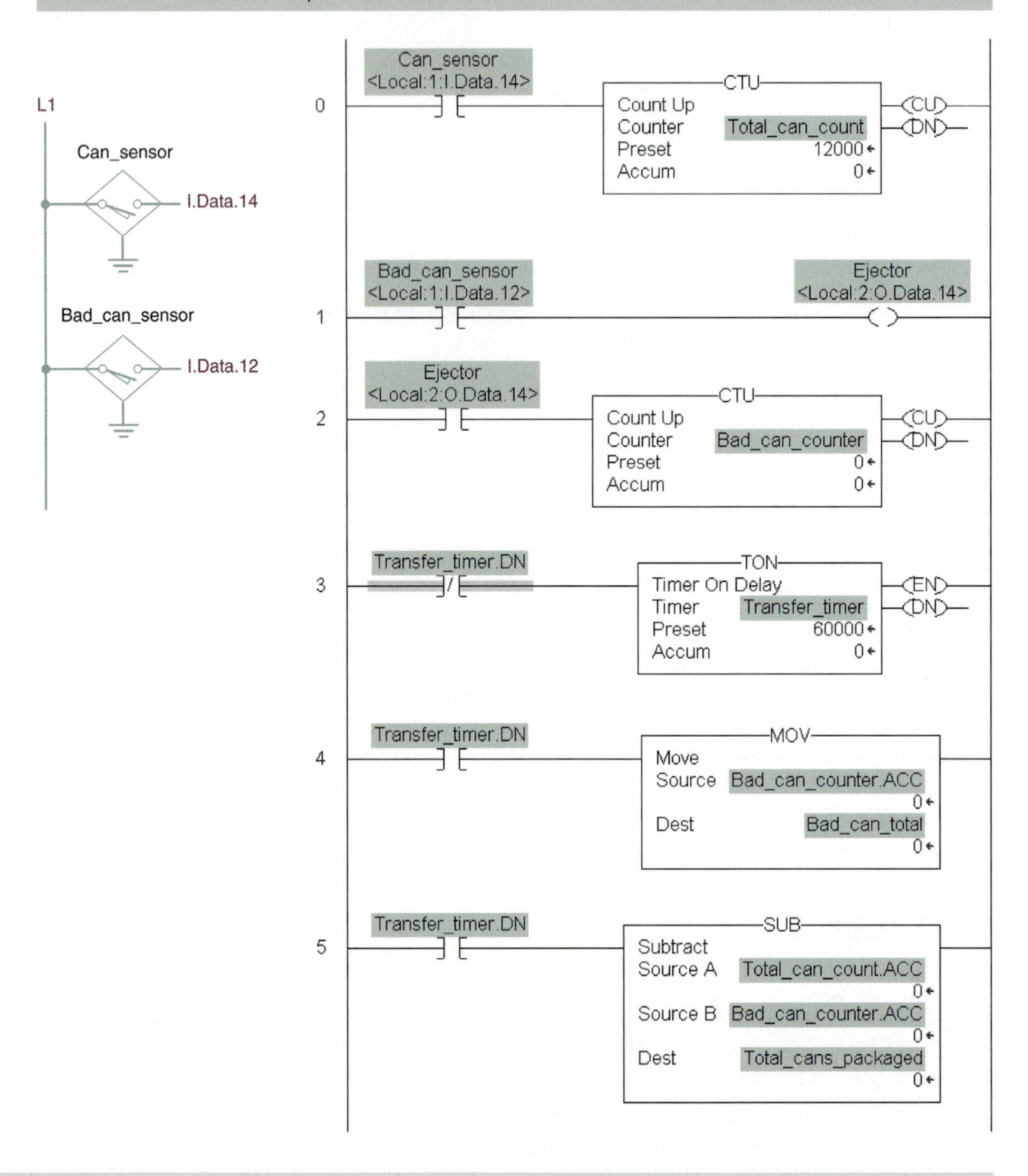

Solution

The empirical design information data types and event triggers includes (Note: the counter and timer instructions are from the ladder logic in Figure 5-27):

- All data values present are integer data.
- The chute 1 total trigger is the Chute 1 counter done (C5:0/DN).
- The chute 2 total trigger is the Chute 2 counter done (C5:1/DN).
- The chute 3 total trigger is the done bit of the last box delay timer (T4:0/DN).
- The data calculate and move trigger is the done bit of the last box delay timer (T4:0/DN).

The standard ladder logic in Figure 6-12(d) is used to add the new total to the previous total and place the answer back into the original register. The following Boolean logic is used to trigger the ADD instruction for the ladder rungs:

Chute 1 adder = C5:0/DN

Chute 2 adder = C5:1/DN

Chute 3 adder = C5:2/DN

Chute 1 and 2 subtotal = Last box delay timer (T4:0/DN)

Total box count = Last box delay timer (T4:0/DN)

Based on this information the ladder solution in Figure 6-19 was developed using an SLC 500 PLC. All three chute total adders (rungs 0, 1, and 2) are configured like the standard in Figure 6-12(d), so the total is continually updated. The OSR is used to make the ADD instructions true for only one scan. The last box delay signal timer (Figure 5-27) done bit is used to add the three chute totals together to determine the total box count. The solution could be simpler since all three chutes get 10 boxes, but the solution illustrated is valid if the box count per chute is changed. If the count would always remain the same, then the T4:0/DN bit could just add 30 (3×10) to the current box total. Study the ladder logic solution and verify that the ladder satisfies the problem description in the example.

6-8 TROUBLESHOOTING MATH AND MOVE LADDER LOGIC

The most difficult programs with math instructions to verify are those with multiple math instructions and that execute at a fast rate. An additional problem in troubleshooting math instruction rungs is to determine if the problem is in the program or if the data coming into the math instructions from the process is bad.

Move instructions offer few operational problems. If a section of the ladder that includes a move instruction is not operating properly, then you can use any of the following suggestions to troubleshoot it along with the math instructions present.

- If PLC rungs with math instructions do not produce the correct results, then first verify that the data from the process is correct by viewing the PLC 5 and SLC 500 dialog boxes for the input image table, integer image table, and floating point image table illustrated in Figure 6-20. If a ControlLogix processor is used the program and control tags are viewed in the dialog box displayed in Figure 6-21. Note that the register values can be displayed in binary, octal, hexadecimal, and decimal format, depending on the type of data present.
- Check the arithmetic status bit S5:0 (see the Section 6-5 subsection *Updates to Arithmetic Status Bits*) to determine if an overflow or divide by zero has occurred.
- Test the sequence of math rungs one rung at a time to verify that each rung is performing properly. Use the temporary end instruction, TND, described in Chapter 4 to halt the ladder execution after each rung.
- Use the suspend instruction, SUS, to verify status values for all registers and bits at critical points in the ladder. The SUS instruction is described in Section 5-8-1.
- Be aware of situations where computed math values are used to update internal PLC memory bits *and* cause the execution of process actions. Since the scan time and internal update time of registers is much faster than most process events, the event may not be able to complete the action before the drive signal for the event is removed. This type of problem is described well in the counter problem described in Section 5-8-2.
- Use the test options on some of the Allen-Bradley SLC 500 models in the single-step, single-scan, or continuous-scan test modes to isolate and run portions of the ladder logic. These tests are described in the next section.

6-8-1 SLC 500 Test Modes

Most of the test modes operate much like the program run mode with the exception of energizing the output field devices. The processor will read input field devices, execute the ladder program, and update the output image table; however, the output field devices are not turned on. The following

FIGURE 6-19: Solution for Example 6-15.

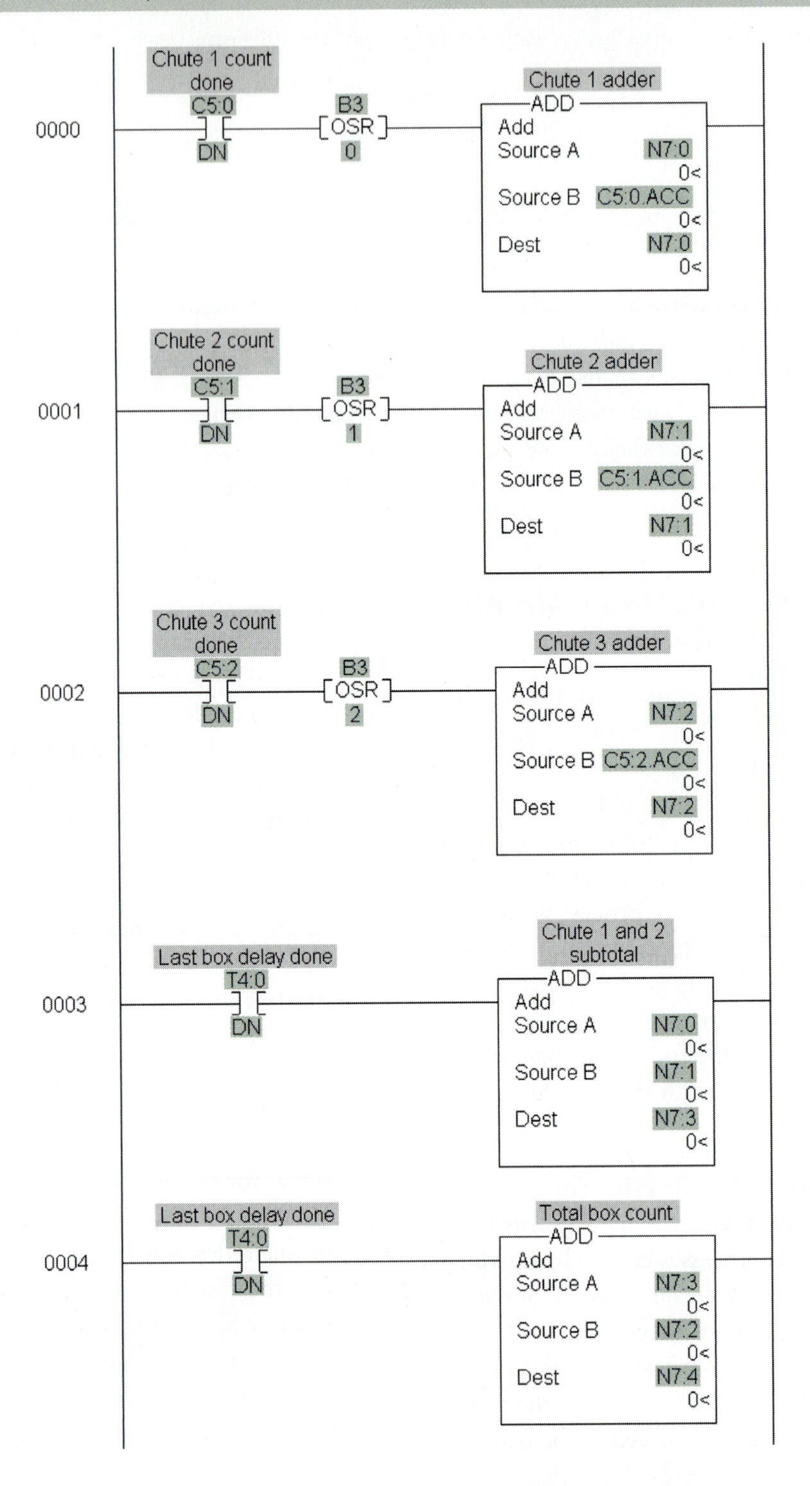

FIGURE 6-20: Input, binary, integer, and floating point data files for PLC 5 and SLC 500.

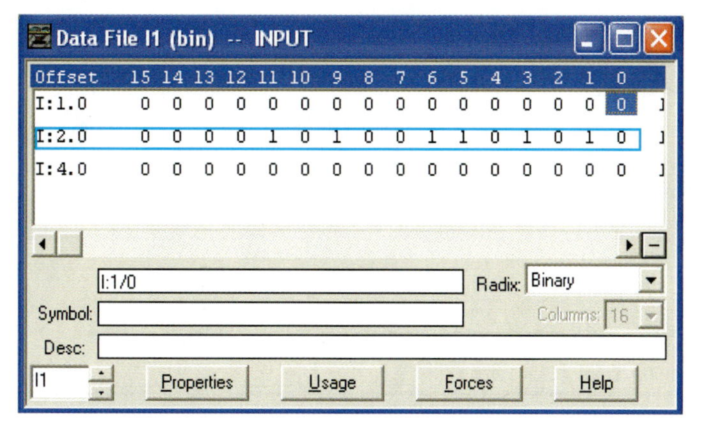

(a) Input image table data files

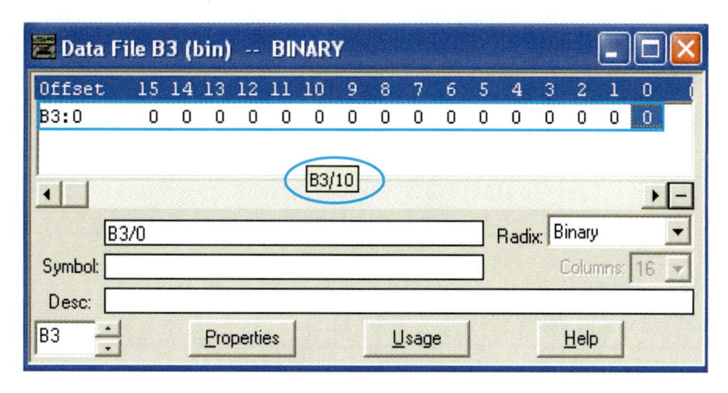

(b) Virtual relay data files

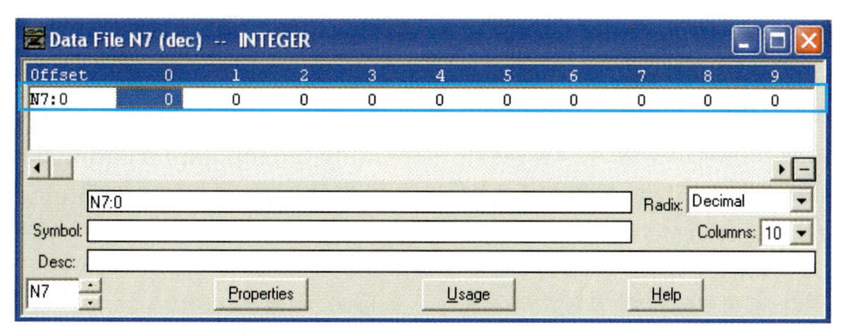

(c) Integer word data files

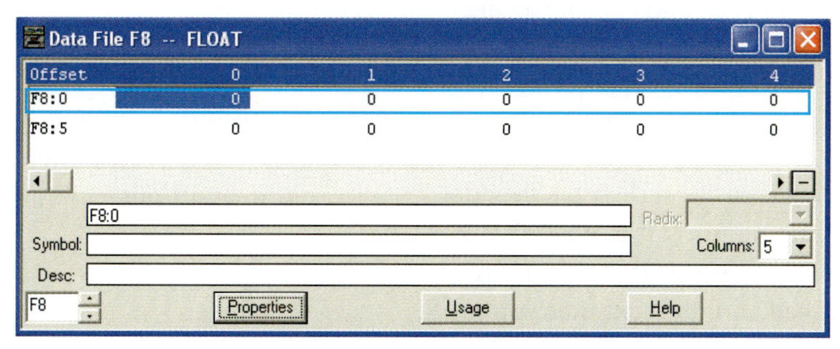

(d) Floating point word data files

FIGURE 6-21: Program Tags dialog boxes—Monitor Tags tab and Edit Tags tab.

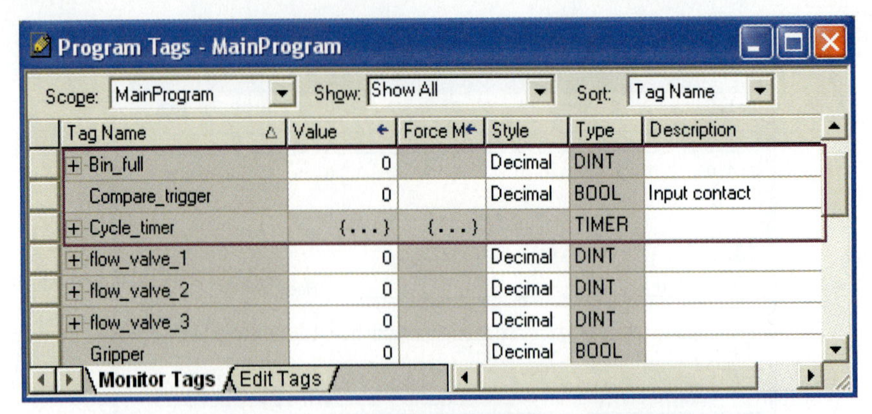

(a) Monitor tags table

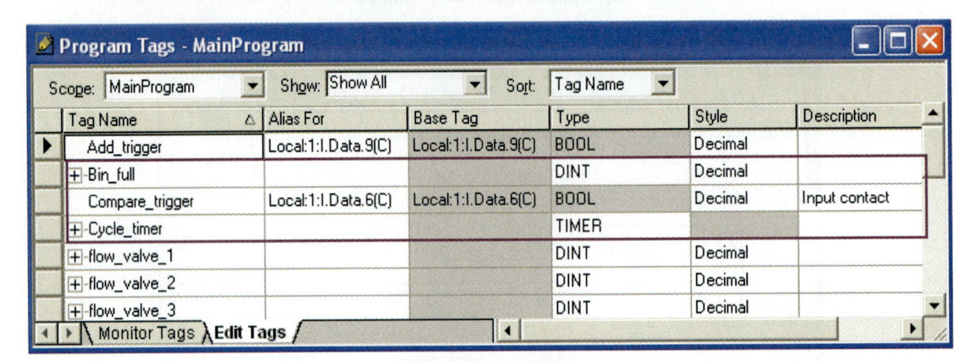

(b) Edit tags table

test modes can be used to test any PLC ladder program in the SLC 500 systems. The modes are selected by clicking on the *Comms* drop-down menu at the top of the RSLogix 500 program screen and then selecting *Mode*. The PLC system must be in the run mode for the options to be active.

Single-Step test mode. The *single-step test mode*, available on SLC 5/02 and higher controllers, initiates the processor to scan and execute a single rung or group of rungs. The mode is set up with parameters in the Test Single Step dialog box. The parameters include:

Current Location: This field displays the current file and rung number; it is not editable. When a single-step scan is enabled, the current location is retained until the next time you call up the dialog box.

Go Breakpoint: A *breakpoint* is established by entering a file and rung number from which the single-step scanning should begin. The scan is started by clicking *Go Single Step*.

Go Single Step: Initiates single-step scanning.

Single-Scan test mode. The *single-scan test mode*, available with all Allen-Bradley controllers, executes a single operating cycle that includes reading the inputs, executing the ladder program, and updating all data without energizing output circuits.

Go Single Scan: Click this button to start a single scan of the ladder program.

Continuous-Scan test mode. The *continuous-scan test mode* is the same as the REM run mode, except output circuits are not energized. This allows you to troubleshoot or test your ladder program without energizing output field devices.

6-9 LOCATION OF THE INSTRUCTIONS

The location of instructions from this chapter in the Allen-Bradley programming software is indicated in Figure 6-22.

FIGURE 6-22: Location of instructions described in this chapter.

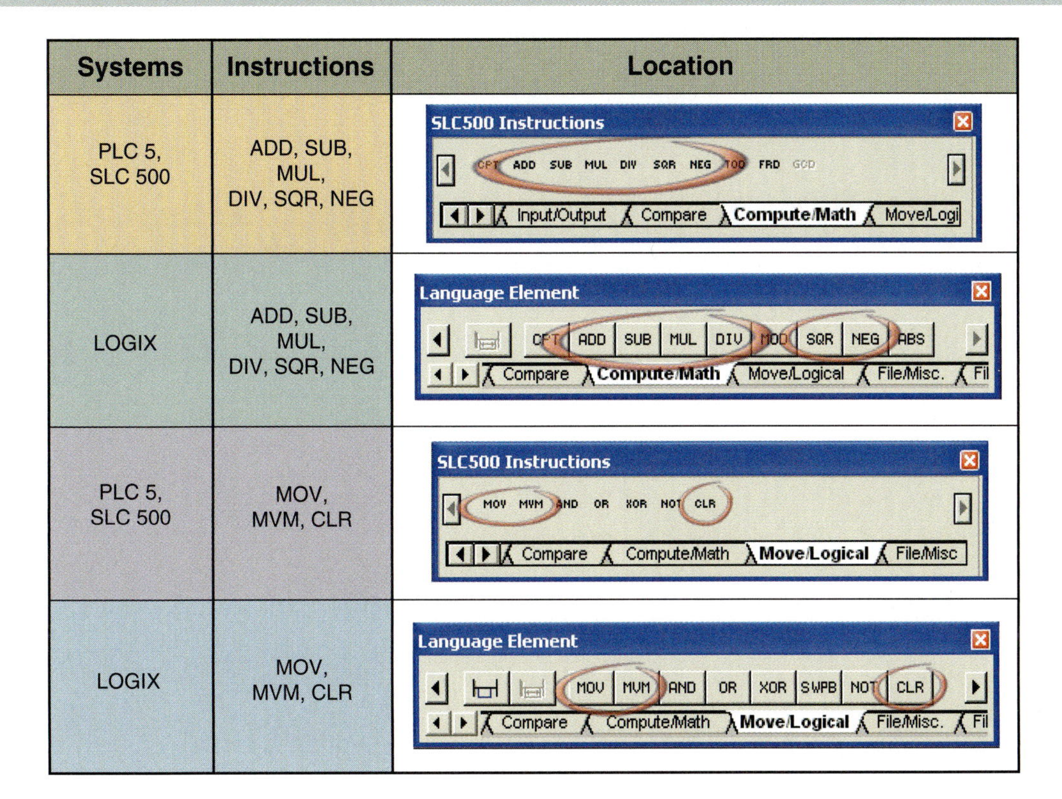

Systems	Instructions	Location
PLC 5, SLC 500	ADD, SUB, MUL, DIV, SQR, NEG	SLC500 Instructions: CPT ADD SUB MUL DIV SQR NEG TOD FRD GCD; Input/Output, Compare, Compute/Math, Move/Logi
LOGIX	ADD, SUB, MUL, DIV, SQR, NEG	Language Element: CPT ADD SUB MUL DIV MOD SQR NEG ABS; Compare, Compute/Math, Move/Logical, File/Misc., Fil
PLC 5, SLC 500	MOV, MVM, CLR	SLC500 Instructions: MOV MVM AND OR XOR NOT CLR; Compare, Compute/Math, Move/Logical, File/Misc
LOGIX	MOV, MVM, CLR	Language Element: MOV MVM AND OR XOR SWPB NOT CLR; Compare, Compute/Math, Move/Logical, File/Misc., Fil

Chapter 6 questions and problems begin on page 567.

CHAPTER

7

Conversion and Comparison Instructions

7-1 GOALS AND OBJECTIVES

There are two principal goals of this chapter. The first goal is to provide the student with information on the binary coded decimal and hexadecimal numbering systems and conversion instructions. The second goal is to show how the PLC's comparison and BCD conversion instructions are used in programs for industrial automation systems.

After completing this chapter you should be able to:

- Explain the binary coded decimal (BCD) and hexadecimal numbering systems.
- Use the convert to BCD (TOD) instruction and the convert from BCD (FRD) instruction.
- Convert numbers between the decimal, binary coded decimal, and hexadecimal systems.
- Describe the operation of the equal to, not equal to, less than, greater than, less than and equal to, greater than and equal to, and limit test instructions in Allen-Bradley PLCs.
- Design and analyze ladder logic programs, which use comparison and BCD conversion instructions.
- Develop automation solutions given an industrial control problem using the comparison and BCD conversion instructions for Allen-Bradley (AB) PLC 5, SLC 500, and ControlLogix PLCs.
- Include comparison instructions in the empirical design process.
- Describe troubleshooting techniques for ladder rungs with comparison instructions.

7-2 BINARY CODED DECIMAL SYSTEM

The *binary coded decimal* (BCD) system is a binary number system with a relatively easy conversion process between the decimal (system used by humans) and binary coded decimal (system used for automation data) number representation. The BCD system uses four binary bits to represent the decimal numbers 0 through 9. The BCD representation of a decimal number is obtained by replacing each decimal digit by its equivalent four-bit binary number. Table 7-1 illustrates the relationship between the decimal, binary, and BCD numbering systems, and Figure 7-1 presents a decimal to BCD conversion example. The BCD number in Figure 7-1 is just the binary equivalent for each decimal digit. Use the data in Table 7-1 to verify the BCD answer in the figure.

A push switch, also called a pushwheel or thumbwheel switch, is often used to input numerical values and set points into a machine

TABLE 7-1 Decimal, binary, and BCD counting comparison.

Decimal	Binary	BCD		
0	0000			0000
1	0001			0001
2	0010			0010
3	0011			0011
4	0100			0100
5	0101			0101
6	0110			0110
7	0111			0111
8	1000			1000
9	1001			1001
10	1010		0001	0000
11	1011		0001	0001
12	1100		0001	0010
13	1101		0001	0011
18	10010		0001	1000
19	10011		0001	1001
20	10100		0010	0000
98	1100010		1001	1000
99	1100011		1001	1001
100	1100100	0001	0000	0000
498	111110010	0100	1001	1000
499	111110011	0100	1001	1001
500	111110100	0101	0000	0000

or process controlled by a PLC. A four-digit push switch is shown in Figure 7-2 with the + button and − button labeled. The push switch is a new version of the older thumbwheel switch where the digits were changed by rotating a geared thumbwheel. The push switch in Figure 7-2 can have 4 (BCD output), 8 (octal output), or 10 (decimal output) SPST (single pole single throw) switches with one terminal from each switch connected to a common terminal. Figure 7-2(b) shows the four SPST switches for a BCD version of the push switch with two BCD digits. The push switch has the decimal number 63 entered, with 6 representing the most significant digit. Note that the four switches for this most significant digit are wired to the four most significant bits of the eight point input module. The switches for the digit 3 connect to the least significant bits of the PLC input.

The push switch output tables in Figure 7-3(a and b) illustrate a 10-position 10 switch decimal and 10-position 4 switch BCD, respectively. The 10-position decimal push switch has 10 SPST switches corresponding to switch values 0 to 9. For example, when the switch has a value of 3, the third SPST switch connecting the common terminal to the output terminal of the switch is closed. The 10-position BCD push switch creates a BCD code for the value dialed into the push switch.

The new push and older thumbwheel switches are examples of a device that converts decimal numbers to BCD numbers and is used as an interface to PLC inputs. Likewise, the PLC output of a BCD number can be interfaced to a seven-segment display, which converts it to a decimal number. PLCs have instructions to convert BCD to binary because push switches produce BCD numbers, but the PLC instructions operate on the

FIGURE 7-1: Decimal to BCD conversion.

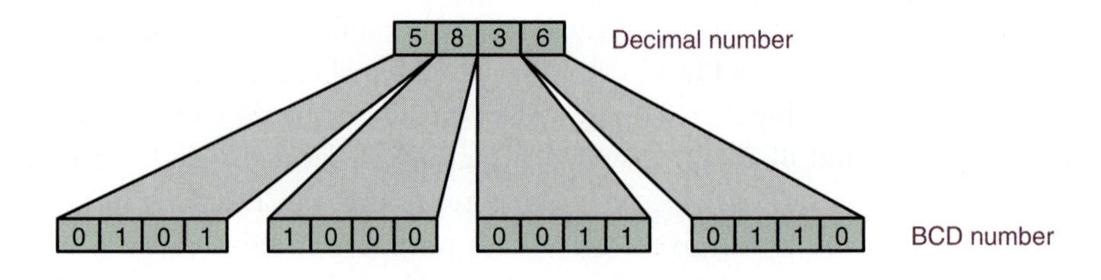

binary value of decimal numbers. Also, the binary to BCD instruction is necessary because seven-segment displays require a BCD output from the PLC. Figure 7-4(a) depicts a four-digit push switch as an input device connected to a PLC and a four-digit, seven-segment display as an output device connected to a PLC. The switch output lines are labeled 8, 4, 2, and 1 for each of the BCD number digits 9234_{BCD}. The BCD bit pattern of 1001 0010 0011 0100 is indicated by the blue and black signal wires (blue closed switch and black open switch). Note that the PLC output wiring, Figure 7-4(b), and seven-segment display input lines are similarly labeled. The Allen-Bradley BCD conversion instructions are covered in the following sections.

7-2-1 Allen-Bradley BCD Instructions and Standard Ladder Logic

The Allen-Bradley BCD conversion instructions are *convert to BCD* (TOD) and *convert from BCD* (FRD). Figure 7-5 illustrates these two instructions. The TOD instruction converts 16-bit integers to a BCD equivalent, and the FRD instruction converts a BCD number to its 16-bit integer equivalent.

Because the BCD number, before and after the conversion, is stored in a 16-bit integer register, the size of the register presents a problem. Note that status bit S:2/1 in Table 7-2 and S:0/1 in Table 7-3 indicates when this problem is present in the PLC 5 and SLC 500 PLCs. The problem arises because the number of bits required to store a number in BCD is larger than the bit total for the same number value stored in binary. An example illustrates this point. A 16-bit integer register with the maximum permitted value of $+32.767_{10}$ would have a bit pattern of 0111111111111111_2 or the first 15 bits for the number and a sign bit. When the same decimal number, $+32.767_{10}$ is converted to BCD, the representation is $0011\ 0010\ 0111\ 0110\ 0111_{BCD}$. As you can see, it takes 20 bits to store the equivalent BCD number. The largest BCD number that a 16-bit integer register (four groups of four bits) can store is 9999. Each of the Allen-Bradley processors handles this problem differently.

PLC 5 BCD Instructions and Standard Logic.

The BCD conversion instructions for the PLC 5 are limited to BCD values in the 0 to 9999

FIGURE 7-2: BCD push switch.

(a) BCD push switch

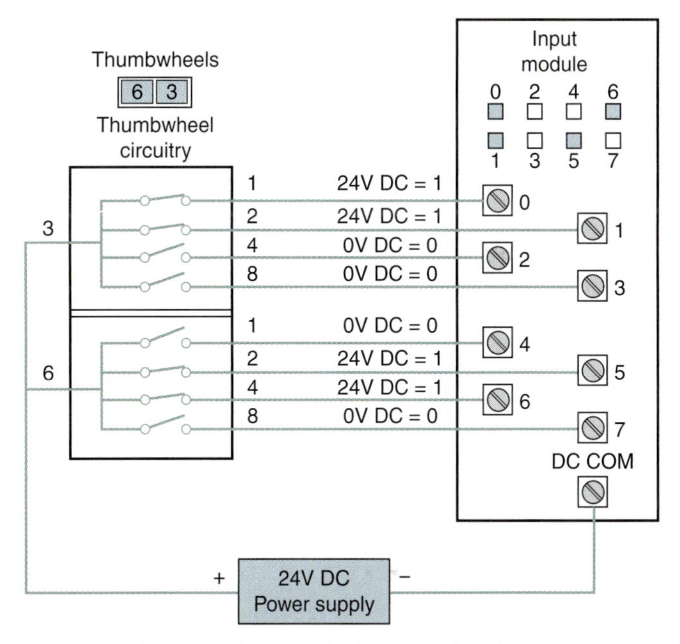

(b) Input wiring for BCD push switch

Courtesy of Cherry Electrical Products.

FIGURE 7-3: Input/Output truth table for 10-position decimal and BCD push witches or thumbwheel switches.

Ten-position decimal

Dial position	Common to: 0	1	2	3	4	5	6	7	8	9
0	●									
1		●								
2			●							
3				●						
4					●					
5						●				
6							●			
7								●		
8									●	
9										●

(a)

Ten-position BCD

Dial position	Common to: 1	2	4	8
0				
1	●			
2		●		
3	●	●		
4			●	
5	●		●	
6		●	●	
7	●	●	●	
8				●
9	●			●

(b)

Rehg and Sartori, Industrial Electronics, *1st Edition, © 2006, Reprinted by permission of Pearson Education, Inc., Upper Saddle River, NJ.*

FIGURE 7-4: Push switch input and seven-segment output.

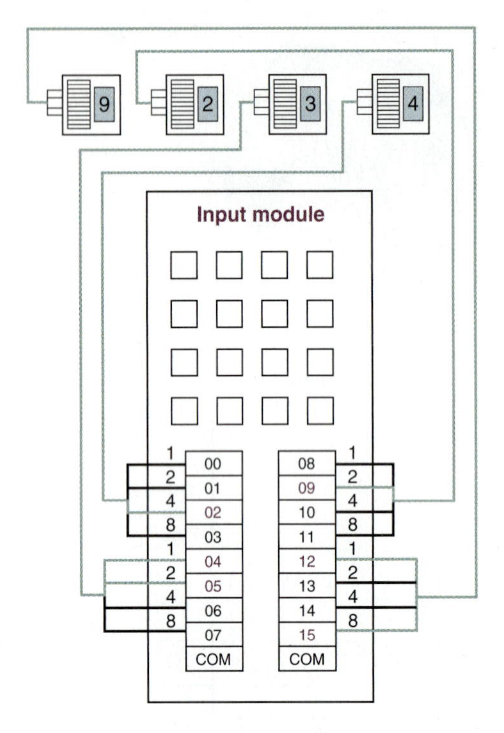

(a) Push switch input with SLC PLC module in slot 2

(b) Seven-segment output with SLC PLC module in slot 3

FIGURE 7-5: TOD and FRD instructions.

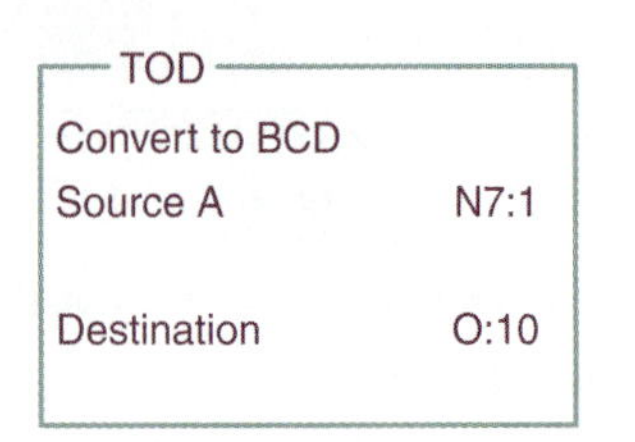

(a) Convert to BCD instruction (TOD)

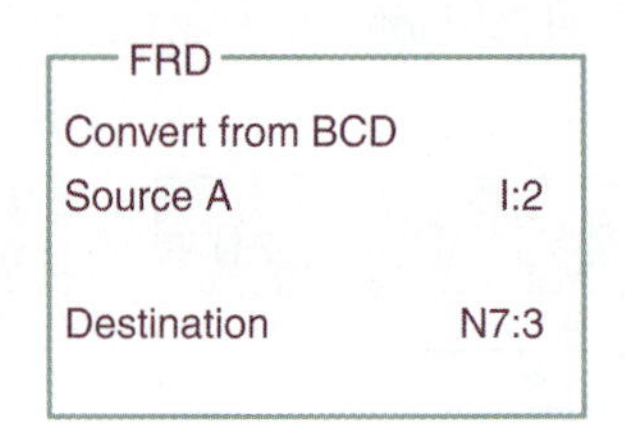

(b) Convert from BCD instruction (FRD)

range. As a result, the integer file bit length is adequate to handle all conversions with both instructions, TOD and FRD. However, the largest binary integer that can be converted to BCD is 10001111101111_2 or the decimal equivalent of 9999. The range of BCD values restricts negative number conversion into BCD as well.

Standard ladder logic rungs for the TOD and FRD instructions are illustrated in Figure 7-6. Read the implementation in the figure.

The status bits for the PLC 5 are listed and described in Table 7-2. A study of the table indicates that BCD values outside the accepted range cause the overflow bit to be set and a minor error to be generated. Zero results cause the S:2/2 status bit to be set to true.

SLC 500 BCD Instructions and Standard Logic.

Model 502 and higher SLC processors execute the BCD instructions in the following modes:

Mode 1: If the BCD source address values are not over 9999 decimal or BCD, then the destination address can be the word address of any integer data file.

Mode 2: If the BCD source address values could reach the integer maximum value of 32,767, then the math register S:13 must be used for the destination address.

In executing the two modes, the SLC processor uses status bits to indicate the conditions

TABLE 7-2 PLC 5 status bits.

Status bit	Name	Description
S:2/1	Overflow (V)	Bit is set if the BCD result is outside the range of 0 to 9999 and a minor error is generated, but is reset otherwise.
S:2/2	Zero (Z)	Bit is set if the result or value after a conversion is negative or zero.

(a) Status bit table for TOD instruction

Status bit	Name	Description
S:2/2	Zero (Z)	Bit is set if the result or value after a conversion is a zero, but is reset otherwise.

(b) Status bit table for FRD instruction

TABLE 7-3 SLC 500 status bits.

Status bit	Name	Description
S:0/1	Overflow (V)	Bit is set if the BCD result is larger than 9999 and minor error bit S:5/0 is also set.
S:0/2	Zero (Z)	Bit is set if the result or value after a conversion is a zero, but is reset otherwise.
S:0/3	Sign (S)	Bit is set if the source word is a negative value, but is reset if the source word is a positive value.

(a) Status bit table for TOD instruction

Status bit	Name	Description
S:0/1	Overflow (V)	Bit is set if the source word is a non-BCD value or if the value is greater than 32,767, but is reset otherwise. If the bit is set the minor error bit S:5/0 is also set.
S:0/2	Zero (Z)	Bit is set if the result or value after a conversion is a zero, but is reset otherwise.

(b) Status bit table for FRD instruction

present in the BCD conversion. Study the status bits in Table 7-3 regarding the operational status of the two SLC 500 BCD conversion instructions.

In mode 1 the SLC operation is similar to the PLC 5 logic described in the standard ladder logic rungs in Figure 7-6. Integer registers are used for source and destination addresses. However, if a negative number is entered into the source address, then the absolute value of the number is generated before the conversion and the sign status bit, S:0/3, is set.

In mode 2 the SLC operation handles BCD numbers over 9999 but not greater than the integer maximum value of +32,767. The conversion of decimal numbers this large into their BCD equivalent requires 20 binary bits (five groups of 4 bits each), so the 16-bit data registers will not work. To facilitate conversions of numbers of this size, the math registers S:13 and S:14 must be used. Also, since the number is larger than 9999, the overflow status bit, S:0/1, is set, and the fault condition must be reset. The standard ladder logic rungs used for the SLC 500 for this type of conversion are illustrated in Figure 7-7.

Some comments on the standard ladder rungs and examples in Figure 7-7 follow:

- When math register S:13 is placed in the source or destination address, both S:13 and S:14 data are used in the conversion.
- The S:5/0 minor error bit is reset in the ladder in Figure 7-7(a) before the end of the scan, so no major error occurs.
- In the TOD instruction in Figure 7-7(a), the converted decimal value is stored as a BCD

FIGURE 7-6: Standard ladder logic for PLC 5 and SLC 500 BCD instructions for BCD values less than 9999.

Application	Standard Ladder Logic Rungs for PLC 5 and SLC 500 BCD Instructions
Convert the binary value of a decimal number to its equivalent BCD value when a trigger is generated by the action of an input field device(s) and the true condition of the input ladder logic. The binary to BCD conversion instruction (TOD) in the PLC 5 system is used to convert a positive binary number with a decimal value of 9999 or less into the equivalent BCD value. The SLC 502 and higher processors use the same standard ladder logic for the TOD instruction to convert a positive or negative binary number with a decimal value of 9999 or less into the equivalent BCD value. Study the example solution illustrated below the ladder rung to understand how the TOD instruction performs when the source decimal values are not greater than 9999_{10}.	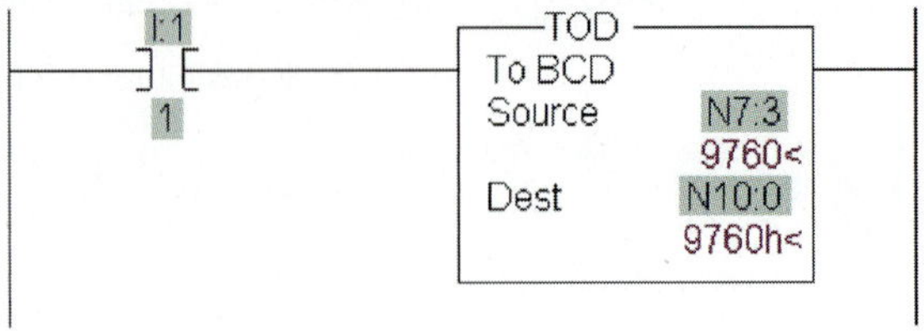 I:1 1 TOD To BCD Source N7:3 9760< Dest N10:0 9760h< 9 7 6 0 **N7:3** **Decimal** 0010 0110 0010 0000 9 7 6 0 **N10:0** **4-digit BCD** 1001 0111 0110 0000 **(a)**
Convert a BCD value to its binary equivalent value when a trigger is generated by the action of an input field device(s) and the true condition of the input ladder logic. The BCD to binary conversion instruction (FRD) in the PLC 5 system is used to convert a positive BCD value of 9999 or less into the binary number equivalent. The SLC 502 and higher processors use the same standard ladder logic for the FRD instruction to convert a positive BCD value of 9999 or less into the binary number equivalent. Study the example solution illustrated below the ladder rung to understand how the FRD instruction performs when the source BCD values are not greater than 9999_{BCD}.	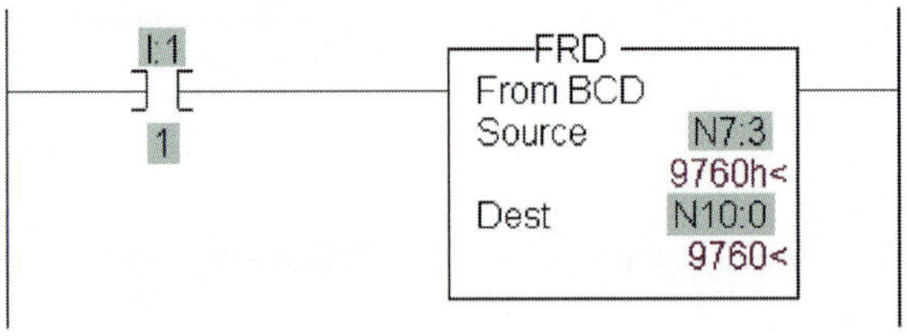 I:1 1 FRD From BCD Source N7:3 9760h< Dest N10:0 9760< 9 7 6 0 **N7:3** **4-digit BCD** 1001 0111 0110 0000 9 7 6 0 **N10:0** **Decimal** 0010 0110 0010 0000 **(b)**

number (00032760) in math registers S:14 (most significant digits 0003) and S:13 (the least significant digits 2760).

- In Figure 7-7(a), the BCD number is moved from the math registers to two outputs: O:3.0, the least significant four digits (2760_{BCD} which is equal to 10080_2), and O:4.0, the most significant digit (3_{BCD} which is equal to 3_2). The output address uses the dot (.) 0 notation indicating that the 16-bit words 3 and 4 are the target of the BCD number move. Also, a masked move (MVM) instruction is used because only the four least significant bits of the S:14 16-bit word are critical (they hold the value 3).
- The math register (S:13 and S:14) is used as the source for the FRD instruction and holds the BCD value of 32,760 in Figure 7-7(d). The 5-digit BCD number $32{,}760_{BCD}$ has a BCD bit pattern of

S:14	S:13			
3	2	7	6	0
0000 0000 0000 0011	0010	0111	0110	0000

The FRD instruction converts the $32{,}760_{BCD}$ to $32{,}760_{10}$, where the decimal value has a binary representation of

0111 1111 1111 1000

Note that when $32{,}760_{10}$ is converted to BCD the BCD value is $32{,}760_{BCD}$ but the bit pattern of the BCD number is not the same as the

FIGURE 7-7: Standard ladder logic for SLC 500 BCD instructions for BCD values greater than 9999.

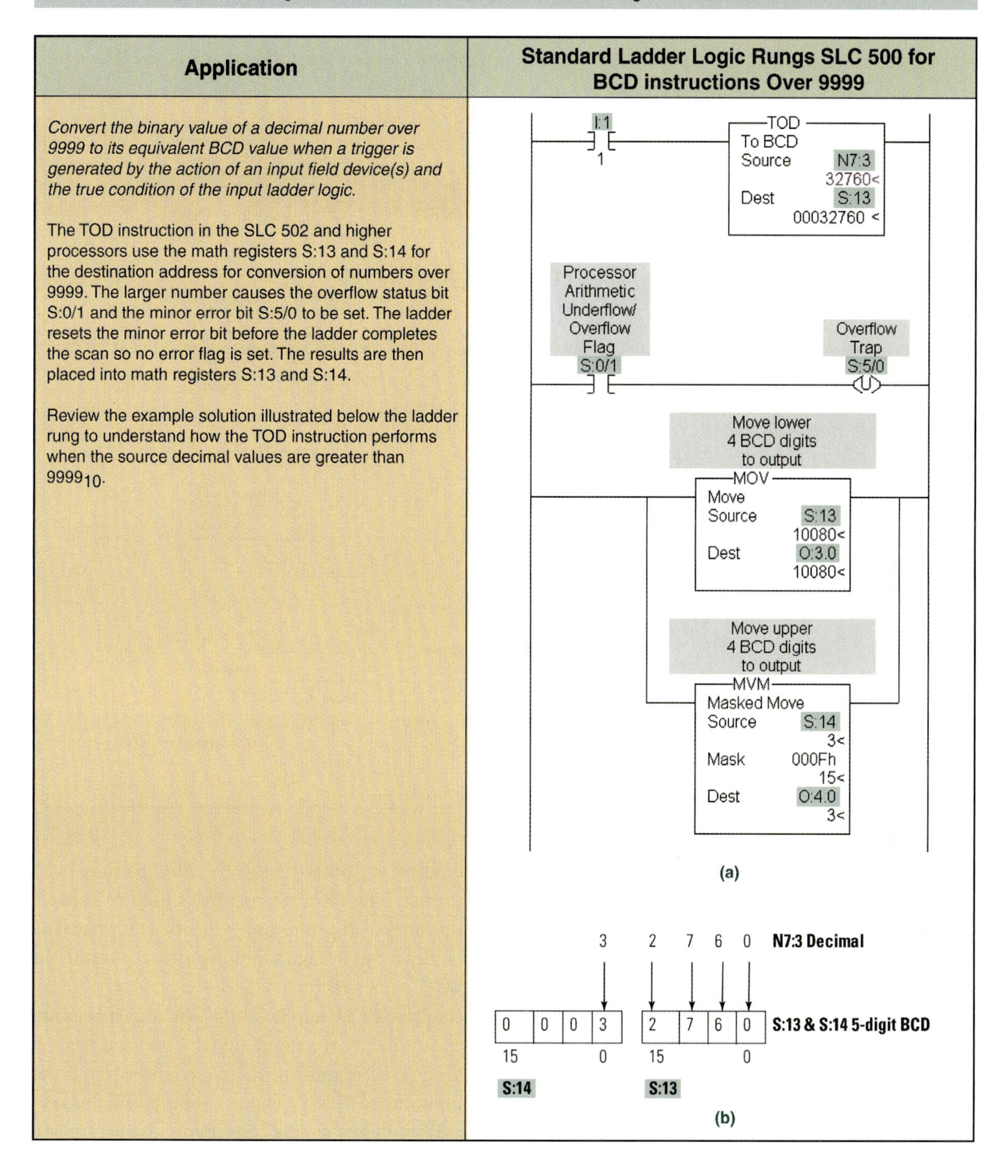

Application	Standard Ladder Logic Rungs SLC 500 for BCD instructions Over 9999
Convert the binary value of a decimal number over 9999 to its equivalent BCD value when a trigger is generated by the action of an input field device(s) and the true condition of the input ladder logic. The TOD instruction in the SLC 502 and higher processors use the math registers S:13 and S:14 for the destination address for conversion of numbers over 9999. The larger number causes the overflow status bit S:0/1 and the minor error bit S:5/0 to be set. The ladder resets the minor error bit before the ladder completes the scan so no error flag is set. The results are then placed into math registers S:13 and S:14. Review the example solution illustrated below the ladder rung to understand how the TOD instruction performs when the source decimal values are greater than 9999_{10}.	(a), (b)

binary representation or bit pattern for the equivalent decimal value. This is a point of confusion when discussing binary coded decimal numbers and regular decimal numbers.

ControlLogix BCD Instructions and Standard Logic. The TOD instruction takes the decimal value between 0 and 99,999,999 stored in the source register and converts it to a BCD value that is stored in the destination register. The allowed source and destination date types are single integer (SINT), integer (INT), and double integer (DINT). The parameters placed into the source include variables in the form of tag names and positive integer values, called *immediate values* by Allen-Bradley. The parameter placed in

FIGURE 7-7: (Continued).

Application	Standard Ladder Logic Rungs SLC 500 for BCD instructions Over 9999
Convert a BCD value over 9999 to its binary equivalent value when a trigger is generated by the action of an input field device(s) and the true condition of the input ladder logic. The FRD instruction in the SLC 502 and higher processors used the math registers S:13 and S:14 for the source address for conversion of numbers over 9999. The overflow bit is only set if a non-BCD number is present in the source address or if the number is greater than 36,767. The S:14 bit is cleared in case the BCD value has only 4 bits (S:14 is not used) and current data in S:14 would make the conversion incorrect. Note that the N7:2 value of 4660_{10} is the decimal equivalent of the BCD number 1234_{BCD}. The FRD instruction registers always display the numbers in BCD. Study the example solution illustrated below the ladder rung to understand how the FRD instruction performs when the source BCD values are greater than 9999_{BCD}.	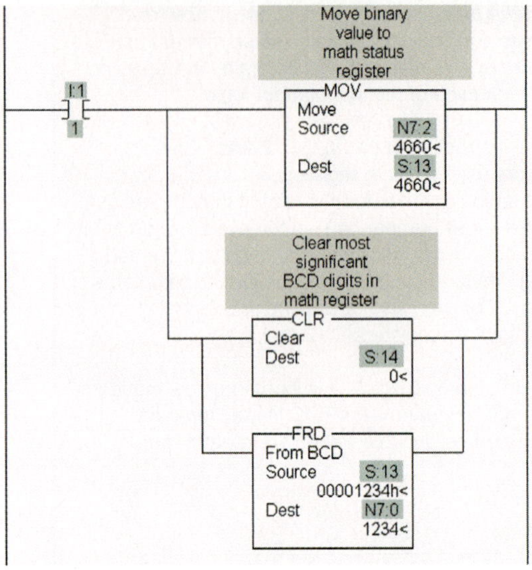 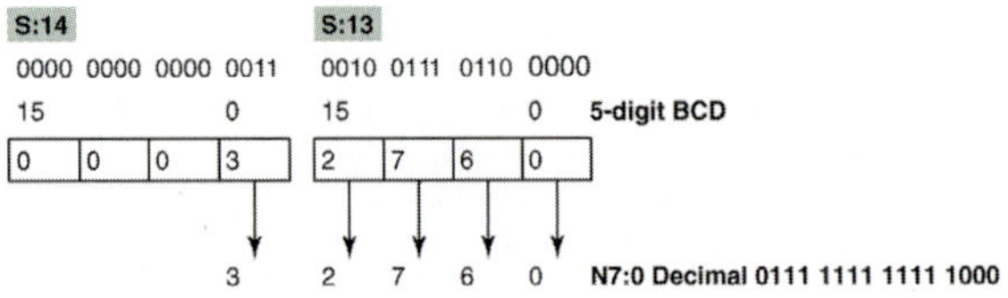 I:1 / 1 Move binary value to math status register MOV Move Source N7:2 4660< Dest S:13 4660< Clear most significant BCD digits in math register CLR Clear Dest S:14 0< FRD From BCD Source S:13 00001234h< Dest N7:0 1234< (c) S:14 0000 0000 0000 0011 (15 … 0) [0 | 0 | 0 | 3] S:13 0010 0111 0110 0000 (15 … 0) [2 | 7 | 6 | 0] 5-digit BCD 3 2 7 6 0 N7:0 Decimal 0111 1111 1111 1000 (d)

the destination is a tag name. Placement of a negative value in the source creates a minor fault and clears the destination.

The standard ladder rung for the ControlLogix BCD TOD instruction is illustrated in Figure 7-8(a). Note that the rung looks like the PLC 5 and SLC 500 rungs except that tags are used in place of memory-specific registers found in the earlier PLCs.

The FRD instruction converts a BCD value in the source register to a decimal value and stores the result in the destination register. The allowed data types and parameters are the same as the TOD instruction, and no fault conditions are present. The standard ladder rung for the ControlLogix BCD FRD instruction is illustrated in Figure 7-8(b).

Standard Logic for BCD with One-Shot (OSR) Instruction. Output field devices used to display process parameters often require that their input data or the output from the PLC be in a BCD format. The FRD instruction is used to convert the internal PLC integer file to a BCD format. The ladder rungs in Figure 7-8(a and b) convert input BCD values to binary and then convert binary results to BCD output values. The conversions are controlled by the input signal *Compare_trigger*. One problem must be addressed, however. If the input instruction in Figure 7-6(a) is true for several seconds, then the conversion instruction will execute on every scan while the instruction is true. Now, if the input data in N7:3 is changing rapidly, then those changes will be passed to the destination, N7:10, and eventually to an output display after every scan, which could make an output display flicker as numbers change.

This situation is solved by placing a one-shot rising (OSR) instruction in the TOD instruction rung. The solution is illustrated in Figure 7-9 in the

FIGURE 7-8: Standard ladder logic for ControlLogix BCD instructions.

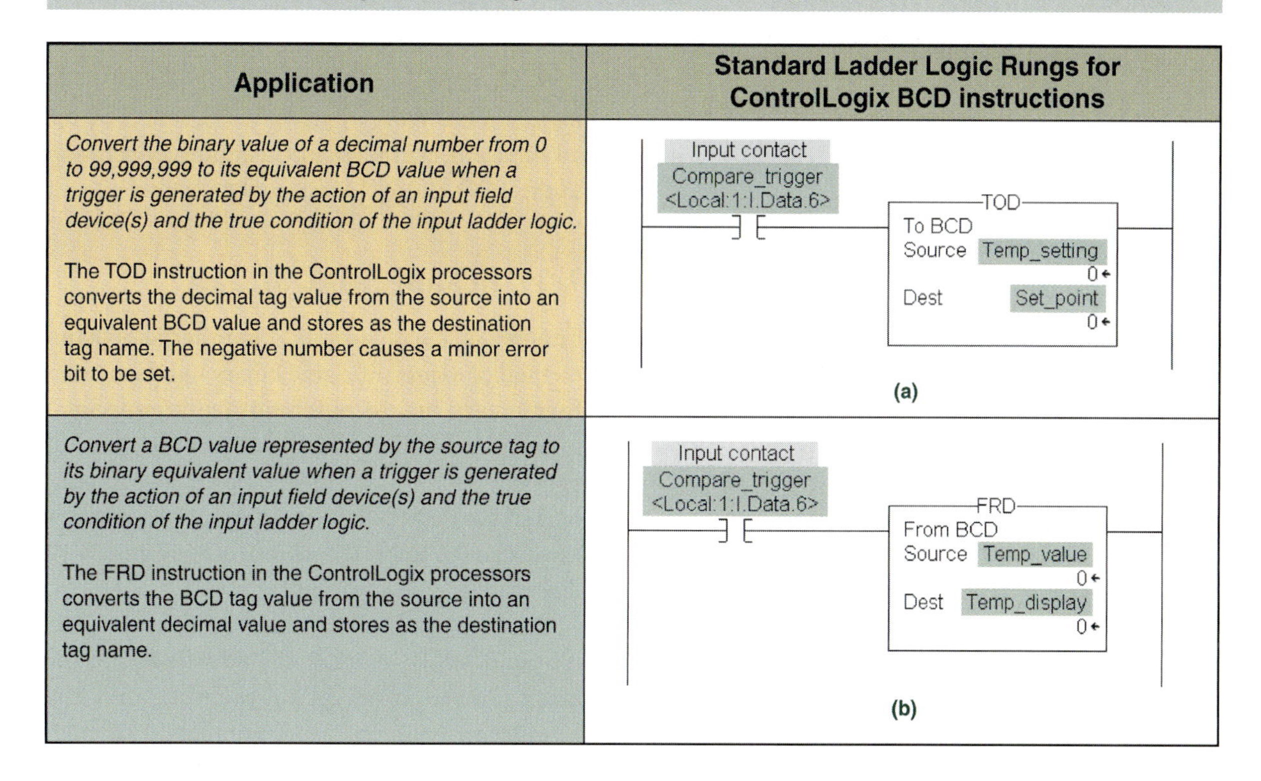

Application	Standard Ladder Logic Rungs for ControlLogix BCD instructions
Convert the binary value of a decimal number from 0 to 99,999,999 to its equivalent BCD value when a trigger is generated by the action of an input field device(s) and the true condition of the input ladder logic. The TOD instruction in the ControlLogix processors converts the decimal tag value from the source into an equivalent BCD value and stores as the destination tag name. The negative number causes a minor error bit to be set.	(a)
Convert a BCD value represented by the source tag to its binary equivalent value when a trigger is generated by the action of an input field device(s) and the true condition of the input ladder logic. The FRD instruction in the ControlLogix processors converts the BCD tag value from the source into an equivalent decimal value and stores as the destination tag name.	(b)

FIGURE 7-9: Standard ladder logic for SLC 500 BCD instructions for BCD values sampled with a one-shot instruction.

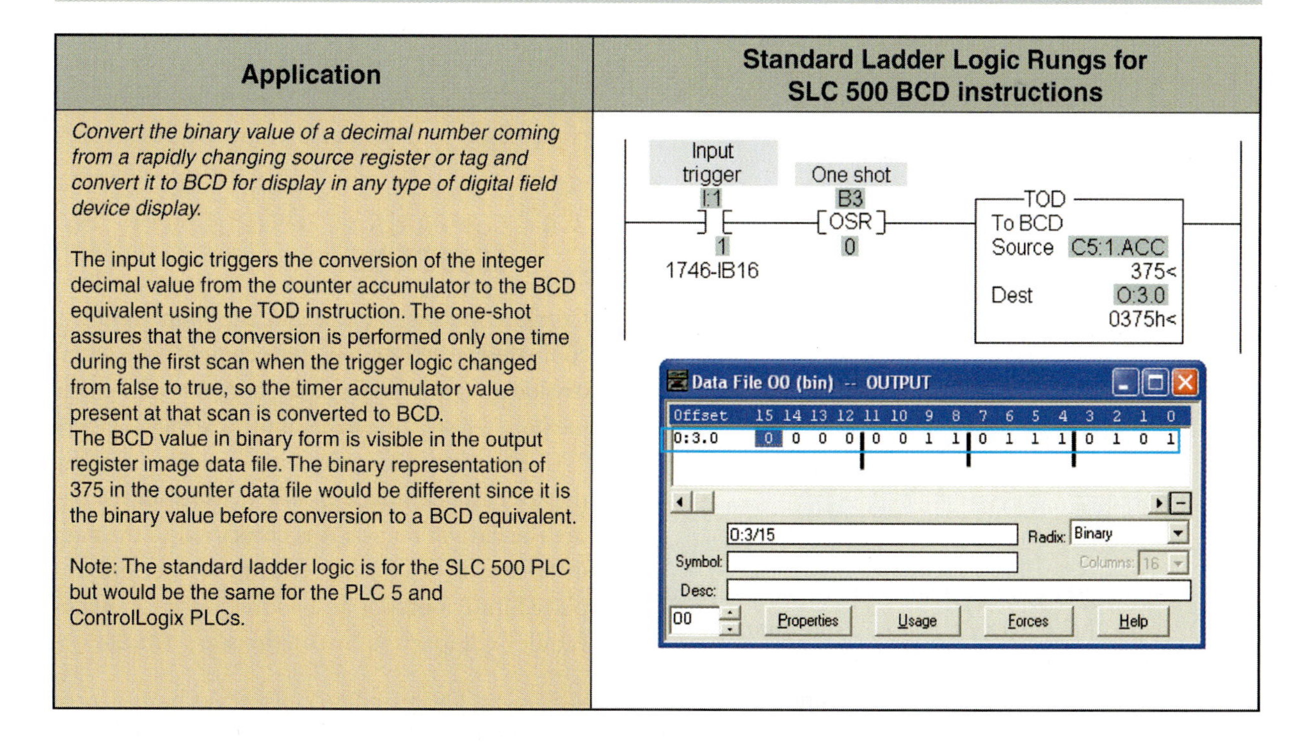

Application	Standard Ladder Logic Rungs for SLC 500 BCD instructions
Convert the binary value of a decimal number coming from a rapidly changing source register or tag and convert it to BCD for display in any type of digital field device display. The input logic triggers the conversion of the integer decimal value from the counter accumulator to the BCD equivalent using the TOD instruction. The one-shot assures that the conversion is performed only one time during the first scan when the trigger logic changed from false to true, so the timer accumulator value present at that scan is converted to BCD. The BCD value in binary form is visible in the output register image data file. The binary representation of 375 in the counter data file would be different since it is the binary value before conversion to a BCD equivalent. Note: The standard ladder logic is for the SLC 500 PLC but would be the same for the PLC 5 and ControlLogix PLCs.	

TABLE 7-4 Counting in binary and hexadecimal number systems.

Binary			Hexadecimal
		0000	0
		0001	1
		0010	2
		0011	3
		0100	4
		0101	5
		0110	6
		0111	7
		1000	8
		1001	9
		1010	A
		1011	B
		1100	C
		1101	D
		1110	E
		1111	F
	0001	0000	10
	1001	1110	9E
	1001	1111	9F
	1010	0000	A0
	1111	1110	FE
	1111	1111	FF
1	0000	0000	100

FIGURE 7-10: Hexadecimal to binary number conversion.

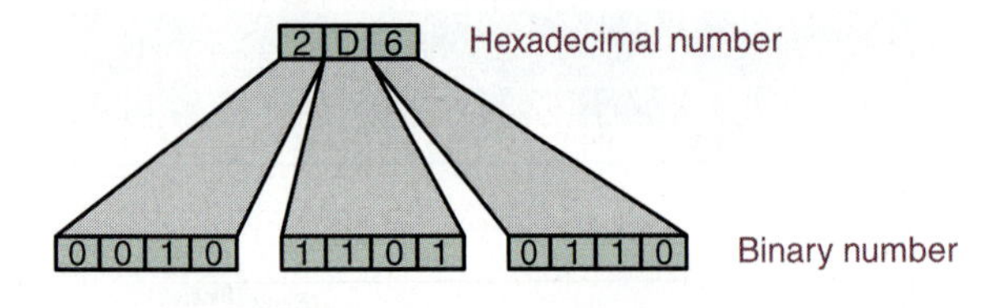

standard logic rung with the OSR instruction used with the TOD instruction. Note that while the input logic can be true for any length of time, the BCD instruction converts the counter accumulator to BCD only once on the rising edge of the input logic. The destination register is an output, O:3.0, that drives a seven-segment display. The next change in the readout occurs when the input logic for the TOD instruction again cycles from false to true.

Study the standard ladder logic for the BCD instruction, Figures 7-6, 7-7, 7-8, and 7-9, for all three Allen-Bradley PLCs until the operation of the TOD and FRD instructions is understood.

7-3 HEXADECIMAL SYSTEM

The *hexadecimal system* is another shorthand method of expressing large binary numbers using four binary bits to represent one hexadecimal number. Table 7-4 illustrates the relationship between binary numbers and hexadecimal numbers. Note that the letters A through F represent the decimal numbers 10 through 15. Figure 7-10 illustrates this relationship.

Whereas the decimal numbering system has multiples of 10 as its weighted values and the octal numbering system has multiples of 8, the hexadecimal system (using numbers 0 through 9 and letters A through F) has multiples of 16. Figure 7-11 illustrates the method to translate a hexadecimal number to a decimal number. Note that the decimal number is the sum of the products. Study the figure until you are familiar with translating hexadecimal numbers to their decimal equivalent. Hexadecimal numbers are used in many PLC instructions, including comparison instructions that are discussed next.

7-4 COMPARISON INSTRUCTION STRUCTURE

Comparison instructions, as the name implies, compare two sets of data—the contents of a memory location or register to the contents of another memory register or a fixed numerical value. Figure 7-12 illustrates the instruction structure used by the six basic comparison instructions. In the figure the data in Source A is compared to the data in Source B. The requirements for parameters placed into Source A and Source B for the Allen-Bradley comparison instructions are as follows:

PLC 5: The parameter placed in both Source A and B can be a numerical value or program constant, an integer data type, or a floating point data type.

FIGURE 7-11: Hexadecimal to decimal conversion.

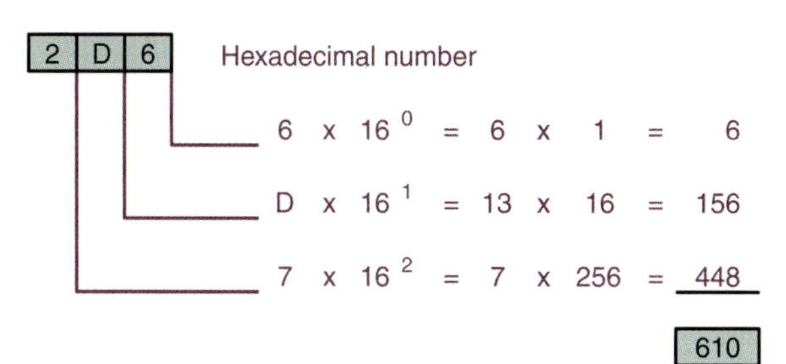

FIGURE 7-12: Comparison instruction.

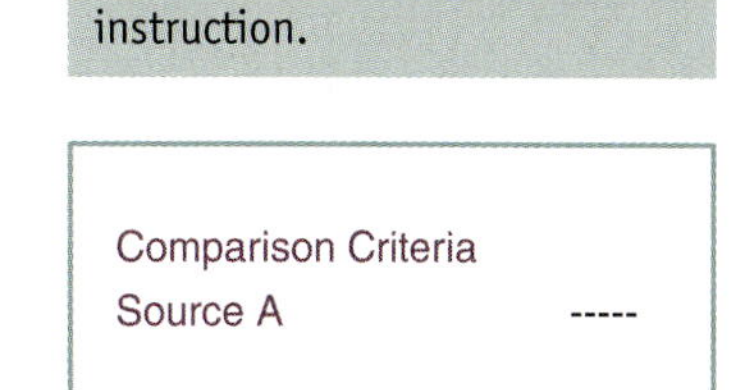

SLC 500: Source A must be a bit or word address and Source B can be either a program constant or a bit/word address. Negative integers are stored in 2's complement.

ControlLogix: The parameter placed in both Source A and B can be any of the following data types: SINT (single integer), INT (integer), DINT (double integer), REAL (real), or string. The parameter can be either an immediately entered program constant or a tag name.

Care must be used when real data types are used in comparison instructions because the decimal component in real variables often causes problems when comparisons like EQU and NEQ are made. All comparison instructions are inputs for the ladder logic and make the rung true or false based on the evaluation of Source A and B parameters using the comparison condition.

Comparison instructions are widely used in industry and have many applications, such as initiating a process when an input is at a proper value, halting a process when an output has reached a prescribed value, and indicating when a process parameter is outside of a tolerance range.

7-5 ALLEN-BRADLEY COMPARISON INSTRUCTIONS

The Allen-Bradley PLCs have a set of comparison instructions available in all three processors. The following comparison instructions, common to all three PLC systems, are described in this section.

- Equal (EQU): evaluates whether two values are equal
- Not equal (NEQ): evaluates whether two values are not equal
- Less than (LES): evaluates whether one value is less than another
- Greater than (GRT): evaluates whether one value is greater than another
- Less than or equal (LEQ): evaluates whether one value is less than or equal to another
- Greater than or equal (GEQ): evaluates whether one value is greater than or equal to another
- Masked equal to (MEQ): evaluates whether one masked value is equal to the same masked area of another value
- Limit test (LIM): tests for values inside of or outside of a specific range

7-5-1 Standard Ladder Logic for EQU, NEQ, LES, and GRT Comparison Instructions

The ladder logic structure used for comparison instructions with Allen-Bradley PLCs is the same. All instructions are part of the input ladder logic and are used individually or in combination with other comparison or input instructions. The PLC 5 and SLC 500 use memory address registers for data, whereas Logix PLCs use tag names. The equal to (EQU), not equal to (NEQ), less than (LES), and greater than (GRT) instructions, illustrated in Figure 7-13, check for a simple comparison between two values. These examples use a variety of data types, Allen-Bradley PLC models, and register/tag examples to show how each would be used with a comparison instruction. Study the standard ladder logic for these comparison instructions in Figure 7-13 before reading the explanations that follow.

Allen-Bradley EQU Instruction and Standard Logic. The Allen-Bradley *equal to* (EQU) instruction compares two sources of data. If the two

FIGURE 7-13: Standard ladder logic for SLC 500 and ControlLogix single comparison instructions.

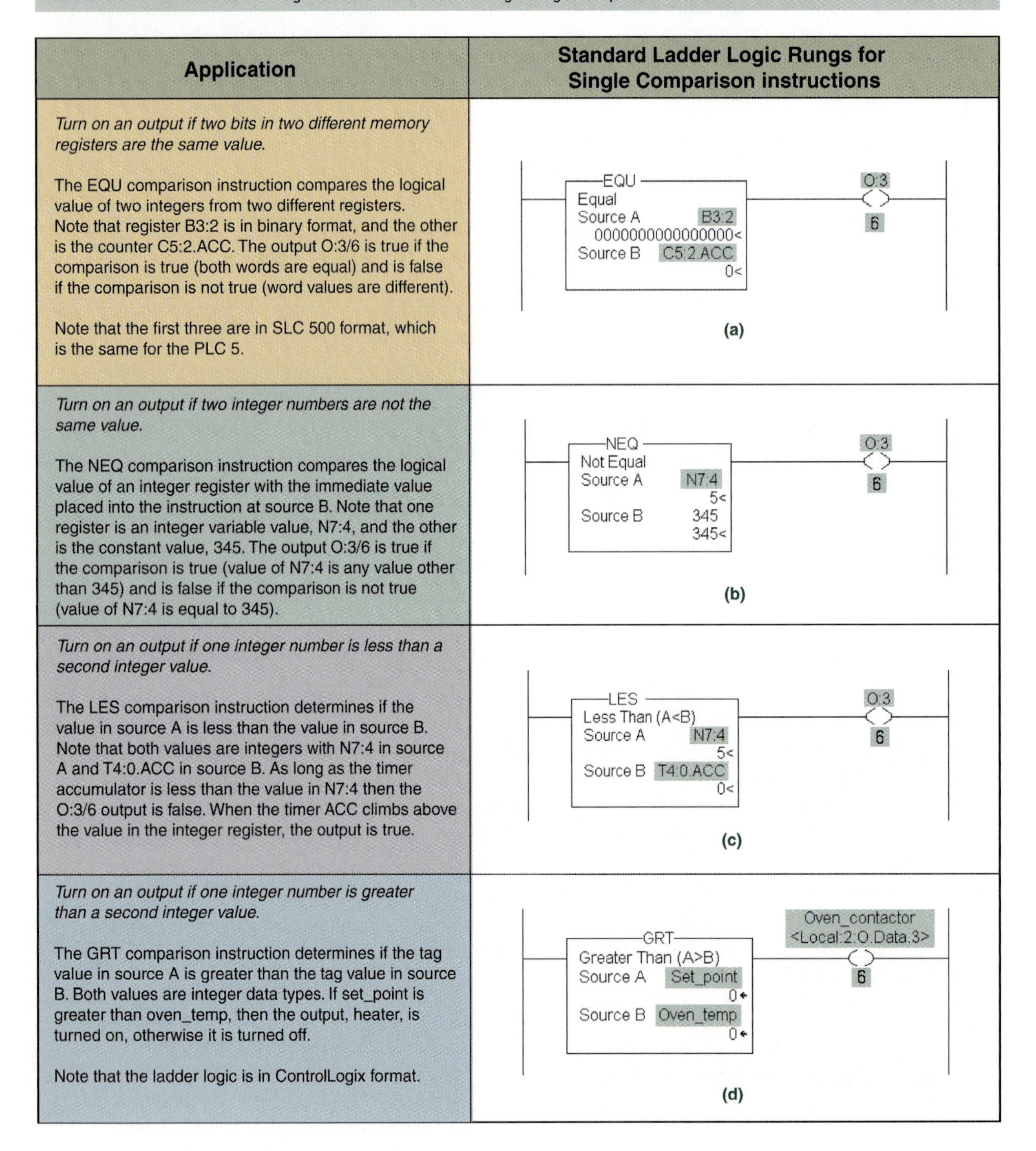

Application	Standard Ladder Logic Rungs for Single Comparison instructions
Turn on an output if two bits in two different memory registers are the same value. The EQU comparison instruction compares the logical value of two integers from two different registers. Note that register B3:2 is in binary format, and the other is the counter C5:2.ACC. The output O:3/6 is true if the comparison is true (both words are equal) and is false if the comparison is not true (word values are different). Note that the first three are in SLC 500 format, which is the same for the PLC 5.	EQU Equal Source A B3:2 0000000000000000< Source B C5:2.ACC 0< O:3 ( ) 6 **(a)**
Turn on an output if two integer numbers are not the same value. The NEQ comparison instruction compares the logical value of an integer register with the immediate value placed into the instruction at source B. Note that one register is an integer variable value, N7:4, and the other is the constant value, 345. The output O:3/6 is true if the comparison is true (value of N7:4 is any value other than 345) and is false if the comparison is not true (value of N7:4 is equal to 345).	NEQ Not Equal Source A N7:4 5< Source B 345 345< O:3 ( ) 6 **(b)**
Turn on an output if one integer number is less than a second integer value. The LES comparison instruction determines if the value in source A is less than the value in source B. Note that both values are integers with N7:4 in source A and T4:0.ACC in source B. As long as the timer accumulator is less than the value in N7:4 then the O:3/6 output is false. When the timer ACC climbs above the value in the integer register, the output is true.	LES Less Than (A<B) Source A N7:4 5< Source B T4:0.ACC 0< O:3 ( ) 6 **(c)**
Turn on an output if one integer number is greater than a second integer value. The GRT comparison instruction determines if the tag value in source A is greater than the tag value in source B. Both values are integer data types. If set_point is greater than oven_temp, then the output, heater, is turned on, otherwise it is turned off. Note that the ladder logic is in ControlLogix format.	GRT Greater Than (A>B) Source A Set_point 0← Source B Oven_temp 0← Oven_contactor <Local:2:O.Data.3> ( ) 6 **(d)**

sources of data are equal, then the instruction is true. Figure 7-13(a) illustrates the standard ladder logic for the EQU instruction. Note that bits are used and that the comparator is true if the bits are the same. If the value of the data in Source A and Source B are not equal, then the output of the instruction is false.

Allen-Bradley NEQ Instruction and Standard Logic. The Allen-Bradley *not equal to* (NEQ) instruction compares two sources of data. If the two sources of data are not equal, then the instruction is true. Figure 7-13(b) illustrates the standard ladder logic for the NEQ instruction. Note that an integer register is compared to a fixed value; if the value in Source A is less than the value in Source B the comparator makes the rung true. If the value of the data in Source A and Source B are equal, then the output of the instruction is false.

Allen-Bradley LES Instruction and Standard Logic. The Allen-Bradley *less than* (LES) instruction compares two sources of data. If the first listed source is less than the second listed source, then the instruction is true. Figure 7-13(c) illustrates the standard ladder logic for the LES instruction. Note that an integer register is compared to a timer accumulator value; if the value of Source A is less than the value of Source B, the comparator makes the rung true. Registers from counters and timers are often used in comparator instructions. If the value of the data in Source A is greater than or equal to the value of the data in Source B, then the output of the instruction is false.

Allen-Bradley GRT and Standard Logic. The Allen-Bradley *greater than* (GRT) instruction compares two sources of data. If the first listed source is greater than the second listed source, then the instruction is true. Figure 7-13(d) illustrates standard ladder logic for the GRT instruction. Note that tag values are compared in this application to demonstrate how a ControlLogix PLC program would look. If the tag name in Source A is less than or equal to the value of the tag in Source B, then the output of the instruction is false.

7-5-2 Standard Ladder Logic for LEQ, GEQ, MEQ, and LIM Comparison Instructions

The less than or equal to (LEQ), greater than or equal to (GEQ), equal to with mask (MEQ), and limit test (LIM) instructions check for two comparison parameters between two values. Read the standard ladder logic for these comparison instructions in Figure 7-14 before reading the explanations that follow.

Allen-Bradley LEQ Instruction and Standard Logic. The Allen-Bradley *less than or equal to* (LEQ) instruction compares two sources of data. If the first listed source is less than or equal to the second listed source, then the instruction is true. Figure 7-14(a) illustrates the standard ladder logic for the LEQ instruction. In this standard circuit example the two integer registers are compared. If the value in Source A is greater than the value of the data in Source B, then the output of the instruction is false.

Allen-Bradley GEQ Instruction and Standard Logic. The Allen-Bradley *greater than or equal to* (GEQ) instruction compares two sources of data. If the first listed source is greater than or equal to the second listed source, then the instruction is true. Figure 7-14(b) illustrates the standard ladder logic for the GEQ instruction. Note that tag values are compared in this application to emphasize the similarity between the Logix implementations of comparison instructions and the older Allen-Bradley PLCs. If the value in Source A is less than the value of the data in Source B, then the output of the instruction is false.

Allen-Bradley MEQ Instruction and Standard Logic. The Allen-Bradley *masked equal to* (MEQ) instruction compares two sources of data that are masked over some of the bits. If those masked areas are equal, then the instruction is true. Figure 7-14(c) illustrates the standard ladder logic for the MEQ instruction. Source A and Source B are replaced with Source and Compare respectively. Therefore, the value in the Source is compared with the value of the Compare parameter. The mask is a hexadecimal number where a 1 in the mask indicates that the corresponding bits from each value are compared, and a 0 indicates that the bits are ignored. If the bits in the Source identified by the mask are equal to the value of the corresponding bits in the Compare parameter, then the output of the instruction is true. If they are not, then the instruction is false.

Allen-Bradley LIM Instruction and Standard Logic. The *limit test* (LIM) instruction is used to set an output when a test value is either inside or outside a lower and higher limit range. The instruction evaluates as follows:

- *The instruction is true if*: The lower limit is equal to or less than the higher limit, and the test parameter value is equal to or inside the limits. Otherwise the instruction is false.
- *The instruction is true if*: The lower limit has a value greater than the higher limit, and the instruction is equal to or outside the limits. Otherwise the instruction is false.

FIGURE 7-14: Standard ladder logic for the SLC 500 double comparison instructions.

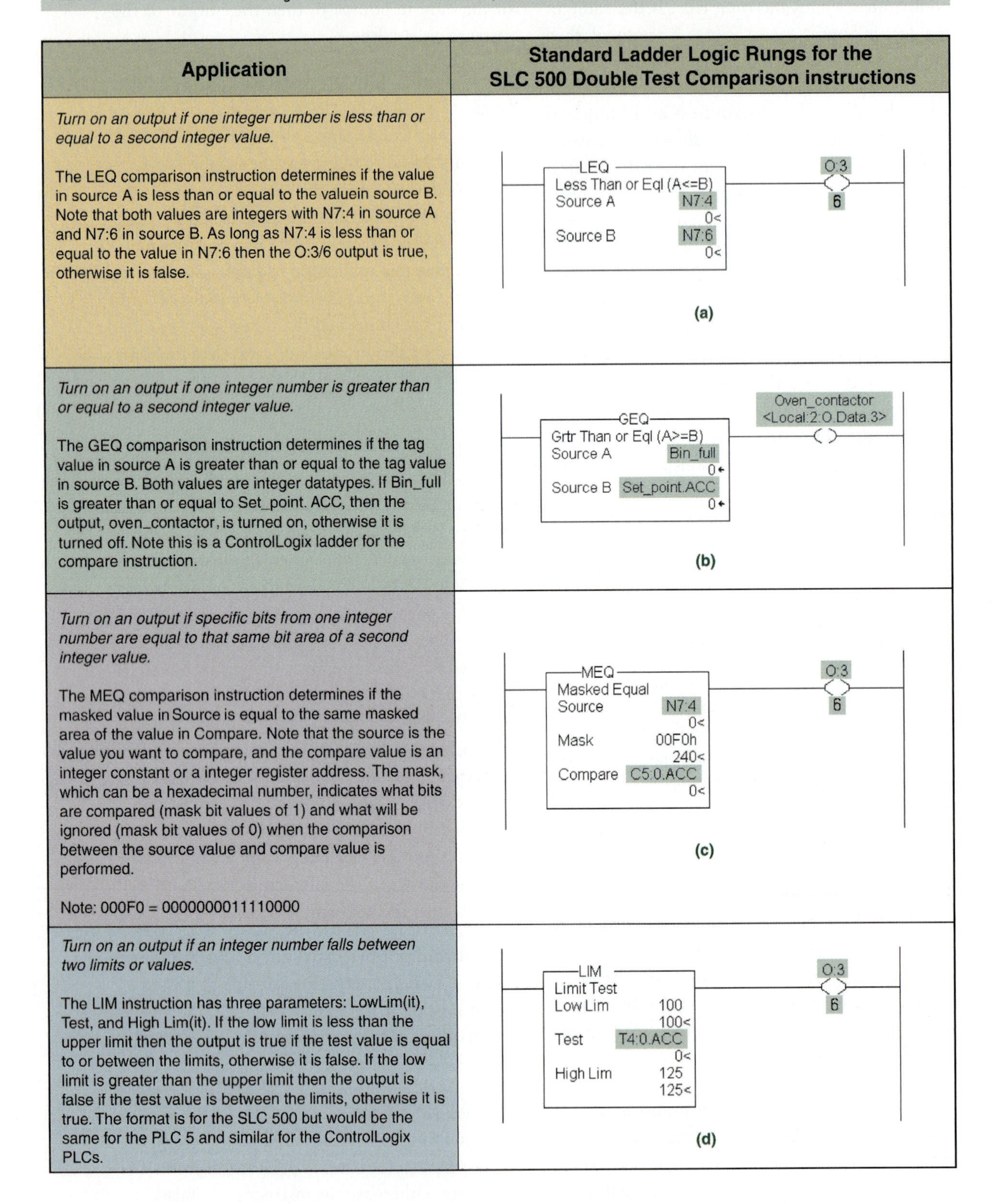

Application	Standard Ladder Logic Rungs for the SLC 500 Double Test Comparison instructions
Turn on an output if one integer number is less than or equal to a second integer value. The LEQ comparison instruction determines if the value in source A is less than or equal to the valuein source B. Note that both values are integers with N7:4 in source A and N7:6 in source B. As long as N7:4 is less than or equal to the value in N7:6 then the O:3/6 output is true, otherwise it is false.	LEQ Less Than or Eql (A<=B) Source A N7:4 0< Source B N7:6 0< O:3 6 **(a)**
Turn on an output if one integer number is greater than or equal to a second integer value. The GEQ comparison instruction determines if the tag value in source A is greater than or equal to the tag value in source B. Both values are integer datatypes. If Bin_full is greater than or equal to Set_point. ACC, then the output, oven_contactor, is turned on, otherwise it is turned off. Note this is a ControlLogix ladder for the compare instruction.	GEQ Grtr Than or Eql (A>=B) Source A Bin_full 0 Source B Set_point.ACC 0 Oven_contactor <Local:2:O.Data.3> **(b)**
Turn on an output if specific bits from one integer number are equal to that same bit area of a second integer value. The MEQ comparison instruction determines if the masked value in Source is equal to the same masked area of the value in Compare. Note that the source is the value you want to compare, and the compare value is an integer constant or a integer register address. The mask, which can be a hexadecimal number, indicates what bits are compared (mask bit values of 1) and what will be ignored (mask bit values of 0) when the comparison between the source value and compare value is performed. Note: 000F0 = 0000000011110000	MEQ Masked Equal Source N7:4 0< Mask 00F0h 240< Compare C5:0.ACC 0< O:3 6 **(c)**
Turn on an output if an integer number falls between two limits or values. The LIM instruction has three parameters: LowLim(it), Test, and High Lim(it). If the low limit is less than the upper limit then the output is true if the test value is equal to or between the limits, otherwise it is false. If the low limit is greater than the upper limit then the output is false if the test value is between the limits, otherwise it is true. The format is for the SLC 500 but would be the same for the PLC 5 and similar for the ControlLogix PLCs.	LIM Limit Test Low Lim 100 100< Test T4:0.ACC 0< High Lim 125 125< O:3 6 **(d)**

The three parameters shown in Figure 7-14(d), Low Lim(it), Test, and High Lim(it), can all be word addresses or tag names. However, if the test is a program constant, then both limits must be word addresses or tag names. If the test is a word address or tag name, then the two other parameters can be word addresses, tag names, or program constants.

7-5-3 Standard Ladder Logic for Multiple Instructions and Hysteresis

Combinations of comparison instructions are often necessary for some automation problems. In addition, it is often useful to have input triggers that include a controlled amount of hysteresis. The standard ladder logic rungs in Figure 7-15 illustrate how comparison instructions are combined and hysteresis is added. Read the description in the standard ladder logic in Figure 7-15 before reading the explanations that follow.

Standard Logic Multiple Compare Instruction. Combinations of comparison instructions are used when multiple input conditions must be tested. Any of the comparison instructions can be used in logical combinations of AND, OR, AND/OR, and OR/AND input logic. The problem description and the logic equation generation will dictate the comparison instructions, the parameters used, and the logic combinations. A trigger instruction to initiate the comparison process is illustrated in OR configuration in Figure 7-15(a) and in the AND configuration in Figure 7-15(b).

Standard Logic Comparison Instructions with Hysteresis. Process applications often require that the comparison process have hysteresis. For example, if a GRT instruction is used for the control of a valve, then the valve action will occur when one parameter exceeds another. However, if the measured parameter has some variation or noise, the valve will be cycled on and off when the input passes the trigger point and the input moves above and below the GRT value. Hysteresis fixes this problem by having the turn on value slightly higher than the turn off value. This means that after the GRT instruction is true, the process parameter must fall to some value below the original trigger point before the valve turns off.

Figure 7-15(c) shows two configurations of comparison with hysteresis. The first uses standard OTE output instructions and the second uses OTL and OTU latched outputs. The latch solution is simpler, but latches are not used for many types of outputs for safety reasons. Analysis of the first solution indicates that both output bits initially are off. This enables the GRT instruction (the B3:0/1 XIO instruction is true and has power flow) and disables the LES instruction because the O:3/0 XIC input instruction is false or has no power flow. When the process parameter is greater than 25, the O:3/0 output is true and the LES instruction is enabled (XIC input instruction has power flow). The O:3/0 sealing instruction assures that this output remains on even when the process parameter is less than 25. When the process parameter falls below 20 the B3:0/1 bit is true, which makes the top rung false and the rung output, B3:0/0, false as well. If the process parameter is an integer, then the hysteresis range is 19–26: 26 to turn on and 19 to turn off.

Review all of the standard logic circuits for the comparison instructions in Figures 7-13, 7-14, and 7-15.

EXAMPLE 7-1

Design a ladder program that sets the upper and lower limits of a laboratory temperature chamber to +/−1/2 percent of the temperature set point. The allowable temperature for the chamber is ambient temperature up to 500° F. The temperature set point is set manually using a thumbwheel or push switch. The push switch is set to 300° F in this example. Include comparison instructions to:

- Turn on a heater and turn on a lamp when the chamber temperature is below the lower limit, and
- Turn off the heater and turn on a lamp when the chamber temperature is above the upper limit.

Solution

The ladder program is illustrated in Figure 7-16. The first through third rungs in the solution set the upper and lower limits by multiplying the temperature set point, N7:0, by the upper and lower limit percent variation. Note that

FIGURE 7-15: Standard ladder logic for the SLC 500 multiple comparison instructions and hysteresis.

Application	Standard Ladder Logic Rungs for Multiple Comparison Instructions and Hysteresis
Turn on an output if one integer number is in a range that is less than or equal to the first reference integer but greater than a second reference integer when the trigger input is received. The LEQ and GRT comparison instructions are used in an OR configuration with an XIC contact to start the comparison evaluation ANDed to the OR circuit. Either of the ORed comparisons AND the trigger contact must be true for the output to be true. Any of the comparisons can be used in this configuration, and the trigger contact is optional.	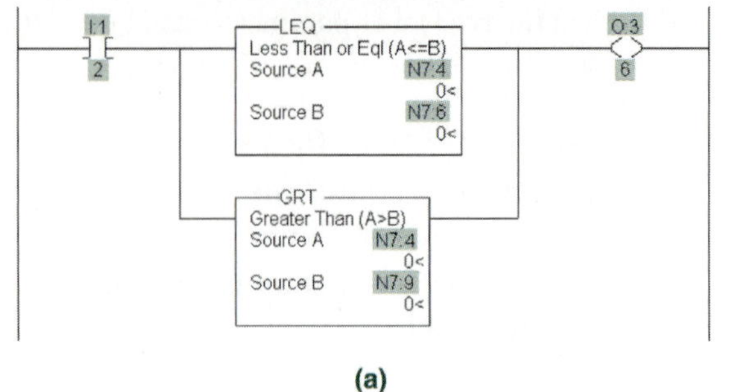 (a)
Turn on an output if one integer number is in a range that is greater than or equal to the first reference integer but less than a second reference integer when the trigger input is received. The GEQ and LES comparison instructions are used in an AND configuration with an XIC contact to start the comparison evaluation. All of the ANDed instructions must be true for the output to be true. Any of the comparisons can be used in this configuration, and the trigger contact is optional. Note the wider ladder rung used a wrap function to display the wider figure.	I:1 2 GEQ Grtr Than or Eql (A>=B) Source A N7:4 0< Source B 240 240< LES Less Than (A<B) Source A N7:4 0< Source B 530 530< O:3 6 (b)
Turn on an output when an integer register reaches a specific value and reset the output when the integer register falls to a value below the higher trigger point. The logic circuit to perform this action includes a GRT and LES comparator along with a B3:0/1 contact to lock the O:3/0 bit on at values above 25. The output bit remains on until the register value drops below 20. The hysteresis band width is set by the values of the upper and lower trigger points in source B. The second ladder logic configuration uses latching and unlatching OTL and OTU instructions to achieve the same results. There are often safety issues when latching outputs are used so use of the second ladder would have to be checked for safety concerns before implementation. Note that the solutions use the SLC 500 format, but the solution would be similar for the PLC 5 and ControlLogix except for the addressing.	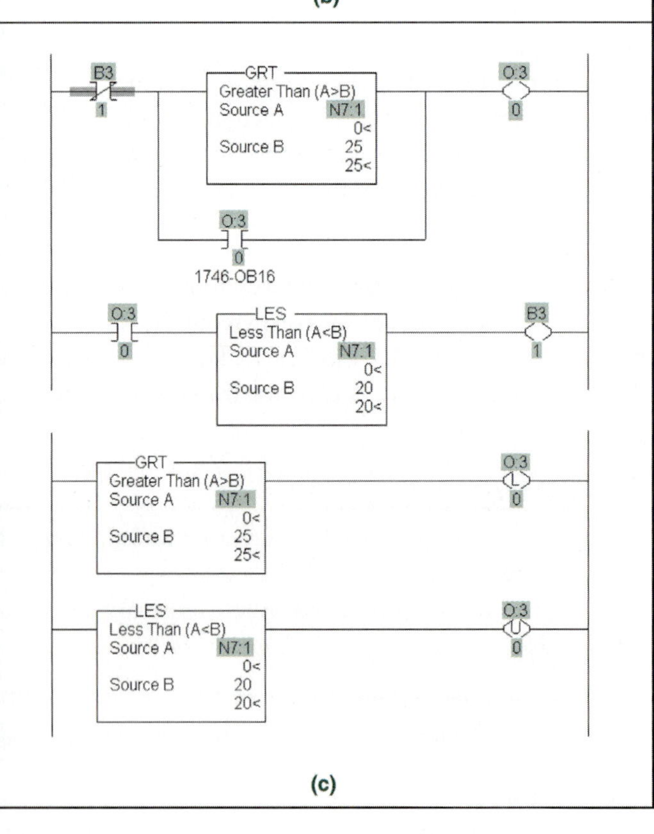 (c)

FIGURE 7-16: Solution for Example 7-1.

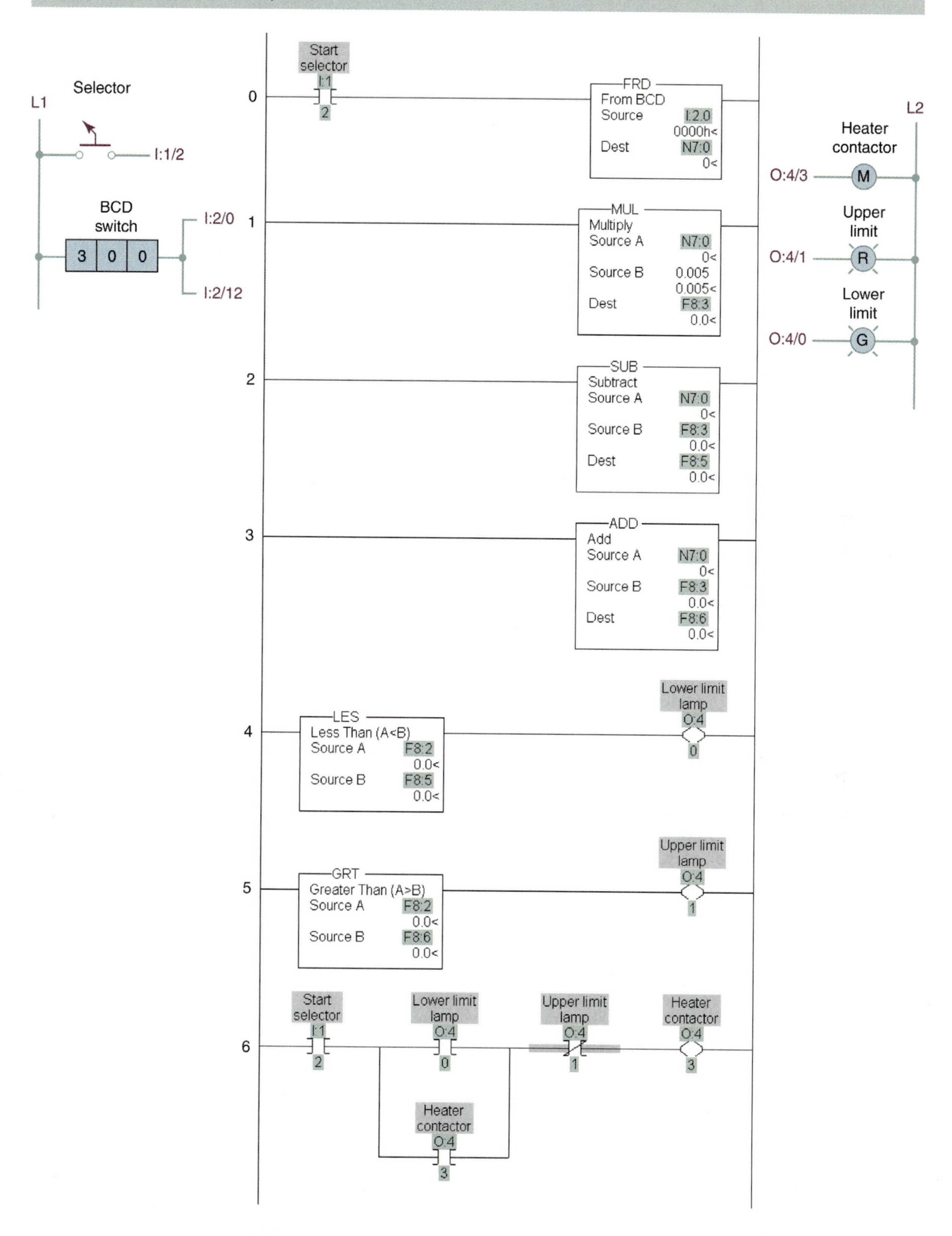

the temperature is set to 300 by a BCD switch. An FRD instruction (rung 0) converts the BCD temperature value to binary and stores the value in N7:0, an integer register. The setpoint range is determined in the second rung and the upper and lower limits are established in rung 2 and 3. Floating point register are necessary because the temperature limits are set to tenths of a degree. The temperature is measured with an RTD and the value scaled to engineering units of degrees ° F. To reduce the complexity of the solution, the RTD input instruction and scaling instruction are not shown. The amended RTD reading is placed in F8:2. An LES instruction compares the RTD reading, Source A (F8:2), to the lower limit set point, Source B (F8:5). Note that Source B of the LES instruction is the same as the destination of the SUB instruction. If the temperature input is less than the lower limit, the heater contactor 0:4/3 is turned on and the lower-limit lamp 0:4/0 is illuminated. The 0:4/3 instruction seals around the 0:4/0 instruction so that the heater remains on after the oven temperature rises above the lower limit and the LES instruction returns to a false state. Also, note that a GRT instruction compares the temperature reading in F8:2, to the upper limit in Source B (F8:6). If the input is greater than the upper limit, the heater solenoid is turned off and upper limit lamp 0:4/1 is illuminated. Study the solution for the oven temperature controller until the use of the BCD and comparison instructions is clear.

7-6 EMPIRICAL DESIGN PROCESS WITH BCD CONVERSION AND COMPARISON INSTRUCTIONS

The empirical design process, introduced in Section 3-11-4, is an organized approach to the design of PLC ladder logic programs. However, the term *empirical* implies that some degree of trial and error is present. As more instructions are added to the design and the process becomes more complex, the empirical design requires more fixes and adjustments. However, the process used for these new instructions is quite similar to that learned in previous chapters.

7-6-1 Adding BCD Conversion Instructions to the Process

The first step in using BCD conversion instructions in PLC ladder designs is to know what parameters are entered for the source and the destination registers and to understand the standard BCD conversion logic circuits illustrated in Figures 7-6, 7-7, 7-8, and 7-9. The size of the BCD number dictates what types of ladder logic are required. In addition, it is important to know when a one-shot instruction is needed to trigger the conversions. Stop now and review all of these if necessary.

BCD Conversion Instructions. The use of BCD instructions implies that process values are required in both the decimal and binary coded decimal (BCD) formats. Review Example 7-1 and notice that the problem statement indicates that the temperature value for the oven is entered from a push switch. Since the push switch has a BCD output a conversion is necessary because the PLC must perform the math in binary. The complete empirical process is listed in Section 3-11-4; the modifications for BCD instructions are as follows:

Step 1: (Write the process description): Include a description of the conversions that must be included; for example, BCD numbers in and BCD values out. Note especially the update rate of the source data in the BCD instructions and the length of time that the BCD rung input logic is true.

Step 2: (Determine the type of BCD instruction to use): Use a TOD instruction for conversion from binary to BCD and a FRD instruction for conversion from BCD to binary. Therefore, FRD is used to input data and TOD is used to output data.

Step 3: (Determine if a one-shot instruction is required): The update time for the data at Source A and the duration of the input logic that triggers the conversion are used to determine if a one-shot instruction is necessary. If the BCD data changes while the input logic in the BCD rung is true, then the value in the destination file will change with every scan. This is often not desired, so a one-shot instruction must be used to make sure that only one conversion occurs at the false to true transition of the input logic.

Step 4: (Determine what type of registers are needed for the size of BCD number present): Use the standard ladder logic in Figures 7-6 and 7-7 to select the correct register for BCD numbers from 0 to 9999 and to select registers for BCD numbers greater than 9999.

7-6-2 Adding Comparison Instructions to the Process

The first step in using comparison instructions in PLC ladder designs is to know what parameters are entered for Source A and Source B registers and to understand the standard comparison logic circuits illustrated in Figures 7-13, 7-14, and 7-15. In addition, identify the type of comparison, for example, simple tests like less than, greater than, or the limit test (Figures 7-13 and 7-14), or more complex checks like less than one value but greater than another [Figure 7-15(a) and (b)]. Also, the need for hysteresis should be determined [Figure 7-15(c)]. Stop now and review all of these if necessary. The complete empirical process is listed in Section 3-11-4; the modifications for comparison instructions are as follows:

Step 1: (Write the process description): Include a complete description of the types of comparisons required for the process, including all the values and registers associated with the process. Identify the outputs controlled by comparison instructions, applications that require hysteresis, and those that have more complex comparison requirements.

Step 2: (Write Boolean equations for all comparison instruction rungs): A Boolean expression is needed to describe each comparison instruction.

7-6-3 Process Tank Design

The following examples demonstrate how to design the control ladder logic using the empirical design process for a relatively complex process control problem. Study the process tank shown in Figure 7-17 and the operator control panel shown in Figure 7-18. After a complete review of the tank layout and operator panel, read the following process description. This is an example of how a process description should be written for the empirical design process.

FIGURE 7-17: Process tank.

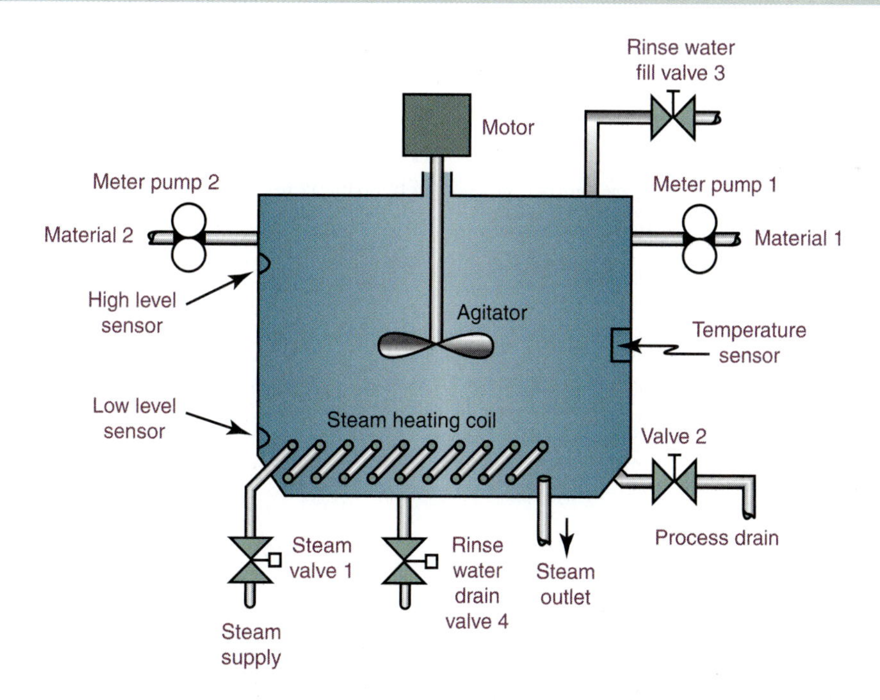

FIGURE 7-18: Process tank operator panel.

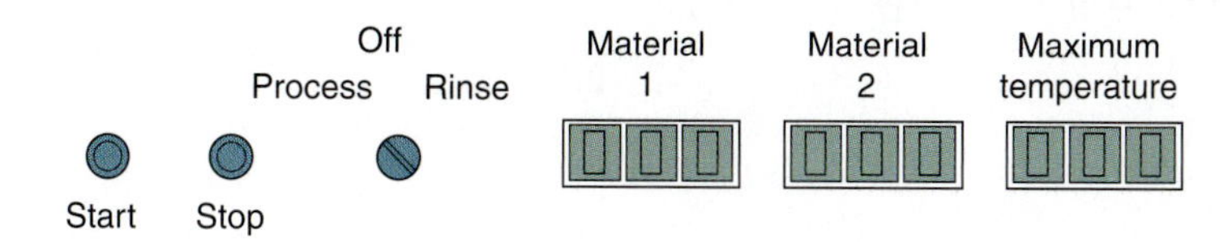

The volume in gallons of two liquid ingredients (material 1 and material 2 in the figure) and the maximum soak temperature are entered on three-digit push switches. The system has a start and stop push button and a two-position (center position off) normally open selector switch to select the rinse and process modes. One seven-segment readout with a three-digit display shows the current process temperature. Indicator lights display the following conditions: fill cycle, drain cycle, water fill valve open, water drain valve open, process drain valve open, steam valve open, and over temperature warning. The two process metering pumps pass 8 ounces of liquid per revolution. The tank has low and high level liquid sensors, and a temperature sensor and transmitter. A mixer, driven by a motor, is used to agitate the process.

When the process selector switch is moved to the process position, process rungs are selected. The process is started with the start push button, which sets a system start bit. Each pump rotates through the correct number of revolutions to put the desired volume of liquid into the tank. The mixer is on whenever the liquid is over the low level liquid sensor. When all liquids are loaded, the steam valve is opened and the process is heated to the temperature set point. When the set point is reached the heater is turned off and the process liquid is drained through the process drain valve. When the low level liquid sensor is not active the system start bit is turned off.

After the process liquid is drained, the system waits for the selector to be placed in the rinse position and the start switch to again be pressed. The tank fills with water and the mixer is on whenever the water is over the low level liquid sensor. When the water reaches the high level liquid sensor, the water is drained through the water drain valve. When the rinse cycle is completed the system returns to the stopped condition.

Complex process problems should be broken into logical solution groups. In this example the solution groups would be fill cycle, process heating and drain cycle, tank rinse cycle, and displays. There may be some overlap between these solutions when outputs are shared in two solution areas. For example, the mixer is used in the fill and rinse cycles, but that problem is addressed as it is uncovered. Working on the solution in these smaller chunks makes it easier to see the total solution. The problems at the end of the chapter include the design of the ladder logic program for displaying process status on the operator panel. The fill cycle, process heating and drain cycle, and tank rinse cycle solutions are addressed in Examples 7-2 and 7-3.

EXAMPLE 7-2—FILL CYCLE

Design the ladder logic used to start the system and complete the fill cycle for the process described above.

Solution

The empirical design information includes:

- All data values present are integer data.
- Gallons data entered with the push switches and FRD instructions are converted to pump revolutions (1/2 gallon per revolution) by multiplying the gallons by 2.
- Outputs are FRD conversion instructions for the volume of each liquid, math instructions, counters for each pump, and motor contactors for each pump.

The following Boolean logic is used for the ladder rungs (logic is the same for both pumps):

- Start bit = start switch (with the start bit used as a sealing instruction) AND NOT stop AND [low level float switch OR start timer timing bit]

  After the start switch is pressed and the tank level is above the low level float switch, the float switch bit determines how long the process cycle and start bit (B3:0/0) are true. However, the low level float switch is initially off and does not come on until some liquid fills the tank, which takes about 80 seconds. Therefore, the start timer (timer timing bit) is used to seal around the low level sensor instruction until the liquid activates the low level sensor. The timer is preset to 120 seconds, so an additional 40 seconds is available to get the tank liquid above the low level point.

- TON timer = start switch (with a sealing instruction of timer timing)

  Note that the start switch is also needed to trigger the start timer input to get the timer active. The timer timing sealing instruction makes sure that the timer continues to increment when the momentary start switch is released.

- Volume FRD instructions = process selected AND start bit AND one shot [Standard logic Figure 7-9(a)]
- MUL instructions (gallons required × 2) = unconditional input [Standard logic Figure 6-12(a)]

  Each pump revolution supplies 0.5 gallons of material; therefore, the multiplication of gallons required times 2 produces the number of pump revolutions necessary to reach the required material amount. This result is placed in an integer register for comparison with the pump revolution counter.

- C5:0 = material pump contactor AND process selected AND pump revolution limit switch

  Every revolution of the pumps increment the pump counters. As a result, the counter ACC tracks the number of half-gallon increments of material placed in the tank.

- Pump contractor = Comparison of C5:0.ACC (pump revolutions) *less than* total number of required pump revolutions AND start bit

  Each pump revolution is 0.5 gallons of material and the total number of revolutions is required gallons times 2.

- Mixed contactor = low level sensor

Based on this information the ladder solution in Figure 7-19 was developed for the Allen-Bradley SLC 500 PLC. The input and output interface is shown for the inputs and output used in this portion of the solution. The sequence for the ladder solution is to:

1. Read in the volume for each liquid in gallons.
2. Calculate the number of pump rotations (0.5 gallons per revolution) to equal the desired material volume (number of half gallon revolutions is equal to the gallons required times 2).
3. Count the pump revolutions.
4. Compare the accumulated pump revolutions with the total required and keep the pump contactors true while the counter ACC value is less than the required value.
5. Start the mixer when the liquid is above the low level float switch.

Compare the solution with the problem statement. Can you identify any rungs or operations that need to be added?

The missing operation is a reset at the start of the process of the Pump 1 volume and Pump 2 volume counter accumulators. They must be zero for a correct comparison. Process statements rarely state control requirements for the instructions used because they may not be known when the process is being defined. The reset of the counters is an operational issue when counters are used, and the designer must recognize when they are necessary. A rung input with the start switch address and counter reset instructions at the output would fix this omission.

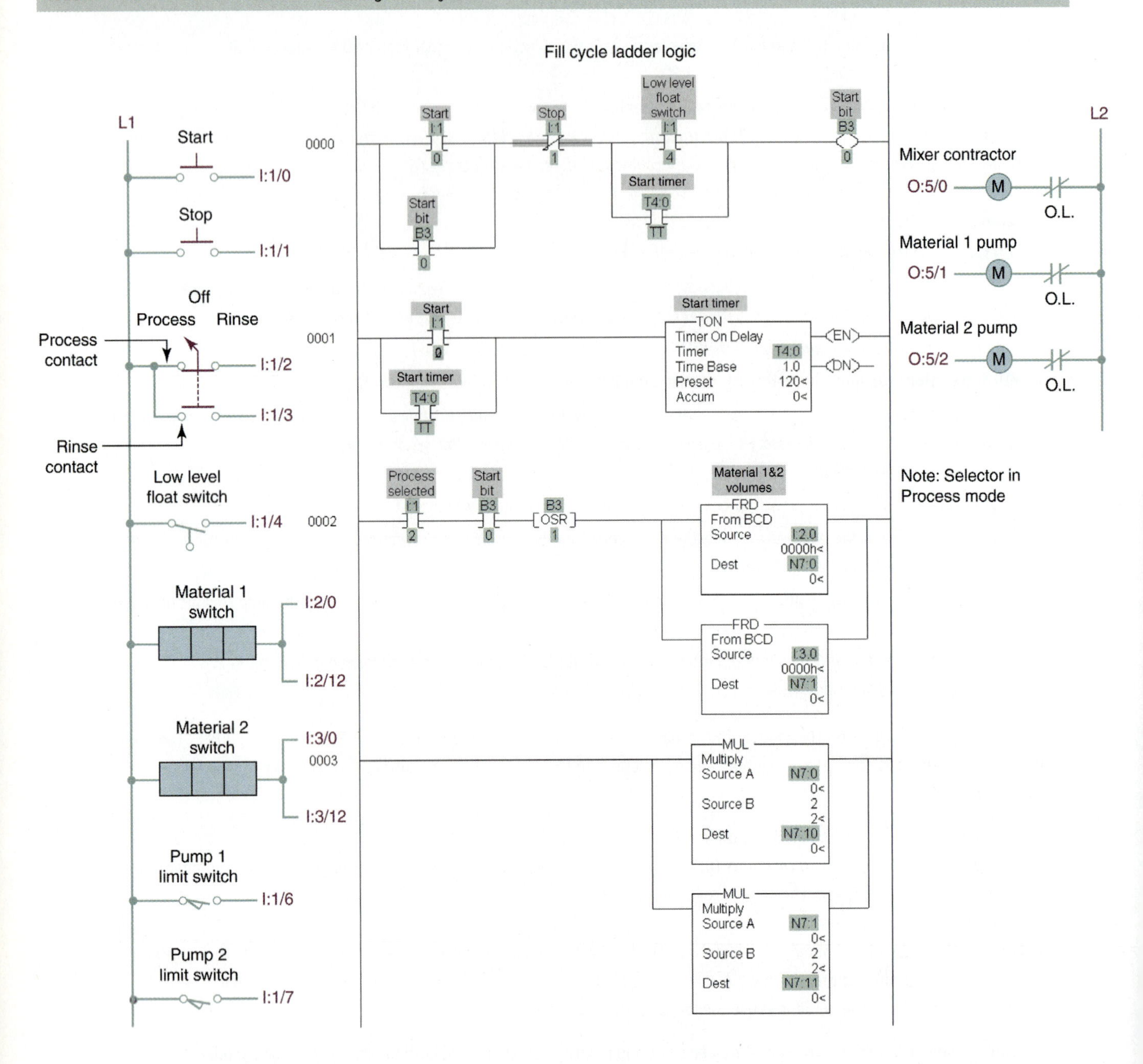

FIGURE 7-19: Process tank control ladder logic—fill cycle.

EXAMPLE 7-3—TEMPERATURE/DRAIN CYCLE

Design the ladder logic used to heat the liquid and the ladder logic for the system drain cycle described in the process statement.

SOLUTION

The empirical design information includes:

- All data values present are integer data.
- The steam valve must open when all liquid has been loaded.

FIGURE 7-19: (Continued).

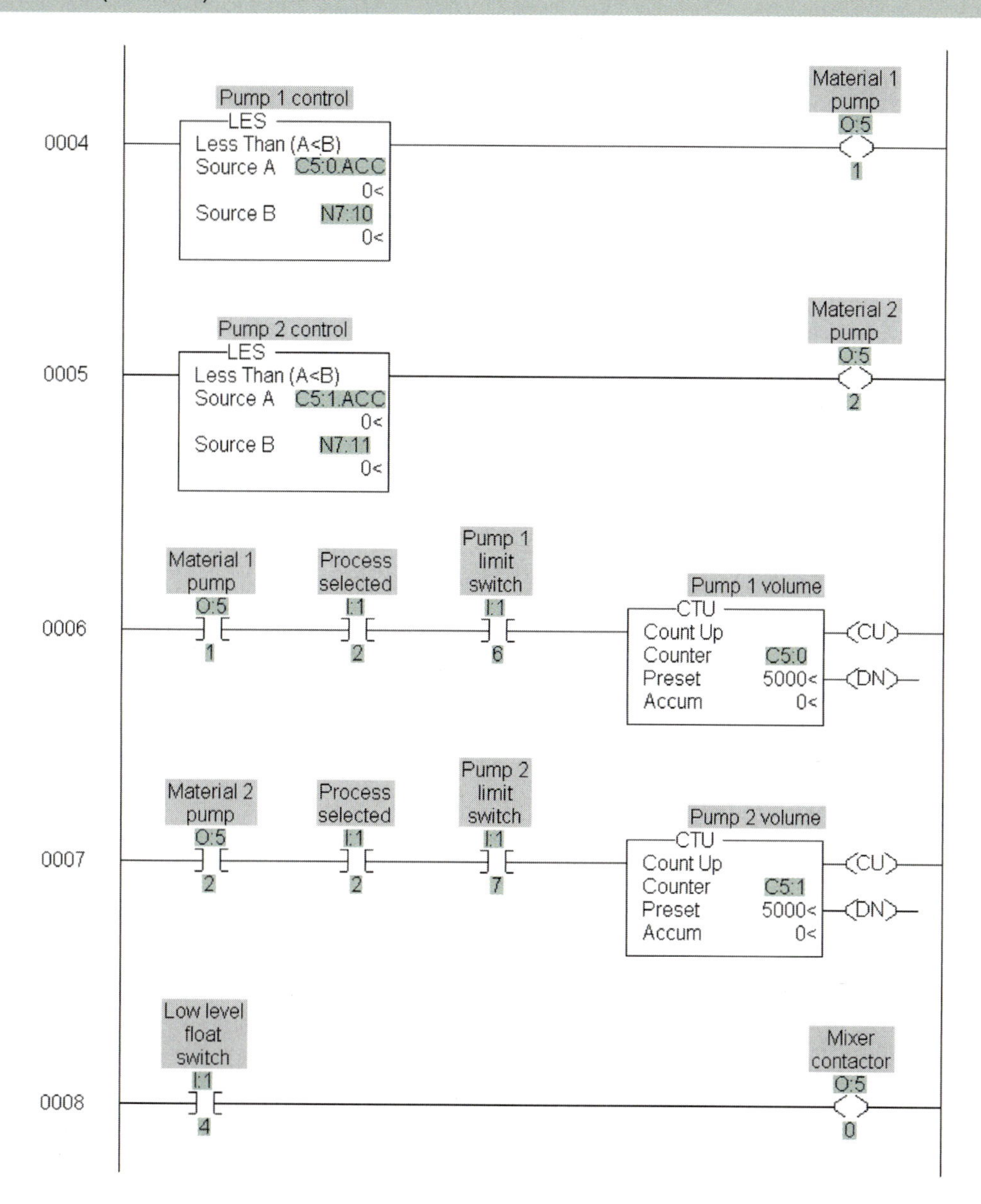

- Outputs are FRD conversion instructions for the heating temperature set point, process drain valve, steam valve, and temperature comparison bit.

The following Boolean logic is used for the ladder rungs:

- Temperature FRD instructions = process selected AND start bit AND one shot [Figure 7-9(a)]
- Maximum temperature bit = GEQ (temp sensor > temp set point) [Figure 7-14(a) and (b)]

  The maximum temperature bit is on when the measured temperature (N7:4) is greater than the temperature set point (N7:3).

- Steam valve = start bit AND process selected AND NOT material 1 pump 1 AND NOT material 2 pump 2 AND NOT process drain valve AND NOT maximum temperature bit

The steam valve is opened when both pumps are off, the process temperature is not at the set point value, the process drain valve is closed, and the start and selected bits are true. The start bit ensures that the steam valve does not turn back on when the drain valve returns to the closed state after the system drains.

- Process drain = process selected AND maximum temperature bit (sealed with XIC instruction using the process drain address) AND low level sensor

The maximum temperature bit opens the process drain valve, but a sealing contact from the output is necessary because the maximum temperature bit will not be true for the entire drain cycle. The low level sensor false condition closes the process drain valve.

Based on this information the ladder solution in Figure 7-20 was developed for the Allen-Bradley SLC 500 PLC. The input and output interface is shown for the inputs and output used in this portion of the solution. The sequence for this part of the ladder solution is to turn on the heater when both pumps are stopped, turn the heater off and the process drain valve on when the liquid reaches the set point temperature, and turn the process cycle and all outputs off when the tank is drained.

FIGURE 7-20: Process tank control ladder logic—temperature/drain cycle.

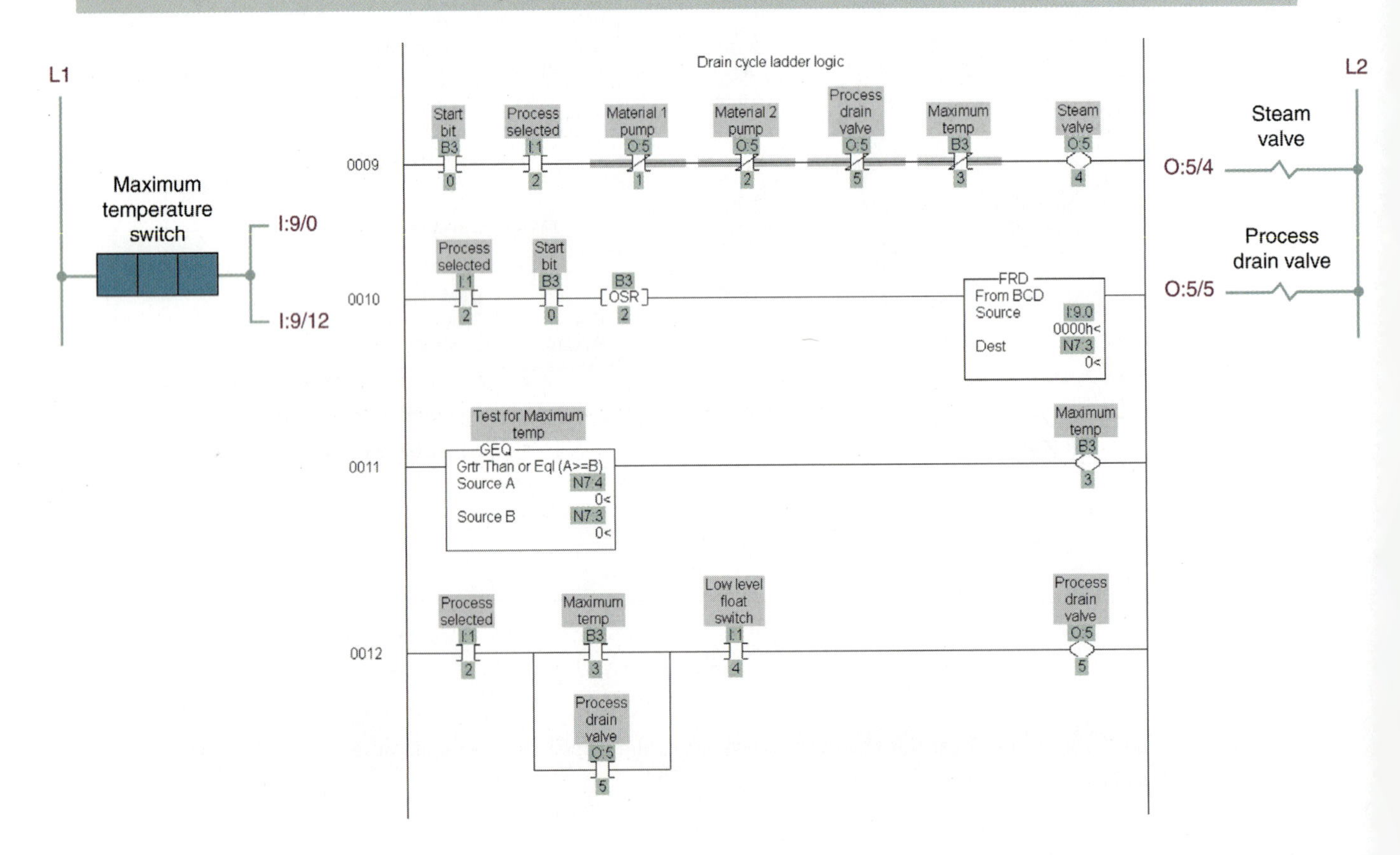

A review of the solution for Examples 7-2 and 7-3 indicates that instructions are often used in rungs to prevent the rung from changing state after the process controlled by the rung is complete. A good example of that is the start bit in the steam valve in rung 9 in Figure 7-20. The steam valve would turn back on when the drain valve closed at the end of the process because all the other instructions would be true at that time. However, the start bit is placed into the rung, B3:0/0, and it opens at the end of the process and keeps the steam valve off or closed.

EXAMPLE 7-4—RINSE CYCLE

Design the ladder logic used to rinse the process tank after each process reaction with the water fill, mix, and flush described in the process statement.

SOLUTION

The empirical design information includes:

- All data values present are integer data.
- The selector switch is in the rinse position and the tank is filled with water and then drained. The mixer is used throughout the cycle.
- Outputs are mixer contactor, water drain valve, and water fill valve.

The following Boolean logic is used for the ladder rungs:

- Mixer contactor = rinse selected AND low level float switch

  The same output in ladder logic should never be referenced or used in two different rungs. The mixer contactor already appears in rung 8 in Figure 7-19; therefore, this rung should be combined with rung 8.

- Rinse water fill valve = start bit AND rinse selected AND NOT high level float switch AND NOT rinse water drain valve

  The false XIO instruction for the high level float switch turns the rinse water fill valve off when the liquid closes the high level float switch contacts. The false XIO instruction for the rinse water drain valve holds the fill valve off until all the liquid is drained and the cycle is terminated with a false start bit.

- Rinse water drain = start bit AND rinse selected AND high level float switch (sealed with rinse water drain valve XIC instruction) AND low level float switch

  The true condition for the high level float switch opens the rinse water drain valve and makes the sealing instruction, O:5/7, true. The seal is needed since the high level float switch is false after the liquid starts to drain. The low level sensor terminates the water drain cycle when the tank is empty and the sensor is not active.

Based on this information, the ladder solution in Figure 7-21 was developed for the Allen-Bradley SLC 500 PLC. The input and output interface is shown for the inputs and output used in this portion of the solution. The sequence for the ladder solution is to turn on the rinse water fill valve, turn on the mixer when liquid is over the low level float switch, turn off the rinse water fill valve and open the rinse water drain valve when the liquid is over the high level float switch, and end the cycle when the water drops below the low level float switch.

Review the solutions for all three parts of the process, Figures 7-19, 7-20, and 7-21. Lessons learned in this solution include:

- The use of a selector switch to select the process under control isolates each solution so that there is no rung interaction between two solutions. This is a good practice if the process permits this type of control.
- Dividing the process into logical sections—a fill cycle, temperature/drain cycle, and rinse cycle—creates three smaller problems instead of one big problem. This makes the problem easier to program and easier to troubleshoot.

7-6-4 Pneumatic Robot Design

The following example is a modification of the ladder logic, Figure 3-44, for the robot control problem in Example 3-14. Study the robot configuration and ladder logic solution in Figure 3-44. After a complete review of the example solution in Chapter 3, read the following modification to the earlier robot operational description.

The two-axis robot for this example does not have end-of-travel sensors to indicate when the axes have reached their limits as did the robot in the Chapter 3 example. To make sure that the robot has finished its

FIGURE 7-21: Process tank control ladder logic—rinse cycle.

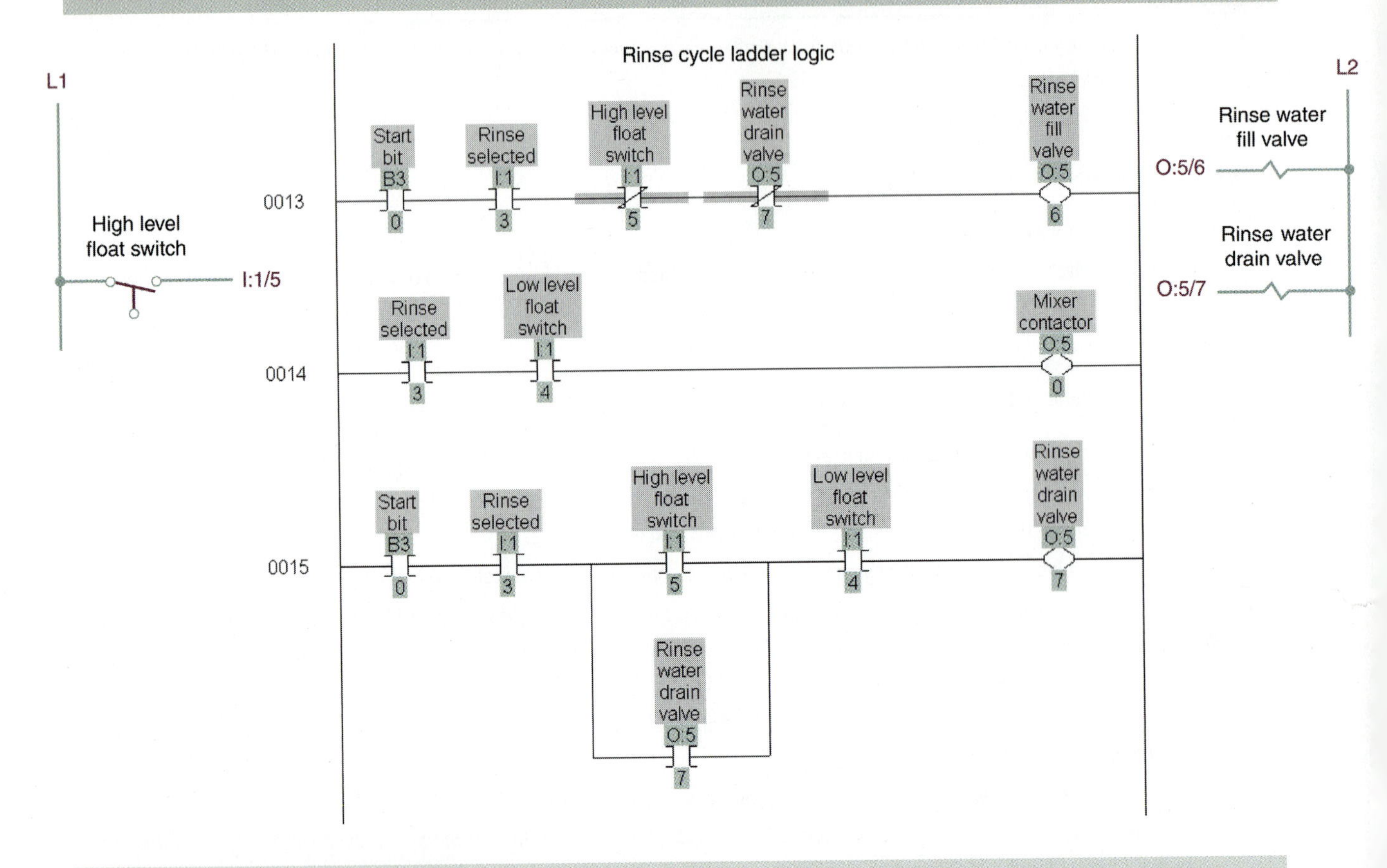

EXAMPLE 7-5—PNEUMATIC ROBOT

Design the ladder logic with ControlLogix ladder logic to control the pneumatic robot described at the start of this section.

Solution

The empirical design information includes:

- A regenerative timer is used to establish the cycle time for the process.
- The timer and comparators are used for control since end-of-travel sensors on the pneumatic actuators are not present.
- The time for each axis move is provided, so the total cycle time for the robot is the sum of each axis move. The moves in Figure 7-22(a) include down, close gripper, up, out, open gripper, in, which results in a total cycle time of 13 seconds.
- Outputs are X-axis valve, Y-axis valve, and gripper valve.

Note that the two-axis pneumatic robot is driven by a set cycle time for all axes, so the optimum speed is not attained. A part is present before the start_selector is active.

The following Boolean logic is used for the ladder rungs:

- Timer = Start selector AND NOT timer done
- X-axis down = Start selector AND part sensor AND timer ACC less than or equal to 4 seconds
- Gripper close = Start selector AND timer greater than 2 seconds but less than or equal to 9 seconds
- Y-axis out = Start selector AND timer equal to 7 seconds but less than or equal to 11 seconds

Based on this information the ladder solution in Figure 7-22(b) was developed for the Allen-Bradley ControlLogix PLC. The input and output interface is shown for the inputs and output used in the solution. The sequence for the ladder solution follows the timing diagram for the robot in Figure 7-22(a).

FIGURE 7-22: Pneumatic robot control ladder logic.

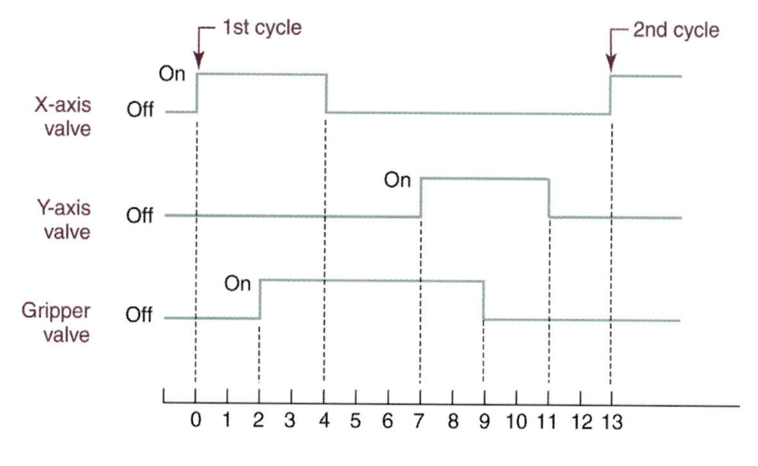

(a) Timing diagram

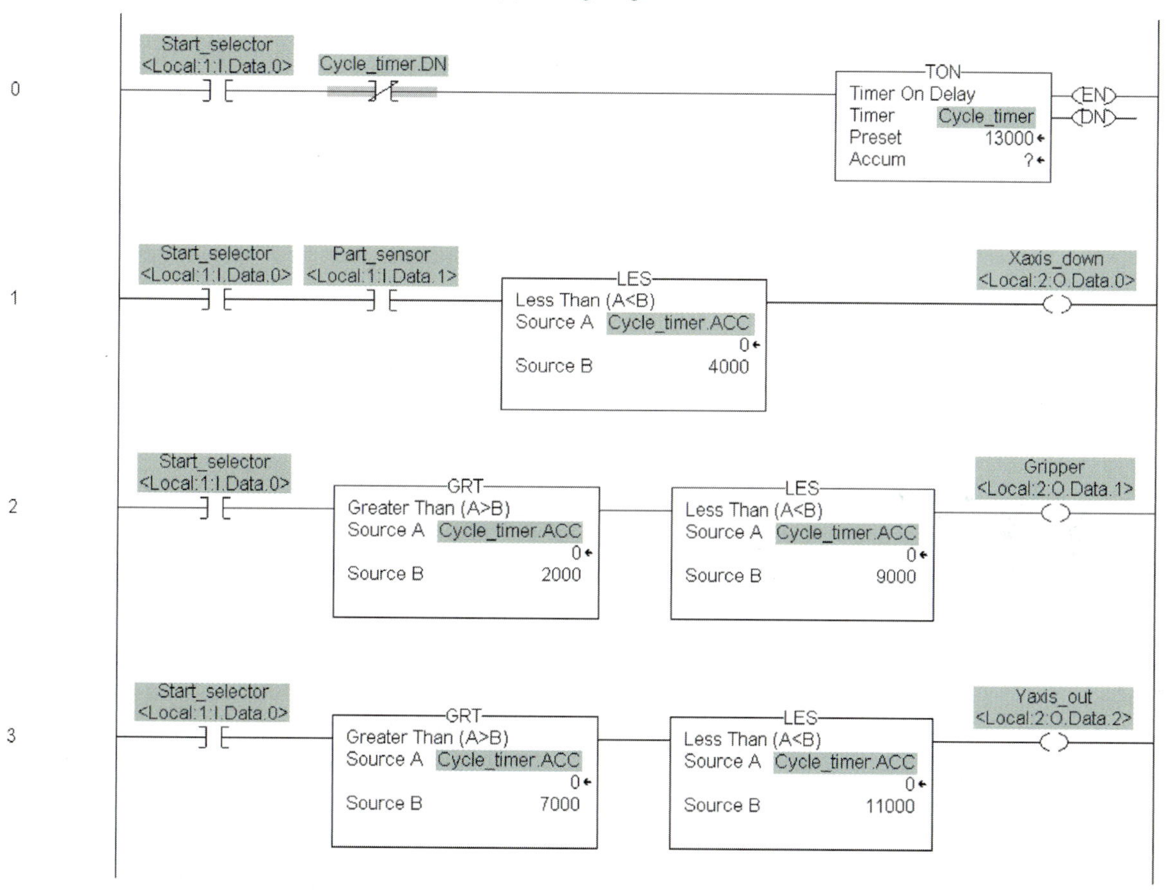

(b) Ladder logic

travel, each axis has a specified time for the travel. A cycle is started when the start selector switch is on AND when the pickup sensor indicates that a part is in the pickup location. Use the following list of actuator times to control the actuator and gripper sequences.

Vertical down – 2 seconds

Vertical up – 3 seconds

Horizontal out – 2 seconds

Horizontal in – 2 seconds

Gripper close – 2 seconds

Gripper open – 2 seconds

The timing diagram for the robot and program is illustrated in Figure 7-22(a).

Compare the solutions for the robot control in Figure 3-44, Figure 4-22, and Figure 7-22. Automation solutions can take numerous forms and still be correct. Some solutions are more efficient than others and some are simpler than others. PLC programmers make the best instruction choice for the application when they have a full understanding of all of the instructions. For example, the solution in Figure 3-44 was the only one possible until timers were learned. Similarly, the solution in Figure 4-22 was the only timer solution possible until comparators were covered. As your knowledge grows, your choice of solutions expands.

7-7 TROUBLESHOOTING BCD CONVERSION AND COMPARISON LADDER LOGIC

The conversion and comparison instructions offer few operational problems. If a section of the ladder that includes these instructions is not operating properly, then you can use any of the following suggestions to troubleshoot.

- If PLC rungs with BCD and comparison instructions do not produce the correct results, then first verify that the data from the process is correct by viewing the PLC 5 and SLC 500 dialog boxes for the input image table, integer image table, and floating point image table illustrated in Figure 6-20. If a ControlLogix processor is used the program and control tags are viewed in the dialog box displayed in Figure 6-21. Note that the register values can be displayed in binary, octal, hexadecimal, and decimal format depending on the type of data present.
- Test the sequence of BCD and comparison rungs one rung at a time to verify that each rung is performing properly. Use the temporary end (TND) instruction described in Chapter 4 to halt the ladder execution after each rung.
- Use the suspend (SUS) instruction to verify status values for all registers and bits at critical points in the ladder. The SUS instruction is described in Section 5-8-1.
- Be aware of situations where comparison instructions are used to make execution decisions and execute process actions since the scan time and internal update time for the instruction is much faster than most process events. This problem is similar to that described in the counter problem discussed in Section 5-8-2.
- Use the test options on some of the Allen-Bradley SLC 500 models in the single-step, single-scan, or continuous-scan modes to isolate and run portions of the ladder logic. These tests are described in Section 6-8-1.

7-7-1 Troubleshooting with the Module Indicators

Review the signal flow diagram in Figure 3-47 and notice that a fault could occur anywhere along the single path from input field device to output actuator. The faults could occur in any of the following locations:

- Input and output wiring between field devices and modules
- Field device/module power supplies
- Input mechanical switch devices
- Input sensors
- Output actuators
- PLC I/O modules
- PLC proccssor

The fault locations are listed in the order of most frequent to less frequent sources of problems. In the middle of the signal flow you have several indicators that are ideal troubleshooting tools. Locating the fault quickly requires observations and measurements at these input and output modules.

Input Module Troubleshooting Analysis. Earlier we discussd using the status lights on the input module for troubleshooting problems with input signals. Now we take a closer look at the information they provide. The input module falls near the center of the signal flow, Figure 3-47, so it is an ideal place to start troubleshooting. Starting in the center supports the "divide and conquer" rule introduced in the troubleshooting procedures in Chapter 1. Study the troubleshooting guide in Figure 7-23(a)

FIGURE 7-23: Troubleshooting discrete input modules.

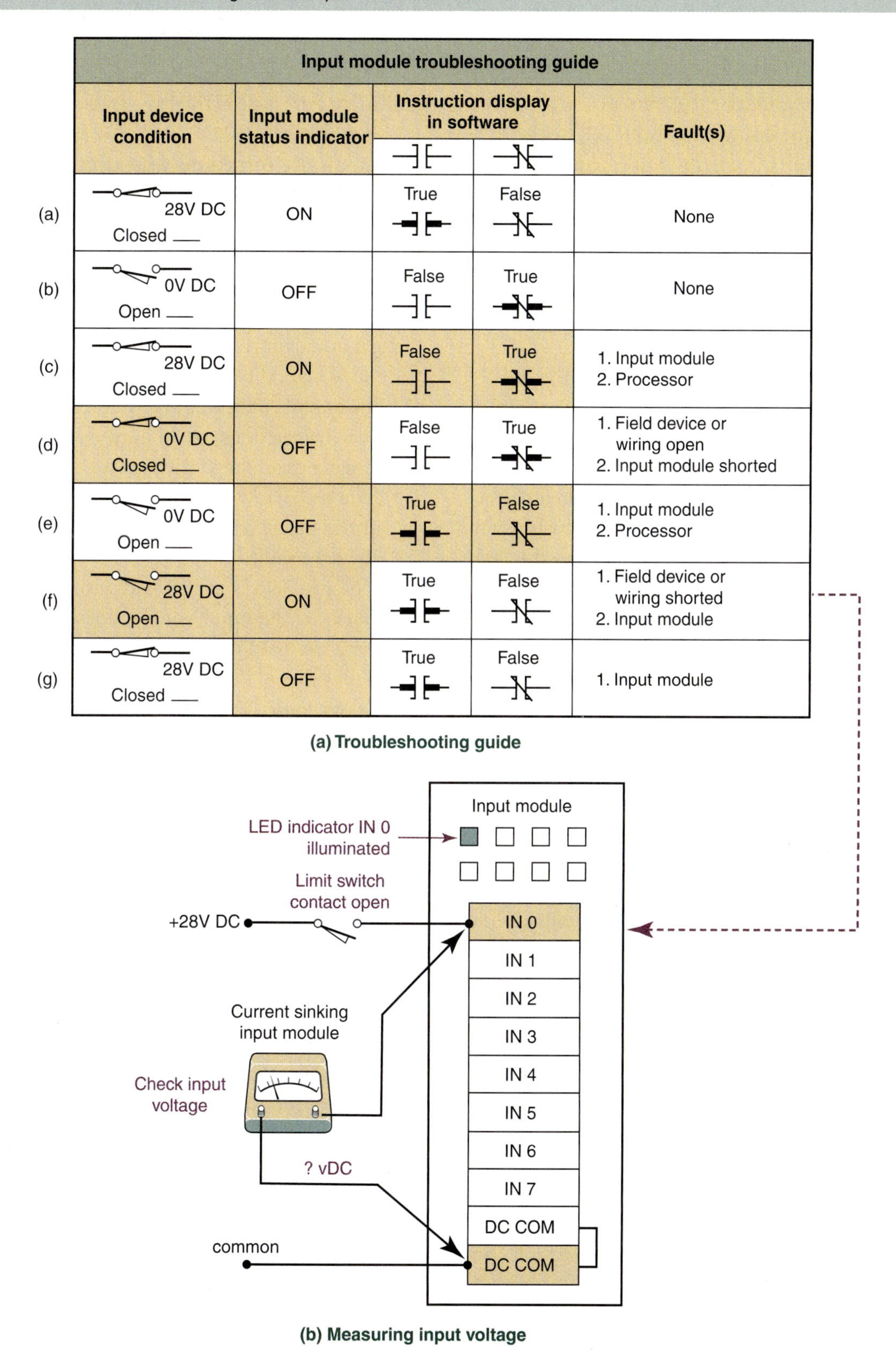

	Input module troubleshooting guide				
	Input device condition	Input module status indicator	Instruction display in software: –] [–	Instruction display in software: –]/[–	Fault(s)
(a)	28V DC Closed	ON	True	False	None
(b)	0V DC Open	OFF	False	True	None
(c)	28V DC Closed	ON	False	True	1. Input module 2. Processor
(d)	0V DC Closed	OFF	False	True	1. Field device or wiring open 2. Input module shorted
(e)	0V DC Open	OFF	True	False	1. Input module 2. Processor
(f)	28V DC Open	ON	True	False	1. Field device or wiring shorted 2. Input module
(g)	28V DC Closed	OFF	True	False	1. Input module

(a) Troubleshooting guide

(b) Measuring input voltage

for discrete input modules before reading a description of each indication. Note that the condition of the input field device and module status light is provided. In addition, the condition (highlighted or true and not highlighted or false) of the instruction displayed by the programming software is shown. The areas where the fault could occur are shaded. Finally, the most likely faults are stated in priority order. Refer to Figure 7-23(a) as you read the following descriptions of each possible fault condition.

a. Correct indications – no fault present.
b. Correct indications – no fault present.
c. Sensor condition, input voltage, and module indicator are correct, but the XIC and XIO ladder instructions have an incorrect indication. The problem is most likely in the input module I/O point, but the fault could be caused by the processor. Since the input module is the most likely cause, replace the module or move the input to another I/O point.
d. The module indicator and the XIC and XIO ladder instructions agree, but not with the state of the field device. The best check is to measure the input voltage at the I/O point as shown in Figure 7-23(b). If the voltage is 0 VDC, then it is either a broken field wire or a bad sensor.
e. The field device status, input voltage, and module indicator all agree, but the XIC and XIO ladder instructions do not. The problem is most likely the input module I/O point. The processor could cause it, but that is less likely.
f. The input voltage, module indicator, and the XIC and XIO ladder instructions agree, but the condition of the field device does not. The fault is most likely a short in the field device or the input wiring. The input module could cause the fault, but that is less likely.
g. The input voltage (28 VDC), the field device, and the XIC and XIO ladder instructions agree, but the module indicator does not. Check for a bad indicator in the input module.

In general, the following items should be remembered during troubleshooting discrete input modules:

- If inputs are fused, then verify that the fuse is not blown.
- If inputs are turned on when the electronic sensor field device driving the input is off, then verify that the sensor off leakage current is not greater than the turn on current for the current sinking input module.
- If the input module is suspected, then replacing it or moving the faulty input to another I/O point is a good test to verify a bad channel.
- If the module indicator and the XIC and XIO ladder instructions are in agreement, then a voltage measurement indicates:
  - The problem is in the input in the module if the input voltage agrees with the field device condition.
  - The problem is in the field wiring or field device if the input voltage does not agree with the field device condition.

Output Module Troubleshooting Analysis. The output module falls near the center of the signal flow, Figure 3-47, so it is another ideal place to start troubleshooting. Study the troubleshooting guide in Figure 7-24(a) for discrete output modules before reading a description of each indication. Note that the condition of the output field device and module status light is provided. In addition, the condition (highlighted or true and not highlighted or false) of the output coil displayed by the programming software is shown. The areas where the fault could occur are shaded. Finally, the most likely faults are stated in priority order. Refer to Figure 7-24(a) as you read the following descriptions of each possible fault condition.

a. Correct indications – no fault present.
b. Correct indications – no fault present.
c. Output instruction and output indicator agree but the field device does not. The problem is most likely that the field wiring is open or the module output circuit is bad. If outputs have individual fuses, then the fuse should be checked.

FIGURE 7-24: Troubleshooting discrete output modules.

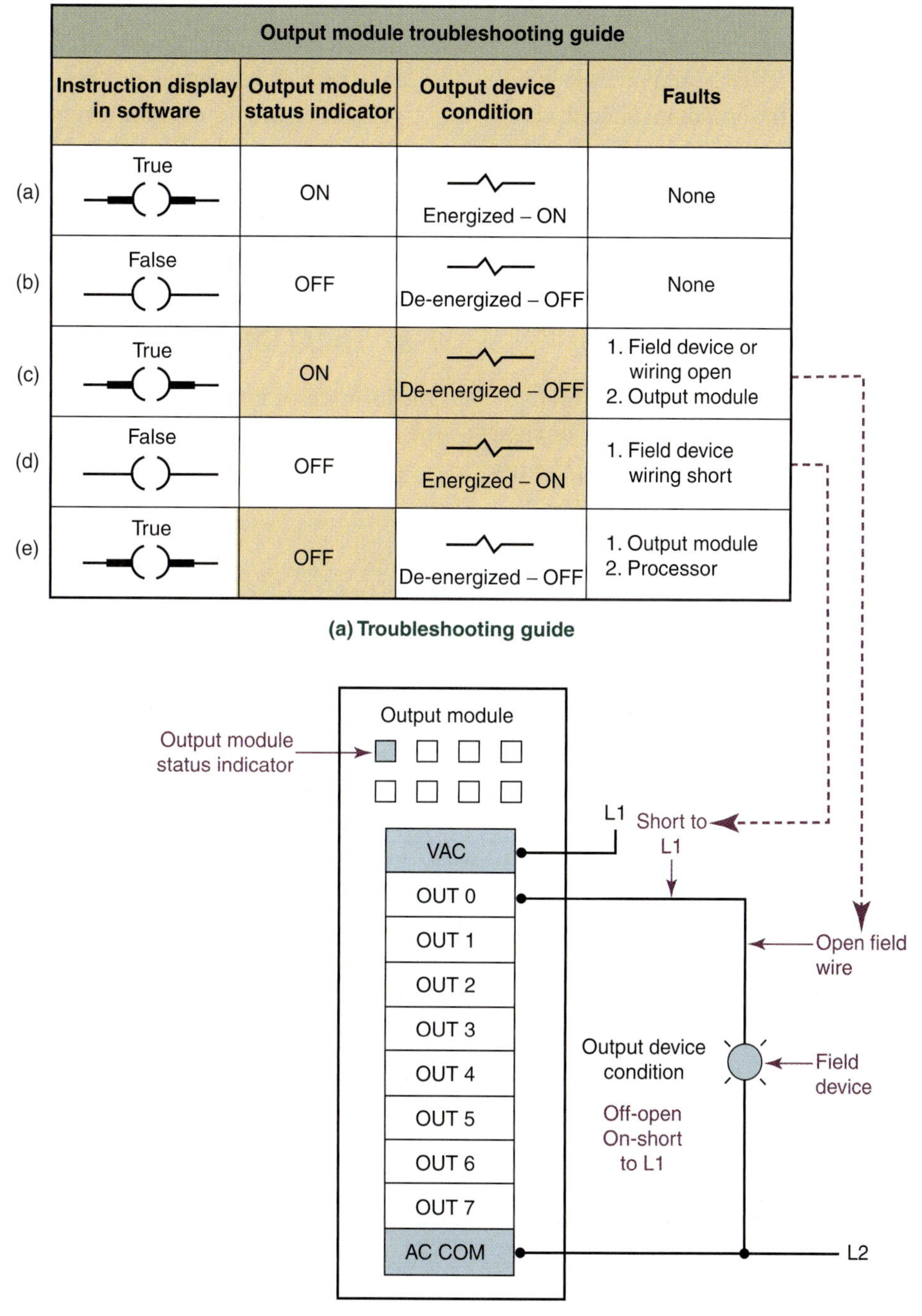

	Output module troubleshooting guide			
	Instruction display in software	Output module status indicator	Output device condition	Faults
(a)	True	ON	Energized – ON	None
(b)	False	OFF	De-energized – OFF	None
(c)	True	ON	De-energized – OFF	1. Field device or wiring open 2. Output module
(d)	False	OFF	Energized – ON	1. Field device wiring short
(e)	True	OFF	De-energized – OFF	1. Output module 2. Processor

(a) Troubleshooting guide

(b) Measuring input voltage

d. Output instruction and output indicator agree but the field device does not. The problem is most likely that the field wiring is shorted to the power line.

e. The field device status and module indicator agree, but the output instruction condition does not. The problem is most likely the output module I/O point. The processor could cause it, but that is less likely.

In general, the following items should be remembered during troubleshooting discrete output modules:

- Many output modules have each I/O or channel fused. In most cases there is a blown fuse indicator that illuminates if the fuse is open when you turn on the output. If an output fails to turn on when the output indicator signals that the output is active, then check the fuse first. Next check for open field wiring using a voltmeter or ohm meter.
- You can use the bit forcing function to make a rung active without running the ladder logic program so that an output fault can be fixed.

7-8 LOCATION OF THE INSTRUCTIONS

The location of instructions from this chapter in the Allen-Bradley programming software is indicated in Figure 7-25.

FIGURE 7-25: Location of instructions described in this chapter.

Systems	Instructions	Location
PLC 5, SLC 500	TOD, FRD	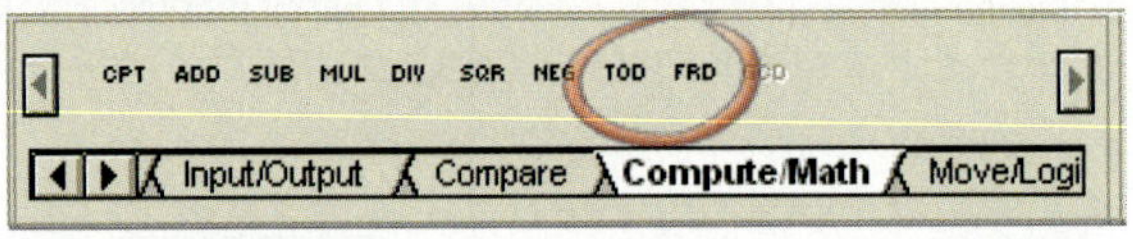 CPT ADD SUB MUL DIV SQR NEG TOD FRD — Input/Output, Compare, **Compute/Math**, Move/Logi
LOGIX	TOD, FRD	 DEG RAD TOD FRD TRN — Trig Functions, Advanced Math, **Math Conversions**, Mc
PLC 5, SLC 500	LIM, MEQ, EQU, NEQ, LES, GRT, LEQ, GEQ	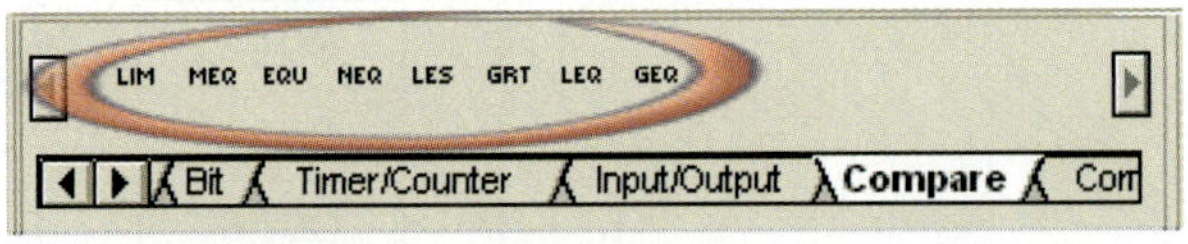 LIM MEQ EQU NEQ LES GRT LEQ GEQ — Bit, Timer/Counter, Input/Output, **Compare**, Con
LOGIX	LIM, MEQ, EQU, NEQ, LES, GRT, LEQ, GEQ	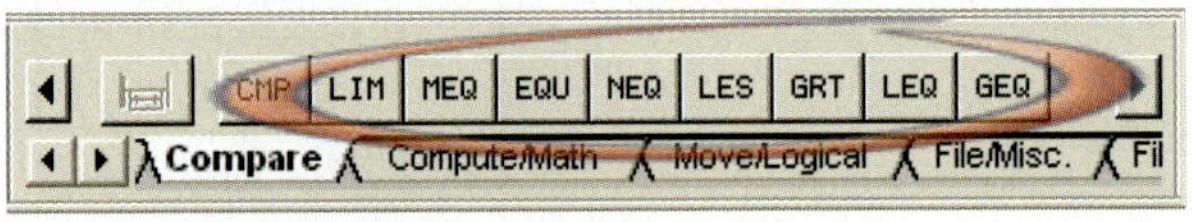 CMP LIM MEQ EQU NEQ LES GRT LEQ GEQ — **Compare**, Compute/Math, Move/Logical, File/Misc., Fil

Chapter 7 questions and problems begin on page 570.

CHAPTER

8

Program Control Instructions

8-1 GOALS AND OBJECTIVES

There are three principal goals of this chapter. The first goal is to provide the student with an overall picture of various program control instructions relative to subroutines and program scan. The second goal is to introduce the Allen-Bradley PLC 5, SLC 500, and Logix systems program control instructions that perform operations such as master control reset, jump, jump to subroutine, and immediate input and output functions. The third goal is to show how the program control instructions are applied to specific PLCs that are used in industrial automation systems.

After completing this chapter you should be able to:

- Explain the function of the program control instructions such as the master control and zone control reset instructions and jump and label instructions.
- Describe the operation of subroutines.
- Explain the function of immediate input and output instructions.
- Draw and describe ladder logic representing applications that use the program control instructions.
- Develop ladder diagram solutions using the program control instructions for the Allen-Bradley PLC 5, SLC 500, and ControlLogix systems.
- Use program control instructions and immediate input and output instructions in the empirical design process.
- Troubleshoot ladder rungs with program control instructions and immediate input and output instructions.

8-2 PROGRAM CONTROL INSTRUCTIONS

Program control instructions direct the flow of operation, as well as the execution of instructions, within a PLC ladder program. When programmed conditions are satisfied, portions of the program can be jumped over or their rungs not scanned so the outputs in these specific program groups or zones remain unchanged. In other words, the program control instructions allow the PLC to efficiently perform user-programmed routines that are executed only when specific automation conditions dictate. The program control instructions alter the program scan time, thereby optimizing total system response.

8-3 ALLEN-BRADLEY PROGRAM CONTROL INSTRUCTIONS

The Allen-Bradley program control instructions that are discussed in this section are as follows:

- The master control reset (MCR) instruction, which is used in pairs, fences in a group of instructions that can be executed or disabled.
- The jump (JMP) and the label (LBL) instructions are used together. When a JMP is enabled, the program jumps to the ladder rung with the LBL instruction and continues the execution of the subsequent instructions. Jumps can be either forward (skip ladder rungs) or backward (rescan ladder rungs).
- The jump to subroutine (JSR), the subroutine (SBR), and the return (RET) instructions are used together in the program. When a JSR instruction is enabled, the program jumps from the main program to the ladder rung in the subroutine program with the SBR instruction. The subroutine is executed until an RET instruction occurs. The RET instruction returns the program to the ladder rung in the main program that follows the JSR instruction rung.

8-3-1 Master Control Reset Instructions

Some output instructions, often called *override instructions*, provide a means of scanning a section of the control ladder when specific input conditions are present. Their use increases program flexibility and efficiency plus offers a reduction in scan time by jumping over portions of the ladder that are not utilized for specific process control situations. In relay logic these instructions are called *master control relays*; in the PLC they are called *master control reset*.

Electromechanical Master Control Relays. A hardwired master control relay (MCR) is used with relay ladder logic to shut down all or a portion of the relay ladder logic by turning on a master control relay with input relay logic. A hardwired master control relay ladder is illustrated in Figure 8-1. Note that rungs 1 and 2 always operate but rungs 4 through 31 operate only if the MCR contactor in rung 1 is active. An active MCR closes the MCR contacts and

FIGURE 8-1: Electromechanical master control relay ladder.

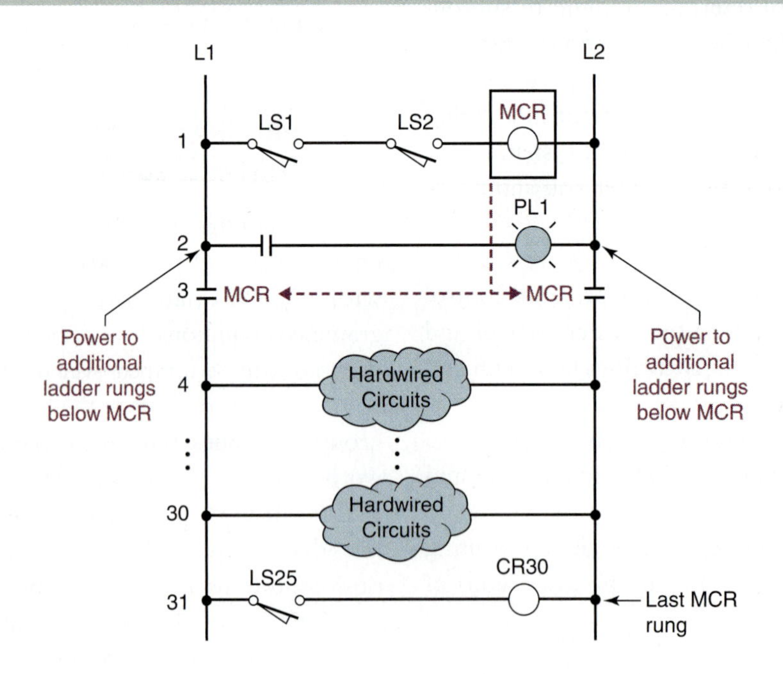

applies power to rungs 4 through 31. There could be additional relay ladder rungs that are not a part of the MCR circuit. These rungs do not rely on the MCR rungs for their power, so they are not controlled by the MCR contacts.

PLC Master Control Reset Instruction and Standard Ladder Logic. The PLC *master control reset* (MCR) output instruction is used in pairs and enables or inhibits the execution of a group or zone of ladder rungs, or it can be used to control the entire ladder logic program. The standard ladder logic for the MCR instruction is provided in Figure 8-2.

The PLC MCR works like its electromechanical counterpart. Two MCR instructions form a fence or zone in a program. Execution or scanning of the PLC instructions within the fence is controlled by the input logic in the rung of the first MCR instruction. The second MCR instruction, which signifies the end of the fence, does not have input logic, so its execution is unconditional. Read the description of the MCR instruction set provided in Figure 8-2. When the MCR instruction is false or de-energized, all non-retentive (non-latched) output instructions within the fence are false, even if their rung inputs are true. All non-retentive rungs are turned off simultaneously, and all retentive rungs remain in their last state. Conversely, when the MCR instruction is true or energized, the rungs within the fence operate as if no fence existed.

The operation of the MCR instruction is demonstrated in the example ladder logic in Figure 8-3. The first rung is labeled *fence start*, which is a rung with a conditional MCR instruction, and the last rung is labeled *fence end*, which is a rung with an unconditional MCR instruction. Note that the MCR instructions do not have an address. The operation of the zone is as follows:

Input I:1/1 is true (On/off switch closed): This causes the MCR instruction to be true, and the rungs within the fence act in accordance with their respective input logic conditions as if no

FIGURE 8-2: Standard ladder logic rungs for master control reset instructions.

Application	Standard Ladder Logic Rungs for Master Control Reset instructions
Control the scanning of all or part of a ladder logic program, disable output instructions, and force all non-retentive outputs to the off state with a single input logic instruction or a combination of instructions. Pairs of MCR output instructions are placed in a ladder logic program to control when the rungs between the instructions are scanned and executed. The MCR instructions are said to "fence" the ladder rungs that they bracket. The logic state (false or true) of the first MCR instructions dictates if the fenced rungs are scanned (true state) or passed over (false state). When the MCR is false, all non-retentive outputs in the fenced rungs are turned off and all retentive outputs retain their last condition. When the MCR is true, the rungs are scanned like the remainder of the ladder logic. The rung with the first MCR instruction is conditional and the second is unconditional as shown in the ladder rungs here.	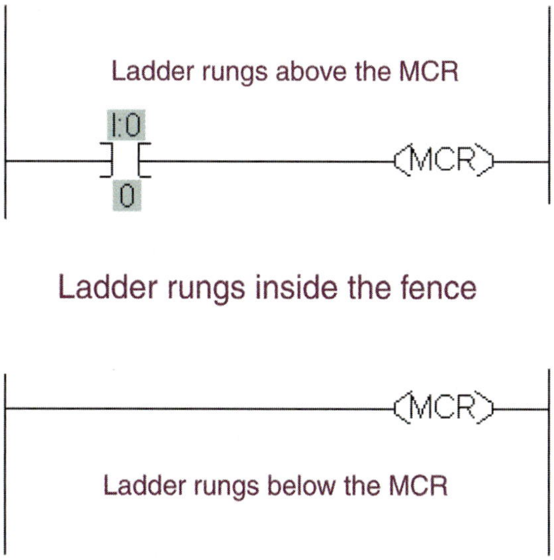 Ladder rungs above the MCR I:0 / 0 —] [— (MCR) Ladder rungs inside the fence (MCR) Ladder rungs below the MCR MCR ladder rungs

FIGURE 8-3: Ladder logic example with MCR instruction.

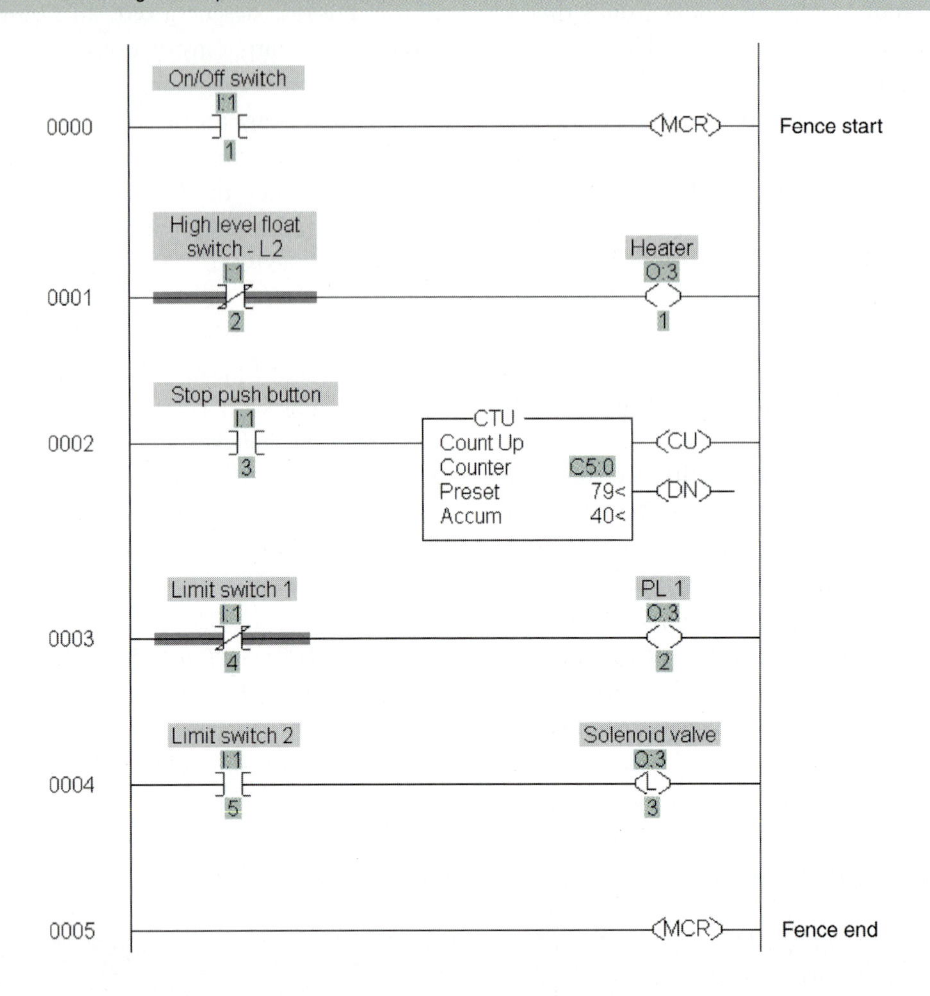

zone exists. With the zone active, input field devices for the ladder in Figure 8-3 turn on the *heater, pilot light*, and *solenoid valve*. The push button drives the *counter* ACC to 40, for example, and at this point, input I:1/1 goes false, and the rungs inside the fence respond as follows.

Input I:1/1 is false (On/off switch open): This causes the MCR instruction to be false, and all rungs within the zone are forced to a false state independent of the condition of the input logic. With the zone inactive, the non-retentive output instructions are false so the *heater* and *pilot light are off* even though their controlling input instructions are true. The *solenoid valve* remains *on* because it is a retentive output instruction and it was on when I:1/1 changed to a false condition. The *counter ACC* remains at 40, even though I:1/3 continues to cycle on and off.

The following comments apply to the MCR operation when the MCR is false:

- The input rung logic does not control the rung output.
- Non-retentive outputs are all false and retentive outputs hold the state they had when the MCR went from true to false.
- The counter accumulator stops incrementing even though the input logic changes state. The current count in the counter accumulator remains. When the MCR becomes active the counter starts from the previous value in the ACC.
- Note that the figure depicts only one MCR zone that includes the entire program. Multiple MCR zones can be constructed within a program that control only parts of the ladder logic program. However, MCR zones are never overlapped or nested.

EXAMPLE 8-1

Design a ladder program using an MCR to initialize a counter and timer preset value with an initialization push button (I:1/2) before a ladder logic program is executed. The counter and timer preset values are located in N7:1 and N7:2, respectively.

SOLUTION

Figure 8-4 shows the solution ladder logic. The initialization push button is pressed, the XIO instruction, B3:0/1, is true (power flow), and the virtual memory bit B3:0/2 is set, which seals the initialization push button and makes the MCR instruction in rung 2 true. A true MCR executes the ladder logic rungs, moving preset values into the counter and timer. It also activates the internal reset memory bit B3:0/1. When the internal reset is true, the XIO instruction, B3:0/1, in rung 0 is false (no power flow) and the MCR becomes false. This makes all of the rungs in the MCR fence false as well. So, the MCR is true for just one scan after the initialization push button is pressed. The program could use the first scan status bit (S:15) instead of the push button so that the process is performed when power is first applied or when power is reapplied after a power failure.

The examination of PLC fault bits is another application of the MCR instruction. When used in this manner, the application places inputs addressed by fault bits in the first MCR rung and disables rungs with outputs that are creating the fault conditions with an MCR fence. The MCR instruction operates the same in the PLC 5, SLC 500, and Logix PLCs, that is, you create an MCR zone using pairs of MCR instructions.

FIGURE 8-4: Ladder solution for Example 8-1.

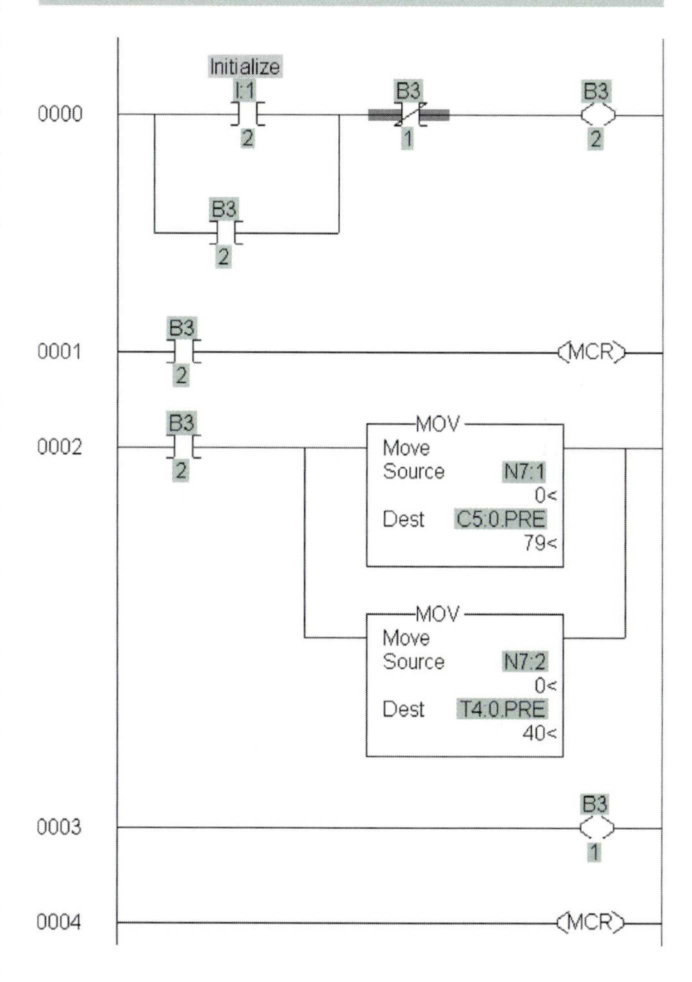

8-3-2 Jump and Label Zone Control Instructions

The *jump* (JMP) output instruction and the input *label* (LBL) instruction are used together and allow the program scan sequence to be changed based on automation system conditions. Some manufacturers use the terms *skip* or *go-to* for their jump instruction. The jump instruction reduces the PLC scan time by jumping over (branching around) instructions that are not relevant to the program's operation at a particular time because of process or operator requirements. The jump instruction steps the program over a group of rungs to the label rung, as illustrated in Figure 8-5.

The *label* (LBL) instruction identifies the destination rung of the *jump* (JMP) instruction. The label instruction reference number must match that of the jump instruction with which it is used. The reference number can be assigned a three-digit address between 000 and 255. The PLC 5 and SLC 500 automatically place Q2: in front of the reference number. The label instruction, which is always true, is the first input instruction in the destination rung. The label is assigned the jump instruction's unique number. Ladder programs are permitted multiple jump instructions with the same number, but that number can be used only

FIGURE 8-5: Jump instruction illustration.

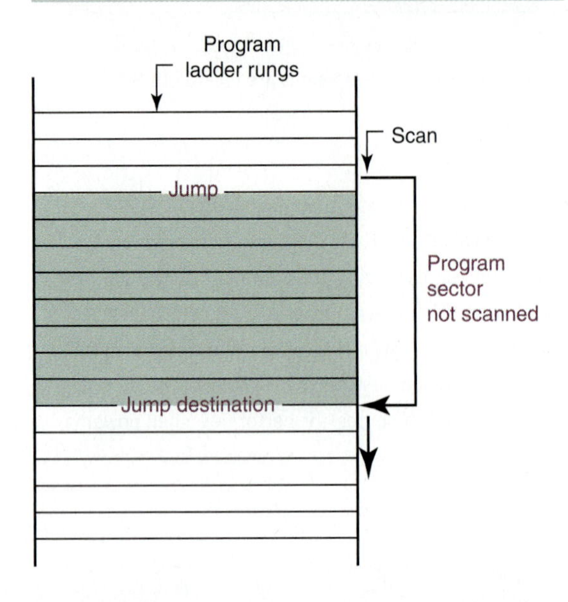

FIGURE 8-6: Ladder logic example with JMP and LBL instructions.

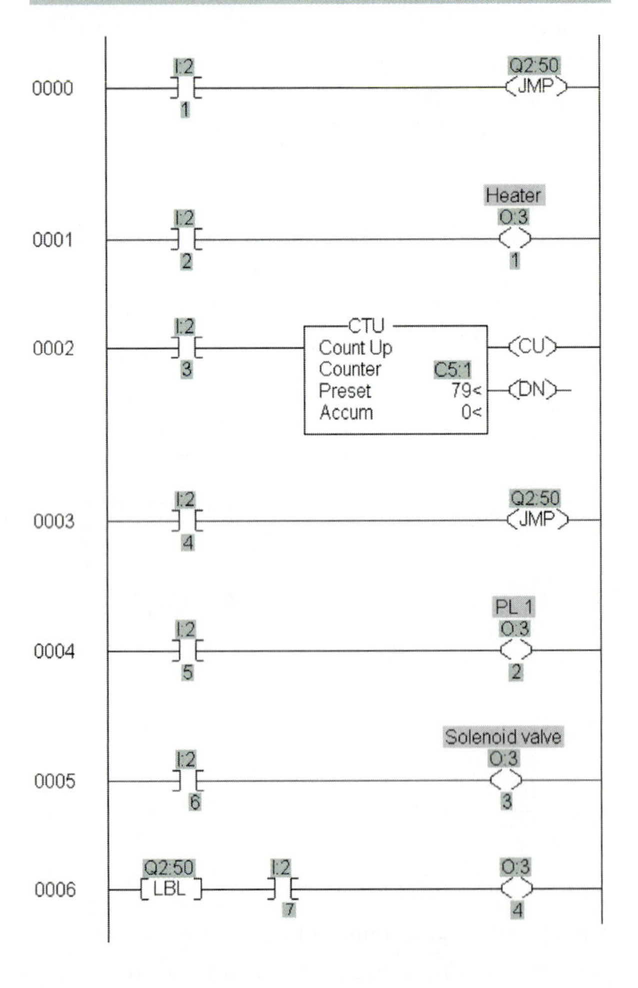

once on a label instruction. Jump instructions located on different rungs can use the same label instruction with the same reference number, as illustrated in Figure 8-6. The operation of the ladder program in the figure is as follows:

- If input I:2/1 is active, then the program jumps to rung 6 and continues the program sequence. Note that the counter in rung 2 will not increment because that rung is not scanned due to the jump. For this reason, the use of timers and counters inside the jump zone should be studied carefully.
- If input I:2/1 is false, then the program executes rungs 1 and 2.
- In rung 3, if input I:2/4 is false, then the program executes rungs 4, 5, and so forth.
- In rung 3, if input I:2/4 is true, then the program jumps over rungs 4 and 5 to rung 6 and continues the program sequence.

Finally, when using jump instructions, the following precautions should be observed.

1. Programming a jump instruction to go backward in the program should only be done with great care. Excessive backward jumps could cause the program to remain in a loop for too long and trigger a watchdog timer fault. The watchdog timer checks to see if the program scan time has exceeded a specific value.
2. Make sure that the LBL instruction is the first input instruction on the rung.
3. Programming a jump instruction into an MCR zone should never be done because the execution of the rungs between the rung with the label instruction and the last rung with the MCR instruction is dictated by the jump instruction and not by the initial MCR instruction.

In the ControlLogix PLC, text labels are used in place of the numeric code required in the PLC 5 and SLC 500 systems. The requirements for the text labels include a unique label name used only once for each jump, and label text with 40 or less characters that can be letters, numbers, and underscores (_).

Standard Ladder Logic for Jump and Label Instructions. There are three configurations of standard ladder logic for the jump and label instructions in the Allen-Bradley PLC systems. Read the description that accompanies each of the three configurations in Figure 8-7.

Figure 8-7(a) illustrates the standard forward jump instruction for the PLC 5 and SLC 500, and Figure 8-7(b) illustrates this for the ControlLogix PLC. Note that a reference number (preceded by Q2:) is used for the PLC 5 and SLC 500, but the jump and label are linked with a jump and label name for the ControlLogix PLC. The forward jump occurs when the input logic in the jump rung is true. The label instruction is placed in the extreme left location in the first rung to be scanned after the bypassed rungs.

Figure 8-7(c) illustrates the backward jump. Note that the label instruction is above (lower rung number) the jump instruction. As the ladder

FIGURE 8-7: Standard ladder logic rungs for jump and label instructions.

Application	Standard Ladder Logic Rungs for Jump and Label Instructions
Jump over a portion of the ladder logic rungs when input field device(s) and the input ladder logic indicate that the rungs shouldnot be scanned. When activated by any combination of input ladder logic instructions, the jump (JMP) instruction in the forward direction forces the ladder program to skip over or not scan that portion of the ladder logic between the JMP instruction and the label (LBL) instruction. The PLC 5, SLC 500, and ControlLogix processor syntax is the same except that jumps are identified by numbers (0 to 255 in PLC 5 and SLC 500) and text names (ControlLogix processor). Note that the PLC 5 and SLC 500 automatically put a Q2: in front of the number. The ControlLogix label can include up to 40 alpha, numeric, or underscore characters.	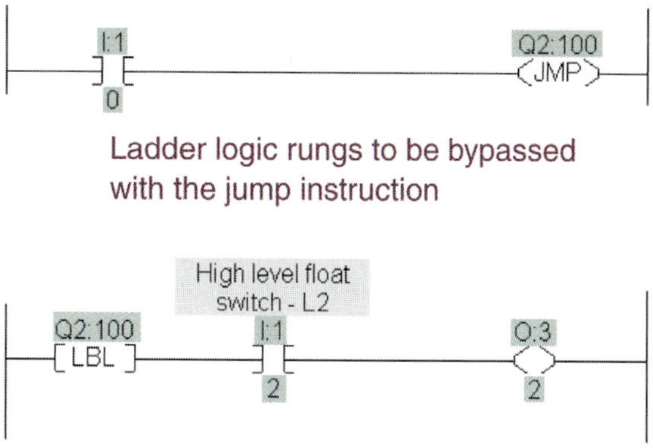 I:1] [0 — Q2:100 (JMP) Ladder logic rungs to be bypassed with the jump instruction Q2:100 [LBL] — High level float switch - L2 I:1] [2 — O:3 ( ) 2 **(a) Forward jump (PLC 5 and SLC 500)** Part_sensor <Local:1:I.Data.1>] [— Sensor_evaluation (JMP) Ladder logic text to be bypassed with the jump instruction Sensor_evaluation [LBL] — Part_sensor <Local:1:I.Data.1>] [— Machine_trigger <Local:2:O.Data.9> ( ) **(b) Forward jump (ControlLogix)**
Jump backward in the ladder logic and execute a portion of the ladder logic additional times when input field device(s) and the input ladder logic indicate that the rungs should be scanned again. The jump (JMP) instruction in the backward direction, when activated by any combination of input ladder logic instructions, forces the ladder program to execute or rescan a portion of the ladder logic that was just scanned. The rungs between the between the label (LBL) instruction and the JMP instruction are rescanned every time the JMP instruction is executed. The PLC 5 and SLC 500 rung format is used in the illustration.	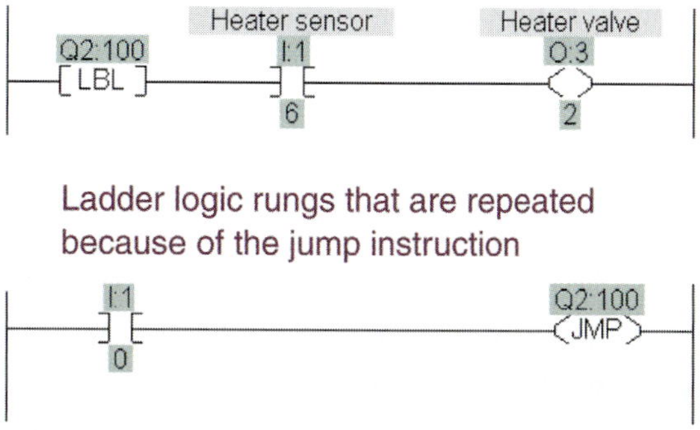 Q2:100 [LBL] — Heater sensor I:1] [6 — Heater valve O:3 ( ) 2 Ladder logic rungs that are repeated because of the jump instruction I:1] [0 — Q2:100 (JMP) **(c) Backward jump**

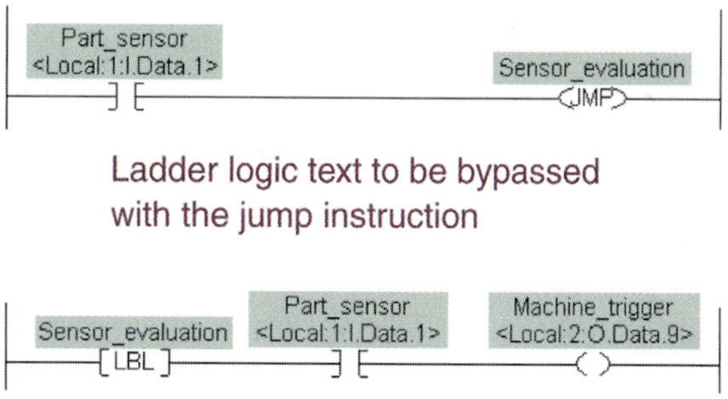

FIGURE 8-7: (Continued).

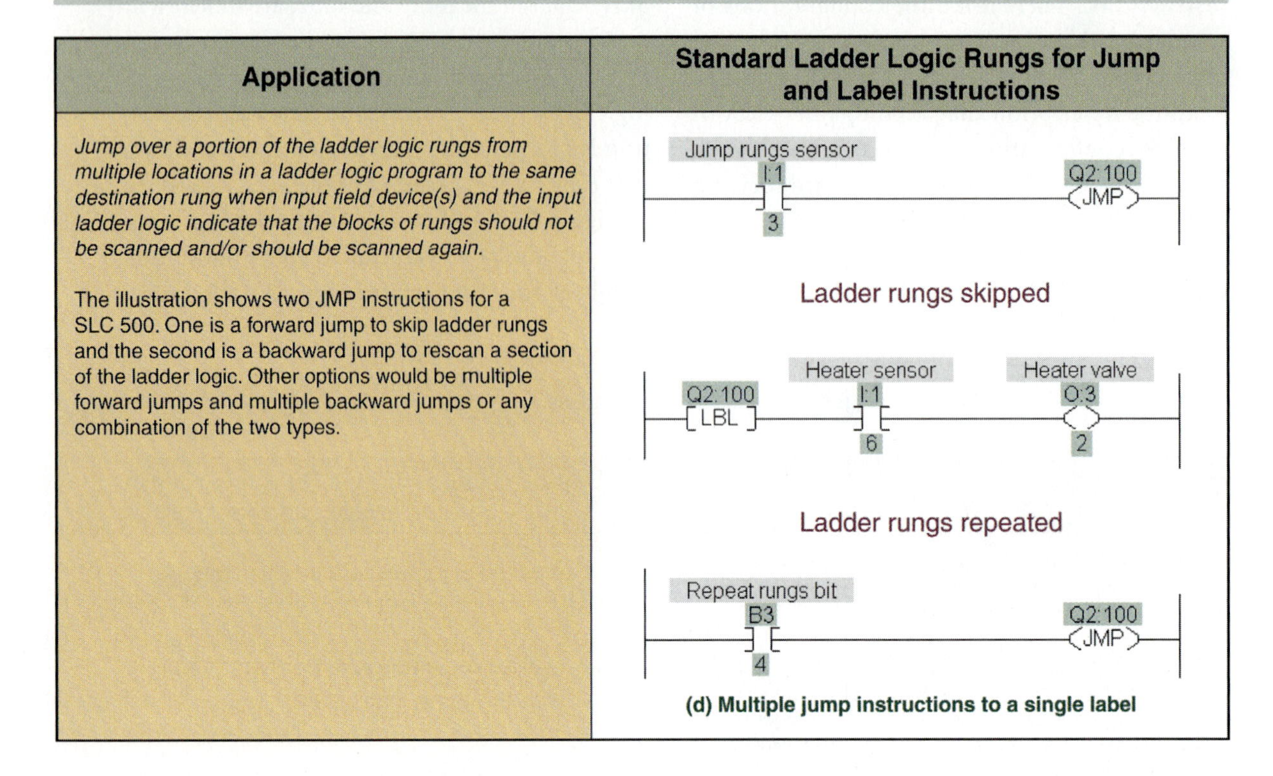

(d) Multiple jump instructions to a single label

scans from top to bottom the label instruction does not create any problems when the rung is scanned since it is always evaluated as true. The backward jump occurs when the input logic in the jump rung is true. The backward jump is used to rescan a block of ladder logic repeated times. The ControlLogix logic implementation is the same except for the difference in the jump and label identification.

The standard configuration for multiple jump instructions to a single label instruction is illustrated in Figure 8-7(d). The figure shows a forward jump and a backward jump, but multiple forward and backward jumps are also permitted.

Jump versus MCR Instructions. A study of the forward jump and MCR instructions indicates a good deal of similarity. Both permit an input logic condition to force the PLC scanner to skip over a block of PLC ladder logic. The primary difference, however, is in how the outputs are handled when the instructions are executed. The MCR sets all non-retentive outputs to the false state (the output image table bits are set to zero) and keeps the retentive outputs in their last state or condition. The jump leaves all outputs in their last condition because the output image table bits are not changed.

8-3-3 Subroutine Instructions

A *subroutine* is a group of ladder logic PLC instructions outside the main ladder program that can be executed with the subroutine instruction. Thus, by using a single subroutine instruction repeated program routines do not have to be duplicated in the main ladder program when the routines must execute repeatedly. For example, when an automated machine has a sequence of rungs that must be repeated numerous times in the machine's cycle, that sequence of rungs could be programmed one time into a subroutine and just called when needed. Variable data can be *passed* to a subroutine when it is called, permitting the subroutine ladder logic to perform mathematical or logical operations on the data. Variable data or results produced in the subroutine can be passed back to the main program in a similar fashion. The subroutine concept is common in all PLCs, but the instruction that moves the program to the subroutine may differ between PLCs. The subroutine instructions in the Allen-Bradley PLCs are the *jump to subroutine* (JSR) output instruction, the

FIGURE 8-8: Jump to subroutine instruction.

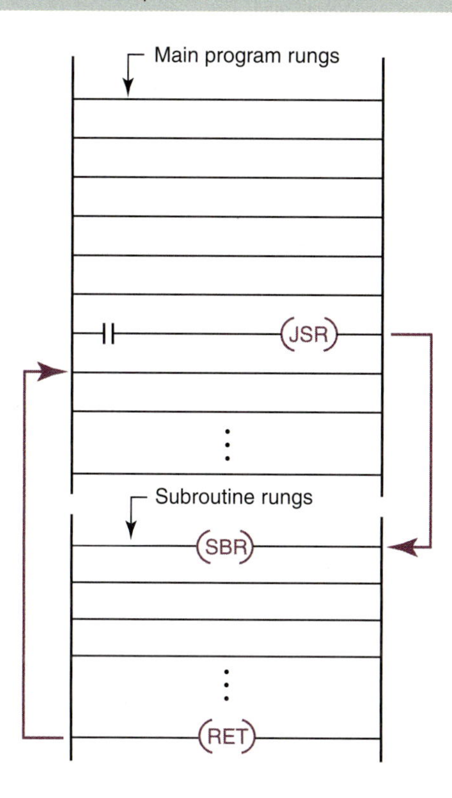

FIGURE 8-9: Nested subroutines.

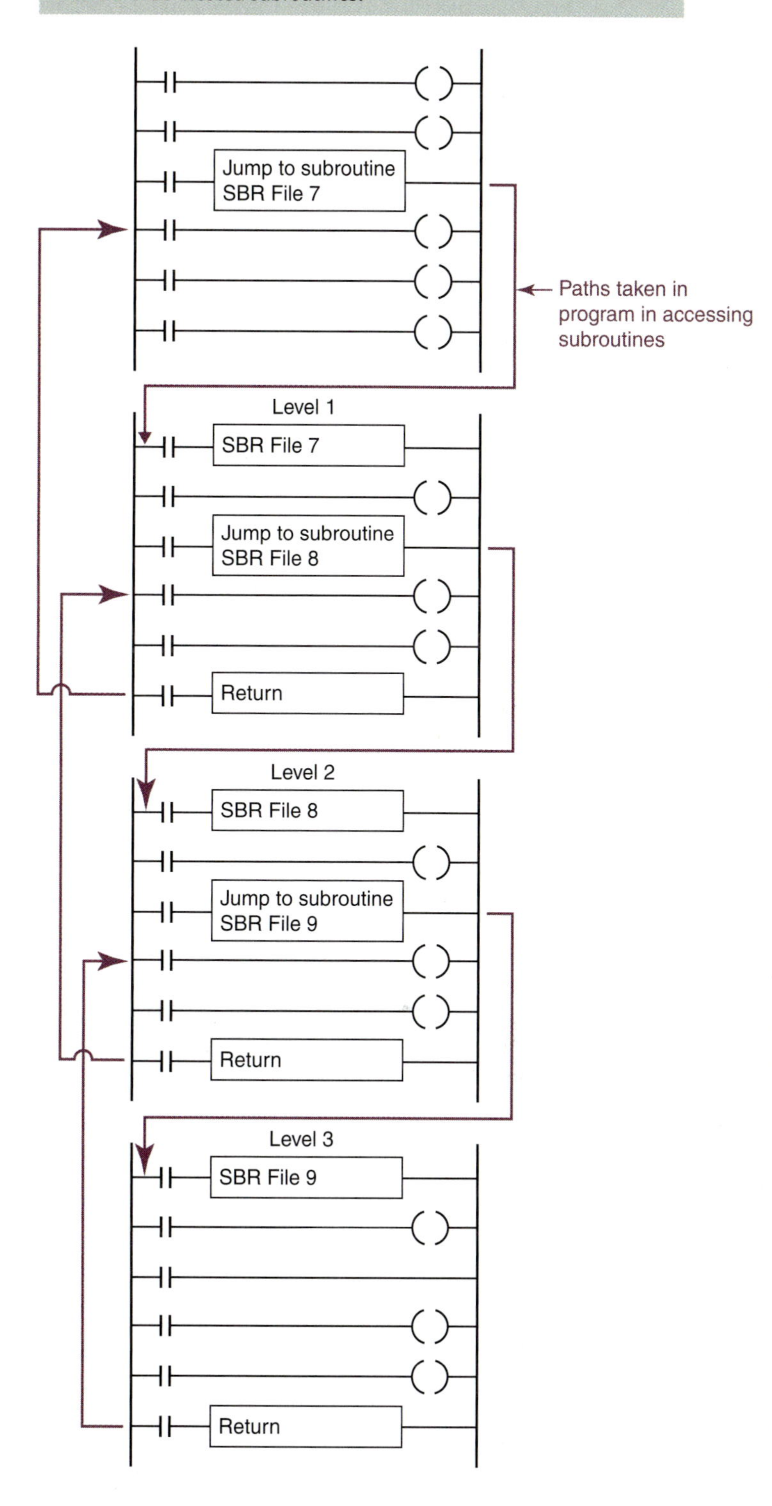

subroutine (SBR) input instruction, and the *return* (RET) output instruction.

Figure 8-8 illustrates the subroutine operation. Observe that the subroutine rungs with the SBR and RET instructions are a separate ladder logic program from the main program. Note that an RET instruction marks the end of the subroutine, similar to the last MCR instruction in an MCR zone. The return instruction is typically in an unconditional rung and returns the ladder program to the rung immediately after the rung containing the JSR instruction. However, if you want to end the subroutine before it completes all the rungs in the subroutine program, then a second conditional RET instruction may be inserted in a rung. Note in Figure 8-8 that the first rung of the subroutine contains the unconditional subroutine instruction SBR, which serves as a marker identifying that subroutine rungs follow.

Subroutines may be *nested*, which allows the programmer to direct the program flow to a subroutine, then to another and to another without returning to the main program each time. For example, sequential mathematical calculations that are needed by an industrial process can be implemented in nested subroutines for use by the main program whenever it's required. Figure 8-9 illustrates the nesting concept. Note that each subroutine return instruction routes the program flow to the previous subroutine, then finally back to the main program. The lines with arrows show the paths taken by the program in accessing the subroutines. Also note that each subroutine is labeled with a level number and a file number. In the Allen-Bradley PLC 5 and SLC 500 systems the subroutines are numbered U:3, U:4, and so on, up to U:255. The ControlLogix system permits the programmer to name the subroutines so that they are identified by the application.

When nesting subroutines, care should be taken because scan time and the scanning rate of the main program may be impacted since the main program is not being scanned while the subroutine is executing. As a result, excessive delays in scanning the main program may cause inputs and outputs not to be assessed at the required time. These delays can be mitigated with the immediate input and output instructions covered later in this chapter.

8-3-4 PLC 5 and SLC 500 Subroutine Instructions

The Allen-Bradley subroutine instructions for the PLC 5 and SLC 500 PLCs are quite similar; however, the PLC 5 has parameter passing options like the ControlLogix PLC. Parameter passing is covered later in this section. The naming procedures used for the PLC 5 and SLC 500 subroutines are the starting point.

PLC 5 and SLC 500 Subroutine Setup. The procedure for selecting and naming the subroutine files in the PLC 5 and SLC 500 systems is illustrated in Figure 8-10. The process starts by right-clicking the Program Files folder in the left file menu, and then selecting New . . . to create a new ladder file [Figure 8-10(a)]. Note that

FIGURE 8-10: PLC 5 and SLC 500 subroutine setup.

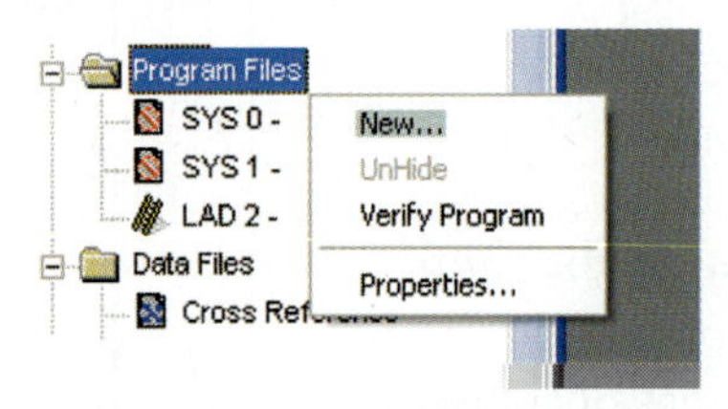

(a) Left file menu – right-click program files so that New... can be selected

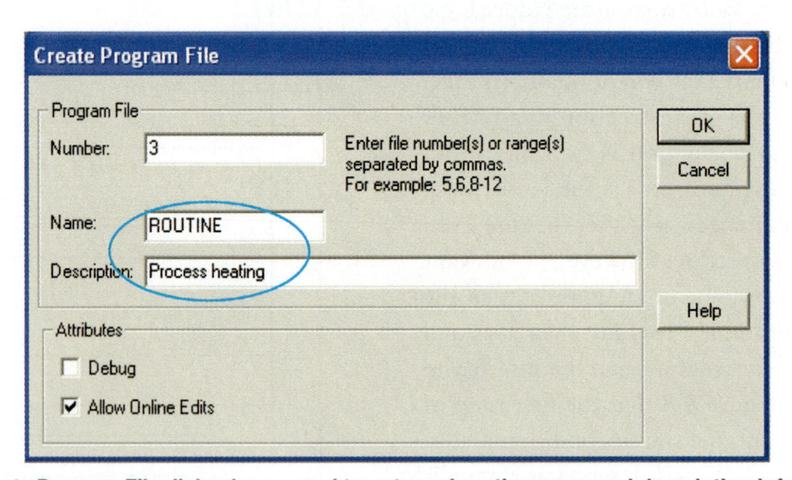

(b) Create Program File dialog box – used to enter subroutine name and description information

(c) Updated Left file menu with new subroutine program file, LAD 3 - ROUTINE

LAD 2 is the default main program ladder file that is always present in this menu.

In the Create Program File dialog box [Figure 8-10(b)], the number (3 to 255) of the desired file is entered. Usually subroutine files are just numbered consecutively. A name and description can be added, but that is optional. In this example file 3 is used with the name ROUTINE. When the dialog box is closed (click OK) the new file is visible in the left file menu [Figure 8-10(c)]. If LAD 2 is double-clicked the main program ladder will open, but if LAD 3 - ROUTINE is selected the subroutine ladder will open. Once they are opened, they can be quickly selected from the tabs at the bottom of the window.

Standard Ladder Logic for the PLC 5, SLC 500, and ControlLogix Jump to Subroutine Instruction. The standard ladder logic for the three configurations of the jump to subroutine (JSR) instruction is illustrated in Figure 8-11. Read the description that accompanies each of the configurations.

Figure 8-11(a) illustrates the standard unconditional subroutine instruction. Note that the subroutine ladders are called U:3, U:4, and U:5. The U: is the required preface and the number indicates the LAD number in the left file menu. Thus, when the first subroutine instruction (rung 0) in Figure 8-11(a) is executed, the ladder LAD 3 is scanned. Since each of the subroutine instructions is unconditional, the ladders in LAD 3, LAD 4, and LAD 5 are executed in that order. This allows a large program to be divided into three logical segments for easier development and maintenance. The naming procedures for the PLC 5 and SLC 500 systems are the same, but the ControlLogix PLC uses a tag to identify the subroutine. The PLC 5 and SLC 500 subroutine numbers are entered by first double-clicking the file number box. Clicking outside the instruction box after entering the number adds the subroutine number to the instruction.

The standard subroutine instruction with controlling input logic is illustrated in Figure 8-11(b). Note that the subroutine ladders are identified in the same fashion, but in this configuration the execution of the subroutines is controlled by the input logic on each subroutine rung. This is the most frequently used configuration for subroutines.

The standard subroutine instruction with controlling input logic and a one-shot instruction is

EXAMPLE 8-2

Implement the process control tank solution in Figures 7-19, 7-20, and 7-21 using the subroutine standard ladder logic configuration in Figure 8-11(a). Review the solution for the tank problem in Chapter 7, and then use unconditional subroutines for execution of the solution.

Solution

The solution to the process tank problem, illustrated in Figure 8-12, is divided into three process solution areas: tank fill and mix (Figure 7-19), material heat and drain (Figure 7-20), and tank rinse (Figure 7-21). Three subroutines are created: Routine_1 (tank fill and mix), Routine_2 (material heat and drain), and Routine_3 (tank rinse). The ladder logic in the main program calls the three unconditional subroutines and the original 15-rung single ladder solution is executed by scanning the three process control ladder sections one after the other.

EXAMPLE 8-3

The PLC program for a production system has 15 fault conditions that stop the production system by latching a bit. Design an operator attention program that will flash a red fault light on the system with a 2-second on, 1-second off duty cycle when these faults occur. If the alert is not acknowledged in 60 seconds by pressing a normally open reset push button, a horn with the same duty cycle is added to the fault warning system. The ladder logic should be placed in a subroutine so it can be called from any one of the 15 locations in the machine control ladder logic where the faults are detected. The system reset should reset the fault indicator as well.

FIGURE 8-11: Standard ladder logic rungs for SLC 500 subroutine instructions.

Application	Standard Ladder Logic Rungs for Subroutine Instructions
Execute all the ladder logic on every scan but isolate blocks of the rungs for better design management and easier editing. The use of subroutines permits a program solution in PLC 5, SLC 500, and ControlLogix to be divided into logical segments. No input logic is present in this standard ladder logic example, so the rung is always active. The result of this unconditional input condition is the execution of the subroutine instruction every scan. As a result, it is like pasting the block of rungs from the subroutine into the main program where the jump to subroutine instruction is located. However, the rungs in the subroutine are isolated for easy editing and maintenance. The tabs at the bottom of the page permit selection of the main program or any of the subroutine ladders. Note: SLC 500 rungs are shown, but the operation would be the same for PLC 5 and ControlLogix.	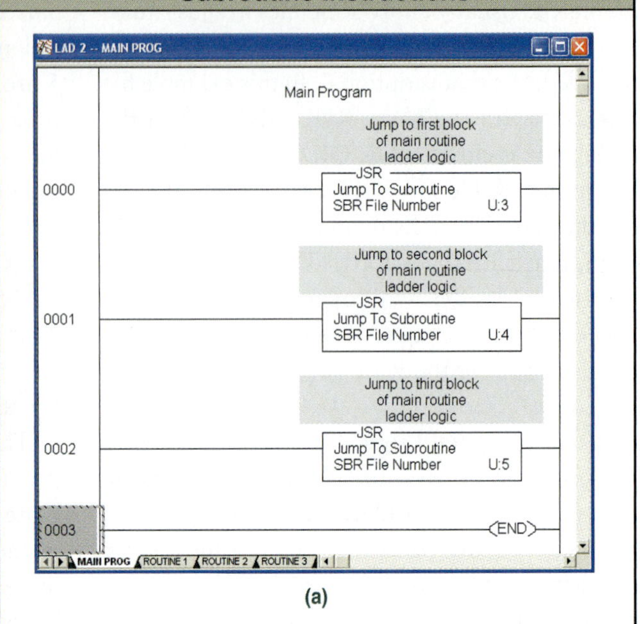 (a)
Schedule the execution of a portion of the ladder logic when one or a combination of input logic instructions make the rung with the subroutine jump instruction active. The input field device(s) and a combination of their input ladder logic instructions in the SLC 500 make the jump to subroutine rung active. As a result, the ladder logic rungs in the subroutine are executed on a schedule and at a rate determined by the field devices, input instructions, or internal virtual memory bits. The tabs at the bottom of the page permit selection of the main program or ladder logic for the subroutine, ROUTINE 1. Note: SLC 500 rungs are shown, but the operation would be the same for PLC 5 and ControlLogix.	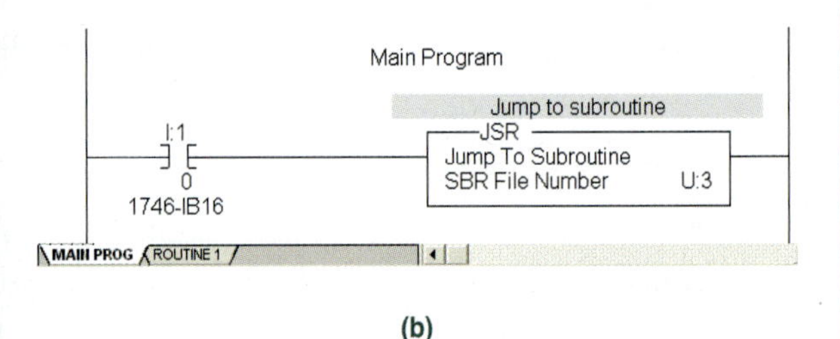 (b)
Execute a portion of the ladder logic for only one scan when an input instruction or a combination of input logic instructions makes the rung with the subroutine jump instruction active. The input field device(s) and the input ladder logic in the jump to subroutine instruction rung determine when the jump rung is active. The one shot assures that the jump to subroutine is performed for only one scan. Note: SLC 500 rungs are shown, but the operation would be the same for PLC 5 and ControlLogix.	(c)

FIGURE 8-12: Solution for Example 8-2.

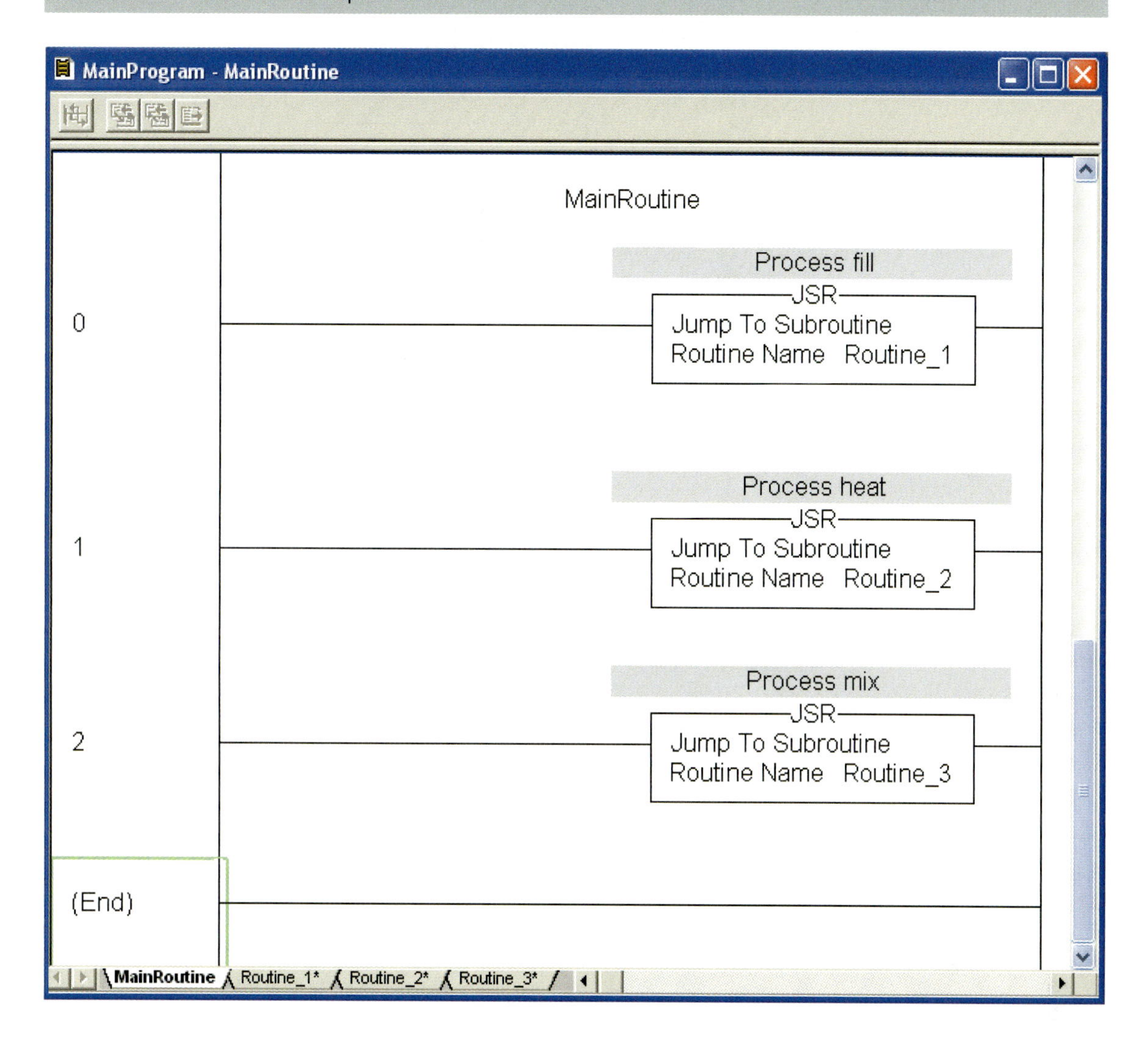

Solution

The operator alert ladder logic program is shown in Figure 8-13. The JSR instruction rung from the machine control ladder logic is shown in Figure 8-13(a) and the subroutine is shown in Figure 8-13(b). Study the ladder solution.

When any production fault is detected the system program [Figure 8-13(a)] latches the retentive bit B3:0/2. That bit is used to call the operator alert subroutine and is reset with the reset push button switch and the unlatch instruction for the B3:0/2 bit in rung 1 of the main program.

The light flashing logic in the subroutine [Figure 8-13(b)] is located in rungs 0 through 2, and the three remaining rungs provide the horn control. The flashing light is produced by two timers with cross-coupled done bits. T4:0 sets the off light time and T4:1 sets the on light time. The done output of T4:0 triggers the light. After T4:0 reaches 1 second, the light comes on (T4:0/DN) and remains on while T4:1 increments for 2 seconds. The output, T4:1/DN, resets the T4:0 timer and the light.

The horn enunciator is cycled on and off by the light timer done bit, T4:0/DN, after the 60-second timer has timed out. The reset for the operator alert also resets all three timers. Verify that the ladder solution satisfies the automation system requirements.

FIGURE 8-13: Ladder solution for Example 8-3.

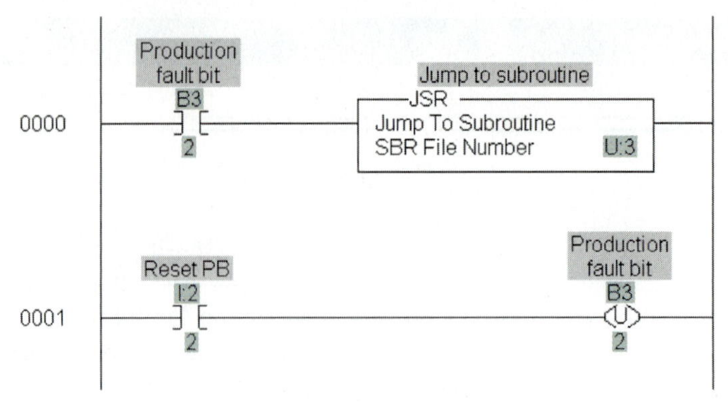

(a) Main routine – LAD 2

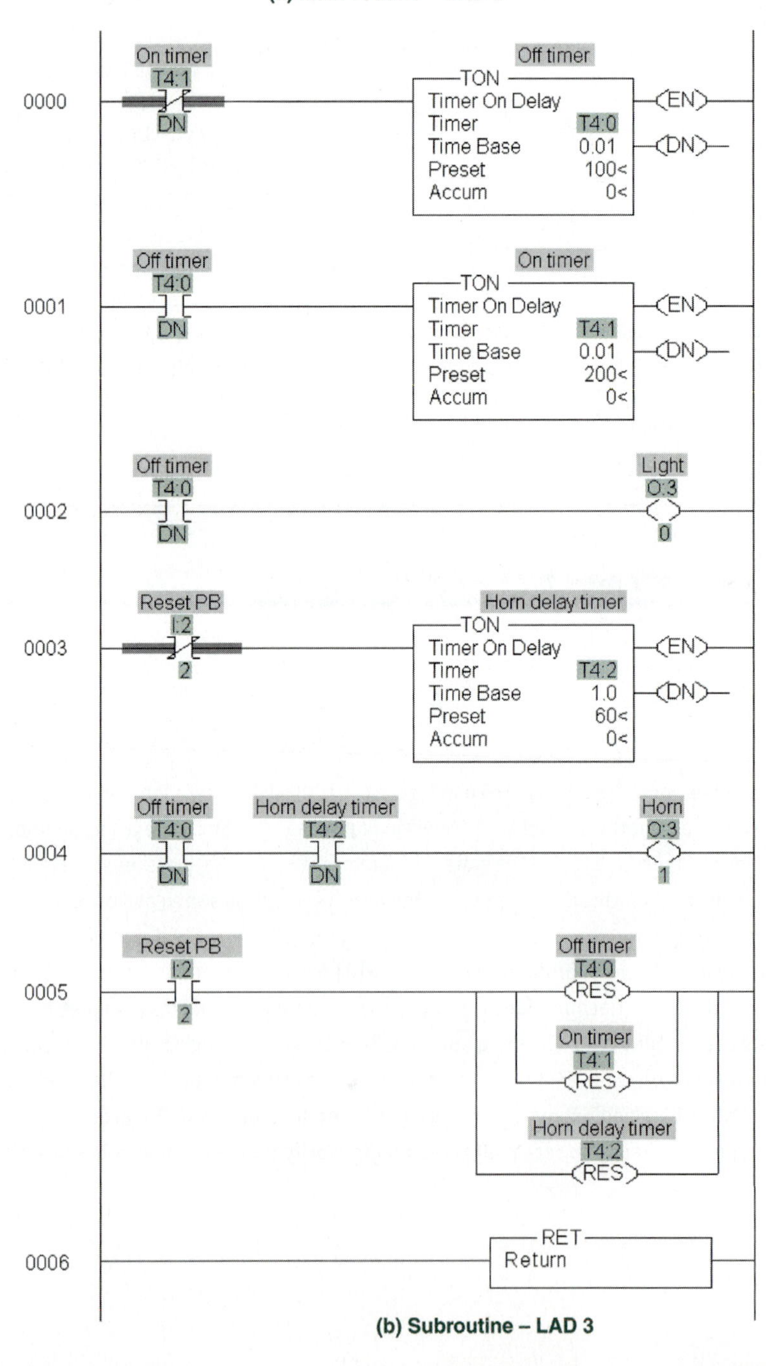

(b) Subroutine – LAD 3

illustrated in Figure 8-11(c). With the one-shot instruction included in this configuration, the subroutine will execute for only one scan after the input logic transitions from false to true. Although used less often, this configuration would be useful for situations where the trigger logic instruction does not return to the false state immediately.

Standard Ladder Logic for Subroutine and Return Instructions. The standard ladder logic for two configurations using the subroutine (SBR) and return (RET) instructions is illustrated in Figure 8-14. Read the description that accompanies each of the configurations in the figure.

Both of the SLC 500 subroutines in Figure 8-14(a) and (b) perform the same, but the subroutine in (b) uses the subroutine (SBR) and return (RET) instructions. When a subroutine is called by a JSR instruction, the program named in the JSR instruction will execute even without the SBR present. However, with the SBR instruction present it is clear that the ladder logic is a subroutine. The END rung present after the last rung of all programs serves the same function as the RET instruction. Although the SBR and RET instructions are recommended for clarity, they are required when parameters are passed using the PLC 5 and ControlLogix subroutine instructions covered later in this section.

The standard ladder logic for a solution with multiple subroutine return rungs is illustrated in Figure 8-14(c). There is no limit on the number of returns that are present in the subroutine. The subroutine in the figure shows two returns. Note that the last return is an unconditional rung, and all returns in the body of the subroutine ladder logic have conditional input logic to trigger the rung.

EXAMPLE 8-4

Modify Example 8-3 as follows:

There are two levels of faults in the production system that trigger the enunciator: level one faults do not require immediate attention; level two faults must be acted upon immediately. The enunciator for level one is the flashing lights only; the enunciator for level two includes both the lights and the horn with no delay for the horn. The trigger for a level one fault is latch bit, B3:0/2, while level two faults are triggered by B3:0/3. Redesign the subroutine so that it handles both types of faults. Reset of both types of faults is required.

SOLUTION

The two-level operator alert ladder logic program is shown in Figure 8-15. The modified JSR instruction rung from the machine control ladder logic is shown in Figure 8-15(a) and the modified subroutine is shown in Figure 8-15(b). Study the ladder solution.

The only change in the jump to subroutine rungs is the addition of the B3:0/3 bit for triggering the subroutine and resetting the operation. The subroutine has the following changes: The 60-second timer is removed, the horn is triggered only by the timer pulse output, and a return (RET) instruction is added between the flashing light and pulsing horn rungs. As a result, a level 1 fault executes only the upper half of the subroutine ladder, and a level 2 fault executes the entire subroutine ladder logic.

8-3-5 PLC 5 and ControlLogix Options for Subroutine Instructions

The Allen-Bradley subroutine instructions for the three PLCs are quite similar; however, the ControlLogix PLC uses a unique naming process plus some additional options. The naming procedure used for a ControlLogix PLC subroutine is illustrated in Figure 8-16.

ControlLogix Subroutine Setup. Study the procedure for selecting and naming ControlLogix subroutine files in the Figure 8-16. The process starts by right-clicking the MainProgram folder in the left file menu, and then selecting New Routine . . . to create a new program file [Figure 8-16(a)]. Note that MainRoutine is the default main program ladder logic file that is always present in this menu.

FIGURE 8-14: Standard ladder logic rungs for SLC 500 subroutine instructions.

Application	Standard Ladder Logic Rungs for Subroutine Instructions
Identify the start and end of a block of rungs in a subroutine (these instructions are optional but recommended). The SBR instruction is used to mark the start of a subroutine and is always ANDed with the input logic instruction in the first rung of the subroutine. This instruction is always evaluated as true. While not required, the instruction is recommended so the start of the subroutine is identified. The RET instruction marks the end of subroutine execution or the end of the subroutine file. It causes the processor to resume execution in the main program file at the instruction following the JSR instruction where it exited the program. An RET instruction is not required, but is recommended for clarity. If it is omitted, the END statement (always present at the end of the subroutine ladder logic) automatically returns program execution back to the rung after the JSR instruction in the calling ladder program. Note: SLC 500 rungs are shown, but the operation would be the same for PLC 5 and ControlLogix.	Subroutine - Routine 1 High level float switch - L2 I:1 2 O:3 1 Additional rungs in routine 1 END **(a)**
Schedule the execution of a subroutine block of ladder logic with multiple returns from the subroutine. The subroutine instructions are the same as in previous standard ladder logic, but the subroutine has multiple return instructions. This permits a return to the calling ladder logic at multiple ladder rungs before the entire subroutine ladder rungs are scanned. Note: SLC 500 rungs are shown, but the operation would be the same for PLC 5 and ControlLogix.	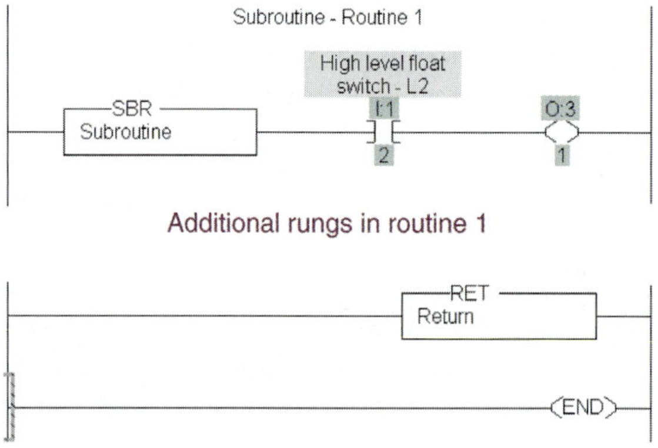 Subroutine - Routine 1 SBR Subroutine High level float switch - L2 I:1 2 O:3 1 Additional rungs in routine 1 RET Return END **(b)**
Schedule the execution of a subroutine block of ladder logic with multiple returns from the subroutine. The subroutine instructions are the same as in previous standard ladder logic, but the subroutine has multiple return instructions. This permits a return to the calling ladder logic at multiple ladder rungs before the entire subroutine ladder rungs are scanned. Note: SLC 500 rungs are shown, but the operation would be the same for PLC 5 and ControlLogix.	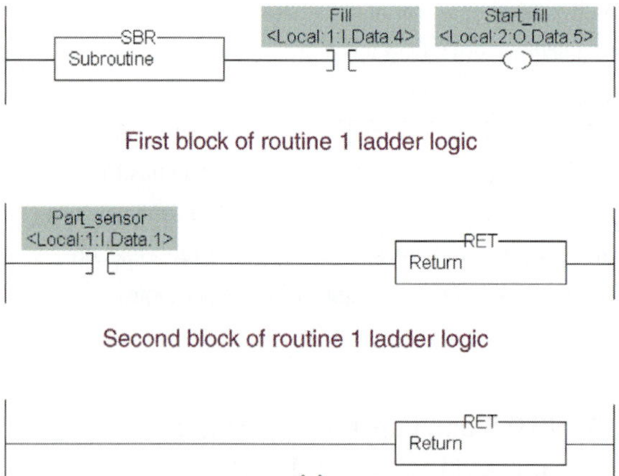 SBR Subroutine Fill <Local:1:I.Data.4> Start_fill <Local:2:O.Data.5> First block of routine 1 ladder logic Part_sensor <Local:1:I.Data.1> RET Return Second block of routine 1 ladder logic RET Return **(c)**

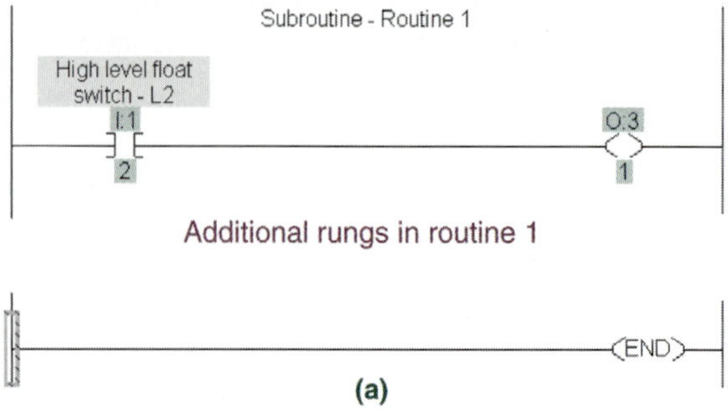

FIGURE 8-15: Ladder solution for Example 8-4.

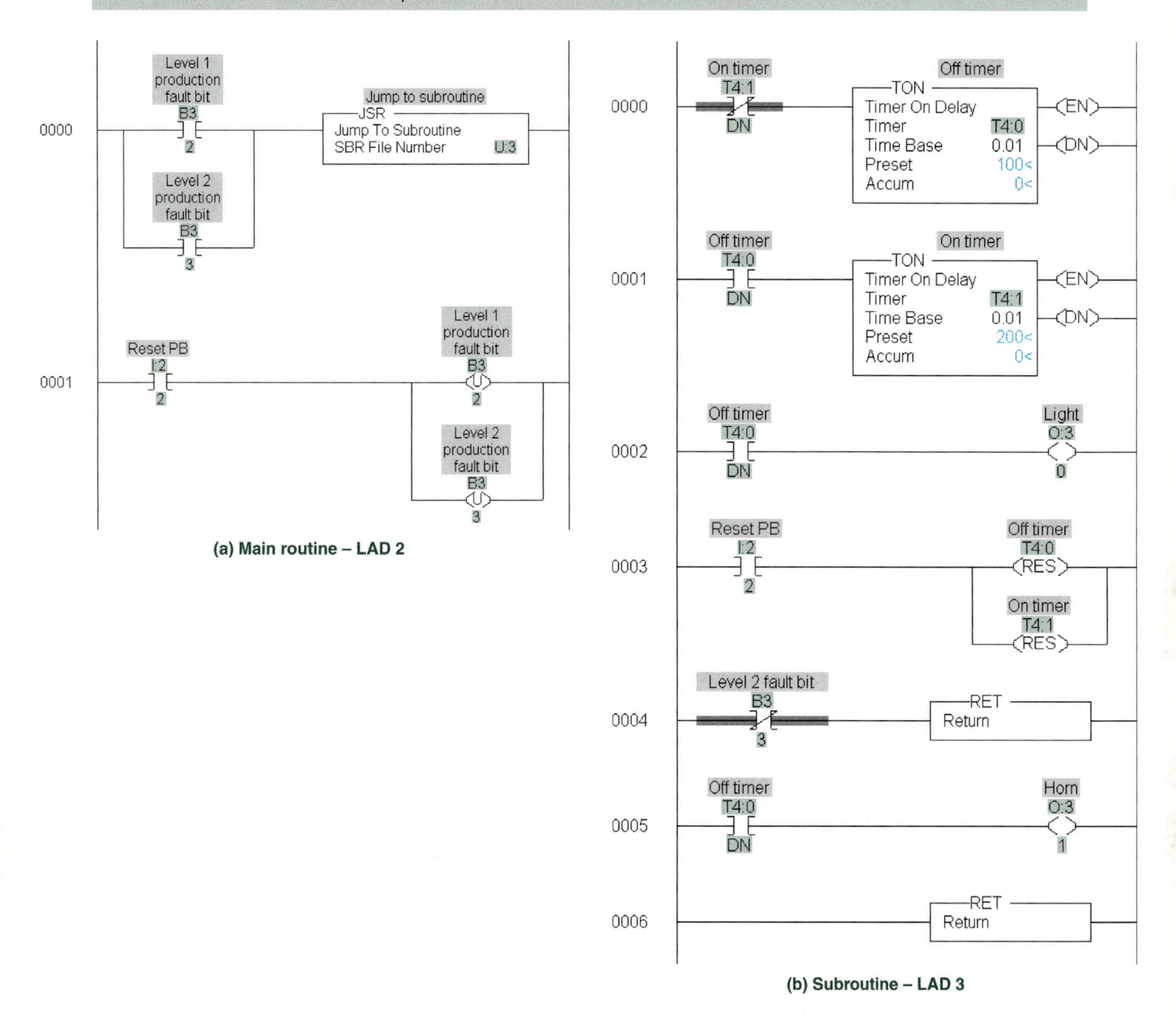

In the New Routine dialog box [Figure 8-16(b)], the routine name (Routine_1) is entered (note spaces are not allowed and underlines are used). A description can be added, and the type of program must be selected [Figure 8-16(c)]. Ladder Diagram is selected in this example. When the dialog box is closed (click OK), the new file is visible in the left file menu [Figure 8-16(d)]. If MainRoutine is double-clicked the main program ladder is opened, but if Routine_1 is selected then that subroutine ladder is opened. Once they are opened, they can be quickly selected from the tabs at the bottom of the window.

PLC 5 and ControlLogix JSR, SBR, and RET Instruction Options. The PLC 5 and ControlLogix instructions for jump to subroutine (JSR), subroutine (SBR), and return (RET) have a parameter passing option. Parameter passing describes a process that moves one or more data values from the main routine to the subroutine at the time that the subroutine is called. Data can

FIGURE 8-16: Selecting and naming ControlLogix subroutines.

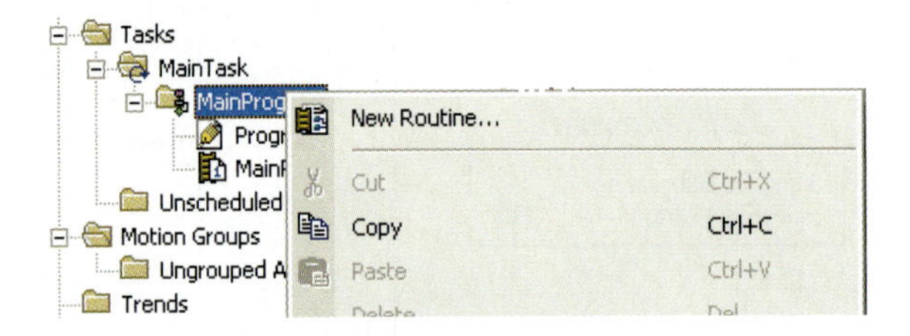

(a) Left file menu – right-click MainProgram so that New Routine... can be selected

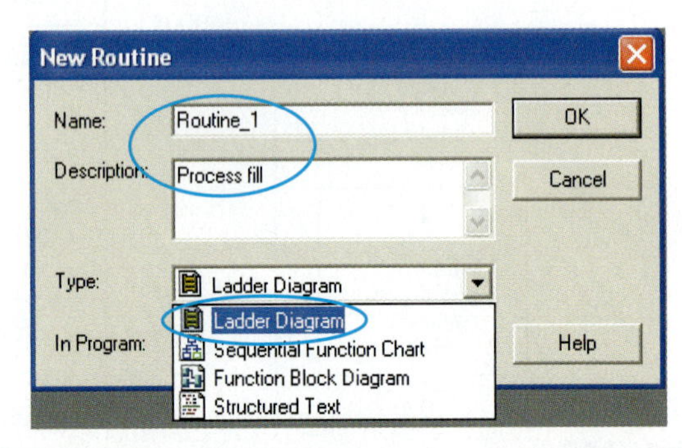

(b) New Routine dialog box – used to enter subroutine name and description information plus to indicate the software language to be used

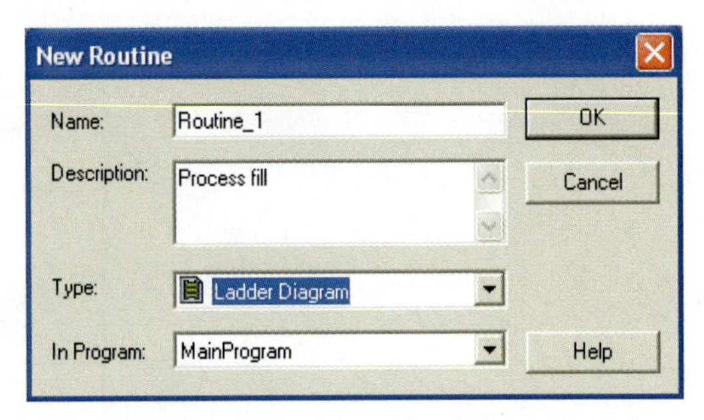

(c) New Routine dialog box with all data present

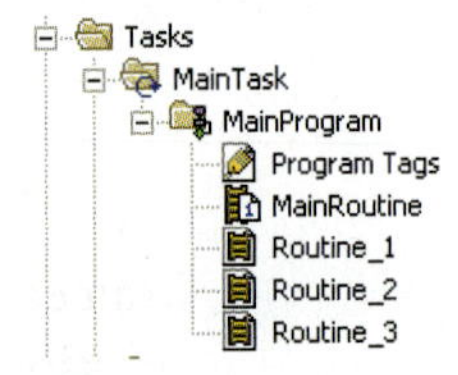

(d) Left file menu with MainProgram elements. Note MainRoutine and three additional ladder files Routine_1, Routine_2, and Routine_3.

also be passed from the subroutine back to the main program. It is important to understand the concept of global and program data values before parameter passing is covered.

Global and Program Data Values. The terms *global* and *program* define where the data values represented by tag names or files can be used and how they are identified. For example, all the data or memory addresses in the PLC 5 and SLC 500 PLCs are global. This means that an input address, such as I:1/0, can be used in LAD 2 and in a subroutine ladder like LAD 3 – ROUTINE, shown in Figure 8-10(c). LAD 2 is the main ladder logic program and LAD 3 – ROUTINE is a separate set of ladder logic rungs used as a subroutine. The value of address I:1/0 is the same whether an input uses this address in the main ladder logic or in any subroutine rung. As a result, if data addresses for bits or words change in one ladder rung they change at every other program location (LAD 2, LAD 3, etc.) where the address is used. Therefore, if you want to use the accumulator value of a counter located in LAD 2 in a comparator in subroutine LAD 3, you just use the counter ACC address.

The ControlLogix PLC, on the other hand, has both global data (controller scoped) and local data (program scoped). The term used by Allen-Bradley for global data is *controller-scoped data.* The controller-scoped tag can be used in all controller programs and it brings the tag value with it. Program-scoped tags, called local variables in computer programming, are restricted to a single program. For example, a program-scoped tag name such as liquid_temp could not be used in a rung in Routine_1. If the value from tag liquid_temp was required in Routine_1, then you could change to a controller-scoped (global) tag or use *parameter passing.*

Parameter Passing in ControlLogix and PLC 5 JSR, SBR, and RET Instructions. The ControlLogix JSR instruction is illustrated in Figure 8-17(a). Note that in addition to the Routine Name parameter there are locations for an input parameter (Input Par) and a return parameter (Return Par). Each parameter is defined as follows:

Routine Name: The routine name parameter is the name of the routine file to be executed. The steps in Figure 8-16 describe how this file is created.

Input Par (JSR and SBR): The input parameter value in the main routine represents process data that must be passed to the subroutine program identified in Routine Name. The data goes into the Input Par tag in the SBR instruction. Main routine input parameters can be fixed values or tag names with Boolean, integer, or floating point data types.

Return Par (JSR and RET): The return parameter value in the return instruction in the subroutine represents process data or subroutine calculations that must be passed back to the main routine program that called the subroutine. The data goes into the Return Par tag in the JSR instruction. The return parameter in the return instruction is a tag name with Boolean, integer, or floating point data types.

FIGURE 8-17: ControlLogix and PLC 5 JSR, SBR, and RET instructions with parameter passing.

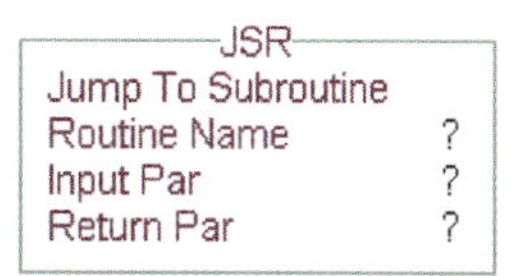

(a) ControlLogix jump to subroutine instruction

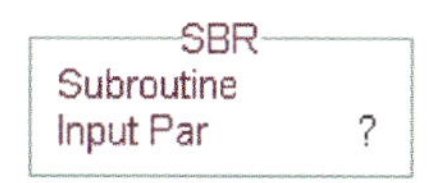

(b) ControlLogix subroutine instruction

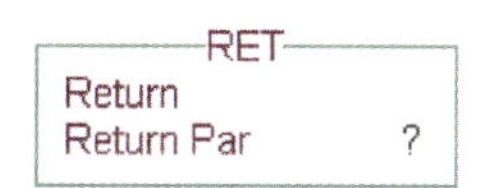

(c) ControlLogix return instruction

The subroutine (SBR) and return (RET) instructions are illustrated in Figure 8-17(b) and (c). Note that the SBR instruction has an input parameter, Input Par, and RET has a return instruction, Return Par. Multiple input and return parameters are allowed. The data passing and operation of the three instructions are illustrated in Figure 8-18.

The following data flow occurs in the figure.

1. The centigrade temperature value in Temp1 (tag in main routine) is transferred to TempC (tag in subroutine).
2. The subroutine ladder calculates the temperature value in Fahrenheit.
3. The temperature value in Fahrenheit in TempF (tag in subroutine) is transferred to Temp2 (tag in main routine).

FIGURE 8-18: Parameter passing for ControlLogix PLC.

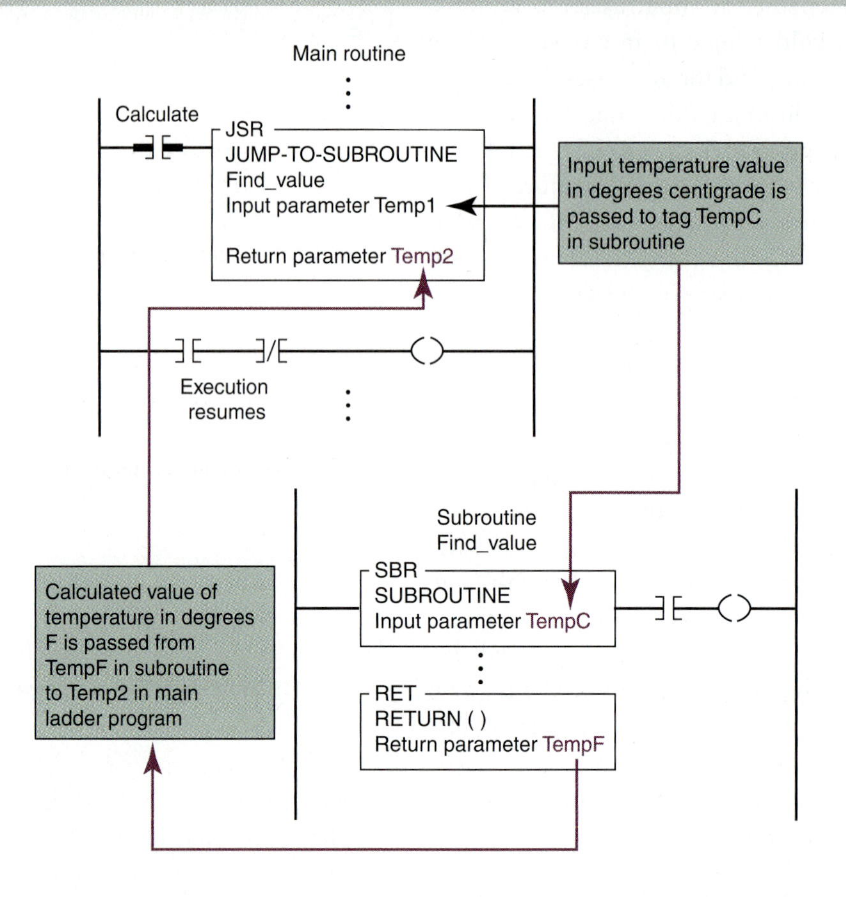

FIGURE 8-19: ControlLogix JSR, SBR, and RET parameter addition and removal.

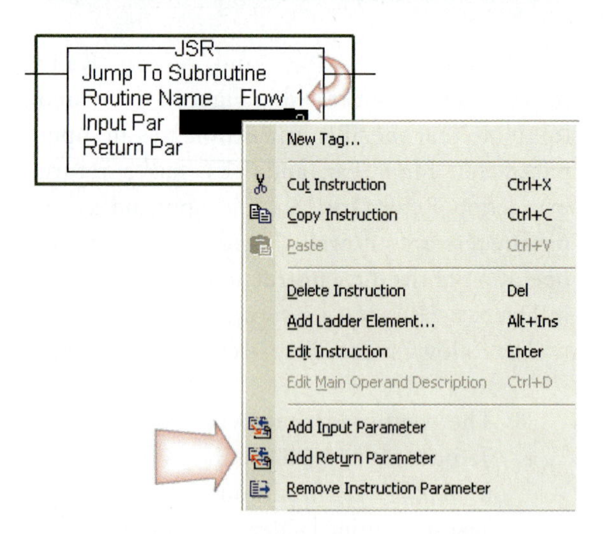

The JSR, SBR, and RET instructions in ControlLogix are used with and without passing parameters. Additional input and return parameters are added or parameters are removed by selecting a parameter location and right-clicking the mouse. Figure 8-19 shows the drop-down menu that appears. Note the add and remove commands and the program name that is present.

If no parameters are present, then the instruction operates like the subroutine instructions in the SLC 500. However, if input and return parameter tag locations are present, then they must have a tag entered. The number of parameters in the JSR instruction must agree with the number in the SBR and RET instructions.

Selection of PLC 5 Parameters. The parameter passing syntax is the same for the PLC 5 as that used in the ControlLogix PLC, but the technique for creating the input and return parameters is different. The first input parameter in the JSR instruction appears when the Enter key is used to input the subroutine number. A new input parameter is generated when the previous parameter is completed and the Enter key pressed. Pressing the Enter key with a blank input parameter box creates a return parameter entry box. Finally, pressing the Enter key with a blank return field cancels the entry mode. Input and return parameters are placed in the SBR and RET instructions by double-clicking the instruction.

Standard Ladder Logic for Parameter Passing in PLC 5 and ControlLogix Subroutine Instructions. The standard ladder logic for the ControlLogix JSR, SBR, and RET instructions using parameter passing is illustrated in Figure 8-20. Read the description of the process in the figure. Note that the subroutine would function correctly for any JSR instruction with any input and return parameter tags located anywhere in the main routine. The ability to have different tags for the data in the main routine and in the subroutine is a major advantage to using parameter passing.

The ControlLogix standard ladder logic for no parameter would require removal of all parameters in the JSR, SBR, and RET instructions. With that accomplished, the ControlLogix instructions have the same form as the SLC 500 instructions. In addition, the variations on the standard ladders presented for the SLC 500 apply to the ControlLogix PLC as well.

FIGURE 8-20: Standard ladder logic rungs for parameter passing in the PLC 5 and ControlLogix subroutine instructions.

Application	Standard Ladder Logic Rungs for Parameter Passing in the ControlLogic Subroutine Instructions
Pass a variety of register values to reusable subroutine code using ControlLogix subroutine instructions. The JSR instruction has the following parameters: tag name of subroutine ladder logic (routine name), input parameter values (Chem1_vol and Chem1_den) to pass, and the return parameter (Chem1_weight) from the subroutine. The SBR instruction in the first rung of the subroutine has the tag that receives the passed parameter value for use in the subroutine. Note that the values from Chem1_vol and Chem1_den are placed in the tags volume and density. Finally, the RET instruction stores the calculated weight value in a tag named Tank_weight. When the return instruction is executed, the value in Tank_weight is transferred to the tag Chem1_weight in the JSR instruction for use in the main routine. The PLC 5 JSR instruction uses the same syntax except a number (U:3 to 255) is used to identify the subroutine in place of the tag name. Also, the input and return parameters can be fixed value Boolean, integers, and floating point numbers, logic element addresses such as N7:2, or logic structure addresses such as C5:0.ACC.	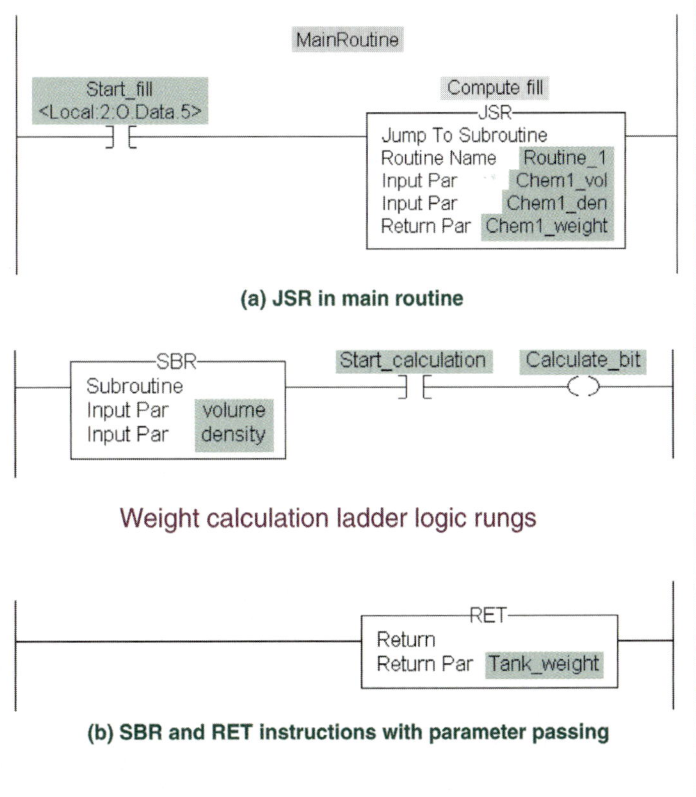 MainRoutine Start_fill <Local:2:O.Data.5> Compute fill JSR Jump To Subroutine Routine Name Routine_1 Input Par Chem1_vol Input Par Chem1_den Return Par Chem1_weight **(a) JSR in main routine** SBR Subroutine Input Par volume Input Par density Start_calculation Calculate_bit Weight calculation ladder logic rungs RET Return Return Par Tank_weight **(b) SBR and RET instructions with parameter passing**

EXAMPLE 8-5

A stack light indicator system for a production area has three lights: green for normal operation, yellow for minor fault, and red for major fault. For each condition the respective light has an on and off duty cycle that is different. Design a ControlLogix PLC ladder logic program using a subroutine that will handle this requirement for stack light control.

SOLUTION

The ControlLogix stack light control ladder logic program is shown in Figure 8-21. Note that five parameters are passed to the subroutine as follows: three Boolean bits to indicate which of the lights should be turned on

FIGURE 8-21: Ladder logic for Example 8-5.

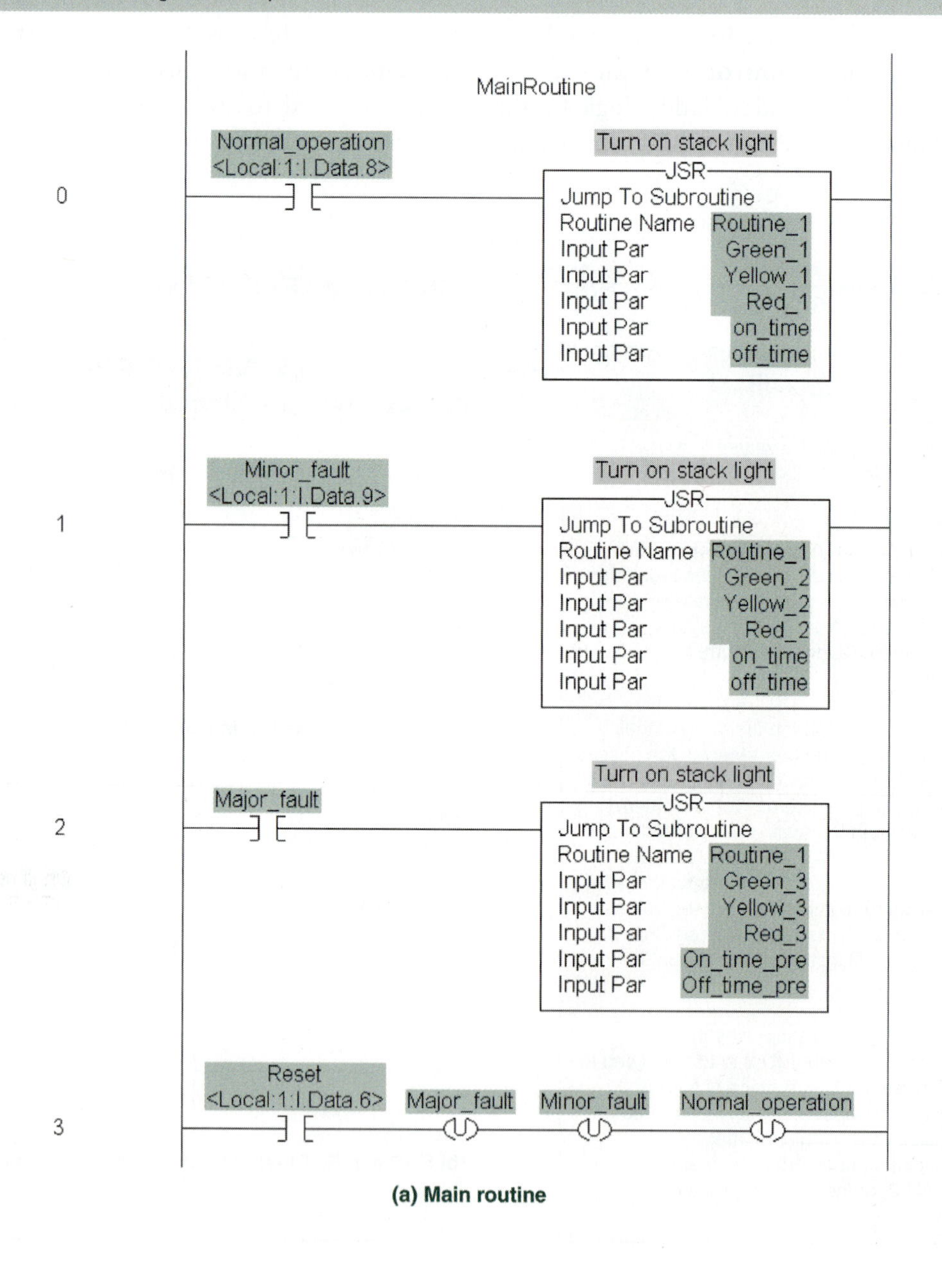

(a) Main routine

and two integers to pass the on and off times for the light. The Input Par bits have different values for each of the conditions. For example, Green_1, Yellow_1, and Red_1 have bits 1, 0, 0 loaded, respectively. So when the *normal_operation* input is true, the bits passed make the green stack light active and the yellow and red inactive. The bit patterns for the minor fault are 0, 1, 0; and for the major fault they are 0, 0, 1. Inside the subroutine, the light indicator bits are received in three tags, Green, Yellow, and Red. They are then used to enable rungs 2, 3, or 4, depending on what light should be on. The timer values for the on and off duty cycle are passed directly into the respective timer preset locations. A manual reset returns all system status bits to false and the system is restarted.

FIGURE 8-21: (Continued).

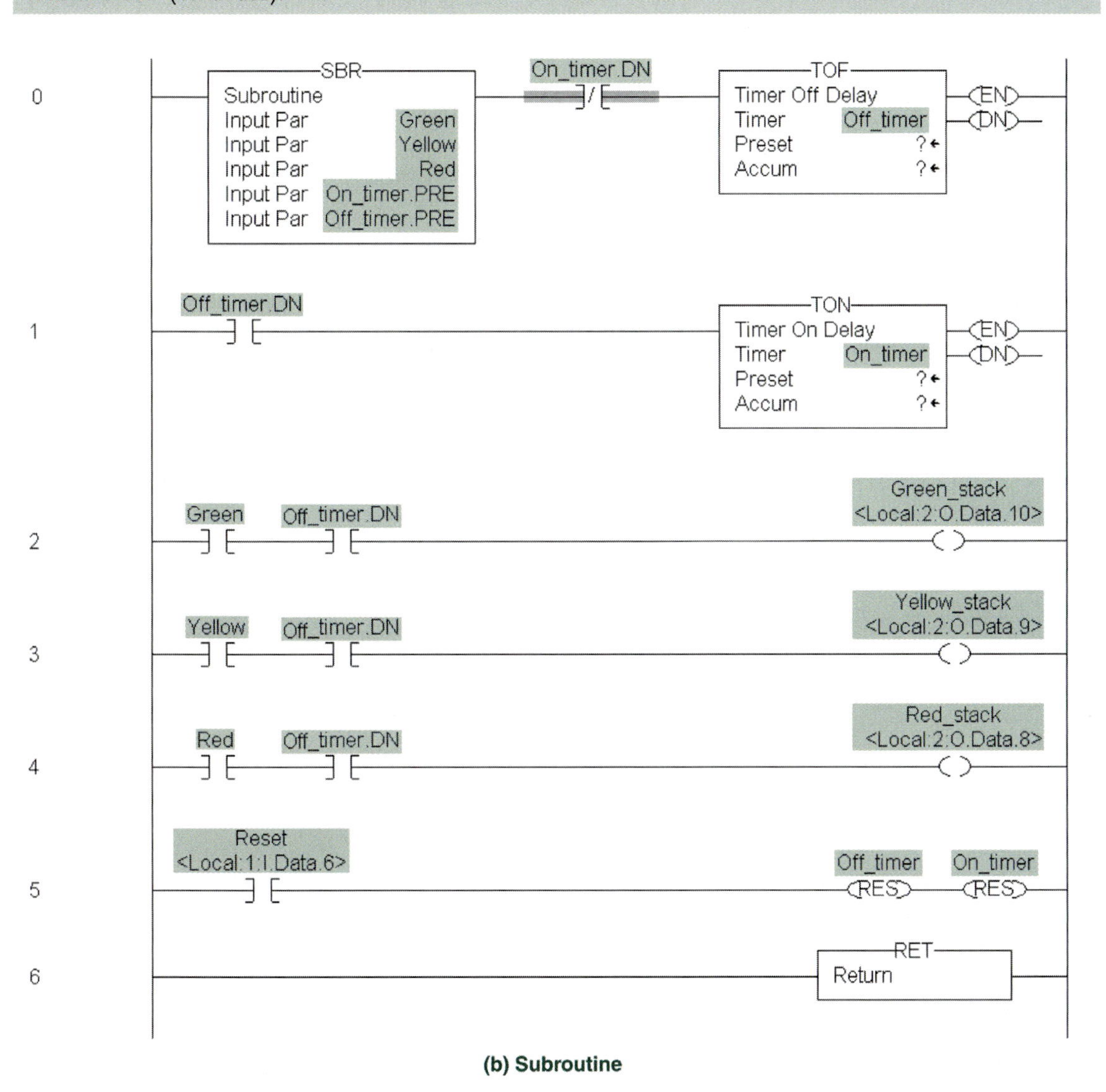

(b) Subroutine

8-4 ALLEN-BRADLEY IMMEDIATE INPUT AND OUTPUT INSTRUCTIONS

When the PLC program scan reaches an immediate input or immediate output instruction, the program flow is interrupted and the input or output data is updated. After the input or output data has been serviced, the program flow is returned to the point of interruption. These instructions are generally used for time-critical I/O data. In the following subsections the format and capability of these instructions are discussed for the PLC 5, SLC 500, and ControlLogix PLCs.

8-4-1 PLC 5 Immediate Input and Output Instructions

The *immediate input* (IIN) and *immediate output* (IOT) instructions are illustrated in Figure 8-22. They can have conditional and unconditional input logic. Note that the IIN and IOT instructions are located in the program scan after the program inputs and outputs have been serviced, noted as housekeeping in the figure. The (y) next to IIN in the figure is the address of the instruction, which is the rack number (two digits) and module group (one digit) of the input device. For example, IIN 014 represents rack 01 and module group 4. For the IIN instruction, the input image table is updated in accordance with the status of the input module, which reads the input field device. Likewise, the (x) next to IOT in the figure is the address of the instruction, which is the rack number (two digits) and module group (one digit) of the output device. For the IOT instruction the output image table is updated immediately without completing the scan in accordance with the status of the output module, which drives the output field device.

The ladder diagrams for the PLC 5 IIN and the IOT instructions are illustrated in Figure 8-23(b) and (d). Note that the number above the IIN and IOT is the address of the instruction and represents the rack number and module group. In Figure 8-23(b), when the program scan reaches the IIN instruction the scan is interrupted and all bits of word 012 in the input image table are updated. In Figure 8-23(d), when the program scan reaches the IOT instruction the scan is interrupted and all bits of word 014 in the output image table are updated.

FIGURE 8-22: Program scan with immediate input and output instructions.

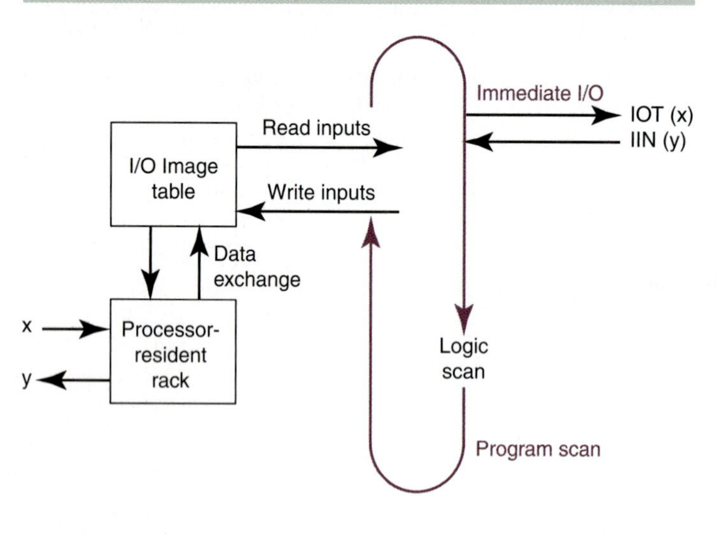

8-4-2 SLC 500 Immediate Input and Output Instructions

In the SLC 500 PLC these I/O instructions are called the *immediate input with mask* (IIM) and the *immediate output with mask* (IOM) instructions. These instructions provide the same function in the program scan as the IIN and the IOT instructions, which were described in Section 8-4-1. However, the mask provides an improvement in capability over the IIN and IOT instructions. The mask specifies the portion of the input or output data to be updated. This allows the programmer to specify which bits of the word are copied from an input module to the input image table or from the output image table to an output module. Figure 8-23(a) illustrates the IIM instruction, and Figure 8-23(c) illustrates the IOM instruction. The IIM instruction operates on the inputs assigned to a specific word of a rack slot, and the IOM instruction operates on the outputs assigned to a specific word of a rack slot. The function of the parameters of the instructions is as follows:

- **Slot:** The slot parameter represents the address of the input or output word and includes the slot and word within the slot that contains the data to update the input or output image tables. In Figure 8-23(a), I:1.0 indicates slot

FIGURE 8-23: Standard ladder logic rungs for immediate input and output instructions.

Application	Standard Ladder Logic Rungs for Immediate Input and Output Instructions
Input a word of data from a field device to a register or tag name without waiting for the completion of the scan cycle. The immediate input instruction in the PLC 5 and the immediate input with mask instruction in the SLC 500 can be implemented with conditional and unconditional input logic. In addition, a one shot can be added to limit the immediate data transfer to just one sample taken in a single scan. The masked version allows the programmer to designate which bits in the sampled input data are transferred to the input register. *Note: The SLC 500 is shown with all three input logic options. The PLC 5 is shown with only one but could have all three as well.*	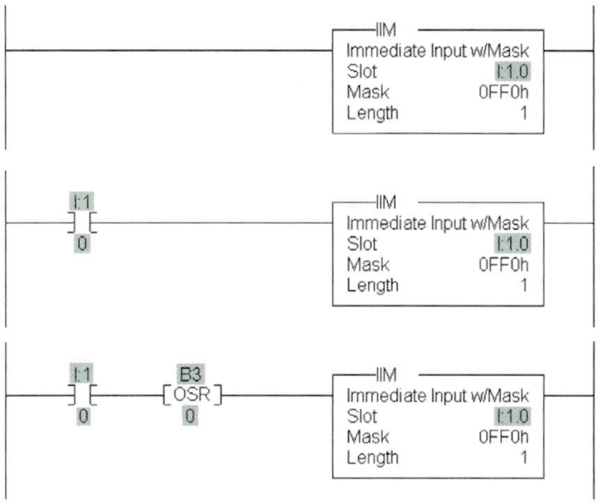 IIM Immediate Input w/Mask Slot I:1.0 Mask 0FF0h Length 1 I:1 0 — IIM Immediate Input w/Mask Slot I:1.0 Mask 0FF0h Length 1 I:1 0 — B3 OSR 0 — IIM Immediate Input w/Mask Slot I:1.0 Mask 0FF0h Length 1 **(a) Immediate input with mask (SLC 500)** 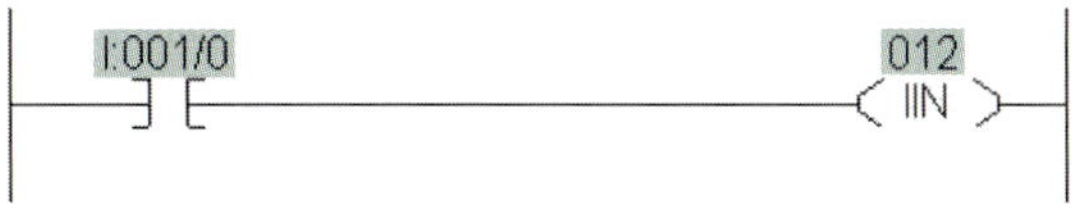 I:001/0 — 012 IIN **(b) Immediate input (PLC 5)**
Output a word of data stored in a PLC register or tag to an output field device without waiting for the completion of the scan cycle. The immediate output instructions in the PLC 5 and ControlLogix plus the immediate output with mask instruction in the SLC 500 can be implemented with conditional and unconditional input logic. In addition, a one shot can be added to limit the immediate data transfer to just one sample taken in a single scan. The masked version allows the programmer to designate which bits in the sampled output register or tag are transferred to the output field device. *Note: All PLCs are shown with one input logic option; all could use any of the three shown on the SLC IIM instruction.*	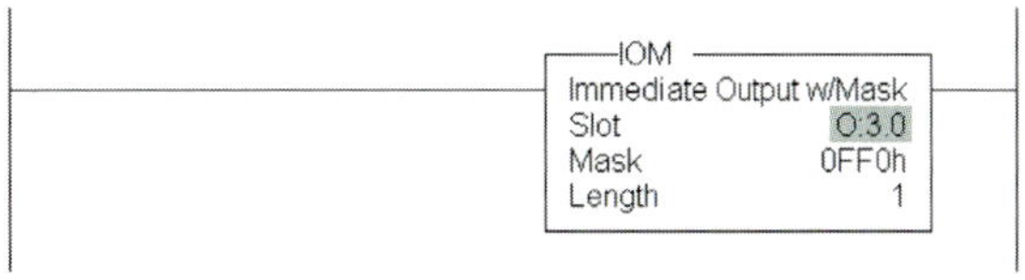 IOM Immediate Output w/Mask Slot O:3.0 Mask 0FF0h Length 1 **(c) Immediate output with mask (SLC 500)** 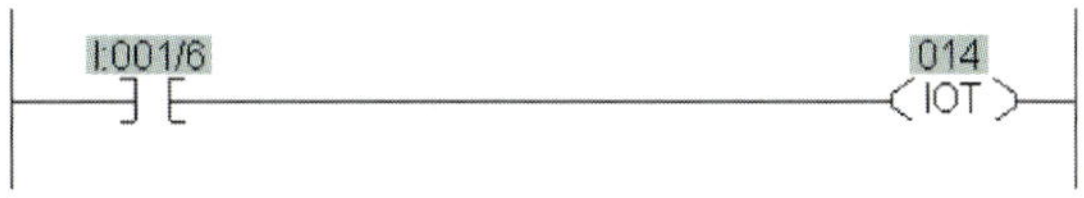 I:001/6 — 014 IOT **(d) Immediate output (PLC 5)** 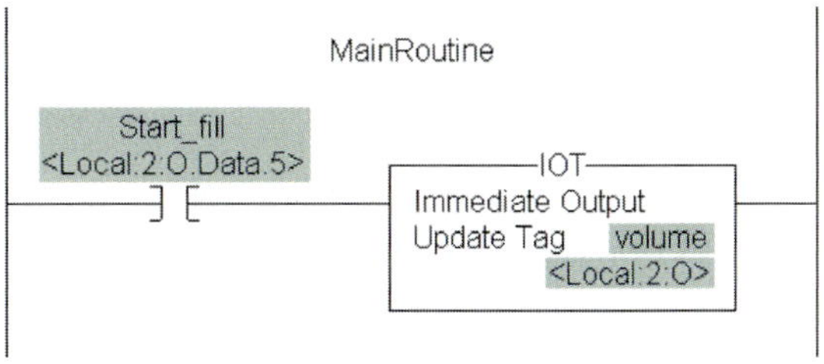 MainRoutine Start_fill <Local:2:O.Data.5> — IOT Immediate Output Update Tag volume <Local:2:O> **(e) Immediate output (ControlLogix)**

1 and word 0. Likewise, in Figure 8-23(c), O:3.0 indicates slot 3 and word 0. Up to 32 words can be associated with the slot depending on PLC capability.

- **Mask:** The mask parameter is typically a hexadecimal number that specifies what bits in the word are 1s and what bits are 0s. Every mask bit location with a 1 will pass that data bit and every 0 in a bit location inhibits the data transfer. Note that the mask in Figure 8-23(a) passes all bits except the four least significant bits and the four most significant bits.
- **Length:** The length parameter specifies the number of words (0 to 32 in some SLC processors) per slot that are transferred.

Use of immediate instruction increases the execution time significantly, so conditional rungs are recommended so that the input instructions control when the immediate instructions are executed. The standard ladder logic for the IOM instruction shows three control possibilities: unconditional, conditional, and conditional with a one-shot. The one-shot would provide the fastest execution time since the IOM instruction would be executed only one time.

8-4-3 ControlLogix Immediate Output Instruction

The immediate output instruction (IOT) for the ControlLogix PLC operates the same as the immediate output instructions for the PLC 5 and SLC 500 PLCs. Figure 8-23(e) illustrates the IOT instruction, which updates the specified output data. Note that when the IOT instruction executes, it immediately transfers the value of the Local:2:0 tag to the output module.

EXAMPLE 8-6

A production assembly system has three robots and numerous other automated assembly machines to produce a variety of products. The PLC program controlling the system has a large number of rungs. Each of the robots has a light curtain to indicate (closure of an NO contact) that an employee has entered the work envelope of the robot while it is running. Design a PLC program that inputs the intrusion data as fast as possible and performs three tasks equally fast:

1. Turn off the invaded robot's servo power (turn on a relay in the robot's control cabinet).
2. Sound a horn in the production area.
3. Light a red warning light over the robot whose light curtain was broken.

Solution

The SLC 500 robot safety system ladder logic program and field device interface is shown in Figure 8-24. Each of the light curtain contacts triggers an immediate input instruction, which moves the word from the input module in slot 4 into the input image table. A mask (0000 0000 0000 0111) is used to block 1's in all bits except the first three locations. Note that the input word decimal value changes due to the terminal to which the curtain contact is attached. Based on input connections, robot 1 is 1_{10} (0000 0000 0000 0001), robot 2 is 2_{10} (0000 0000 0000 0010), and robot 3 is 4_{10} (0000 0000 0000 0100). Three equal to comparison instructions check for the values of 1, 2, and 4, and then cause the appropriate robot action and alarms. The operational sequence includes:

1. Load the following constant mask values in the IOM instruction in rungs 2, 3, and 4. Put 43 hex (0000 0000 0100 0011) into IOM in rung 2, 4C hex (0000 0000 0100 1100) in rung 3, and 70 hex (0000 0000 0111 0000) into IOM in rung 4.
2. Use an unconditional MOV instruction to put all 1's into the output image table for 0:3.0.
3. The output (0:3.0) with all 1's present is moved through a mask with desired output bit patterns for each robot. The image table for 0:3.0 has all 1's present, but the outputs are only true where there is a 1 in the mask. So the masks control what outputs are turned on (0043 for robot 1, 004C for robot 2, 0070 for robot 3).

Mask bits 0, 2, and 4 control the servo stop relays for robots 1, 2, and 3 because the relays are connected to terminals 0, 2, and 4, respectively. Similarly, mask bits 1, 3, and 5 control the red intrusion indicator light because terminals 1, 3, and 5 are used for the lights for robots 1, 2, and 3, respectively. Since the warning horn field device is wired to terminal 6 (bit 03.0/6) for all three robots, each mask has a 1 in that bit location.

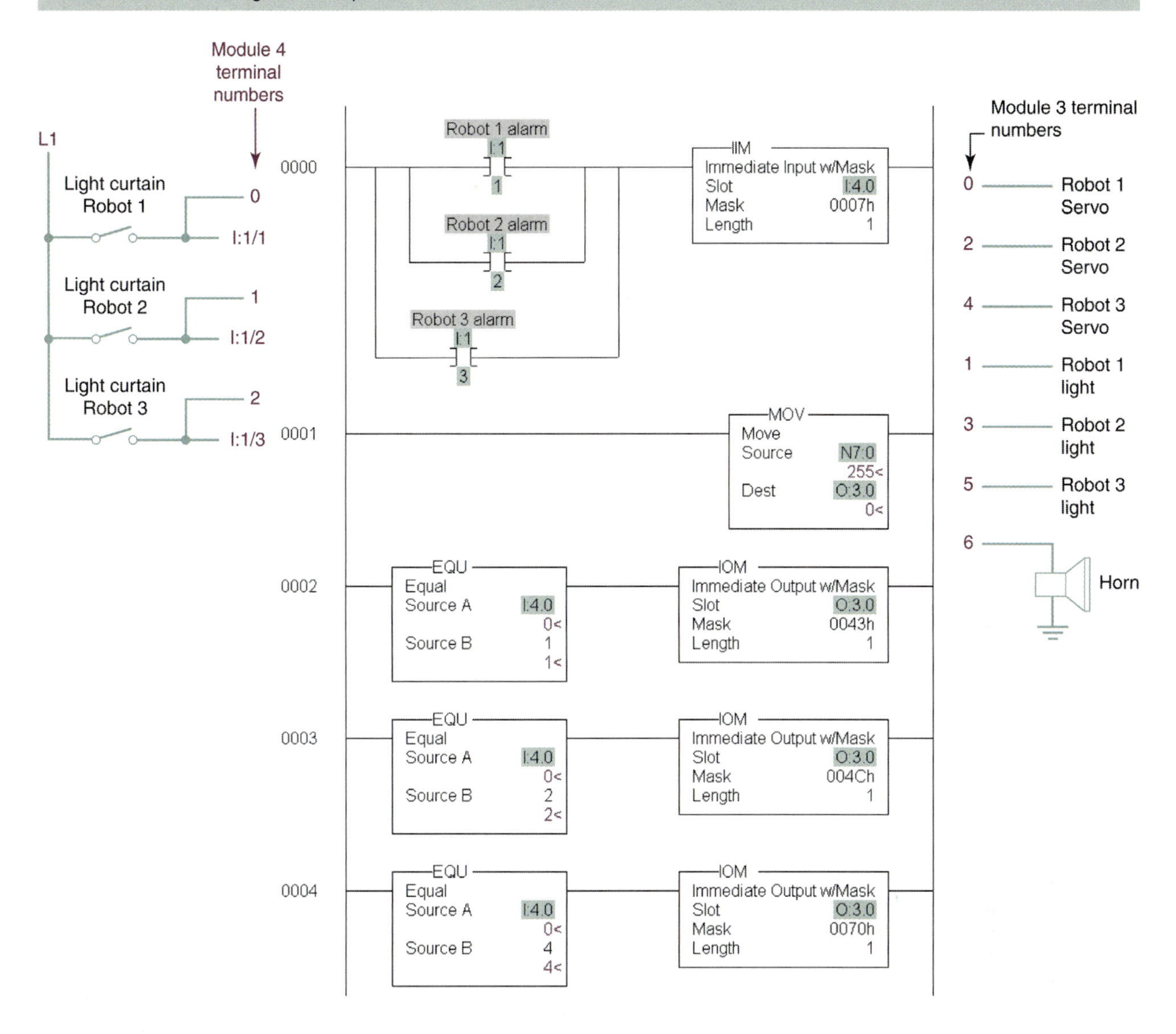

FIGURE 8-24: Ladder logic for Example 8-6.

8-5 EMPIRICAL DESIGN PROCESS WITH PROGRAM CONTROL INSTRUCTIONS

The empirical design process, introduced in Section 3-11-4, is an organized approach to the design of PLC ladder logic programs. However, the term *empirical* implies that some degree of trial and error is present. As more instructions are added to the design and the process becomes more complex, the empirical design requires more fixes and adjustments. However, the process used for these new instructions is quite similar to that learned in previous chapters.

8-5-1 Adding Program Control Instructions to the Process

The first step in using program control instructions in PLC ladder designs is to know when to use master control reset (MCR); jump (JMP) and label (LBL); jump to subroutine, subroutine, and return (JSR, SBR, and RET); and immediate inputs (IIN and IIM) and immediate outputs (IOT and IOM). In addition, there are a number of standard logic configurations for each, so selecting the proper configuration adds another degree of challenge to the empirical design. The following discussion addresses some of the issues associated with these instructions.

MCR Instruction. The master control reset instruction offers numerous advantages and features for a control solution. The MCR instruction sets up a fence, called the MCR zone, around an entire ladder logic program or around a block of ladder rungs. The MCR zone is either scanned and executed or bypassed. Use this instruction when you want to:

- Control how frequently an entire program or a set of rungs in a program is scanned.
- Force a group of non-retentive outputs to the false state and freeze the state of retentive outputs in their current state.
- Force a group of outputs to the false state and ignore the condition of the input ladder logic.
- Pause the operation of a counter so that the accumulator stops incrementing even though the input logic changes state. The current count in the counter accumulator remains. When the MCR becomes active the counter starts from the previous value in the ACC.

JMP and LBL Instructions. The jump and label instructions offer advantages and features for a control solution similar to that found in the MRC instruction. The JMP and LBL instructions identify blocks of ladder rungs, called JMP zones, that are passed over and not scanned when the input logic for the JMP instruction is true. The JMP zone is either scanned and executed or bypassed. Use this instruction when you want to:

- Control how frequently an entire program or a set of rungs in a program is scanned.
- Retain the state of all outputs in the JMP zone when the ladder block is not scanned.
- Jump from multiple points in a ladder program to a common label rung.
- Jump either forward or backward in a ladder program.
- Pause the operation of counters and timers so the accumulator stops incrementing even though the input logic changes state. The current value in the accumulators remains. When the JMP zone becomes active the counters and timers start from the previous value in the ACC.

JSR, SBR, and RET Instructions. The jump to subroutine, subroutine, and return program control instructions are used most often to change the flow of the ladder logic. There are numerous configurations and many benefits from using these instructions that allow the program to be directed to other ladder logic programs called subroutines. Use these instructions when you want to:

- Organize a program into smaller blocks of ladder logic.
- Isolate frequently used ladder logic blocks in a subroutine and execute them from numerous locations in the main ladder logic.
- Pass data to another program and execute algorithms containing that data.
- Return results from another program that can be used in the main routine.
- Reduce scan time and ladder rung count by separating the control logic into blocks that can be called from the main routine only when necessary.

IIN, IOT, IIM, and IOM Instructions. The immediate instructions change the program flow normally associated with traditional input and output instructions. The immediate instructions permit faster data measurement since the movement of the data value does not have to wait until the usual input or output update point in the scan cycle. Use these instructions when you want to:

- Move field device data to the input image table without waiting for the scan cycle to reach the image table update point.
- Interrupt the scan process and input a data word immediately after the logic rung with the IIN or IIM instruction is scanned.
- Interrupt the scan process and move a data word immediately after the logic rung with the IOT or IOM instruction is scanned.

- Interrupt the scan process and move only the masked portion of a data word immediately after the logic rung with the IIM instruction is scanned.

8-6 TROUBLESHOOTING PROGRAM CONTROL INSTRUCTIONS IN LADDER LOGIC

The program control instructions offer few operational problems. If a section of the ladder that includes these instructions is not operating properly, then you can use any of the following suggestions to troubleshoot.

- Use the always false (AFI) instruction in the input logic of MCR and JMP rungs to eliminate the MCR zone and JMP zone ladder rungs until the proper operation of the main part of the routine can be verified. The AFI sets its rung-condition-out to false; in other words, the AFI disables all the instructions on its rung.
- Use the temporary end (TND) instruction, described in Chapter 4, or the suspend (SUS) instruction, described in Section 5-8-1, to terminate an MCR, JMP, or JSR instruction immediately after the branch to examine the input logic conditions that caused the branch to occur. This is useful when branches occur under the wrong process conditions.
- Troubleshoot portions of a ladder not functioning properly by using the single-step mode. Put the SLC 500 PLC into the single-step mode (select the test mode instead of the run mode after a program is downloaded) and select Execute Step once to run one scan and then a second time to execute the first rung. Subsequent clicks on Execute Step runs the next rung in the ladder.
- Use breakpoints in the single-step mode to execute the ladder down to a breakpoint inserted into the ladder logic. After selecting test mode and single step, select set end rung. The file number and rung number are entered to indicate where the breakpoint should be established. Every time the Enter key is pressed, one scan is executed with the processor stopping at the breakpoint.

In addition, verify that:

- MCR zones are never overlapped or nested.
- MCR zones and JMP zones do not overlap.
- Backward jumps do not cause a scan time greater than the watchdog timer. Backward jumps increase scan time when rungs are rescanned. If excessive backward jumps are performed and the watchdog timer is exceeded (maximum value is about 2.5 seconds), then the processor has a major fault condition.
- JSR and SBR instructions have the same number of input parameters.
- JSR and RET instructions have the same number of return parameters.
- The LBL and the SBR instructions are the first input instructions on the rung.
- The scope and data types of the tags are consistent with scope and data present in their use.

8-7 LOCATION OF THE INSTRUCTIONS

The location of instructions from this chapter in the Allen-Bradley programming software is indicated in Figure 8-25.

FIGURE 8-25: Location of instructions described in this chapter.

Systems	Instructions	Location
PLC 5, SLC 500	MCR, JMP, LBL, JSR, SBR, RET	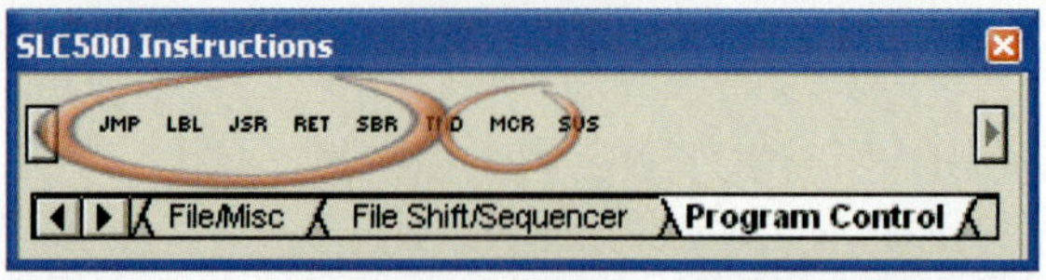 SLC500 Instructions: JMP LBL JSR RET SBR TND MCR SUS — File/Misc, File Shift/Sequencer, **Program Control**
LOGIX	MCR, JMP, LBL, JSR, SBR, RET	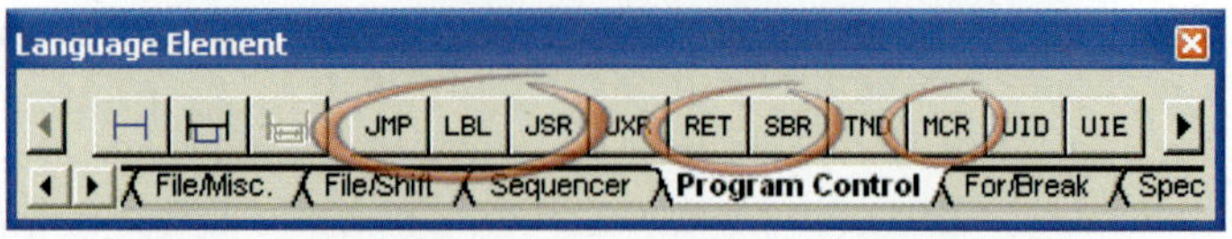 Language Element: JMP LBL JSR JXR RET SBR TND MCR UID UIE — File/Misc., File/Shift, Sequencer, **Program Control**, For/Break, Spec
PLC 5	IIN, IOT	 PLC5 Instructions: BTR BTW IIN IOT MSG CIO IDI IDO — User, Bit, Timer/Counter, **Input/Output**, Compare
SLC 500	IIM, IOM	 SLC500 Instructions: BTR BTW IIM IOM SVC MSG IIE IID RMP RPI REF — User, Bit, Timer/Counter, **Input/Output**, Compare
LOGIX	IOT	 Language Element: MSG GSV SSV IOT — Favorites, Bit, Timer/Counter, **Input/Output**, Compare

Chapter 8 questions and problems begin on page 574.

CHAPTER

9

Indirect and Indexed Addressing

9-1 GOALS AND OBJECTIVES

There are two principal goals of this chapter. The first goal is to provide the student with the overall concept of addressing. The second goal is to show how the addressing modes are applied to specific Allen-Bradley PLCs that are used in industrial automation systems.

After completing this chapter you should be able to:

- Explain the concept of direct, indirect, and indexed addressing.
- Describe how to combine addressing modes.
- Apply the indirect and indexed addressing modes to Allen-Bradley PLC 5 and SLC 500 PLCs.
- Create a multidimensional array for the ControlLogix PLC.
- Apply the indirect addressing mode to the ControlLogix PLC using arrays.
- Troubleshoot the indirect and indexed addressing modes.

9-2 ALLEN-BRADLEY ADDRESSING MODES

This chapter explores the addressing modes—direct, indirect, and indexed—used in the Allen-Bradley PLCs. An *addressing mode* is the means by which the PLC selects the *data* that is used in an instruction. The addressing mode is determined by how you specify the instruction's *operand*. The terms data, operand, and addressing mode are defined as follows:

- *Data* are numerical values that are used in computations. For example, if the PLC has the value 3 in a memory location and the value 4 in another location, and there is an instruction to add the values in these two locations, then there are two data values involved: 3 and 4.
- *Operands* are the symbols in an instruction. Again, if the instruction were to add the two memory locations, then the data would be the same, but the operands would be the symbols for the locations.
- *Addressing mode* describes the relationship between the operands and the data, that is, how we use the operands to get the correct data.

9-2-1 Direct Addressing

In *direct addressing*, the memory address of the data is supplied with the instruction. Figure 9-1 illustrates this addressing mode. Note that the address (2112) in the instruction points directly to the address that contains the data (345).

FIGURE 9-1: Direct addressing mode.

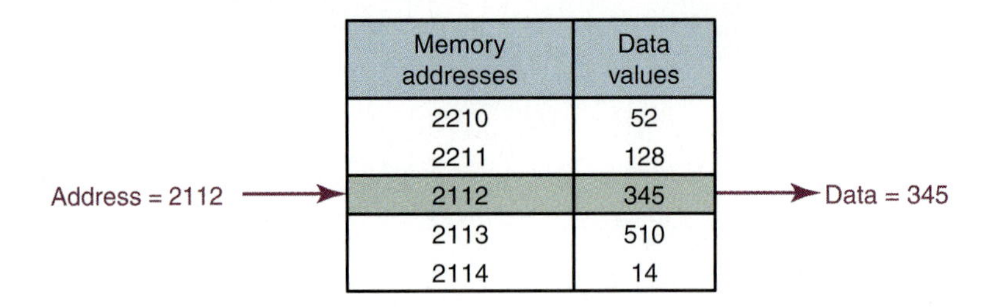

FIGURE 9-2: Indirect addressing mode.

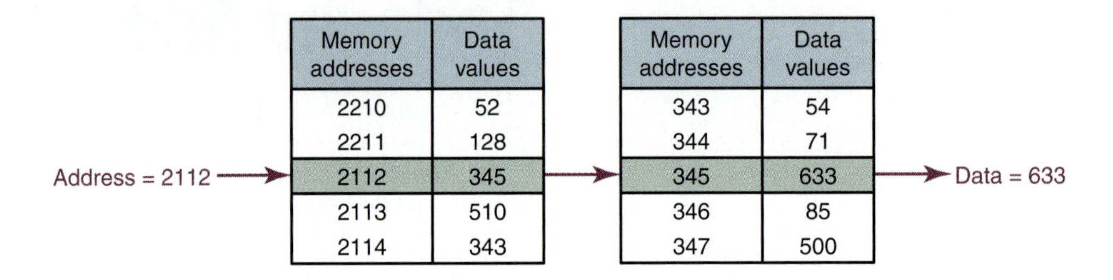

9-2-2 Indirect Addressing

In *indirect addressing,* the address in the instruction serves as a reference point and does not point directly to the data location. In other words, the instruction's memory address contains the address of a memory location, which either contains the address of the operand or specifies another location that contains the effective address of the data. This is confusing, so the example in Figure 9-2 helps to illustrate this addressing mode. Note that the address of the instruction points to a memory location that contains another address, 345, which contains the data. Memory data is a general term because it could be the value used for a timer preset value, the setpoint for a temperature comparator, or in this case the address for the location of the actual data.

An everyday example of indirect addressing is Internet domain names. When you point to an address, such as www.jimandglenn.com, you are really pointing to an entry in a Domain Name Service (DNS) table, which maps the domain name to a specific numeric address representing a physical server on the network. This is a simple two-stage indirection that allows people and documents to use more or less meaningful names instead of meaningless Internet address numeric strings. The cost of this indirection is that the DNS table entries must be maintained as servers' addresses change and as new servers and domain names come online. This cost is more than offset by the value of meaningful names and by eliminating the need for the documents that point to a given server to have to be updated every time the server's numeric address changes.

9-2-3 Indexed Addressing

Indexed addressing is an addressing mode for referencing a memory location that is the original memory address plus the value stored in an index register. In other words, the content of the index register is added to the original address of the instruction. Indexed addressing is useful to access elements in an array or list for the purpose of averaging a set of values in the array. The address in the instruction does not change, but the value in the index register is incremented, thus sequentially accessing array locations one by one. Figure 9-3 illustrates indexed addressing. Note that the address of the instruction (2112) is added to the index (23), thus forming the effective address of 2135 (2112 + 23), which contains the data.

Indirect addressing is preferred over indexed addressing because indirect addressing allows for a different pointer for each location where you want to use this "variable" addressing method.

FIGURE 9-3: Indexed addressing mode.

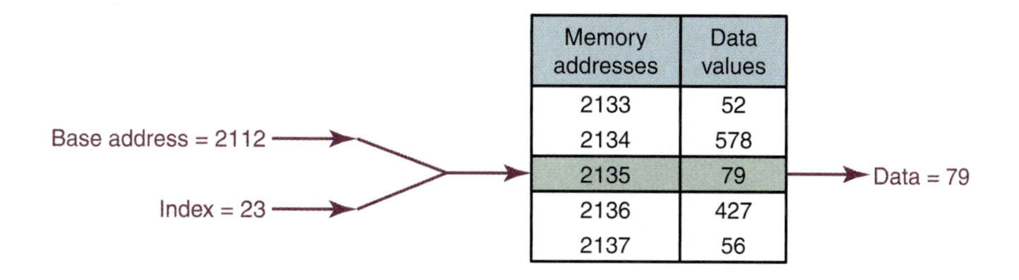

Since indexed addressing requires you to take your pointer and load it into a fixed "offset" location of S:24, there is one location for the pointer of all indexed addresses. Indexing also requires one more instruction than indirect since the pointer needs to be moved to S:24.

9-2-4 Indexed Indirect Addressing

Indexed and indirect addressing modes can be combined. One technique is called *pre-indexed addressing*, which is a combination addressing method that incorporates indexed addressing prior to the use of indirect addressing. Another technique is called *post-indexed addressing*, which is a combination addressing method that incorporates indexed addressing after the use of indirect addressing. Figure 9-4 illustrates the two combination addressing modes. In Figure 9-4(a) the result of the pre-indexed addressing mode supplies the address 3000 for the indirect addressing mode, which yields the data value of 77. Figure 9-4(b) shows the results when the index in Figure 9-4(a) is incremented by 1. In Figure 9-4(c) the result of the post-indexed addressing mode supplies the address 1533 for the indexed addressing mode, which yields the data value of 25. Figure 9-4(d) shows the results when the index in Figure 9-4(c) is incremented by 1. Now, before you continue, review the concepts of the indirect and indexed addressing modes until you are completely familiar with them because the next subsections discuss the addressing mode syntax for various Allen-Bradley PLCs.

9-2-5 PLC 5, SLC 500, and Logix Systems Syntax

The syntax for the addressing modes in the AB PLCs is slightly different, but the addressing concept is the same. First of all, direct addressing mode is the mode that has been used in the examples of the PLC instructions in the first eight chapters. That is, when N7:1 is used as the address in an instruction, it is a direct address. The data located in address N7:1 is the data used in the instruction.

PLC 5 and SLC 500 Indirect Addressing. Allen-Bradley indirect addresses use brackets, [ ], to hold an address where the data is located. The brackets indicate that the logical address is being used in an indirect addressing mode. Figure 9-5 illustrates this addressing mode. Source A in the equal to comparison instruction uses indirect addressing for the location of the data file. The integer register has the indirect address, N7:2, enclosed in brackets. The actual data address is built by using the data value (5) at the indirect location, N7:2, to form the indirect address N7:5, which holds the actual data value of 13. Note that 5 is not the value of the data but rather an offset or the value placed in the address N7:[5]. This offset forces the file to point to address N7:5, which contains the data. Since the value in N7:[N7:2] or N7:5 is equal to 13, the output of the comparison instruction is true.

Figure 9-6 illustrates an indirect address similar to Figure 9-5, but in this case the indirect address is T4:0.ACC. The indirect address, T4:0.ACC, contains a 7. This offset of 7 points to the address that contains the data, which is N7:7. In other words, the value in N7:7, 10, is the number that is used in the comparison instruction, and thus the output of the comparison instruction is false. In this example, the address changes as the timer accumulator increments toward the preset value.

The indirect address can be used for a file number, word number (element + *subelement*), and

FIGURE 9-4: Indexed indirect addressing mode.

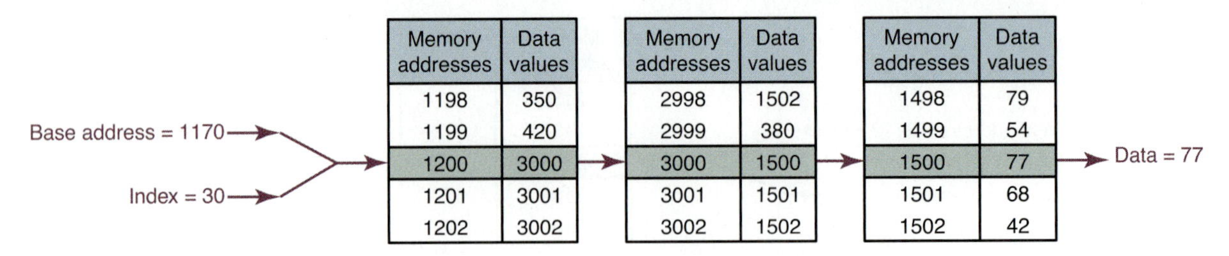

(a) Pre-indexed addressing method with an index of 30

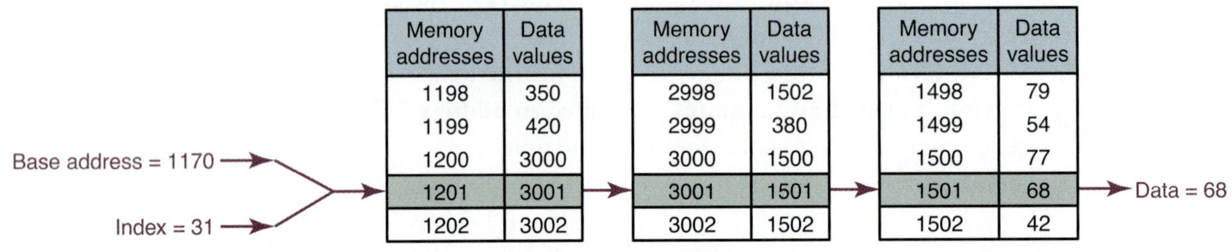

(b) Pre-indexed addressing method with an index of 31

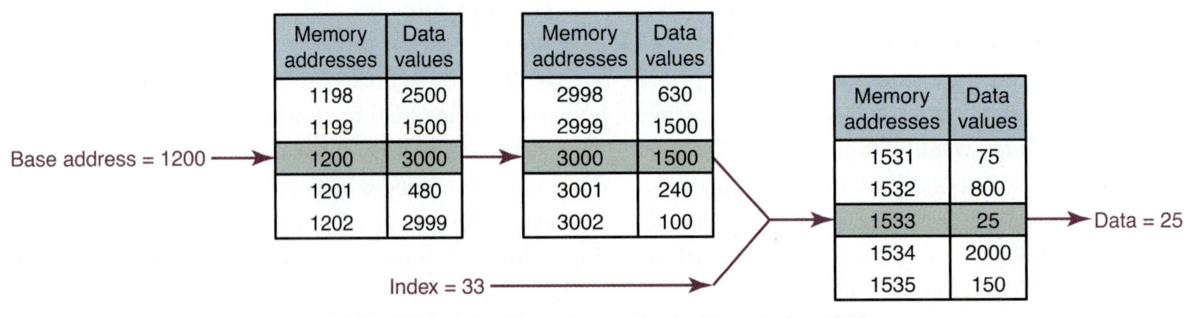

(c) Post-indexed addressing method with an index of 33

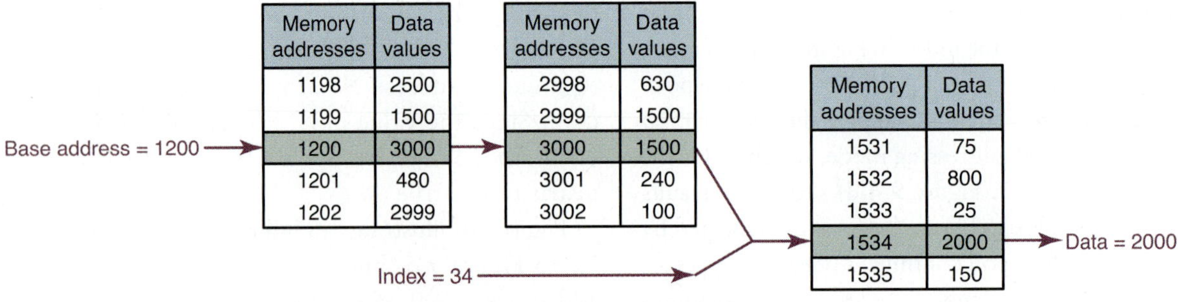

(d) Post-indexed addressing method with an index of 34

bit number. The indirect address(es) can be in one or all of these fields since the address is specified to the word or bit level. The indirect address is always enclosed in square brackets ([ ]). Any part of a bit or word address can use indirect addressing notation to point to the actual location of the data. Table 9-1 lists a number of examples of indirect addressing for the SLC 500 system.

PLC 5 and SLC 500 Indexed Addressing. The indexed address in an instruction consists of a prefix, the pound symbol (#), and the logical address referred to as the base address. The index value to be added to the base address is stored in the status file, word S:24. A move instruction is used to place either a positive or negative index value in S:24. However, when specifying indexed addresses make

FIGURE 9-5: Indirect addressing in a comparison instruction where the instruction is true.

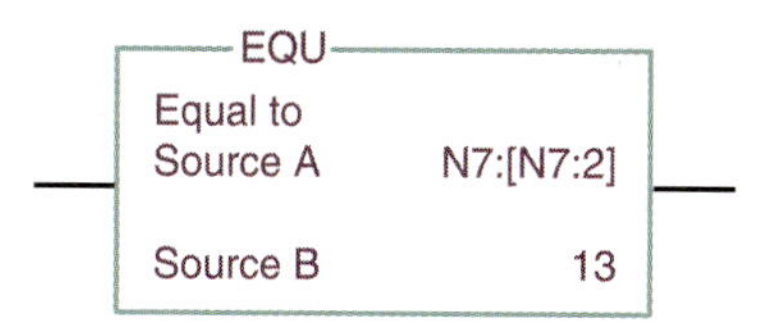

Data file

Addresses	Values
N7:0	1
N7:1	400
N7:2	5
N7:3	32
N7:4	250
N7:5	13
N7:6	42

FIGURE 9-6: Indirect addressing mode in a comparison instruction where the instruction is false.

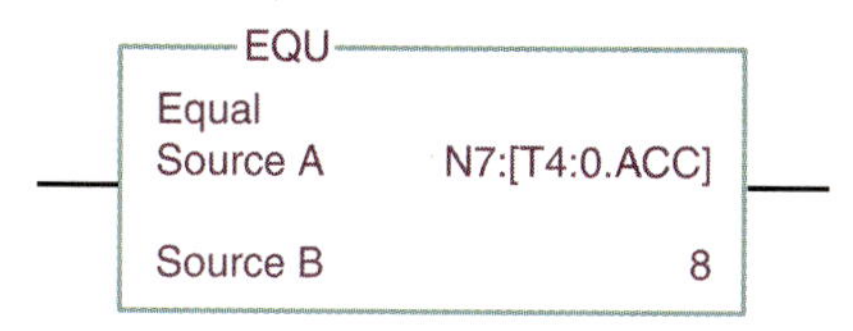

T4:0.ACC = 7

Data file

Addresses	Values
N7:0	100
N7:1	250
N7:2	6
N7:3	5
N7:4	0
N7:5	0
N7:6	0
N7:7	10
N7:8	8

FIGURE 9-7: Move with mask instruction using indexed addressing.

If S:24 has a value of 6, then the addresses are:

Location	Base addresses	Indexed addresses
Source	N7:8	N7:14
Destination	N7:13	N7:19

sure the value in S:24 does not cause the indexed address to exceed the file boundary. For example, if N7 contains 20 elements and S:24 contains a value of 10, then #N7:15 refers to (N7:15 + 10) which is outside the file boundary. This is referred to as *crossing a file boundary* and causes a fault. Figure 9-7 illustrates a move with mask instruction using indexed addresses. Note that when S:24 has a value of 6, the indexed Source base address, #N7:8, would actually reference location N7:14(8 + 6 = 14) for the source and #N7:13 would reference N7:19(13 + 6 = 19) for the destination.

ControlLogix Arrays. An array is a data structure that allocates a contiguous block of memory to store a specific data type as a table of values. The term *contiguous block* means memory locations that are adjacent to each other. Tag arrays in the Logix systems can have one, two, or three dimensions. Figure 9-8 illustrates and describes these three types of arrays. Read the description in the figure before continuing.

A one-dimensional array is used to store a series of values for use in a process problem. For

TABLE 9-1 Indexed Addressing Examples

Valid Address	Variable	Explanation
N7:[C5:8.ACC]	Word number	The word number is the accumulated value of counter 8 in file 5.
B3/[I:0.6]	Bit number	The bit number is stored in input word 6.
N[N7:4]:[N9:3]	File and word number	The file number is stored in integer address N7:4 and the word number in integer address N9:3.
S10:[N7:5].1	Element number	The element number is stored in N7:5.
I:[N7:0].1/4	Slot number	The slot number is stored in N7:0.

FIGURE 9-8: One-, two- and three-dimensional arrays.

Array types	Stores data in this structure	Description
One dimension		Single or one-dimensional arrays are called flat arrays since there is a single column of data just one layer deep. The data is stored in one long list of consecutive or contiguous memory locations. This is the most commonly used type of array structure.
Two dimension		Double or two-dimensional arrays are also called flat arrays for files because there is just one layer. The data is stored in multiple long columns, which are placed next to each other. This is the next most commonly used type of array structure.
Three dimension		Three-dimensional arrays are not flat files since there are a number of two-dimensional arrays stacked on top of each other.

example, if a process had five different temperatures for a heating operation, then each of the five values would be saved in an element of a one-dimensional array with five elements. The ladder logic could recall any of the five temperatures by referencing the array element with the desired value. In another application, the process recipe values could be stored in a tag array and then recalled during the execution of the process control program.

A two-dimensional array is often just an extension of the one-dimensional array. If the one-dimensional array column is used to hold the numerical quantities for each ingredient of a recipe, then multiple columns in a two-dimensional array could hold ingredient quantities for other recipes. For example, if a process has four recipes with 10 ingredient values in each, then a 10 by 4 two-dimensional array would hold the data.

The previous two-dimensional array analog can be applied to a three-dimensional array. Assume that the 10 ingredients and four recipes are used for process A, but the automation area has two processes, A and B. The first two-dimensional array holds the quantities for process A, the second holds the quantities for process B. Therefore, the first dimension is for ingredient quantities, the second dimension has the quantities for different recipes, and the third dimension is the quantities for the two processes. The result is a 10 by 4 by 2 array where element 1,1,1 is the quantity of ingredient 1, for recipe 1, and process A, and element 6, 3, 2 is the quantity of ingredient 6, recipe 3, and process B.

Figure 9-9 illustrates an array tag (Tag name column), the double integer with dimensions (Data type column), and the value of each dimension (Dimension columns). The valid subscript ranges for the examples in Figure 9-9 are as follows:

- One-dimensional array with 7 elements and a valid subscript range: DINT[x], where x = 0 to 6
- Two-dimensional array with 4 and 5 elements in each dimension and a valid subscript range: DINT[x,y], where x = 0 to 3, y = 0 to 4
- Three-dimensional array with 2, 3, and 4 elements in each dimension and a valid subscript range: DINT[x,y,z], where x = 0 to 1; yx = 0 to 2, z = 0 to 3

FIGURE 9-9: One-, two-, and three-dimensional array parameters.

Tag name	Data type	Dimension 0	Dimension 1	Dimension 2	Total elements
one_d_array	DINT[7]	7	0	0	7
two_d_array	DINT[4,5]	4	5	0	20
three_d_array	DINT[2,3,4]	2	3	4	24

FIGURE 9-10: Setting up ControlLogix arrays.

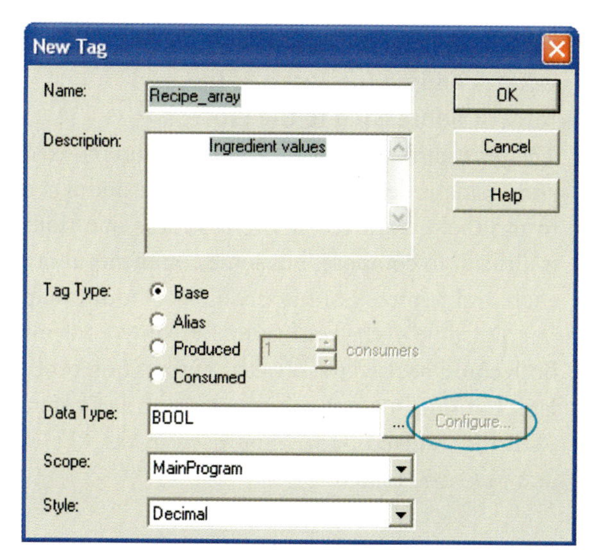

(a) New tag dialog box

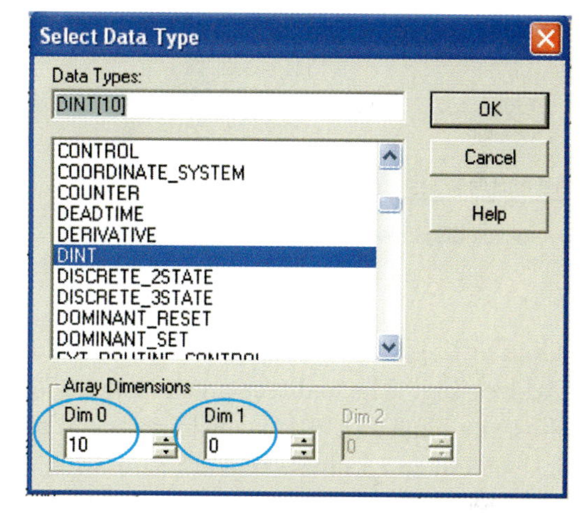

(b) Select data type dialog box with DINT selected

(c) Final data type displayed

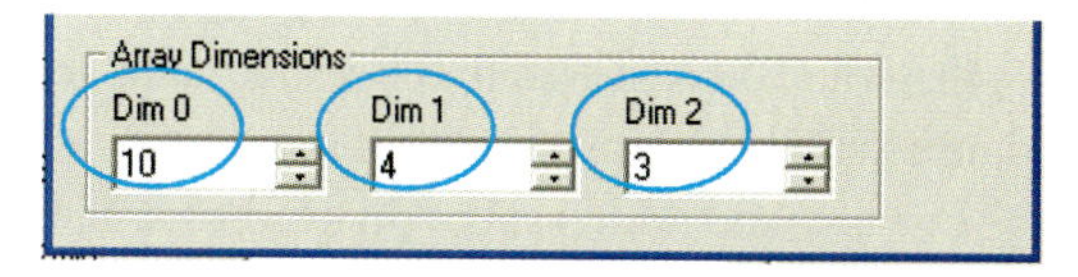

(d) Three array dimensions added

Note that the maximum range value is always one less than the dimension since subscript values start with zero.

Setting Up ControlLogix Arrays. Arrays are set up in ControlLogix by first creating a tag name and then defining or declaring the data type. Figure 9-10(a) shows the New Tag dialog box with the new tag and its description displayed. Note that the default value of Boolean is present, but arrays in ControlLogix should be either DINT or REAL data types. DINT is selected in Figure 9-10(b), and the array dimension boxes appear at the bottom. Each higher dimension is grayed out until you enter values into the lower dimension. The illustrations in Figure 9-10(c) and (d) show the results of entering the three-dimensional array.

FIGURE 9-11: Indirect addressing example for ControlLogix PLC.

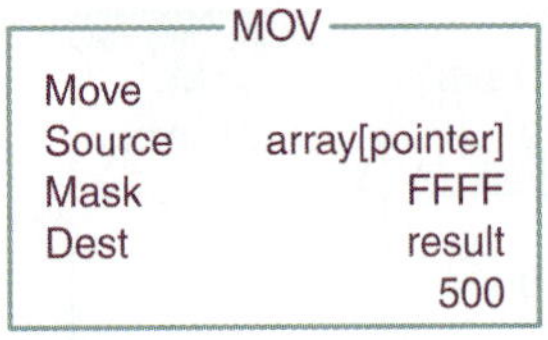

```
MOV
Move
Source      array[pointer]
Mask                  FFFF
Dest                result
                       500
```

(a) Move instruction with indirect addressing using a one-dimensional array

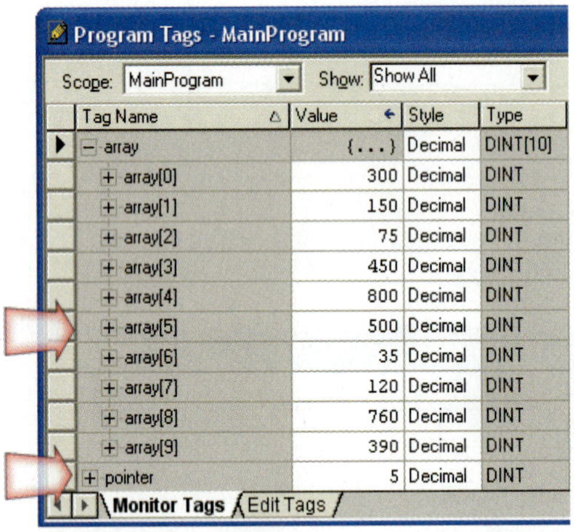

Program Tags - MainProgram

Scope: MainProgram Show: Show All

Tag Name	Value	Style	Type
− array	{...}	Decimal	DINT[10]
+ array[0]	300	Decimal	DINT
+ array[1]	150	Decimal	DINT
+ array[2]	75	Decimal	DINT
+ array[3]	450	Decimal	DINT
+ array[4]	800	Decimal	DINT
+ array[5]	500	Decimal	DINT
+ array[6]	35	Decimal	DINT
+ array[7]	120	Decimal	DINT
+ array[8]	760	Decimal	DINT
+ array[9]	390	Decimal	DINT
+ pointer	5	Decimal	DINT

Monitor Tags / Edit Tags

(b) Array values displayed in Program Tags Window

ControlLogix PLC Indirect and Indexed Addressing. The indirect addressing mode is illustrated in Figure 9-11, which includes a move instruction and a nine-element, one-dimensional array. Note that in the move instruction the indirect address is enclosed in brackets with the tag name pointer. By inserting a number between 0 and 9 in the pointer, data is moved from the array to the result. For example, if the pointer is set at 5, then the 500 is moved to the result tag in accordance with the array shown in the figure.

For indexed addressing you need to create an indexing mechanism because there is no index register in the ControlLogix processor. Many of the instructions that are used for functions requiring indexing use a control tag and are discussed in Chapter 10. A control tag is a unique tag that is created for each instruction instance. The control tag has both length and position fields for control of indexing and is used in place of the S:24 register in the PLC 5 and SLC 500 PLCs. The ControlLogix PLC cannot perform all the indexing maneuvers that the PLC 5 and SLC 500 PLCs can, but its advantages are not having all index instructions using the same address to store the offset, and the ability to use user-defined arrays.

9-3 EMPIRICAL DESIGN PROCESS WITH INDIRECT AND INDEXED ADDRESSSING

The empirical design process, introduced in Section 3-11-4, is an organized approach to the design of PLC ladder logic programs. However, the term *empirical* implies that some degree of trial and error is present. As more instructions are added to the design and the process becomes more complex, the empirical design requires more fixes and adjustments.

9-3-1 Adding Indirect and Indexed Addressing to the Process

The most difficult aspect of indirect and indexed addressing is determining which to use and when to use them. A direct answer to these questions is difficult to compose, but some comments about each and some example problems should help remove some of the confusion. In many problems both could be used to solve the automation problem, but one is usually more efficient. Let's start with the indexed addressing used in the PLC 5 and SLC 500 systems.

Indexed Addressing Used in PLC 5 and SLC 500. The indexed addressing has a fixed offset stored in the status bit S:24. That address is added to the base address to determine the new address location. The problem is that the same offset value must be used with every indexed base address. Example 9-1 illustrates this point.

Indirect Addressing Used in ControlLogix PLC. The indirect addressing places an address within an address to indicate the location of that address parameter. In the PLC 5 and SLC 500 systems, the imbedded address changes an address parameter of the base address to determine the new address location. In the ControlLogix PLC, indirect addressing is linked to the use of DINT arrays. Example 9-2 illustrates this point.

EXAMPLE 9-1

The process tank design problem in Section 7-6-3 (Figures 7-17 through 7-21) has been modified as follows:

The material 1 and 2 quantity inputs by BCD push switches have been removed and receipts have been stored in memory for the needed quantities. The number of the receipt to run is entered with a BCD push switch and stored in N7:15 when an NO load push button is pressed. The material values are transferred by the same push button. The first three of the process receipts are listed in Table 9-2. Note that the offset listed in the table is located in S:24. Design an indexed addressing solution for this problem.

SOLUTION

Review the process tank problem statement and the solution to tank problem in Chapter 7 before reading the solution to this modification.

The modified system ladder logic is presented in Figure 9-12(a) and the integer register values are illustrated in Figure 9-12(b). In the Chapter 7 solution, the material 1 and 2 quantities are placed into N7:0 and N7:1 and then multiplied by 2 with the results going into N7:10 and N7:11.

In this modification, the process starts with the load push button that loads the receipt number from the BCD switch into N7:15. The same load push button makes three comparison instructions active and the receipt number makes a EQU comparison instruction true, which loads the correct index offset into the S:24 register. Next, the indexed values from the receipt locations in N7:30 through N7:35 are multiplied by 2 and then placed into N7:10 and N7:11. From this rung forward the solution is unchanged from that illustrated in Figures 7-20 and 7-21.

TABLE 9-2 Receipt parameters for Example 9-1

	Receipt 1		Receipt 2		Receipt 3		
Parameters	**Values**	**Address location**	**Values**	**Address location**	**Values**	**Address location**	**Base address**
Material 1 (gallons)	**50**	**N7:30**	**75**	**N7:32**	**60**	**N7:34**	**N7:0**
Material 2 (gallons)	15	N7:31	25	N7:33	45	N7:35	N7:1
Offset		30		32		34	

EXAMPLE 9-2

A color sensor is used to check the color of products moving down a conveyor. The sensor returns two types of signals: a discrete signal to indicate that a part has been detected and a color value signal, which is an analog value based on the color sensed. In order to get a more reliable color value, the sensing ladder logic should sample the color 10 times and find the average value. Design a ladder logic program for collecting the color data.

SOLUTION

The ladder logic solution and array for storing the data is presented in Figure 9-13. The solution has four rungs to sample the 10 color values. The rungs for averaging the values are not shown. The ladder action starts when the color_switch is true, indicating that a part is present. The data is stored in a one-dimensional array, Color_sample[Array_index.ACC]. The tag is found in the move instruction in rung 2, Figure 9-13(a), and in the Monitor

FIGURE 9-12: Solution for Example 9-1.

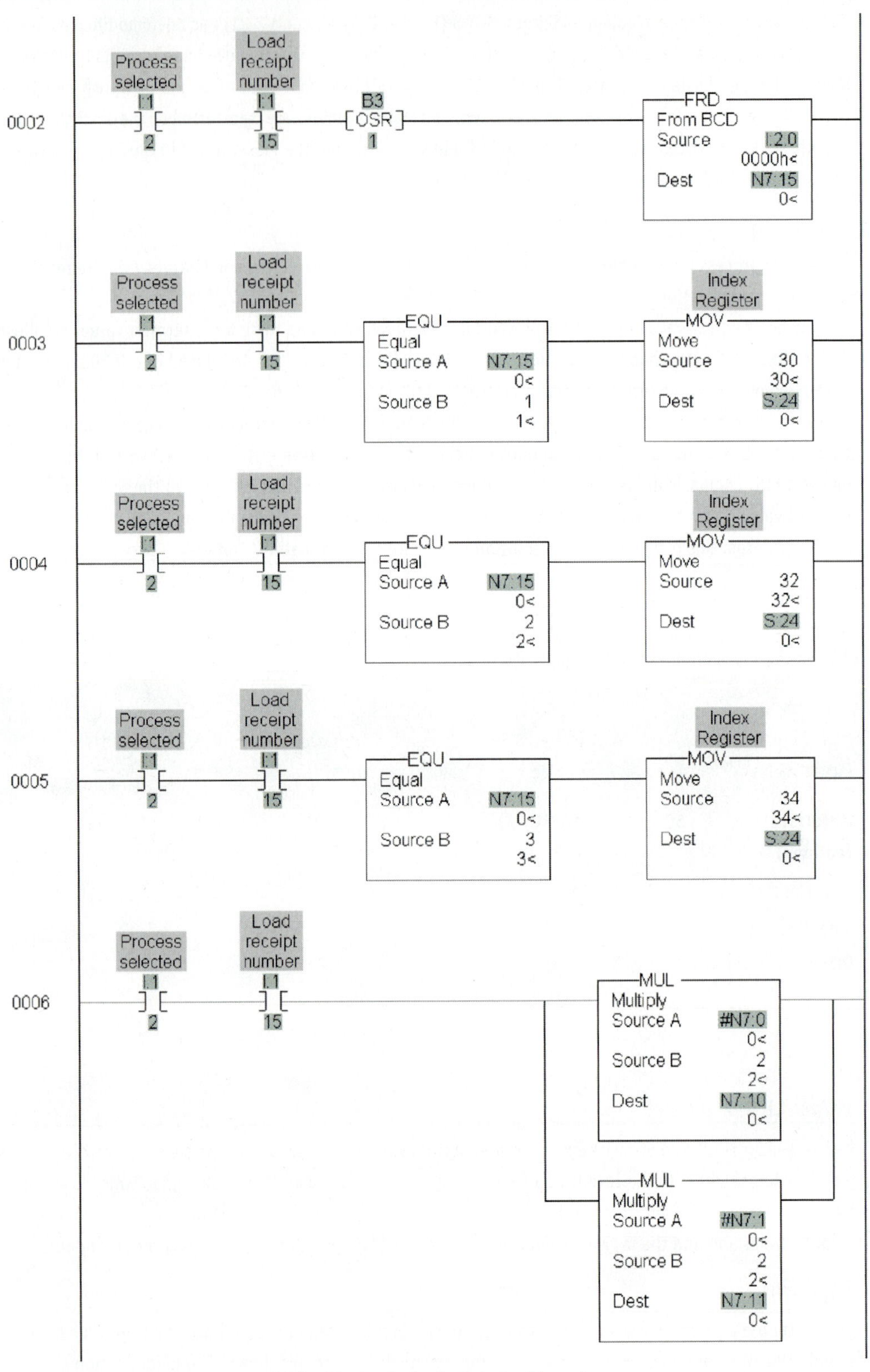

(a) Ladder logic for indexed addressing

FIGURE 9-12: (Continued).

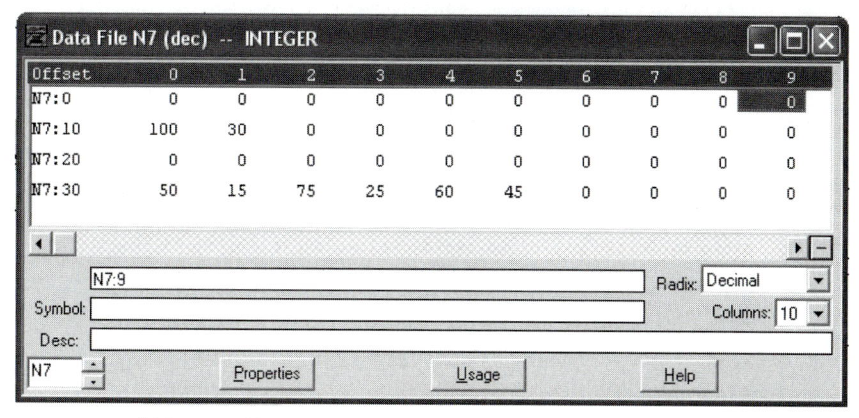

Data File N7 (dec) -- INTEGER

Offset	0	1	2	3	4	5	6	7	8	9
N7:0	0	0	0	0	0	0	0	0	0	0
N7:10	100	30	0	0	0	0	0	0	0	0
N7:20	0	0	0	0	0	0	0	0	0	0
N7:30	50	15	75	25	60	45	0	0	0	0

(b) Integer files with stored data values for indexed addressing

Tags tab of the Program Tags window, Figure 9-13(b). Note in the window that the database has the first three values stored and the fourth value is visible in the sensor tag, Color_data, for the color sensor.

A move instruction moves the value from the input image table, Color_data, every time the half-second timer's done bit becomes true. The timer triggers the counter, which acts as the index for the data array. The array is indexed by the tag Array_index.ACC, which is the accumulator of the counter. When the counter reaches 10 the counter done bit inhibits the timer and counter and the logic waits for the Color_switch to cycle off and reset the system.

FIGURE 9-13: Solution for Example 9-2.

0
Part present
Color_switch
Array index value
Array_index.DN
0.5 sec sample rate
Sample_timer.DN
0.5 sec sample rate
TON
Timer On Delay
Timer Sample_timer
Preset 500
Accum ?
EN
DN

1
Array index value
Array_index.DN
0.5 sec sample rate
Sample_timer.DN
Array index value
CTU
Count Up
Counter Array_index
Preset 10
Accum ?
CU
DN

2
0.5 sec sample rate
Sample_timer.DN
Array index value
Array_index.DN
single_move
ONS
MOV
Move
Source Color_data
<Local:3:I.Ch0Data>
0.0
Dest Color_sample[Array_index.ACC]
0

3
Part present
Color_switch
Array index value
Array_index
RES
0.5 sec sample rate
Sample_timer
RES

(a) Ladder logic for indirect addressing

FIGURE 9-13: (Continued).

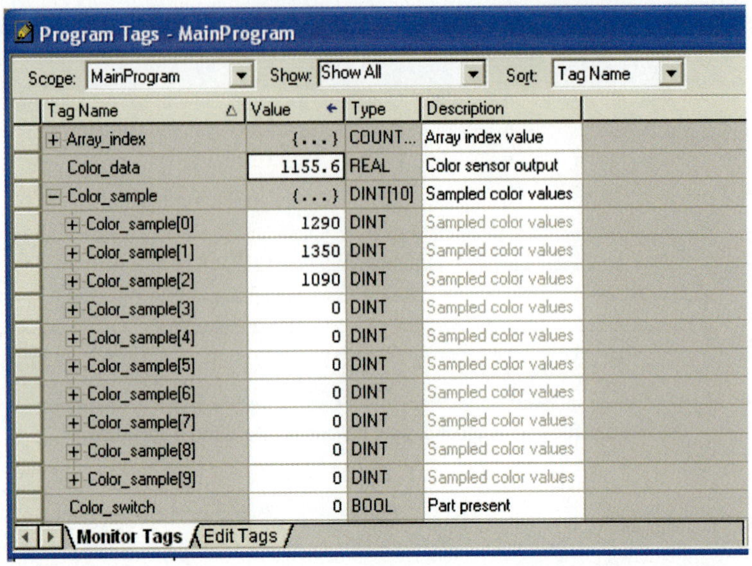

Program Tags - MainProgram

Scope: MainProgram | Show: Show All | Sort: Tag Name

Tag Name	Value	Type	Description
+ Array_index	{...}	COUNT...	Array index value
Color_data	1155.6	REAL	Color sensor output
− Color_sample	{...}	DINT[10]	Sampled color values
+ Color_sample[0]	1290	DINT	Sampled color values
+ Color_sample[1]	1350	DINT	Sampled color values
+ Color_sample[2]	1090	DINT	Sampled color values
+ Color_sample[3]	0	DINT	Sampled color values
+ Color_sample[4]	0	DINT	Sampled color values
+ Color_sample[5]	0	DINT	Sampled color values
+ Color_sample[6]	0	DINT	Sampled color values
+ Color_sample[7]	0	DINT	Sampled color values
+ Color_sample[8]	0	DINT	Sampled color values
+ Color_sample[9]	0	DINT	Sampled color values
Color_switch	0	BOOL	Part present

Monitor Tags / Edit Tags

(b) Integer files with stored data values for indirect addressing of an array

9-4 TROUBLESHOOTING INDIRECT AND INDEXED ADDRESSING IN LADDER LOGIC

Addressing modes are not instructions but can be used with most of the PLC instructions. The major cause of problems with indirect and indexed addressing is the selection of the offset or pointer values. The pointers redirect the integer or floating point file to a new location in memory. The problem occurs when the redirected location is outside the data file boundary. In Allen-Bradley PLCs the boundaries for integer and floating point files are declared or set by the user. Indexed or indirect addresses that fall outside the file boundaries cause a fault condition. Problems in the PLC 5 and SLC 500 occur in indexed addressing, which uses the S:24 register to store the pointer index value. Some program instructions store values in this S:24 register and will overwrite the indexed pointer number placed there. A good practice is to move the index value for the pointer to the S:24 register just ahead of the rung where the indexing is performed.

When PLC programs are developed, some problems result from the complexity of the process problem to be solved, and others result from programming techniques that produce ladder rungs that are not a correct solution. If indirect or indexed addressing modes are present in the rungs where the process control solution is not working, then troubleshooting those rungs is far more difficult. Troubleshooting indirect addressing is more difficult than troubleshooting indexed addressing. Use the following guidelines when troubleshooting indirect and indexed addressing.

- Verify that pointers do not point to addresses outside the memory boundaries.
- Use the temporary end (TND) instruction to stop scanning at points in the ladder where addressing indexes can be verified.
- Use single-step options to scan one rung at a time to analyze how the pointers are changing the addressing flow.
- The value displayed in an instruction beneath an indexed address, such as #N7:5, is the value in N7:5, not the indexed value in N7:[5 + S:24]. To view the indexed value, add an unconditional rung to move the indexed value to another integer register for display.
- This guideline could be listed in the empirical design area as well. When designing an indexed or indirect addressing mode solution with a large database containing many elements, start with just a small data set. For example, if you can make the ladder operate with just a few receipts and a small number of ingredients, then you can easily expand it for the larger data set.

Chapter 9 questions and problems begin on page 578.

PART 2

Advanced PLC Instructions and Applications

GOALS

The goal for the second half of the text, Chapters 10 through 17, is to expand your knowledge of PLCs into the use of the more sophisticated PLC instructions and introduce some of the new programming and networking technology. In Part 1 a subset of the PLC instructions was covered, but that subset is used to write a majority of the ladder logic used in manufacturing automation. However, PLCs are being used for an increasing number of control requirements, and as they move into these new areas additional advanced instructions are required. Using these more complex instructions correctly is a goal of this section.

The PLC is a participant in the trend toward distributed control and use of information networks in factory automation. Therefore, another goal of Part 2 is to introduce the networks and subnetworks that permit PLC control to be distributed across the factory and across the Internet.

The final goal for Part 2 is to introduce the programming standard IEC 61131-3 with multiple programming language styles. As PLCs take on more complex control tasks, standard ladder logic is not sufficient for the task. The languages defined by the standard take PLC programming to the information system level. As PLCs move into this new territory, so must the men and women who install, program, and maintain them. Therefore, the general goal of Part 2 is to prepare you for the future of PLC control.

OBJECTIVES

After completing Part 2 you should be able to:

- Write a program using the four language formats supported by Allen-Bradley that are defined in the IEC 61131-3 programmable logic controller standard.
- Describe a control solution using sequential function charts techniques.
- Apply PLC instructions for data handling, shift register, and sequencer functions to the solution of control problems.
- Apply analog input and output hardware and the PID function to the solution of process control problems.
- Develop a PLC solution using network technology at three information technology levels.
- Apply sound troubleshooting and documentation techniques to systems under PLC control.
- Select industrial networks for factory floor applications.

CAREER INSIGHTS

The following insight provides some information on the jobs available to students who master the contents of Part 2 of the text. Some of the terms used in this description may not be familiar to you at this time but will be when the material in Part 2 is covered. Read the following paragraphs now to get a general feel for the additional opportunities in PLCs, then read it again after completing Part 2 to see where you could work in the PLC and automation control area. If you enjoy the technology and problems covered in this text, then there are numerous paths that you can follow in PLCs and manufacturing automation.

In Part 1 a number of paths were identified for students who enjoy PLCs and factory automation hardware and software responsibilities. The concepts learned in the second half of this text open up similar paths, but at a higher technical level. Although standard ladder logic is still the primary programming tool used in North America, the more structured languages defined in IEC 61131-3 will work their way into PLC solutions. The new standard has languages that are graphical and text based. Learning the graphical is not a stretch for current ladder logic programmers; however, the development of PLC programs in the text-based languages requires skills learned in computer science. Some of the more advanced PLC instructions for motion control, process control, and data manipulation require greater understanding of related technologies such as servo and stepper motor control, closed loop feedback concepts, and databases. These advanced instructions require a broader educational preparation. As a result, the path for students interested in PLC program development diverges into several interesting directions.

PLC technology has embraced the movement toward distributed control through the use of vendor-specific networks and international network standards such as Ethernet/IP, DeviceNet, and ControlNet. Ethernet/IP (Ethernet Industrial Protocol) is a high-level network for linking industrial systems, whereas DeviceNet and ControlNet are subnetworks for linking devices and controllers. Knowledge in these communication technologies opens additional career doors for students who wish to work in distributed control using PLC systems. If you enjoy this technology, it is a wonderful career area.

CHAPTER

10

Data Handling Instructions and Shift Registers

10-1 GOALS AND OBJECTIVES

There are three principal goals of this chapter. The first goal is to provide the student with the overall concept of how bits, words, and files are handled in a PLC. The second goal is to discuss the data transfer and manipulation by logical instructions, the file-arithmetic-logic instruction, and shift registers, including first in, first out and last in, first out operations. Also, the copy and fill instructions are included. The third goal is to show how data handling and shifting instructions are applied to Allen-Bradley PLC 5, SLC 500, and Logix PLCs.

After completing this chapter you should be able to:

- Explain the concept of how data is handled by bits, words, and files.
- Describe the logical instructions of and, or, and exclusive or.
- Describe the arithmetic-file-logic function.
- Explain the concept of shift registers.
- Describe the operation of first in, first out and last in, first out shifting functions.
- Describe the operation of copy and fill instructions.
- Describe the data handling and manipulation instructions for the Allen-Bradley PLC 5, SLC 500, and Logix systems.
- Develop ladder logic solutions using data handling and manipulation instructions for the Allen-Bradley PLC 5, SLC 500, and Logix systems.
- Include data handling and manipulation instructions in the empirical design process.
- Describe troubleshooting techniques for ladder rungs with data handling and manipulation instructions.

10-2 DATA HANDLING

Data handling includes the movement and manipulation of data by arithmetic and logical operations. This data handling is accomplished on *bits, words,* and *files*, where words are also referred to as registers and files as tables, blocks, or arrays. The relationship between bits, words, and files is illustrated in Figure 10-1. Note that bits are represented as 1s or 0s, a word is a group of bits, and a file is a group of words. Bit patterns in words and word patterns in files are discussed in the following subsections.

10-2-1 Bit Patterns in Words

The bit pattern in one word or register can be transferred to another word or register as shown in Figure 10-2. Note that in Figure 10-2(a), word 40 contains all 0s and word 43 contains a bit

FIGURE 10-1: Bits, words, and files.

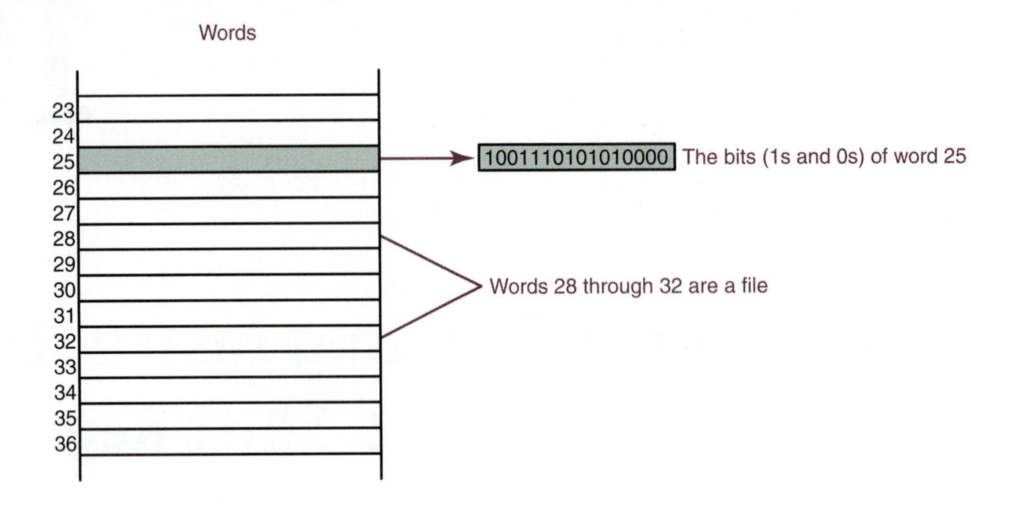

FIGURE 10-2: Bit patterns in words.

0000	0000	0000	0000	Word 40
–	–	–	–	Word 41
–	–	–	–	Word 42
1010	0011	0001	1011	Word 43

(a) Initial contents of words 40 and 43

1010	0011	0001	1011	Word 40
–	–	–	–	Word 41
–	–	–	–	Word 42
1010	0011	0001	1011	Word 43

(b) Contents of words 40 and 43 after the transfer

pattern of 1s and 0s. In Figure 10-2(b), the contents of words 40 and 43 are the same because the contents of word 40 were replaced or written over with the contents of word 43. Words 41 and 42 were unchanged. Even if the contents of word 40 were a pattern of 1s and 0s, its contents would still be replaced with the contents of word 43. Recall that an individual bit(s) within a word or register can be changed with the use of a mask, as discussed in Section 6-5-6.

10-2-2 Word Patterns in Files

The pattern of words that comprises a file has distinct *starting* and *ending* locations. In some PLC applications the entire sequence of words is transferred from one location to another. This *file-to-file transfer* is used when the words in one file represent a set of data that is used many times within a process, but must remain unchanged after each operation in the process. Because the data within the file is manipulated in a process, a second file is needed to handle the data changes. In addition to file-to-file transfers, *word-to-file* and *file-to-word* transfers are used in processes driven by a PLC. Figure 10-3 illustrates these three types of transfers. The *file-arithmetic-logic* (FAL) instruction is used to accomplish these transfers.

10-3 ALLEN-BRADLEY DATA TRANSFER AND MANIPULATION INSTRUCTIONS

The Allen-Bradley data transfer and manipulation instructions allow the user to operate on data with *logical, arithmetic,* and *shifting functions,* which handle single- or multiple-word data. The Allen-Bradley PLCs have data transfer and manipulation instructions available in all three processors. The following instructions, common to all three PLC systems, are described in this section.

- **Bitwise and (AND):** performs a bit-by-bit logical *AND* operation.
- **Bitwise or (OR):** performs a bit-by-bit logical *OR* operation.
- **Bitwise exclusive or (XOR):** performs a bit-by-bit logical *EXCLUSIVE OR* operation.

FIGURE 10-3: Word and file transfers.

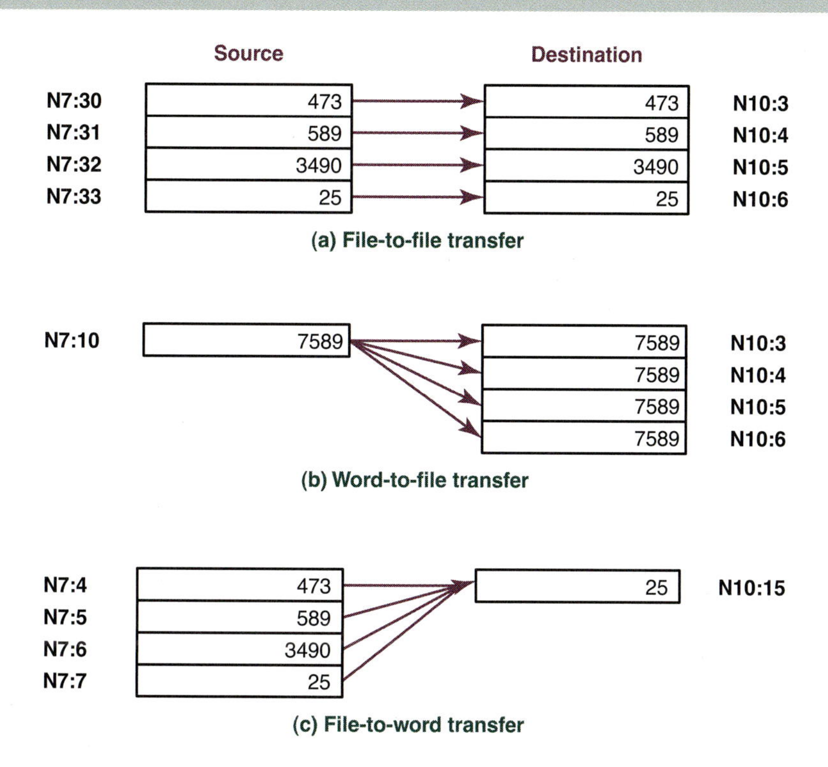

- **File-arithmetic-logic (FAL):** performs copy, arithmetic, logic, and function operations on the data stored in files or arrays.
- **Bit shift left (BSL) and Bit shift right (BSR):** shifts the bits within a word or specified bits in an array one position to the left or right, respectively.
- **FIFO load (FFL) and FIFO unload (FFU):** loads data into a file and unloads data from a file, respectively. When used in pairs, they perform the FIFO function—first in, first out.
- **LIFO load (LFL) and LIFO unload (LFU):** loads data into a file and unloads data from a file, respectively. When used in pairs, they perform the LIFO function—last in, first out.
- **Copy (COP):** copies data files from one register location to another, maintaining the sequential sequence of the data.
- **Fill (FLL):** copies the contents of one register location to a specified number of other register locations.

10-3-1 AND, OR, and XOR Instructions

The *AND, OR,* and *XOR* instructions are logical instructions. They are illustrated in Figure 10-4(a) for the SLC 500 system, but these instructions are also available in the PLC 5 and ControlLogix systems. These logical instructions perform a bit-by-bit logical operation using the values in Sources A and B and storing the result of the operation in the destination. Figure 10-4(b) illustrates the truth table for each of these logical operations. Figure 10-4(c) illustrates an AND, OR, and XOR operation on two identical words. The AB requirements for parameters placed into Sources A and B are:

PLC 5 and SLC 500: The parameter placed in both Source A and Source B can be a program constant or a bit/word address.

ControlLogix Systems: The parameter placed in both Source A and Source B can be any of the following data types: SINT (single integer), INT (integer), or DINT (double integer). The parameter used can be either an immediately entered program constant or a tag name.

Standard Ladder Logic for AND, OR, and XOR Instructions. The standard ladder logic illustrating how the AND, OR, and XOR instructions are

FIGURE 10-4: AND, OR, and XOR instructions.

AND
Bitwise AND
Source A
Source B
Dest

OR
Bitwise Inclusive OR
Source A
Source B
Dest

XOR
Bitwise Exclusive OR
Source A
Source B
Dest

(a) AND, OR, and XOR instructions

AND Instruction

Source A	Source B	Destination
0	0	0
1	0	0
0	1	0
1	1	1

OR Instruction

Source A	Source B	Destination
0	0	0
1	0	1
0	1	1
1	1	1

XOR Instruction

Source A	Source B	Destination
0	0	0
1	0	1
0	1	1
1	1	0

(b) AND, OR, and XOR truth tables

	AND Instruction	OR Instruction	XOR Instruction
Source A	1 0 1 0 1 0 1 0	1 0 1 0 1 0 1 0	1 0 1 0 1 0 1 0
Source B	1 1 1 0 0 0 1 1	1 1 1 0 0 0 1 1	1 1 1 0 0 0 1 1
Destination	1 0 1 0 0 0 1 0	1 1 1 0 1 0 1 1	0 1 0 0 1 0 0 1

(c) AND, OR, and XOR operations

used in a ladder rung appears in Figure 10-5. Compare the ladder implementations with the descriptions and truth tables in Figure 10-4. Note that the AND instruction can be used for a masked move requirement because the destination bit pattern is the same as Source A for every bit that is a 1 in Source B. Also, 0's in Source B produces 0's in the destination. The XOR instruction works like a compare instruction, comparing each pair of bits and placing a 1 in the destination bit location when the bits are not the same, such as a 1 and 0 or 0 and 1. The OR instruction performs the logical OR between corresponding bits.

The use of the AND for a masked move instruction is illustrated in Figure 10-5(a). The mask is 0FF0h, so the middle eight bits are all ones ($0000\ 1111\ 1111\ 0000_2$). Source A has 3D49h and the mask eliminates the 3 and 9, but passes the D4. That passes the middle hex values (D4h or $1101\ 0100_2$) from Source A to the destination.

The OR instruction produces a 1 in the destination register for every pair of bits with a 1 in either or both of the pair bits. The XOR instruction is a comparator that produces a 0 in the destination if the pairs of compared bits are equal (both ones or both zeros). Use the truth tables in Figure 10-4 to verify the destination values in Figure 10-5 for each instruction.

10-3-2 File-Arithmetic-Logic (FAL) Instruction

The *file-arithmetic-logic* (FAL) instruction is used to transfer data from one file to another and perform arithmetic and logic operations on files. The Allen-Bradley FAL instruction, available on the PLC 5 and ControlLogix PLCs, is illustrated in Figure 10-6.

The terms within the FAL instruction of Figure 10-6 are as follows:

- **Control:** the address of the control word, which provides the information necessary to execute the instruction.
- **PLC 5:** the address is in the control area R of processor memory and the default file 6. The operation and format for the control data file is described in the next section.

FIGURE 10-5: Standard ladder logic rungs for SLC 500 AND, OR, and XOR instructions.

Application	Standard Ladder Logic Rungs AND, OR, and XOR Instructions
Perform a bit by bit AND operation between two registers or perform a masking operation while moving a value from one register to another. The bitwise AND instruction in PLC 5, SLC 500, and ControlLogix ANDs each bit of the Source A and Source B words and places the ANDed bit operation into the Destination register. The instruction can also be used as a masked move operation, which in not available in the PLC 5 and ControlLogix PLCs, by placing the data value to be masked and moved in Source A and placing the mask into Source B. The result is placed in the Destination. Note: SLC 500 rungs are shown, but the operation would be the same for PLC 5 and ControlLogix.	Start I:1 0 AND Bitwise AND Source A N7:2 3D49h< Source B N7:5 0FF0h< Dest T4:2.PRE 0D40h< Note: middle 8 bits are passed so the destination has 0D40h after the AND **(a)**
Perform a bit by bit OR operation between two registers. The bitwise OR instruction in PLC 5, SLC 500, and ControlLogix logically ORs each bit of the Source A and Source B words and places the result of that OR operation into the correspond bit in the Destination register. Note: SLC 500 rungs are shown, but the operation would be the same for PLC 5 and ControlLogix.	I:1 5 OR Bitwise Inclusive OR Source A N7:4 00C8h< Source B N7:8 0050h< Dest N7:1 00D8< **(b)**
Perform a bit by bit exclusive OR (XOR) operation between two registers to identify the bits in the two words that are not equal. The bitwise XOR instruction in PLC 5, SLC 500, and ControlLogix determines if the respective data bits in two registers are the same. If the two words are equal the result is zero for all bits. All bit patterns that are not equal will result in a 1 placed in that bit location in the Destination register. Note: SLC 500 rungs are shown, but the operation would be the same for PLC 5 and ControlLogix.	C5:2 DN XOR Bitwise Exclusive OR Source A T4:0.ACC 0082h< Source B N7:3 01C2h< Dest N7:11 0140h< **(c)**

- **ControlLogix:** the control address would be a tag such as control_1 with a data type of control.
- **Length:** the number that represents the length of the file.
  PLC 5 – a number from 1 to 1000
  ControlLogix – DINT (double integer)
- **Position:** the number that specifies the word location within the file. It typically starts at zero and indexes to one less than the file

FIGURE 10-6: FAL instruction.

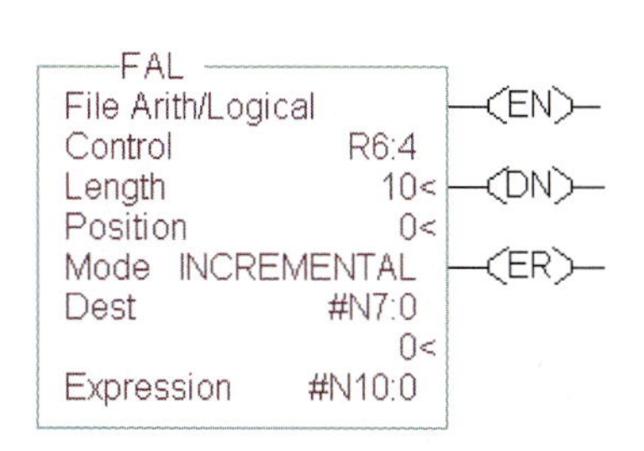

length, although you can start at a number other than zero.

- **Mode:** there are three modes: In the *all* mode a complete file is transferred in one scan, in the *incremental* mode one element of data is operated upon for each time the FAL ladder rung instruction goes from false to true, and in the *numeric* mode a decimal number (1 to 1000) indicates the number of elements operated on per scan. The instruction remains true until all words are transferred in block size indicated.
- **Destination:** the address of the location for the result of the operation.
- **Expression:** the address that specifies the source of the file and the type of arithmetic or logical operation to be performed.

Control Word Addressing. This instruction marks the first use of *control words* or the *control data file* to provide the information necessary to execute instructions. The control data file, illustrated in Table 10-1(a), has three-word elements where word 0 is the status word, word 1 indicates the length of stored data, and word 2 indicates position. The control register is used with the following instructions: bit shift, FIFO shift register, LIFO shift register, and sequencer instructions in all Allen-Bradley PLC models. In addition, it is used in the file-arithmetic-logic (FAL) instruction in the PLC 5 and ControlLogix PLCs.

The addressable bits and words from the control data file are listed in Table 10-1(b) and the control elements or bits are in Table 10-1(c).

The addressing used for control data files in the PLC 5 and SLC 500 is described in Table 10-2.

Control data words in ControlLogix PLCs are tags with a data type of control. The control structure in ControlLogix is the same three-word sequence used in PLC 5 and SLC 500 and illustrated in Table 10-1. The addressing of bits and words from the three-word structure is similar as well, but tags are used with element and bit identifiers.

TABLE 10-1 Control word bit and word structure.

15	14	13	12	11	10	09	08	07	06	05	04	03	02	01	00	Word
EN	EU	DN	EM	ER	UL	IN	FD				Error Code					0
Length of Bit Array or File (LEN)																1
Bit Pointer or Position (POS)																2

(a) Control data file words

Addressable Bits	Addressable Words
EN = Enable	LEN = Length
EU = Update Enable	POS = Position

(b) Addressable bits and words

DN = Done

EM = Stack Empty

ER = Error

UL = Unload

IN = Inhibit

FD= Found

(c) Control bits

TABLE 10-2 Control data file addressing format.

Format	Explanation		
	R	Control file	
	f	File number	
	:	Element delimiter	Number 6 is the default file. A file number between 10 and 255 can be used if additional storage is required.
	e	Element number	Ranges from 0 to 255. These are three-word elements. See Table 10-1.
Rf:e.s/b	Rf:e	Explained above.	
	.	Word delimiter	
	s	Indicates word	
	/	Bit delimiter	
	b	Bit	
Examples:			
R6:2		Element 2, control file 6. Address bits and words by using the format Rf:e.s/b	
R6:2/15 or R6:2/EN		Enable bit	
R6:2/14 or R6:2/EU		Unload Enable bit	
R6:2/13 or R6:2/DN		Done bit	
R6:2/12 or R6:2/EM		Stack Empty bit	
R6:2/11 or R6:2/ER		Error bit	
R6:2/10 or R6:2/UL		Unload bit	
R6:2/9 or R6:2/IN		Inhibit bit	
R6:2/8 or R6:2/FD		Found bit	
R6:2.1 or R6:2.LEN		Length value	
R6:2.2 or R6:2.POS		Position value	
R6:2.1/0		**Bit 0 of length value**	
R6:2.2/0		**Bit 0 of position value**	

Standard Ladder Logic for the FAL Instruction. The standard ladder logic in Figure 10-7 describes how the FAL instruction copies a single element to an array, copies an array to an array, or copies an array to a single element. These three file transfers were introduced in Figure 10-3. The instruction in Figure 10-7 uses the ControlLogix PLC format, which includes arrays and array operations. You may want to review the array material in Chapter 9 if you are not familiar with their operation.

First, the *word-to-file transfer* using the FAL instruction is illustrated in Figure 10-7(a). Note that the mode is *incremental*, that is, with each enabling of the instruction, the contents of the expression are moved into the destination. The first time Color_switch is true, the position value is 0 and the contents of tag value_1 are moved into the first element of a two-dimensional tag, array_2[0,Control_4.pos]. On the second transition of the input instruction, the contents of value_1 are moved into the second element of the

FIGURE 10-7: Standard ladder logic rungs for ControlLogix FAL instructions.

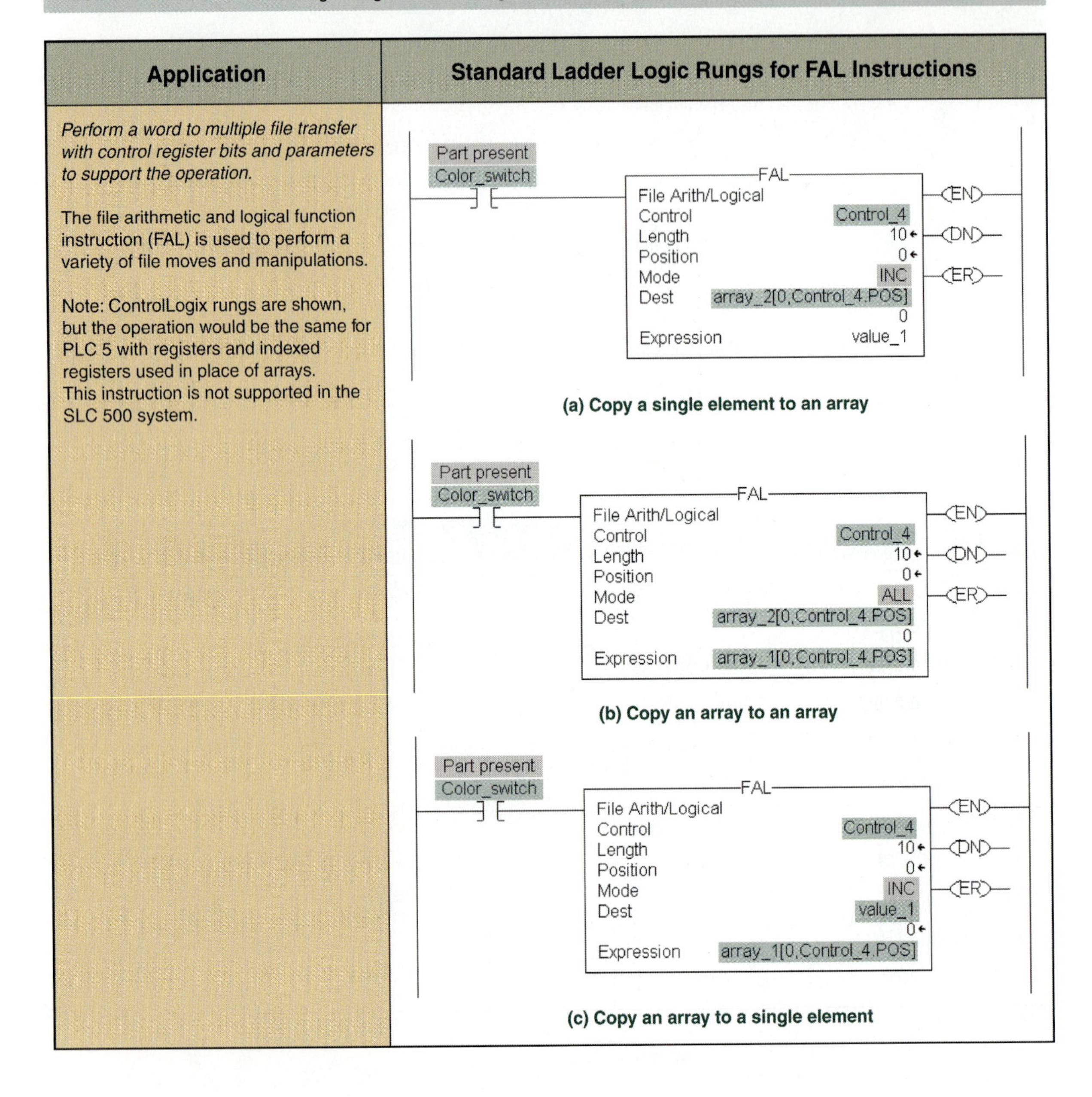

two dimensional array. After these two transitions, values are in array locations Array_2[0,0] and Array_2[0,1]. Eventually, the FAL instruction copies value_1 into the first 10 elements or positions of the two dimensional array. Therefore, the values are copied to the ten positions from Array_2[0,0] to Array_2[0,9].

Next, the *file-to-file transfer* using the FAL instruction is illustrated in Figure 10-7(b). Note that the FAL instruction is in the ALL mode, which means that all data are transferred in the first program scan when the FAL instruction is enabled. The data moves all of the elements from Array_1 into the same position within Array_2. In this case Array_1[0,0] to Array_1[0,9] are moved to Array_2[0,0] to Array_2[0,9]. After this transfer, for example, Array_1[0,5] has the same value as Array_2[0,5].

The third transfer, the *file-to-word transfer*, is illustrated in Figure 10-7(c). It is similar to the word-to-file transfer except that the FAL instruction transfers the data from a sequence of array elements into a single tag. Each time the FAL instruction is enabled, it copies the current value of array_1 to value_1. The FAL instruction uses the incremental mode, so only one array value is copied each time the Color_switch instruction is enabled. The next time the instruction is enabled, the instruction overwrites value_1 with the next value in array_1.

The FAL instruction has a control data register; therefore, in the ControlLogix system the tag used to identify the control data register must be a control data type. Figure 10-7(d) illustrates how the Control_4 tag is assigned a control data type.

PLC 5 FAL Instruction. Figure 10-8 illustrates an arithmetic operation using the FAL instruction in the PLC 5 system. When the input instruction I:001/0 is true, the FAL instruction is enabled, and the FAL instruction adds the

FIGURE 10-7: (Continued).

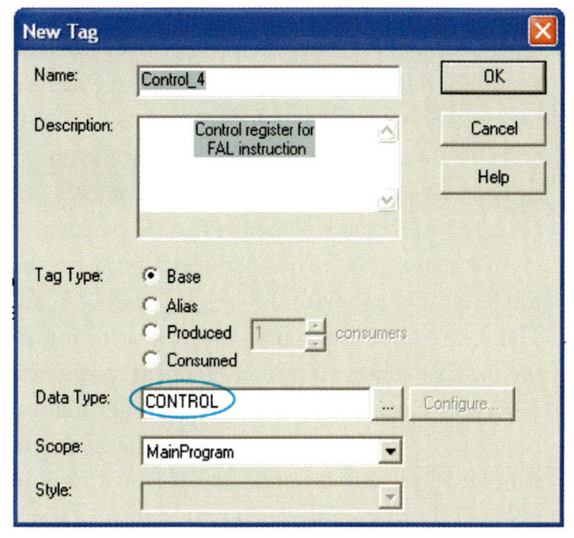

(d)

FIGURE 10-8: FAL operation in a PLC 5 processor.

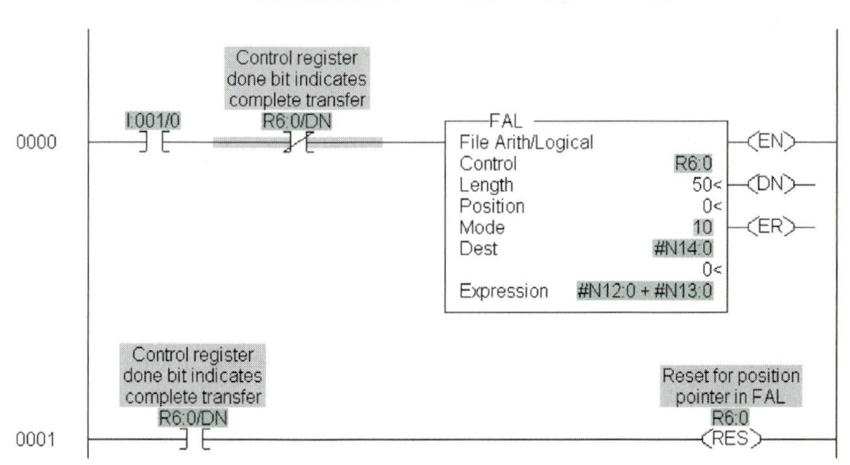

(a) PLC 5 FAL ladder logic

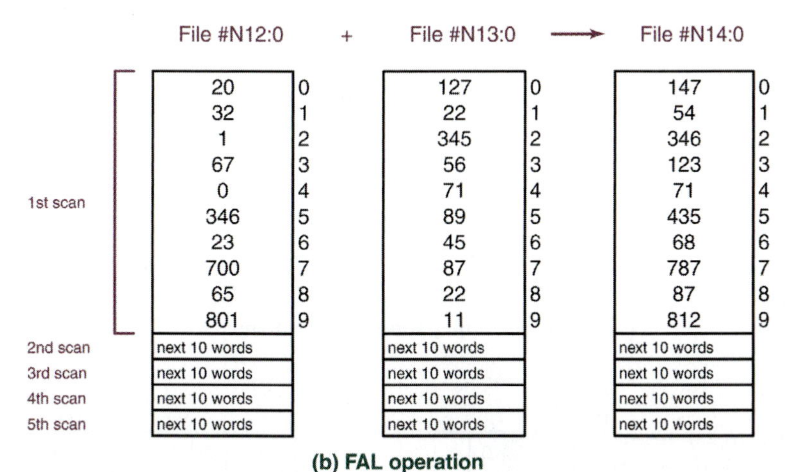

(b) FAL operation

50 integer values from registers starting at #N12:0 to the corresponding 50 values in registers starting at #N13:0 and then copies the 50 integer sums to registers starting at #N14:0. The # symbol indicates an indexed address. (Section 9-2-5). The # is placed before the word (#NR:D) address to indicate that the word is automatically indexed by 1 after each operation (N12:0, N12:1, N12:2, etc.). The operation is performed in five scans (10 words per scan) because the mode is numeric with 10 operations per scan. The expression block indicates the addition of the first 50 values starting at #N12:0 to the first 50 values starting at #N13:0, and the destination block indicates that the result of the addition should be placed into the first 50 positions in #N14:0. For example, using the data in Figure 10-8(b), N12:5 (346) is added to N13:5 (89) and the result is stored in N14:5 (435). After the transfer starts, input I:001/0 cannot stop the data transfer. Note that the control register done bit, R6.0/DN, is used to reset the pointer in the control register using the standard reset instruction. The example also illustrates that the R6.0/DN bit can be used with an XIO instruction (rung 0) to make the FAL instruction false after the 50 operations. Applications would use either the reset instruction or this inhibit bit for control of the FAL, but not both.

10-3-3 Shift Registers

The *shift register* is a logical operator that permits a register to move its contents to the right or to the left. The shift register, or *bit shift register* as it is often called, serially shifts a bit to the adjacent bit location through a register or group of registers. The concept of bit shifting to the right is illustrated in Figure 10-9. Note that at the most significant bit (MSB) end of the register there are two inputs—data and clock. Figure 10-9(a) shows the register before any shifting has taken place. Figure 10-9(b) shows the register after a data bit equal to 1 has been clocked into the register. Note that each bit within the register has been moved to the right, and the bit at the far right of the register has been lost. Figure 10-9(c) shows the register after a data bit equal to 0 has been clocked into the register and all bits have shifted to the right. Finally, shifting left is the same concept as shifting right but with the data and clock at the least significant bit (LSB) end of the register.

When a group of registers is involved in the shifting operation, they are connected as shown in Figure 10-10. Note that the clock is connected to all the registers. The data line is at the MSB end of the first register, the LSB of the first register is connected to the MSB of the second, and the LSB of the second is connected to the MSB of the third. The LSB of the third register is lost in this right-shifting operation.

Another shifting operation is rotate. *Rotate* is an operation of a shift register where the LSB connects to the MSB for a right rotate and the MSB connects to the LSB for a left rotate. Figure 10-11 illustrates the right rotate. Note that instead of losing the LSB, it replaces the MSB. Figure 10-11(a)

FIGURE 10-9: Shift registers.

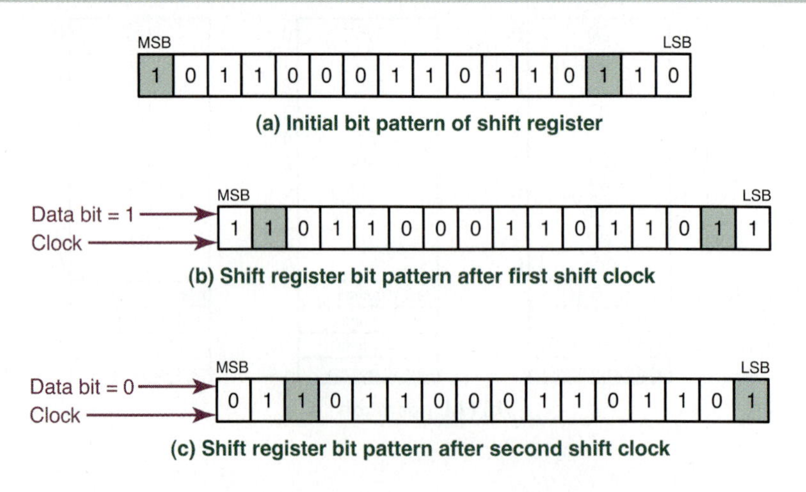

FIGURE 10-10: Multiple register shift registers.

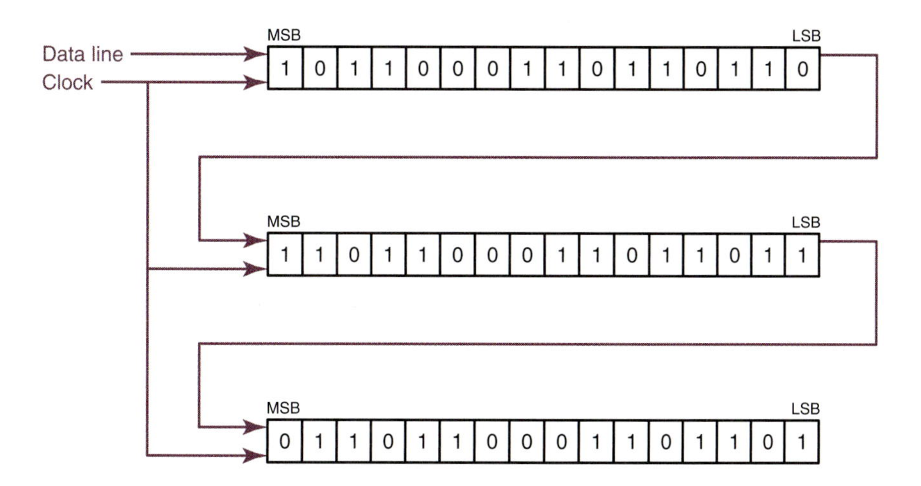

FIGURE 10-11: Right rotate shift registers.

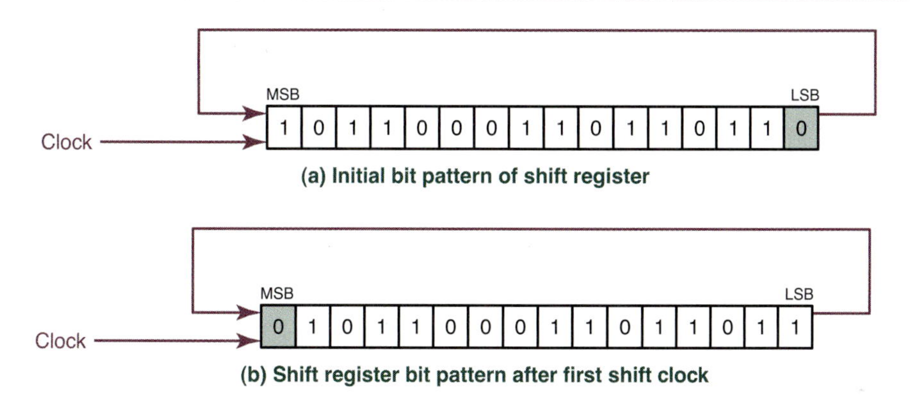

shows the register before the shift, and Figure 10-11(b) shows the register after the shift.

The following example illustrates how the shift register is used in an automated ice-cream cone conveyor system. In the process a PLC controls the following four actions:

1. Verify that the cone is not broken.
2. Put ice cream inside the cone.
3. Add sprinkles on the ice cream.
4. Add peanuts on the top.

If the cone is broken, we obviously don't want to add ice cream and the other items. Therefore, we have to track the bad cone down our process line so that we can tell the machine not to add each item. A sensor monitors cone quality, and if the sensor is on, then the cone is good; if it's off, then the cone is broken. Figure 10-12(a) depicts the automated process, and Figure 10-12(b) depicts the operation of the PLC shift register that drives the process. Refer to the figure as you read the process functional description.

The automated process in Figure 10-12 depicts only the first four steps of the entire operation. These four steps function as follows:

Step 1: Sensor S1 monitors the first cone. Since it's a good cone, a 1 is shifted into register bit 00.

Step 2: S1 senses the second cone, which is broken, and a 0 is shifted into bit 00. Note that the 1 from bit 00 is shifted into bit 01, thus turning on M1, which puts ice cream in the first cone.

Step 3: S1 senses the third cone, which is good, and a 1 is shifted into bit 00. Note that the 0 from bit 00 is shifted into bit 01, which inhibits M1 from putting ice cream into the broken

FIGURE 10-12: Shift register application.

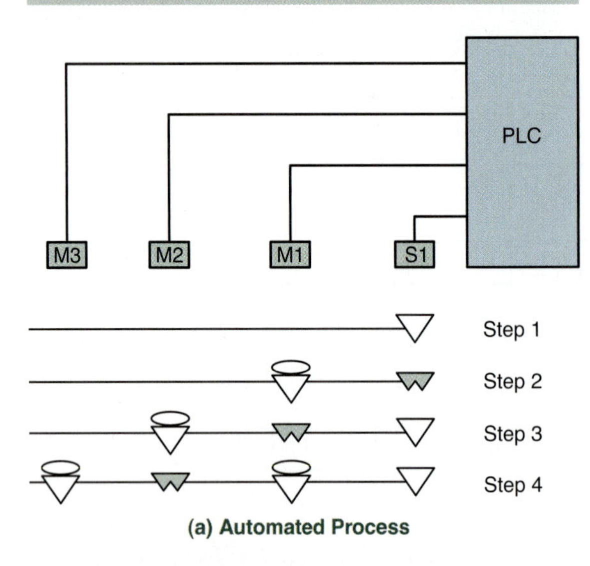

(a) Automated Process

03	02	01	00	Bits
0	0	0	1	Step 1
0	0	1	0	Step 2
0	1	0	1	Step 3
1	0	1	1	Step 4

(b) Shift register contents for steps 1 through 4

cone. Also, the 1 in bit 01 is shifted into bit 02, thus turning on M2, which adds sprinkles on the first cone.

Step 4: S1 senses the fourth cone, which is good, and a 1 is shifted into bit 00. Note that the 0 from bit 01 is shifted into bit 02, which inhibits M2 from adding sprinkles. Also, the 1 in bit 00 is shifted into bit 01, thus turning on M1, which puts ice cream into the third cone. Also, the 1 in bit 02 is shifted into bit 03, thus turning on M3, which adds peanuts to the first cone.

Review Figure 10-12 and reread the description until you are familiar with shift register operation in an automated process. Shift registers are most commonly used in processes that involve conveyor systems, labeling, or bottling applications.

Bit Shift Left (BSL) and Bit Shift Right (BSR) Instructions. These two shift register instructions (BSL and BSR) allow you to shift a specific number of bits either to the left or right. The positions of shifted bits in the shift register are identified by a word/bit address or a tag for a one-dimensional array. The BSL and the BSR instructions are shown in Figure 10-13(a) for PLC 5 and SLC 500 and in Figure 10-13(b) for ControlLogix. The terms within the instructions have the following meaning:

- **File** (PLC 5 and SLC 500): the address of the bit array that is to be shifted. The bit array is a contiguous collection of 16-bit words from one word to the file maximum.
- **Array** (ControlLogix): an array tag specifying the array to be shifted.
- **Control:** the address of the operational structure, which consists of the following:
  PLC 5 and SLC 500: the operational structure is as follows:

Word	Contents
Control	Bit 15 (EN) is a 1 when the instruction is enabled
	Bit 13 (DN) is a 1 when the bits have shifted
	Bit 11 (ER) is a 1 when the length is negative
	Bit 10 (UL) stores the state of the bit that was shifted out of the range of bits
Length	The bit length of the file
Position	The current position of the bit pointed to by the instruction

ControlLogix: the operational structure is as follows:

Mnemonic	Data Type	Description
.EN	BOOL	Indicates the instruction is enabled
.DN	BOOL	Indicates the bits have shifted
.UL	BOOL	Stores the state of the bit that was shifted out of the range of bits
.ER	BOOL	Indicates the length is negative
.LEN	DINT	Indicates the number of array bits to shift

FIGURE 10-13: BSL and BSR shift register instructions.

BSL
Bit Shift Left
File #B3:2
Control R6:1
Bit Address I:1/2
Length 12
(EN)
(DN)

BSR
Bit Shift Right
File #N7:4
Control R6:2
Bit Address I:1/2
Length 15
(EN)
(DN)

(a) BSL and BSR instructions for the Allen Bradley PLC 5 and SLC 500m

(b) BSL and BSR instructions for the Allen Bradley ControlLogix

- **Bit address** (PLC 5 and SLC 500): the address of the data that is to be shifted.
- **Source bit** (ControlLogix): a tag identifying the bit to be shifted.
- **Length:** the number of bits in the register or array to shift.

The operation of the BSL instruction is illustrated in Figure 10-14(a). When the rung goes from a false to true transition, the EN bit is set and one left shift is generated. Note that the input bit address is I:1/5, which is typically a discrete input field device, such as a sensor, and the input status, which is 0, is shifted into bit position B3:12/0. All bits of the file are shifted one position to the left. Note that B3:13/10 (also called address B3:12/26) is the last bit of the 27-bit file and is shifted into the unload bit R6:0/UL. The state of the UL bit prior to the left shift is lost.

The operation of the BSR instruction is illustrated in Figure 10-14(b). When the rung goes from a false to true transition, the EN bit is set and one right shift is generated. In this second example, the input bit address is I:1/6, and the status of the field device, which is 1, is shifted into bit position B3:31/4 (or B3:30/20), which is the first bit in the 21-bit file. All bits of the file are shifted one position to the right. Note that the last bit is B3:30/0, which is the last bit of the file, and it is shifted into the unload bit R6:1/UL. Likewise, the state of the UL bit prior to the right shift is lost.

Bit shift registers often are used to track parts moving on a conveyor or transfer system. In some applications the locations on the conveyor are associated with bits in the shift register. The locations on the conveyor that hold parts have a 1 in the shift register and those that are empty contain a 0. Operations on the parts as they move along the conveyor are triggered by the 1s present in the shift register.

Standard Ladder Logic for BSL and BSR Instructions. The standard ladder logic for shift register operation is illustrated in Figure 10-15. Note that in Figure 10-15(a) the BSL instruction has input logic that is used to make the rung active. Therefore, the input logic must transition from false to true for every shift required in the BSL instruction.

Figure 10-15(b) illustrates the ControlLogix version of the BSR instruction. Tags are used for all parameters and an array is used for the shift register. The BSR and BSL instructions have similar formats to those demonstrated in the standard ladder logic in Figure 10-15.

The standard ladder logic in Figure 10-15(c) illustrates one of the most common applications for the bit shift register instructions, the tracking of parts on a synchronous conveyor. The conveyor layout and ladder logic are illustrated, and a complete description is provided. Study the two illustrations as you read the description of the operation. Example 10-1 applies these concepts to an industrial automation problem.

FIGURE 10-14: Shift register operation.

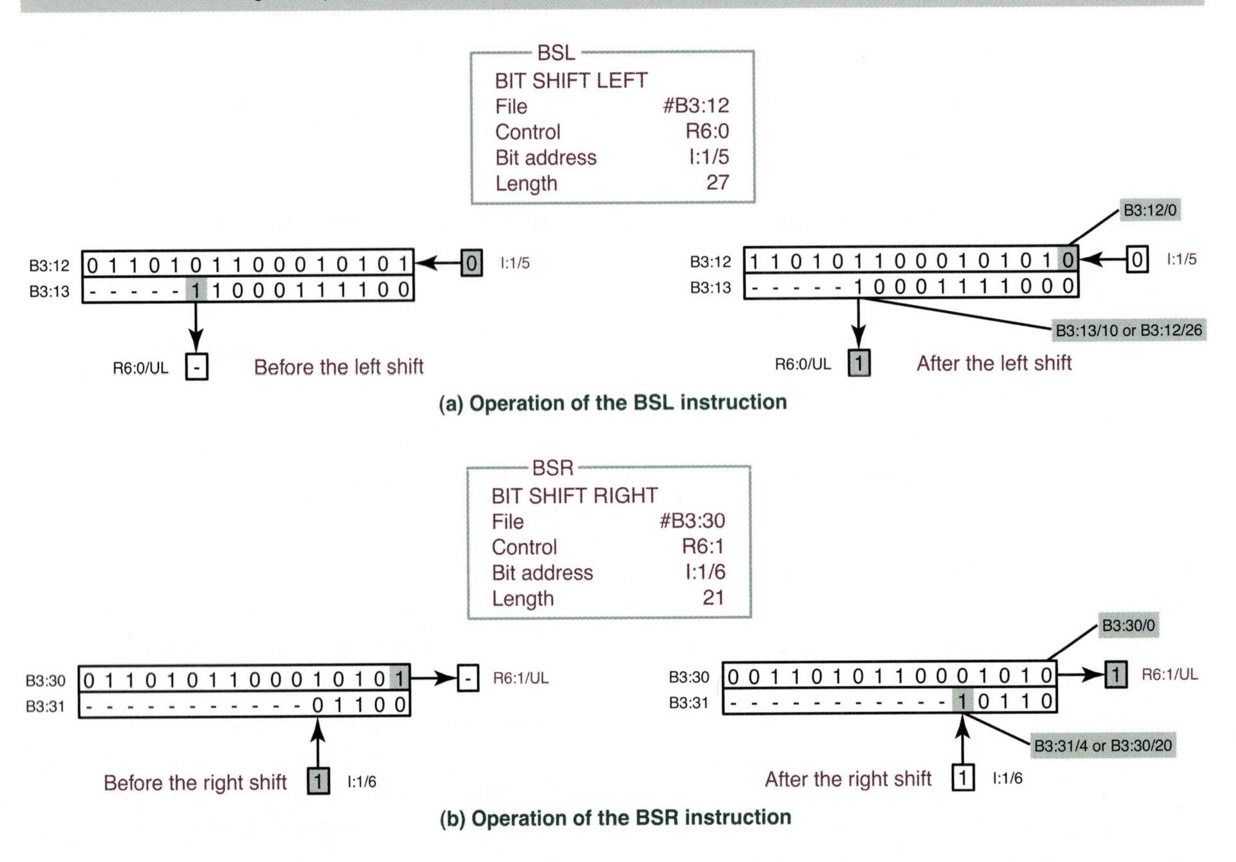

(a) Operation of the BSL instruction

(b) Operation of the BSR instruction

FIGURE 10-15: Standard ladder logic rungs for SLC 500 BSL and BSR instructions.

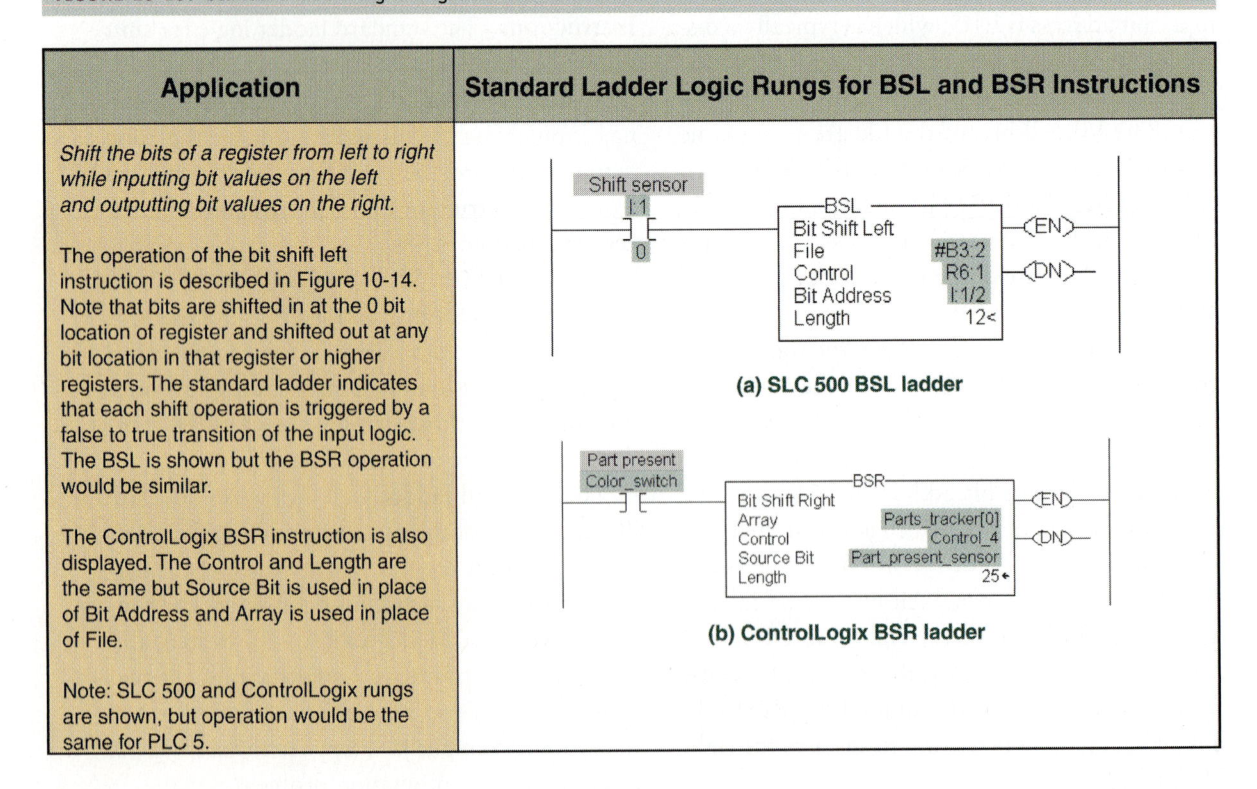

Application	Standard Ladder Logic Rungs for BSL and BSR Instructions
Shift the bits of a register from left to right while inputting bit values on the left and outputting bit values on the right. The operation of the bit shift left instruction is described in Figure 10-14. Note that bits are shifted in at the 0 bit location of register and shifted out at any bit location in that register or higher registers. The standard ladder indicates that each shift operation is triggered by a false to true transition of the input logic. The BSL is shown but the BSR operation would be similar. The ControlLogix BSR instruction is also displayed. The Control and Length are the same but Source Bit is used in place of Bit Address and Array is used in place of File. Note: SLC 500 and ControlLogix rungs are shown, but operation would be the same for PLC 5.	Shift sensor I:1 0 — BSL Bit Shift Left, File #B3:2, Control R6:1, Bit Address I:1/2, Length 12 — (EN) (DN) (a) SLC 500 BSL ladder Part present Color_switch — BSR Bit Shift Right, Array Parts_tracker[0], Control Control_4, Source Bit Part_present_sensor, Length 25 — (EN) (DN) (b) ControlLogix BSR ladder

FIGURE 10-15: (Continued).

Application	Standard Ladder Logic Rungs for BSL and BSR Instructions
Track parts on a synchronous conveyor so that specific parts can be selected for a quality check or corrective action. This standard ladder logic is used frequently to track parts that are equally spaced on a synchronous conveyor system. Every part over a portion of the conveyor is assigned a bit from the shift register. The conveyor and bits corresponding to conveyor positions are illustrated at the right. In this example the shift register length is 11 so 11 parts could be followed as they pass through this product window. For example, part quality could be checked by a sensor and passed to the shift register by input I:1/2. If that input is high (bad part) when the part is at position 0, then a 1 is entered into the shift register at the zero bit location. A second sensor at input I:1/12 shifts all register data left one bit as the parts move to the next location. When the bad part's bit is shifted 11 times, its bit value of 1 is in the 10th bit location in the register and the part is in the rejection location on the conveyor (one shift to bring it into the shift register from I:1/2 and 10 to move to location 10). The logic in the second rung on the right is used to eject the bad part that was detected at position 0. The part ejector can be located at any position along the conveyor that corresponds with a bit in the shift register. A second alternative is to place the ejector at location length plus one (11 + 1 = 12 for this ladder) and use the unload control bit (UL) to trigger the ejector. Several examples are provided so you can learn this important application for shift registers. The BSL is shown but the BSR could be used in this standard logic as well. Also, the SLC 500 rungs are shown, but the operation would be the same for PLC 5 and ControlLogix.	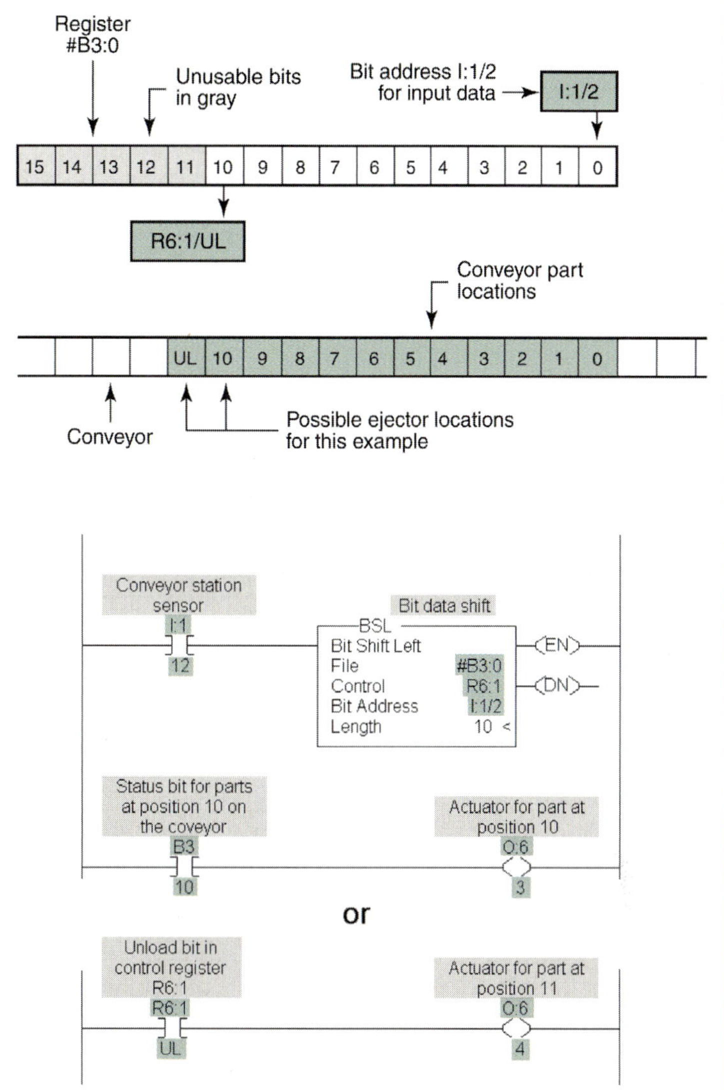 (c)

EXAMPLE 10-1

The packing testing system in Figure 10-16(a) uses a proximity sensor (S2) to look through a box of material and determine if a metal part has been packaged. Design a ladder program using a shift register instruction to eject any box from the line at the ejector station that does not have the part present. A photoelectric thru-beam sensor (S1) detects every passing box.

SOLUTION

The solution for the problem, shown in Figure 10-16(b), follows the standard ladder logic concept from Figure 10-15(c) but with a 0 indicating that the location has a problem. The proximity sensor, S2, detects if the metal part is in the box through the box

FIGURE 10-16: Quality system for packaging testing system.

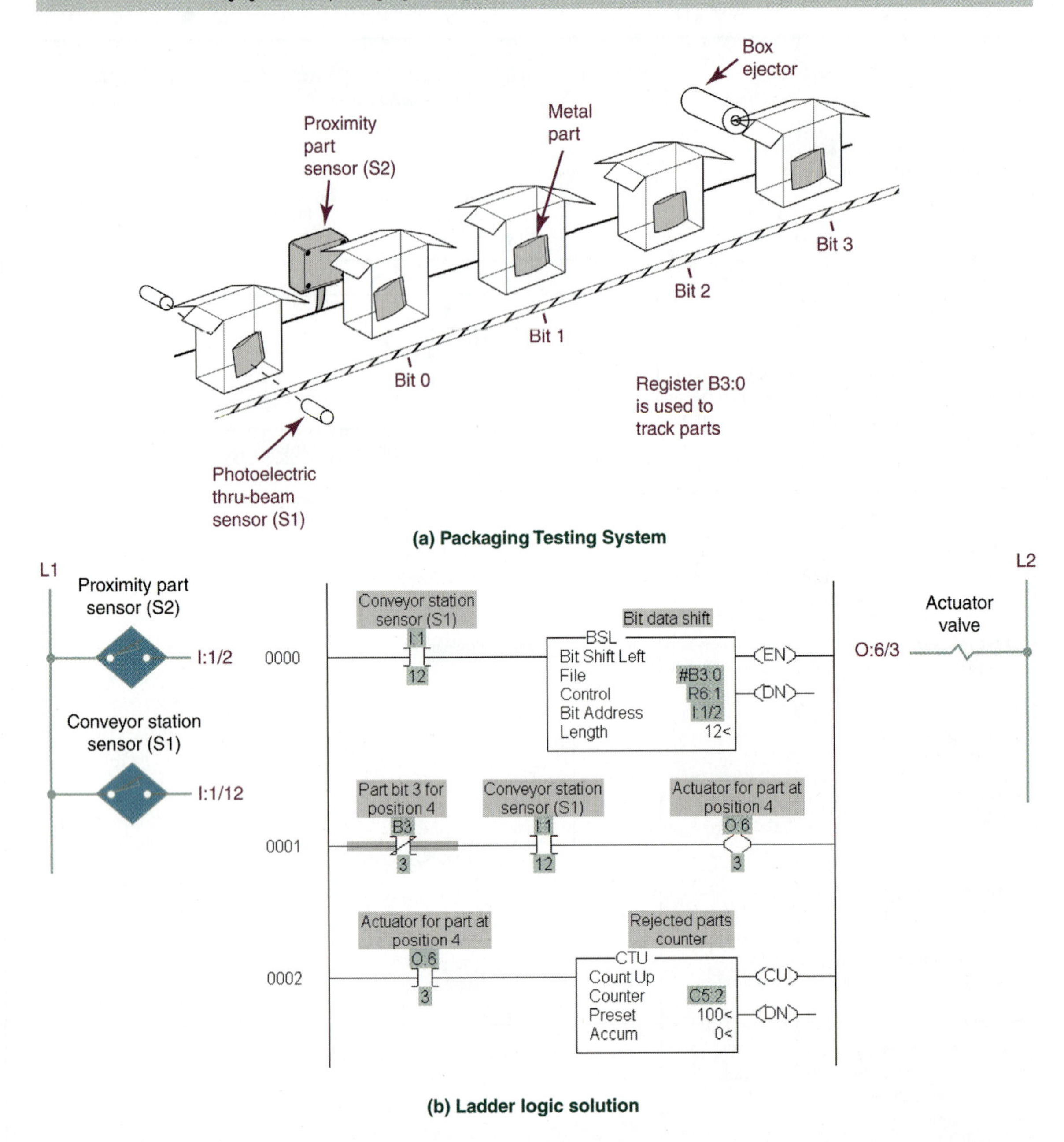

material. If it is present it places a 1 in the corresponding register bit; if it is absent it places a 0 in that register bit. All boxes without a part inside are ejected at position 4 based on the shift register bit 3 value. Good boxes have a 1 in their shift register location and bad boxes have a 0. Therefore, rung 1 uses an XIO instruction with the bit 3 address and an XIC instruction with the conveyor station sensor (S1) address. When a part is in the ejection position, sensor S1 is true and rung 1's logic state is controlled by the XIO (-] / [-) instruction and the value of bit 3 (B3:0/3 address). So if bit 3 has a 1 (good box) the rung is false and the actuator is off, but if bit 3 has a 0 (bad box) the rung is true and the actuator ejects the box. Rung 2 has a counter that counts the number of bad boxes. Figure 10-17(b) shows the register after one shift with a bad box detected and Figure 10-17(c) shows the register after four shifts with bit 3 holding the 0 for the bad box. Figure 10-17 shows the shift registers for four consecutive shifts after a bad box is detected.

10-3-4 First In, First Out (FIFO) Function

The *first in, first out (FIFO)* function is a word shift operation, which is similar to the bit shift operation that was discussed in Section 10-3-3. Word shifting provides a simple method of loading data into a file, usually called the stack. As in the bit shift register, two inputs are used in the FIFO or word shift operation—load and unload, which operate independently of each other. Figure 10-18 illustrates this operation. Note that the *load input* enters the data word into the next available word at the bottom of the stack from a source location, and the *unload input* removes the data word from the top of the stack into a destination location. Figure 10-18(a) and (b) shows the status of the stack before and after the load; Figure 10-18(c) and (d) shows the status of the stack before and after the unload. The data is shifted into the stack in the order in which it is received. Therefore, the first word shifted in will be the first word shifted out. For a practical example, think of a queue at a grocery store checkout or at a movie theater ticket window.

FIGURE 10-17: Bit movement in B3:0 register for good and bad boxes.

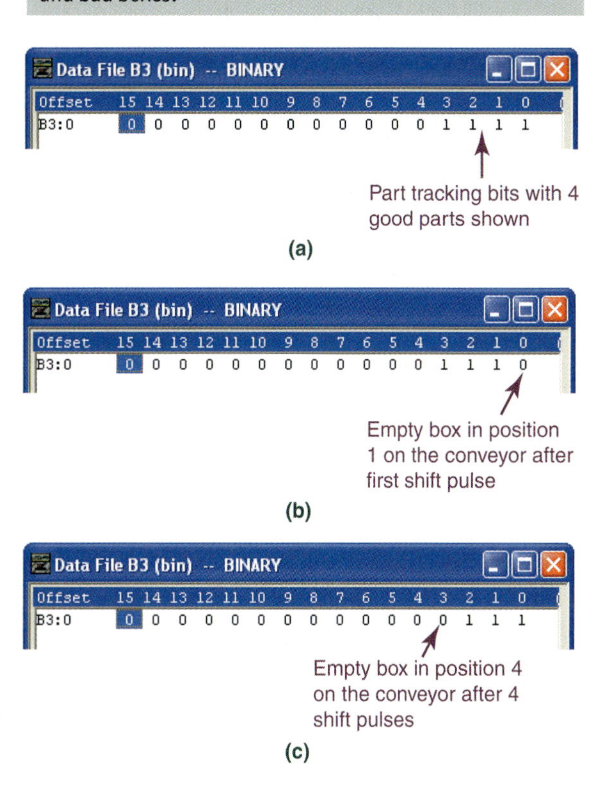

FIGURE 10-18: Word stacks FIFO load and unload operations.

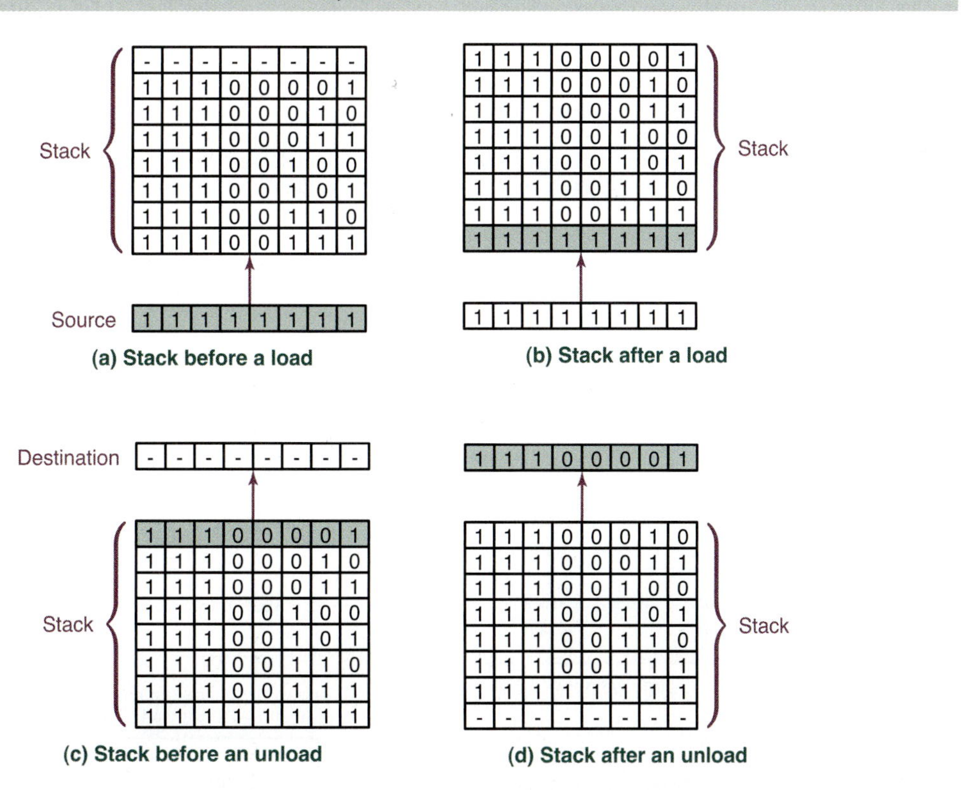

The first person in line is serviced first and leaves, then the next person, and so forth. The people enter and leave the queue at different times, but the first one in line is the first one out.

The FIFO function has many applications. One application of the FIFO function is a process where different parts are pulled from inventory to be used in production. Each part is assigned a unique code, which is loaded into a FIFO stack, and parts are pulled in the order prescribed by the stack. This type of inventory control ensures that oldest part in the inventory is used first. Another use of the FIFO function is the storage and retrieval of data that is synchronized to an external movement of parts on a conveyor or transfer machine. The FIFO function is also useful when keeping values that are obtained from a process in a *moving window* situation such as the process shown in Figure 10-19. Figure 10-19(a) illustrates the temperature profile as a function of time, and Figure 10-19(b) shows that the desired window from time t_0 to t_1 is kept in a FIFO stack. In this application the temperature values are sampled at regular time intervals and the stack load sample values are listed from the top to the bottom, with the first value in at the top of the stack. The data is unloaded from the top registers, so the first in is the first out. The stack has sufficient words to hold the samples over the specified time period t_1 to t_0 Thus, the stack always contains the most recent t_0 through t_1 temperature values.

FIFO Instructions. To program the FIFO function a pair of instructions is used—*FIFO load* (FFL) and *FIFO unload* (FFU). The FFL instruction loads data into a file from a source, and the FFU instruction unloads data from a file to a destination. The FFL and the FFU instructions are shown in Figure 10-20(a). Note that for the FFL and the FFU instructions, the FIFO addresses the same file, the control address is the same, and the length and position are the same. These common addresses/numbers are required so that the instructions operate on the same FIFO file. The terms within these instructions have the following meaning:

- **Source:** the address location of the data word that is the next value loaded onto the FIFO stack by the FFL instruction at the location contained in the pointer.
- **Destination:** the address location of the next data word that is unloaded from the FIFO stack by the FFU instruction.
- **FIFO:** the *indexed* address of the stack or an array tag. The same FIFO address is used for the FFL and FFU instructions.

FIGURE 10-19: Temperature monitoring process using a FIFO stack.

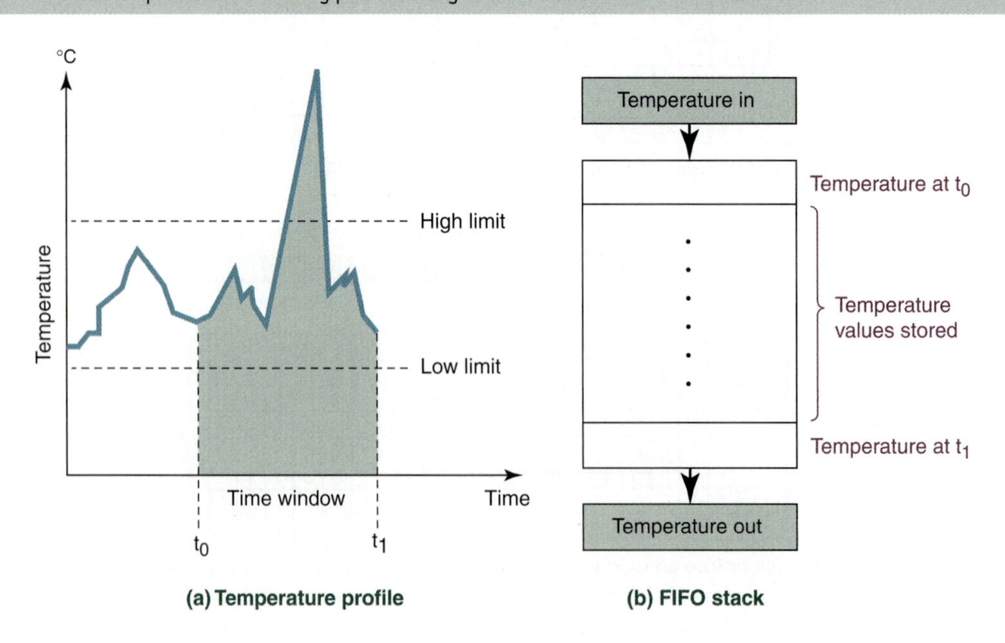

FIGURE 10-20: FIFO load and unload example.

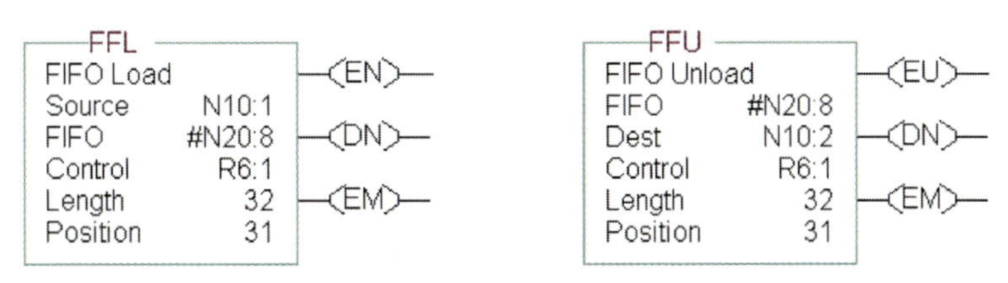

(a) FFL and FFU instructions for the Allen Bradley PLC 5 and SLC 500

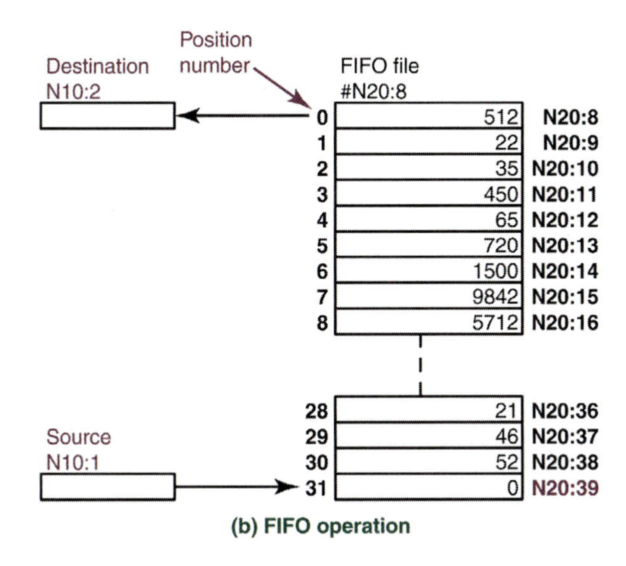

(b) FIFO operation

- **Control:** the address of the command structure for the operation.

PLC 5 and SLC 500 have the following control structure:

Bit	Function	Description
15	EN load enable	Indicates if the FFL instruction is enabled
14	EU unload enable	Indicates if the FFU instruction is enabled
13	DN done bit	Indicates the FIFO stack is full. Note that when the DN bit is set, the transfer of data from the source to the stack is inhibited.
12	EM empty bit	Indicates the FIFO stack is empty, that is, all the data has been transferred to the destination. Note that if the FFU instruction is activated after the EM bit is set, then zeros are transferred to the destination.

Logix systems have the following operational structure:

Mnemonic	Data Type	Description
.EN	BOOL	Indicates if the FFL instruction is enabled
.EU	BOOL	Indicates if the FFU instruction is enabled
.DN	BOOL	Indicates the bits have shifted
.EM	BOOL	Indicates that the FIFO stack is empty
.ER	BOOL	Indicates length is negative
.LEN	DINT	Indicates number of elements to shift
.POS	DINT	Identifies where the next load/unload will occur

- **Length:** the number of words/elements in the stack.
- **Position:** the stack pointer, which is the stack address where the next source word/element will be loaded.

FIFO Load (FFL) Instruction. The PLC 5 and SLC 500 FIFO operation is depicted in Figure 10-20(b). Note that the length of the stack is 32 from N20:8 through N20:39. With each activation of the FFL instruction, the data from source N10:1 is loaded into the FIFO stack at the position indicated by the position parameter in the instruction. The position indicator is normally set to 0 for the start of the FIFO instruction, and the indicator value increases by 1 with every entry of data (false to true transition of FFL instruction) until it advances to the last file dictated by the length parameter. In this example, on the first activation of the FFL instruction the source data is loaded into N20:8, on the next activation it is loaded into N20:9, and so forth. When the stack is full, the DN (done) bit is set to 1. Note in Figure 10-20 that the position parameter is 31, so the stack has just one open element. If that element is loaded the done bit is set.

FIFO Unload (FFU) Instruction. When rung conditions change from false to true, the FFU enable bit (EU) is set. This unloads the contents of the element at stack position 0 into the destination, N10:2 in Figure 10-20. All data in the stack is shifted one element toward the position zero, and the stack element that contained the last data value is zeroed. The position value then decrements to point at this recently zero stack element. The FFU instruction unloads an element at each false to true transition of the rung, until the stack is empty. Applying this to the example in Figure 10-20, with each activation of the FFU instruction, the data from the starting address (N7:8) is loaded into destination N10:2, and all data are shifted one position toward the starting address. When all the data have been shifted out of the stack the EM (empty) bit is set to 1.

10-3-5 Last In, First Out (LIFO) Function

The *LIFO* function shifts words as does the FIFO function, but reverses the order of the data, outputting the last word received first and the first word received last. As a result, words can be added to the LIFO stack without disturbing the words already loaded on the stack. The load and unload of the LIFO stack operates similarly to that of the load and unload of the FIFO stack, only the last word in the LIFO stack is the first word that is unloaded from the stack. Figure 10-21(a) and (b) shows the status of the stack before and after

FIGURE 10-21: Word stacks LIFO load and unload operations.

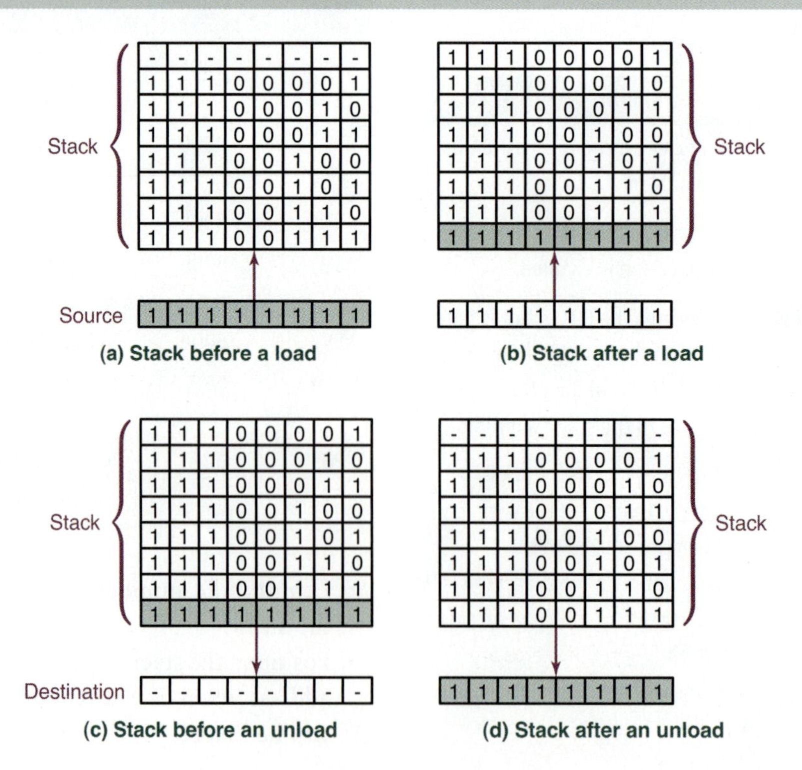

the load. Figure 10-21(c) and (d) shows the status of the stack before and after an unload. Typically the words are loaded onto the stack, then unloaded at a different time. For a practical example, picture a magazine parts feeder used in an automated process that works as follows. The first part placed into the feeder compresses the spring that is attached to the bottom of the feeder. When the second part is placed on top of the first, it further compresses the spring as the first part is pushed deeper into the magazine's case. This continues until the magazine is full. The first part in is at the bottom of the magazine and the last part in is at the top. This is just like the bullet magazines used on automatic pistols and rifles. In this situation the last part in is the first part removed. As in a FIFO operation, the load and unload of the stack occurs at different times in the LIFO operation.

LIFO Instructions. The LIFO operation is depicted in Figure 10-22 has two instructions: *LIFO load* (LFL) and *LIFO unload* (LFU). Note that the length of the stack is 32 from N20:8 through N20:39, with position numbers 0 through 31. With each activation of the LFL instruction, the data from source N10:1 is loaded into the LIFO stack, and the position is advanced by 1 toward the highest stack element position. With the stack empty and the LFL instruction activated, the position pointer is at 0 and the first entry goes into the stack at position 0. The position pointer is incremented in preparation for the next LFL instruction activation. At any point, an LFU instruction could become active. If this occurs the last data entered is pulled from the stack (position pointer = position pointer − 1). After the data is removed the position pointer is de-cremented by 1 to be ready for the next LFL or LFU instruction. When the stack is full, the DN bit is set to 1. With each activation of the LFU instruction, the last data entered is pulled and placed into destination file N10:2. When all the data have been shifted out of the stack the EM bit is set to 1.

The LIFO example in Figure 10-22 shows the stack with one open element. If an LFL instruction is executed, then position location 31 will be filled with the data from the source register, N10:1. However, if an LFU instruction is executed next, then the position pointer is changed to 30 (31-1) to point to last data entered so it can be moved to the destination register, N10:2.

FIGURE 10-22: LIFO load and unload example.

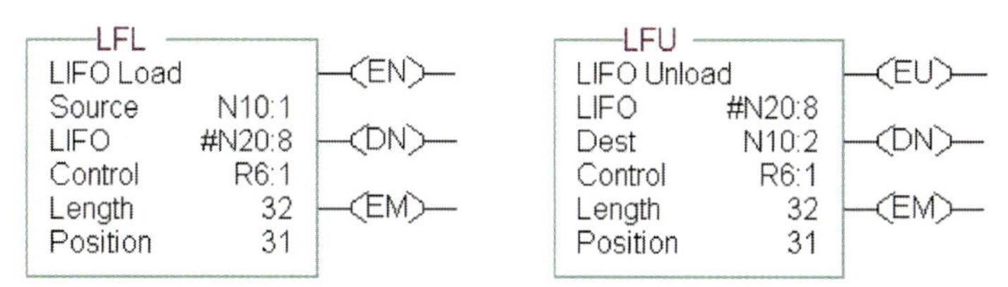

(a) LFL and LFU instructions for the Allen Bradley PLC 5 and SLC 500

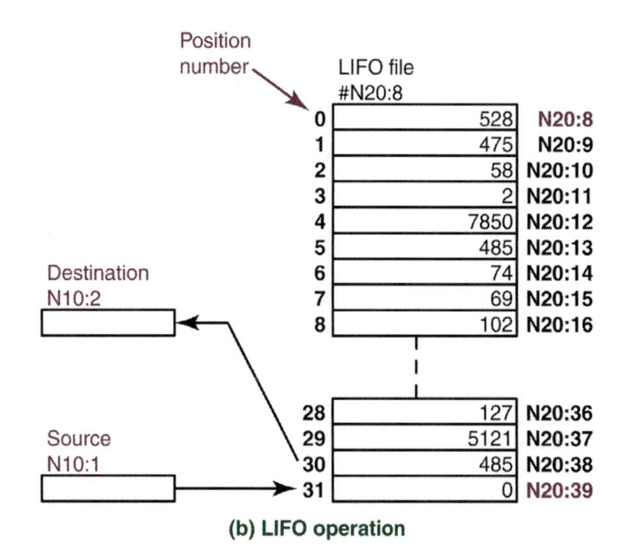

(b) LIFO operation

To program the LIFO function LFL and LFU instructions are used. The LFL loads data into a stack file from a source, and the LFU unloads data from a stack file to a destination. The LFL and the LFU instructions are shown in Figure 10-22(a). Note that for the LFL and the LFU instructions, the LIFO addresses the same file, the control address is the same, and the length and position are the same. These common addresses/numbers are required so that the instructions operate on the same LIFO file. The terms within these instructions have the same meaning as the FIFO instructions described in Section 10-3-4.

ControlLogix FIFO and LIFO Instructions. The ControlLogix PLC operation with FIFO and LIFO functions like the corresponding instructions in the PLC 5 and SLC 500 except

FIGURE 10-23: FIFO and LIFO instruction illustration.

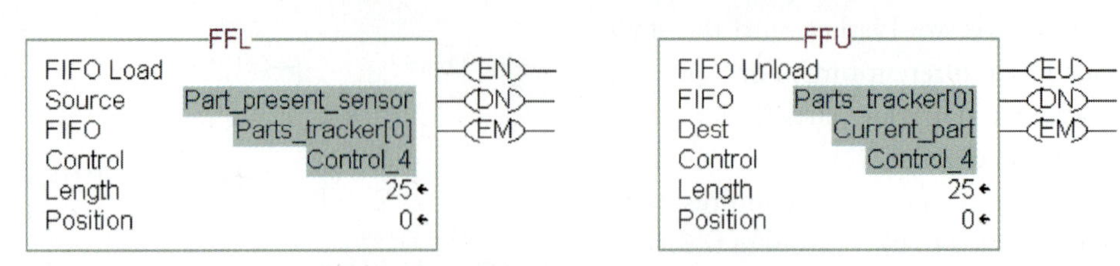

(a) ControlLogix FFL and FFU instructions

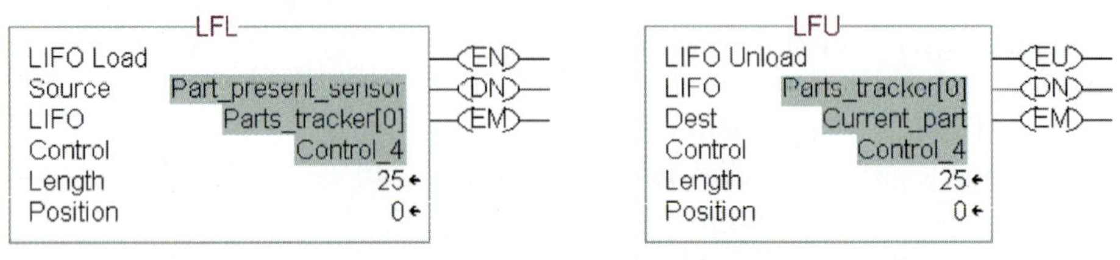

(b) ControlLogix LFL and LFU instructions

that arrays are used for the sequential files and tags are used to hold the parameter data. The instructions for ControlLogix are shown in Figure 10-23. Note that the instructions have the same parameters as the PLC 5 and SLC 500, but that you must use tags and arrays in the parameter definitions.

10-3-6 Copy and Fill Instructions

The copy (COP) and fill (FLL) instructions are word transfer instructions without any control bits. Therefore, bits are not available to indicate that the transfer has occurred and parameters, like length, are not present to indicate how many files to transfer. They are, however, useful for moving a number of data files (COP instruction) from one location to another or moving a single data file (FLL instruction) to a specified number of other data files.

COP Instruction. In the *copy instruction* the destination file type determines the number of words that are transferred from the source to destination. For example, if the destination file type is a counter and the source file type is an integer, three integer words are transferred for the counter file type since the counter is a three-word instruction. This provision is necessary because a control register is not present to specify the length for the transfer. However, if you need an enable bit, program an output instruction (OTE) in parallel with the COP instruction using an internal bit as the enable bit indicator. The PLC 5 and SLC 500 instruction format is illustrated in Figure 10-24(a) and the process for the transfer is illustrated in Figure 10-24(b).

Enter the following parameters when programming this instruction:

- *Source* is the address of the file you want to copy. You must use the file indicator (#) in the address so that indexing of the address occurs.
- *Destination* is the starting address where the instruction stores the copy. You must use the file indicator (#) in the address so that indexing of the address occurs.
- *Length* is the number of elements in the file you want to copy where maximum length is based on the destination file type. If the destination file type is three words per element (timer or counter), you can specify a maximum length of 42. If the destination file type is one word per element, you can specify a maximum length of 128 words.

All elements are copied from the source file into the destination file each time the instruction is executed. Elements are copied in ascending order.

FIGURE 10-24: Copy instruction illustration.

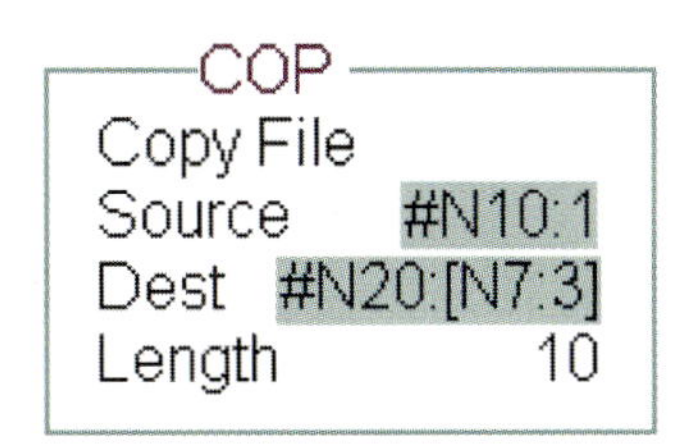

(a) PLC 5 and SLC 500 copy instruction

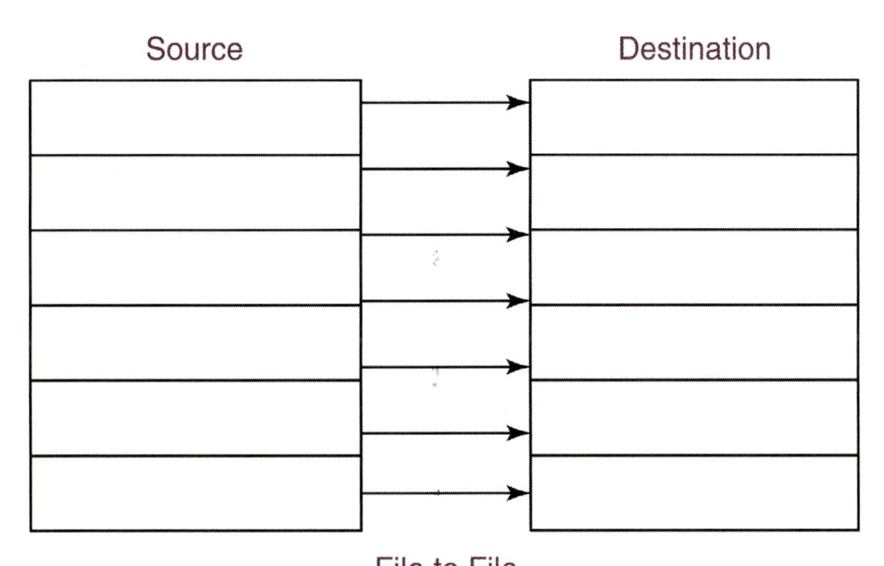

(b) Copy instruction data transfer

Fill Instruction. The fill (FLL) instruction loads a sequence of file elements or registers with either a program constant or a value from another file register address. The PLC 5 and SLC 500 instruction format is shown in Figure 10-25(a) and the process for the transfer is illustrated in Figure 10-25(b).

The following parameters are required for programming this instruction:

- *Source* is the program constant or element address and does not need a pound sign (#) file indicator because it is not indexed.
- Floating point and string values are supported for higher versions of the SLC systems.
- *Destination* is the destination starting address of the file to be filled and must be indexed, so the indicator (#) in the address is required.
- *Length* is the number of elements in the destination file that will be filled.
- The source value, typically a constant, fills the specified destination file each scan the rung is true. Elements are filled in ascending order.

ControlLogix Copy and Fill Instructions. The ControlLogix PLC operation with copy and fill instructions is like the PLC 5 and SLC 500 except that arrays are used for the sequential files and tags are used to hold the parameter data. In addition, the FLL instruction will not write past the end of an array. If the length is greater than the total number of elements in the destination array, the FLL instruction stops at the end of the array, but no major fault is generated.

For best results, the source and destination should be the same data type. If you mix data types for the source and destination, the data type

FIGURE 10-25: Fill instruction illustration.

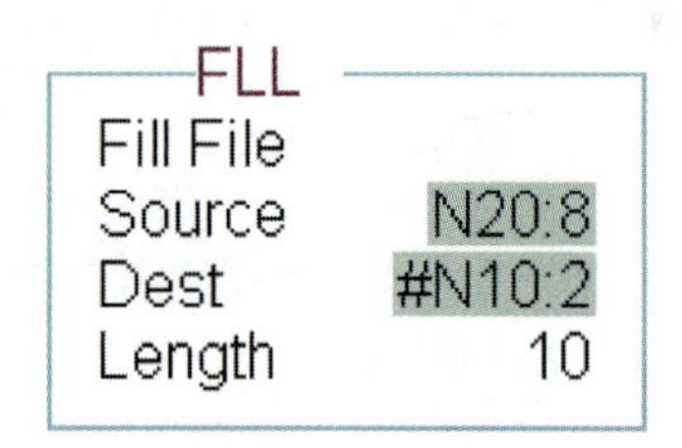

(a) PLC 5 and SLC 500 fill instruction

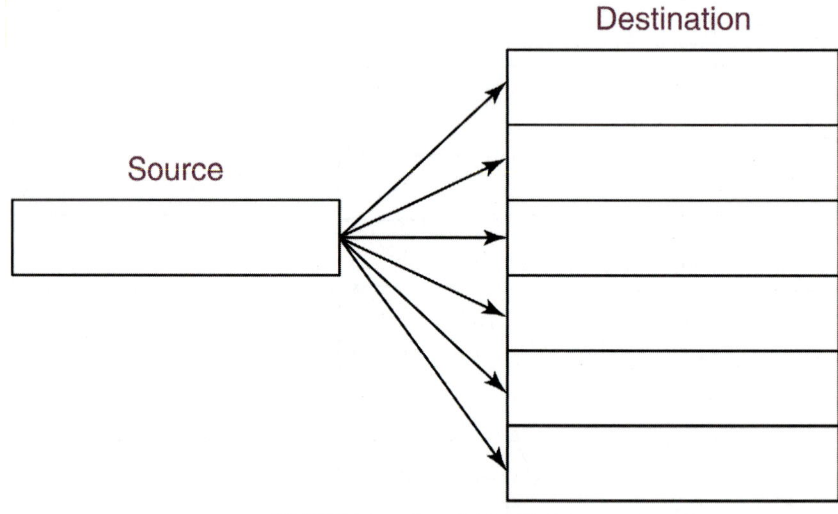

(b) Fill instruction data transfer

of the destination elements are used and they are filled with converted source values.

Standard Ladder Logic Rungs for PLC 5 and SLC 500 COP and FLL Instructions. The standard ladder logic used for COP and FLL are illustrated in Figure 10-26. The input logic is not unique but the addressing of the source and destination need some explanation. The source is a standard indexed address with the number symbol (#) placed before the starting address location. The source address starts at location N10:1 and increments to N10:10 due to the value 10 in the length parameter.

The destination is an addressing mode not discussed previously; it is called *indexed indirect addressing*. A combination of indexed and indirect, it permits the starting register location for the indexed address (#N20:) to be defined by the value in another address (N7:3). First the indirect address is resolved, and then the indexed final address is implemented. For example, if N7:3 has a value of 15, then the indexed address is #N20:15.

This addressing scheme is ideal for data lookup tables. Assume that three sets of five operational parameters needed for a process are saved in address locations starting at address N10:0. The three sets of data would look like that displayed in Table 10-3.

The starting register for each set is 0, 5, and 10 for sets 1, 2, and 3, respectively. Assume that the source address is #N10:[N7:3] and the destination is #N20:5. If 0, 5, or 10 is placed into N7:3, then the needed block of parameters is moved to the process locations starting at #N20:5. After an operator enters the set number into N7:3, the COP instruction picks the corresponding data and copies it to the process program file locations.

FIGURE 10-26: Standard ladder logic rungs for SLC 500 COP and FLL instructions.

Application	Standard Ladder Logic Rungs for COP and FLL Instructions
Used in applications that require the movement of data words between registers and arrays without the support of a control data file to indicate the status of the operation. *Use COP to copy blocks of data from one set of registers to a second set in one PLC scan. Use FLL to copy a program constant or register contents from a single register or tag address to multiple register addresses or an array.* The copy instruction transfers blocks of data words from the indexed source address to the indexed destination address using the length parameter to indicate how many words to copy. Note that the destination address is an indexed indirect addressing mode. In this mode, the indirect address is solved first (the value in N7:3 is to identify the file number). For example, if N7:3 holds a value of 20, then the indirect address is N20:20. That value is indexed by one for ten locations, so N20:20 through N20:29 for the indexed part of the address. This is an an example where indexed addressing with an offset is often used. The fill instruction fills a block of words with the value from a single word. Indexed addressing is used in the example. Note: SLC 500 rungs are shown, but the operation would be the same for PLC 5 and ControlLogix PLCs. Also, the ControlLogix processors have a synchronous copy instruction, CPS.	Copy data bit I:1 12 COP Copy File Source #N10:1 Dest #N20:[N7:3] Length 10 (a) Fill data bit B3 10 FLL Fill File Source N20:8 Dest #N10:2 Length 10 (b)

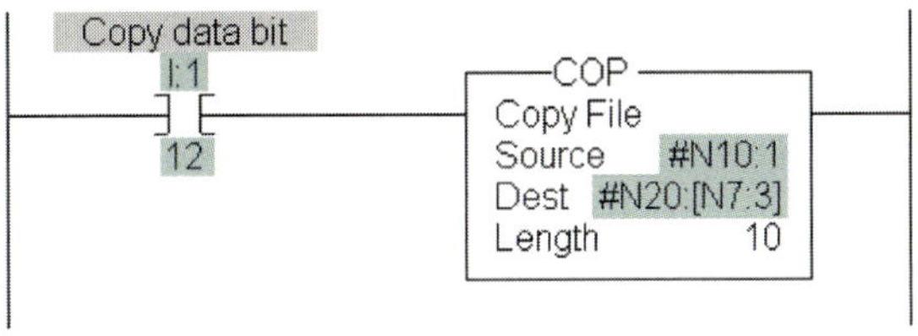

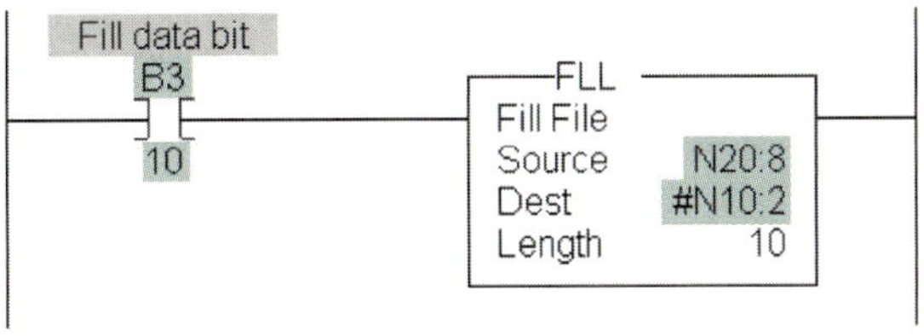

TABLE 10-3 Data for indexed indirect address.

Set	Address	Data
1	N10:0	1030
1	N10:1	58
1	N10:2	24526
1	N10:3	2345
1	N10:4	1450
2	N10:5	2030
2	N10:6	75
2	N10:7	31526
2	N10:8	2300
2	N10:9	2000
3	N10:10	3030
3	N10:11	158
3	N10:12	33526
3	N10:13	2745
3	N10:14	3450

10-4 EMPIRICAL DESIGN PROCESS WITH BIT AND WORD OPERATION INSTRUCTIONS

The empirical design process, introduced in Section 3-11-4, is an organized approach to the design of PLC ladder logic programs. However, the term *empirical* implies that some degree of trial and error is present. The instructions introduced in this chapter are used for special problem applications. Therefore, the use of these instructions in the design of ladder logic focuses on recognizing the types of problems where these instructions are required. Suggested application areas for the bit and word manipulation instructions follow.

AND, OR, and XOR Instructions: Used in applications that require:

- Specific bits in a word to be masked or changed to 0 while the rest of the word remains the same. An AND instruction with 1s for the mask or pass bits is used for this type of masking.

- Specific bits in a word to be changed to 1s while the rest of the word remains the same. An OR instruction with 1s for the mask bits is used for this type of masking.
- A comparison of two register values to identify the bits that are the same in both registers. XOR is used for this type of comparison.

BSL and BSR Instructions: Used in applications that require:

- A sequence of bits in a word to be sequentially shifted to a new location within the word.
- Tracking of a condition or part from one location on a conveyor to another location. In most cases, the parts and conveyor movement must be synchronous.

FAL, FFL, FFU, LFL, and LFU Instructions: Used in applications that require the movement of data words between registers and arrays with the support of a control data file to indicate the status of the operation.

- Use FAL for applications where movement of data must be completed in one scan or in successive scans. Note that this instruction is not available on the SLC 500 PLC.
- Use FFL and FFU for applications where movement of data is performed one data word at a time when the first data entered is the first data that must be pulled out.
- Use LFL and LFU for applications where movement of data is performed one data word at a time when the last data entered is the first data that must be pulled out.

COP and FLL Instructions: Used in applications that require the movement of data words between registers and arrays without the support of a control data file to indicate the status of the operation.

- Use COP to copy blocks of data from one set of registers to another in one PLC scan.
- Use FLL to copy a program constant or register contents from a single register or tag address to multiple register addresses or an array.

EXAMPLE 10-2

The color sensor in Example 6-11 has an output timing diagram in Figure 6-14.

In Example 6-11 the color value is valid for 0.5 seconds; the trigger starts 5 milliseconds after the color value is valid and is true for 0.7 seconds. The color sensor is modified as follows: the color value is valid for 6 seconds; the trigger pulse starts 5 milliseconds after the color value is valid and is true for 7 seconds.

- Review Example 6-11 problem statement and the solution in Figure 6-14 and 6-15.
- Modify the color sensor sampling design by using the new sensor operation values and an FAL instruction and ControlLogix software to copy 10 values of color data into an array when the color_switch is active.

Solution

The three-rung solution is illustrated in Figure 10-27.

- Rung 0 is a 0.5-second regenerative timer activated by the color_switch tag. The timer done bit is true for one scan and false for 0.5 seconds. The timer starts when the color_switch tag is true.
- Rung 1 is triggered for one scan every half second. When the FAL is true it copies the data from Color_data to the array Color_sample[Control_4.POS]. Since the FAL mode parameter is set for incremental (INC), one data value is copied every time the FAL is true. The copies will occur until 10 array elements are full (length equals 10). The XIO instruction, with a Control_4.DN address, makes the input logic false after the tenth sample is copied (FAL is done). Note that the array Color_sample[Control_4.POS] is a tag with an indirect addressing mode. With every true transition of the FAL the position value is incremented, which gives the array an indirect operation.
- Rung 2 resets the FAL instruction (position is reset to 0) when the color_switch tag is false.

Compare the solution in Figure 6-15 with that in Figure 10-27. Note how much more efficient the FAL solution is.

FIGURE 10-27: Solution for Example 10-2.

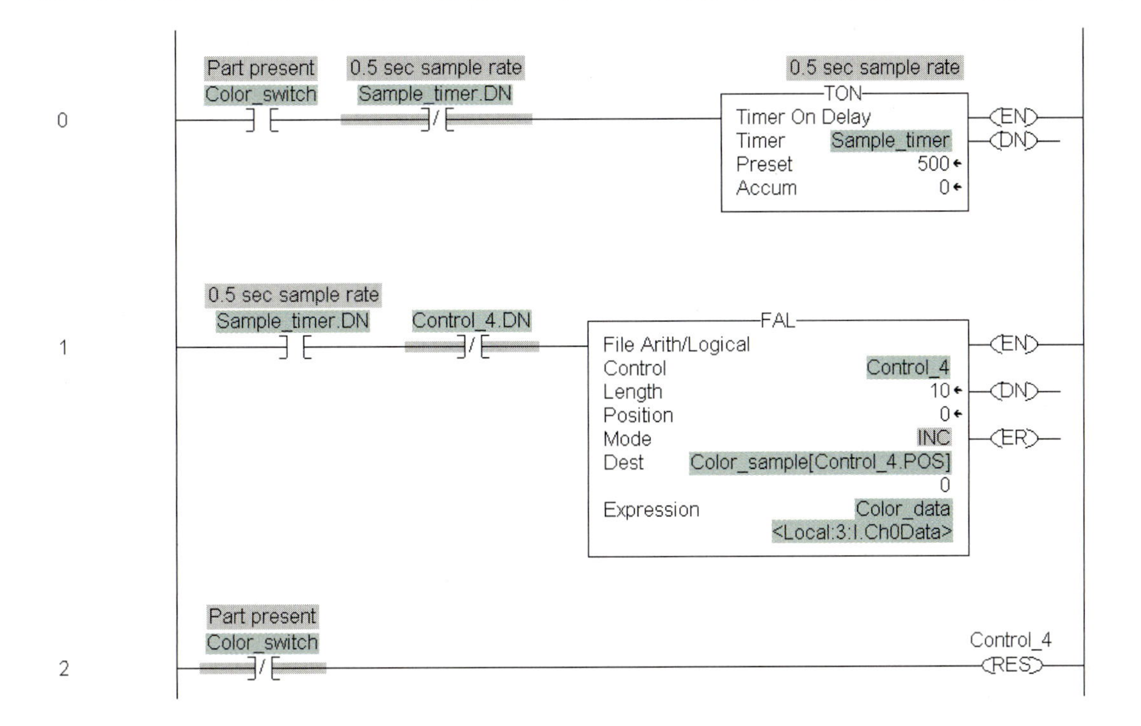

EXAMPLE 10-3

A machine makes 12 bits of data available at an output one bit at a time and provides a contact closure to indicate when each data bit is present. Design a ladder that captures the serial data in a register starting at B3:10 and transfers all bit values to a register at N10:0 when all 12 bits are captured.

Solution

The three-rung solution is illustrated in Figure 10-28.

- Rung 0 uses a BSR instruction to move each bit into register B3:10. The data valid instruction, I:1/2, makes the BSR output true when the machine indicates that the data at I:1/3 is valid.
- Rung 1 has a COP instruction to copy the data from register B3:10 to N10:0. The input is true when the control data word's done bit is true. This indicates that all 12 bits have been stored.
- Rung 2 has a reset instruction activated by the same control done bit to reset the control data registers for another cycle.

FIGURE 10-28: Solution for Example 10-3.

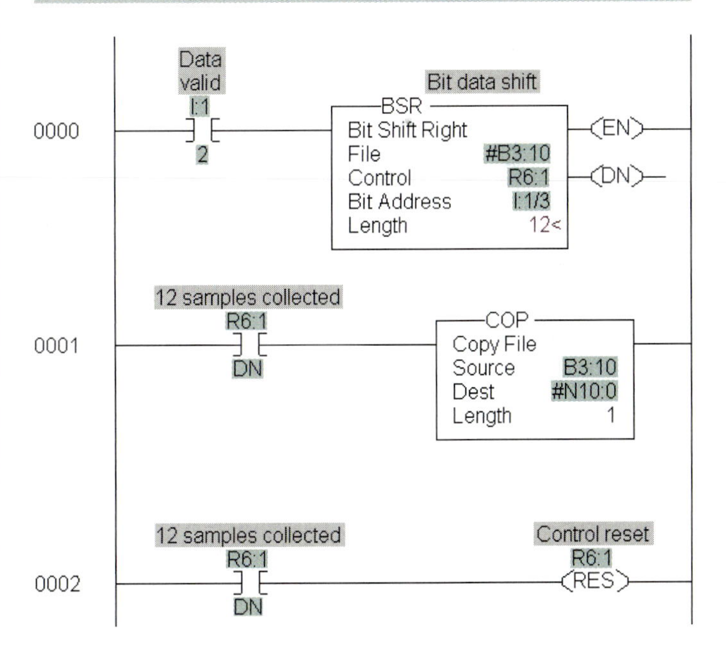

EXAMPLE 10-4

Example 9-1 modified the process tank design problem in Section 7-6-3 (Figures 7-17 through 7-21) so that a receipt number could be entered by an operator to move the receipt values using indexed addressing. Review that solution in Figure 9-12.

Modify the process tank design again using an algorithm to calculate the indirect address offset from the receipt number, and use the COP instruction with indexed indirect addressing to load the number of half-gallon quantities of materials 1 and 2 into the process ladder logic.

SOLUTION

The five-rung solution is illustrated in Figure 10-29.

Rung 0 loads the receipt number from a thumbwheel switch using the BCD to decimal conversion instruction to put the receipt number into N7:15.

Rungs 1 through 3 convert the receipt number into an offset that points to the starting address for each set of receipt parameters. The algorithm is:

$$\text{Starting register number} = [(\text{RN} \times \text{NOI}) - \text{NOI}] + \text{SA}$$

where RN is the receipt number, NOI is the number of ingredients, and SA is the starting address register number for the ingredients block of memory. In this example the starting address for the ingredients is N7:10, and there are three receipts with two ingredients in each. For example, if receipt 2 is desired, then:

$$\text{Starting register number} = [(2 \times 2) - 2] + 10 = 12$$

Note that the offset value of 12 is stored in N7:20, which indicates that receipt 2 data starts at N7:12.

Rung 4 copies the indexed receipt data from #N7:[N7:20] where #N7:[N7:20] is #N7:12 since N7:20 holds the value 12. The receipt data is copied to #N10:10 for use by the process.

This approach has many advantages. Most important is that the number of rungs of ladder logic does not increase as the number of receipts and ingredients gets larger

10-5 TROUBLESHOOTING DATA HANDLING INSTRUCTIONS AND SHIFT REGISTERS IN LADDER LOGIC

The data handling and shift register instructions use indexed, indirect, and indexed indirect addressing modes plus arrays to perform the data movement required in the instructions. As a result, all of the major causes of problems with addressing modes and array operations described in the last chapter would apply here as well.

If problems occur in rungs with data handling and shift register instructions using indirect or indexed addressing modes, use the following guidelines when troubleshooting these rungs.

- Verify that pointers do not point to addresses outside the memory boundaries.
- Use the temporary end (TND) instruction to stop scanning at points in the ladder where addressing indexes can be verified and data transferred checked.
- Use single-step options when available to scan one rung at a time to analyze how the data is changing and where the flow is incorrect.

These guidelines could be listed in the empirical design areas as well. When designs include data handling and shift register instructions with

FIGURE 10-29: Solution for Example 10-4.

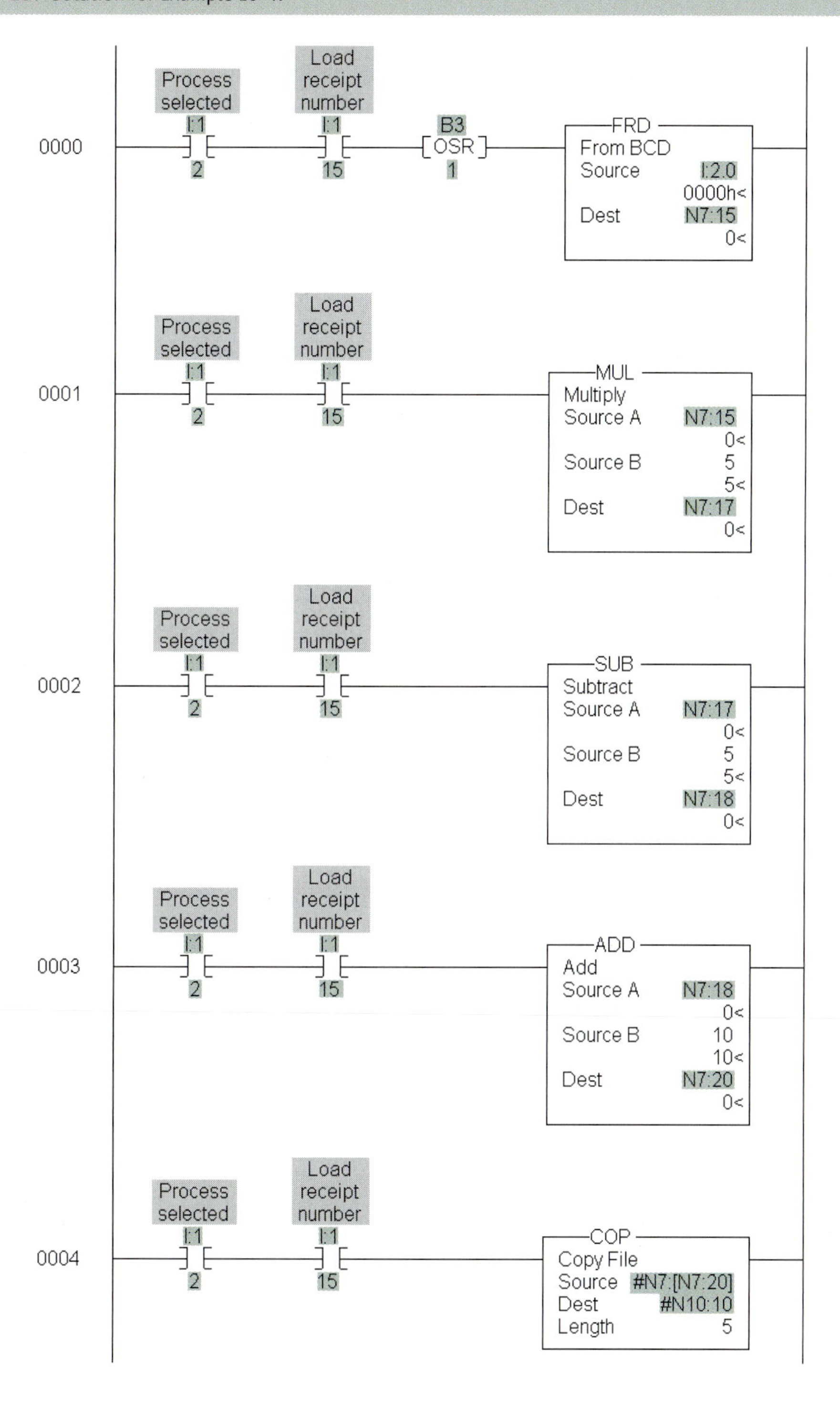

large data blocks or arrays, start with just a small data set. For example, if you can make the ladder operate with a FIFO instruction with a length of 3 you can make it work with a length of 100. So start with a small set and work up to the larger value.

10-6 LOCATION OF THE INSTRUCTIONS

The location of instructions from this chapter in the Allen-Bradley programming software is indicated in Figure 10-30.

FIGURE 10-30: Location of instructions described in this chapter.

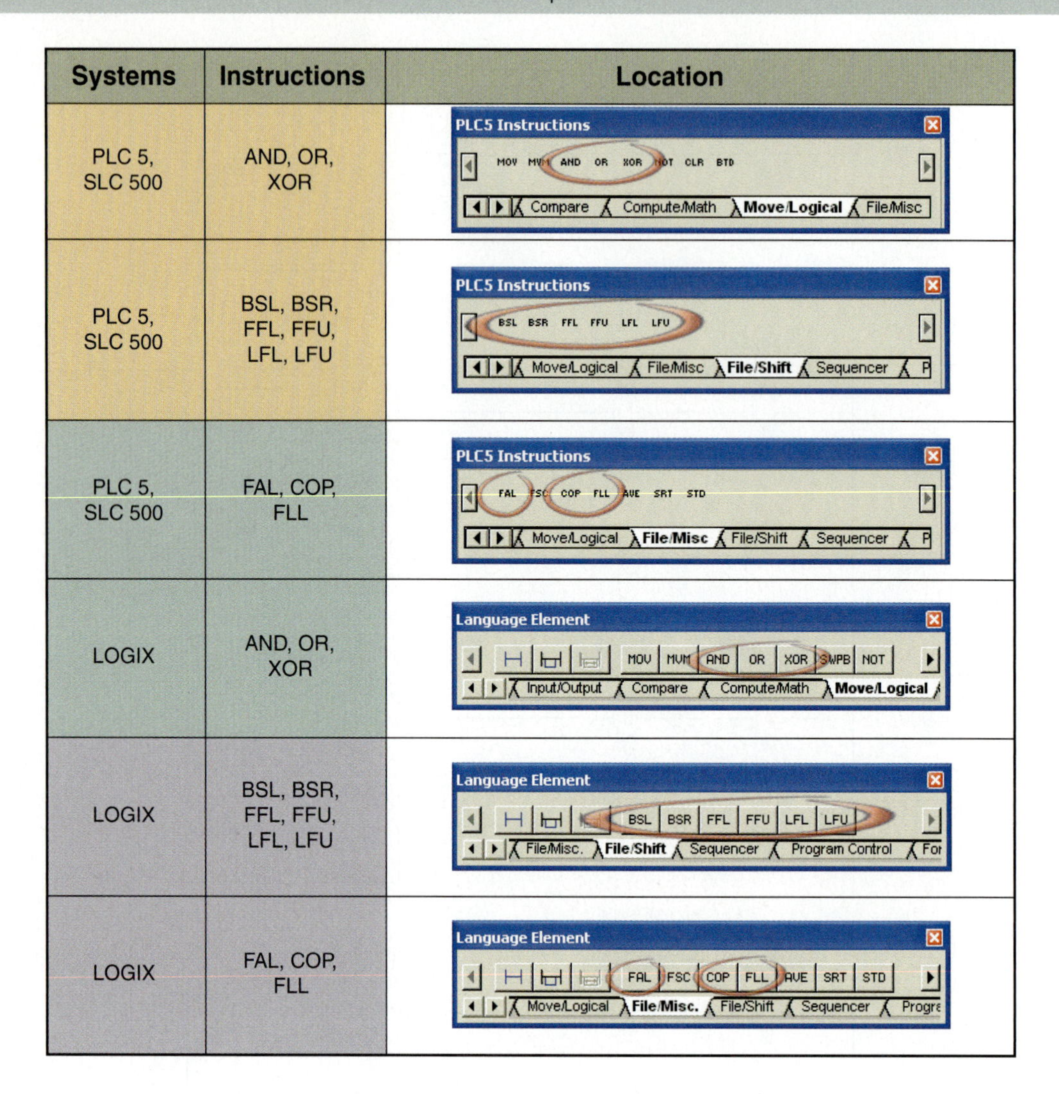

Systems	Instructions	Location
PLC 5, SLC 500	AND, OR, XOR	PLC5 Instructions: MOV MVM AND OR XOR NOT CLR BTD — Compare / Compute/Math / Move/Logical / File/Misc
PLC 5, SLC 500	BSL, BSR, FFL, FFU, LFL, LFU	PLC5 Instructions: BSL BSR FFL FFU LFL LFU — Move/Logical / File/Misc / File/Shift / Sequencer / P
PLC 5, SLC 500	FAL, COP, FLL	PLC5 Instructions: FAL FSC COP FLL AVE SRT STD — Move/Logical / File/Misc / File/Shift / Sequencer / P
LOGIX	AND, OR, XOR	Language Element: MOV MVM AND OR XOR SWPB NOT — Input/Output / Compare / Compute/Math / Move/Logical
LOGIX	BSL, BSR, FFL, FFU, LFL, LFU	Language Element: BSL BSR FFL FFU LFL LFU — File/Misc. / File/Shift / Sequencer / Program Control / For
LOGIX	FAL, COP, FLL	Language Element: FAL FSC COP FLL AVE SRT STD — Move/Logical / File/Misc. / File/Shift / Sequencer / Progra

Chapter 10 questions and problems begin on page 580.

CHAPTER

11

PLC Sequencer Functions

11-1 GOALS AND OBJECTIVES

There are three principal goals of this chapter. The first goal is to provide the student with an introduction to the operation and function of electromechanical sequencing devices. The second goal is to introduce the basic PLC sequencer function and timing. The third goal is to show how the programmable sequencer instructions are applied to specific PLCs that are used in industrial automation systems.

After completing this chapter you should be able to:

- Identity and describe the operation and function of electromechanical sequencing devices.
- Describe the basic PLC sequencer function.
- Develop and describe the ladder diagram for the operation of a PLC sequencer with timing.
- Describe the technique of cascading sequencers.
- Develop ladder logic solutions using sequencer instructions for Allen-Bradley PLC 5, SLC 500, and ControlLogix PLCs.
- Troubleshoot ladder rungs with sequencer instructions.

11-2 ELECTROMECHANICAL SEQUENCING

Sequencing refers to a predetermined step-by-step process that accomplishes a specific task. A familiar application of sequencing events to achieve a specific task is the operation of a dishwasher or a clothes washer. Each washer sequences through multiple steps that perform a fixed routine of actions to complete its specific task. Simple sequencing is performed by electromechanical *drum switches*. A three-position, six-electrical-terminal drum switch is shown in Figure 11-1. The operator handle on the switch shown in Figure 11-1(a) is in the off position. Rotating the handle to the left creates the reverse contact condition, and moving it to the right creates the forward contact condition. These contact conditions are depicted in Figure 11-1(b).

Drum switches are frequently used for reversing the direction of AC and DC motors, or for controlling multispeed motors with two to four speed settings. Applications include directional control of motors for overhead hoists and doors, and speed control of motors used in blowers, mixers, pumps, and special machine tools. Figure 11-2 illustrates

FIGURE 11-1: Drum switch and contacts.

(a)

Handle end

Reverse	Off	Forward
1—2	1 2	1 2
3—4	3 4	3 4
5—6	5 6	5—6

Internal switching

(b)

Courtesy of Square D/Schneider Electric.

FIGURE 11-2: Drum switch switching three-phase voltage to a motor.

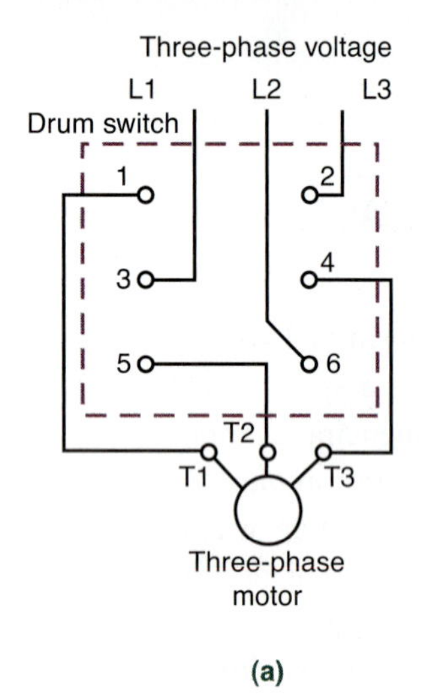

(a)

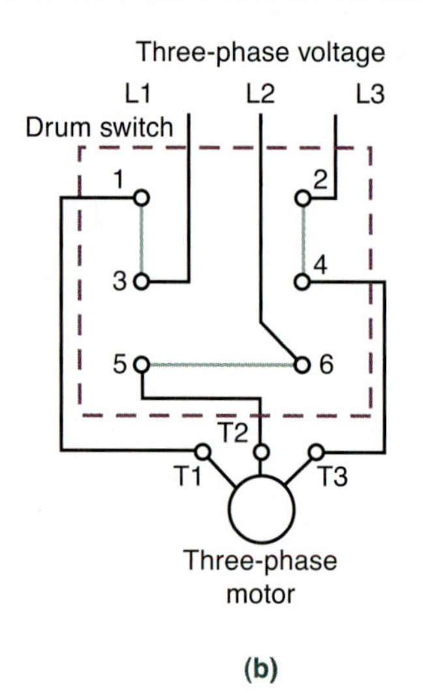

(b)

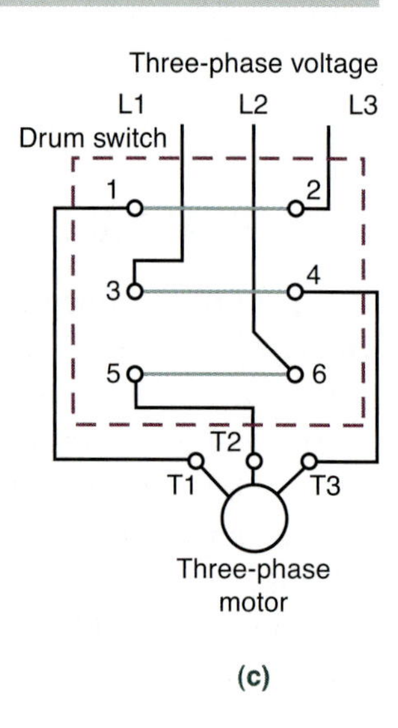

(c)

Rehg and Sartori, Industrial Electronics, *1st edition. © 2006, reprinted by permission of Pearson Education, Inc., Upper Saddle River, NJ.*

how a drum switch can be used to operate a three-phase motor in a rotational switching application. Note that a three-phase motor must have any two of the three-phase input lines switched to reverse the direction of rotation of the motor. The three-phase motor and input voltages are wired to the switch, as illustrated in Figure 11-2(a). When the switch handle is in the forward position, the switch makes the connections illustrated in Figure 11-2(b) as follows: L1 is connected to T1, L2 to T2, and L3 to T3. When the switch is in the reverse position [Figure 11-2(c)] lines L1 and L3 are reversed, so L1 is connected to T3 and L3 is connected to T1.

Another electromechanical sequencing operation is the lighting in a movie theater, where a motor-driven timing cam typically controls switches that set the theater illumination. This cam-operated sequencer is commonly known as a timer. The lighting sequence is shown in Table 11-1. Note that the four steps in the lighting sequence are as follows:

1. The five banks of lights are on high brightness, H, for the cleaning crew.
2. The five banks of lights are on medium brightness, M, during the pre-show time, allowing patrons to find seats.
3. The five banks of lights are on low brightness, L, for the showing of the previews.
4. The five banks of lights are off for the showing of the movie.

TABLE 11-1 Lighting sequence for theater illumination.

	Brightness level				
Step	Bank 1	Bank 2	Bank 3	Bank 4	Bank 5
1	H	H	H	H	H
2	M	M	M	M	M
3	L	L	L	L	L
4	Off	Off	Off	Off	Off

TABLE 11-2 Sequencer with seven steps and four outputs.

Step	01	02	03	04
1	0	1	0	0
2	0	0	0	1
3	1	1	0	0
4	1	1	0	1
5	0	1	1	1
6	1	0	1	1
7	1	0	0	0

11-3 BASIC PLC SEQUENCER FUNCTION

The basic *PLC sequencer function* provides the capability to program many steps and provide multiple outputs. Table 11-2 illustrates an output sequencer function that has seven steps and four outputs. Note that the steps are labeled 1 through 7, with each step having a specific pattern. For example, step 3's pattern is 1100. The outputs are labeled O1, O2, O3, and O4, and an output is on when a 1 is in its column and off when a 0 is in its column. The sequencer steps through the seven patterns, turning on the outputs as dictated by the pattern. Other sequencers change the output sequence bit pattern after comparing an input word with bit patterns stored in sequencer registers. Several types of sequencer instructions are discussed in the following sections.

11-4 ALLEN-BRADLEY SEQUENCER INSTRUCTIONS

The Allen-Bradley PLCs provide a group of sequencer instructions that can be used individually or in pairs. The operation of these instructions and specific PLC applications are addressed in the subsequent subsections. The instructions are as follows:

- **Sequencer output (SQO):** an output instruction that uses a file or an array to control various output devices (PLC 5, SLC 500, and ControlLogix).
- **Sequencer input (SQI):** an input instruction that compares bits from an input file or array to corresponding bits from a source address. The instruction is true if all pairs of bits are the same (PLC 5 and ControlLogix).
- **Sequencer compare (SQC):** an output instruction that compares bits from an input source file to corresponding bits from data words in a sequence file. If all pairs of bits are the same, then a bit in the control register is set to 1 (PLC 5 and SLC 500).
- **Sequencer load (SQL):** an output instruction that functions like a word-to-file or file-to-file transfer (PLC 5, SLC 500, and ControlLogix).

11-4-1 PLC 5 and SLC 500 SQO and SQC Sequencer Instruction Structure

The SQO and SQC instructions are shown in Figure 11-3; the terms *file*, *mask*, *control*, *length*, and *position* are used in all instructions. The terms *destination* and *source* are used in SQO and SQC sequencers, respectively. These parameters used in sequencer instructions are defined as follows:

- **File:** is the starting address for the registers in the sequencer file, and you must use the indexed file indicator (#) for this address.
- **Mask:** is the code or bit pattern through which the sequencer instruction moves source data to the destination address. In the mask bit pattern, a 1 bit passes values (1 or 0) from source to destination and a 0 mask bit inhibits the data flow and puts a 0 in the bit location in the destination. Use a mask register or file name if you want to change the mask pattern under program control. Place an *h* behind the parameter to indicate that the mask is *hexadecimal* number notation or a *B* to indicate *binary* notation. *Decimal* notation is entered without any indicator.

FIGURE 11-3: Sequencer parameters for SLC 500.

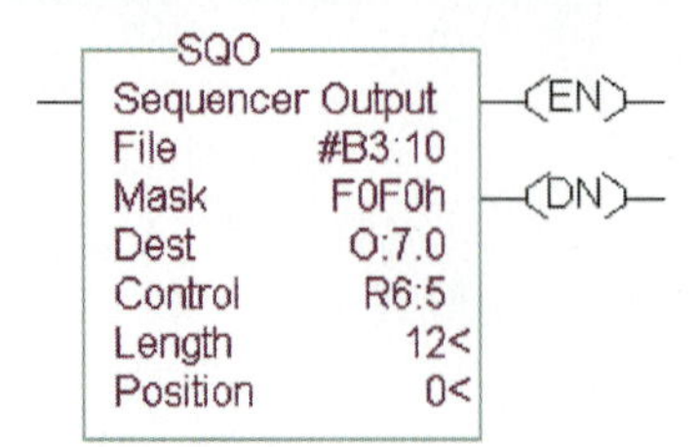

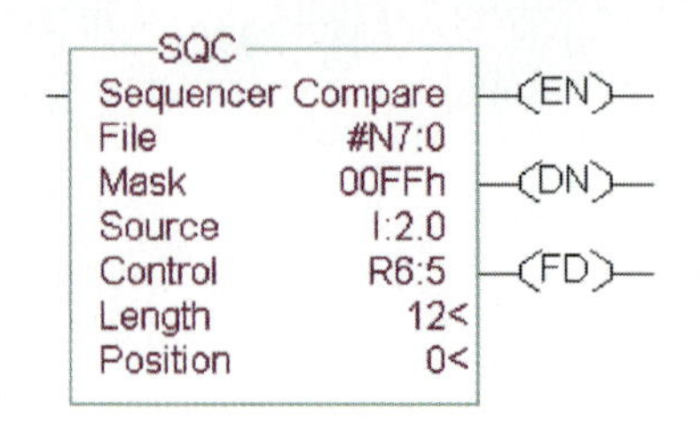

- **Source:** is the address of the input word or file for an SQC from which the instruction obtains data for comparison to its sequencer file.
- **Destination:** is the address of the output word or file for an SQO to which the instruction moves data from its sequencer file.
- **Control:** is the address that contains parameters with control information for the instruction and discrete outputs to indicate sequencer instruction results and status. The general control register file is described in Chapter 10 in Tables 10-1 and 10-2. Review that information if necessary.

  The control file address is in the control area R of processor memory and the default file is 6. The first of three control words have status bits including:

  - **Bit 8:** the *found bit* (FD) is only used for the SQC instruction. When the status of all non-masked bits in the source address match those of the corresponding reference file word, the FD bit is set. This bit is assessed each scan of the SQC instruction while the rung is true.
  - **Bit 11:** the *error bit* (ER) indicates a negative position value or a negative or zero length value.
  - **Bit 13:** the *done bit* (DN) is set by the SQO or SQC instruction after it has operated on the last word in the sequencer file. It is reset on the next false to true rung transition after the rung goes false.
  - **Bit 15:** *the enable bit* (EN) is true when the rung goes from false to true and is used to indicate that the SQO or SQC instruction is enabled or active.
  - Note in Figure 11-3 that the control bits present for each sequencer instruction are shown on the right side of the instruction box. If they are true the bit initials are highlighted.
- **Length:** is the number of steps of the sequencer file starting at position 1. Since the file parameter starts at position 0, the number of words in the file is length plus one.
- **Position:** is the word location or step in the sequencer file from/to which the instruction moves data. The position pointer is incremented by 1 before action is taken by the sequencer instructions. Therefore, if the pointer is at 0, then after the first false to true transition of the sequencer rung the pointer is a 1 and points at the first word in the sequencer file. A position value that points past the end of the programmed file causes a runtime major error.

11-4-2 PLC 5 and SLC 500 SQO Instruction Operation

The *SQO instruction* is available in all Allen-Bradley PLCs, and the operation in the PLC 5 and SLC 500 processors is the same. The operation of the SQO during the first scan is different from the operation for all subsequent scans.

SQO Instruction Operation during the First Scan. When the processor is first changed from the program to the run mode, the operation of the SQO instruction is dependent upon the condition of the rung (true or false) at the time of the first scan. The two conditions are:

- If the rung is *true* at the time of the first scan and the position parameter is 0, then the data in the memory location at position 0 is transferred through the mask to the destination register and the position pointer is incremented by 1.

- If the rung is *false* at the time of the first scan and the position parameter is 0, then the data in the memory location at position 0 is NOT transferred to the destination register and the position pointer is NOT incremented.

SQO Instruction Operation after the First Scan. After completion of the first scan, the SQO instruction operates as follows:

- When the rung has a false to true transition the position pointer is incremented by 1 and the data in the memory location at the new position is moved through the mask to the destination file.
- If the position memory location data changes while the rung is true, the SQO continues to update the destination file on every scan.
- When the rung goes false the output is no longer updated, and the SQO instruction waits for the next false to true rung transition.

After the first scan, every time the SQO rung is true the instruction increments the position number by 1 and moves the sequencer file for that position number through a mask to the destination file. Bits passing through a 1 in the mask are placed in the destination register (these are the *unmasked* bits) and those encountering a 0 in the mask are blocked (these are the *masked* bits). For example, the first time the input logic is true after a false first scan, the SQO instruction increments the position pointer from 0 to 1 and moves the data word from the first position in the file through the mask to the destination. On each subsequent true logic condition, data is moved from the second position, from the third position, and so forth until all word locations of the file have been moved to the destination location. When the position parameter reaches the value in the length parameter (all words have been moved), the done bit in the control register is set. On the next false to true transition of the rung with the done bit set, the position pointer is reset to 1. A simple illustration of how the SQO moves data from the file to the destination is illustrated in Figure 11-4 for the SLC 500. Figure 11-4(a) depicts the sequencer

FIGURE 11-4: Sequencer moving data from a file to an output.

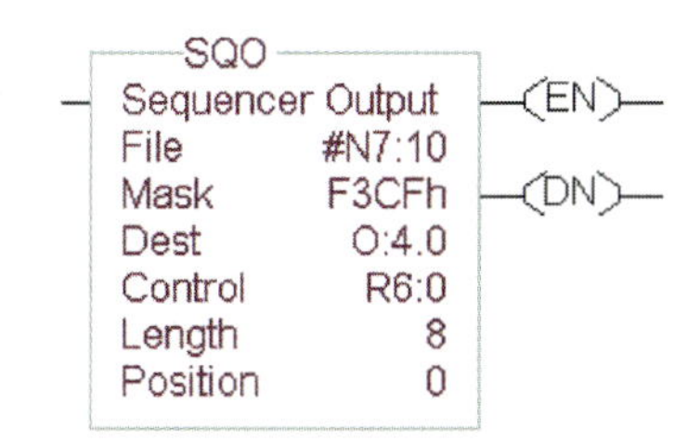

(a) Sequencer instruction for SLC 500

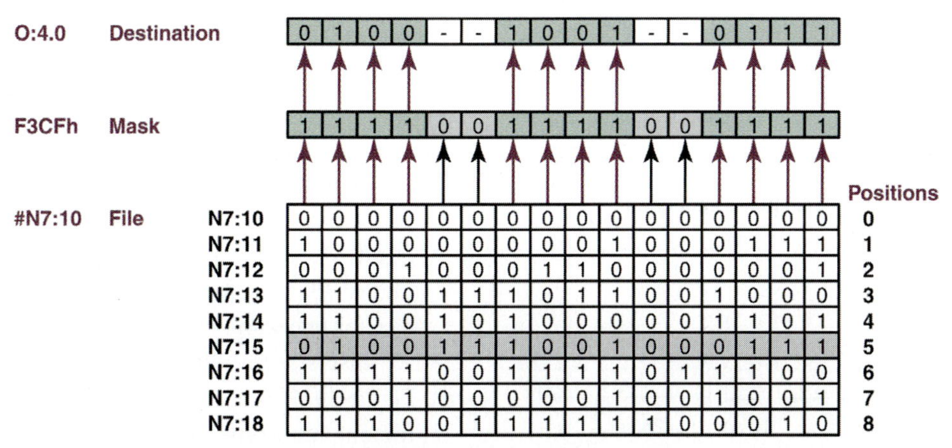

(b) Sequencer data movement for SLC 500 with the position pointer at 5

instruction where the file address is #N7:10, the mask is F3CFh, the destination is O:4.0, the length is 8, and the starting position is 0. Note that the length parameter indicates the file length starting with the data in position 1.

Figure 11-4(b) illustrates how the data flows when the sequencer file executes. Note that the data in position 5, which is file N7:15, moves through the mask to the destination location, O:4.0. The arrows in the figure indicate the unmasked bits that are passed through the mask and into the destination location. The dashes in the bits of the destination register indicate that those bits remain unchanged in the destination location during the sequencing. In many applications, timed pulses are used to automatically increment the sequencer through the positions in the file.

11-4-3 ControlLogix SQO Sequencer Instruction

The operation of the ControlLogix SQO instruction, shown in Figure 11-5, is similar to the PLC 5 and SLC 500 description. The operation during the first scan and for all subsequent scans is the same as that described in the previous section for the PLC 5 and SLC 500. Therefore, when this SQO instruction is enabled the first time after a false first scan, it first increments the position value from 0 to 1, passes the data from the File tag array_dint[1] through the mask 16#0F0F, and stores the result in the destination tag called value_1. Notice that the ControlLogix processor puts the radix and the # in front of the value to indicate the radix of the displayed number. The Control tag is control_1, the length is 10 (array_dint[1] to array_dint[10]), and the starting position value starts at 0. When the pointer reaches the value in the length parameter, the done bit in the control file is set. The first false to true transition of the rung after the done bit is set causes the position parameter to reset to 1 for the next sequence through the file array. Note that a tag is used for the destination location in place of the memory file parameter used in the PLC 5 and SLC 500 PLCs. Also, a one-dimensional array is used to hold the output word sequence in place of an integer memory file with the indexed mode (#) symbol. The array parameter tag is entered with a zero index, for example, array_dint[0]. The array index is set by the value in the position parameter.

FIGURE 11-5: Sequencer SQO instruction for the ControlLogix PLC.

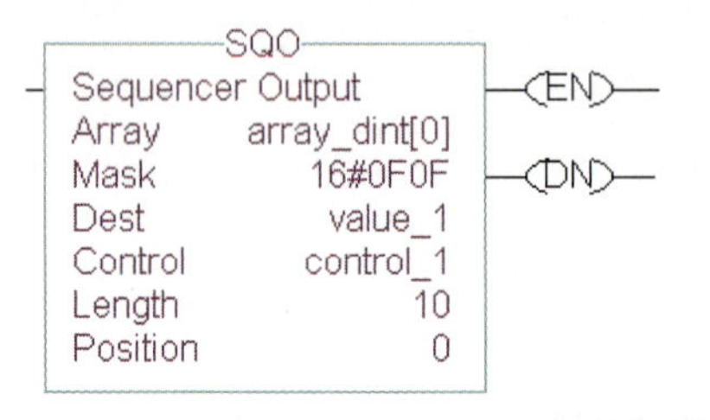

FIGURE 11-6: Pulses used to control SQO instruction.

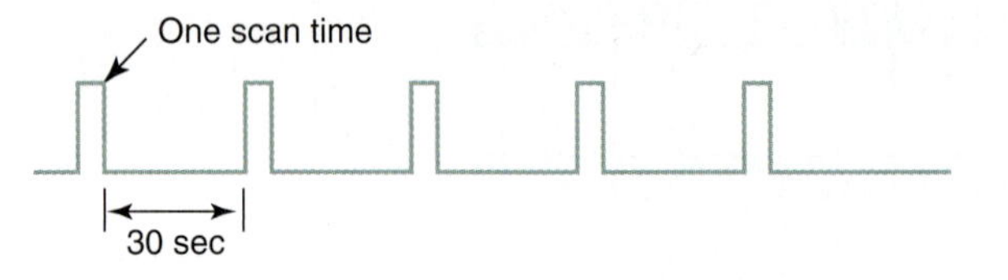

(a) Constant duty cycle pulse sequence

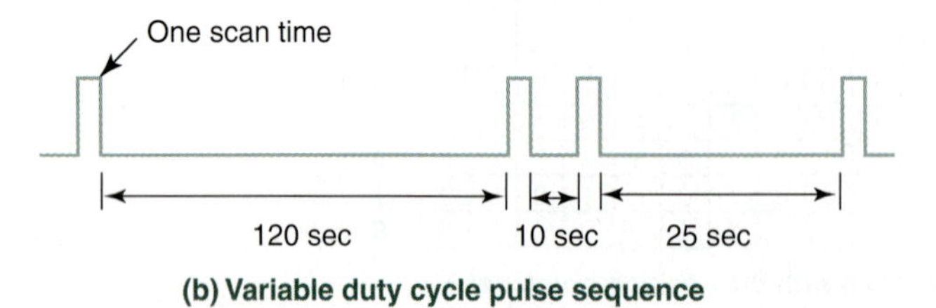

(b) Variable duty cycle pulse sequence

Standard Ladder Logic for Time- and Event-Driven SQO Instructions. The SQO instructions are often used to force a sequential machine to move through a series of steps. The SQO instruction can be stepped to the next sequencer word by two types of triggers:

- A series of *timed pulses*
- A sequence of *external events*

When timed pulses are used, you have the following two options, which are illustrated in Figure 11-6:

- A pulse stream with a constant duty cycle, Figure 11-6(a); for example, pulses that are true for one scan time and false for 30 seconds.
- A pulse stream with a variable duty cycle, Figure 11-6(b); for example, each pulse is true for one scan time, but false for 120 seconds for the first pulse, false for 10 seconds for the

second pulse, and so forth with the off time dictated by the control requirements for the process machine.

The SQO instruction rung is false during the first scan for all timer-driven applications. Therefore, the data in the word in position 0 is not transferred and the position pointer remains at 0 until the first transition from the timer.

When external events are used to trigger the SQO instruction, each external event moves the next sequencer word to the destination file and a different set of outputs are turned on. If a robot is loading parts into a machine, then a sensor in the parts feeder indicating that a part is present would be the event that moves the robot to pick up the part. These events are called "transition conditions" because they indicate a transition from one machine state (certain outputs on) to a new machine state (other outputs on). Again, two options are available for the SQO solution when events are used to load words into the destination register. They are:

1. A machine that has less than four states or sequences with one or two transition conditions to indicate the next state.
2. A machine with numerous states or sequences and/or one with multiple transition conditions to indicate the next state.

The SQO instruction rung is usually false during the first scan for event-driven applications. Therefore, the data in the word in position 0 is not transferred and the position pointer remains at 0 until the first transition from the timer.

The standard ladder logic for the SQO instruction with a fixed duty cycle timer and an instruction with few machine states and transitions are illustrated in Figure 11-7. Standard ladder logic

FIGURE 11-7: Standard ladder logic rungs for SLC 500 SQO instructions.

Application

A machine or process with a small number of machine or process states that are active for a fixed time period needs a PLC controller.

The SQO instruction is used with a pulse input pictured in Figure 11-6(a). The standard ladder logic has two rungs. The first has the timer done bit for the input logic and an SQO instruction for the output. The second rung has a regenerative timer with a preset time of 30 seconds and a 1 second time base. The timer is started by a field device with an address of I:1/1 and reset by its done bit. The done bit also increments the SQO instruction to the next output word. The process has three states: state one outputs are on for 90 seconds, state two outputs are on for 60 seconds, and state three outputs are on for 120 seconds. Since the sequencer is incremented every 30 seconds, it takes three words for state one, two words for state two, and four words for state three. This is a total of nine words. The outputs are the same for each group of words associated with a specific state. For example, the first three words are the same since they turn on state one outputs. The next two are the same and the final four are the same. The first word is in N7:21 and the last is in N7:29 so the length is nine and the starting position is 0. Three bits of O:3.0 are used for outputs, bits 0, 6, and 12. So the destination is O:3.0, control is R6:0, and the mask is 001000001000001B or 1041h. Outputs 0 and 12 are on for state 1, 6 for state 2, and all three bits for state 3. The word values are shown below the ladder logic.

Note: The operation would be the same for PLC 5 and ControlLogix PLCs.

Standard Ladder Logic Rungs for SQO Instructions

SQO increment
T4:0
DN

SQO
Sequencer Output
File #N7:20
Mask 1041h
Dest O:3.0
Control R6:0
Length 9
Position 0
(EN)
(DN)

Start
I:1
1

Timer reset
T4:0
DN

TON
Timer On Delay
Timer T4:0
Time Base 1.0
Preset 30
Accum 0
(EN)
(DN)

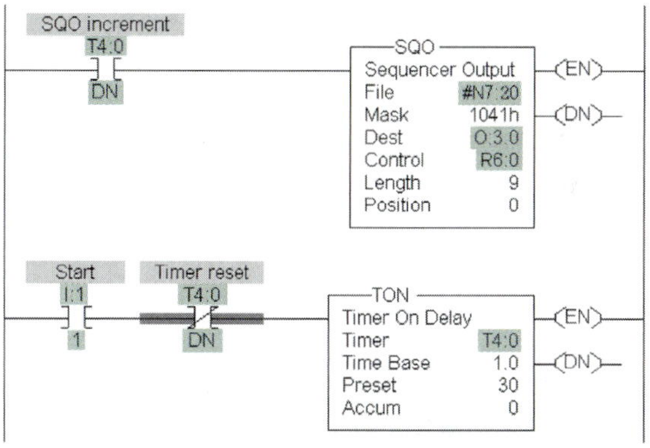

Offset	15	14	13	12	11	10	9	8	7	6	5	4	3	2	1	0
N7:20	0	0	0	0	0	0	0	0	0	0	0	0	0	0	0	0
N7:21	0	0	0	1	0	0	0	0	0	0	0	0	0	0	0	1
N7:22	0	0	0	1	0	0	0	0	0	0	0	0	0	0	0	1
N7:23	0	0	0	1	0	0	0	0	0	0	0	0	0	0	0	1
N7:24	0	0	0	0	0	0	0	0	0	1	0	0	0	0	0	0
N7:25	0	0	0	0	0	0	0	0	0	1	0	0	0	0	0	0
N7:26	0	0	0	1	0	0	0	0	0	1	0	0	0	0	0	1
N7:27	0	0	0	1	0	0	0	0	0	1	0	0	0	0	0	1
N7:28	0	0	0	1	0	0	0	0	0	1	0	0	0	0	0	1
N7:29	0	0	0	1	0	0	0	0	0	1	0	0	0	0	0	1

(a)

FIGURE 11-7: (Continued)

Application	Standard Ladder Logic Rungs for SQO Instructions
A machine or process with a small number of machine or process states and a limited number of unique external event signals needs a PLC controller. The machine has three states with an output word for each state starting at N7:1. The destination is O:3.0, the length is 3, and the starting position is 0. The output bits are 1, 3, 5, 8, and 12 for state one; 2, 3, 9, and 10 for state two; and 3, 4, 5, 8, and 12 for state three. That requires a mask of 0001011100111110B or 173Eh and uses control R6:0. The standard ladder logic has one rung, which consists of an SQO instruction with input ladder logic including the four input field device sensors. The input logic is [(S1 AND S2) OR S3 OR S4] where (S1 AND S2) triggers the state 1 outputs, S3 triggers state two outputs, and S4 triggers state 3 outputs. After state three outputs are moved, the position is equal to the length. When the state one outputs return to a true state, the position resets to 1 and the outputs in N7:1 are moved to the destination. The word values are shown below the ladder logic. Note: SLC 500 rungs are shown, but the operation would be the same for PLC 5 and ControlLogix.	See ladder diagram and word values below.

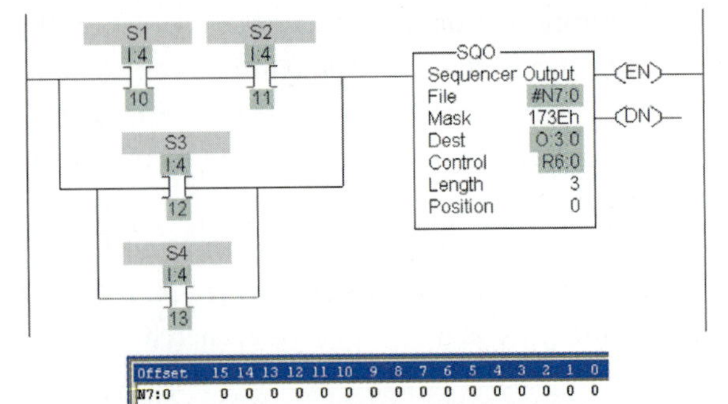

Offset	15	14	13	12	11	10	9	8	7	6	5	4	3	2	1	0
N7:0	0	0	0	0	0	0	0	0	0	0	0	0	0	0	0	0
N7:1	0	0	0	1	0	0	0	1	0	0	1	0	1	0	1	0
N7:2	0	0	0	0	0	1	1	0	0	0	0	0	1	1	0	0
N7:3	0	0	0	1	0	0	0	1	0	0	1	1	1	0	0	0

(b)

for the other two options is provided in the sections that follow. Study the standard ladders and read the descriptions in the figure.

The standard ladder logic for the SQO instructions demonstrates how a timer, Figure 11-7(a), or a set of input transitions conditions, Figure 11-7(b), are used to control a machine with data words stored in consecutive integer files. PLC sequencers are often used to control automatic industrial equipment that operates in a fixed and repeatable timed sequence. The PLC sequencer makes programming these applications much simpler. For example, the off/on operation of 12 lamps can be controlled using a PLC sequencer with one rung of a ladder diagram. By contrast, the equivalent PLC solution regular ladder logic would require 12 rungs (one for each lamp) in the ladder diagram.

When a fixed time base is used, data words are repeated to set the on time required for each output state. In the standard ladder logic example illustrated in Figure 11-7(a), the regenerative timer has a preset of 30 seconds so sequencer words are delivered every 30 seconds. The first, fourth, and sixth data words are repeated (3, 2, and 4 sets of identical words) in order to get the accumulated time required for the outputs in each state. The first state is on for 90 seconds (3 identical output words each on for 30 seconds), the second for 60 seconds, and the third for 120 seconds. Match the mask bits to the desired output bits and notice that every output bit has a 1 in the mask and all other bits are 0. This permits the programmer to use the other output bits for another output requirement without affecting this sequencer rung.

The event-driven sequencer in Figure 11-7(b) uses an *OR* configuration with four transition conditions in the three *OR* paths. Any one of the three paths can make the SQO rung true. So as each event occurs, that OR branch is true and the sequencer is stepped through the three output words to control the machine. S1 AND S2 trigger the word in position 1, S3 triggers the word in position 2, and S4 triggers the word in position 3.

Both of the standard circuits use SLC 500 logic rungs, but the operation would be the same for the PLC 5 and ControlLogix PLCs with the following modifications. The PLC 5 input addressing in Figure 11-7(b) would follow the rack/group/bit format, and the ControlLogix ladder would use arrays and tags in place of memory addresses.

EXAMPLE 11-1

Use a sequencer to control a dishwasher cycle that has the following timed steps in its operation sequence.

1. The pre-rinse timing cycle
2. The soap release
3. The wash timing cycle
4. The post-rinse timing cycle
5. The drying timing cycle

Table 11-3 shows the sequential steps for the dishwasher. Note that the sequence numbers are in the first column, the step numbers are in the second column, and the sequencing patterns are in columns 3 through 7. The sequencing pattern columns are labeled with the dishwasher actions, which are soap, fill, wash, drain, and dry. A 1 in these pattern columns indicates that the dishwasher action is on. Conversely, a 0 in these columns indicates that the dishwasher action is off. For example, in step 6, which is the second step in sequence 3, the wash cycle is on, and all other actions are off. Before continuing, study the table until you're familiar with all the sequences, steps, and patterns.

SOLUTION

The ladder diagram that implements the dishwasher steps is illustrated in Figure 11-8. Note that start and stop switches are located in rung 0, enabling or disabling the dishwasher operation. In the SQO sequencer, the length is 11. The mask is 001F hex (0000 0000 0001 1111 binary), which allows the first five bits from the sequencer file to be passed to the destination location. The timer T4:1 is an on-delay timer that is preset to 30 seconds. The description of the ladder diagram representing the dishwasher operation is as follows:

1. The momentary start switch input I:1/0 is turned on, which activates the start bit, output B3:0/0, thus energizing all the instructions that reference this virtual control relay.
2. The XIC instruction (B3:0/0) seals the start switch and activates the timer rung.

TABLE 11-3 Sequential steps for a dishwasher.

1 Machine Sequence	2 Sequencer Step	3 Soap	4 Fill	5 Wash	6 Drain	7 Dry	
	0	0	0	0	0	0	N7:10
1	1	0	1	0	0	0	N7:11
1	2	0	0	1	0	0	N7:12
1	3	0	0	0	1	0	N7:13
2	4	1	0	0	0	0	N7:14
3	5	0	1	0	0	0	N7:15
3	6	0	0	1	0	0	N7:16
3	7	0	0	0	1	0	N7:17
4	8	0	1	0	0	0	N7:18
4	9	0	0	1	0	0	N7:19
4	10	0	0	0	1	0	N7:20
5	11	0	0	0	0	1	N7:21

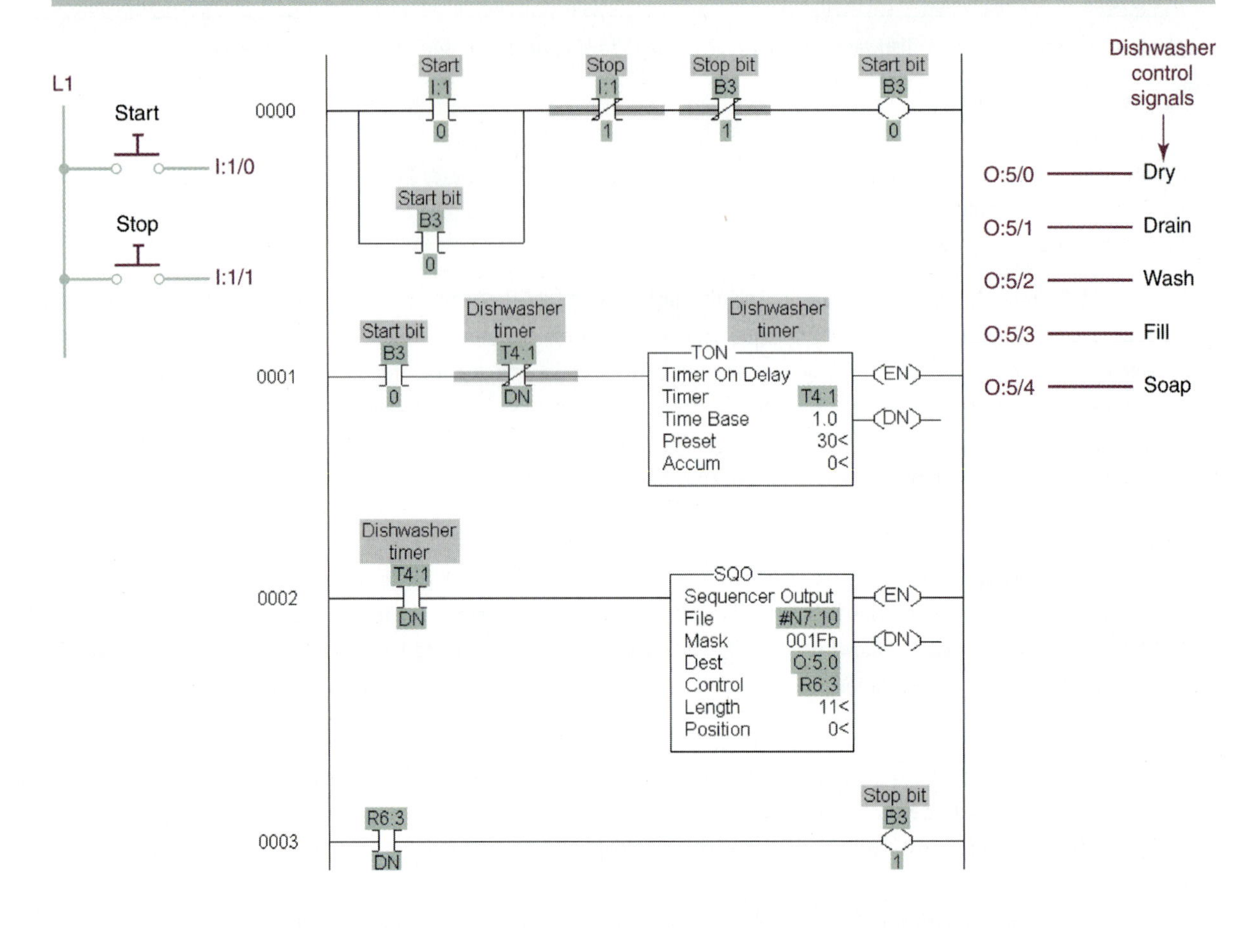

FIGURE 11-8: Ladder diagram for dishwasher example.

3. After 30 seconds the timer DN bit goes from false to true for one scan, which advances the sequencer from step 0 to step 1. Step 0 is the starting position pointing to file N7:10, step 1 is pointing to N7:11, and step 11 points to N7:21. The first five bits in these 11 files hold the 0 and 1 combinations from columns 3 to 7 in Table 11-3.
4. In step 1 the sequencer moves the word located in N7:11 to O:5.0, the output file to control dishwasher functions.
5. The timer DN bit deactivates the timer, which causes the DN bit to go from true to false, which reactivates the timer after one scan.
6. The previous three operational steps (3, 4, and 5) are repeated for sequencer steps 2 through 11. This moves files N7:12 through N7:21 to the output file (O:5.0) as each sequencer step is executed.
7. When the sequencer reaches step 11, the sequencer's control register (R6:3) done bit (DN) is activated, which activates the stop bit, output B3:0/1, and all instructions with that reference.
8. The binary bit (3:0/1) in rung 0 is true so the XIO instruction is false and the start bit is false, which makes the timer rung false.
9. The ladder is now ready for another activation of the start switch.

Each step in Table 11-3 is incremented every 30 seconds. The same time increment for every step is impractical because some steps in the dishwasher cycle require more than 30 seconds for the task. Assume that the fill and the drain cycles require 60 seconds each; the three wash cycles require 60 seconds, 180 seconds, and 120 seconds, respectively; and the dry cycle requires 240 seconds. Because all the cycle times are multiples of 30 seconds, they can be implemented by increasing the number of steps for each pattern to accommodate the required times. This implementation is shown in Table 11-4 and is loaded into registers N7:11 through N7:43.

TABLE 11-4 Dishwasher steps with non-uniform sequence times.

Machine Sequence	Sequencer Step	Soap	Fill	Wash	Drain	Dry	
	0	0	0	0	0	0	N7:10
1	1	0	1	0	0	0	N7:11
1	2	0	1	0	0	0	N7:12
1	3	0	0	1	0	0	N7:13
1	4	0	0	1	0	0	N7:14
1	5	0	0	0	1	0	N7:15
1	6	0	0	0	1	0	N7:16
2	7	1	0	0	0	0	N7:17
3	8	0	1	0	0	0	N7:18
3	9	0	1	0	0	0	N7:19
3	10	0	0	1	0	0	N7:20
3	11	0	0	1	0	0	N7:21
3	12	0	0	1	0	0	N7:22
3	13	0	0	1	0	0	N7:23
3	14	0	0	1	0	0	N7:24
3	15	0	0	1	0	0	N7:25
3	16	0	0	0	1	0	N7:26
3	17	0	0	0	1	0	N7:27
4	18	0	1	0	0	0	N7:28
4	19	0	1	0	0	0	N7:29
4	20	0	0	1	0	0	N7:30
4	21	0	0	1	0	0	N7:31
4	22	0	0	1	0	0	N7:32
4	23	0	0	1	0	0	N7:33
4	24	0	0	0	1	0	N7:34
4	25	0	0	0	1	0	N7:35
5	26	0	0	0	0	1	N7:36
5	27	0	0	0	0	1	N7:37
5	28	0	0	0	0	1	N7:38
5	29	0	0	0	0	1	N7:39
5	30	0	0	0	0	1	N7:40
5	31	0	0	0	0	1	N7:41
5	32	0	0	0	0	1	N7:42
5	33	0	0	0	0	1	N7:43

Sequence 1 encompasses six sequencer steps, which are two steps for the fill cycle, two steps for the wash cycle, and two steps for the drain cycle. Because each step is 30 seconds, the fill cycle, the wash cycle, and the drain cycle each take the required 60 seconds. The dry cycle encompasses eight steps in sequence 5, which is 240 seconds, the required drying time. Review all the steps in the table until you are confident that you understand the implementation of dishwasher operation with different cycle times.

The ladder logic in Figure 11-8 can be used for this more complete solution to the dishwasher problem by just changing the value of the length parameter in the SQO instruction from 11 to 33. With the data from Table 11-4 loaded into the files N7:11 through N7:43, the new dishwasher times are implemented.

EXAMPLE 11-2

A magazine parts feeder is shown in Figure 11-9. The feeder pushes a cylindrical part from the bottom of the stack to a pickup point for a robot. Three NO sensors are shown: S1 indicates when a part is in the pick up position, S2 indicates that there is a part in the feeder ready to be inserted, and S3 indicates that the push cylinder is retracted. When the start switch is closed, the system operates as follows:

1. The pusher cylinder valve (O:3/0) is turned on so that a part is pushed into the pickup point for the robot. The valve actuates only when the piston is fully retracted (S3 closed), when there is a part in the magazine (S2 closed), and no part is at the pickup point (S1 open). The feeder part pusher holds the part against the stop pins until the robot has acquired the part.
2. The robot input (O:3/1) is turned on when a part is at the pickup point to signal the robot program to acquire the part. The robot has two outputs: I:4/13 Part Acquired and I:4/8 Gripper Clear. The Part Acquired contacts close when the gripper closes on the part and remain closed until the Gripper Clear pulse returns to a false state, and the Gripper Clear pulse occurs when the part is above the pickup point.

Develop a SQO sequencer ladder program that meets this control description.

Solution

The ladder diagram that implements the parts feeder is illustrated in Figure 11-10(a) and the data used in the sequencer is shown in Figure 11-10(b). Position 1 data (valve on) is present at the output when logic path Start AND NOT S1 AND S2 AND S3 is true. Position 2 data (robot pickup signal plus valve on) is present at the output when logic path S1 AND NOT Part Acquired is true. Position 3 data (all outputs off) is present when logic path Gripper Clear is true. The sequencer returns to position 1 when the part pusher returns to the retracted location and the logic path Start AND NOT S1 AND S2 AND S3 is true again.

FIGURE 11-9: Magazine round parts feeder.

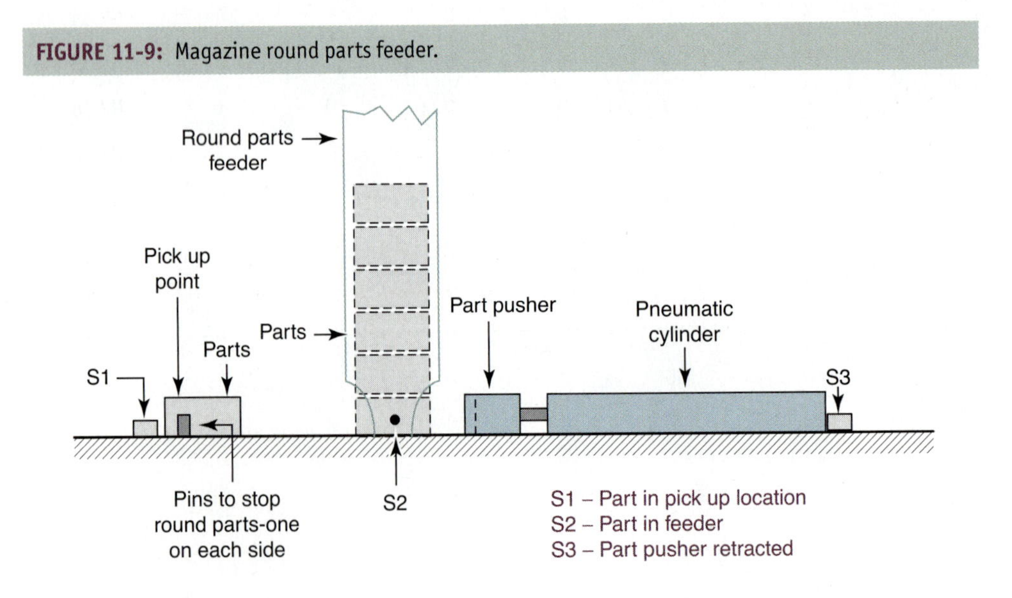

FIGURE 11-10: Ladder diagram for magazine parts feeder.

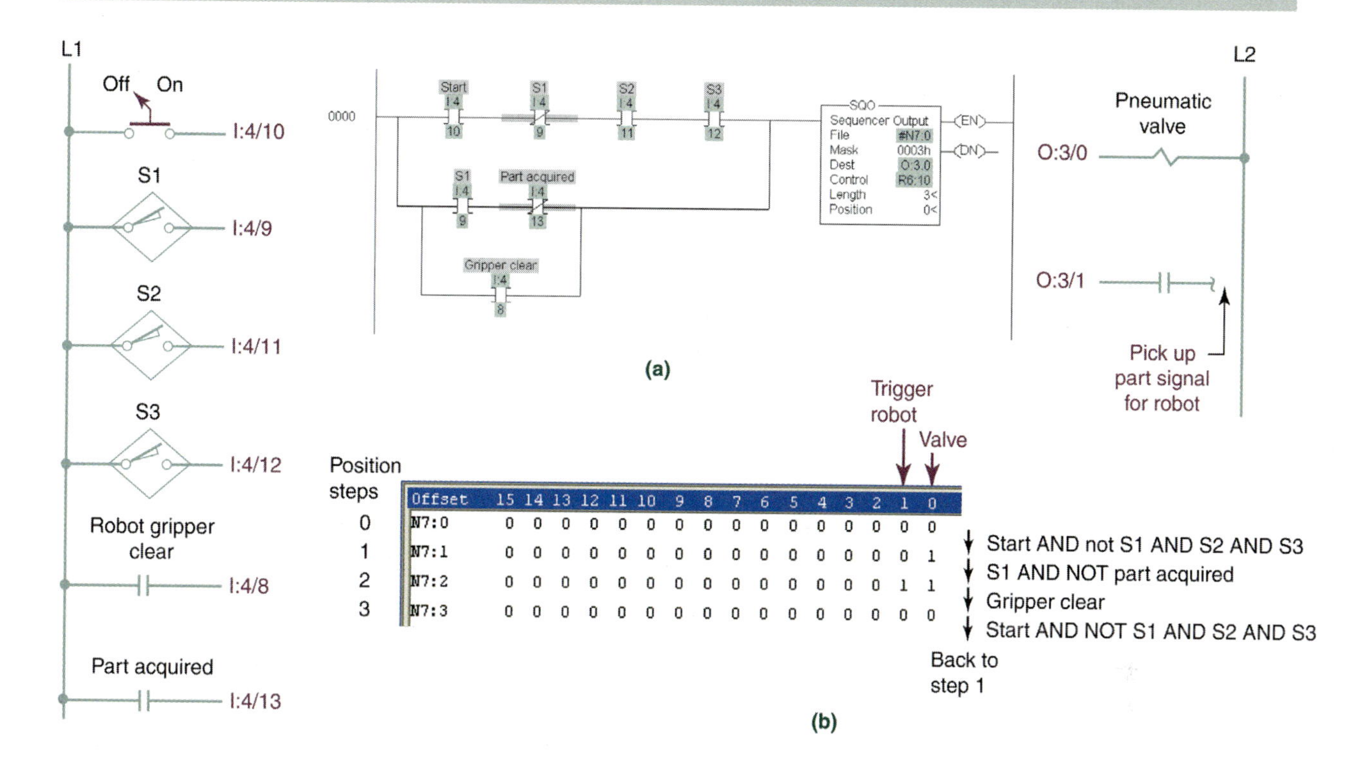

11-4-4 PLC 5 and ControlLogix Sequencer Input (SQI) Instruction

The *SQI instruction* is an *input* instruction available in PLC 5 and ControlLogix. The SQI is like an equal comparison instruction but a mask is added, and multiple word combinations can be compared using this instruction. Figure 11-11(a) illustrates this instruction for the PLC 5 and ControlLogix PLCs, and Figure 11-11(b) illustrates its operation using PLC 5 parameters. Note that the location of the *destination* parameter in the SQO instruction just studied (Figures 11-4 and 11-5) is replaced by the *source* parameter in the SQI instruction. The PLC 5 uses file registers and the ControlLogix uses tags and arrays. The source contains the address of the input data word to be compared against the sequencer file words. The initial position value is 0 and the length is the number of files used for the comparisons.

Figure 11-11(b) illustrates how the SQI instruction compares the input data in I:003 through the mask F3CFh with the data in the sequencer file N7:10 through N7:18. The specific data in the sequencer file used in the comparison is identified by the pointer in the position parameter. In Figure 11-11(b), the data at position 5 matches the unmasked input data, so the SQI instruction is true, thus making the rung true and also making any output instruction(s) on the rung true. When the unmasked source bits do not match the sequencer file word, the instruction is false and the rung is also false. In the SLC 500 the SQC performs a similar function that is covered in a later section.

The SQI Control Register. The SQI instruction has a control register like the SQO but eliminates the done bit. The SQI instruction does not automatically increment the position parameter on a false to true transition at its input like the SQO instruction. When the SQI is paired with an SQO instruction with identical control tags, the position pointer is incremented by the SQO instruction for both. If an SQI instruction is used alone, then the position pointer must be changed by other instructions such as an ADD and a MOV instruction in order to select another input. The logic required to use only an SQI instruction is illustrated in Figure 11-12. Note that the SQI instruction can be used to turn on an output when the input word tag, value_1, matches a word in the sequencer array, array_1[0]. The output instruction(s) can be placed in rung 0 with the

FIGURE 11-11: Sequencer comparing data between an input and a file word.

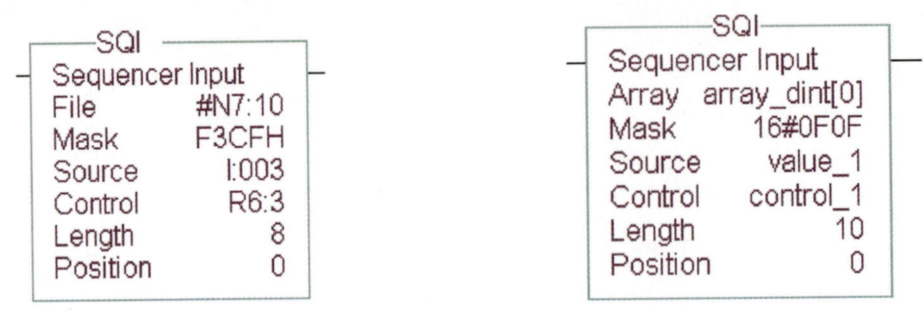

(a) Sequencer instruction for PLC 5 and ControlLogix

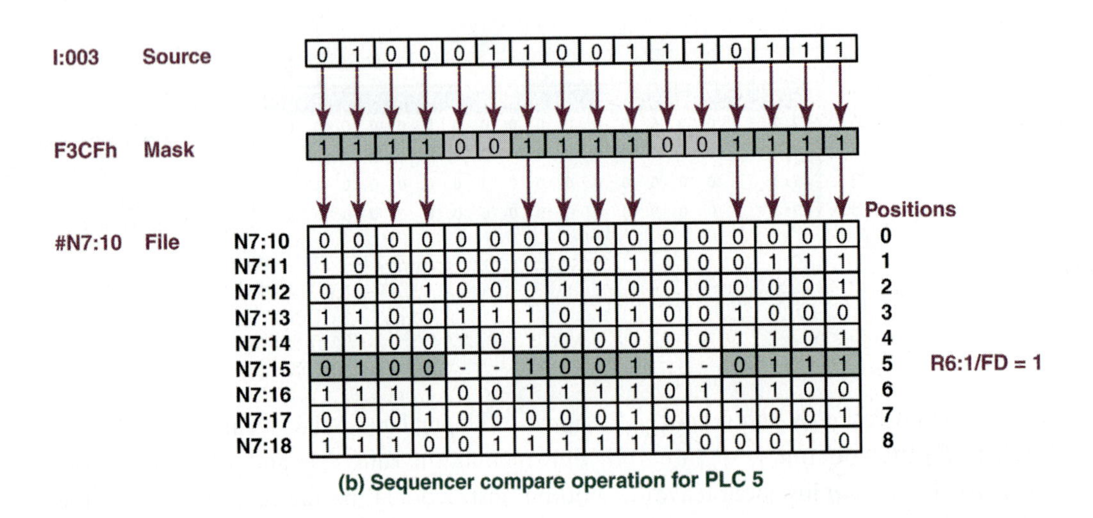

(b) Sequencer compare operation for PLC 5

FIGURE 11-12: Ladder logic for SQI instruction.

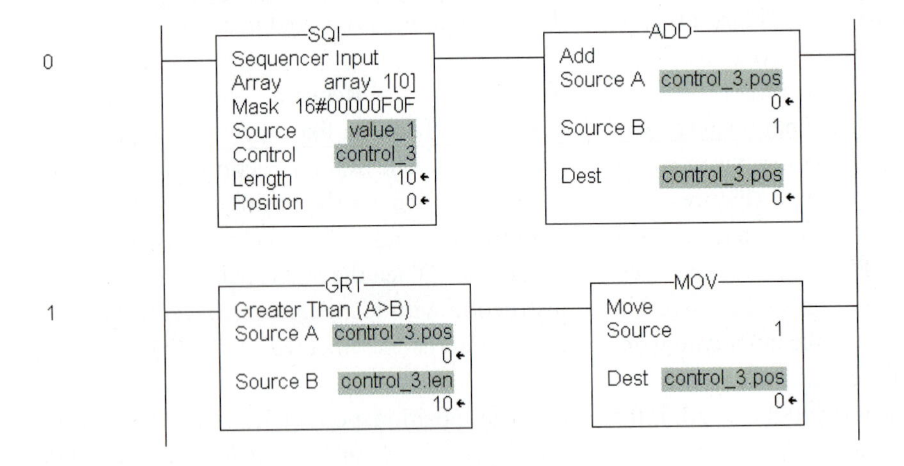

ADD instruction, which adds a 1 to the current pointer, control_3pos. In the next rung, a GRT instruction checks to see if the pointer, control_3pos, is greater than the length, control_3len. If it is, then the pointer is reset to 1 by the output MOV instruction.

Standard Ladder Logic for Paired SQI and SQO Instructions. In the PLC 5 and ControlLogix PLCs, the SQI and SQO instructions can be used in pairs as illustrated for the ControlLogix PLC in the standard ladder logic in Figure 11-13. Note that the control address, length, and

FIGURE 11-13: Standard ladder logic rungs for paired ControlLogix SQI and SQO instructions.

Application	Standard Ladder Logic Rungs for Paired ControlLogix SQI and SQO Instructions

Application

A machine or process that operates in a sequential fashion must execute its operational sequence based on a changing set of discrete input field devices

The paired SQI and SQO instruction ladder described in the text is illustrated in this standard ladder logic configuration. There are two versions shown, with the first being an unconditional execution and the second triggered by an external bit. In the unconditional configuration, the SQI tests the masked input word against the array value selected by the position pointer on every scan. The second ladder rung (b) only makes the test when the input instruction before the SQI is true.

Parameter requirements include:

- The control, length, and position parameters must be the same in both instructions.
- The array tag in the array parameter must have the zero element of the array specified as shown in these ladders.
- The source and destination parameter must be input and output tags, respectively.
- The position parameter is set to zero and the length is the total number of values in the arrays, not counting position zero in the arrays.

The instruction pair operates as follows:

- The SQI source word, value_in, is passed through the mask and all 1's in the mask pass their respective bits and 0's block the transfer.
- If the unmasked SQI source value, value_in, matches the array word, array_SQI, the SQI instruction is true and the rung is active.
- If the rung is true, then the SQO increments the position pointer and moves the selected array word, array_SQO, through a mask to the destination location, value_out.
- When the position pointer equals the length parameter, the last data file is moved to the output array and the control file done bit is set.
- On the next false-to-true transition of the rung with the done bit set, the position pointer returns to position 1.
- The control file error bit is set if the position value is negative.

Sample array files are illustrated.

Note: The operation would be the same for PLC 5.

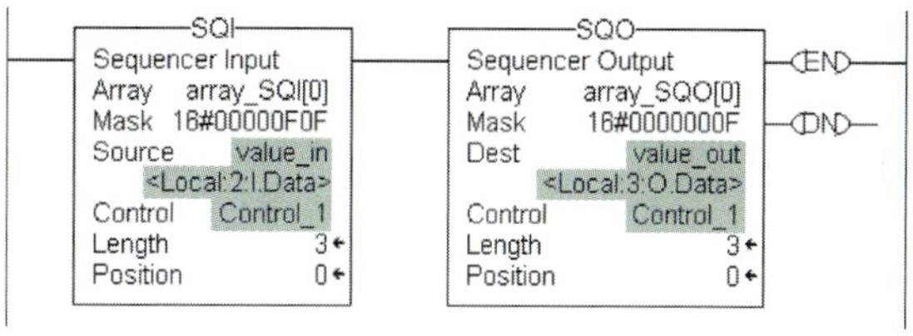

Standard Ladder Logic Rungs for Paired ControlLogix SQI and SQO Instructions

SQI: Sequencer Input; Array array_SQI[0]; Mask 16#00000F0F; Source value_in <Local:2:I.Data>; Control Control_1; Length 3; Position 0
SQO: Sequencer Output; Array array_SQO[0]; Mask 16#0000000F; Dest value_out <Local:3:O.Data>; Control Control_1; Length 3; Position 0; (EN) (DN)

(a)

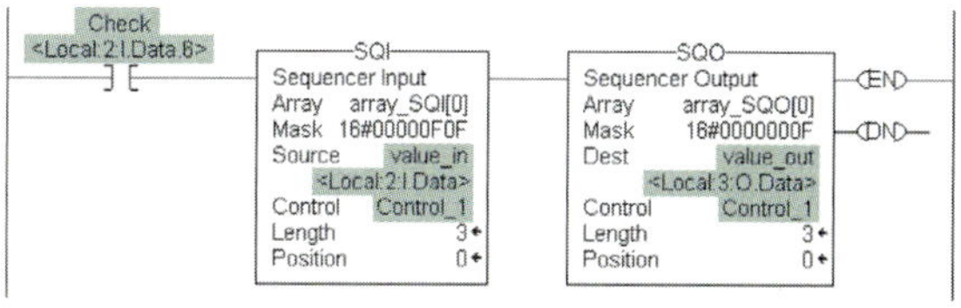

Check <Local:2:I.Data.6>
SQI: Sequencer Input; Array array_SQI[0]; Mask 16#00000F0F; Source value_in <Local:2:I.Data>; Control Control_1; Length 3; Position 0
SQO: Sequencer Output; Array array_SQO[0]; Mask 16#0000000F; Dest value_out <Local:3:O.Data>; Control Control_1; Length 3; Position 0; (EN) (DN)

(b)

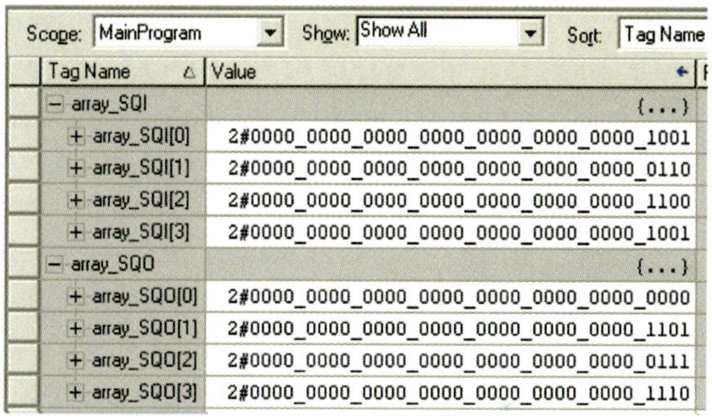

Scope: MainProgram Show: Show All Sort: Tag Name

Tag Name	Value
− array_SQI	{...}
+ array_SQI[0]	2#0000_0000_0000_0000_0000_0000_0000_1001
+ array_SQI[1]	2#0000_0000_0000_0000_0000_0000_0000_0110
+ array_SQI[2]	2#0000_0000_0000_0000_0000_0000_0000_1100
+ array_SQI[3]	2#0000_0000_0000_0000_0000_0000_0000_1001
− array_SQO	{...}
+ array_SQO[0]	2#0000_0000_0000_0000_0000_0000_0000_0000
+ array_SQO[1]	2#0000_0000_0000_0000_0000_0000_0000_1101
+ array_SQO[2]	2#0000_0000_0000_0000_0000_0000_0000_0111
+ array_SQO[3]	2#0000_0000_0000_0000_0000_0000_0000_1110

(c)

position parameters are the same. Also, the index on the sequencer array in the SQI instruction must be set to zero (array_SQI[0]). The array for the SQO must be set to a zero index as well. The inputs for the SQI instruction are in the source tag, value_in, and outputs for the SQO instruction are in the destination tag, value_out.

The operation of the SQO instruction during the first scan depends on the status of its rung during that scan. The standard ladder logic in Figure 11-13 assumes that the SQI instruction makes the rung false for the first scan, so the SQO instruction is false, which causes no action or change in the SQO during the first scan. If the standard ladder in Figure 11-13(a) is used, then the SQI is forced to a false condition for the first scan by making the number in value_in different from the value of the number in array_SQI[0] when the first scan occurs. If the standard ladder in Figure 11-13(b) is used, then the SQI is forced to a false condition for the first scan by making the start input (XIC instruction) false when the first scan occurs. The start input could be a system start selector switch that is turned on after the PLC program is downloaded. Another option is to set the start input to the first scan status bit (true for first scan only) address and make the instruction an XIO type.

When used in pairs with the same control register parameter, the position pointer of both instructions is changed whenever the position in the SQO instruction is incremented. Restated, when the SQI instruction finds a match between the unmasked bits in the input tag and the SQI array value, the rung becomes true. This causes the SQO to increment the position parameter by 1 and move the value in the SQO array at this index or pointer location to the output tag. Because they share the same control tag name, the pointers of the SQI and SQO instructions are the same, so they work in unison as they step through their array values. This programming technique provides input and output sequencers that function in unison, resulting in the activation of a selected set of outputs when a set of specific inputs is present. The inputs used for control must all be connected to the same input word and the controlled outputs must be in the same output word. Read the description of the SQI and SQO pair in Figure 11-13.

The organization of the array data for the SQI and SQO instructions is illustrated in Figure 11-14(a) and described as follows:

- The input word that represents the input sensors and switches and the desired output words for each input are displayed in the first table in Figure 11-14(a). Note the input bit pattern and the desired output bit pattern.
- The output bit patterns are shown in the second table in Figure 11-14(a) in the order necessary to make the sequencer operate properly. Note that all zeros (0000) are loaded into the 0 element of the array, the last output (1101) is loaded in element 1, and the first output (0111) is loaded into element (2). All other output words are loaded in the array in their sequential order after the first output word.

With the arrays loaded, the operation of the standard ladder logic for the SQI and SQO pair, as shown in Figure 11-14, can be evaluated:

- **First scan:** The standard ladder in Figure 11-13(b) is used with the start input representing the system start selector switch. The switch is off for the download of the program, so during the first scan the SQO rung is false. This keeps the position value at 0 and no word is moved to the destination address.
- **Step 0:** When the start switch is turned on, the all zeros in value_in match the all zeros in the SQI array at position 0, so the rung goes true and the SQO increments the position parameter to 1 and moves the SQO array value of 1101 to the output file.
- **Step 1:** With the position parameter at 1, the value_in is changed to 0110, which matches the position 1 value in the SQI array. This forces the rung true and the SQO increments the position parameter to 2 and moves the SQO array value of 0111 to the output file.
- **Step 2:** With the position parameter at 2, the value_in is changed to 1100, which matches the position 2 value in the SQI array. This forces the rung true and the SQO increments the position parameter to 3 and moves the SQO array value of 1110 to the output file. Now the length

FIGURE 11-14: SQI/SQO array values and matches.

Process input value and desired output value

value_in	value_out
0000	0000
1001	1101
0110	0111
1100	1110

Sequencer array value organization

Position	SQI array array_SQI	value_in	SQO array array_SQO	value_out
0	1001	1001	0000	
1	0110	0110	1101	1101
2	1100	1100	0111	0111
3	1001	1001	1110	1110

(a)

FIGURE 11-14: (Continued).

First scan occurs with the rung false (start switch off). When the rung goes true (start switch on and Value_in equal to 1001) the pointer moves to 1 and moves 1101 into value_out.

Position	SQI array value	value_in	Rung condition	SQO array value	value_out	Operational steps
0	1001	1001	False to True	0000		
1	0110			1101	1101	1
2	1100			0111		
3	1001			1110		

Value_in is changed to 0110 and the sequencer increments to position 2 and moves 0111 to value_out.

Position	SQI array value	value_in	Rung condition	SQO array value	value_out	Operational steps
0	1001			0000		
1	0110	0110	False to True	1101		
2	1100			0111	0111	2
3	1001			1110		

Value_in is changed to 1100 and the sequencer increments to position 3 and moves 1110 to value_out.
With the position and length parameters equal, the done bit is set, and the sequencer position will reset back to 1 with the next false to true change in the rung.

Position	SQI array value	value_in	Rung condition	SQO array value	value_out	Operational steps
0	1001			0000		
1	0110			1101		
2	1100	1100	False to True	0111		
3	1001			1110	1110	3

Value_in is changed to 1001 and the sequencer increments to position 1 and moves 1101 to value_out.

Position	SQI array value	value_in	Rung condition	SQO array value	value_out	Operational steps
0	1001			0000		
1	0110			1101	1101	1
2	1100			0111		
3	1001	1001	False to True	1110		

(b)

and position value are the same, so the done bit in the control register is set to 1.

- **Step 3:** With the position parameter at 3, the done bit is set to 1 and the value_in is changed to 1001, which matches the position 3 value in the SQI array. This forces the rung true and the SQO does not increment the position parameter but resets it to 1, and the value of the SQO file at the 1 position is moved to the output file.

The sequencer continues to cycle through the outputs as the input values change. Note that the input word pattern that matches the output word pattern is not in the same element number in the SQI and SQO arrays. This is necessary because a match between an SQI input and an element in the SQI array cause the SQO to increment the position parameter *before* selecting the SQI array element to move to the output. As a result, the two arrays are off by one position for every move.

Standard Ladder Logic for Variably Timed SQO Instructions. The standard ladder logic described in Figure 11-7 illustrated how the SQO instructions are used for machines that have timed machine cycle elements. In the previous ladder logic standard and Example 11-1, variation in times among machine cycles was handled by having a constant timer pulse width and using multiple sequencer file words. Review this process and example before you continue.

The standard ladder logic in Figure 11-15 uses two SQO instructions and a TON timer. One is used to move the sequencer output files to the destination address for control of the process, and

FIGURE 11-15: Standard ladder logic rungs for SLC 500 timed SQO instructions.

Application

A machine or process that operates in a sequential fashion must execute its operational sequence based on a set of step times that are not the same.

The machine has three states with an output word for each state stored in memory. The true output bits for each state are 1, 3, 5, 8, and 12 for state one; 2, 3, 9, and 10 for state two; and 3, 4, 5, 8, and 12 for state three. The first state should be on for 100 seconds, the second for 50 seconds, and the third for 112 seconds.

The system requires two SQO instructions: one for the machine outputs (Output SQO) and one for the times at each state (State timer SQO). Both of the SQOs have R6:10 for control, 3 for length, and 0 for position. The other configuration for the SQOs includes:

The Output SQO (b) is:

- The three output words for states 1, 2, and 3 are stored in files N7:1, N7:2, and N7:3 so the file parameter is #N7:0.
- The mask is 0001011100111110B or 173Eh.
- The destination file is O:3.0.

The State timer SQO (c) is:

- The three output word values for the states are 100, 50, and 112 (seconds). They are stored in files N7:11, N7:12, and N7:13 so the file parameter is #N7:10.
- The mask is 00FFh.
- The destination file is T4:1.PRE.

The standard ladder logic (a) has two rungs and operates as follows:

- The regenerative timer cycles the two SQO instructions through their three states. The timer has an output waveform that looks like Figure 11-6(b) with the off time set by the values moved into the timer PRE parameter (T4:1.PRE) by the State timer SQO instruction.
- In position 1 the Output SQO moves the proper true bits through the mask to O:3.0. The timer has 100 seconds loaded into the timer PRE parameter, and state 1 is on for 100 seconds.
- In position 2 the outputs are on for 50 seconds. Output SQO moves the output bits and State timer SQO loads the correct on time to T4:1. Position 3 works in a similar fashion.

The output and timer data tables are shown below the ladder. SLC 500 rungs are shown, but the operation would be the same for PLC 5 and ControlLogix.

Standard Ladder Logic Rungs for SLC 500 Timed SQO Instructions

T4:1 DN

SQO
Sequencer Output
File #N7:0
Mask 173Eh
Dest O:3.0
Control R6:10
Length 3<
Position 0<
(EN)
(DN)

SQO
Sequencer Output
File #N7:10
Mask 00FFh
Dest T4:1.PRE
Control R6:10
Length 3<
Position 0<
(EN)
(DN)

T4:1 DN

TON
Timer On Delay
Timer T4:1
Time Base 1.0
Preset 0<
Accum 0<
(EN)
(DN)

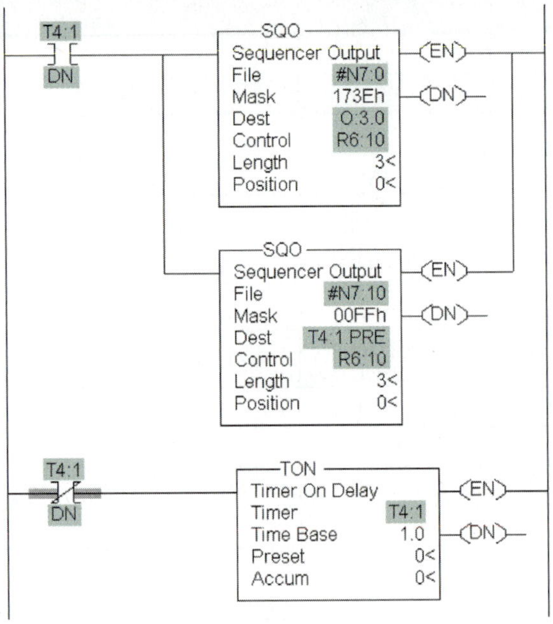

(a)

N7:0	0	0	0	0	0	0	0	0	0	0	0	0	0	0	0	0
N7:1	0	0	0	1	0	0	0	1	0	0	1	0	1	0	1	0
N7:2	0	0	0	0	0	1	1	0	0	0	0	0	1	1	0	0
N7:3	0	0	0	1	0	0	0	1	0	0	1	1	1	0	0	0

(b)

N7:10	0
N7:11	100
N7:12	50
N7:13	112

(c)

the second is used to move the required time for that machine cycle to the timer preset location. The ladder logic in Figure 11-15 serves both as a standard and as an example of how this type of control is accomplished. Read the description for the application in Figure 11-15.

The timer is a regenerative TON instruction with a done bit that is true for one scan time and false for the time value placed into the preset parameter. Every time the timer done bit is true, the position pointers on both of the SQO instructions increment to the next location. The timer preset value SQO moves the next timer preset value to T4:0.PRE and output SQO moves the next combination of true output bits to the destination, O:3.0. The outputs remain active for the preset time moved into the timer. After the values for the three position steps are moved to the output, the position parameter is reset to 1 and the process repeats.

EXAMPLE 11-3

Develop ladder logic to control traffic lights for north/south and east/west lanes at an intersection using SQO instructions. The east/west red light time should be 25 seconds, the north/south red light time should be 35 seconds, and the yellow should be on for 5 seconds for all directions. The green plus yellow times must correspond to the red time values.

SOLUTION

The ladder and timing diagrams for the traffic light controller are illustrated in Figure 11-16(a), the sequencer output file is shown in Figure 11-16(b), and the timer preset file is shown in Figure 11-16(c). A start switch is added for control. Note that

FIGURE 11-16: Ladder solution for Example 11-3.

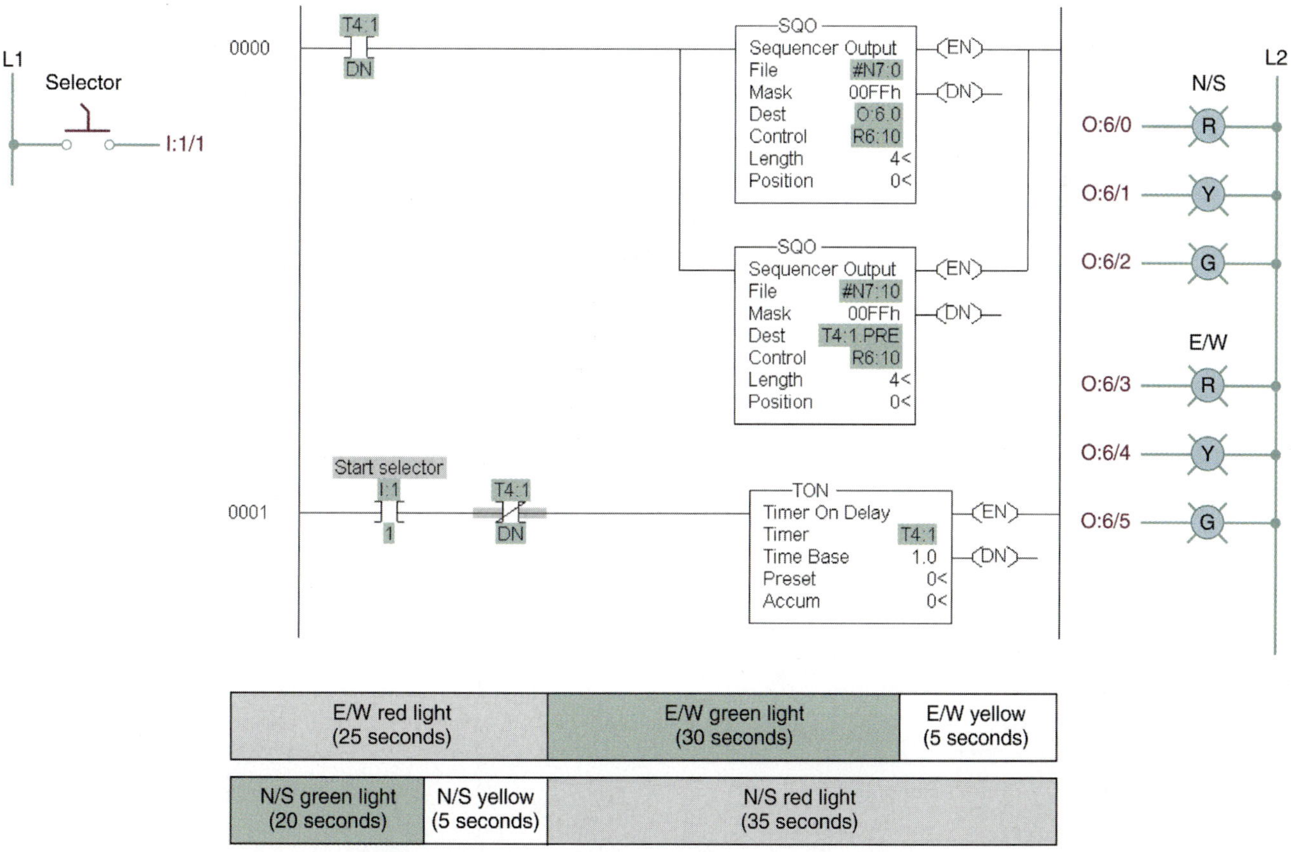

L1 Selector I:1/1

0000 T4:1 DN

SQO Sequencer Output File #N7:0 Mask 00FFh Dest O:6.0 Control R6:10 Length 4< Position 0< (EN) (DN)

SQO Sequencer Output File #N7:10 Mask 00FFh Dest T4:1.PRE Control R6:10 Length 4< Position 0< (EN) (DN)

0001 Start selector I:1 1 T4:1 DN

TON Timer On Delay Timer T4:1 Time Base 1.0 Preset 0< Accum 0< (EN) (DN)

L2 N/S O:6/0 R O:6/1 Y O:6/2 G E/W O:6/3 R O:6/4 Y O:6/5 G

E/W red light (25 seconds)	E/W green light (30 seconds)	E/W yellow (5 seconds)

N/S green light (20 seconds)	N/S yellow (5 seconds)	N/S red light (35 seconds)

(a) Ladder and timing diagram for intersection traffic light

the position, length, and control on both SQO instructions are the same. The sequencer files and output destination files are different, so the SQOs operate in parallel but with different data and different target destinations.

Start with the position parameter at 0 and let the timer done bit become active for the first time. The position parameter is incremented to 1 and data at that location in both sequencer files is moved to the destination address. Output bits 0 and 5 are turned on for O:6.0, and the 30-second value is moved to the timer preset, T4:1.PRE. The timer restarts with a 30-second delay, which keeps output bit 0 (N/S red) and bit 5 (E/W green) on for 30 seconds. When the done bit occurs after the 30 seconds, the sequencers move to position 2. The timer preset is changed to 5 seconds, and the data word here keeps the N/S red light on again (O:6.0 bit 0), while the E/W light changes to the yellow light (O:6.0 bit 4). Each time the timer done bit is true, the next data words (timer preset value and output bit on combination) are moved to the destination addresses. When the position parameter equals the length parameter, the position parameter returns to 1 on the next execution of the timer. This sequence continues until the start switch is false.

FIGURE 11-16: (Continued).

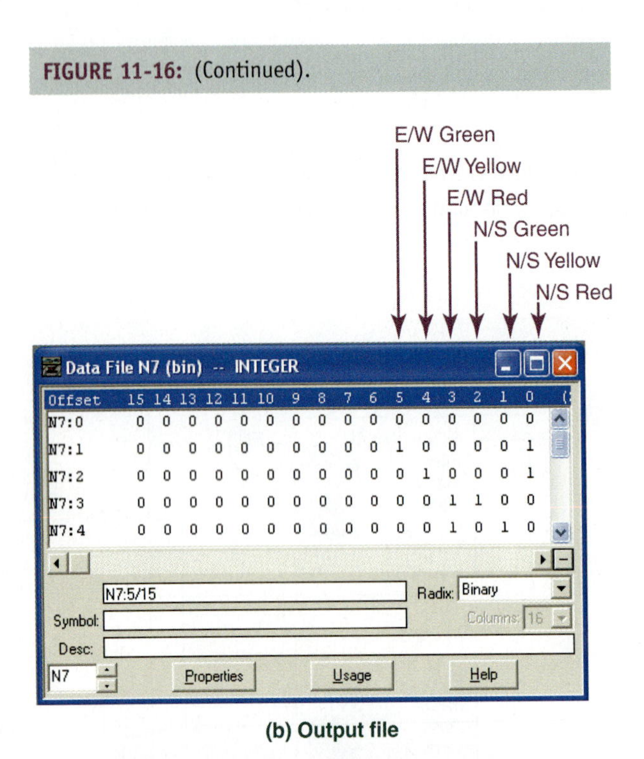

(b) Output file

Data File N7 (dec) -- INTEGER

Offset	0	(Symbol) Description
N7:0	0	
N7:1	30	
N7:2	5	
N7:3	20	
N7:4	5	

N7:5 Radix: Decimal
Symbol: Columns: 1
Desc:
N7 Properties Usage Help

(c) Timer preset file

11-4-5 Sequencer Compare (SQC) Instruction

Figure 11-17 illustrates the SLC 500 *SQC instruction*, which is similar to the PLC 5 and ControlLogix SQI instruction. The instruction parameters are the same as those used for the SQI instruction, but the differences between the two instructions include:

- The SQC is an output instruction in the SLC 500, whereas the SQI is an input instruction in the other two PLCs.
- Unlike the SQI instruction, the SQC instruction increments the position parameter when its input sees a false to true transition.
- The SQC instruction has an additional status bit—the *found bit* (FD). When the source pattern matches the sequencer file word pattern, the FD bit is a 1. It is a 0 under all other conditions.

The SQC instruction can be used as a diagnostic tool to determine if a machine has the correct switch/signal bit values in a register to start a production run. The control file found bit (FD) is

FIGURE 11-17: Sequencer SQC instruction for SLC 500.

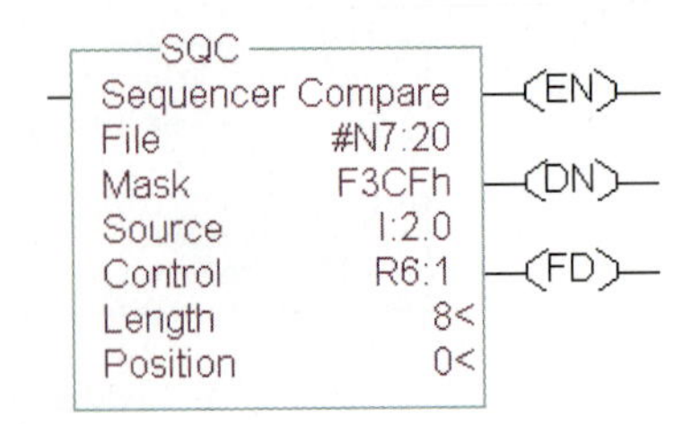

used to activate an indicator to let an operator know that the machine is ready for production.

The SQO and SQC instructions are also used as a pair, as illustrated in the standard ladder logic in Figure 11-18. This is similar to the SQO and the SQI pair for the PLC 5 and ControlLogix PLCs. In this case the FD bit in the control word is used to make sure that the pair of instructions increment in unison. In other words, if input I:2 matches the SQC file word, then the SQC instruction sets the FD bit, which in turn enables the SQO function. Read the description of the operation in Figure 11-18.

11-4-6 Sequencer Load (SQL) Instruction

The *SQL instruction* transfers data from a source address into a sequencer file on every false to true transition of the SQL instruction rung. It's a word-to-file transfer without a mask and is illustrated in Figure 11-19. With a length of 8, the 8 values or bit patterns of source I:4.0 are moved into file #N7:11 to #N7:18. In the figure, the SQL is at position 4 so the 4th value of the input bit pattern is being moved into file #N7:14 (see highlight). On each false to true transition at its input, the SQL instruction position is incremented by 1. When the instruction has reached position 8, the length value, the DN bit is set. On the next false to true transition, the instruction recycles to position 1 and the value of input I:4.0 is moved into that location. The SQL instruction loads data into position 0 only if the instruction was true during the first PLC scan.

The most common application for the SQL instruction is for loading data registers for the SQC and SQI instructions. Loading a large number of input conditions for a large number of machine steps is tedious and prone to errors. The robot in Figure 11-20 is a good example. Note that each axis and the gripper have a sensor at the extremes of travel. Ladder logic using the SQL instruction could read the outputs of every sensor and move the sensor bit patterns into a contiguous set of sequencer word files. The procedure includes manually moving the robot to the home position and pressing a push button to make the SQL instruction true. With all the sensor outputs connected to the same input word, the SQL ladder logic moves the value of all sensors to an integer register. Next, the robot arm is moved to the first position and the sensor values are loaded. This

FIGURE 11-18: Standard ladder logic rungs for SLC 500 SQC and SQO instruction pairs.

Application	Standard Ladder Logic Rungs for SLC 500 SQC and SQO Instruction Pairs
A machine or process that operates in a sequential fashion must execute its operational sequence based on a set of discrete input conditions using an SLC 500 PLC. The SLC 500 does not have an SQI instruction like that found on the PLC 5 and ControlLogix. The sequencer compare, SQC, instruction performs the same function as the SQI but is an output instruction. The SQC can be paired with an SQO so that input words are used to send selected output words to the output destination address. Since both the SQC and SQI are output instructions, the SQC has a control bit, called Found (FD), that is used to link the two instructions. Ladder logic linking the two instructions using the found bit is illustrated on the right. Note that both instructions have the same control register, length, and position values. The input logic on both instructions is the NOT found bit. When the words in I:2.0 are the same as the 8 register words in N7:21 to N7:28, the found bit is true and the SQO instruction outputs the corresponding output words from N7:11 to N7:18 to the destination file.	Found bit R6:1 FD Data compare SQC Sequencer Compare File #N7:20 Mask F3CFh Source I:2.0 Control R6:1 Length 8< Position 0< (EN) (DN) (FD) Found bit R6:1 FD SQO Sequencer Output File #N7:10 Mask F3CFh Dest O:5.0 Control R6:1 Length 8< Position 0< (EN) (DN)

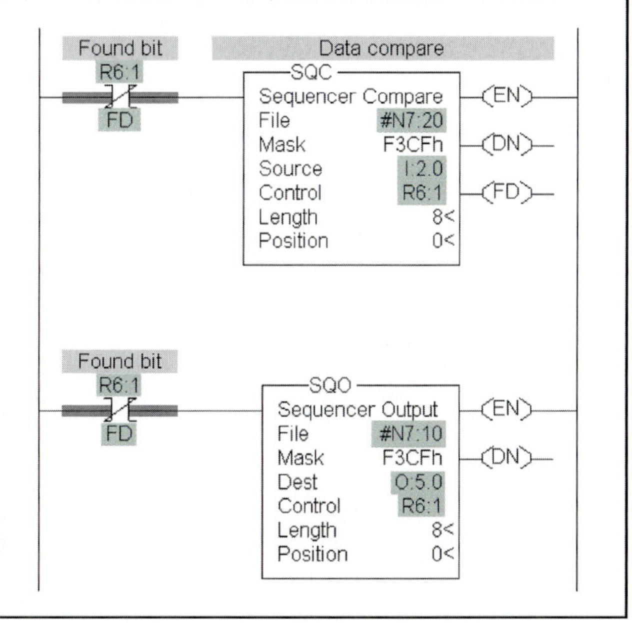

FIGURE 11-19: Sequencer load (SQL) instruction.

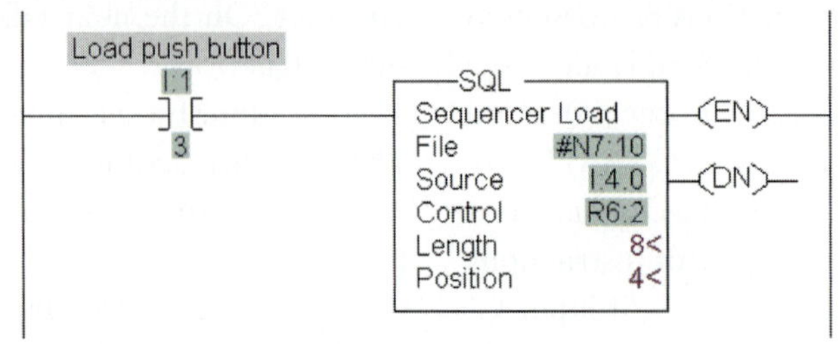

(a) Sequencer instruction

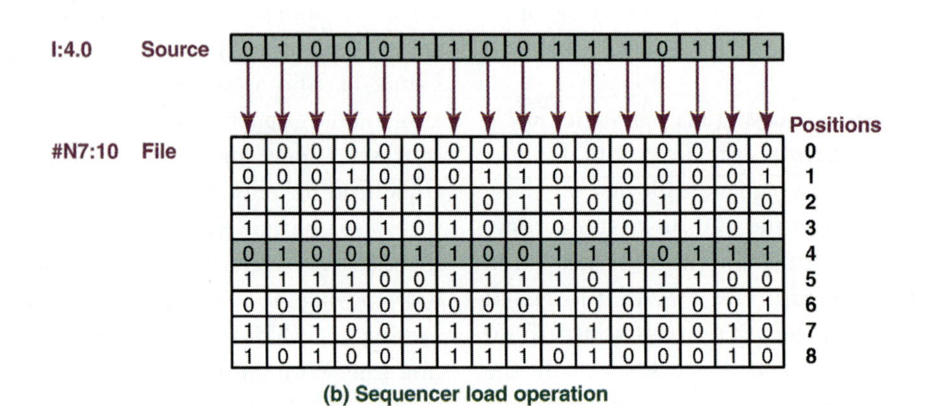

(b) Sequencer load operation

FIGURE 11-20: Two-axis robot part loader.

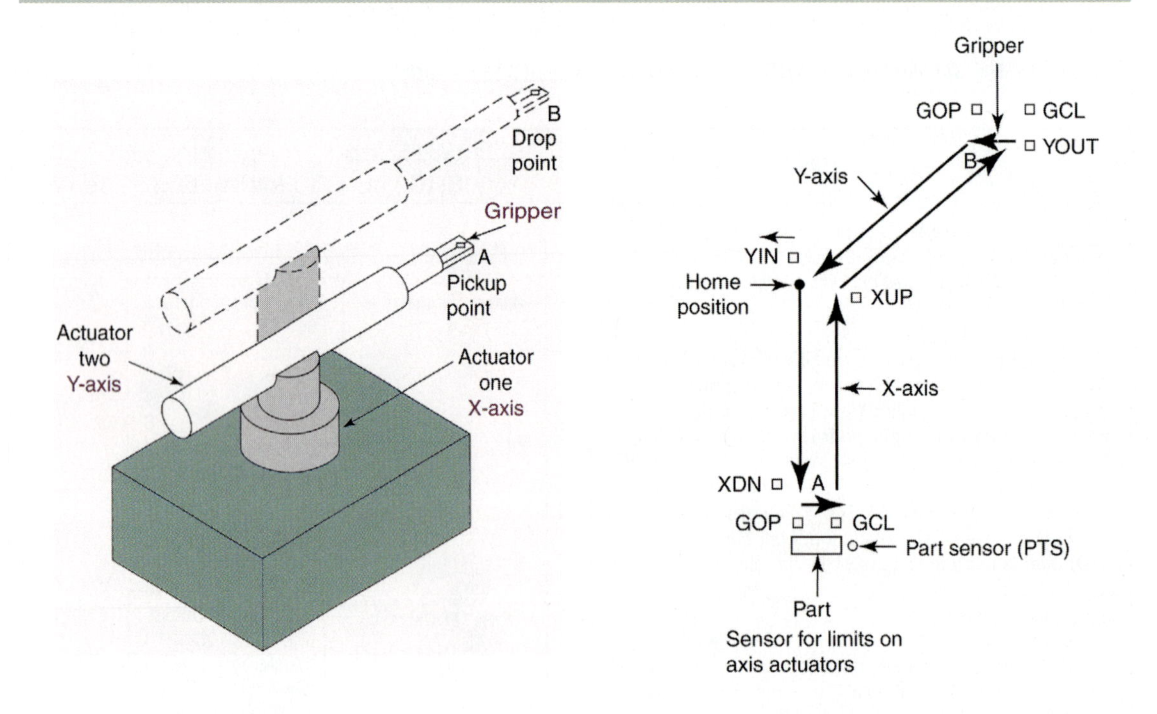

continues for every position in a complete robot motion cycle. The result is a complete sequencer file for use in an SQI/SQO pair or an SQC/SQO pair.

11-5 CASCADING SEQUENCERS

The technique of *cascading*, or *chaining* as it's called in some literature, provides a greater number of output bits than a single sequencer can provide with 16-bit words. Figure 11-21 depicts two 16-bit output sequencers, which are chained to achieve a 20-bit output. Figure 11-21(a) illustrates the pattern output for each step. Note that all 16 output bits of the top SQO are active and only 4 bits of the lower SQO are active, with the other 12 bits (bits 5 through 16) always zero. The active bits provide a 20-bit output. For example, the 20-bit output pattern for step 3 is 00110001001100001010. Figure 11-21(b) shows the ladder diagram of the two sequencers that share the same control, length, and mask. Because the sequencers have common control, length, and mask parameters, they are always on the same step and sequence together through the 15 patterns. However, the destination files are chained by using consecutive output files O:5.0 and O:6.0. The mask in the second sequencer could be 000Fh since only the first four bits are used.

11-6 EMPIRICAL DESIGN PROCESS WITH SEQUENCER INSTRUCTIONS

The empirical design process, introduced in Section 3-11-4, is an organized approach to the design of PLC ladder logic programs. However, the term *empirical* implies that some degree of trial and error is present. In some chapters the instructions covered required numerous rungs. As a result of these more complex solutions more fixes and adjustments were necessary as the programs were

FIGURE 11-21: Cascaded sequencers.

Step	Lower SQO Bits 16 - 13	Lower SQO Bits 12 - 9	Lower SQO Bits 8 - 5	Lower SQO Bits 4 - 1	Top SQO Bits 16 - 13	Top SQO Bits 12 - 9	Top SQO Bits 8 - 5	Top SQO Bits 4 - 1
1	0000	0000	0000	1101	0001	0010	1000	1000
2	0000	0000	0000	1000	0001	1000	0001	0011
3	0000	0000	0000	0011	0001	0011	0000	1010
4	0000	0000	0000	1010	1111	1010	0011	1000

Note: Only 4 of the 15 steps are shown.

(a) Cascaded sequencers pattern output

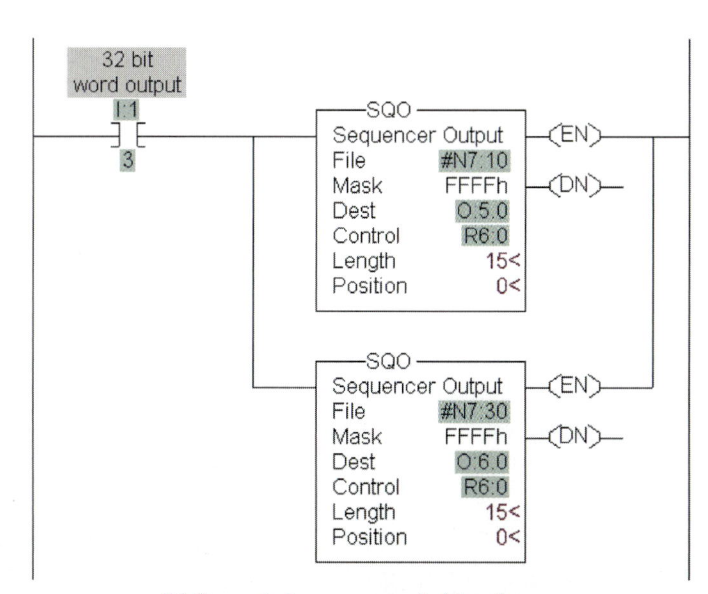

(b) Cascaded sequencers ladder diagram

debugged. However, the instructions added in this chapter use few rungs, so if the word data files are set up properly, the instruction parameters are selected properly, and the large amounts of data are entered correctly, then these rungs should require a minimum of change during initial tests.

11-6-1 Adding Sequential Instructions to the Process

The first step in using sequencer instructions is to know which instructions to use and in what combinations. The instructions available for the SLC 500 are SQO, SQC, and SQL; for the PLC 5 and ControlLogix they are SQO, SQI, and SQL. They operate the same in each PLC model except for the variable naming and memory identification differences mentioned frequently between the ControlLogix and the SLC 500 and PLC 5 processors. The guidelines in Table 11-5 can be used to identify the sequencer solution for a sequential process where the machine has a sequence of steps or states that are executed in order during every machine cycle.

In all the control requirements where an SQO instruction is used, the discrete control outputs are stored in a sequence of words. The PLC 5 and SLC 500 use integer or bit type memory registers and the ControlLogix uses arrays. Therefore, the deciding factor on which configuration to choose is based on the type of input used to trigger the move through the output data.

TABLE 11-5 Sequencer instructions for control requirements.

Processor/Sequencer Instructions	Control Requirements	Standard Ladder Logic Solution
All processors SQO	Each step or state in the sequence is active for a fixed time period.	(a) Standard logic in Figure 11-7(a) or in Figure 11-15 can be used for this requirement. If there are a small number of machine states and several have the same time, then the solution in Figure 11-7(a) can be used. If there are many machine states and the times for each vary widely, then the solution in Figure 11-15 is used. In general, the solution in Figure 11-15 is used for all timer-based sequencers.
All processors SQO	The sequencer input is driven by a small set of steps or states that are triggered from a set of external events.	(b) Use the standard logic in Figure 11-7(b).
SLC 500 SQC and SQO	Each sequencer output condition is triggered by a set of input bits from external field devices or other type of source register.	(c) Use the standard logic in Figure 11-18.
PLC 5 and ControlLogix SQI and SQO	Each sequencer output condition is triggered by a set of input bits from external field devices or other type of source register.	(d) Use the standard logic in Figure 11-12.
All processors SQL	Load a word data file with input field device bit patterns.	(e) Use the standard logic in Figure 11-19.

EXAMPLE 11-4

Determine the combination of sequencer instructions for all three processors that are required for programming the following automation situations. Use Table 11-5.

1. The robot in Figure 11-20 when the robot does not have end-of-travel sensors on the actuators, and the times for each movement are 2 seconds for the gripper and 2 seconds for each linear actuator movement.
2. The robot in Figure 11-20 when sensors are present at the end of travel for each actuator and for the gripper.
3. The tank control problems in Figures 1-17 and 1-19(b).
4. The process rinse cycle in Figure 7-21.
5. A process that has seven states or steps with eight outputs per state. The states must be on or active for the following times in seconds: 33, 150, 3, 45, 67, 300, and 1 for states 1 through 7, respectively.
6. The robot in Figure 11-20 needs the SQC or SQI sequencer file or array loaded with input data for each machine state.

SOLUTION

The automation problems require the following solutions.

1. Use the solution in Table 11-5(a) with the standard solution from Figure 11-7(a).
2. Use the solution in Table 11-5(c) and (d), depending on the type of processor present.
3. Use the solution in Table 11-5(b) with the standard solution from Figure 11-7(b).
4. Use the solution in Table 11-5(c) and (d), depending on the type of processor present.
5. Use the solution in Table 11-5(a) with the standard solution from Figure 11-15.
6. Use the solution in Table 11-5(e).

11-7 TROUBLESHOOTING SEQUENCER INSTRUCTIONS

The major causes of problems with sequencer ladder rungs are incorrect data in the sequencer data files and arrays and improper entry of parameter values in the instructions. If a sequencer does not operate properly, use the following guidelines when troubleshooting the sequencer rungs.

- Verify all the data in the sequencer file(s).
- Verify the instruction parameters, especially the length value. In applications where the sequencer instruction is false during the first PLC scan, the 0 position is not included in the length value.
- If an SQO instruction is used with a timed input, then replace the timed input with a manual switch or forced input instruction to manually step through the sequencer file while all the sequencer outputs and machine actions are verified.
- If an SQC or SQI instruction is used to drive an SQO instruction, then add an input to force the SQO instruction to manually step through the sequencer file while all the machine inputs, sequencer outputs, and machine actions are verified.
- This guideline could be listed in the empirical design areas as well. When troubleshooting a sequencer problem with a large database having many elements, start with just a small data set. For example, if you can make the ladder operate with just a few machine inputs and sequencer outputs, then you can easily expand it for the larger data set.

11-8 LOCATION OF THE INSTRUCTIONS

The location of instructions from this chapter in the Allen-Bradley programming software is indicated in Figure 11-22.

FIGURE 11-22: Location of instructions described in this chapter.

Systems	Instructions	Location
PLC 5	SQI, SQO, SQL	 PLC5 Instructions: SQI SQO SQL SDS DFA — Move/Logical, File/Misc, File/Shift, Sequencer, P
SLC 500	SQC, SQO, SQL	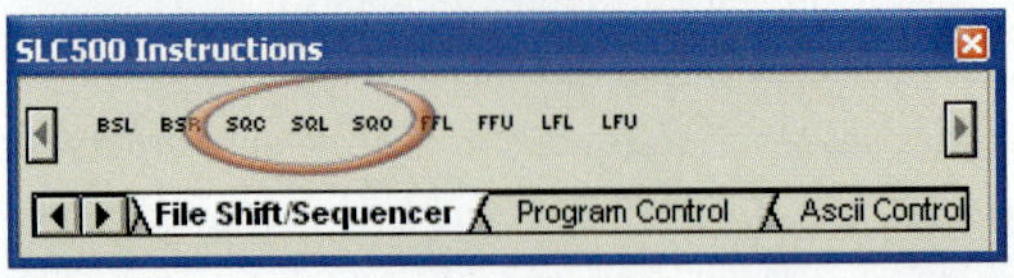 SLC500 Instructions: BSL BSR SQC SQL SQO FFL FFU LFL LFU — File Shift/Sequencer, Program Control, Ascii Control
LOGIX	SQC, SQO, SQL	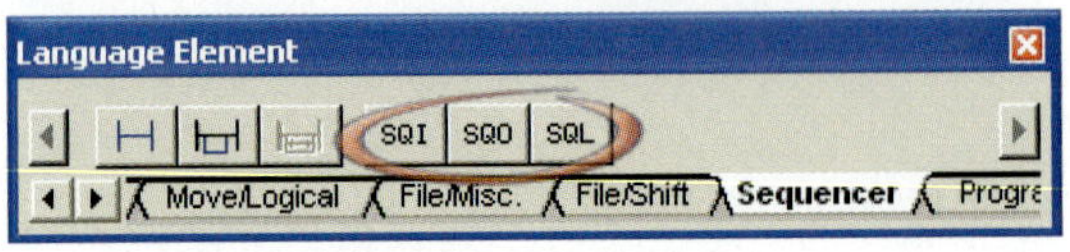 Language Element: SQI SQO SQL — Move/Logical, File/Misc., File/Shift, Sequencer, Progra

Chapter 11 questions and problems begin on page 582.

CHAPTER

12

Analog Sensors and Control Systems

12-1 GOALS AND OBJECTIVES

There are three principal goals of this chapter. The first goal is to provide the student with an introduction to the operation and function of analog sensors. The second goal is to describe the analog I/O modules and their operation for the Allen-Bradley PLC 5, SLC 500, and Logix systems. The third goal is to provide the basics of the closed-loop control system operation, including system attributes and PLC applications.

After completing this chapter you should be able to:

- Describe the operation and function of analog devices such as temperature, pressure, flow, and position sensors.
- Name and describe the components of a vision system and describe the operation.
- Select the appropriate type of vision and illumination system given the parameters for a sensing application.
- Describe the analog I/O modules for the Allen-Bradley PLC 5, SLC 500, and Logix systems.
- Explain the generalized closed-loop block diagram, and state the purpose of each of the blocks.
- State the characteristics that differentiate between an effective and an ineffective control system.
- List the general closed-loop control modes and explain how each acts to correct the system error.
- Describe PLC proportional closed-loop process control.

12-2 ANALOG SENSORS

Analog sensors, in contrast to discrete sensors, measure a range of input conditions and generate a range of output values. In analog temperature measurement, for example, the analog sensor typically produces an output that varies from 1 to 5 volts or 4 to 20 milliamperes when exposed to a temperature range of 08 to 1008 Celsius. Most sensors have a direct and linear relationship between the input condition and the output response. However, some sensors have inverse input and output relationships and others produce a nonlinear output from a linear change in the input. Analog sensors measure the common process parameters such as temperature, pressure, flow, level and position. These sensors and vision systems are discussed in the following subsections.

12-2-1 Temperature Sensors

Temperature sensors are classified based on their output as follows:

- **Resistance as a function of temperature:** resistance temperature detectors and thermistors
- **Voltage as a function of temperature:** thermocouples and solid-state temperature sensors

A *resistance temperature detector* (RTD) is a temperature-sensing device that detects a change in the resistance of a metal as a function of temperature. The most common metals used in RTDs are listed in Table 12-1 along with their temperature range and their resistance coefficient. Note that platinum has the widest temperature range and lowest coefficient of resistance. This large temperature range makes it the most popular metal for RTDs.

TABLE 12-1 Common materials used in RTDs.

RTD Material	Temperature Range In °C	Resistance Coefficient In Ohms/°C
Copper	−151 to +149	0.0042
Nickel	−73 to +149	0.0067
Platinum	−184 to +815	0.0039
Tungsten	−73 to +276	0.0045

The resistance coefficient is the amount of resistance change for each degree of temperature change. The RTD's change in resistance must be converted to a change in voltage or current so it can be usable. Figure 12-1 depicts a two-terminal RTD in a bridge circuit and a differential amplifier connected to points X and Y of the bridge. Note that the RTD is represented as the variable resistor R_T. If the resistance of the RTD is at room temperature, the potential difference at the bridge output is zero volts, and the differential amplifier output is zero volts. If the temperature of the RTD changes, the resistance changes and the bridge is no longer balanced, producing a potential difference across the bridge at X and Y and causing an output voltage on the differential amplifier. In addition to the two-terminal RTD as depicted in the bridge circuit, three-terminal and four-terminal RTDs are available and used in applications that require high accuracy.

Thermistors are electronic components that exhibit a large change in resistance with a change in the device temperature. The symbol for a thermistor is the same as for an RTD, the variable resistor symbol. The word thermistor is a contraction of the words thermal resistor. Depending on the type selected, thermistors have either a negative temperature coefficient (NTC) or positive temperature coefficient (PTC) of resistance. NTC thermistors exhibit decreasing electrical resistance with increases in environmental temperature and

FIGURE 12-1: Bridge circuit with two-terminal RTD.

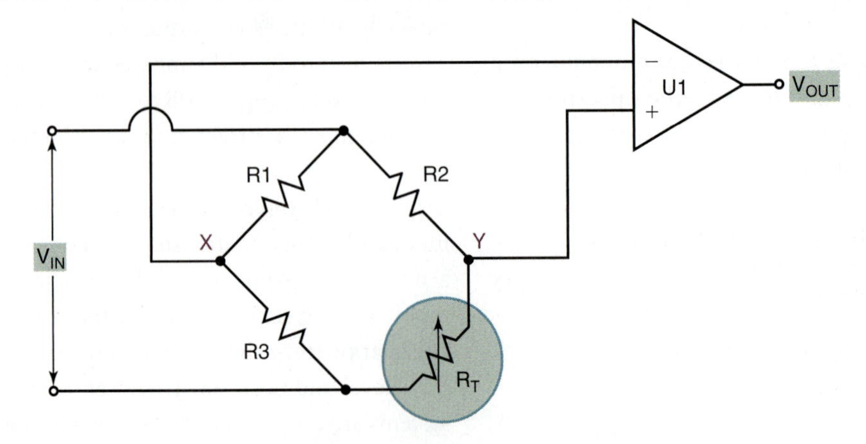

Rehg and Sartori, Industrial Electronics, *1st Edition, © 2006, Reprinted by permission of Pearson Education, Inc., Upper Saddle River, NJ.*

increasing electrical resistance with decreases in environmental temperature, and PTC thermistors exhibit increasing electrical resistance with increases in environmental temperature and decreasing electrical resistance with decreases in environmental temperature. Figure 12-2 illustrates the resistance-temperature graph of both the NTC and PTC thermistors.

Thermocouples are temperature-sensing devices that produce a small voltage in the millivolt range as a function of temperature. Thermocouples are constructed using two dissimilar metal wires, which are listed in Table 12-2 along with the applicable temperature range and an assigned letter designating the thermocouple type.

The operation of the thermocouple can be traced to a discovery made by Thomas Seebeck in the early 1800s known as the Seebeck effect. The discovery was that if two wires made from dissimilar metals are connected at both ends, making two junctions, then when one end is heated a small amount of current will flow through the wires. Figure 12-3(a) illustrates the Seebeck effect. Figure 12-3(b) shows a voltmeter replacing one of the junctions where the millivolt level voltage is proportional to the amount of heat at the remaining junction.

With the hookup as in Figure 12-3(b), the connection between the thermocouple wire and the voltage lead forms another junction of dissimilar metals; thus a small voltage is produced, which is opposite to the voltage produced by the thermocouple junction, causing an erroneous voltage reading. However, if this newly created junction, called the *reference junction*, is at zero volts, then the voltmeter will read the correct voltage. In the late 1800s the reference junction was called the *cold junction* because it was immersed in an ice bath, which held the temperature to 0° Celsius, thereby holding the voltage to zero volts. Rather than the ice bath, a more convenient technique called a *reference* block, which is shown in Figure 12-4, uses junction temperature compensation to keep the block at a predetermined temperature. The extension wires in the

FIGURE 12-2: Resistance-temperature curves for thermistors.

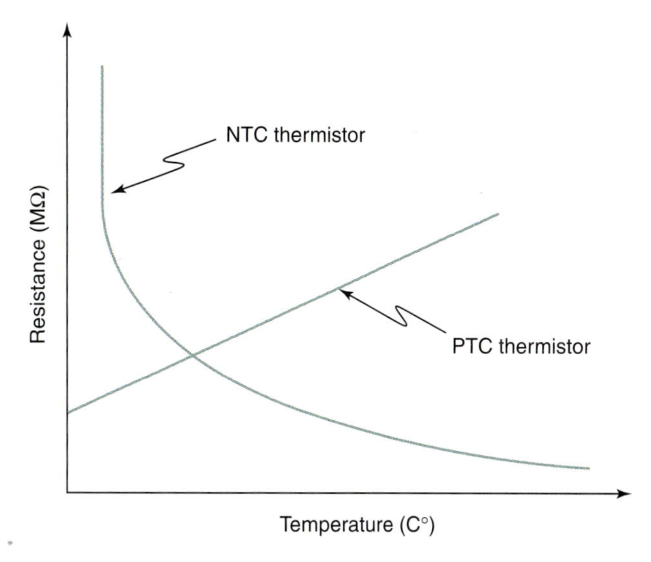

Rehg and Sartori, Industrial Electronics, *1st Edition, © 2006, Reprinted by permission of Pearson Education, Inc, Upper Saddle River, NJ.*

TABLE 12-2 Thermocouple materials.

Type	Thermocouple Materials	Temperature Range (°C)
B	Platinum & 6% or 30% Rhodium	0 to +1800
E	Chromel & Constantan	−190 to +1000
J	Iron & Constantan	−190 to +800
K	Chromel & Alumel	−190 to +1370
R	Platinum or Platinum & 13% Rhodium	0 to +1700
S	Platinum or Platinum & 10% Rhodium	0 to +1765
T	Copper & Constantan	−190 to +400

FIGURE 12-3: Thermocouple junction connections—Seebeck effect.

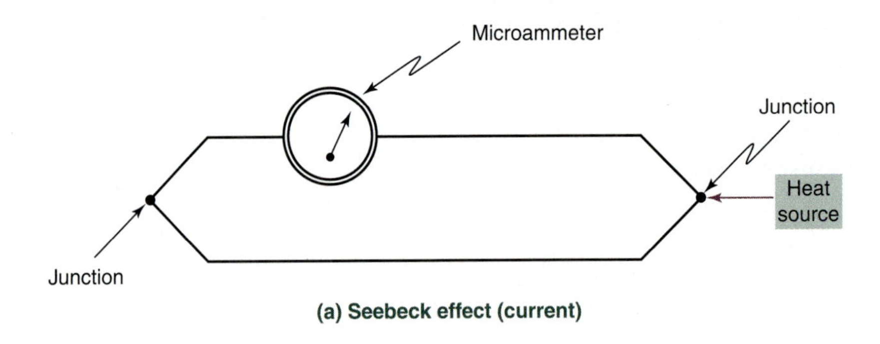

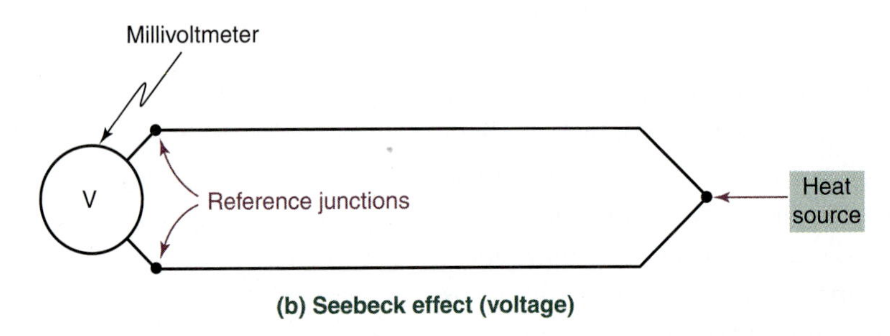

Rehg and Sartori, Industrial Electronics, *1st Edition, © 2006, Reprinted by permission of Pearson Education, Inc., Upper Saddle River, NJ.*

FIGURE 12-4: Thermocouple reference block.

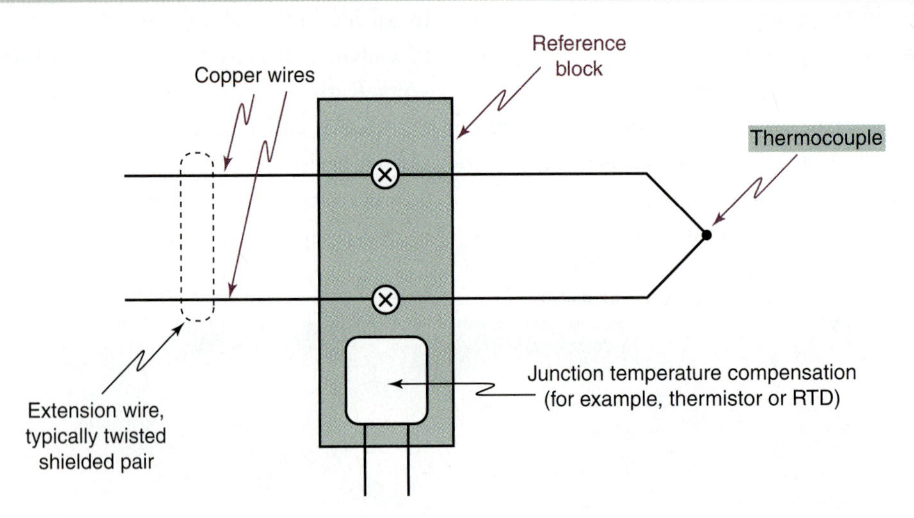

Rehg and Sartori, Industrial Electronics, *1st Edition, © 2006, Reprinted by permission of Pearson Education, Inc., Upper Saddle River, NJ.*

figure are standard copper conductors that are twisted, and sometimes shielded, and connected to an analog input of a PLC.

Solid-state temperature sensors are integrated circuits (ICs) whose output is linearly proportional to a temperature scale, typically Celsius. Designers who have embedded these ICs in cellular phones usually include one or more sensors in the battery pack; notebook computers might have four or more sensors for checking temperatures in the CPU, battery, AC adapter, and heat-generating assemblies. Although they

TABLE 12-3 Temperature sensor comparisons.

Temperature Sensor	Advantages	Disadvantages
RTD	Very stable and very accurate	Expensive, requires external power supply and is self-heating
Thermistor	Fast reacting to temperature changes	Non-linear in parts of its response curve, limited temperature range and fragile
Thermocouple	Self-powered, rugged, operates over large temperature range	Non-linear in parts of its response curve, requires temperature compensation, least sensitive
IC Temperature Sensor	Most linear, most inexpensive	Limited temperature range, requires external power supply and is self-heating

cannot always replace the traditional temperature sensors—resistance temperature detectors, thermistors, and thermocouples—IC temperature sensors offer many advantages. For instance, they require no linearization or reference-junction compensation. In fact, they often provide reference-junction compensation for thermocouples. They generally provide better noise immunity through higher-level output signals, and some provide logic outputs that can interface directly to a PLC.

Finally, a *comparison of temperature sensors* is shown in Table 12-3, which lists the advantages and disadvantages of the RTD, the thermistor, the thermocouple, and the IC temperature sensor.

12-2-2 Pressure Sensors

Pressure sensors are devices that detect the force exerted by one object on another. Before we discuss the various pressure sensors, it's important to understand the terms that are used in the discussion. First, *pressure* is defined as the amount of force applied to an area, where pressure in expressed in pounds per square inch (psi). Pressure in liquids and gases is referred to as *hydraulic pressure*, but in the case of gases pressure may not be a constant, but a variable. The change in an object's shape when force is applied is called *stress* or *strain*. Force applies to many situations and is basically a push or a pull. Pressure sensors are typically classified based on their operational characteristics as follows:

- **Direct pressure sensors:** pressure is a function of the amount of deformation of an object.
- **Deflection pressure sensors:** pressure is a function of the amount of deflection of an object.
- **Differential pressure sensors:** pressure is a function of the difference in two measurements.
- **Piezoelectric and solid-state sensors:** pressure is a function of the deformation/deflection of a crystal or electronic component.

Direct pressure sensors measure *strain*, which is the amount of deformation of a body due to an applied pressure or force. More specifically, strain is defined as the fractional change in length—either positive (tensile) or negative (compression). It is measured by a *strain gage*, which consists of a very fine wire or metallic foil arranged in a grid pattern. When the grid pattern in Figure 12-5 is placed in parallel with the pressure it maximizes the amount of metallic wire or foil subjected to strain. The grid is bonded to a thin backing, called the carrier, which is attached directly to the test specimen. Therefore, the strain experienced by the test specimen is transferred directly to the strain gage, which responds with a small linear change in electrical resistance. This change in resistance is based on the principle that the thinner the wire the higher the resistance. So in a strain gage, the applied pressure elongates the wire, making it thinner, and thus increases its resistance.

FIGURE 12-5: Metal strain gage bonded to a carrier.

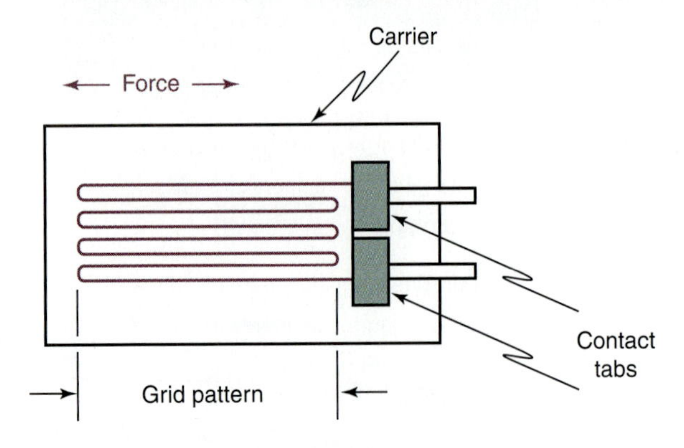

Rehg and Sartori, Industrial Electronics, *1st Edition, © 2006, Reprinted by permission of Pearson Education, Inc., Upper Saddle River, NJ.*

FIGURE 12-6: Active and dummy gages in a bridge circuit.

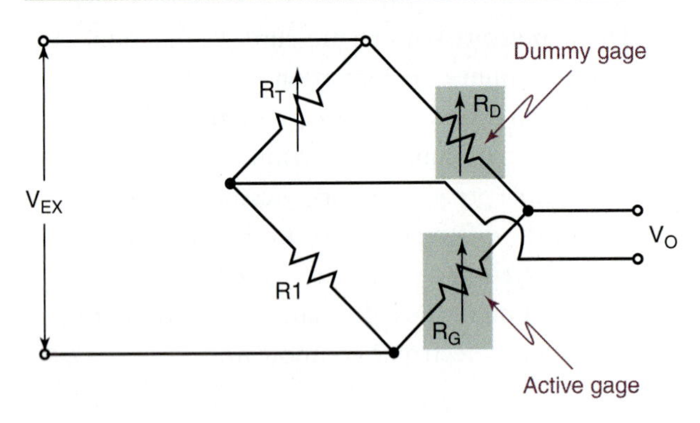

Rehg and Sartori, Industrial Electronics, *1st Edition, © 2006, Reprinted by permission of Pearson Education, Inc., Upper Saddle River, NJ.*

In order to make measurements of such small changes in resistance, strain gages are almost always used in a bridge circuit configuration, as depicted in Figure 12-6. Note that the figure shows two strain gages—an active gage and a dummy gage. The dummy strain gage is added to compensate for temperature changes, which impact the small resistance changes in strain gage. The dummy gage is attached perpendicular to the line of force, thus the force has no impact, but temperature changes occur in both gages and are canceled out since the gages are in opposite legs of the bridge. The balancing of the bridge by R_T compensates for resistor variations and for preload conditions.

Deflection pressure sensors such as the bourdon tube, diaphragm, and bellows consist of the following basic segments:

FIGURE 12-7: Primary segments of deflection-type pressure sensors.

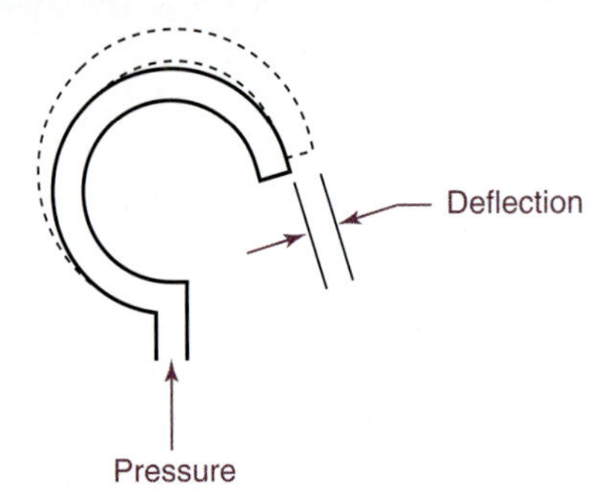

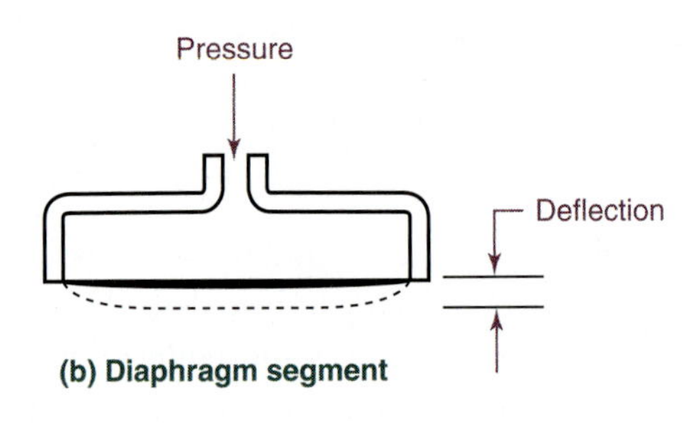

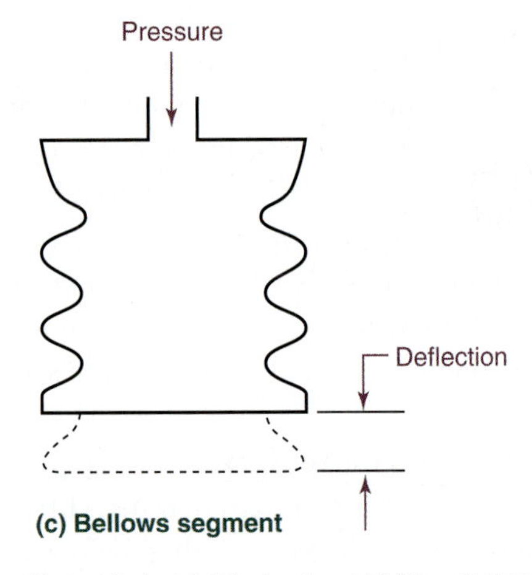

Rehg and Sartori, Industrial Electronics, *1st Edition, © 2006, Reprinted by permission of Pearson Education, Inc., Upper Saddle River, NJ.*

- A primary segment, which converts the applied pressure into a proportional displacement. (Figure 12-7 illustrates examples of the primary segments.)
- An intermediate segment, which converts the displacement into a change of an electrical component.
- A transmitter, which converts the change of the electrical element into a signal usable by a PLC.

The *bourdon tube* is a deformed hollow metal tube opened at one end and sealed at the other. The tube is available in various shapes, such as the shape of the letter C as illustrated in Figure 12-7(a), a spiral, or a helix. Bronze, steel, or stainless steel is typically used for the tube. The operation is as follows: The liquid whose pressure is being measured enters the tube at the mechanically constrained open end. As pressure increases, the tube tends to straighten, providing a displacement, which is proportional to the pressure.

The *diaphragm* illustrated in Figure 12-7(b) is a flexible membrane, which is made of rubber for low-pressure measurements and metal for high-pressure measurements. The diaphragm is mounted in a cylinder, creating a space on both of its sides. One space is open to the atmosphere and the other is connected to the pressure source to be measured. When pressure is applied, the diaphragm expands into the open space, and the amount of movement is proportional to the pressure being applied.

The *bellows* is a thin sealed metal cylinder with corrugated sides like the pleats of an accordion; it is illustrated in Figure 12-7(c). When pressure is applied to the bellows, the bellows expands and opens the pleats, producing a displacement that is proportional to the applied pressure.

Differential pressure sensors are modified diaphragm sensors where the diaphragm is located halfway between the ends of the cylinder. When pressure is applied to both ends the diaphragm will move toward the end with the lowest pressure. Figure 12-8 illustrates a differential pressure sensor, where P1 and P2 are the two pressure inlets of the sensor. A sensing device, whose output is sent to a PLC, detects the movement of the diaphragm, which is the difference between the two pressures.

Piezoelectric pressure sensors operate on the piezoelectric effect discovered in the late 1800s by Jacques and Pierre Curie. The piezoelectric effect occurs when pressure is applied to a crystal; the crystal deforms and produces a small voltage, which is proportional to the deformation.

S*olid-state sensors* derive measurements from pressure exerted on one side of a diaphragm. They differ from electromechanical sensors in that rather than consisting of several discrete components, these sensors have all their electrical and mechanical components built into a single piece of silicon. A small deflection of the diaphragm causes implanted resistors to exhibit a change in ohmic value. The sensor converts this change into a voltage that can be easily interpreted as a continuous and linear pressure reading. In addition, a laser-trimmed resistor network on the device provides temperature compensation and calibration.

FIGURE 12-8: Differential pressure sensor.

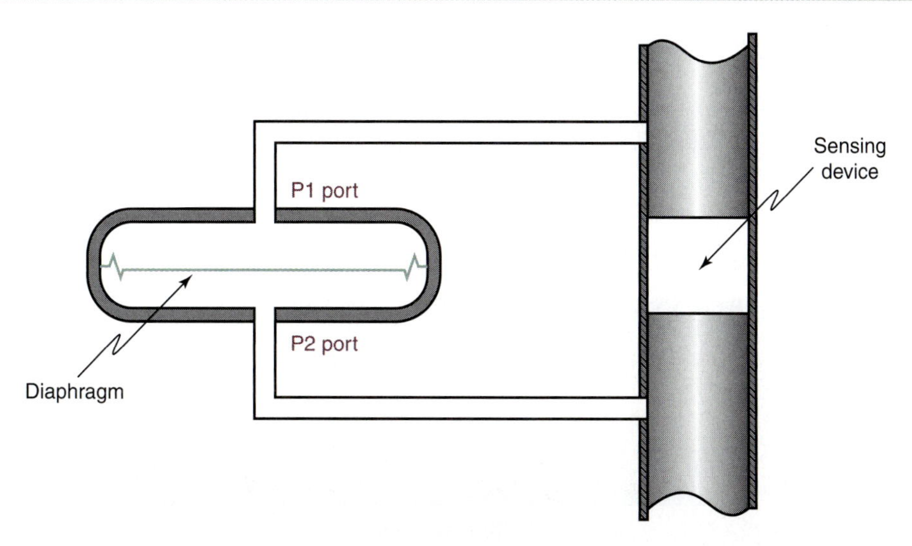

Rehg and Sartori, Industrial Electronics, *1st Edition, © 2006, Reprinted by permission of Pearson Education, Inc., Upper Saddle River, NJ.*

12-2-3 Flow Sensors

Sensing the flow of liquids and gases is an important measurement in industrial processes. The measurement of flow rate indicates how much fluid is used or distributed in a process and it is frequently used as control variable, which aids in maintaining the efficiency of a process. Flow rate is also used to indicate a fault such as improper flow of fluid through a pump or dysfunctional drain operation. Two broad categories of flow sensors are as follows:

- **Intrusive:** those sensors that disturb the flow of the fluid that they are measuring.
- **Non-intrusive:** those sensors that do not disturb the flow of the fluid that they are measuring.

However, before we discuss flow sensors, let's look at some of the terms and formulas that are used in industrial processes involving the measurement of flow. *Flow rate* is defined as the volume of material passing a fixed point per unit of time, whether the material is a solid, liquid, or gas—for example, lumber being conveyed into a sawmill or hydraulic fluid moving through a piping system. In this text, we'll be discussing the flow rate of fluids—liquids and gases. The flow rate of a fluid, flowing through a pipe, can be expressed in the following terms: $Q = V \times A$, where Q is the flow rate of the fluid through a pipe, V is the velocity of the fluid, and A is the cross-sectional area of the pipe.

Two more terms used to describe fluid flow are *laminar* and *turbulent*. They are depicted in Figure 12-9 and defined as follows:

- **Laminar flow:** fluid flows rather smoothly parallel to the walls of the pipe.
- **Turbulent flow:** fluid flows down the pipe, but swirls within the flow.

FIGURE 12-9: Laminar flow and turbulent flow.

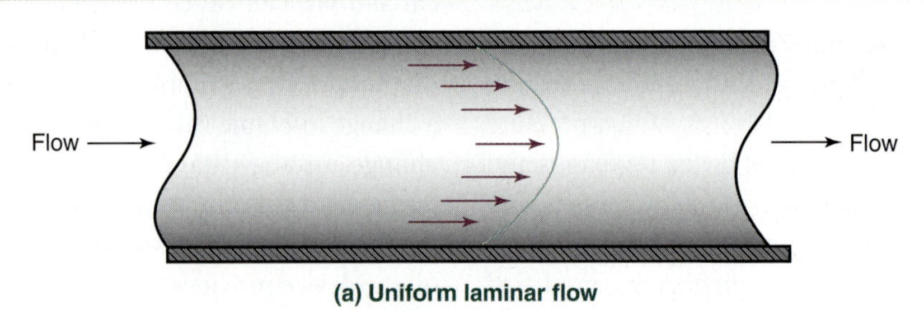

(a) Uniform laminar flow

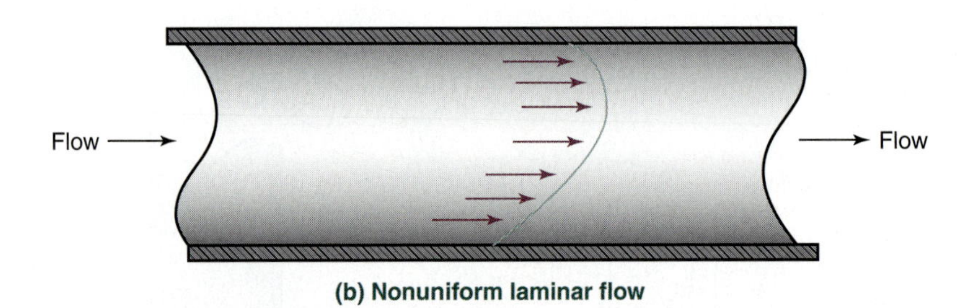

(b) Nonuniform laminar flow

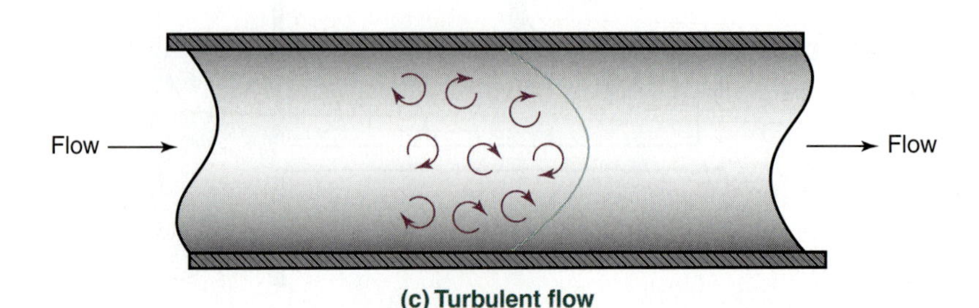

(c) Turbulent flow

The flows in Figure 12-9(a) and (b) depict laminar flow, but the shape of the flow in Figure 12-9(b) is non-uniform relative to the pipe walls, which indicates an imbalance in the frictional characteristics of the walls. Figure 12-9(c) depicts turbulent flow, illustrating the swirling action within the flow.

In the following subsections flow sensors are classified by their method of measurement because this is applicable to industrial applications. The *measurement methods* of flow sensors are classified as follows:

- **Differential:** the flow rate is determined by calculating the pressure difference (or pressure drop) as the fluid flows through an obstruction such as an orifice plate.
- **Velocity:** the flow rate is determined based on the velocity of the fluid as it passes through a turbine.
- **Displacement:** the flow rate is determined by measuring all the fluid used.
- **Mass:** the flow rate is determined based on the total volume of the fluid that passes through the sensor.
- **Visual:** the flow rate is determined by a visual reading of a graduated scale.

Differential Pressure Flow Sensors. Flow sensing from differential pressure (or pressure drop) is based on the fact that the difference in the pressure measurements on both sides of physical restriction in the flow of a fluid is proportional to the square of the flow rate. In other words, the *flow rate* is proportional to the square root of the pressure difference and is expressed as follows:

$$Q = k(P1 - P2)^{1/2}$$

Where Q is the flow rate

k is a constant specific to the type of restriction, in gallons per minute (gpm)

P1 is the pressure in front of the restriction in pounds per square inch (psi)

P2 is the pressure behind of the restriction in pounds per square inch (psi)

There are several types of devices that use this differential pressure relationship to flow rate. The method of generating a differential pressure differs in each of these devices, but all use the differential pressure sensor that was discussed in a previous section of this chapter to generate the output signal. The following is a brief overview of some of devices that are available in the marketplace.

- **Orifice plate:** An orifice plate is a washer-shaped device that is installed in a piping system, and the flow rate is determined from the measurements of pressure in front of and behind the plate. Figure 12-10 shows a diagram of an orifice plate installed in the pipe

FIGURE 12-10: Orifice plate flow sensor.

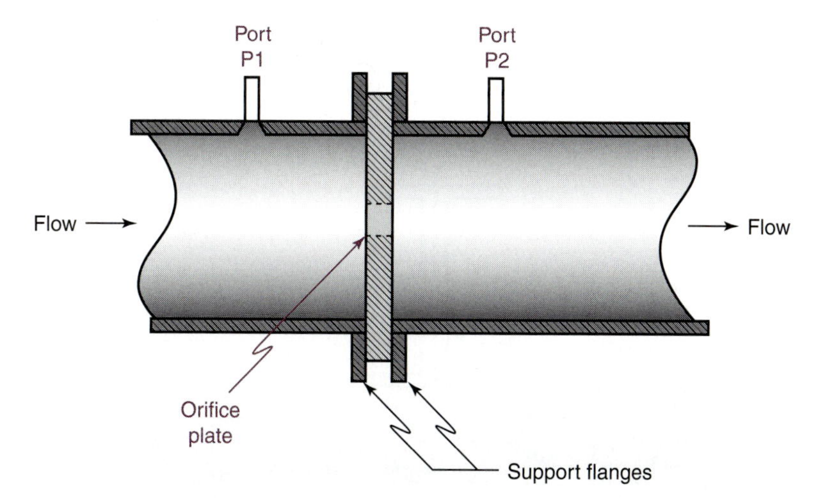

Rehg and Sartori, Industrial Electronics, *1st Edition, © 2006, Reprinted by permission of Pearson Education, Inc., Upper Saddle River, NJ.*

with the hole shown as dashed lines. The hole in the orifice plate is accurately drilled to a specific size, which impacts the constant in the flow rate formula. The flange plates are part of the mounting assembly for the orifice plate and provide easy access to inspect or install the plate. Note that the pressure measurement ports are labeled P1 and P2 and are sampled by a differential pressure sensor for flow rate determination.

- **Venturi:** A venturi is a flow-measuring device that consists of a gradual contraction followed by a gradual expansion within a fluid-carrying pipe. Figure 12-11 illustrates a venturi. Where the pipe is narrowed, the flow is slightly restricted, creating a difference in pressure between the flow entering the narrow area and leaving the narrow area. The high-pressure port P1 is used to sample the pressure as the fluid enters the narrow area, and the low-pressure port P2 is used to sample the pressure as the fluid exits the narrow area. In larger venturies additional ports are used to provide a means to determine an average pressure. A venturi has no installed parts and causes much less disturbance in the fluid flow than an orifice plate.
- **Pitot tube:** A pitot tube is a flow-measuring device consisting of two tubes placed in the fluid flow that sense two pressures—impact pressure and static pressure. Note that pitot is pronounced *pea toe*. Figure 12-12 illustrates

FIGURE 12-11: Venturi flow sensor.

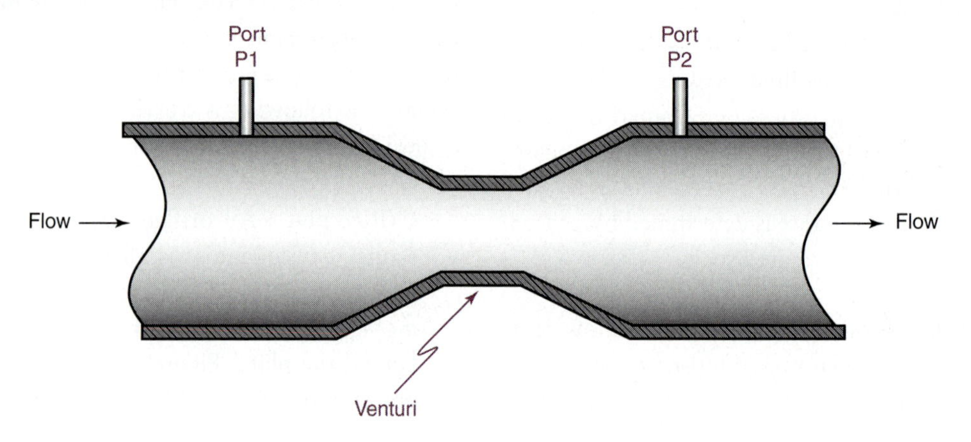

Rehg and Sartori, Industrial Electronics, *1st Edition, © 2006, Reprinted by permission of Pearson Education, Inc., Upper Saddle River, NJ.*

FIGURE 12-12: Pitot tube flow sensor.

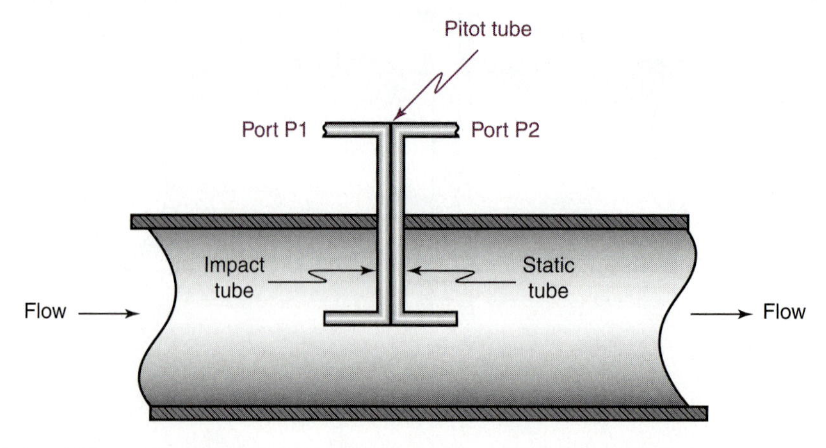

Rehg and Sartori, Industrial Electronics, *1st Edition, © 2006, Reprinted by permission of Pearson Education, Inc., Upper Saddle River, NJ.*

FIGURE 12-13: Flow nozzle flow sensor.

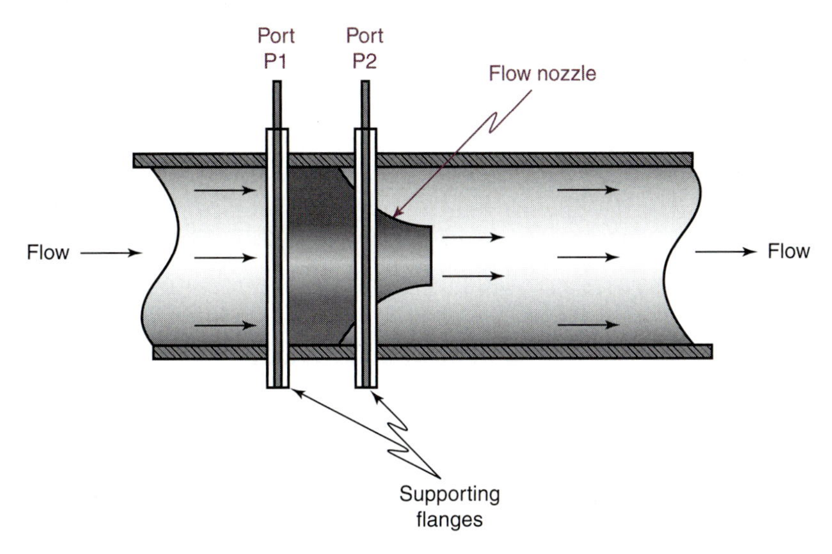

Rehg and Sartori, Industrial Electronics, *1st Edition, © 2006, Reprinted by permission of Pearson Education, Inc., Upper Saddle River, NJ.*

pitot tubes installed in a pipe. The pitot tubes are mounted inside of each other, with the inside tube, which faces the fluid flow, measuring the impact pressure. The second tube, which faces away from the fluid flow, measures the static pressure. The higher pressure is at the impact tube, whereas the lower pressure is at the static tube, and the difference between the two pressures are used to determine flow rate.

- **Flow nozzle:** A flow nozzle is a narrowing spout installed inside a piping system to obtain a pressure differential. It is illustrated in Figure 12-13. The nozzle is generally made of stainless steel, but it can also be made from other corrosion-resistant materials. A flow nozzle is typically installed between pipe flanges, but also is available with a machined ring on the outside diameter of the nozzle, allowing it to be welded between two sections of piping with the nozzle itself entirely within the pipe. Flow nozzles are often used as measuring elements for air and gas flow in industrial applications. Ports P1 and P2 are used to obtain the pressure differential measurements used in determining the flow rate.
- **Elbow:** An existing elbow in the piping system can be used to obtain a differential pressure; Figure 12-14 illustrates this technique. As the fluid flows through the elbow, it exhibits a slight pressure differential. The fluid that flows near the outer radius of the elbow has a slightly higher pressure than the fluid flowing near the inside radius. The pressure ports are labeled P1 and P2; these are the higher-pressure port and lower-pressure port, respectively. Depending on the radius of the elbow, the distance traveled by the fluid could be relatively farther on the outer radius than on the inner. In that case, two ports on both sides of the elbow can be used to obtain an average of each pressure, resulting in a more accurate pressure reading. The pressure difference is used to determine the flow rate.

FIGURE 12-14: Elbow flow sensor.

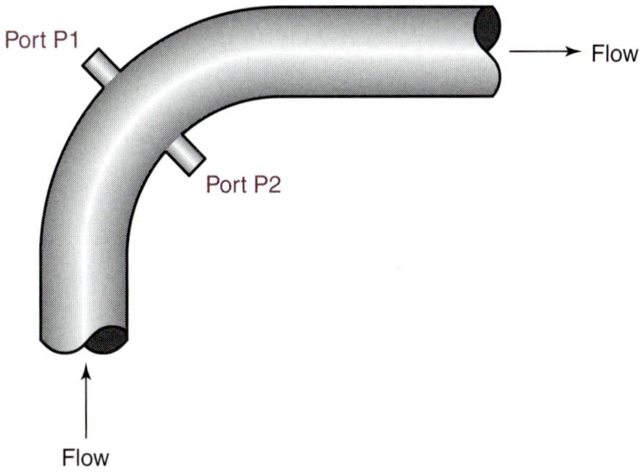

Port P1

Flow

Port P2

Flow

Rehg and Sartori, Industrial Electronics, *1st Edition, © 2006, Reprinted by permission of Pearson Education, Inc., Upper Saddle River, NJ.*

Velocity Flow Sensors. *Velocity flow sensors* are devices that measure the fluid flow rate based on changes in flow velocity. They include the paddlewheel, turbine, vortex, electromagnetic, and ultrasonic flow sensors.

- **Paddlewheel flow sensor:** This flow sensor is installed in a pipe so that the flowing fluid causes its paddlewheel to rotate. Figure 12-15 shows a paddlewheel flow sensor installation. The sensor is mounted in the pipe so that only a portion of the paddle extends into the flow. The paddlewheel flow sensor is mounted in a straight run of piping where the flow is laminar. The flow causes the paddlewheel to rotate, and its revolutions are converted to an electronic output. Generally a magnet is mounted to the paddles so that as they spin, a magnetic sensor converts the rotation of spin to electronic pulses. The sensor output is typically available as a square wave whose frequency is a function of the flow rate or as a variable voltage, but some sensors have a battery-powered electronic display mounted on the sensor for easy viewing.
- **Turbine flow sensor:** There are many different manufacturing designs of turbine flow sensors, but in general they are all based on the same principle—if a fluid moves through a pipe and strikes the vanes of a turbine, the turbine will start to spin and rotate. Figure 12-16 illustrates the turbine flow sensor, whose operation is similar to the paddlewheel where the rate of spin is measured to calculate the flow. High-quality jewel bearings and nickel-tungsten carbine turbine that are used for long life and low friction. The flange fitting ensures correct depth placement of the turbine. The rotation of the rotor is typically detected by a magnetic field and converted to square wave pulses. The number of square wave pulses is proportional to the volume of liquid passing through the turbine, and the flow rate Q can be expressed as follows:

$$Q = V/T = (k \times N)/T$$

Where V is the volume of liquid
k is the turbine constant in cm^3
N is the number of pulses
T is the time interval when the pulses are counted

- **Vortex flow sensor:** The vortex flow sensor includes a non-streamlined object (a bluff body) placed in the fluid flow, which creates vortices (whirlpools or eddies) in a down-

FIGURE 12-15: Paddlewheel flow sensor.

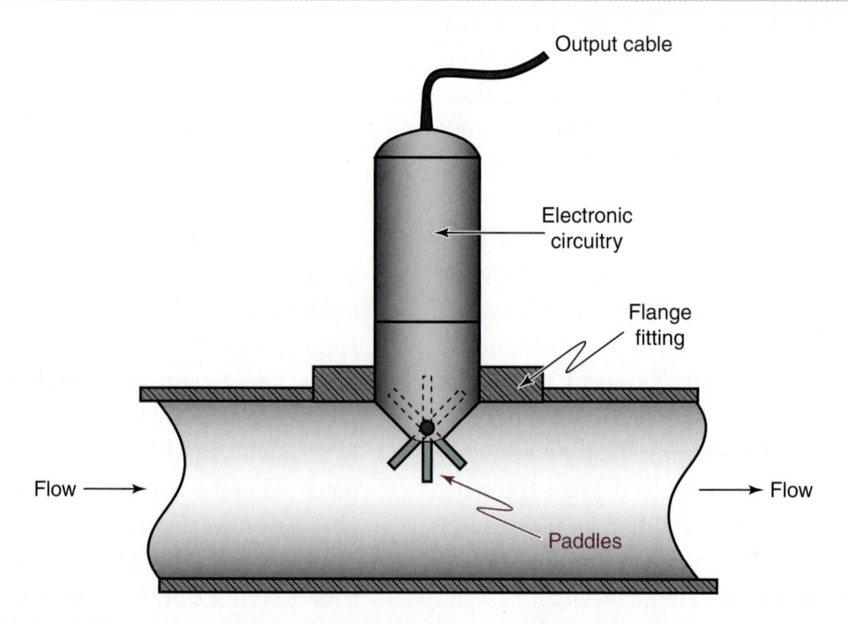

FIGURE 12-16: Turbine flow sensor.

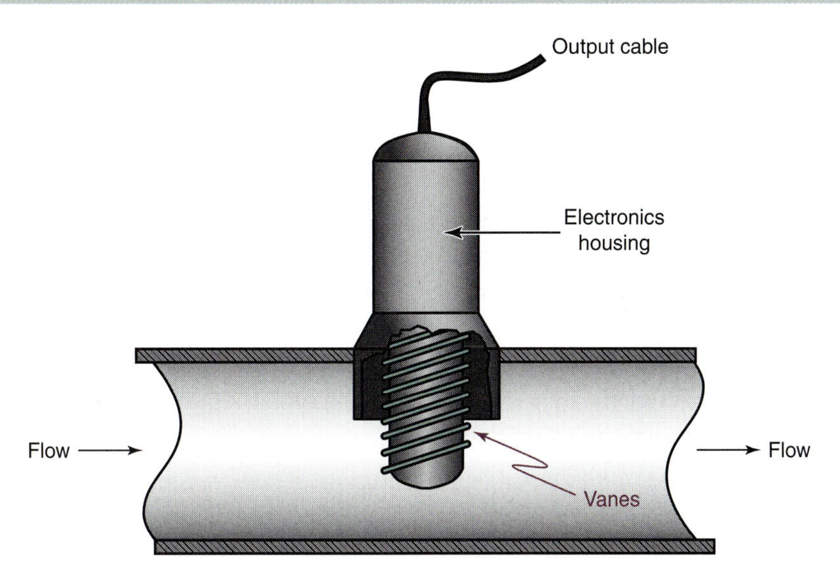

Rehg and Sartori, Industrial Electronics, *1st Edition, © 2006, Reprinted by permission of Pearson Education, Inc., Upper Saddle River, NJ.*

FIGURE 12-17: Vortex flow sensor.

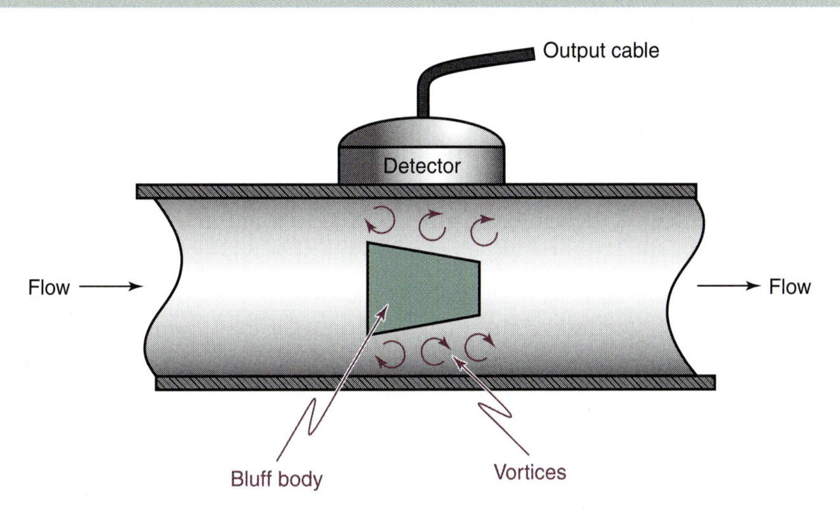

Rehg and Sartori, Industrial Electronics, *1st Edition, © 2006, Reprinted by permission of Pearson Education, Inc., Upper Saddle River, NJ.*

stream flow. When fluid flow strikes the bluff body, a series of vortices are produced or *shedded*. Figure 12-17 illustrates the vortices around the bluff body in the fluid flow. Vortex shedding is the instance where alternating low-pressure zones are generated in the downstream. On the side of the bluff body where the vortex is being formed, the fluid velocity is higher and the pressure is lower. As the vortex moves downstream, it grows in strength and size, and eventually detaches itself. This is followed by the forming of a vortex on the other side of the bluff body. The alternating vortices are spaced at equal distances. The bluff body is shown as a trapezoid, but other shapes also work as long as the bluff body has a width that is a large enough fraction of the pipe diameter so that the entire flow participates in the shedding. Vortex shedding frequency is directly proportional to the velocity of the fluid in the pipe, and therefore to volumetric flow rate. The shedding frequency is independent of fluid properties such as density, viscosity, conductivity, etc., except that the flow must be turbulent for vortex shedding to occur.

- **Electromagnetic flow sensor:** The electromagnetic flow sensor measures the electrical charges in flowing fluid, which are proportional to the fluid velocity. Its operation is based on Faraday's Law, which states that the voltage induced in a conductor moving through a magnetic field is proportional to the velocity of that conductor. Figure 12-18 illustrates the electromagnetic flow sensor. The flow sensor has a small insertion depth (in the neighborhood of 0.2 inches), thus it creates no pressure drop. A set of coils creates a magnetic field, and as the fluid flows through the magnetic field it acts like an electrical conductor and becomes electrically charged. A set of electrodes detects and measures the electrical charge. The electrical charge is converted to velocity from which the flow rate can be determined. Electromagnetic flow sensors are used widely in urban and wastewater systems and in industrial applications where non-intrusive flow sensors are needed, such as measuring the flow rate of chemicals, heavy sludge, paper and pulp stock, mining slurries, and acids.
- **Ultrasonic flow sensor:** The operation of the ultrasonic flow sensor is based on the motion of a sound source and its effect on the frequency of the sound. This phenomenon was observed and described in the mid-1800s by Christian Johann Doppler as *the frequency of the reflected signal is modified by the velocity and direction of the fluid flow.* If a fluid is moving towards a sensor, then the frequency of the returning signal increases, and if the fluid is moving away from a sensor, then the frequency of the returning signal decreases. The frequency difference is equal to the reflected frequency minus the originating frequency and is used to calculate the fluid flow rate. Figure 12-19 depicts an ultrasonic flow sensor that clamps onto a pipe and contains transmitting and receiving components. The bubbles or suspended solids in the fluid reflect the transmitted signal back to the receiver, which is a Doppler meter. The Doppler meter detects frequency differences between the transmitted frequency and the reflected frequency. This type of flow sensor does not insert an obstruction in the piping system, which makes it very portable. The ultrasonic flow sensor can be used as a troubleshooting tool for detecting improper flow rates, as backup to an already installed flow sensor, or to check existing meters in a number of locations. A handheld microprocessor-based converter is also available, which provides various functions such as local graph-

FIGURE 12-18: Electromagnetic flow sensor.

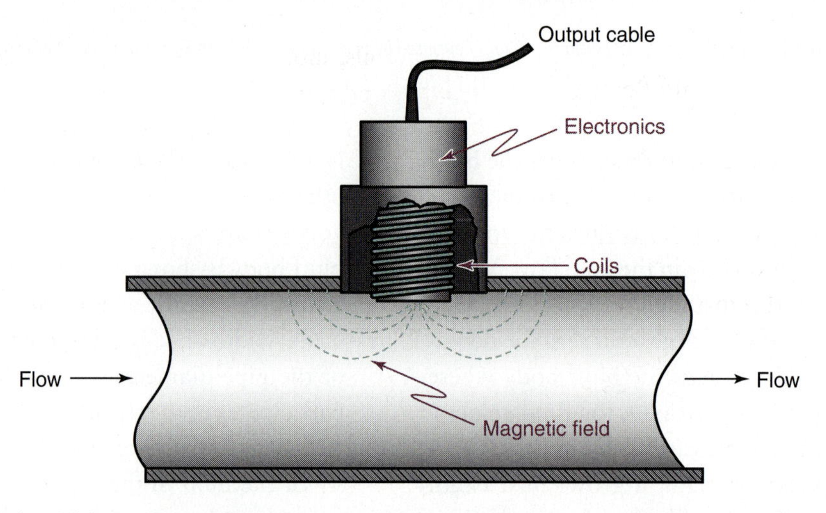

Rehg and Sartori, Industrial Electronics, *1st Edition, © 2006, Reprinted by permission of Pearson Education, Inc., Upper Saddle River, NJ.*

FIGURE 12-19: Ultrasonic flow sensor.

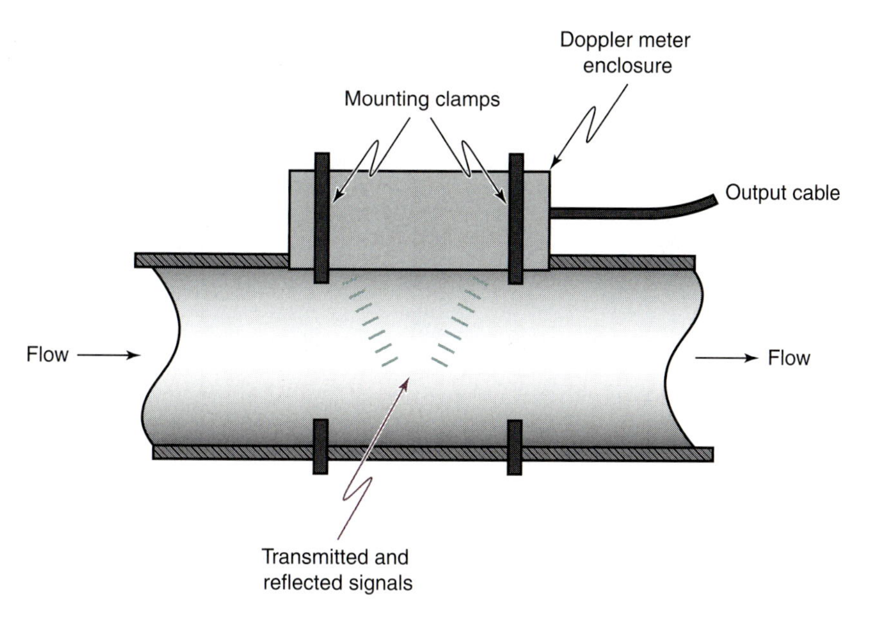

Rehg and Sartori, Industrial Electronics, *1st Edition, © 2006, Reprinted by permission of Pearson Education, Inc., Upper Saddle River, NJ.*

ics display, an integral keypad for calling up menus for flow data, trend displays, and setting up site parameters.

Displacement Flow Sensors. The *displacement flow sensor* measures fluid flow by precision-fitted rotors as flow-measuring elements where known or fixed volumes are displaced between the rotors. The rotation of the rotors is proportional to the volume of the fluid being displaced. The number of rotations of the rotor is counted by an integral electronic pulse transmitter and converted to volume and flow rate. The displacement rotor construction is accomplished in several ways:

- **Piston pumps:** the fluid flowing through the piston is a known quantity, thus the flow rate is a function of the number of piston strokes.
- **Oval gears:** a fixed volume of fluid passes through two rotating, oval-shaped gears with synchronized close-fitting teeth for each revolution, thus shaft rotation can be counted to obtain specific flow rates.
- **Nutating disks:** these moveable disks are mounted on a concentric sphere located in a spherical chamber, and the pressure of the liquid passing through the chamber causes the disks to rotate in a circulating path without rotating about its own axis. Each disk rotation traps a known amount of fluid and the number of disk revolutions is counted, thus the flow rate can be calculated.
- **Rotary vanes:** these vanes are equally divided, rotating impellers inside a casing. The impellers are in continuous contact with the casing, and a fixed volume of liquid is swept through as the impellers rotate. The revolutions of the impeller are counted and flow rate can be determined.

The displacement flow sensor may be used for all relatively nonabrasive fluids such as heating oils, lubrication oils, polymer additives, animal and vegetable fat, printing ink, and freon.

Mass Flow Sensors. *Mass flow sensors* directly measure the flow rate of fluids, especially gases, and tend to be highly accurate—they are off less than 0.1 percent. Two such sensors, the thermal mass flow sensor and the Coriolis mass flow sensor, are discussed in this section.

- **Thermal mass flow sensor:** The thermal mass flow sensor operates independently of density, pressure, and viscosity. This sensor uses a heated sensing element isolated from the fluid

FIGURE 12-20: Thermal mass flow sensor.

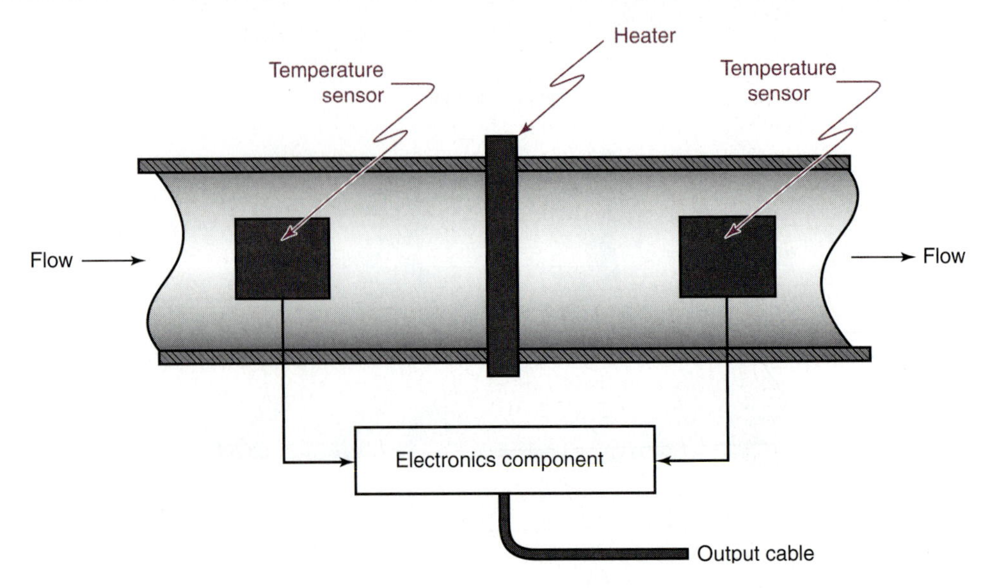

Rehg and Sartori, Industrial Electronics, *1st Edition, © 2006, Reprinted by permission of Pearson Education, Inc., Upper Saddle River, NJ.*

flow path where the flow stream conducts heat from the sensing element. The conducted heat is detected with thermal sensors, and the temperature difference between the sensors is directly proportional to the mass flow rate. Figure 12-20 illustrates a thermal mass flow sensor. Note that one temperature sensor is on the upstream side of the heater and the other sensor is on the downstream side. One type of temperature sensor used in thermal mass flow sensing is the *thermopile*, which consists of thousands of thermocouples gathered in a band of plastic or tape. The electronics component of the thermal mass flow sensor includes a flow analyzer, temperature compensation, and signal conditioner.

FIGURE 12-21: Coriolis flow sensor.

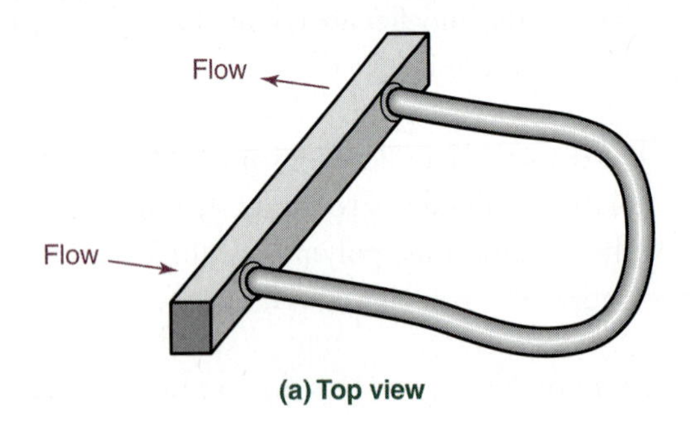

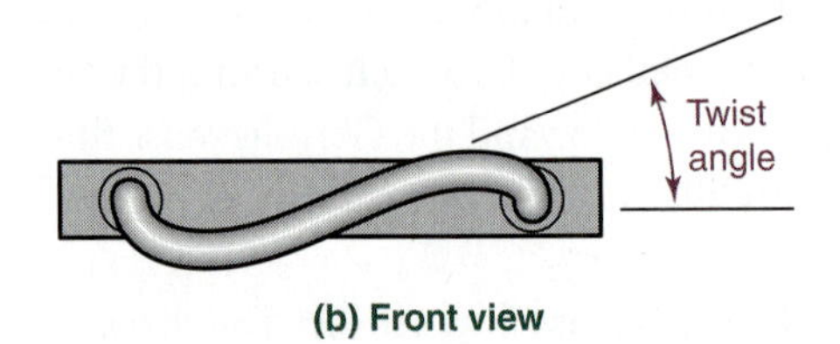

Rehg and Sartori, Industrial Electronics, *1st Edition, © 2006, Reprinted by permission of Pearson Education, Inc., Upper Saddle River, NJ.*

- **Coriolis mass flow sensor:** The Coriolis flow sensor uses a vibrating U-shaped tube to measure fluid flow rate based on an inertial force phenomenon as described in the mid-1800s by Gustave-Gaspard Coriolis and named the Coriolis effect. The fluid to be measured runs through a U-shaped tube, shown in Figure 12-21, that is caused to vibrate in an angular harmonic oscillation. The fluid flowing through the tube causes the tube to deform as in a twist, and the amount of twist is directly proportional to the fluid flow. Detection of the amount of twist is done by sensors, which are mounted near the tube and drive a signal conditioner whose output is typically a variable voltage. This voltage is used to determine the fluid flow rate. The Coriolis mass flow sensor does not require temperature compensation, which is an advantage over the thermal mass flow sensor. In the dredging and other industries with slurries and highly corrosive and/or dirty liquids, using the inline Coriolis mass flow sensor is not practical at high velocities due to the wear on the tube.

Visual Flow Sensor. In some applications, a visual indication of flow rate rather than an electrical signal is more appropriate. The *visual flow sensor*, called a variable area flow meter or a rotameter, consists of a vertically oriented glass or plastic tube with a larger end at the top and a metering float, which is free to move within the tube. Figure 12-22 illustrates the visual flow sensor. Fluid flow causes the float to rise in the tube as the upward pressure of the fluid overcomes the effect of gravity. In other words, gravity pulls the float down and the fluid flow pushes it up. The float rises until the annular area between the float and tube increases sufficiently to allow a state of dynamic equilibrium between the upward pressure and the downward gravity factor. The tube is graduated in appropriate flow units, and the height of the float indicates the flow rate with a typical accuracy of 1 percent of full-scale. This type of flow indicator is used in industrial applications where water is used as a cooling agent in the process, and the flow is adjusted manually until the float is at the desired graduated mark on the sensor.

FIGURE 12-22: Visual flow sensor.

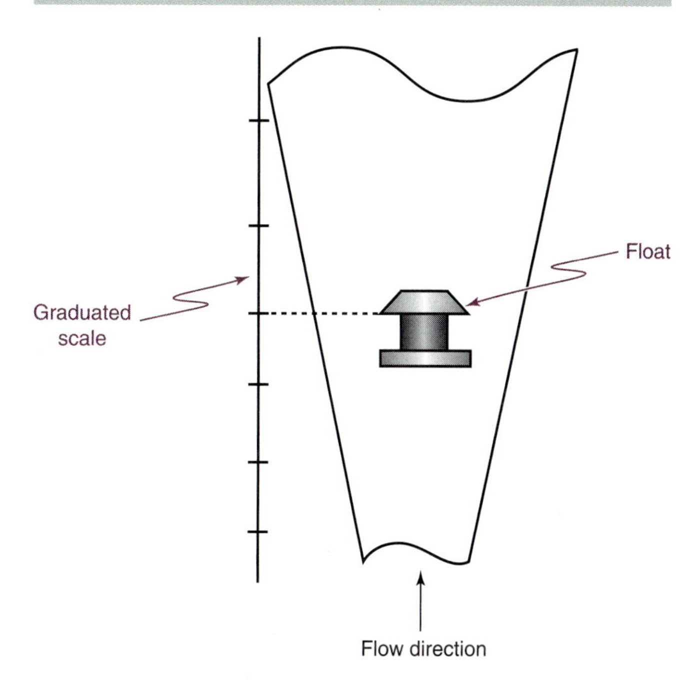

Rehg and Sartori, Industrial Electronics, *1st Edition, © 2006, Reprinted by permission of Pearson Education, Inc., Upper Saddle River, NJ.*

12-2-4 Position Sensors

Position sensors measure both linear and rotary motion and distance. In this section we will discuss three position sensors—the linear and rotary potentiometers and the linear variable differential transformer.

Linear and rotary potentiometers are sensors that produce a resistance output proportional to the displacement or position. *Linear potentiometers* are essentially variable resistors, which are either wire-wound or conductive plastic and either rectangular or cylindrically shaped. The resistance element is excited with a voltage, and the output voltage is ideally a linear function of the input displacement, thus providing a voltage as a position measurement. *Rotary potentiometers* are sensors that produce a resistance output proportional to a rotational position; they operate the same as a linear potentiometer.

The *linear variable differential transformer* (LVDT) is a position sensor consisting of a transformer with a movable core, as shown in Figure 12-23. The transformer has a primary winding and two identical secondary windings that are wound around a hollow tube containing the movable core, and the movable core is attached to the item whose position is to be measured.

The output voltage of secondary A is greatest when the core is opposite it; conversely, secondary B's output is the greatest when the core is opposite it. If the core is equidistant between the two cores, the output voltage is zero, which is referred to as the null position of the sensor. This is shown in Figure 12-23(a). In between, the output voltage varies from zero volts to maximum volts. In the null position an equal voltage is induced into both secondary A and secondary B, but since they are wired so the voltages are opposed they cancel each other out, resulting in a zero output. When the core aligns with secondary A as shown in Figure 12-23(b), the output is in phase with the input voltage. Conversely, with the core aligned with secondary B, the output is out of phase with the input voltage, as depicted in Figure 12-23(c).

FIGURE 12-23: Linear variable differential transformer (LVDT).

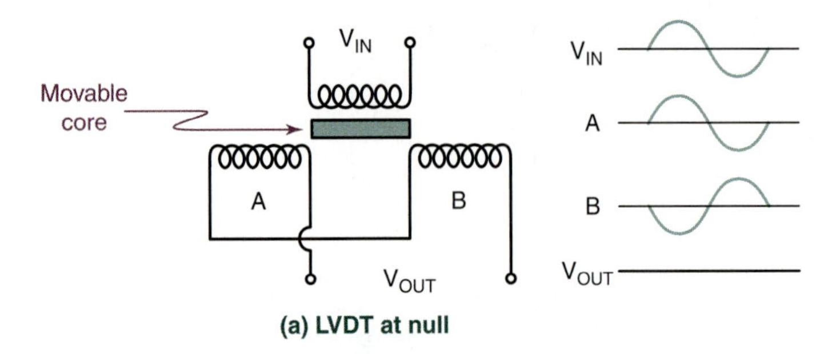

(a) LVDT at null

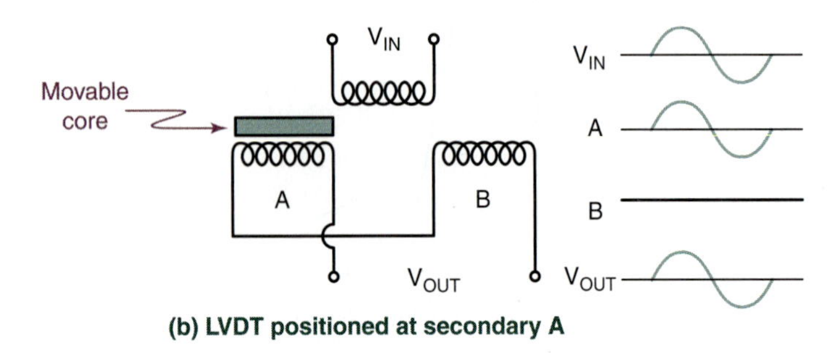

(b) LVDT positioned at secondary A

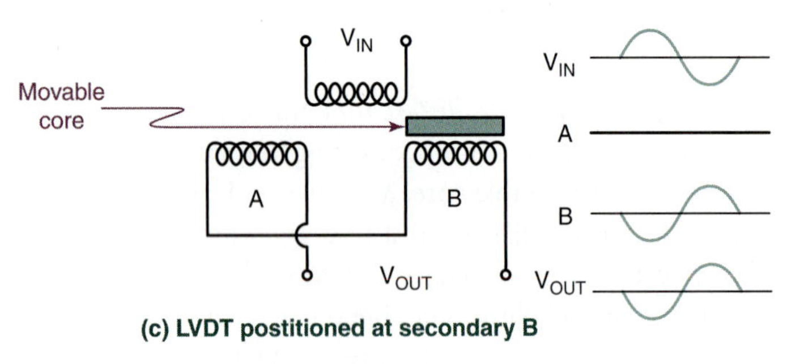

(c) LVDT postitioned at secondary B

Rehg and Sartori, Industrial Electronics, *1st Edition, © 2006, Reprinted by permission of Pearson Education, Inc., Upper Saddle River, NJ.*

The electronics that interface with the LVDT sensor combine information on the phase of the output with information on the magnitude of the output, and provide a DC output voltage indicating where the core is relative to its electrical zero position.

12-2-5 Vision Systems

Vision systems are growing at a rapid rate in manufacturing automation because system cost has dropped, system capability and intelligence has significantly increased, and the size of the camera and camera controllers has plummeted. The applications fall into five general categories: part identification, part location, part orientation, part inspection, and range finding.

- **Part identification:** Vision systems store data for different parts in memory and use the data to distinguish between parts as they enter the work cell. The system can learn the characteristics of different parts and identify each part from its two-dimensional image. Based on the vision image and analysis, the system triggers actions on the part in the production system. Vision systems are also taking over many of the functions previously performed by bar code systems in reading bar codes and extracting product data from the production database.
- **Part location:** Vision technology allows the automation system to locate randomly placed parts entering the work cell. The vision system measures the X and Y distances from the center of the camera coordinate system to the center of the randomly placed part. This application is used most often by robot systems to direct the robot gripper to the location of the part to be moved.
- **Part orientation:** Part orientation and part location are linked because every part must be gripped in a specified manner by the robot gripper. The vision system supplies the orientation information and data that are used to drive the robot gripper into the correct orientation for part pickup. Many part orientation parameters, both measured and calculated from measured data, are provided by the vision system for use in automated part handling. Orientation is used when robots are not present to verify that parts or products are oriented properly for automated assembly.
- **Part inspection:** Vision systems are used to check parts for dimensional accuracy (for example, the diameter of a part) and geometrical integrity (for example, the number of holes). The parts are measured by the camera, and the dimensions are calculated; at the same time, the vision system checks the parts for any missing features or changes in the part geometry.
- **Range finding:** In some applications the system uses two or more cameras to measure the X, Y, Z location of the part. This technique is also used to measure and calculate the cross-sectional area of parts.

Vision Systems Operation and Applications. The block diagram of a vision system shown in Figure 12-24 illustrates the architecture used to implement vision technology. In this configuration the vision system is totally stand-alone; however, some robot manufacturers build the vision system directly into the robot controller. The basic vision system components shown include the following: one or more cameras, a camera controller, interface circuits and systems for the camera and work-cell equipment (PLC), discrete inputs for synchronizing sensors, power supply, a flat-screen LCD display, a programming console, and a lighting system for the parts.

Vision systems operate by processes images captured electronically on either CCD (charge-coupled device) or CMOS (complementary metal oxide semiconductor) cameras. Both types of cameras use a solid-state array of light-sensitive cells, called pixels, deposited on an integrated circuit substrate. Each cell or pixel is a small light-sensitive transistor whose output is a function of the intensity of light striking its surface. The systems come in two basic configurations: *imaging* arrays and *linear* arrays. Figure 12-25 shows how each is organized. In Figure 12-25(a), a 1024 by 1280 image array camera with an LED light source attached is capturing numeric and bar code information on a moving conveyor. In Figure 12-25(b), a linear array of 6144 pixels is scanning in the data from a high-speed conveyor one line at a time. Higher scan rates are achieved with this type of

FIGURE 12-24: Vision system components.

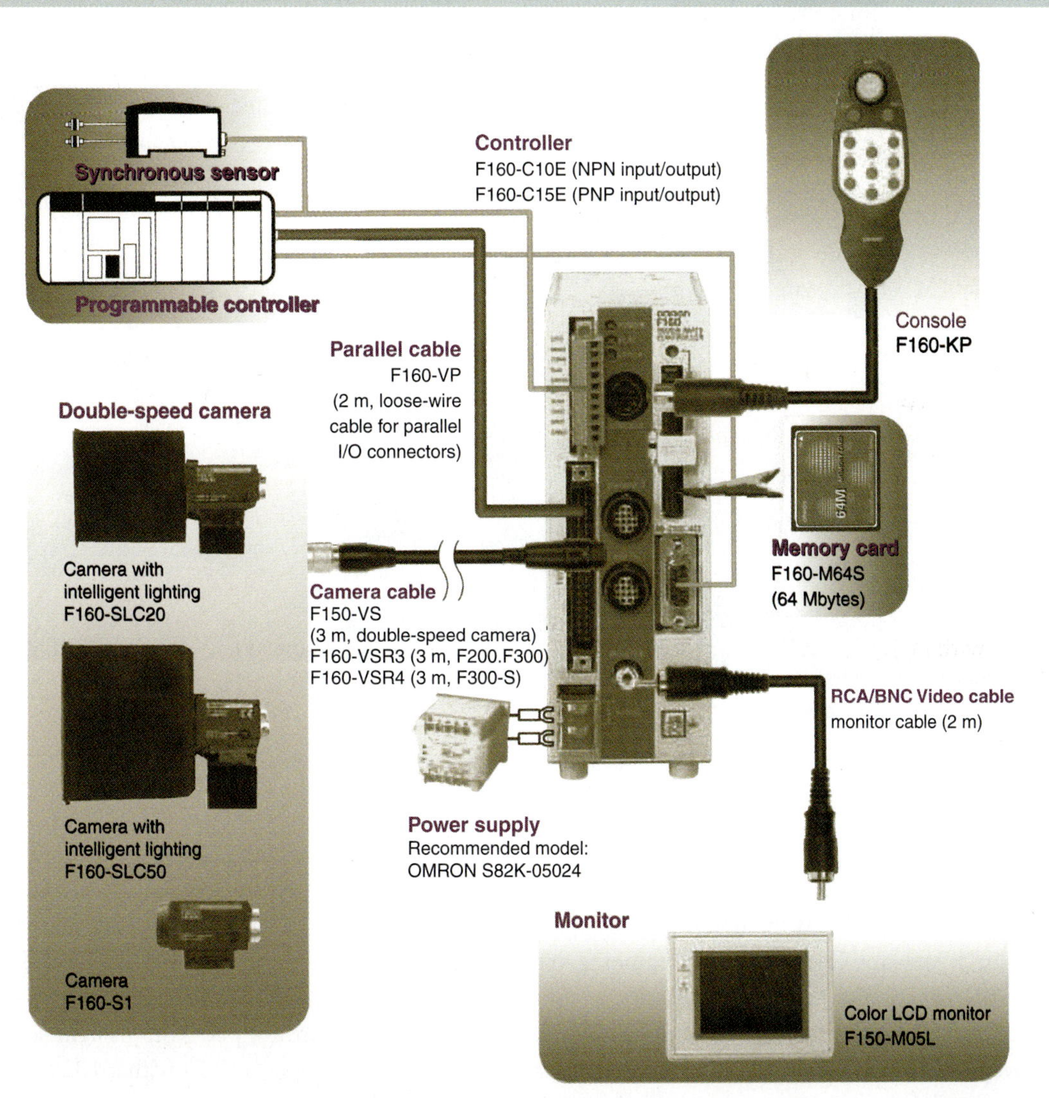

Rehg and Sartori, Industrial Electronics, *1st Edition, © 2006, Reprinted by permission of Pearson Education, Inc., Upper Saddle River, NJ.*

FIGURE 12-25: Image and linear array vision cameras.

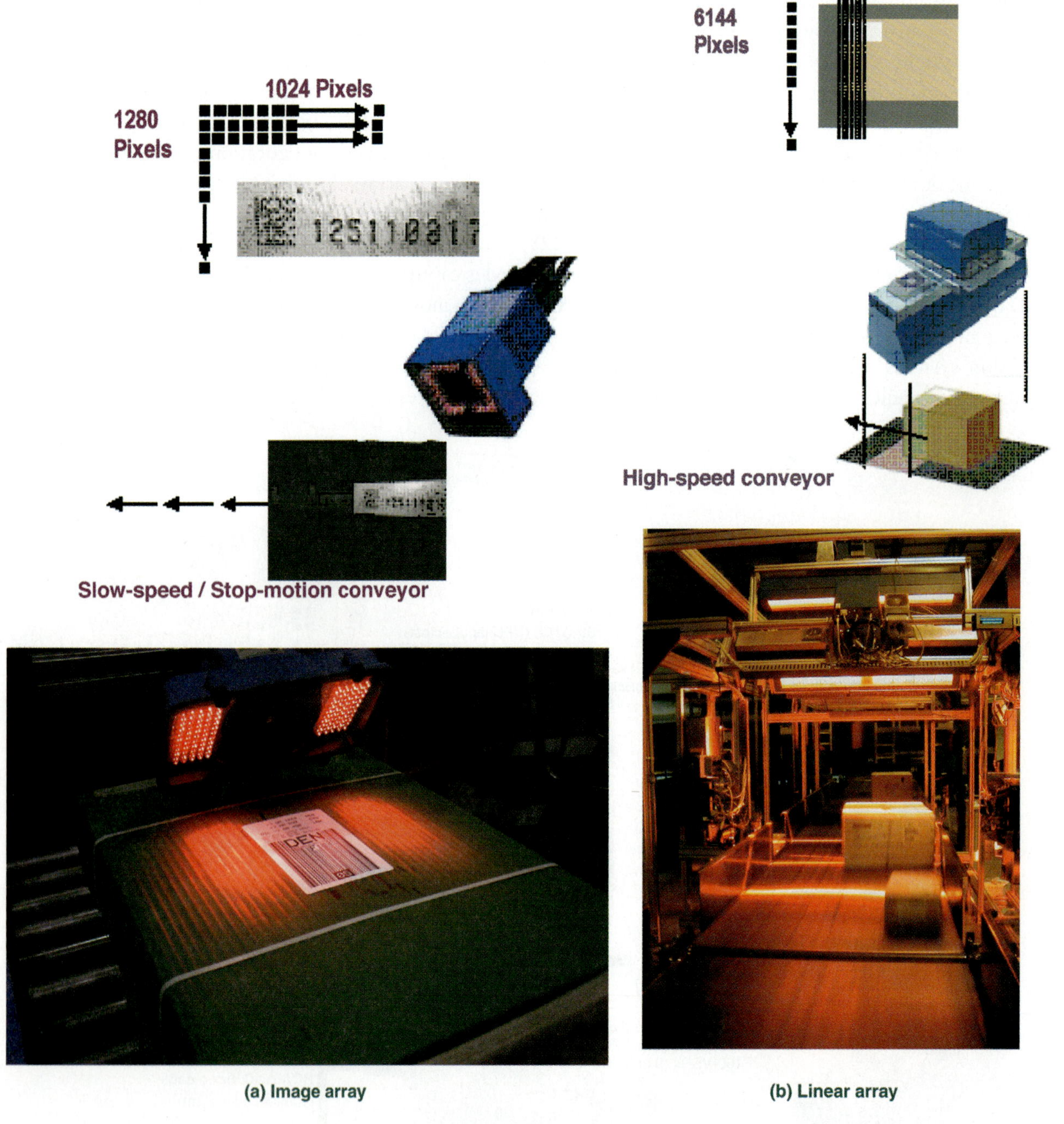

Rehg and Sartori, Industrial Electronics, *1st Edition, © 2006, Reprinted by permission of Pearson Education, Inc., Upper Saddle River, NJ.*

system since the system processes only a small part of the image in each scan. The linear arrays perform a line scan, whereas the imaging arrays perform an area scan. Charge storage cameras are accurate, rugged, and have good linearity.

The unit of measurement in a vision system is the *gray scale,* the parameter measured is *light intensity,* and the measurement element is the *pixel.* The vision system lens focuses light from the part onto the CCD or CMOS light-sensitive surface. The two-dimensional surface, Figure 12-25, is divided into small regions or picture cells called pixels. The resolution of the vision system is directly proportional to the number of pixels on the light-sensitive surface. For example, a CCD with a 1024 by1280 array will have a higher resolution than a CCD with a 128 by 128 array of pixels. Each pixel builds up charge proportional to the amount of light reaching the surface through the camera lens. This charge value is an analog signal that is digitized into a discrete binary value by the central vision processor. With no light present, the pixel is turned *off*, but when light reaches a *saturation level*, the pixel is full *on*. Between those two extremes are shades of gray that cause the pixel to be excited to a partially *on* condition. The number of excitation states between *off* and *on* is called the gray scale. Current systems have gray scales of 256. The gray scale of a pixel is the numerical representation of brightness for one small spot on the part. The cameras in most vision systems conform to the RS-170 standard, which requires that images are acquired at a rate of 30 frames per second with even and odd line interlacing.

The vision system captures an image when a trigger signal from a sensor indicates that the part is in position for an image capture by the camera. When the image is captured, the system transfers the image to a central processing unit for analysis of the vision data. The type of analysis depends on the vision algorithms present and the requirements present in the vision application.

Four typical applications for a vision system are illustrated in Figure 12-26. In application

FIGURE 12-26: Vision system applications.

Product sorting
Sort boxed product by size labels and inspect seams

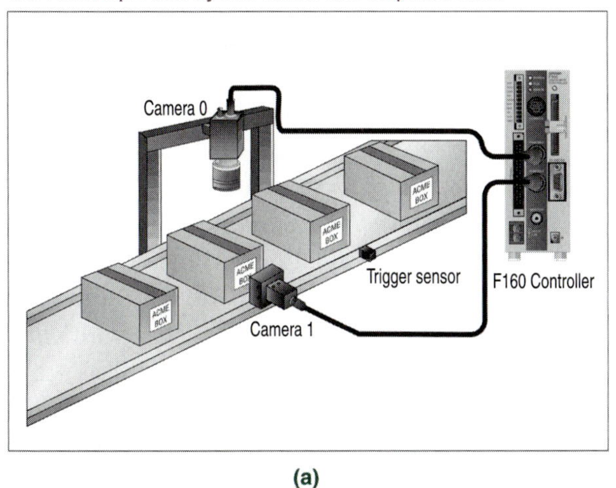

(a)

Optical character recognition
Pill presence/absence and Lot/Date code confirmation on blister pack

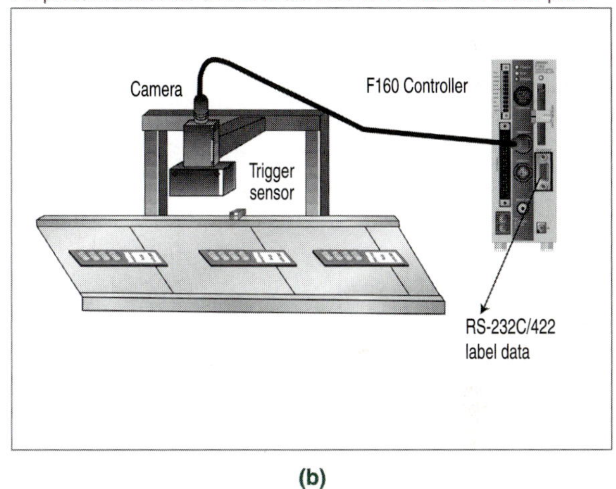

(b)

High speed bottle inspection
Inspect dimensions, conformance and date code

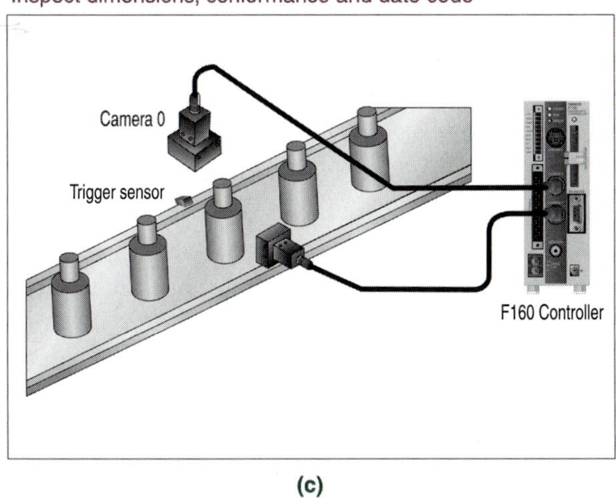

(c)

Position reference
Identify random, odd-shaped product positioning on a conveyor

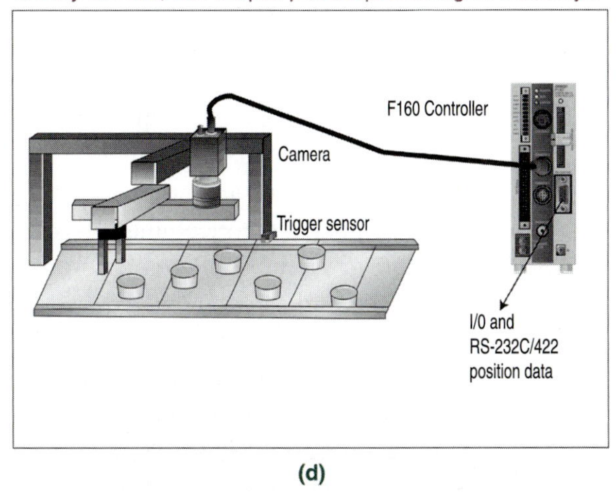

(d)

Rehg and Sartori, Industrial Electronics, *1st Edition, © 2006, Reprinted by permission of Pearson Education, Inc., Upper Saddle River, NJ.*

(a) a trigger sensor is used to synchronize the image capture with the time when a box is present. Note that the boxes are evenly spaced and the next box on the conveyor triggers the inspection. Camera 0 checks box size and a closed top seam, while camera 1 reads the label. In application (b), a vision system inspects blister packs to verify that pills are present and checks the lot and date code. Label data is sent from the vision system by serial transmission to a product data management system. Application (c) uses two cameras to check dimensions, conformance, and the date code on bottles moving by at high speed. Again, a trigger sensor synchronizes the image capture with a passing bottle. Application (d) uses a single camera to locate the X and Y position of randomly placed parts on a conveyor belt. The data is passed over a serial link to a robot controller that changes the robot arm's pickup point based on the data.

Vision Systems Classifications. Vision systems have three basic classifications—*binary systems, gray scale,* or *color processing.*

- *Binary systems*, the lowest in cost and easiest to use, do basic visual inspections. They operate on an image made up of black and white pixels and calculate results by counting the pixels to determine the shape or size of an object. Binary systems allow inspection of silhouettes, profiles, and outlines and are best suited for area measurement and sizing. On the other hand, this type of system performs poorly when light contrast is low or variable and cannot detect imperfections in product appearance.
- *Gray scale* imaging typically processes images using 256 levels of black and white. The most popular gray scale image-processing technique compares the incoming image with a template or model. The correlation between the image and template is used to decide acceptance or rejection of the part. Gray scale processing is more accurate than binary in dimensional measurements because of a technique called sub-pixel processing. Using this process, an edge is defined to an accuracy of less than one pixel; as a result, surface imperfections, scratches, textures, and shadows can be detected. However, it is more expensive and in some applications has longer processing times.
- *Color processing* uses information from the red, blue, and green color spectra to detect and differentiate shades of color relevant in food industry and pharmaceutical applications. Finding blemishes in food and checking the color of pills in blister packs requires a color vision system. Color processing is more expensive than the two other types of visual systems, but it offers unique advantages for applications that justify the cost.

Vision System Configurations. Three vision system configurations are available: *stand-alone, PC-based, and VME-based.* While all three offer some degree of modularity for customization, *stand-alone systems*, like the system in Figure 12-24, have features that make them well suited for factory automation. These vision "black boxes" use Application Specific Integrated Circuit (ASIC) or Systems on a Chip (SoC) technology in the vision electronics that gives them accelerated processing power. As a result, they are the fastest systems available. In addition, stand-alone systems work seamlessly with other factory automation devices, such as programmable logic controllers, photoelectric and proximity sensors, and radio frequency identification systems, to enhance integration of the shop floor production and data tracking. Finally, they work over a wide range of temperatures, vibrations, and electrical interference.

A *PC-based system* generally consists of a microcomputer with a processing board that includes dedicated vision ASICs. Use of the computer's CPU lowers the system price and allows you to take advantage of advances in microprocessor technology. Although lower in cost, standard PC vision systems cannot operate in harsh factory environments. Industrial PCs are available, but the higher cost reduces the advantage offered by the PC-based vision approach.

VME-based systems, an IEEE bus standard used in reliability industrial inspection applications, combine a robust operating system with a flexible user platform. Vision system and computer components are connected via a pin-and-socket connection to the bus, which improves signal transfer rates and is less affected by the industrial environment. However, these systems have a higher cost.

Each configuration offers performance at three levels, which are described as follows:

- **High performance, fixed position:** Applications at this level include scanning parcels using a single or multiple cameras with a 560 frames per minute capture rate. Package spacing is typically 8 inches with conveyors up to 75 inches across. Good field of view (FOV) is achieved with cameras set at 36 inches from the target objects.
- **Mid-range performance, fixed position:** Applications at this level are found in warehouse identification and bar code reading requirements, and also in general sorting of items on moving conveyors. Single or multiple cameras are used with a 350 frames per minute or less capture rate. The depth of field (DOF) coverage is typically 25 inches.
- **Low performance:** At the lowest level, the objects are stationary, indexed (move and then stopped), or moving at a slow speed. The applications are usually manufacturing lines and warehouse related. The frame rate is 250 frames per minute or less, the camera has a fixed focus lens, and FOV is small—usually in the 1- to 2-inch range.

Vision System Programming Methods. The programming options for vision systems range from no programming requirements for special-purpose turnkey systems for a single application to a highly flexible programming environment using Visual Basic or C++. In some systems a menu interface is used, which integrates an internal program with a point-and-click selection mechanism. This method is easy to use and offers a moderate degree of flexibility. In other systems a proprietary programming language must be used, which requires a high level of programming skills and knowledge of vision algorithms. The console in Figure 12-24 is used to program the stand-alone system for Omron.

Lighting for Vision Applications. Although lighting is an important consideration in every vision application, it is frequently overlooked. In many cases a greater effort on the lighting and optics problems would result in a less sophisticated and lower cost vision system. The objective of lighting is to create a high-contrast image that clearly distinguishes the features to be captured. Proper lighting can:

- Shorten the image analysis and recognition time.
- Increase the number of captured images in a fixed time.
- Improve the quality of the captured data.
- Reduce the shadows and glare that cause analysis problems.

Vision System Light Sources and Configurations. The light source chosen for the vision application is critical for a successful project. The major types of sources include *incandescent* and *quartz halogen, fluorescent, light-emitting diodes (LEDs), xenon flash,* and *lasers*. Each of the sources has unique properties such as the location of the source on the *electromagnetic radiation spectrum, operational life, heat generated, level of illumination energy, coherent structure,* and *compatibility* with human operators in the work cell. Significant characteristics of the frequently used sources include:

- **Incandescent and quartz halogen:** These commonly used light sources include devices ranging from standard household bulbs with reflectors to high-power quartz halogen lamps. The quartz halogen sources are often used with fiber-optic bundles to pipe the light onto specific locations on the part. Two major disadvantages are bulb life and high generated heat. However, they have an advantage of little decay of light intensity over time.
- **Fluorescent:** The reduced infrared energy (heat) produced by this type of illumination, plus the extended life of the tube, make it more efficient than the incandescent and quartz halogen source. The natural diffused light produced by a fluorescent source is preferred for vision applications with highly reflective parts.

- **LED:** Arrays of LEDs are arranged in a variety of configurations, including circular and linear sources to light objects. The LED has a long life and produces little heat.
- **Xenon flash:** When used as a *strobe light* source, the xenon flash tube is an important light source in vision applications. Xenon strobe light is used when the vision system must capture the image of a moving part. The 5- to 200-microsecond flash of light illuminates the part at one point in its motion. Since CCD vision cameras are temporary storage devices, the image of the part created by the flash is stored by the camera and the part features are scanned into the vision system memory. In addition, xenon tubes produce light with frequencies from a broad part of the spectrum and have high-intensity levels.
- **Lasers:** The laser is an important source because it produces coherent light that does not disperse as it travels from the source to the target. As a result, the diameter of the laser beam at the target is very close to the beam size leaving the laser. In *structured light* applications, lasers are used to place a thin line of light across the object. The light follows the contours of the part so the vision camera can capture the shape of the part.

Vision System Lighting Techniques. Two design decisions are associated with vision lighting:

1. What type of light source should be chosen?
2. How should the light be positioned with respect to the object?

In general, most lighting applications fall into two categories: front lighting and backlighting. In front lighting the object is lit from the top, and the angle of the lights is changed to achieve different desired effects. One problem in vision is specular reflection off of a specular surface. This is a surface that reflects a light ray in a single direction from any given point on the surface. If a specular surface is present, then the lighting system attempts to minimize the specular reflections. Backlighting provides an outline of the object because the light is placed behind or below the object. Figure 12-27 illustrates a number of lighting techniques and lists attributes and typical applications. Study it carefully until the many different lighting options are understood. The lighting schemes in Figure 12-27 would have to use a strobe source if the objects are moving rapidly past the camera.

Vision System Optics. Optics or lenses, like lighting, are crucial to obtaining effective operation of vision systems. A variety of lenses are available, from low-cost lenses designed for the surveillance industry to high-quality, high-resolution lenses necessary for precise measurement and for detecting defects on small parts. Identifying the required field of view is important in selecting a lens. It is a question of how much of the object needs to be in the image. The selection of a lens has the following steps.

1. Define the field of view (FOV) as illustrated in Figure 12-28(a). Note that the FOV does not have to include the entire object or product but only the area of interest.
2. Determine the distance from the camera to the viewed object, called the *setting distance*. Keep this distance flexible so that the lens choice is not restricted.
3. Find the approximate setting distance on the vertical axis of the graph in Figure 12-28(b), then find the FOV on the horizontal axis, and find the recommended lens where these coordinates intersect. The "t" value is the length of the lens extension tube, which is placed between the lens and the camera as shown in Figure 12-28(b). The extension is necessary to allow the lens to focus on the object at that setting distance.

12-2-6 Troubleshooting Analog Sensors

With the resistance temperature detectors (RTDs) and thermistors, an ohmmeter check is a simple way to validate that the device is operational, but the device must be removed or isolated from the circuitry. However, before isolating the device, make sure the leads are intact and not broken. With the ohmmeter connected to the device, heat the device and verify

FIGURE 12-27: Lighting options for vision systems.

Type of lighting	Lighting configuration	Description	Application	Light type
Directional front lighting	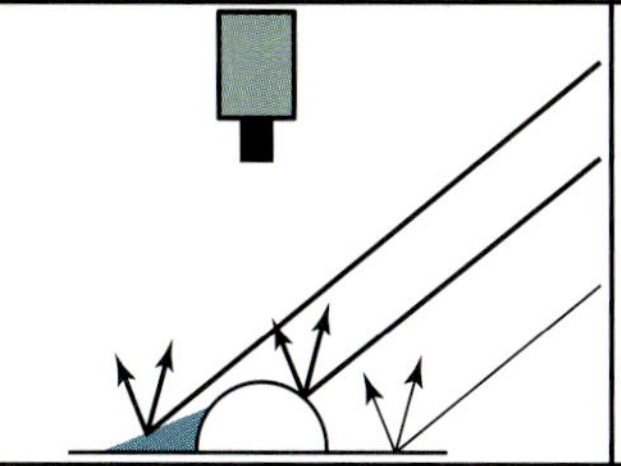	• Similar to oblique but light is at 30 degree or higher angle • The contrast between the object and the background is reduced	Extracts greater information from the flat surfaces of objects	Spot lights or ring lights
Diffuse front lighting	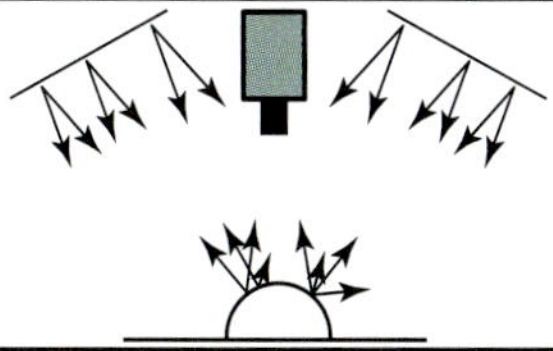	• Large area of uniform illumination • Eliminates shadows and specular reflections • De-emphasizes three-dimensional part characteristics	Applications where the shadow elimination is critical	Dome lights
Polarized lighting	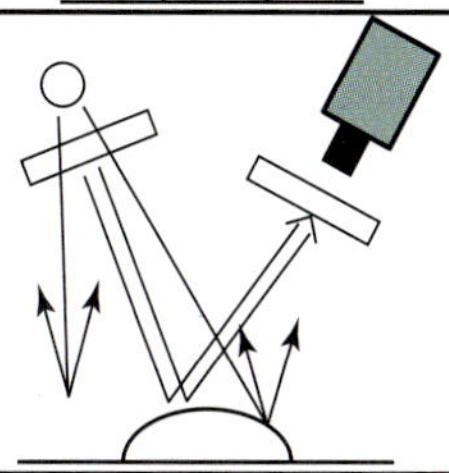	• Polarized light is used and a polarizing filter, called an analyzer, is placed on the camera • Total elimination of specular reflections from any viewing angle are possible when light and camera polar filters are at right angles to each other	Applications where specular surfaces are present	Polarized lights
Fiber optic or LED ring lighting	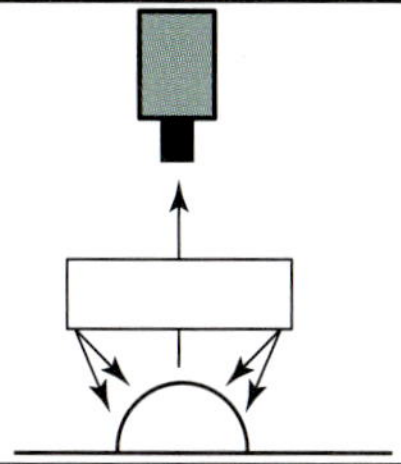	• Ring lights provide intense shadow-free on axis lighting • Can be combined with polarizer and analyzers if specular reflections are a problem	Useful for images of highly reflective objects	Fiber optic and LED ring lights
Fiber optic near in-line lighting	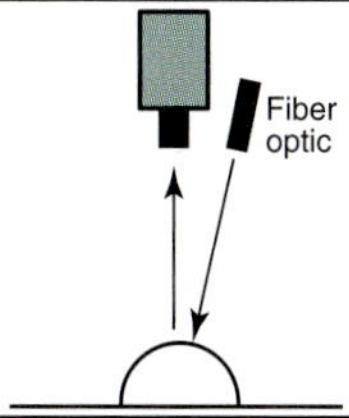 Fiber optic	• Light source is nearly aligned with viewing angle of camera • Difficult to use with shiny object having specular surfaces • Eliminates shadows from details on object surface	Useful for viewing raised details on the object without shadows to obscure the details	Spot light or fiber optic spot
Oblique lighting	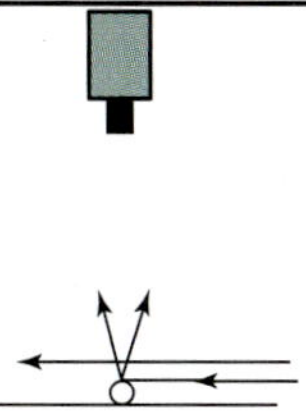	• Called extreme directional lighting • Lighting has low angle of incidence to the surface being illuminated • For example, 20° angle is best for textured surface • Edges are illuminated and surfaces are black	Useful for showing surface texture or edge details on objects	Linear arrays or spots
Structured lighting	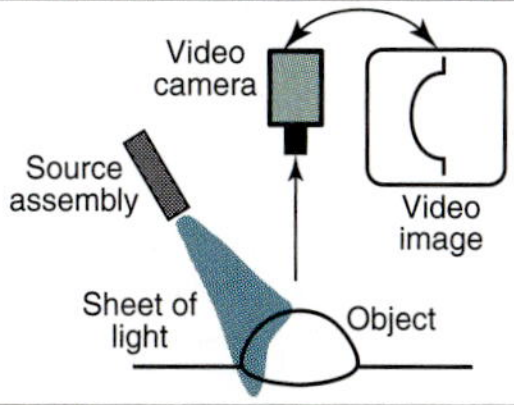 Video camera Source assembly Video image Sheet of light Object	• A laser is often used to create structured light line or beam • A structured light line when falling onto an irregular surface shows all the contours of the surface in the capture image	Useful for surface feature and contour extraction	Fiber optic line light or laser

Rehg and Sartori, Industrial Electronics, *1st Edition, © 2006, Reprinted by permission of Pearson Education, Inc., Upper Saddle River, NJ.*

FIGURE 12-27: (Continued).

Type of lighting	Lighting configuration	Description	Application	Light type
Diffuse back lighting		• Provides the highest contrast and is easiest to set up • Object is between camera and diffused light source • Object appears as a black outline against a white background	Useful for emphasizing object edges for dimensional checks	Back light surfaces
Directional back lighting		• Like diffused back lighting but uses a collimator to make light waves parallel to each other	Used to produce shapely defined shadows of object details	Collimated light source
Dark field illumination lighting		• Cone of light is directed from behind transparent object • Object scatters light to camera to make transparent object visible • Edges of the transparent object are bright against a black background	Used to back light transparent objects that would not be visible with just diffused back lights	Dark field lights

FIGURE 12-28: Lens selection guide.

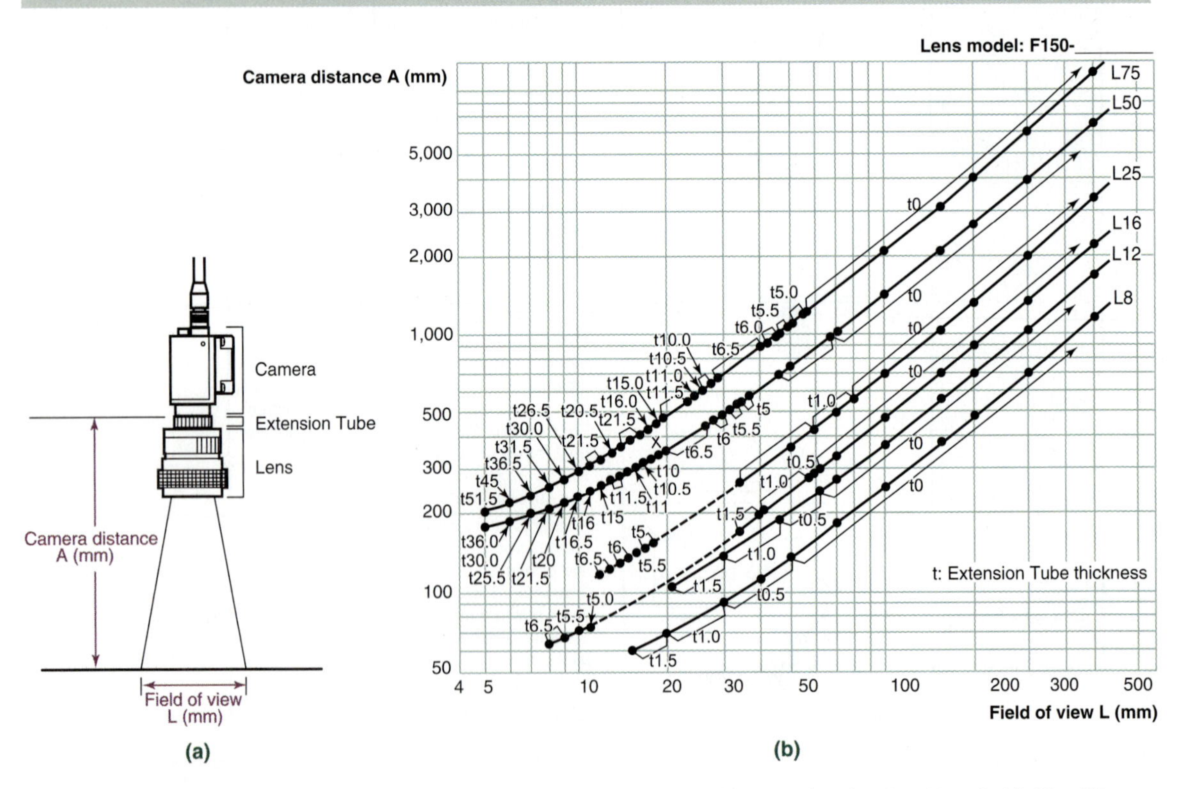

Rehg and Sartori, Industrial Electronics, *1st Edition, © 2006, Reprinted by permission of Pearson Education, Inc., Upper Saddle River, NJ.*

that the resistance changes. With an RTD the resistance should increase, but only by a few ohms. With a negative temperature coefficient thermistor the resistance should decrease by many ohms, much more so than with an RTD. The manufacturer's data sheet, which specifies device characteristics over temperature, should be reviewed. If the resistance of the device does not change or if its value is always infinity, then the device is defective and must be replaced.

With thermocouples an accurate millivoltmeter is needed to measure the thermocouple output as a function of temperature. Most thermocouple manufacturers provide thermocouple testers, including a millivoltmeter, power supply, and reference junction compensation. You simply connect the thermocouple to the tester and select the temperature range, and the tester provides the measurement.

Strain gages act like resistance sensors in a bridge circuit. If the input voltage to the bridge is correct, then a physical load should be applied to validate the operation of the device. The manufacturer's data sheet specifies the range of loads that the device operates. Choose a value in the midrange and measure the output voltage, which is expressed as Output voltage = Maximum output voltage × test load/maximum load.

EXAMPLE 12-1

Validate the operation of a strain gage that can operate up to a maximum pressure of 600 pounds. Its output is 3 mV per volt of input voltage, which is set at 8 V.

SOLUTION

Recalling:

Output voltage = Max output voltage × test load/maximum load

Choose the test load at 300 pounds

Substituting Output voltage = (3 mV/V × 8 V) × 300 lb/600 lb

Output voltage = 12 mV

12-3 ANALOG MODULES AND FIELD DEVICES INTERFACING

Analog modules are used to interface with field devices so that the PLC can operate on digital values. Analog input modules convert analog inputs, typically from the sensors that were discussed in Section 12-2, to a digital value using an analog-to-digital (ADC) converter. Analog output modules convert digital data to analog outputs, typically to control elements such as valves and heating coils, using a digital-to-analog (DAC) converter. A typical analog I/O interface in an automated process is illustrated in a simple block diagram in Figure 12-29 where the PLC controls the heating of the liquid in a tank as a function of the temperature. The PLC controls a heating element via an analog output module and accepts the output of a temperature sensor via an analog input module.

FIGURE 12-29: Analog I/O interface block diagram.

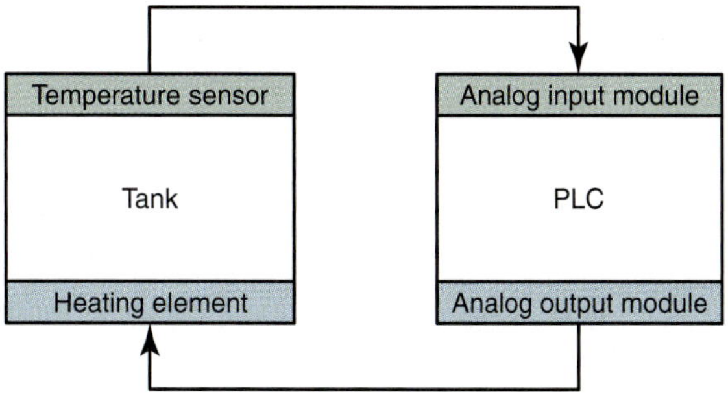

Figure 12-30 depicts an 8-channel analog input module interfaced to thermocouples. Note that cold junction compensation (CJC) transducers are part of the module and are required to ensure accurate thermocouple readings. Each output of the thermocouples is a varying DC voltage in the millivolt range, which is proportional to the temperature. This voltage is digitized via an ADC and read by the processor on command from the PLC program. Note also that the wiring is shielded

FIGURE 12-30: Thermocouple connections to analog input module.

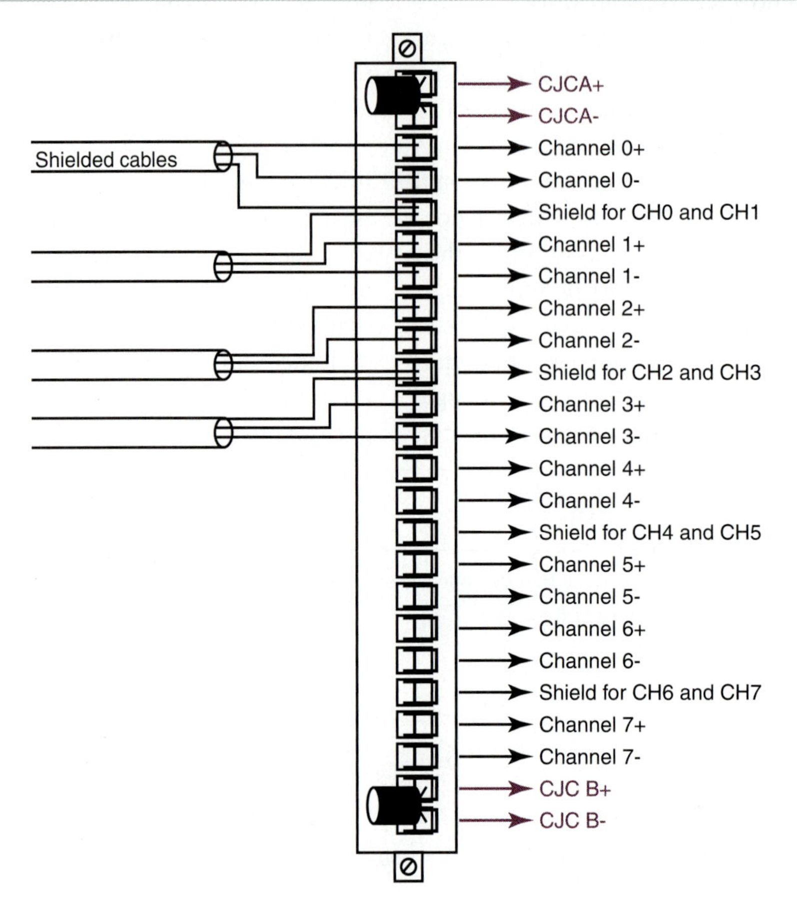

Courtesy of Rockwell Automation, Inc.

to protect the low-level voltage signal from outside electrical noise signals, which can cause errors in the reading of the millivolt signals.

12-3-1 Analog Input and Output Data

The two basic types of analog input modules are voltage sensing or current sensing. *Voltage-sensing* input modules accept either *unipolar* or *bipolar* data. If the field device outputs 0 V to +10 V then the unipolar modules would be used, whereas the bipolar analog module would be used if the field device outputs −10 V to +10 V. The bipolar module accepts both negative and positive input data. *Current-sensing* input modules typically accept analog data over the range of 4 mA to 20 mA, but can accommodate signal ranges of −20 mA to +20 mA. The analog input and output data are specified relative to operational parameters, which impose limitations on the PLC modules. The following is a list of typical I/O specification parameters, including a brief description.

- **Number of I/O:** this indicates the quantity of inputs or outputs that can be interfaced to the analog module. Some PLC types provide combination modules, which accept both inputs and outputs.
- **Data Format:** the representation of analog data that is used by the PLC. The data is represented in BCD, natural binary, or floating point.
- **Voltage Range:** the magnitude and type of voltage signal that is accepted.
- **Current Range:** the magnitude and type of current signal that is accepted.

- **Backplane Current Load:** this value indicates the amount of current that the module requires from the PLC backplane. The backplane power supply supplies all the analog modules' current requirements.
- **Step Response:** this value specifies the time necessary for the signal to go to typically 95 percent of its final value.
- **Update Period:** this value specifies how often the I/O signal is updated by the PLC.
- **Resolution:** this number is expressed in bits and indicates how accurately the analog value can be represented digitally. The higher the number the more accurate is the analog value.

12-3-2 PLC 5, SLC 500, and Logix Options

The Allen-Bradley family of PLCs provides a collection of analog I/O modules to interface with various field devices. The I/O modules use a DAC or ADC to interface the analog signals to data-table values, giving the ladder logic direct access to I/O values. Analog modules also provide a high level of resolution for accurate control in a broad range of automated process applications. The analog modules for the PLCs provide the following options.

- **Input modules:** 2, 4, 8, or 16 channels per module with either differential, high-speed differential, or single-ended options and input ranges of +/−10 V or +/−20 mA.
- **Output modules:** 4, 6, or 8 channels per module with either differential, high-speed differential, or single-ended options.
- **Combination modules:** typically 2 input channels and 2 output channels available per module.
- **RTD and thermocouple modules:** uniquely designed modules for RTD inputs or thermocouple inputs with 4, 6, or 8 channels per module.
- **Special modules:** specific function modules that are used for a singular task such as stepper motor module, encoder module, high-speed counter module, and servo module. Additional specific modules continue to be developed as applications demand.

12-4 CLOSED-LOOP CONTROL SYSTEMS

Control systems, in which the PLC is a main component, can be classified as open-loop and closed-loop. *Closed-loop* control systems are self-regulating and eliminate many of the disadvantages present in open-loop control, such as sensitivity to disturbances and the inability to correct for these disturbances.

The major difference between open-loop and closed-loop systems is the addition of the feedback loop from the output back to the input in the closed-loop system. The insertion of this loop necessitates the addition of two other elements, an *output transducer* and a *summing junction* between the input transducer and controller. The output transducer is a device that is used to convert the measured value of the system output into a physical form that is consistent with the input requirements of the control system. The transducer is a combination *sensor* and *signal conditioner* that passes a sample of the *actual output* (controlled variable) back to the input to be compared with the *desired output* (set point). The gain of the transducer is often labeled with the letter H and has a value or gain of 1 in many feedback systems.

The signal conditioning function of the output transducer ensures that the feedback signal has the proper units for comparison with the output of the input transducer in the summing junction. The units of output and input transducers are dictated by the controller, which incorporates the *summing junction* function into its input circuits. Most controllers accept either 4 to 20 milliamps, 1 to 5 volts or −10 to +10 volts for the 0 to 100 percent range of these parameters.

12-4-1 Direct-Acting and Reverse-Acting Controllers

The summing junction that compares the process output with the set point has signs associated with the inputs that are *opposite*, since this is a *negative feedback* system. If the set point (SP) summing input is *plus* and the feedback process variable (PV) is *minus*, then the error (E) or actuating signal is SP − PV. If the signs are reversed, then the error changes to PV − SP. Process characteristics dictate which type of error equation should be used. As an example, two process tanks are illustrated in Figure 12-31. Figure 12-31(a) has a fill pipe controlled by a fill valve, and a drain pipe with back pressure from downstream processes that creates a non-uniform drain flow. Figure 12-31(b) shows the opposite situation. A drain valve controls outflow from the tank, and a fill pipe provides liquid at a non-uniform rate from upstream processes. The controller in each case must maintain the level in the tank for variations in the drain [Figure 12-31(a)] and for variations in the input flow rate [Figure 12-31(b)].

The control sequence for the tank in Figure 12-31(a) is as follows: The level in the tank falls (PV decreases) because the drain flow increases suddenly, the error value must increase so that the controller output can increase, and the input valve is opened more to increase the input flow. The controller is set to the *reverse-acting* mode (Error = SP − PV) because the controller output needed to rise when the PV decreased. Study the tank and control block diagram in Figure 12-31(a) until the reverse-acting operation is clear.

FIGURE 12-31: Direct and reverse error and process relationship.

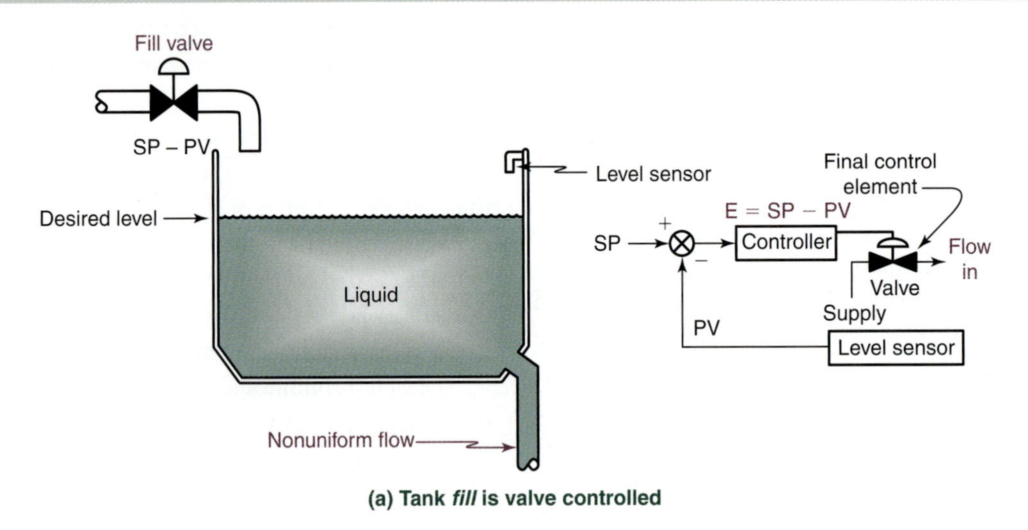

(a) Tank *fill* is valve controlled

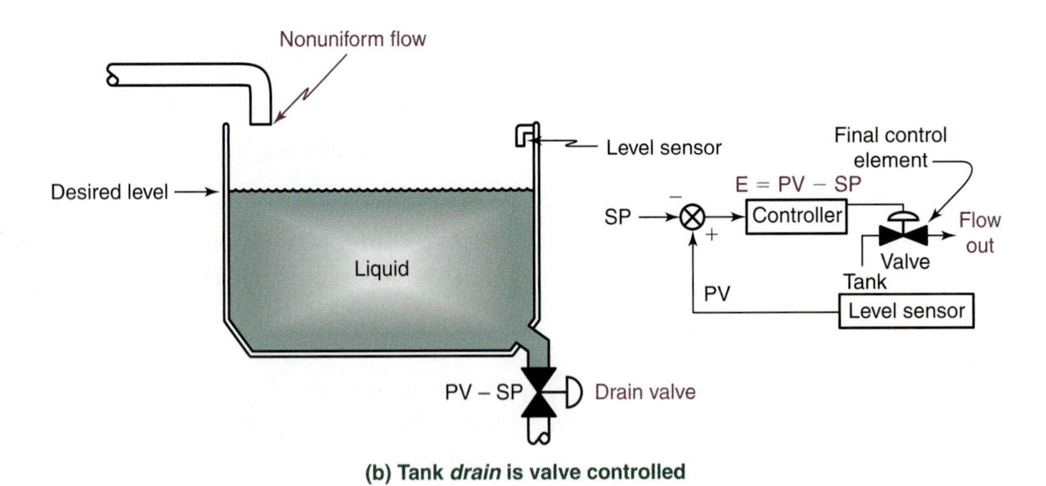

(b) Tank *drain* is valve controlled

Rehg and Sartori, Industrial Electronics, *1st Edition, © 2006, Reprinted by permission of Pearson Education, Inc., Upper Saddle River, NJ.*

The control sequence for the tank in Figure 12-31(b) is as follows: The level in the tank falls (PV decreases) because the input flow decreases suddenly, the error value must decrease so that the controller output can decrease, and the drain valve is closed more to decrease the outlet flow. The controller is set to the *direct-acting* mode (Error = PV − SP) because the controller output needed to decrease when the PV decreased. Study the tank and control block diagram in Figure 12-31(b) until the direct-acting operation is clear.

12-4-2 Analysis of Closed-Loop Systems

Let us take a closer look at closed-loop operation by analyzing the cruise control system, which is illustrated in Figure 12-32.

The closed-loop system has the controller gain (G_2) set at 150 and a summing junction to compare the output, S, with the set point, R. This reverse-acting system will cause the output, S, to increase when the process variable (PV) decreases. The 150 value for the controller gain was not calculated for this model, but selected as a starting point. In industry the gain is sometimes calculated theoretically, but most often it is determined experimentally on a functioning system. Determining optimum controller parameters is discussed later in this chapter.

The next step in the analysis is to develop the steady-state output equation.

Equation 1

$$S = 5 \times S_0$$
$$S = 5(R_0 - G_0)$$

Equation 2

$$R_0 = 150(R - S)$$
$$R_0 = 150R - 150S$$

Substitute equation 1 into equation 2 and solve for S.

$$S = 5(R_0 - G_0)$$
$$S = 5(150R - 150S - G_0)$$
$$S = 750R - 750S - 5G_0$$
$$S + 750S = 750R - 5G_0$$
$$751S = 750R - 5G_0$$
$$S = 0.999R - 0.007G_0$$

FIGURE 12-32: Closed-loop block diagram for the cruise control.

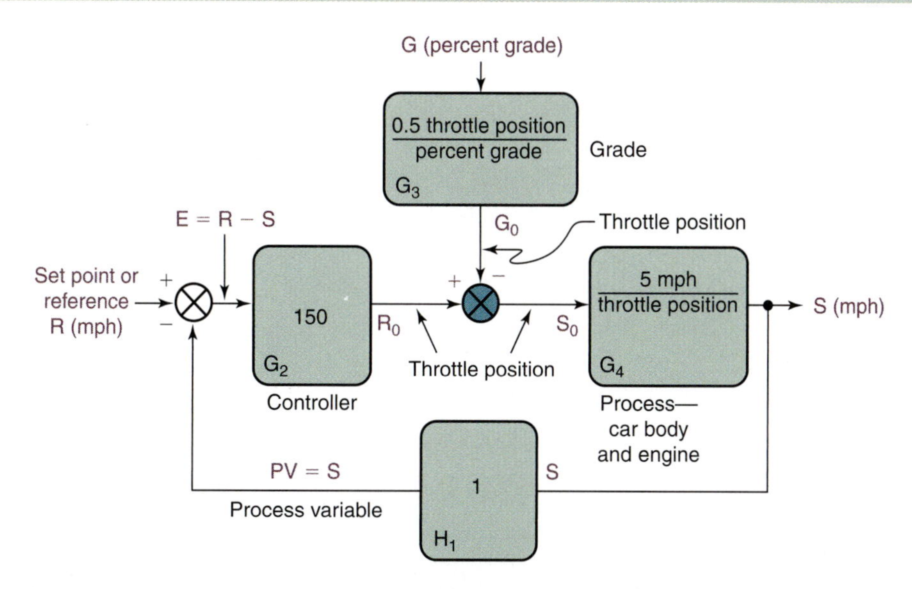

EXAMPLE 12-2

Determine the steady-state output for the closed-loop cruise control system for a reference speed of 55 mph and grades of 0, 1, 10, and 20 percent.

SOLUTION

Evaluate for grade of 0 percent:

$$S = 0.999R - 0.007G_0$$

$$S = 0.999 \times 55 \text{ mph} - 0.007 \times 0$$

$$S = 54.945 \text{ mph}$$

Evaluate for grade of 1 percent:

$$S = 0.999 \times 55 - 0.007 \times 1$$

$$S = 54.938 \text{ mph}$$

Evaluate for grade of 10 percent:

$$S = 0.999 \times 55 - 0.007 \times 10$$

$$S = 54.875 \text{ mph}$$

Evaluate for grade of 20 percent:

$$S = 0.999 \times 55 - 0.007 \times 20$$

$$S = 54.805 \text{ mph}$$

As you can observe from the calculations, the error is very small as long as the controller gain is large (150). The following table compares errors for the open- and close-loop cruise control system for different grades and process gains.

The table illustrates the significant improvement that feedback provided. Consider the following observations.

- At grades starting at 1 percent, the open-loop system had an error of 4.55 percent; above that grade the errors increased dramatically.
- The largest error for the closed-loop system was 0.35 percent for a grade of 20 percent.
- Process gain is difficult to establish accurately and may not be constant, so the effect of process gain change is important to consider. Changes in the process gain had little effect on the closed-loop system, but significant impact on the open-loop implementation.
- The closed-loop system output error (0.10 percent) when all disturbances are zero is a part of the steady-state error inherent in all feedback systems.
- The insensitivity of output speed to disturbances and changes in process gain are due in part to the high controller gain (150). However, if the closed loop gain is too large the system could become unstable.

Control System	Grade	Process Gain	Output Speed	Percent Error
Open-Loop	0	5	55	0%
Closed-Loop	0	5	54.95	0.10%
Open-Loop	1	5	52.5	4.55%
Closed-Loop	1	5	54.94	0.11%
Open-Loop	2	5	50	9.09%
Closed-Loop	10	5	54.88	0.23%
Closed-Loop	20	5	54.81	0.35%
Open-Loop	0	4	44	20.00%
Closed-Loop	0	4	54.89	0.18%
Open-Loop	1	4	42	23.64%
Closed-Loop	1	4	54.90	0.20%

In summary, closed-loop systems have the advantage of greater accuracy than open-loop systems, plus they are less sensitive to noise, disturbances, and changes in the environment. In addition, transient response and steady-state error are relatively easy to control with a gain adjustment in the loop or the addition of compensation to the controller. However, closed-loop systems are more complex and expensive than open-loop systems, and require some control system knowledge to efficiently implementation and maintain.

12-4-3 Load Change—Process Disturbance

The feedback system is called upon to bring the process output to the correct operational level when the input set point is changed or when a disturbance occurs. The term *disturbance* has been used frequently and appears on the block diagram in Figure 12-33. The two disturbances in the figure, Disturbance 1 and Disturbance 2, represent change imposed on one of the control elements and on the process, in that order. The type of change that is imposed most often is Disturbance 2, a forced change in the process's controlled variable, which includes a non-uniform input or output flow rate in a level control process or a non-uniform input fluid temperature in a tank temperature control process. The controlled variables, liquid level and process mixture temperature, have a step-type change in the controlled parameter due to a disturbance. As a result, this type of forced change is called a *process load change*.

The *process load* is best described by a study of a home heating system. The loss of heat through the insulation in the walls, floor, and ceiling of a home is replaced by the heat generated by the furnace. The heat lost is changing continuously as the temperature and other climatic factors on the outside go through many step changes. The control system must balance the system load (heat lost) by regulating the rate at which energy is added back into the system through a change in the final control element (gas valve in the furnace) and eventually in the process variable (PV), which is the temperature in the home. The *load* is a disturbance that acts directly on the process, and the load on a process is always evident in the process variable. In feedback systems any change is evident in the process variable (PV) and system error. Controller action then causes a corresponding change in the setting of the final controlling element, which changes the process variable and corrects for the initial load change. In an ideal situation, the control action causes the manipulated variable to match the increased load placed on the process so that the controlled variable remains at the desired value.

FIGURE 12-33: Closed-loop system.

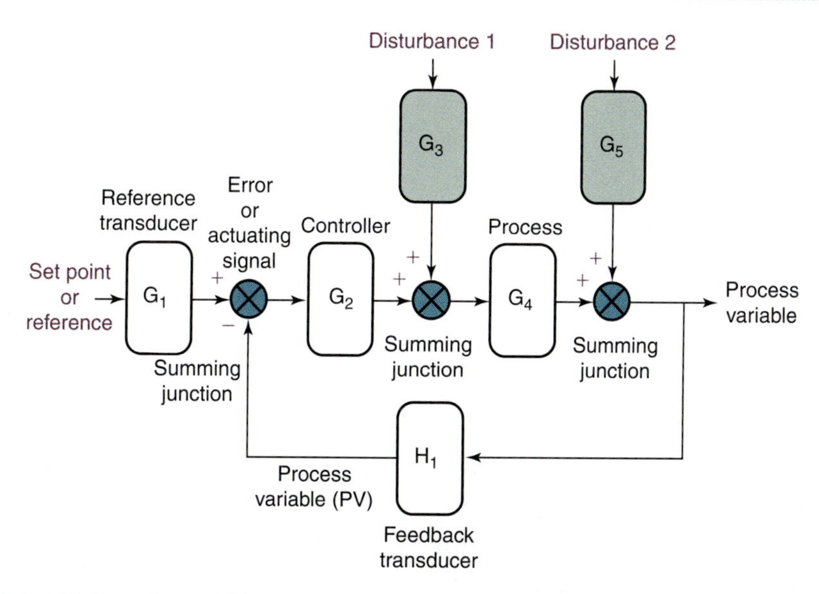

Rehg and Sartori, Industrial Electronics, *1st Edition, © 2006, Reprinted by permission of Pearson Education, Inc., Upper Saddle River, NJ.*

As a result, a good place to check for a system load is at the manipulated variable. Therefore, load is defined as follows:

> *The load on a control system is measured by the value of the manipulated variable required by the process to eliminate the effect of the load.*

The load on a control system is never constant because disturbances that affect the controlled variable are all capable of causing a load change. When a process system must correct for load change, a self-correcting closed-loop system is necessary because it automatically makes the necessary change in the manipulated variable. Most often, uncontrolled system variables and parameters are the root cause of load changes. For example, changes in ambient conditions (temperature, humidity, pressure, power sources), the quality of the material used in the process, and flow rates for materials would require intervention by the control system.

12-5 ATTRIBUTES OF AN EFFECTIVE CONTROL SYSTEM

Control systems are *dynamic* because they respond to an input stimulus with a transient response before reaching a steady-state response that has an output level near the desired input set point. More dynamic input stimulus needs better control systems for the output to track changing input reference. To achieve this level of control, system analysis and design must address three major control objectives: producing the desired *transient response*, reducing *steady-state error*, and achieving *stability*. In addition, designers must also consider the trade-off between *performance and cost* and *system robustness*.

12-5-1 Transient Response

Transient response is an important control attribute because all physical systems have inertia that resists *change*, and change is the very thing that a control system tries to achieve. In some applications the rate at which the output responds to a change in the set point or to a load change is not as critical as others. For example, an aircraft autopilot system must react much faster to a change in direction due to a gust of wind than a process soup kettle must react to a new temperature setting. In another example, the control system for a construction crane can be only as fast as the safe non-destructive movement of the large crane structure permits. Therefore, the application dictates the level of response, and the control system delivers an output that is most appropriate for the application. What is clear from our previous discussion of control systems is that the change in set point or the load change occurs and the control system's effort to conform to the new set point or to correct for the load change occurs some time later. The first objective of good control system is to respond to the demand for change as quickly as possible.

12-5-2 Response to Change

The waveforms illustrated in Figure 12-34 show a typical response from a temperature controller on a process mix tank. The tank mixes and heats several liquids in a continuous process. The load change occurs when one of the ingredients start to come in at a higher temperature than normal. The waveforms include the load change [Figure 12-34(a)], process variable (PV) [Figure 12-23(b)], and controller output (CO) [Figure 12-34(c)], which is the steam valve control signal. A study of the PV and CO curves indicates:

- The process change starts when the load change occurs, Figure 12-34(a).
- There is dead-time delay before the PV begins to react to the load change.
- The mix tank liquid increases in temperature, which forces the steam control valve to close and remove heat from the tank.
- The tank temperature then becomes too low and the steam valve opens to correct that problem.
- Oscillations continue until the steady-state condition is reached and the PV is constant again.
- The ideal PV response is indicated by the dashed line. To achieve this, MV would have to be at a closed position indicated by the dashed line on the MV graph.
- The steady-state response was not able to fully compensate for the increase in tank temperature, so a constant error, called the steady-state error or offset, is present.
- Study the figures and the analysis until the interaction between the curves is clear.

FIGURE 12-34: Response to a load charge.

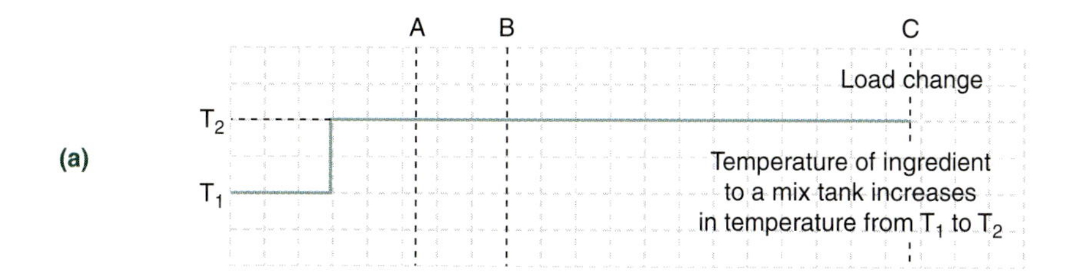

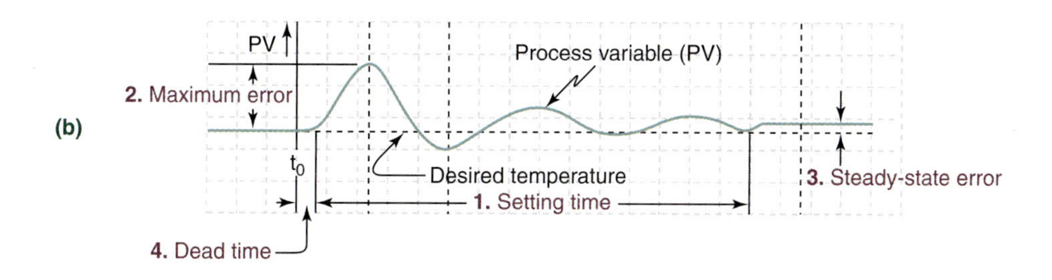

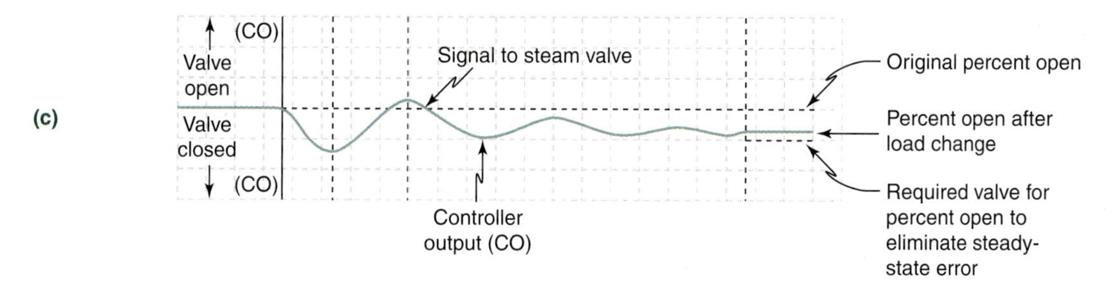

Rehg and Sartori, Industrial Electronics, *1st Edition, © 2006, Reprinted by permission of Pearson Education, Inc., Upper Saddle River, NJ.*

In order to achieve optimum performance, four attributes of the waveform must be reduced:

1. The length of the settling time
2. The maximum value of the response error
3. The residual or steady-state error
4. The dead time

The first two attributes are addressed in the transient analysis of the response, the third is addressed in the steady-state analysis, and the last, dead time, is the most difficult to correct and often cannot be fixed because it is inherent in the process. Some issues associated with dead time are addressed later in this section.

12-5-3 Controller Response and Damping

The first transient response objective of good control is to reduce the settling time required for the system to reach a steady-state value. The second is to reduce the maximum error. There are three distinct types of response curves: *overdamped, critically damped,* and *underdamped.* Figure 12-35 illustrates how a control system with each type would react to step set point change. In each graph the relative value of the rise time, the time required for the waveform to go from 10 to 90 percent of the final value, is reported. In addition, the relative value of the damping coefficient, a number between 0 and 1, is also listed. Some observations:

- The overdamped controller has a very large rise time and damping coefficient (0.8 or 0.9), whereas the rise time and damping coefficient (0.4 or 0.5) of the underdamped controller is very small.
- If fast response is required then a slightly underdamped response is necessary, but you

FIGURE 12-35: Three types of transient response damping to a step change in the set point.

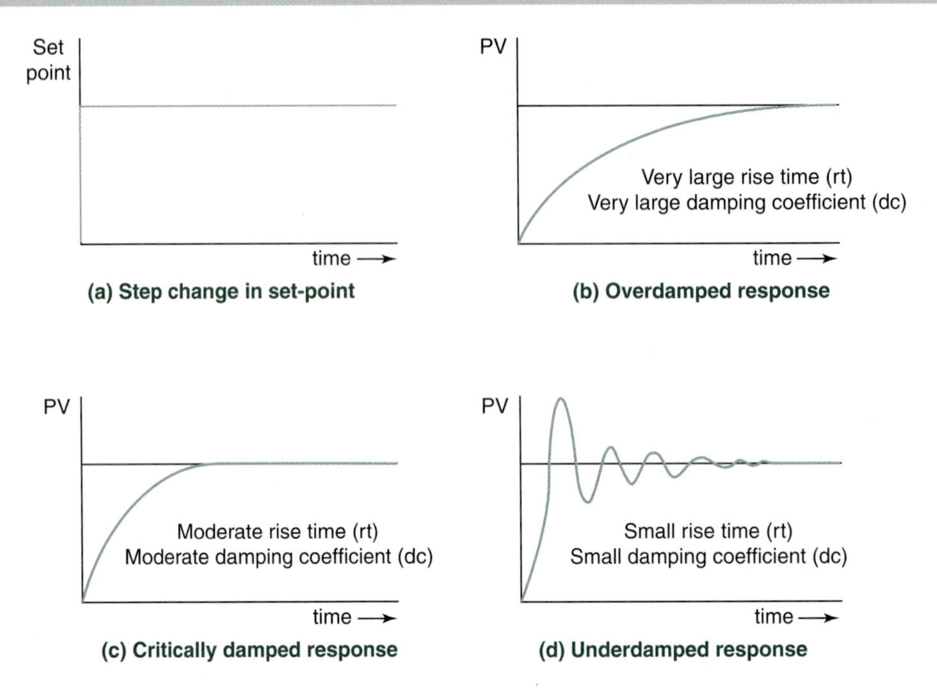

Rehg and Sartori, Industrial Electronics, *1st Edition, © 2006, Reprinted by permission of Pearson Education, Inc., Upper Saddle River, NJ.*

have to live with the larger maximum error and longer settling time. The oscillations present while the response settles to a final value are called *ringing*.

- The critically damped response is between the other two response extremes, with a moderately fast rise time produced by a damping coefficient in the 0.6 to 0.7 range but no overshoot.
- The damping and the controller response are directly related to controller gain. Therefore, increases in controller gain change the damping from overdamped to critically damped and finally to underdamped.

The optimum response is either critically damped or slightly underdamped with the optimum value dictated by the control requirements of the process. All design is a series of trade-offs, as the discussion on the three responses makes very clear. The optimum response for a given application requirement is a compromise.

12-5-4 Transient Response Options

The transient response form is dictated by the application requirements; however, for most applications a compromise between the *best* possible *rise time*, *lowest* possible *overshoot* (another term for *maximum error*), and the *shortest* achievable *settling time* is a good choice. Design of controllers and feedback systems is a complex process. It is easy to find entire texts devoted to a discussion on one element in the process. However, as a starting point for an understanding of a good response consider the following four responses: *quarter amplitude decay, peak percentage overshoot, minimum integral of absolute error,* and *critical damping*. These are frequently used in industrial applications with less complex control requirements and are illustrated in Figure 12-36, along with the step change in set point used to test them. Each is described as follows:

- **Quarter amplitude decay** [shown in Figure 12-36(b)]: The level of ringing is specified such that each successive positive peak value is one-fourth of the value present in the previous positive peak. Control system parameters are adjusted until the desired results are obtained. Quarter amplitude decay is used frequently, easy to implement, and a good trade-off between rise time, overshoot, and settling time.

FIGURE 12-36: Workable transient response options.

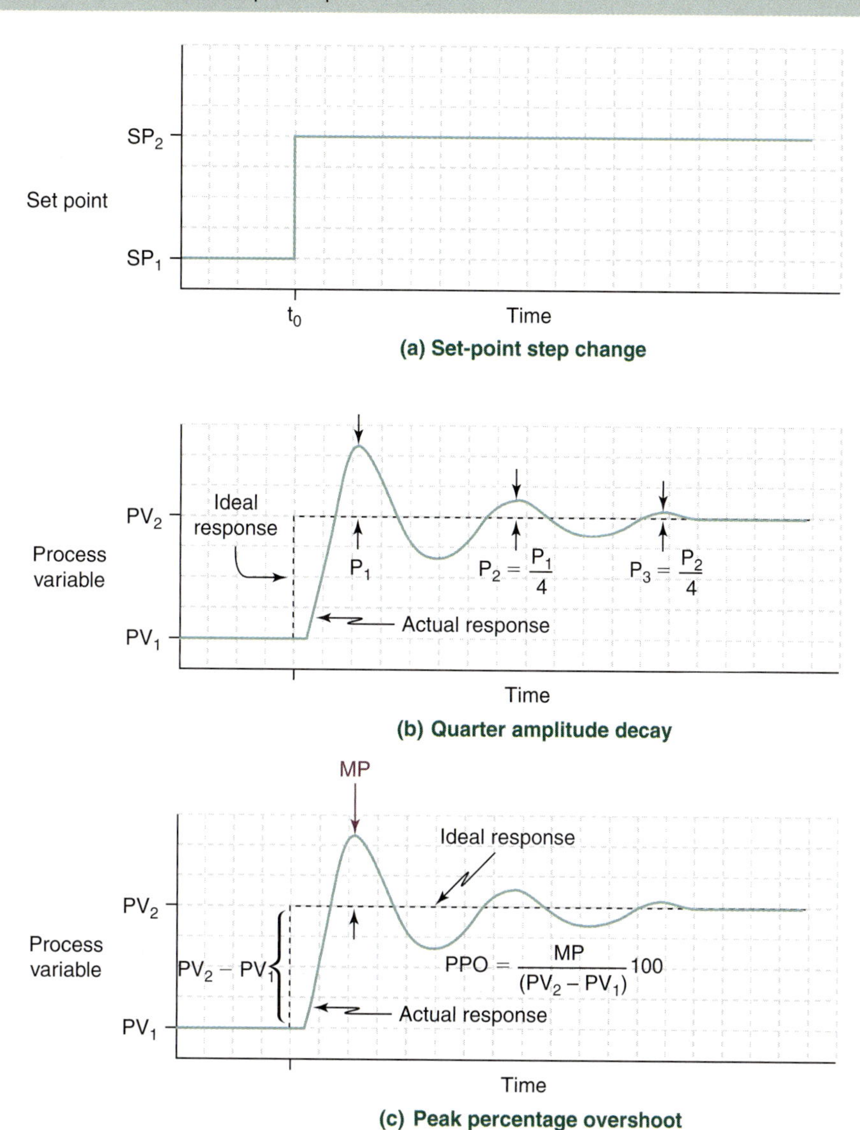

(a) Set-point step change

(b) Quarter amplitude decay

(c) Peak percentage overshoot

Rehg and Sartori, Industrial Electronics, *1st Edition, © 2006, Reprinted by permission of Pearson Education, Inc., Upper Saddle River, NJ.*

- **Peak percentage overshoot** [shown in Figure 12-36(c)]: Peak percentage overshoot (PPO) specifies the maximum peak (MP) overshoot as a percentage of the step change. The formula for MP is:

$$MP = \frac{PPO \times (PV_2 - PV_1)}{100}$$

where the peak percentage overshoot is selected for an anticipated change in PV and the MP is calculated. Control system parameters are adjusted until the desired results are obtained. Like the quarter amplitude decay, this response form is easy to implement, and allows for rise time adjustments as a function of overshoot limits.

- **Minimum integral of absolute error** [shown in Figure 12-36(d)]: The basis for this response option is to keep the integral of the error (E) or the shaded area in the figure to a minimum. The shaded area represents the distance between the ideal response and the actual controlled variable. If that area is minimized, then the PV approaches the ideal response. This technique is used when the process has been modeled and a numerical

FIGURE 12-36: (Continued).

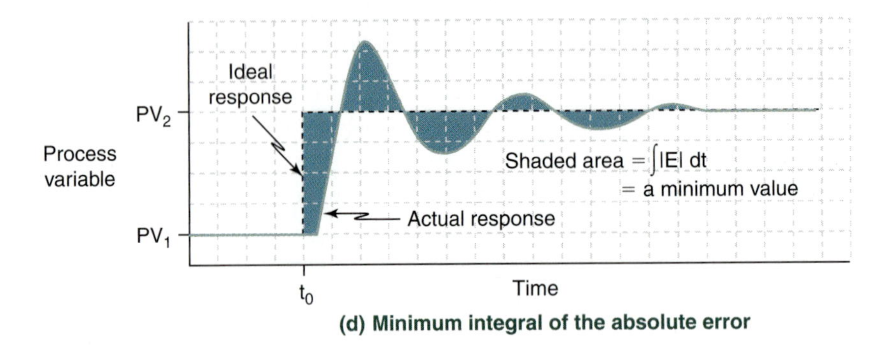

(d) Minimum integral of the absolute error

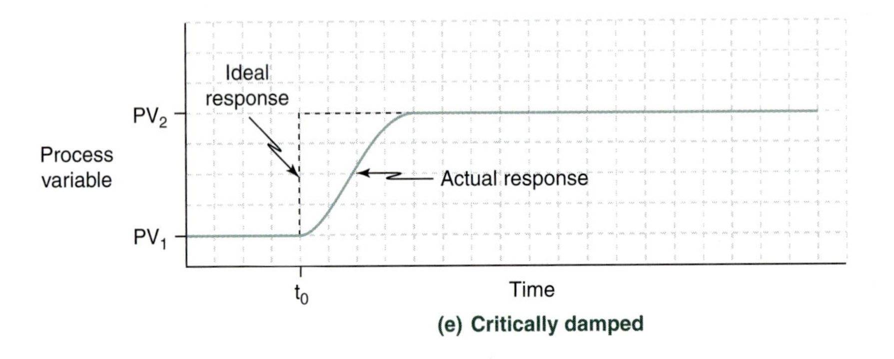

(e) Critically damped

representation of the control system can be used for the integration process.

- Critical damping [shown in Figure 12-36(e)]: Critical damping is selected when overshoot cannot be tolerated. Critical damping is the minimum damping that will bring the PV to the desired value with the best rise time with no ringing.

12-5-5 Steady-State Response

The second major control objective is to reduce the steady-state error that is often present in the steady-state response. Figure 12-34 illustrates the response element that remains after the transients have decayed to zero. If the steady-state response were always equal to the new set point value after a reference change or equal to the previous controlled variable output after a load change, then this discussion of steady-state response would end here. However, what often occurs is a steady-state error or offset in the PV after a change in the set point or load. First let's explore why steady-state error or offset is present, then we will look at corrections that are used.

12-5-6 Understanding Steady-State Error

The cause of steady-state error is often hard to understand for students studying feedback control for the first time. In the model section earlier, the closed-loop cruise control system analysis showed that the output of the cruise control was not exactly equal to the anticipated output as a result of steady-state error. Turn back and review those calculations.

From an analytical standpoint, steady-state error is present because some error or actuating signal is necessary to move the controlled variable closer to the desired output after a set point or load change. This concept is fairly abstract when described as text, so it is often easier to demonstrate the condition that produces steady-state error. Note the mechanical level control system and the results in Figure 12-37. Study the diagram.

The outlet flow from the vessel is 5 gallons per minute (gpm). The input valve is manually opened and the tank is allowed to fill to the 75 percent level [Figure 12-37(a)]. At that point the percent open value of the input valve is adjusted until it is also 5 gpm, so the level remains constant

FIGURE 12-37: Input flow and outlet flow changes in a tank control system.

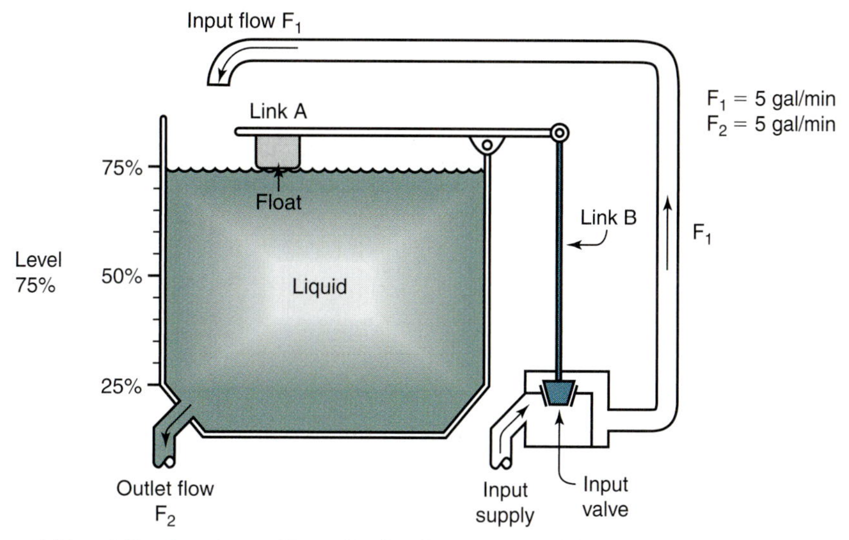

(a) Input flow is set equal to outlet flow by operator with level at 75 percent

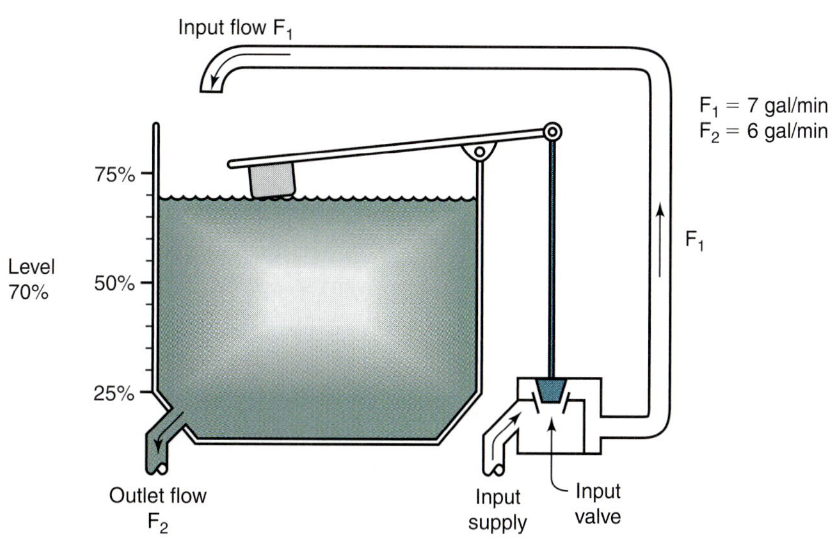

(b) Increased outlet flow to 6 gpm—level drops to 70 percent—input flow rises to 7 gpm

Rehg and Sartori, Industrial Electronics, *1st Edition, © 2006, Reprinted by permission of Pearson Education, Inc., Upper Saddle River, NJ.*

at 75 percent. This may take some adjustment until the operator achieves this exact setting. At that point the length of link B is adjusted so that link A allows the float to rest on the liquid surface. The system is now in automatic control, and the mechanical control system keeps the liquid level constant. Let's explore a load change scenario.

1. Due to process problems the outlet flow is increased to 6 gpm in a step load change. The increased outflow causes the liquid level to fall to 70 percent [Figure 12-37(b)]. Link A at the float follows the float to the lower liquid level. This raises link B and increases the flow in the input valve to 7 gpm, which is an overcorrection.
2. The increased input (7 gpm) allows the level to recover from this load change response, and the liquid level rises and

FIGURE 12-37: (Continued).

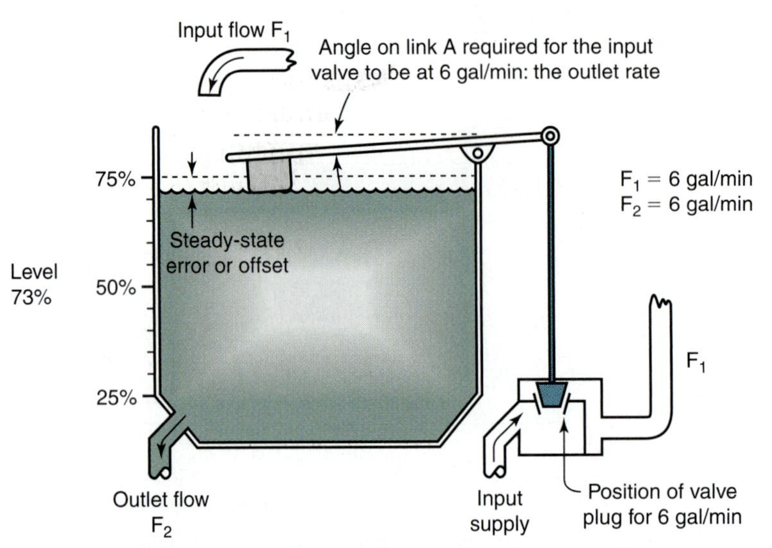

(c) Input flow and outlet flow are 6 gpm but level is at 73 percent

achieves a new equilibrium level of 73 percent [Figure 12-37(c)]. At this point the flow from the input valve is 6 gpm, which is the same as the new outlet flow.

After the load change the system only returned to the 73 percent level. Why is the level not at 75 percent, and why is link A at a slight angle compared to the initial system condition? The answer to these questions is the key to understanding why steady-state error is present and must be addressed by the controller. One response to the question is:

> *If the float was at the 75 percent level, then links A and B would put the input valve at a position for a 5 gpm flow as in the initial setup. Since this flow would not balance the new outlet flow of 6 gpm, the valve must be in the 6 gpm position. The 6 gpm input flow is only achieved when the liquid and float are at 73 percent, not 75 percent. That difference is the steady-state error.*

The 2 percent difference between the desired level of 75 percent and the system response of 73 percent after a load change is the *steady-state error*. Another response is:

> *With this mechanical controller, the error is necessary to drive the final control element (input valve) into a 6 gpm position so that the system is balanced. If the error was 0, then the final control element flow would not match the new flow after the load change.*

The response waveform for the load change that triggered the system response is illustrated in Figure 12-38(a). Following the disturbance of a 1 gpm increase in the outlet flow, the input flow rate, Figure 12-38(b), initially rises to 7 gpm and then reaches equilibrium at 6 gpm, the input rate. When the system comes into balance, distinguished by no change in the level, the input and output flow rates must be equal. The response of the controlled variable is illustrated in Figure 12-38(c). Note that the level drops rapidly at first due to the step increase in outlet flow to 6 gpm. The falls to 70 percent and starts to level off only when the input flow exceeds the new outlet flow of 6 gpm. As the input flow reaches the 7 gpm rate, the liquid level starts to climb toward the final value of 73 percent. The input flow falls to the 6 gpm flow necessary for a balance system when the level reaches 73 percent. Spend time studying Figures 12-37 and 12-38 and thinking about the two explanations just presented until the concept of steady-state response is clear.

FIGURE 12-38: Response curve for 1 gpm flow load change.

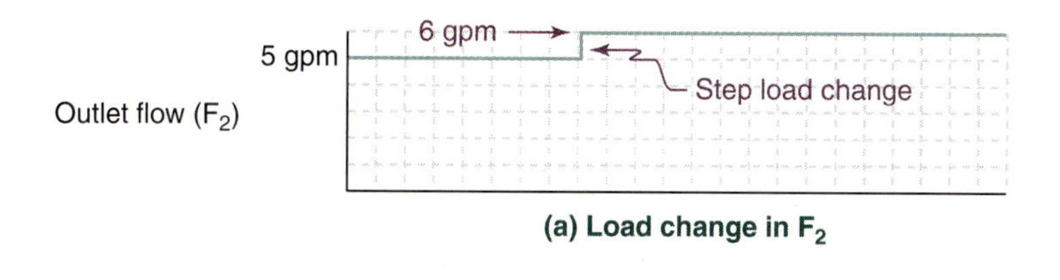

(a) Load change in F_2

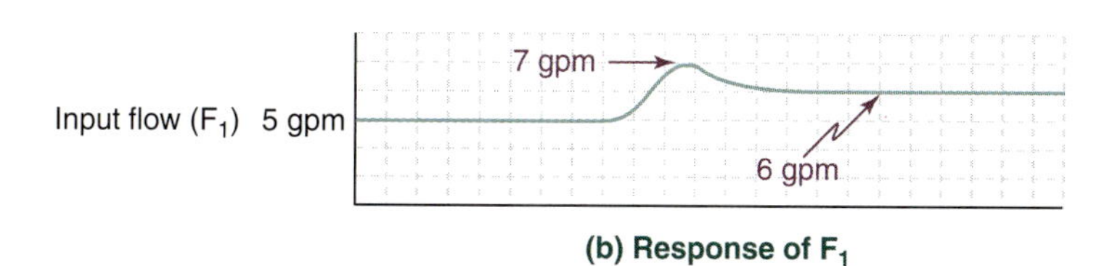

(b) Response of F_1

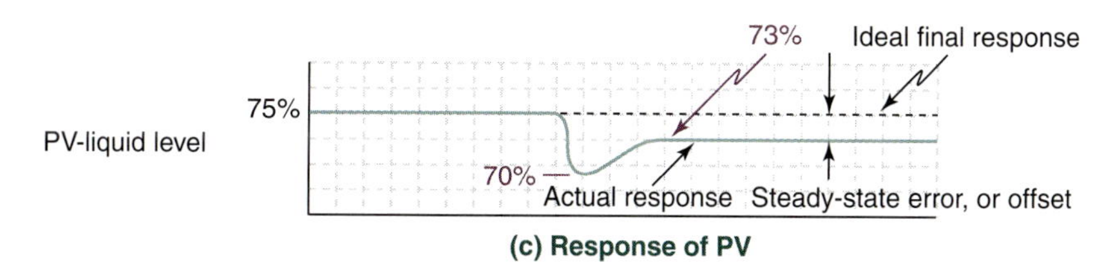

(c) Response of PV

Rehg and Sartori, Industrial Electronics, *1st Edition, © 2006, Reprinted by permission of Pearson Education, Inc., Upper Saddle River, NJ.*

EXAMPLE 12-3

The system outlet flow is at the 6 gpm rate as illustrated in Figure 12-37(c). The system outlet experiences 1 gpm decrease as a step load change. Determine the steady-state output for the system.

Solution

The system change that results from the decrease in load is as follows:

1. The outlet now has a 5 gpm flow, whereas the input is still at the 6 gpm rate.
2. The greater input flow causes the liquid level to rise to 78 percent as the tank fills more rapidly than it drains.
3. The level increase raises the float and lowers link B, which closes the input valve, reducing the input flow to 4 gpm.
4. The decrease in input flow rate (4 gpm) allows the level to recover from this load change reaction, and the liquid level drops from 78 percent to an equilibrium level of 75 percent.
5. As the level falls to the 75 percent level, the input flow approaches 5 gpm, the same as the new outlet flow.
6. The system is back in balance and the steady-state error is zero.

No steady-state error occurs with the load change back to the initial system flows since the system was manually set up to control the level at 75 percent. No system error need be present to position the input valve for 5 gpm since that is the initial control condition.

12-5-7 Correction for Steady-State Error

A correction for the steady-state error is hidden in the load change analysis just completed. To help locate the solution, study the more detailed analysis of the system load change in Figure 12-39. The following important observations are made.

Figure 12-39(a) Observations (System setup – system balanced)

- The desired level is 75 percent and a liquid level at that percentage produces an outlet flow into the downstream process of 5 gpm.

FIGURE 12-39: Analysis of system steady-state error.

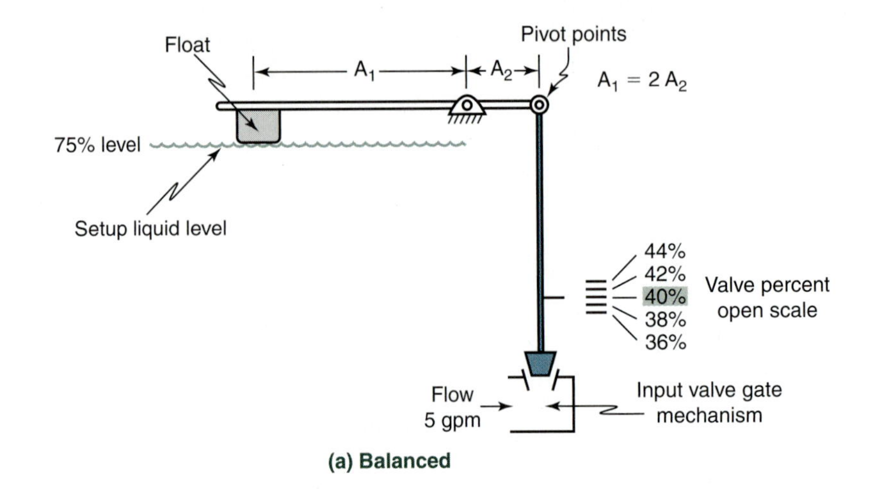

(a) Balanced

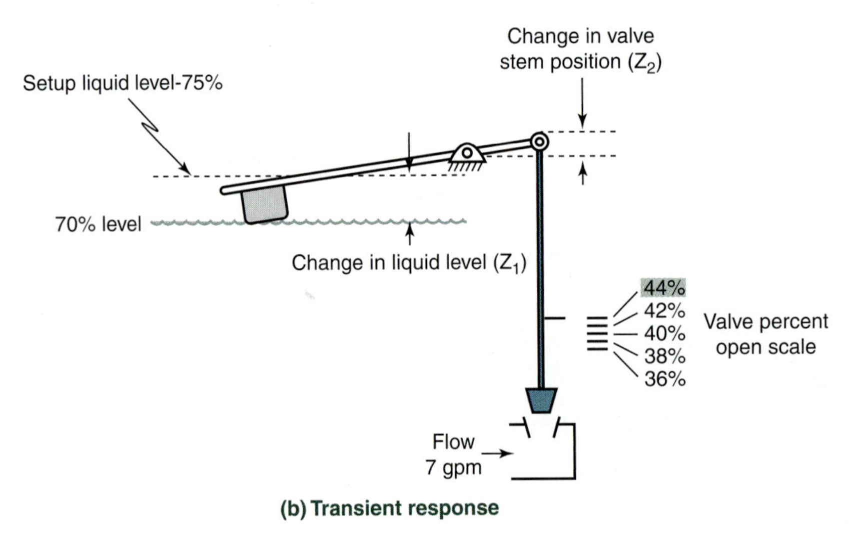

(b) Transient response

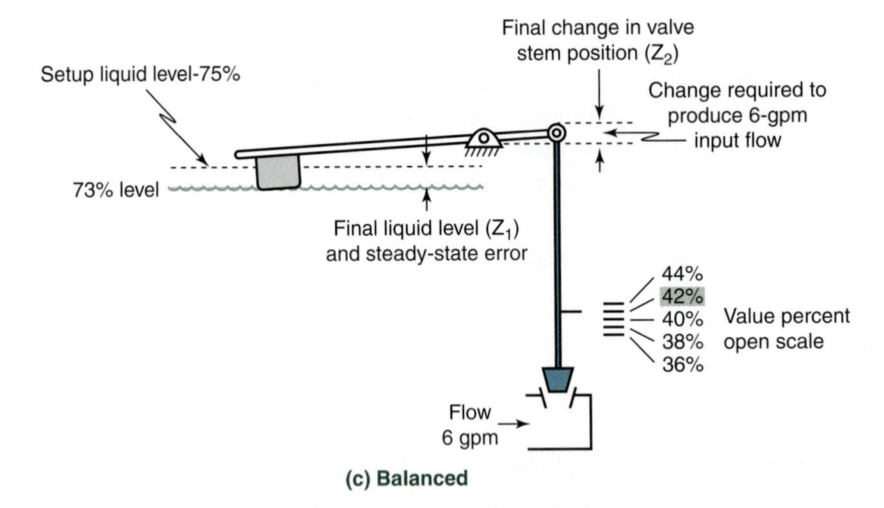

(c) Balanced

Rehg and Sartori, Industrial Electronics, *1st Edition, © 2006, Reprinted by permission of Pearson Education, Inc., Upper Saddle River, NJ.*

- To maintain a 75 percent level link B is adjusted so that the input valve is set at 40 percent open, which produces an input flow of 5 gpm.
- The system is now balanced.

Figure 12-39(b) Observations (Reaction to disturbance – transient response)

- The change in outlet flow from 5 gpm to 6 gpm causes the liquid level to drop initially to 70 percent.
- The change in the float and liquid level is indicated as Z_1. The fulcrum produces a corresponding change in the position of link B that is indicated as Z_2. Since A_2 is twice as long as A_1 (link A lengths on each side of the pivot), the Z_2 change is only half of the change present at Z_1.
- The rise in link B by the Z_2 value opens the input valve to 44 percent.
- The increased input flow of 7 gpm results from the valve opening of 44 percent and is a transient response to the increase in outlet flow.
- In this stage of the response the system is searching for a liquid level which will hold the input valve at a percent open value that will balance input flow with the new output flow of 6 gpm.
- The link B value, Z_2, is changing continuously as the system moves to a new balance condition.

Figure 12-39(c) Observations (New outlet and input flows – system balanced)

- The feedback system for the level control (links A and B) determines that a liquid level of 73 percent will produce a value of Z_2 that opens the input valve to 42 percent.
- A 42 percent valve opening produces an input flow of 6 gpm, which balances the new outlet flow of 6 gpm.
- The difference in the current level (73 percent) and the desired level (75 percent) is 2 percent; this represents the system steady-state error that results from the load change.
- The important insight at this point is that as long as the system is in balance, the valve *must stay at 42 percent* to produce the required flow of 6 gpm, so the *current value of Z_2 cannot change.*

It is not clear what must be held fixed for the system to reach a new balance after the load change. You might ask the question:

> *What could be changed in Figure 12-39(c) that would allow the liquid level to rise closer to 75 percent while keeping the value of Z_2 unchanged?*

The answer to that question leads to one solution for steady-state error.

At this point it should be clear that steady-state error is present in control systems. There is more than one solution for steady-state error, but the one usually considered first is illustrated in Figure 12-40. The figure shows the level control problem after the load change disturbance [Figures 12-37(c) and 12-39(c)] when the system balance has returned with an input and outlet flow of 6 gpm. At this point the steady-state error is 2 percent (75% desired level − 73% actual level = 2% error).

The position of link B is fixed with the Z_2 value setting the input valve at 42 percent open and an input flow of 6 gpm. Link A is divided into two distances: the center of the float to the fulcrum location (A_1) and the fulcrum to the connection to link B (A_2). Since Z_2 cannot change the angle of link A is also fixed. Therefore, if the float was moved toward the fulcrum, then the liquid level could rise and the 6-gpm input valve position is not affected. So changing the position of the float from position P_1 (73 percent liquid level) to P_2 allows the tank level to increase to 73.6 percent, a reduction of 0.6 percent in the steady-state error. If the float was repositioned again to P_3, then the liquid could rise to the 74.3 percent level, which is only a 0.7 percent steady-state error. To understand what is changing in the control system to allow this improvement in steady-state error, the elements present in the tank process must be related to a block diagram.

The diagram in Figure 12-41 represents the tank control process. Compare the blocks and terms in the figure with the system in Figure 12-37 while you review the procedure used to set the initial level at 75 percent. Link B was adjusted to achieve the desired level; therefore, it is the set point. The float determines the current level, so it is the feedback position sensor. The controller is link A since it determines how much change should occur in the final control element when the set level and the actual level are not equal. The controller summing junction is the difference with a positive error for drops in tank level that would increase

FIGURE 12-40: One technique for reducing steady-state error.

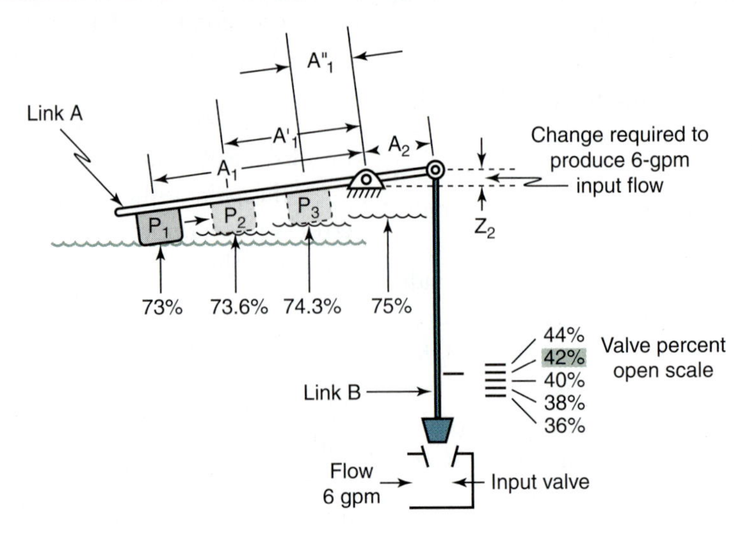

Rehg and Sartori, Industrial Electronics, *1st Edition, © 2006, Reprinted by permission of Pearson Education, Inc., Upper Saddle River, NJ.*

FIGURE 12-41: Block diagram for the level control tank in Figure 12-37.

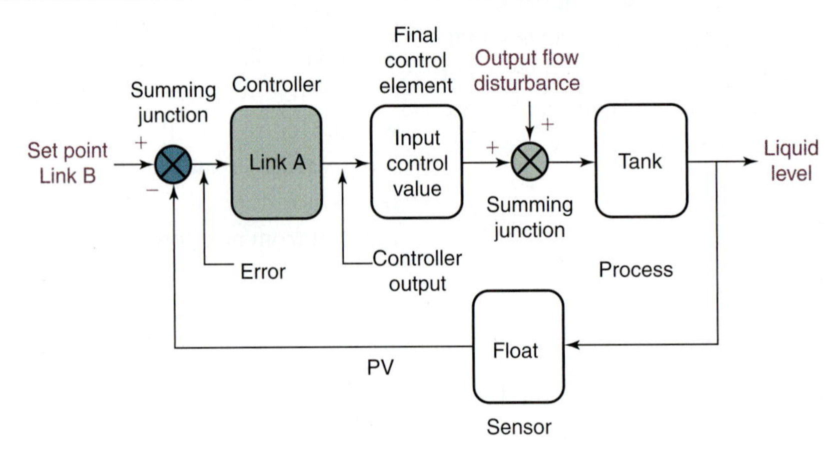

Rehg and Sartori, Industrial Electronics, *1st Edition, © 2006, Reprinted by permission of Pearson Education, Inc., Upper Saddle River, NJ.*

the opening on the input valve, the final control element. The disturbance summing junction is the sum of the balanced flow rate and the change in flow represented by a load change. Since it was a change in the location of the float along link A that improved the steady-state error, analysis of link A provides the reason for this reduction in error.

Link A is a lever with the requisite fulcrum. As the float moves from position P_1 to position P_2 the lengths of the link elements on the left side of the fulcrum changed. These changes are illustrated in Figure 12-42. Every lever has an input and output end as illustrated in Figure 12-42(a). The change in input and output is the difference between the original position of the lever (horizontal) and the final position (some rotation about the fulcrum). In this case the input is on the left and the output is on the right. Using the relationship between sides of similar triangles, the mechanical advantage or gain is the ratio of y over x, as shown.

Apply the information from the lever analysis in Figure 12-42(a) to the position changes of the float in the tank level control problem. Note that as the float moved closer to the fulcrum the gain increased from 0.5 to 2. Earlier we determined

FIGURE 12-42: Controller analysis for gain.

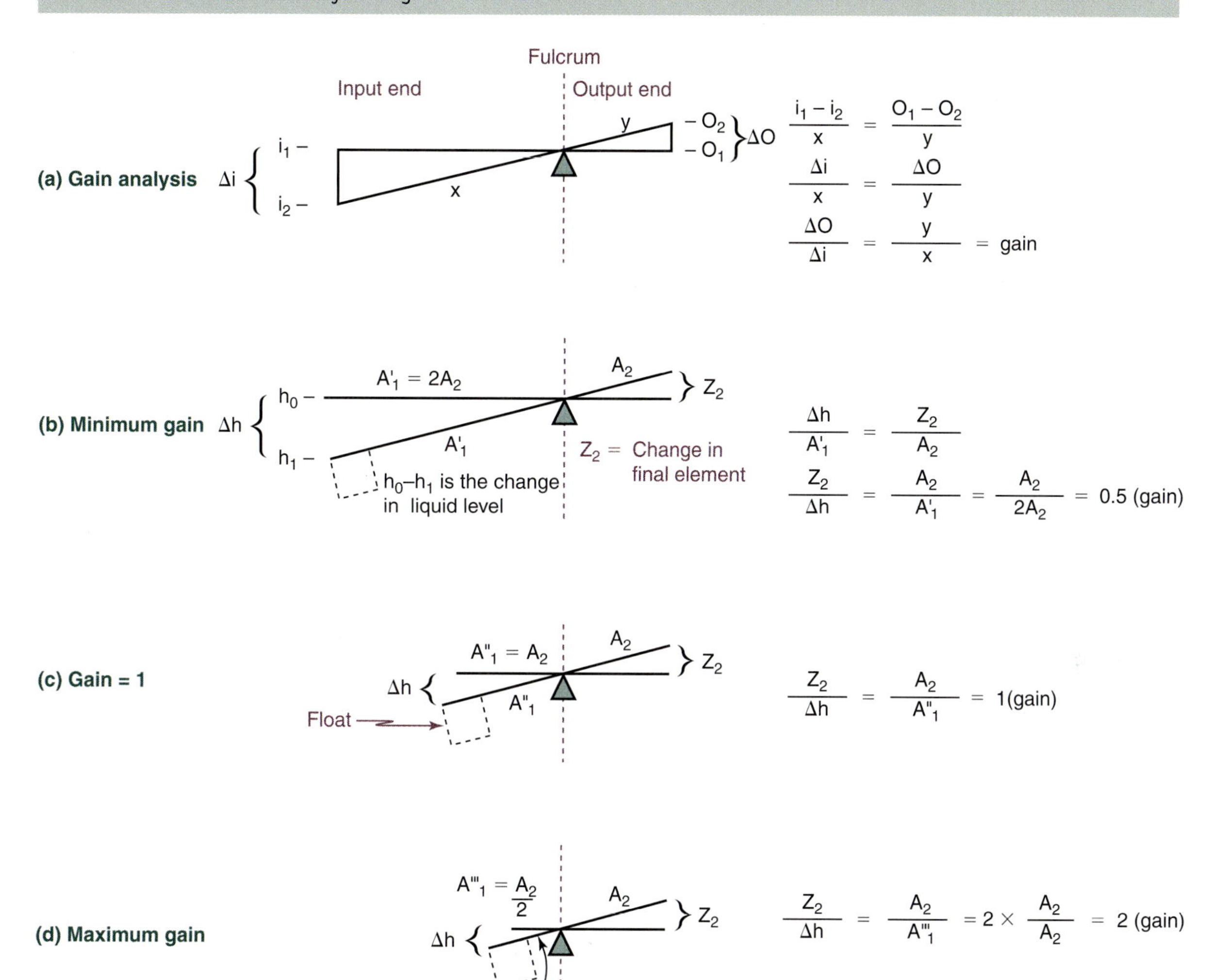

Rehg and Sartori, Industrial Electronics, *1st Edition, © 2006, Reprinted by permission of Pearson Education, Inc., Upper Saddle River, NJ.*

that as the float moved closer to the fulcrum the steady-state error decreased. Therefore, we observe that control system steady-state error and gain have an inverse relationship.

An increase in the controller system gain decreases the steady-state error.

The control system gain includes the cumulative gain of all the control system functional blocks except the process block. For example, an increase in the final element gain reduces the steady-state error just as does an increase in the controller gain. However, the easiest place to change the gain is in the controller.

12-5-8 Controller Gain Side Effects

The decrease in the steady-state error as a result of an increase in controller gain causes problems in other parts of the system response. Analysis of the system during the transient phase of the response to a load change in Figure 12-39(b) illustrates the problem. Assume that the float is located at P_2, so the controller gain is changed from 0.5 to 1, and the system response is modified to incorporate the increase in gain by changing Figure 12-39(b) to the illustration in Figure 12-43. Compare the two figures until the effect on the valve position is understood.

FIGURE 12-43: Change in transient response for higher controller gain.

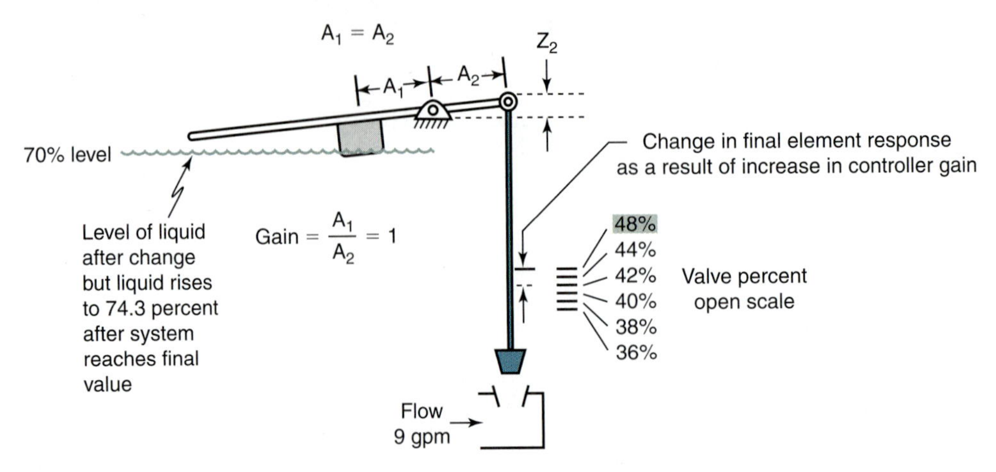

Rehg and Sartori, Industrial Electronics, *1st Edition, © 2006, Reprinted by permission of Pearson Education, Inc., Upper Saddle River, NJ.*

With gains of 0.5 and 1, the initial drop in the tank level after the load change is about equal at 70 percent. Also, the input and outlet flows balance at 6 gpm. However, two things are different. The steady-state level error drops from 2 to 1.4 percent with the higher gain, and the value of Z_2 the input valve transient change to correct for the disturbance, is twice as large when the gain is 1. The larger change in Z_2 causes the input value to open to 48 percent, which causes the input flow to jump to 9 gpm. This strong response reduces the amount that the level will fall, but it also causes the level to recover to over the 74 percent mark before settling down the steady-state value of 73.6 percent. Increased controller gain benefits steady-state response by reducing error and improving response rise time; however, the transient response becomes less optimum because of a higher maximum error and longer settling times. The ideal controller gain is a compromise built on the following criteria:

- The level of steady-state error that can be tolerated.
- The minimum rise time required.
- The maximum value of overshoot.
- The maximum allowed settling time.

If a compromise cannot be reached, then a priority for the criteria must be established for determining the controller gain.

12-5-9 Steady-State Error Correction with Bias

A second technique to reduce steady-state error is to use a *bias* or *offset* input to the controller. The functional block diagram for the tank level controller in Figure 12-41 has a bias or offset input added in Figure 12-44. Compare the two figures.

The function of the bias or offset input is to add enough signal to the final control element so that the steady-state error left in the system is removed. For example, in the tank control problem the bias input would replace the signal generated by the position of the float below the 75 percent fill level. If the bias increased the flow in the input valve, then the tank level would increase. Increased tank level would reduce the steady-state error and the amount of signal from the controller used to keep the input flow at 6 gpm. When the bias provides all the signal necessary for the input valve to deliver 6 gpm, then the error into the controller and its output can be zero. At this point the steady-error falls to zero.

The problem with this solution is that it is only good until another set point or load change occurs. At that point the bias may make the situation worse or at least must be changed every time a set point or load change happened.

12-5-10 Stability

Improving the transient response and reducing steady-state error is of little value if the system is

FIGURE 12-44: Controller bias for steady-state error reduction.

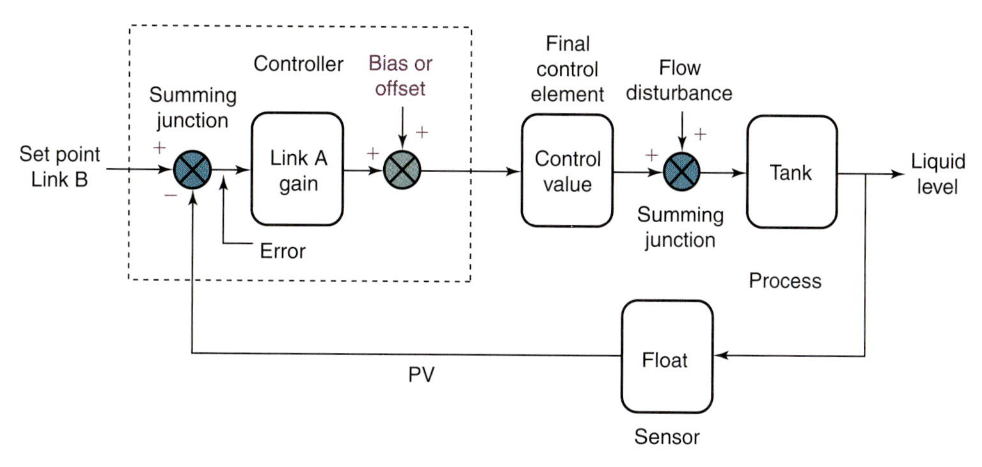

Rehg and Sartori, Industrial Electronics, *1st Edition, © 2006, Reprinted by permission of Pearson Education, Inc., Upper Saddle River, NJ.*

FIGURE 12-45: Oscillatory system responses.

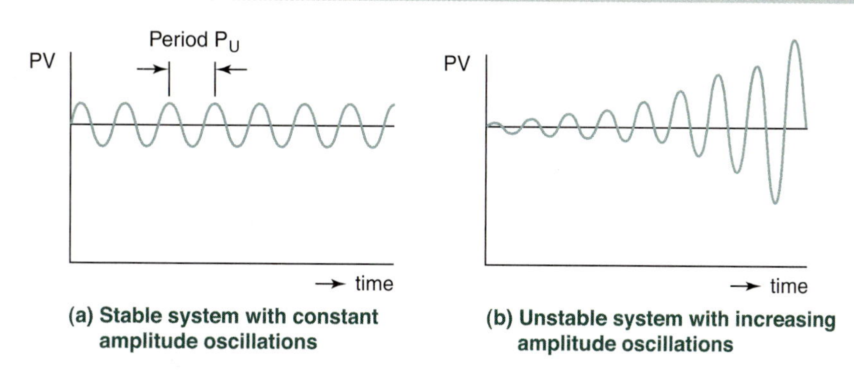

(a) Stable system with constant amplitude oscillations

(b) Unstable system with increasing amplitude oscillations

Rehg and Sartori, Industrial Electronics, *1st Edition, © 2006, Reprinted by permission of Pearson Education, Inc., Upper Saddle River, NJ.*

not *stable. Stability* is linked to the *total* system response, which is the sum of the *natural* and *forced* system responses. The natural system response is what a system does after a disturbance or set point change; it must either die out to zero or oscillate for a stable system. In some systems, the natural response to a change is to initiate an oscillation with increasing larger peak values. The responses shown in Figure 12-45 illustrate two of these conditions. The first is called a stable system with a *constant amplitude oscillation response;* the second is called an unstable system with an *increasing amplitude oscillation response.* For many processes, the first condition causes no system damage and is used in tuning processes, but the second can be destructive for the process system. In each case the damping ratio is zero, so the system is not damped.

The total system response, as illustrated in Figure 12-46, includes the dead-time delay, transient portion, and steady-state portion. The total response is the sum of the natural system response and the forced response (the step change in Figure 12-46). In a stable system, the natural system response dies out after a disturbance or change in set point to leave only the forced response (the new step level). That is the case in Figure 12-46 but not in Figure 12-45. The natural response describes the way the system acquires and dissipates energy. The input to the system does not affect the form or nature of the natural response since it is dependent only on the system parameters. However, the system input directly affects the form or nature of the forced response. The total response illustrated in Figure 12-46 is what

FIGURE 12-46: System response curve.

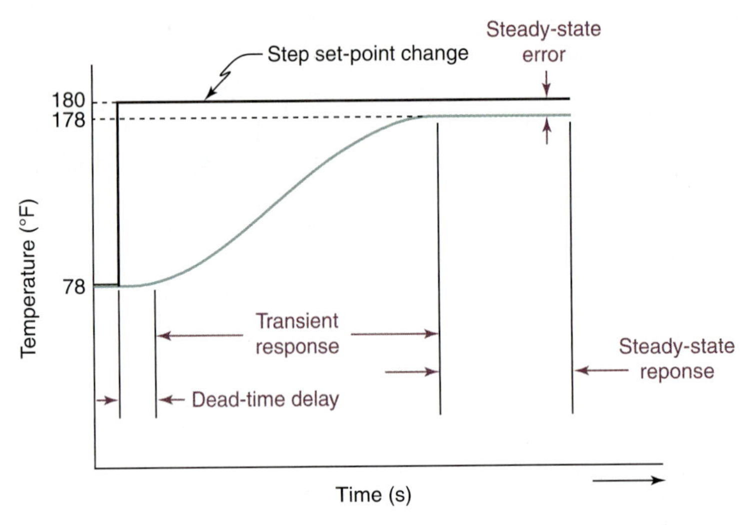

Rehg and Sartori, Industrial Electronics, *1st Edition, © 2006, Reprinted by permission of Pearson Education, Inc., Upper Saddle River, NJ.*

you actually see, and it is the combination of the natural and forced responses that has been divided into three components, dead-time delay, transient response, and steady-state response.

The natural response of a control system will approach zero or oscillate for a stable system. However, if the natural response oscillations grow without bound rather than diminishing to zero the system is *unstable*. In this case, the natural response is much greater than the forced response and control has been lost for the system. This condition is called *instability* and can cause the physical device to self-destruct if limit stops are not part of the design.

The goal is to design stable control systems. To achieve this goal the system's natural response must decay to zero or oscillate as time approaches infinity. The transient response is usually dominated by the system's natural response. Therefore, the system transient response will die out as time increases if the natural response decays to zero. What remains then is only the steady-state response, which is dominated by the forced response or input. The desired transient and steady-state response can be established during the design process for a stable system.

Two other important considerations must be considered in the design of a control system: cost constraints and robust design. Designers must consider the economic impact of control system design based on budget allocations and competitive pricing.

A second important consideration is robust design. Components that set system parameters and establish the transient response, steady-state errors, and stability can change over time after the actual system is built. As a result, the performance of the system changes and no longer fulfills the design goals. To solve this problem, designers try to design a robust system where the system is not sensitive to component and parameter changes.

Closed-Loop Process Control. The closed-loop control systems described in this chapter are used in every type of process and in every process industry. The development of a control system to drive these closed-loop requirements is the topic of Chapter 14. The following statements summarize the operation found in closed-loop systems.

- Disturbances to the controller or process are frequently present.
- Final control elements are continuous devices with outputs from 0 to 100 percent.
- Controller output values are the product of the controller gain (K) and the input error (SP − MV or MV − SP).

- The controller output changes the final control element so that the controlled variable moves closer to the desired set point.
- A continuous proportional control system has zero steady-state error only at the control variable and set point values established at system startup.
- If a disturbance or set point change occurs in a system with proportional control, then some steady-state error will result.
  - The steady state error and the proportional controller gain have an *inverse* relationship (increase gain and reduce steady-state error).
  - Transient response maximum peak and settling time have a *direct* relationship with controller gain (increase gain and increase overshoot and ringing).
  - Steady-state error cannot be completely eliminated with increases in controller gain.

12-6 TROUBLESHOOTING THE PROPORTIONAL GAIN CONTROLLER

Proportional gain controllers used in closed-loop processes involve many intricate machines and systems. As a result, it is difficult to provide a troubleshooting process for every conceivable process system. However, here are some general procedures and suggestions that can help in locating problems.

1. **Know the process:** before you can troubleshoot the proportional gain controller you must first understand the process inside and out—how it operates and how disturbances affect the process.
2. **Know the control system:** it is equally important to know the control system used in the process. It is important to separate controller issues from system component failure problems.
3. **Process versus control problems:** three situations may be present if the process product is defective:
   1. The process may be operating correctly, but the control system may be malfunctioning.
   2. The control system may be operating correctly, but the process may be malfunctioning.
   3. The control system and the process may be malfunctioning. This is generally ruled out because multiple failures are not as common as single failures.

Chapter 12 questions and problems begin on page 586.

CHAPTER

13

PLC Standard IEC 61131-3 Function Block Diagrams

13-1 GOALS AND OBJECTIVES

The goals for this chapter are to provide an overview of IEC 61131-3 graphic languages and to provide the Allen-Bradley implementation of the IEC 61131 Function Block Diagram language.

After completing this chapter you should be able to:

- Describe the difference between Ladder Logic, Function Block Diagram, and Sequential Function Chart graphic languages.
- Define the components of the Allen-Bradley Function Block Diagram (FBD) language.
- Describe the process used to build an FBD program.
- Create a function block diagram program for automation process problems using the Allen-Bradley implementation.

13-2 PLC STANDARDS

The first standard for PLCs was published in 1978 by the *National Electrical Manufacturers Association* (NEMA). However, the rapid growth in PLCs across national boundaries demanded a broader standard. In 1979 the *International Electrotechnical Commission* (IEC) established a working group to look at the complete standardization of PLCs. The PLC standard, called IEC 1131 (changed later to IEC 61131), has the following parts.

- **Part 1: General information** establishes the definitions and identifies the principal characteristics relevant to the selection and application of programmable controllers (PLCs) and their associated peripherals.
- **Part 2: Equipment requirements and tests** specify equipment requirements and related tests for programmable controllers and their associated peripherals.
- **Part 3: Programming languages** define, as a minimum set, the basic programming elements; syntactic and semantic rules for the most commonly used programming languages, as well as major fields of application; and applicable tests and means by which manufacturers may expand or adapt those basic sets to their own programmable controller implementations. Five PLC languages are defined.
- **Part 4: User guidelines** is a technical report providing general overview information and application guidelines of the standard for the end user of programmable controllers.
- **Part 5: Messaging service specification** defines the data communication between programmable controllers and other electronic

systems using the Manufacturing Message Specification (MMS), according to International Standard ISO/IEC 9506.
- **Part 6: Reserved for future use.**
- **Part 7: Fuzzy control programming** defines basic programming elements for fuzzy logic control as used in programmable controllers.
- **Part 8: Guidelines for the application and implementation of programming languages** provides a software developers guide for the programming languages defined in Part 3.

Part 3 of the standard, *Programming Languages for Programmable Controllers*, was released in 1993 and specifies the standards for PLC software. The IEC 61131-3 standard provides a very specific and detailed definition of PLC configuration, programming, and data storage. The programming standard specifies five languages that PLC vendors should support if their controller is specified as IEC 61131 compatible. The languages are:

- Ladder Diagrams (LD)
- Sequential Function Chart (SFC)
- Function Block Diagram (FBD)
- Structured Text (ST)
- Instruction List (IL)

Ladder logic has been described in detail in the first 11 chapters of this text; the concepts learned there will make learning the other languages easier. This chapter covers the programming format for Function Block Diagrams, one of the three graphic languages. However, it is important to define all the languages in the standard.

13-2-1 IEC 61131-3 Standard Languages

A general description of each language follows.

- **IL (Instruction List)—Textural:** A low-level, text-based language that uses mnemonic instructions like those used in machine code in microprocessors. Consequently, IL is a powerful language but it has many disadvantages, such as a cryptic nature. In addition, Allen-Bradley does not support this language. It is not a good fit for most applications, and as a result it is used infrequently in the United States.
- **ST (Structured Text)—Textural:** A high-level, text-based language, such as BASIC, C, or PASCAL, with commands that support a highly structured program development and the ability to evaluate complex mathematical expressions. Structured text is covered in Chapter 15.
- **LD (Ladder Diagram)—Graphical:** The 61131-3 graphical language is based on traditional relay ladder logic. LD is used for logic operations involving only Boolean variables (e.g., True and False). The language elements include only input instructions (i.e., XIO and XIC) and outputs (i.e., OTE, OTL, and OUT) that are placed on a ladder rung to produce the required logic. If timers, counters, or other special functions are required, then function blocks with those functions are included. Most vendors do not support the LD part of the standard since they already have a fully functional ladder logic format for their PLC systems.
- **FBD (Function Block Diagram)—Graphical:** Another graphical language where the basic programming elements appear as blocks. They are linked together to form a final control circuit. The blocks manipulate or operate on the data that flows from input to output connections. Data can be Boolean, integer, real, or text with other types also supported. A large library of standard function blocks is available. Function block diagrams are covered in this chapter.
- **SFC (Sequential Function Chart)—Graphical:** Based on the French standard, GRAFCET, this is a graphical language used most often for sequential control problems. The basic language elements are steps or states that perform associated actions on the process and transitions used to move from the current step or state to next. The actions and transitions are defined in terms of the four other languages. As a result, the SFC language is often used for the basic sequential program flow with FBD, LD, and ST programs embedded into the SFC structure.

Two of the languages, Instruction List and Structured Text, are textural, so commands and their arguments are entered from a keyboard, one line at a time. Three, Ladder Diagram, Function Block Diagram, and Sequential Function Chart, are graphical. A graphical editor is used to build the program on the computer screen of the programming device.

Compliance. Many PLC vendors support the 61131-3 standard; however, the degree of compliance with the standard varies greatly. Three levels of compliance are currently defined:

- **Base level compliance:** defines an essential core of the 61131 standard and the necessary features of each supported language. A product can be certified as base level compliant in one, several, or all of the languages. Base level compliance includes a small number of basic data types and a restricted set of standard functions.
- **Portability level compliance:** defines a larger set of required features and must incorporate an import/export tool to exchange 61131 software with other portability level compliant systems.
- **Full compliance:** full compliance indicates complete implementation of the 61131 standard and fully interchangeable programs.

Allen-Bradley implements the standard on their Logix family of PLCs, so all programming is performed with the RSLogix 5000 software. Achieving interchangeability of programs between different vendor PLCs is a long way off, but support for the standard has increased the programming options on PLCs. Since the standard is still relatively new and the task to achieve compliance is large, the majority of current vendors do not completely implement all the features specified in the standard.

13-3 FUNCTION BLOCK DIAGRAM (FBD)

The *Function Block Diagram (FBD)* language is used to program automation problems as a set of interconnected graphical function blocks. For example, to add two values you select an addition block. If you want a timer, then you select a timer block. Block inputs and outputs are connected by lines. This is very similar to the signal flows depicted in electronic circuit diagrams. A typical signal flow for an FBD is illustrated in Figure 13-1. Note that the function blocks are

FIGURE 13-1: Example FBD circuit.

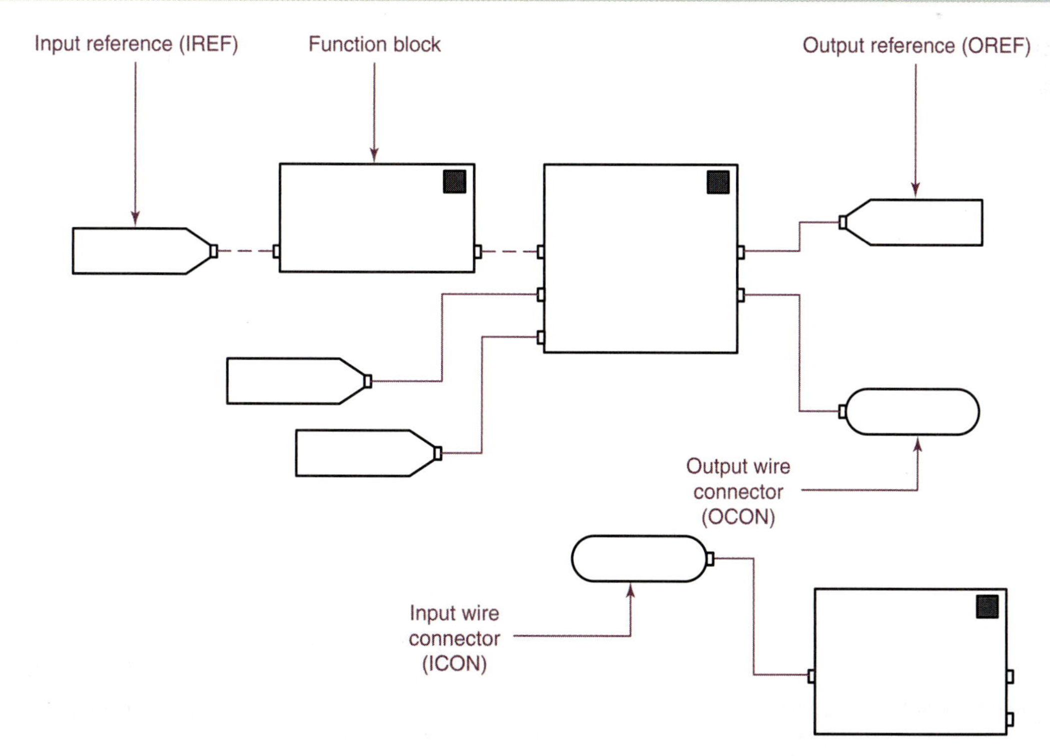

Rehg and Sartori, Industrial Electronics, *1st Edition, © 2006, Reprinted by permission of Pearson Education, Inc., Upper Saddle River, NJ.*

linked together to complete a circuit that satisfies a control requirement. The inputs, placed into an *input reference symbol*, are on the left and the outputs, placed into an *output reference symbol*, are on the right. The input and output tag names are placed inside the *IREF* and *OREF* symbols. A jump between segments of the FBD is represented by the *OCON* (output wire connector) and *ICON* (input wire connector) symbols.

FIGURE 13-2: Example function block and standard displays.

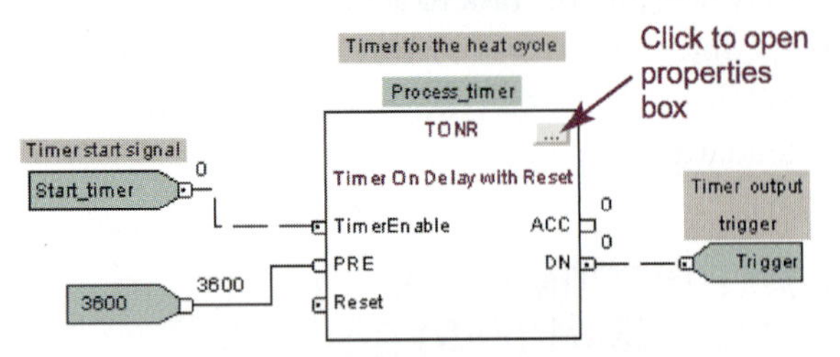

(a) TONR type timer default display

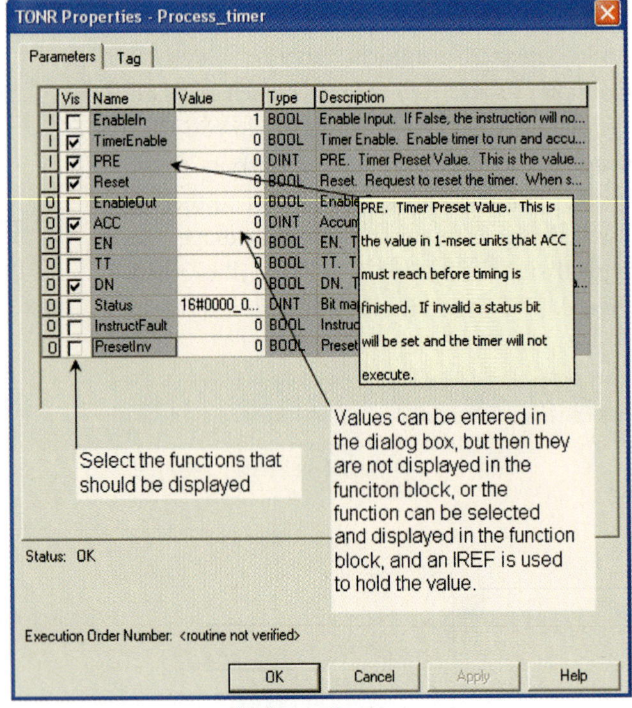

(b) Properties box

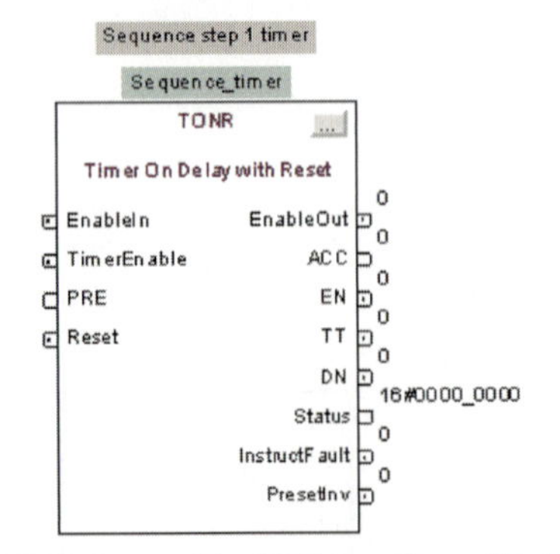

(c) TONR timer with all I/O functions displayed

The line type of the link between function blocks indicates what type of signal is present. A dashed line, as shown in Figure 13-2(a), indicates a Boolean signal path (e.g., 0 or 1) and a solid line indicates an integer or real value (e.g., 527 or 21.7) signal path. An FBD circuit is analogous to an electrical circuit diagram or system block diagram where links and wires depict electrical signal paths between components.

A function block is depicted as a rectangular block, as shown in Figure 13-2(a), with inputs entering from the left and outputs exiting on the right. The function block type is always shown within the block, a tag name for the block is placed above it, and the names of the FBD inputs (left side of the block) and outputs (right side of the block) are shown within the block. The default view of all boxes has some but not all of the input and output parameters visible when the box is placed into the program. The FBD properties box, used to set the visibility option of input and output parameters, is displayed by left clicking the selection button at the upper right of the FBD. The selection button, indicated in Figure 13-2(a), is clicked to produce the timer properties box displayed in Figure 13-2(b).

The properties box has two tabs: *Parameters* and *Tag*, in the upper left corner. The Parameter tab, which is displayed in the figure, shows a list of I/O functions that are available, the parameter value, the data type, and a description. The full description for a truncated listing is displayed by placing the cursor on the description in the box. Parameters with a check in the check box in front of the parameter are listed inside the FBD block when it is placed in the program. Parameter values can be entered directly into the dialog box. For example, the timer preset value could be entered into the dialog box in the Value cell after the PRE cell [see Figure 13-2(b)]. A second option would be to select the preset parameter so it appears in the FBD box and then use an IREF to hold the value [see Figure 13-2(a)]. All parameter values are displayed in the properties box and a select few are displayed in the FBD block in the program.

The Tag tab of the properties box permits the tag properties to be adjusted and descriptions to be entered. A timer function block with all input and output parameters displayed is illustrated in Figure 13-2(c). Note how much larger the block is when all input and outputs are displayed. The display for all function blocks is adjusted using this process. Study the block in Figure 13-2(a) to see a typical FBD with inputs and outputs. This is how a typical block appears in the FBD programming environment for the Allen-Bradley ControlLogix 5000 processor.

The table in Figure 13-3 describes how to choose FBD elements to build an FBD control circuit like that in Figure 13-1. Note that IREF and OREF symbols are used to define tags for the input and output values. The OCON and ICON symbols allow for breaks in the links between blocks to make the diagrams less cluttered and easier to read.

Each function block uses a *tag or variable* to store configuration information, status information, and parameter values for the function block instruction. When a function block instruction is added in the RSLogix 5000 software FBD language, a tag is automatically created for the block. This default tag name can be edited by double-clicking on it and changing it to a name that reflects the function of the block in the control program. In addition, the programmer must create a tag name or assign an existing tag for all IREFs and OREFs used with FBD inputs and outputs. The table in Figure 13-4 outlines the format for an FBD tag name, which is created like the tags in the ControlLogix ladder logic programs.

13-3-1 Signal Flow Types, Execution Order, and Data Latching

There are two types of signal data flows in an FBD: Boolean and real values. A dashed line indicates that the data is a discrete (1 or 0) or Boolean type. A solid line indicates that the data is an integer or real number.

Execution order is dictated by the signal flow and impacted by the *data latching* process covered next. *Signal flow* inside the FBD blocks is always from inputs on the left to the outputs on the right side. Outside the FBD blocks, signal flow moves from function block to function block in a left-to-right fashion, so it flows from a block output on the left to a second block input on the right.

If a feedback path is present, then the signal flows from the output of a block on the right back toward the left to the input of a block on

FIGURE 13-3: Choosing FBD elements.

If you want to:	Then use a(n):
Supply a value from an input device or tag	Input reference (IREF)
Send a value to an output device or tag	Output reference (OREF)
Perform an operation on an input value or values and produce an output value or values	Function block
Transfer data between function blocks when they are: • Far apart on the same sheet • On different sheets within the same routine	Output wire connector (OCON) and an input wire connector (ICON)
Disperse data to several points in the routine	Single output wire connector (OCON) and multiple input wire connectors (ICON)

Rehg and Sartori, Industrial Electronics, *1st Edition, © 2006, Reprinted by permission of Pearson Education, Inc., Upper Saddle River, NJ.*

FIGURE 13-4: Tag name generation for RSLogix 5000.

For a(n):	Specify:
Tag	*tag_name*
Bit number of a larger data type	*tag_name.bit_number*
Member of a structure	*tag_name.member_name*
Element of a one-dimension array	*tag_name[x]*
Element of a two-dimension array	*tag_name[x,y]*
Element of a three-dimension array	*tag_name[x,y,z]*
Element of an array within a structure	*tag_name.member_name[x]*
Member of an element of an array	*tag_name[x,y,z].member_name*

where:

x is the location of the element in the first dimension.

y is the location of the element in the second dimension.

z is the location of the element in the third dimension.

Rehg and Sartori, Industrial Electronics, *1st Edition, © 2006, Reprinted by permission of Pearson Education, Inc., Upper Saddle River, NJ.*

the left. Execution of each function block follows the signal flow path described in the graphic program and is generally from left to right. However, execution is affected by the data latching strategy utilized.

Data latching refers to how the PLC verifies that the data present at the input to an FBD is valid so that the algorithm within the function block can process the data and produce an output. When an IREF is used to specify input data for a function block instruction (see Figure 13-1), the data in that IREF is latched (held at the current value) for the scan of the function block routine. The controller updates all IREF data at the beginning of each scan. When the IREF tag is used at multiple inputs, each input latches the same data. The FBD is evaluated from left to right and top to bottom. Outputs are fixed at the value from the previous scan and not permitted to change until all inputs are latched.

13-3-2 Feedback Loops

When a wire from an output pin of the block is connected back to an input pin of the same block, a feedback loop around a block is created. Figure 13-5(a) illustrates this concept. The loop contains only a single block, so execution order does not matter because the input pin uses an output that the block produced on the previous scan.

If a group of blocks are in a loop, as in Figure 13-5(b), then the controller cannot determine which block input to latch and execute first because it cannot resolve the loop. The question marks above the function blocks as shown in Figure 13-5(b) indicate that the order of execution is not clear. The problem is resolved by placing an Assume Data Available indicator mark, shown in Figure 13-5(c), at the input of the function block that should be executed first. In Figure 13-5(c), the input for block 1 uses data from block 3 that was produced in the previous scan. The problem is that the output from block 3 is dependent on the input to block 1. The Assume Data Available indicator defines the data flow within the loop and indicates what data should be latched and which block is executed first in the loop. To place the indicator, first select the interconnect wire by left-clicking on it, then right-click on it. The dialog box lets you place the indicator at the input where the selected wire is attached.

FIGURE 13-5: Feedback in function blocks.

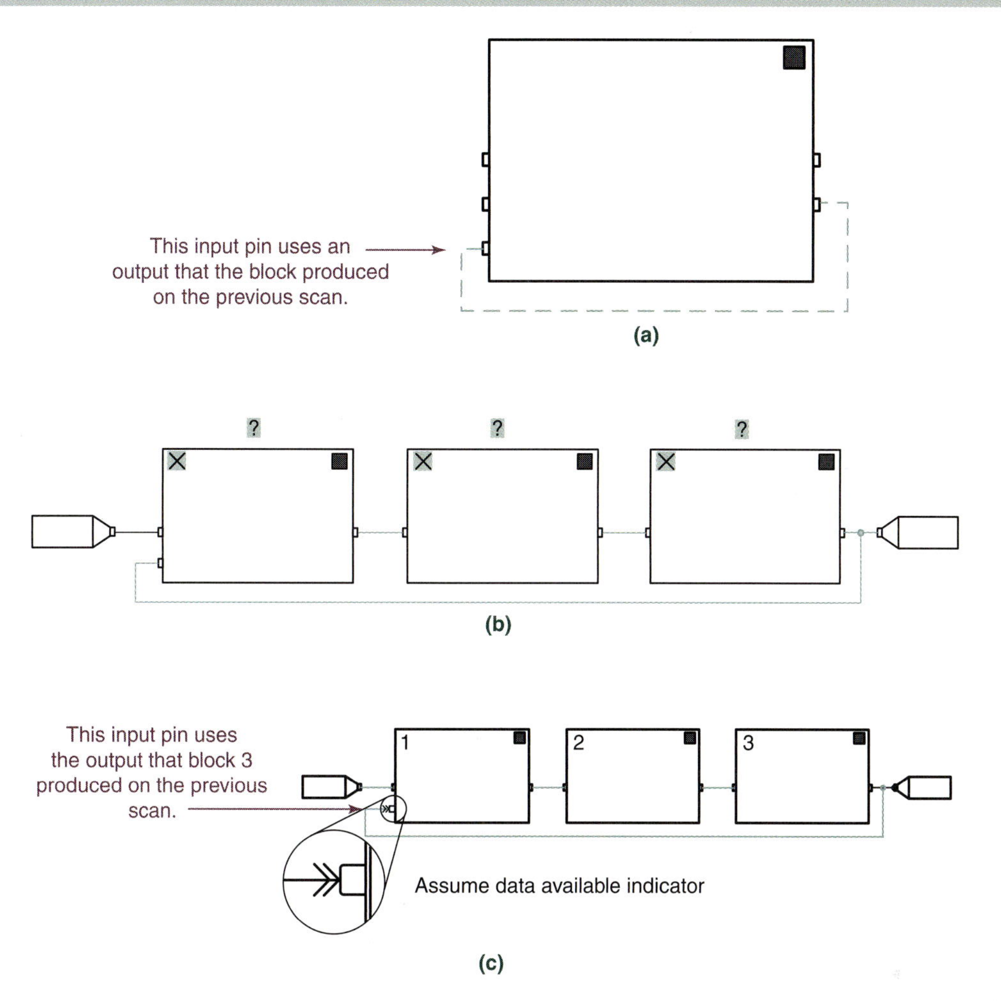

Rehg and Sartori, Industrial Electronics, *1st Edition, © 2006, Reprinted by permission of Pearson Education, Inc., Upper Saddle River, NJ.*

13-3-3 Function Block Diagram Program Development Sequence

The sequence used to develop an FBD routine for a control problem is as follows:

1. Develop the control strategy required for the problem. In ladder logic this meant to develop the Boolean logic for each output, considering the need for special instructions such as timers, counters, comparators, math functions, and one-shots; selecting the standard ladder logic for the application; and making a first cut at the automation solution. The FBD solution process is same, except that the solution must be visualized in function blocks instead of ladder rungs. The FBD standard block solutions, provided in Section 13-4, are useful because they illustrate how the ladder logic translates to FBD.
2. Choose the function block elements. A list of the blocks that are needed is generated. Be liberal here as you drag the blocks onto the solution window since it may be better to have a wider variety of blocks initially considered for the solution. Unused blocks can be deleted from the solution if they are not used.
3. Choose a tag name for each function block and a tag name or constant for each input (IREF) and output (OREF).

FIGURE 13-6: Setup for FBD program development.

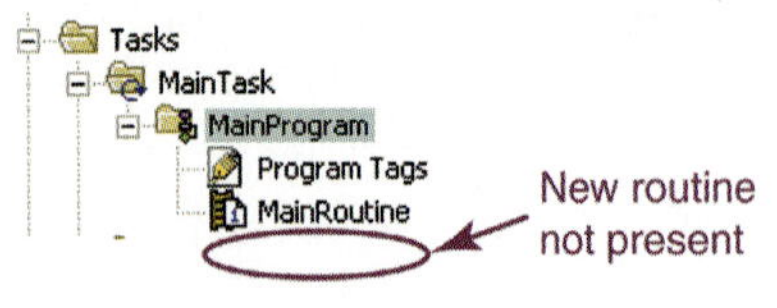

(a) Open status of file menu

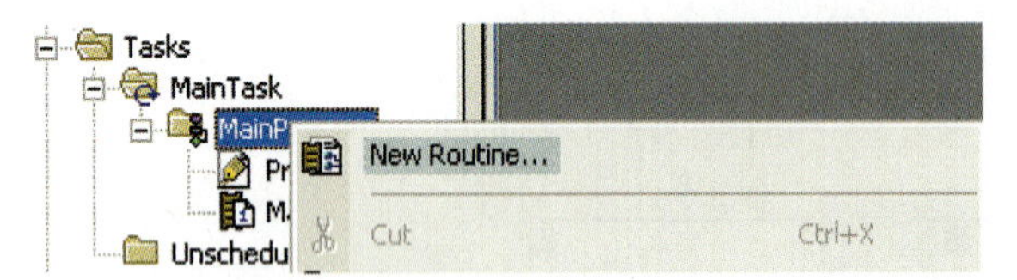

(b) Right-click MainProgram for selection of New Routine...

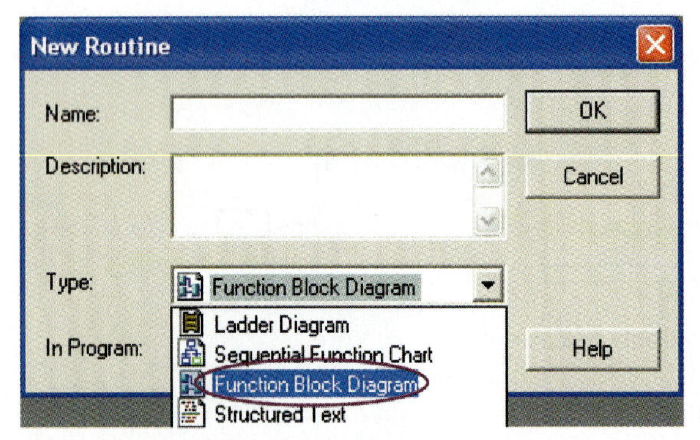

(c) New Routine dialog box to select the routine for the program

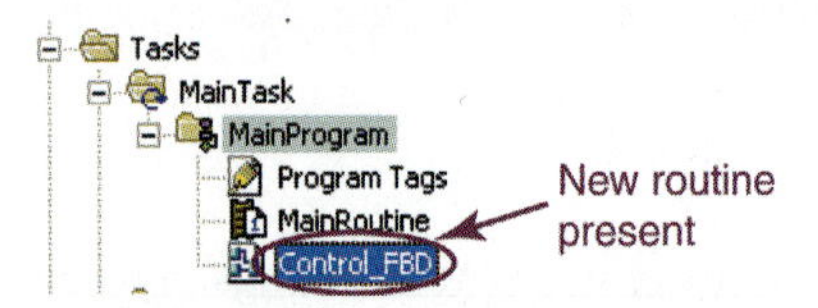

(d) New routine named Control_FBD is now in the file list under MainProgram

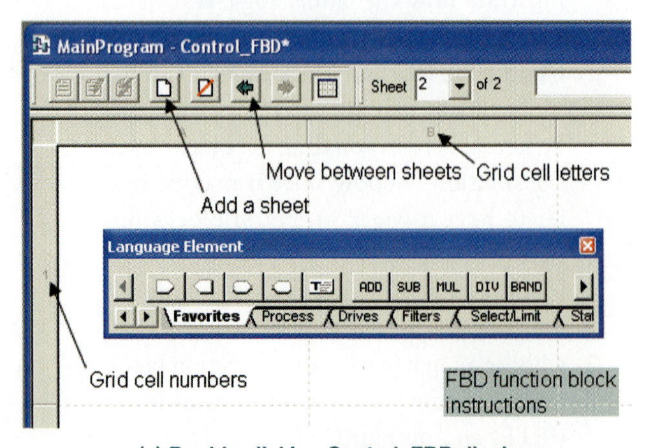

(e) Double clicking Control_FBD displays the FBD graphic development window

4. Define the order of execution. Establish the order of the signal flow and group the blocks accordingly. Determine what calculation is performed first and how that result is used in the next set of blocks.
5. Identify any connectors and feedback loops.
6. Define program/operator control. Determine what field devices are required or available and how they will be used in the automation solution.
7. Add a function block element. Start with the blocks at the far left of the signal flow. These are the first blocks to interface with the input field devices. Configure the blocks by making the input and output parameters needed for the solution visible inside the block.
8. Build the signal flow from left to right.
9. Connect blocks with an OCON and ICON that are on different sheets.
10. Verify the routine.

An example problem later in the chapter will demonstrate how these steps are implemented. Another requirement is to open a programming session in RSLogix 5000 for use in programming an FBD solution.

The sequence used to open an FBD session is as follows:

1. Locate the MainProgam file list on the left of the screen [see Figure 13-6(a)].
2. Right-click on the MainProgram file to open the pop-up menu and then select the New Routine menu item [see Figure 13-6(b)].
3. The New Routine dialog box is opened and the drop-down selection list is expanded in the Type text window [see Figure 13-6(c)]. All four of the 61131-3 standard languages supported by Allen-Bradley are displayed. Select the Function Block Diagram entry and tag named for the program plus a description.
4. The new FBD program named Control_FBD is now listed under MainProgram [see Figure 13-6(d)].
5. Left-clicking Control_FBD twice opens the graphic development window for

FBD programs [see Figure 13-6(e)]. Note in the figure that the Language element window has been pulled onto the graphic development grid. FBD instructions are selected from this window, just as the ladder logic was earlier. The grid has cells identified by letters and numbers to help locate ICON and OCON jump links. Extra sheets can be added when the current sheet is full using the extra sheet icon. Movement between sheets is provided by the left and right arrows indicated.

The MainRoutine is always a ladder logic program in RSLogic 5000, and all other routines are called from the MainRoutine. So the MainRoutine will have one unconditional rung with a jump to subroutine (JSR) calling Control_FBD. The FBD program will execute from the JSR instruction. No subroutine or return from subroutine instruction in the FBD is necessary.

13-3-4 Allen-Bradley RSLogix 5000 FBD Programming

The Logix 5000 language FBD has a complete range of function blocks. The blocks are organized into 14 groups and the numerous instructions for each group are listed in Figure 13-7. Read through the instruction list to see the range of instructions available and to see which instructions are the same as those used in the ladder logic programming. The

FIGURE 13-7: Function blocks supported by RSLogix 5000.

Group Name	Function Block Name and Symbol
Advanced Math Instructions	Natural Log (LN) Log Base 10 (LOG) X to the Power of Y (XPY)
Bit Instructions	One Shot Rising with Input (OSRI) One Shot Falling with Input (OSFI)
Compare Instructions	Limit (LIM) Mask Equal To (MEQ) Equal To (EQU) Not Equal To (NEQ) Less Than (LES) Greater Than (GRT) Less Than or Equal To (LEQ) Greater Than or Equal To (GEQ)
Computer/Math Instructions	Add (ADD) Subtract (SUB) Multiply (MUL) Divide (DIV) Modulo (MOD) Square Root (SQR/SQRT) Negate (NEG) Absolute Value (ABS)
Drive Instructions	Pulse Multiplier (PMUL) S-Curve (SCRV) Proportional + Integral (PI) Integrator (INTG) Second-Order Controller (SOC) Up/Down Accumulator (UPDN)
Filter Instructions	High Pass Filter (HPF) Low Pass Filter (LPF) Notch Filter (NTCH) Second-Order Lead Lag (LDL2) Derivative (DERV)

(continued)

FIGURE 13-7: (Continued).

Group Name	Function Block Name and Symbol
Math Conversion Instructions	Degrees (DEG) Radian (RAD) Convert to BCD (TOD) Convert to Integer (FRD) Truncate (TRN)
Move/Logical Instructions	Masked Move with Target (MVMT) Bitwise And (AND) Bitwise Or (OR) Bitwise Exclusive Or (XOR) Bitwise Not (NOT) Bit Field Distribute with Target (BTDT) Boolean AND (BAND) Boolean OR (BOR) Boolean Exclusive OR (BXOR) Boolean NOT (BNOT) D Flip-Flop (DFF) JK Flip-Flop (JKFF) Set Dominant (SETD) Reset Dominant (RESD)
Process Control Instructions	Alarm (ALM) Scale (SCL) Enhanced PID (PIDE) Ramp/Soak (RMPS) Position Proportional (POSP) Split Range Time Proportional (SRTP) Lead-Lag (LDLG) Function Generator (FGEN) Totalizer (TOT) Deadtime (DEDT) Discrete Two-State Device (D2SD) Discrete Three-State Device (D3SD)
Program Control Instruction	Jump to Subroutine (JSR) Subroutine (SBR) Return (RET)
Select/Limit Instructions	Select (SEL) Enhanced Select (ESEL) Selected Summer (SSUM) Selected Negate (SNEG) Multiplexer (MUX) High/Low Limit (HLL) Rate Limiter (RLIM)
Statistical Instructions	Moving Average (MAVE) Moving Standard Deviation (MSTD) Minimum Capture (MINC) Maximum Capture (MAXC)
Timer/Counter Instructions	Timer On Delay with Reset (TONR) Timer Off Delay with Reset (TOFR) Retentive Timer On with Reset (RTOR) Count Up/Down (CTUD)
Trigonometric Instructions	Sine (SIN) Cosine (COS) Tangent (TAN) Arc Sine (ASN) Arc Cosine (ACS) Arc Tangent (ATN)

instruction format for the RSLogix 5000 controller is documented in reference manuals available in Adobe Acrobat pdf file format from the CD accompanying the text or from the Allen-Bradley Web site. Access the 61131 programming manuals and review the operation for some instructions from the following groups: timers, counters, mathematic instructions, comparators, logical operations, flip flops, and one-shots.

13-4 EMPIRICAL DESIGN WITH FUNCTION BLOCK DIAGRAMS

The Function Block Diagram program development sequence is listed in the previous section. After you are familiar with the operation of the commonly used function blocks, following these design steps will yield good design results. However, to learn the program structure for an FBD solution it best to study FBD equivalents for the ladder logic rungs used in similar control problems.

13-4-1 Standard Function Block Control Solutions

The following standard function block programs indicate how function block solutions compare to common ladder logic rungs and standard ladder logic diagrams.

FBD Standard for Unconditional and Conditional Logic. The comparison between unconditional and conditional ladder logic and the FBD equivalent is illustrated in Figure 13-8(a)

FIGURE 13-8: FBDs for standard ladder logic with conditional XIC and XIO logic rungs and unconditional rungs.

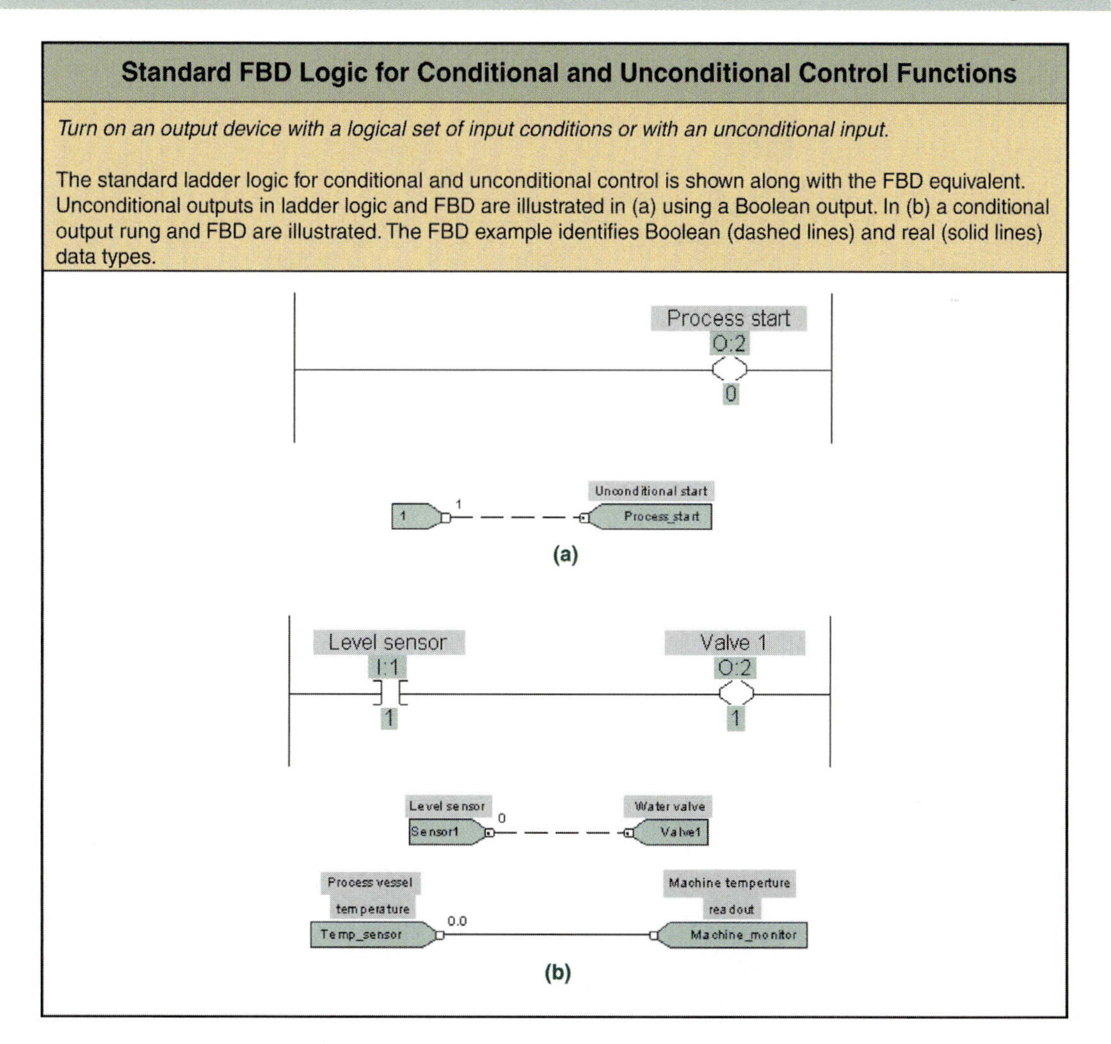

FIGURE 13-8: (Continued).

Standard FBD Logic for Combinational Logic

Turn on an output device for a logical combination of multiple input instructions.

The development of combinational logic in ladder rungs is simply AND/OR or OR/AND instructional groupings, Figure 13-8(c). When implementing the same logic in FBD, Figure 13-8(d), Boolean logic gates like BAND, BOR, BXOR, AND NOT are used to develop the combinational logic. The Boolean logic gates are used since the inputs and output are either on or off. Note that an XIO instruction in the ladder logic requires that a NOT function block is used in the FBD to produce the same logical output. The bitwise AND, OR, and XOR blocks (not shown here) are used with words and each bit in the word is compared with respective bits of other inputs using the logic of the function.

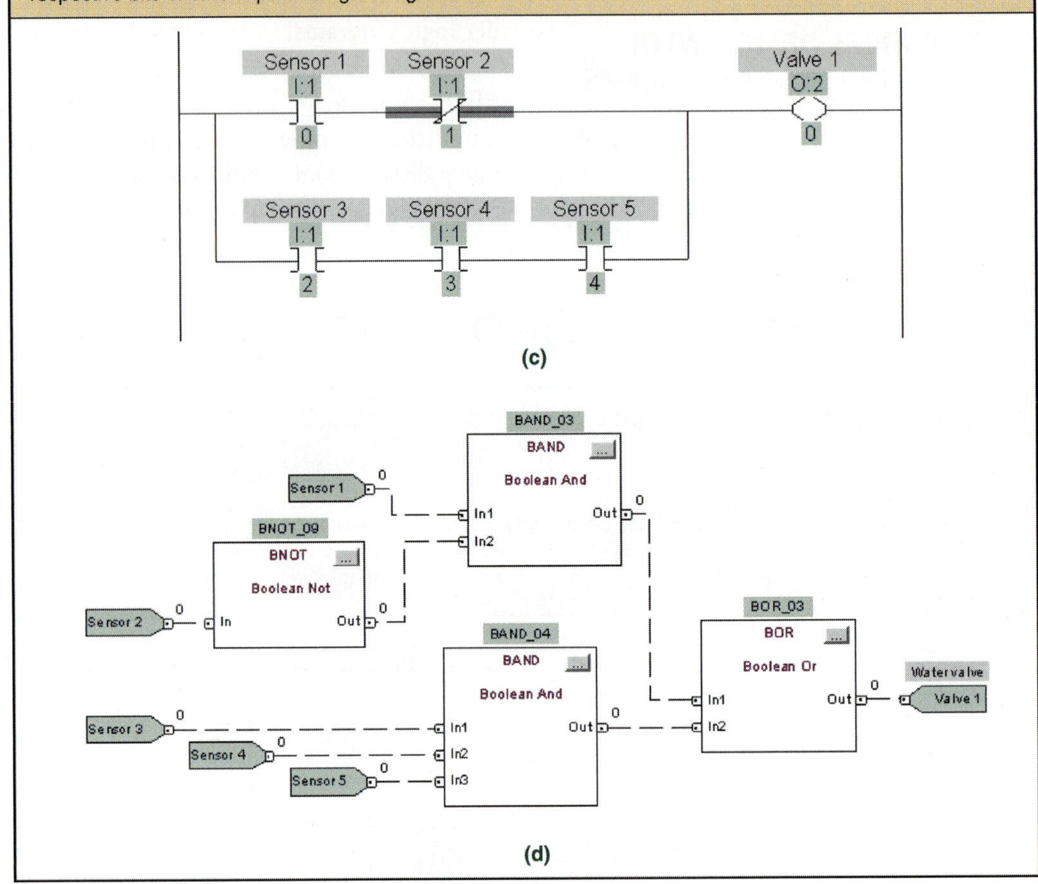

and (b). Read the description of the operation included in the figure. It is important to match the data types for the conditional logic. If the IREF tag is a Boolean, integer, or real data type, then the OREF tag must have the same data type.

FBD Standard for Combinational Conditional Logic. The comparison between combinational logic for ladder logic inputs and their FBD equivalent is illustrated in Figure 13-8(c) and (d). Read the description of the operation included in the figure. Combinational logic is achieved in FBD through the use of Boolean logic functions such as Boolean ANDs, ORs, and NOTs.

FBD Standard for Timers. The comparison between timer instructions for standard ladder

FIGURE 13-9: Standard ladder logic for TONR timers in FBDs.

Standard FBD Logic for TONR Timers

Turn on an output device for a set time period when a maintain contact input device changes states.

The standard ladder logic for the TON timer in Figure 4-12(b) is shown below along with the FBD equivalent. The input for both is a NO maintain contact selector switch. If the switch is closed, then the FBD timer is enabled and the EN and TT bits (not shown) are true because tag Maintain_start is true. When the ACC equals the preset value of 30 seconds (30000 x 0.001) the done bit is set and output tag Timer_done1 is true. An output can also be taken from the TT bit (not shown) on the FBD timer block to get a timed pulse output.

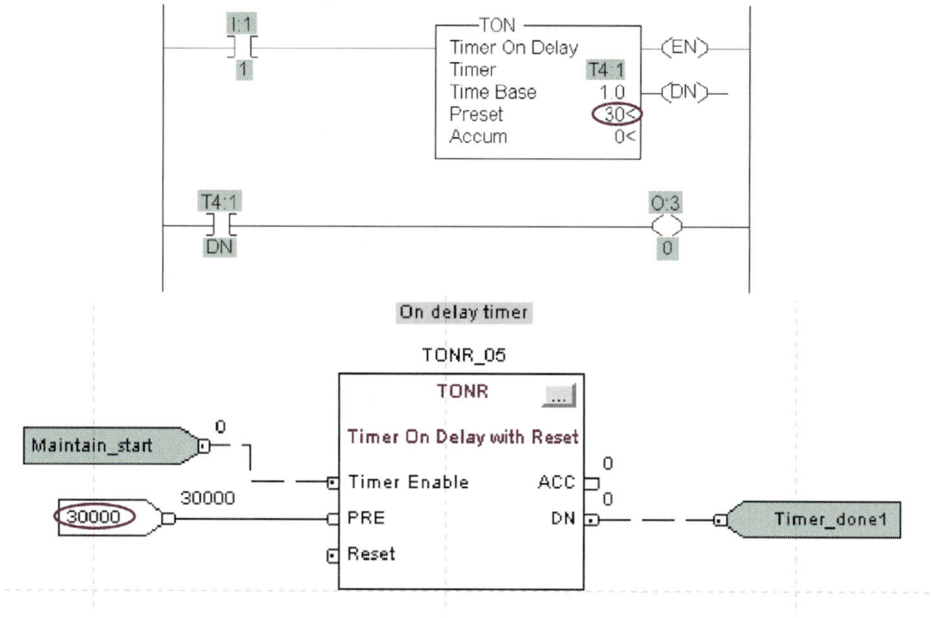

(a)

logic inputs and their FBD equivalent is illustrated in Figures 13-9 and 13-10, respectively. The ladder logic uses the TON symbol and the FBD uses the TONR symbol for a *time on* timer. Read the description of the operation included in the figures. The timer instructions in FBD operate the same as the timers in the ladder logic. However, in FBD you can enter operational parameters, such as timer preset, into the properties box for the timer or as a fixed IREF value connected to the timer preset input.

FBD Standard for Counters. The comparison between counter instructions for standard ladder logic inputs and their FBD equivalent is illustrated in Figures 13-11 and 13-12, respectively. The ladder logic software has two instructions, CTU and CTD, to handle the count up and count down applications. In the function block software, there is just one instruction, which handles both functions. The function block, CTUD, has a CU enable for counting up and a CD enable for counting down. The operation of the FBD counter for counting up and counting down applications is illustrated in the standard circuits in Figure 13-11. The standard FBD in Figure 13-12 illustrates how an up/down counter is implemented with the function block counter

FIGURE 13-9: (Continued).

Standard FBD Logic for TONR Timers

Turn on an output device for a set time period when a momentary contact input device changes states.

The standard ladder logic for the TON timer in Figure 4-12(c) is shown along with the FBD equivalent. The input field device is a NO push button switch. If the switch is closed, then the tag Momentary_start is true and the TT bit is true. The active TONR_11.TT bit seals the input contacts and makes the output tag Timer_done4 true as well. The TT bit and output tag are true for the preset time value of 30 seconds. Note that the preset value is not visible since it was set in the properties box for the TONR timer. It can be viewed by double clicking the properties expansion button in the top right corner of the FBD instruction. An output can also be taken from the DN bit on the FBD timer block to get an on delayed timed output. Note that an assumed data available symbol is placed at the In2 terminal of the feedback loop.

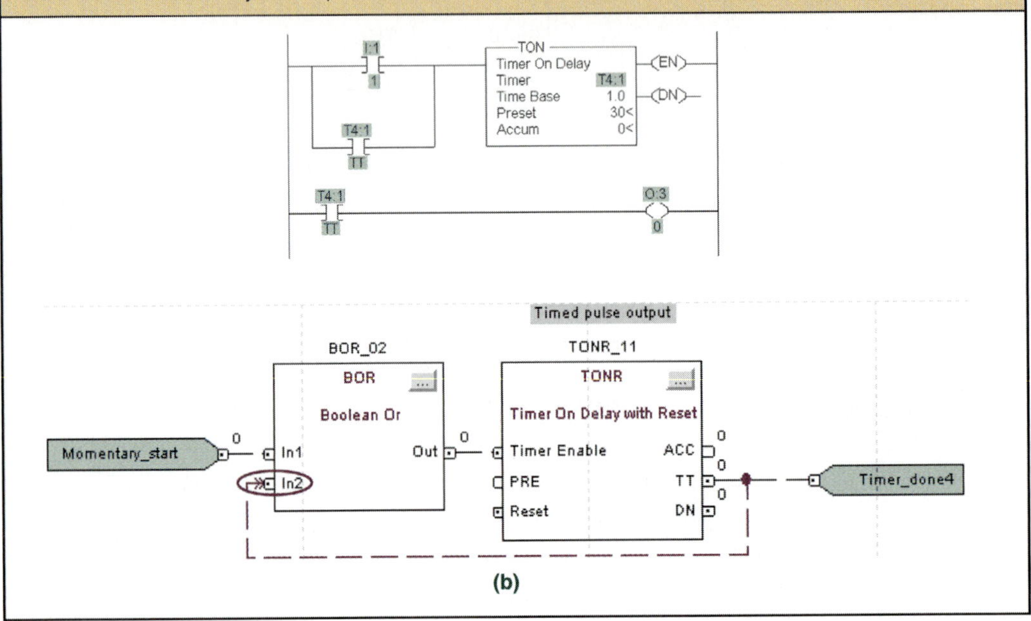

(b)

instruction. Read the description of the operation included in the figures.

FBD Standard for Latch Outputs. The comparison between output latch instructions for standard ladder logic inputs and their FBD equivalent is illustrated in Figure 13-13. The ladder logic software has two instructions, OTL and OTU, to handle applications that require a retentive output. Function block software uses binary element (flip-flop) function blocks to create retentive outputs. There are four types of binary elements: D type flip-flop, JK type flip-flop, set dominant (SETD) flip-flop, and reset dominant (RESD) flip-flop. The set dominant and reset dominant types are used to create the retentive outputs. The operation of the FBD retentive outputs in control applications is illustrated in the standard circuits in Figure 13-13. Read the description of the operation included in the figure.

FBD Standard for One-Shots and Math Blocks. The comparison between one-shot and math instructions for standard ladder logic inputs and their FBD equivalent blocks is illustrated in

FIGURE 13-10: Standard FBD logic for TONR pulse generators.

Standard FBD Logic for TONR Pulse Generators

Turn on an output device with a pulse sequence where one half of the duty cycle has a variable time and the other has a time equal to the scan time of the ladder.

The standard ladder logic for the TON timer in Figure 4-12(e) is shown along with the FBD equivalent. The input field device is a NO maintain contact selector switch. If the switch is closed, then the tag Maintain_start is true along with the EN bit. As a result, the timer is incrementing the ACC and the TT bit is true. When the ACC equals the preset (5 seconds), the timer DN bit is true. This makes the In2 input to the BAND instruction false so the BAND output is false and the timer resets, which makes the DN bit true for only one scan, and the timer starts the timing process again.

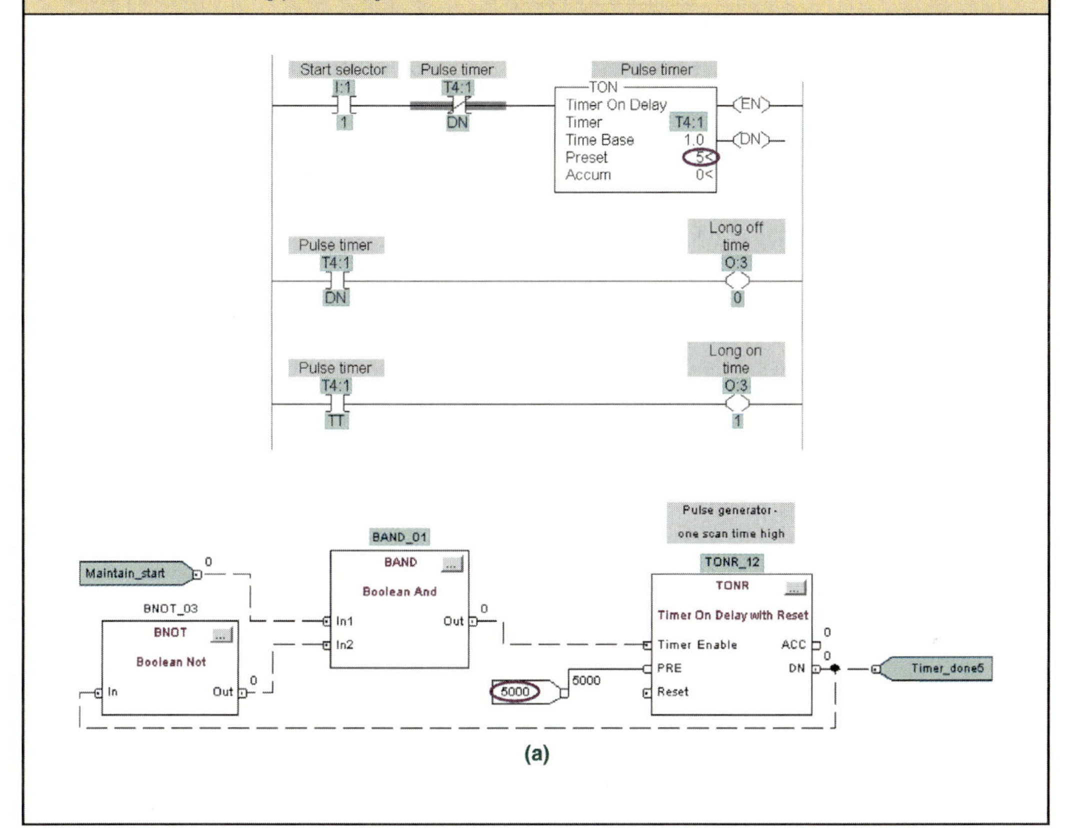

(a)

Figure 13-14. The one-shot is used in control programs to limit the execution of the output to one scan of the program. This is useful when an instruction or block executes every scan when the input is true, but the control requires just one execution. The one-shot output is true for one scan and the add block has its enable pin connected to the one-shot output. Therefore, the add block is enabled for only the first scan after the input goes high. This configuration is used most often with math and move blocks and could have a one-shot falling block used in place of the one-shot rising. Read the description of the operation included in the figure.

FIGURE 13-10: (Continued).

Standard FBD Logic for TONR Pulse Generators

Turn on an output device with a pulse sequence where both halves of the duty cycle have a variable time.

The standard ladder logic for the TON timer in Figure 4-12(f) is shown below along with the FBD equivalent. The input field device is a NO maintain contact selector switch. If the switch is closed, then the tag Maintain_start is true along with the EN bit of timer TONR_13. As a result, the timer is incrementing the ACC and the TT bit is true. When the ACC equals the preset (2 seconds), the timer DN bit is true. This enables the timer TONR_14 and it starts moving toward its 4-second preset value. When TONR_14's ACC equals 4 seconds its DN bit transitions to a 1, and this causes the first timer, TONR_13, to reset because the done bit is inverted. Note that the up time for the pulse is determined by the preset value of TONR_14 since it determines how long the done bit of TONR_13 is held on after its DN bit becomes true.

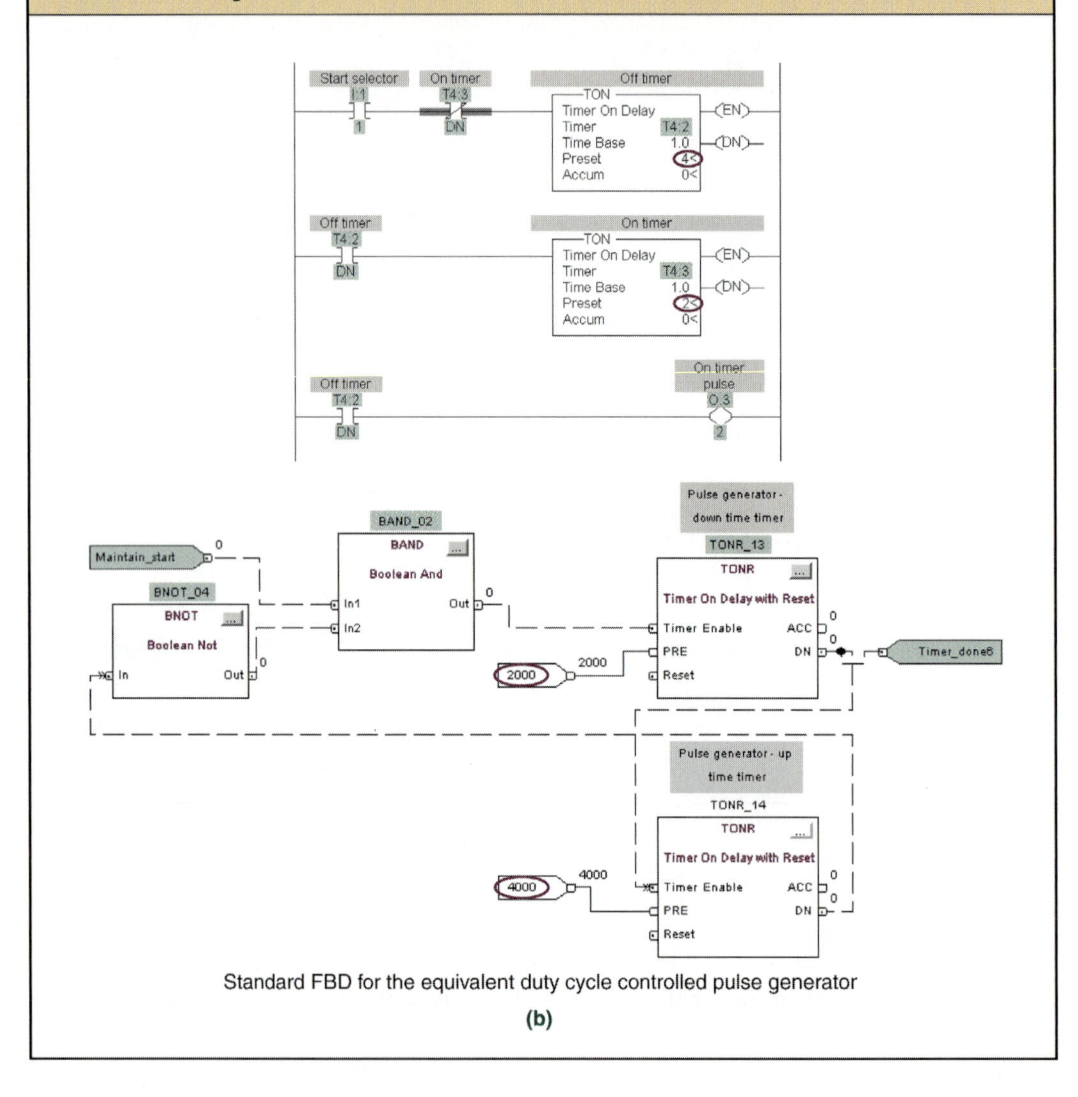

Standard FBD for the equivalent duty cycle controlled pulse generator

(b)

FIGURE 13-11: Standard ladder logic for counters in FBDs.

Standard FBD Logic for Counters

Turn on an output field device after a preset number of false to true transitions of the input logic or turn off an output field device after a preset number of counter inputs.

The standard ladder logic for the CTU counter in Figure 5-10(a) is shown along with the FBD equivalent. The function block software has a combination up and down counter, which can be used as an up counter, down counter, or as a combined up/down counter. If the CU enable input goes from false to true the accumulator increments up by one. If the CD enable input has that transition then the accumulator decrements down by one. So if both input are actively controlled, then the function moves the accumulator in response to both up and down active inputs. The preset value can be entered in the properties dialog box or by an IREF input as shown below. The reset input is used to reset the accumulator and the BNOT function block produces a NOT done output. If the CD enable is used then the FBD is a count down counter.

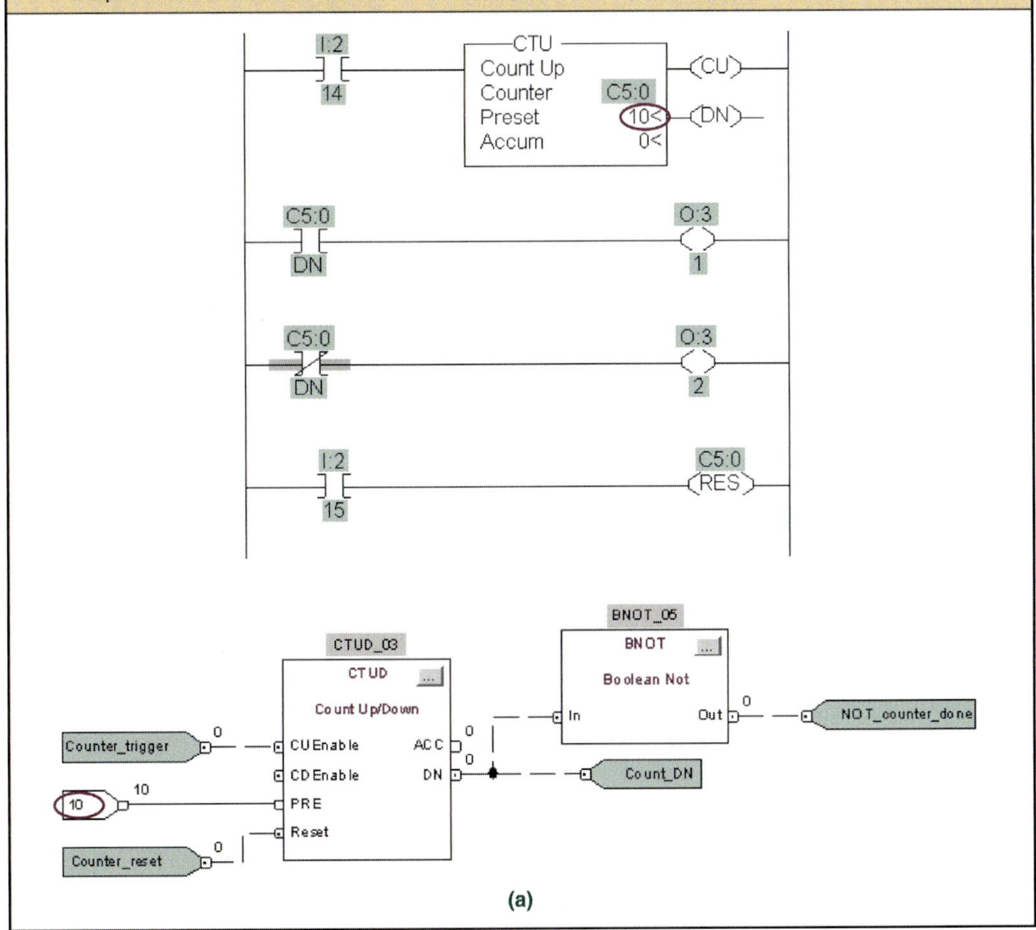

(a)

FIGURE 13-11: (Continued).

Standard FBD Logic for Counters

Turn on an output field device after a preset number of false to true transitions of the input logic or turn off an output field device after a preset number of counter inputs. Hold the ACC value at the PRE value when the ACC value equals the PRE value.

The standard ladder logic for the CTU counter in Figure 5-10(b) is shown below along with the FBD equivalent. The FBD implements this ladder standard by using a Boolean AND function block to combine the counter trigger with the counter NOT done bit. When the counter ACC is less than the preset value, the feedback bit is a 1 and counter trigger passes to the CU enable input, causing the counter ACC to increment with every input transition. When the done bit is true, the feedback is a 0 so all future input triggers are inhibited until the counter is reset. Note that an assumed data available symbol is added as shown. If the CD enable is used then the FBD is a count down counter.

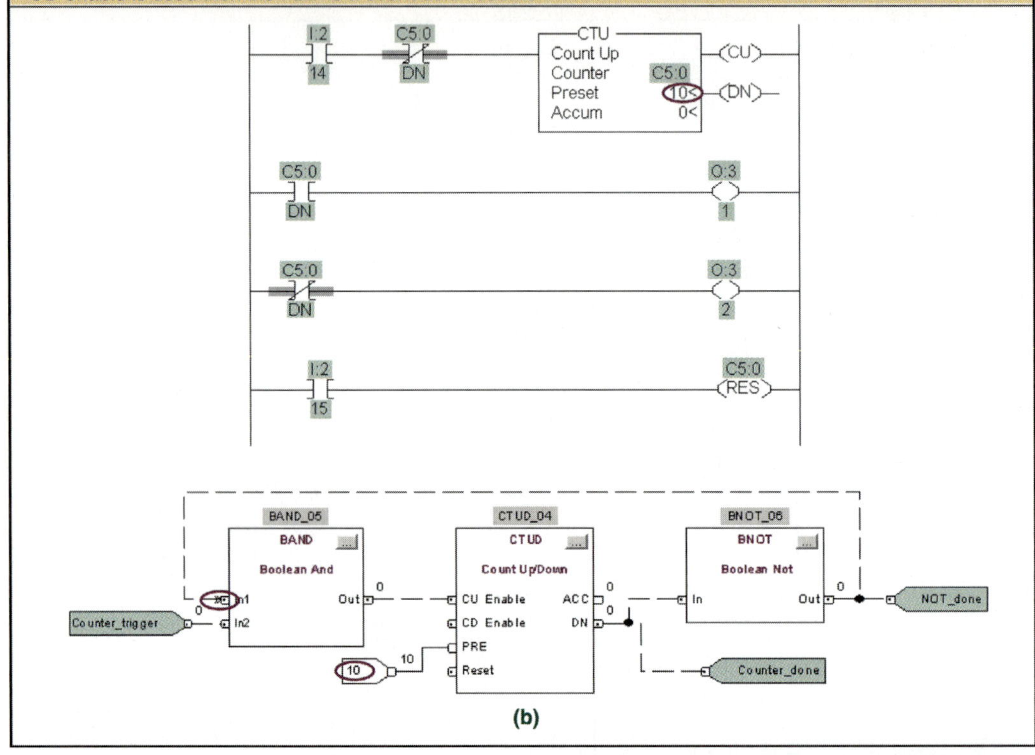

(b)

EXAMPLE 13-1

Design an FBD solution for the process tank problem illustrated in Figure 13-15. The tank should have 25 percent of one ingredient and 75 percent of a second by weight. It should be heated to 180 degrees and then mixed for 30 minutes. Mixing should continue until the tank has drained.

Solution

Some other parameters need to be established. Total weight is 1000 pounds and the FBD solution steps are used to solve this process problem.

1. Develop the control strategy required for the problem.

   The control will use the following logic:

   Valve 1 = start switch AND tank liquid weight less than 250 pounds on fill cycle only
   Valve 2 = start switch AND tank liquid weight greater than 250 pounds but less than or equal to 1000 on fill cycle only
   Mixer on = tank liquid weight greater than 499 pounds

FIGURE 13-12: Standard ladder logic for an up/down counter in FBDs.

Standard FBD Logic for Up/Down Counters

Increment an accumulator tag for every false to true transition of an up count signal and decrement the same tag for every false to true transition of a down count signal. Turn on (or turn off) an output field device after the accumulator tag is equal to or greater than a preset value.

The standard ladder logic for the CTU plus CTD counters in Figure 5-10(d) is shown along with the FBD equivalent. The function block software has a combination up and down counter, which can be used as an up counter, down counter, or as a combined up/down counter. The standard FBD has a tag to signal count up, Up_count_trigger, and a tag to signal count down, Down_count_ trigger. The accumulator tag, CTUD_08.ACC, increments when the up trigger is active, and decrements when the down trigger is active. The done bit is true when the accumulator tag has a value of 10 or higher.

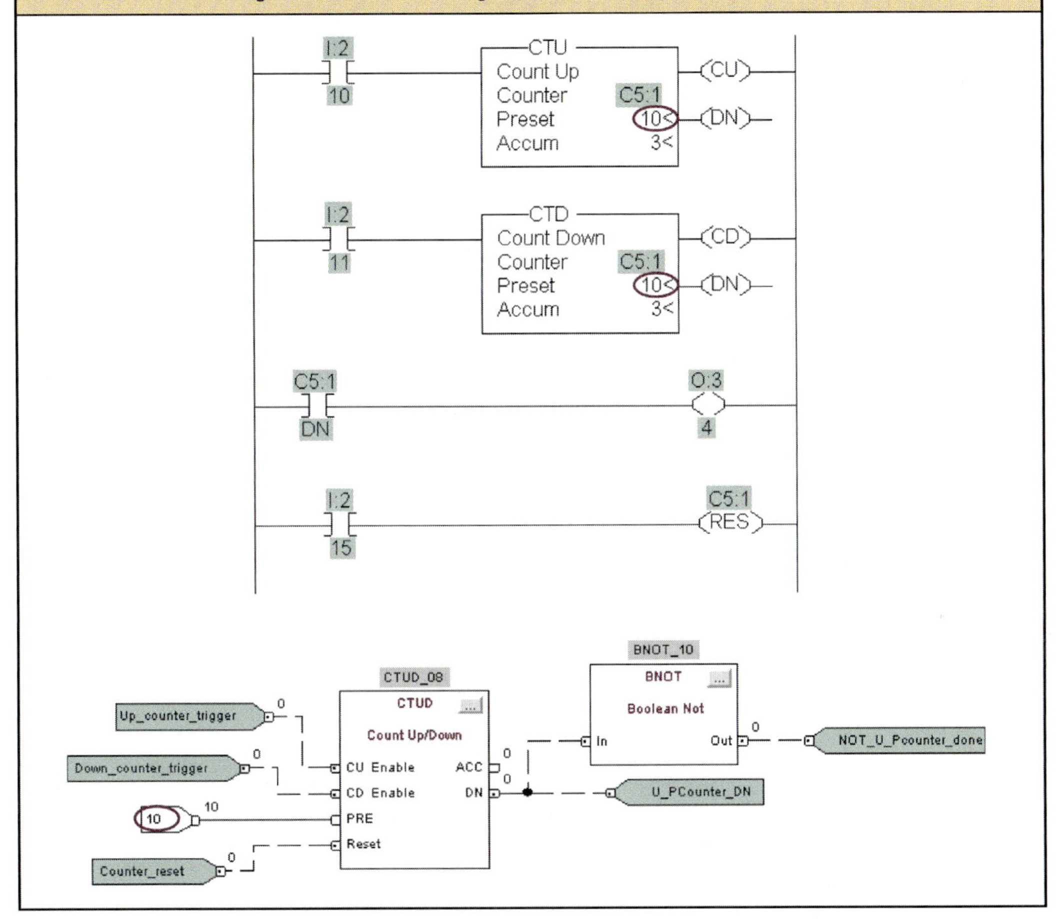

Mixer off = tank liquid weight equal to 0 pounds

Heater on = tank liquid weight equal to or greater than 750 pounds

Heater off = 30 minutes after tank full and temperature at 180 degrees

Valve 3 = tank liquid weight equal to 1000 pounds on drain cycle AND liquid temperature equal to 180 degrees for 30 minutes

2. Choose the function block elements.

   Choosing FBD elements is an iterative process. You start with an initial set and then modify the list as you progress through the design. The list for this design includes the elements in the table in Figure 13-16.

3. Choose a tag name for an element and for each input and output.

FIGURE 13-13: Standard ladder logic for retentive outputs in FBDs.

Standard FBD Logic for Retentive Outputs

Latch and unlatch an output on with an input signal.

The standard ladder logic for the OTL and OTU instructions from Figure 3-32 is shown along with the FBD equivalent. The function block software uses a Set Dominant block, but a Reset Dominant could be used as well. The SETD block makes the output true when the set input is true regardless of the condition on the reset input. The block is reset by a true on the reset input, but only when the set input is false. The RESD block will reset the output when the reset input is true regardless of the condition on the set input. In RESD case the set input only sets the output when the reset input is false. The NOT output is also available.

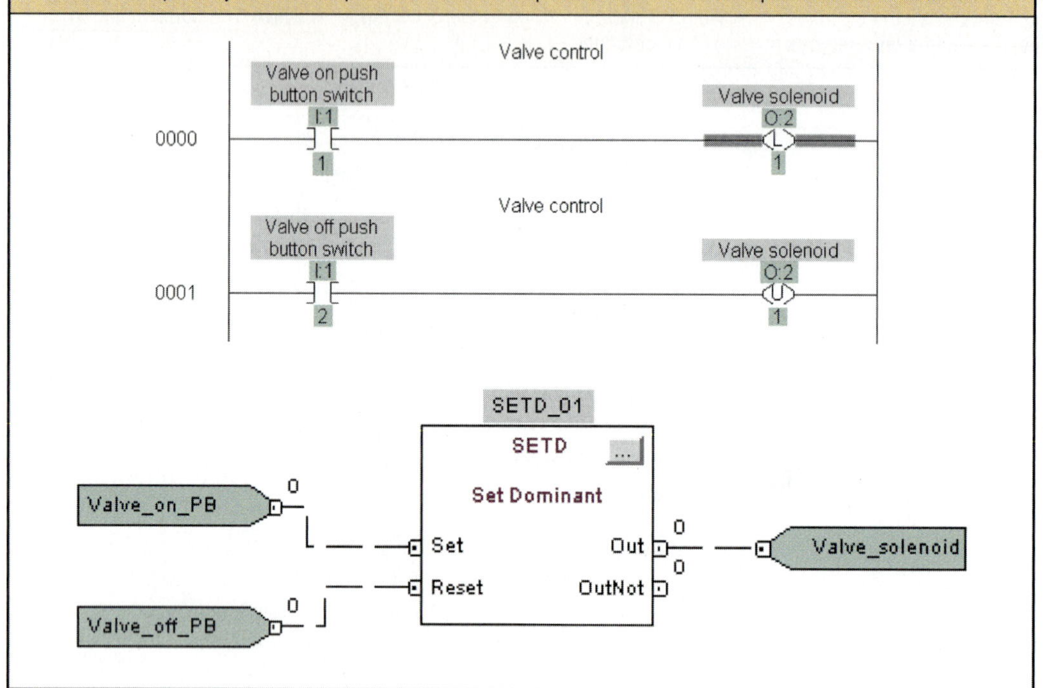

The tag names are variables that hold the data for the process. They should be as descriptive as possible using the fewest words. The tag names include the names of the data values (IREF and OREF) and names for the FBD blocks. Both are listed in Figure 13-17.

4. Define the order of execution.
5. Identify any connectors and feedback loops.
6. Define program/operator control.
7. Add a function block element.
8. Connect elements.
9. Verify that all tags and constant values for IREFs have been made and all outputs, OREFs are defined as well.

Connect blocks with an OCON and ICON that are on different sheets. Steps 4 through 7 are addressed as the function blocks are placed into the programming area in the software. Since it is a design, the development of the FBD program has much iteration, and there is more than one plausible solution. One possible solution is illustrated in Figures 13-18 and 13-19.

At first glance, the FBD solution seems complex and difficult to interpret, but with a little practice it is read more easily than a ladder diagram. The solid lines represent integers and real values, and the dashed lines Boolean or discrete on/off bit values. Input variable values are on the left and output values are on the right.

FIGURE 13-14: Standard ladder logic for one-shots and arithmetic blocks in FBDs.

Standard FBD Logic for One Shots and Arithmetic Blocks

Turn on an output instruction or block like a math or move instruction for only one scan after an input field device(s) is active.

The standard ladder logic for the one shot and addition instruction in Figure 6-12(c) is shown along with the FBD equivalent. The input field device(s) determines when the output is active and the one shot assures that the addition is performed for only one scan. In applications that use blocks that are triggered every scan it often is important to execute those instructions for only one time. The one shot makes the instruction look like one that is triggered by a rising edge (false to true) on the input. While the addition instruction is used in this example, any of the other math instructions (SUB, DIV, MUL, OR SQR) or the move instruction (MOV) could be substituted. Also, the one shot rising block is used, but a one shot falling block could be used to trigger on a falling input edge (true to false input change).

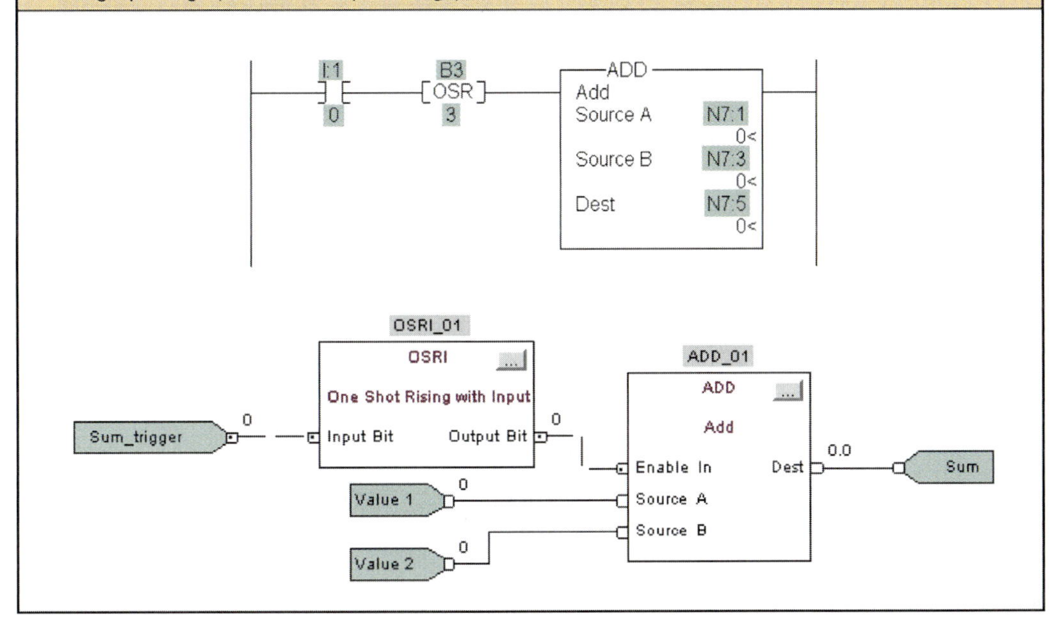

FIGURE 13-15: Process tank for Example 13-1.

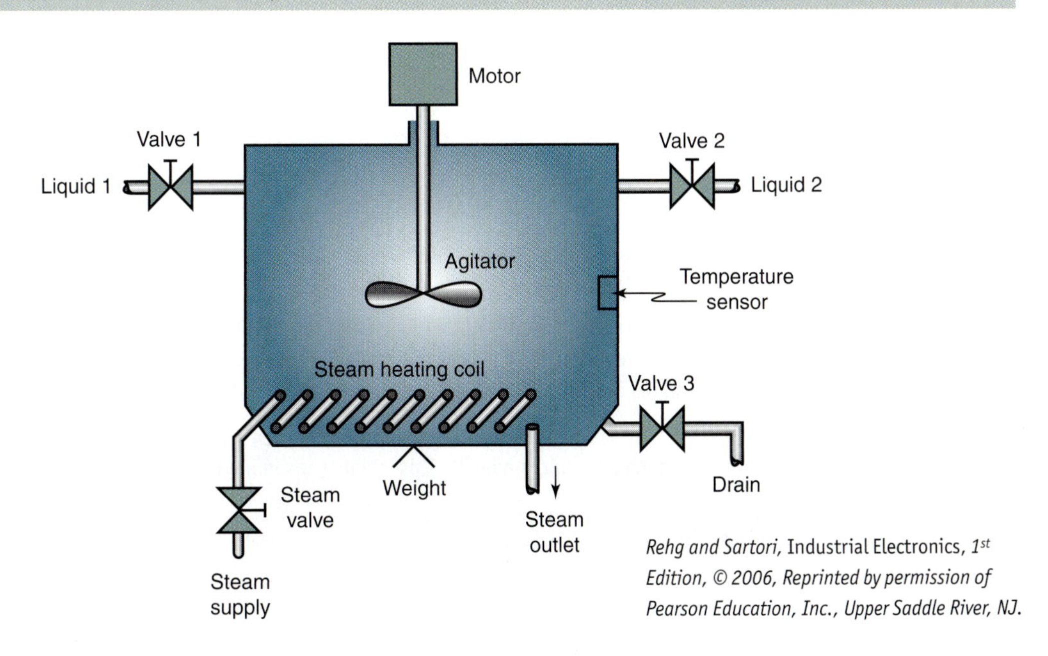

Rehg and Sartori, Industrial Electronics, *1st Edition, © 2006, Reprinted by permission of Pearson Education, Inc., Upper Saddle River, NJ.*

FIGURE 13-16: List of function block elements for Example 13-1.

Element	Description
IREF	Input variable or value
OREF	Output variable or value
ICON	Input connection point for a jumper
OCON	Output connection point for a jumper
OSRI	One Shot Rising with Input—one shot element that changes the output state from 0 to 1 and back to 0 when the input goes from 0 to 1
RESD	Reset Dominant—A set/reset latch that sets the Out pin high when set is high and sets the Out pin low when reset is high. If both inputs are high the reset determines the output
SETD	Set Dominant—A set/reset latch that sets the Out pin high when set is high and sets the Out pin low when reset is high. If both inputs are high the set determines the output
BAND	Boolean AND—An AND gate
BOR	Boolean OR—An OR gate
EQU	Equal To—Comparator that sets the output high when inputs are equal
LES	Less Than—Comparator that sets the set output high when A < B
GRT	Greater Than—Comparator that sets the set output high when A > B
GEQ	Greater Than or Equal To—Comparator that sets the set output high when A ≥ B
TONR	Timer On Delay with Reset—Timer that set the done bit high when the accumulator is equal to the preset values—accumulator increments when the enable input is high

Rehg and Sartori, Industrial Electronics, *1st Edition, © 2006, Reprinted by permission of Pearson Education, Inc., Upper Saddle River, NJ.*

FIGURE 13-17: Tag names, tad data types, and tag descriptions for Example 13-1.

Tag Name	Tag data type	Tag description
Heater	BOOL	Output variable
Mixer	BOOL	Output variable
Start	BOOL	Input variable
Stop	BOOL	Input variable
Valve_One	BOOL	Output variable
Valve_Three	BOOL	Output variable
Valve_Two	BOOL	Output variable
Liquid_weight	DINT	Input variable
Temperature	DINT	Input variable
Latch_Heater	DOMINANT_RESET	Function blocks
Latch_mixer	DOMINANT_RESET	Function blocks
Start_latch	DOMINANT_RESET	Function blocks
Heater_unlatch	DOMINANT_SET	Function blocks
Latch_tank_full	DOMINANT_SET	Function blocks
Timer_start_latch	DOMINANT_SET	Function blocks
Timer_start_AND	FBD_BOOLEAN_AND	Function blocks
Valve_one_AND	FBD_BOOLEAN_AND	Function blocks
Valve_three_AND	FBD_BOOLEAN_AND	Function blocks
Valve_two_AND	FBD_BOOLEAN_AND	Function blocks
Start_unlatch	FBD_BOOLEAN_OR	Function blocks
Temp_at_180	FBD_COMPARE	Function blocks
Weight_at_0	FBD_COMPARE	Function blocks
Weight_at_1000	FBD_COMPARE	Function blocks
Weight_over_249	FBD_COMPARE	Function blocks
Weight_over_500	FBD_COMPARE	Function blocks
Weight_over_750	FBD_COMPARE	Function blocks
Weight_under_249	FBD_COMPARE	Function blocks
Start_oneshot	FBD_ONESHOT	Function blocks
Timer_30min	FBD_TIMER	Function blocks

Rehg and Sartori, Industrial Electronics, *1st Edition, © 2006, Reprinted by permission of Pearson Education, Inc., Upper Saddle River, NJ.*

Wire interconnections are indicated with OCON (pointing right) and ICON (pointing left) symbols and the word *Jumper_x* inside. The numbers below each wire connector are the page coordinates for the other end of the connection.

10. Verify the routine.

The routine is verified by the Rockwell software to uncover any format errors in the programming process, then it must be verified by the designer to ensure that it solves the design requirements for the process. The design verification requires the designer to go through the following steps:

a. Assume that the start switch is off and the tank is empty for the first system scan—mark the input values before the scan and the output values after the scan for each FBD function block. This is time consuming but necessary to verify that the solution produces the correct results.

b. Assume the start switch is pressed—mark the input values before the scan and the output values after the scan for each FBD function block.

c. Assume the liquid just passed the 250 point—mark the input values before the next scan and the output values after the next scan for each FBD function block.
d. Continue through the operational sequence, noting the change in inputs and outputs at each stage of the process operation.
e. Verify that the operation recorded conforms to the desired operation stated at the start of the design.

EXAMPLE 13-2

Study the mixer control routine in step 1 of Example 13-1 and verify that the system executes correctly using the FBD solution in Figure 13-18 (bottom two blocks in the figure).

FIGURE 13-18: Valve 1, valve 2, and mixer FBD for Example 13-1.

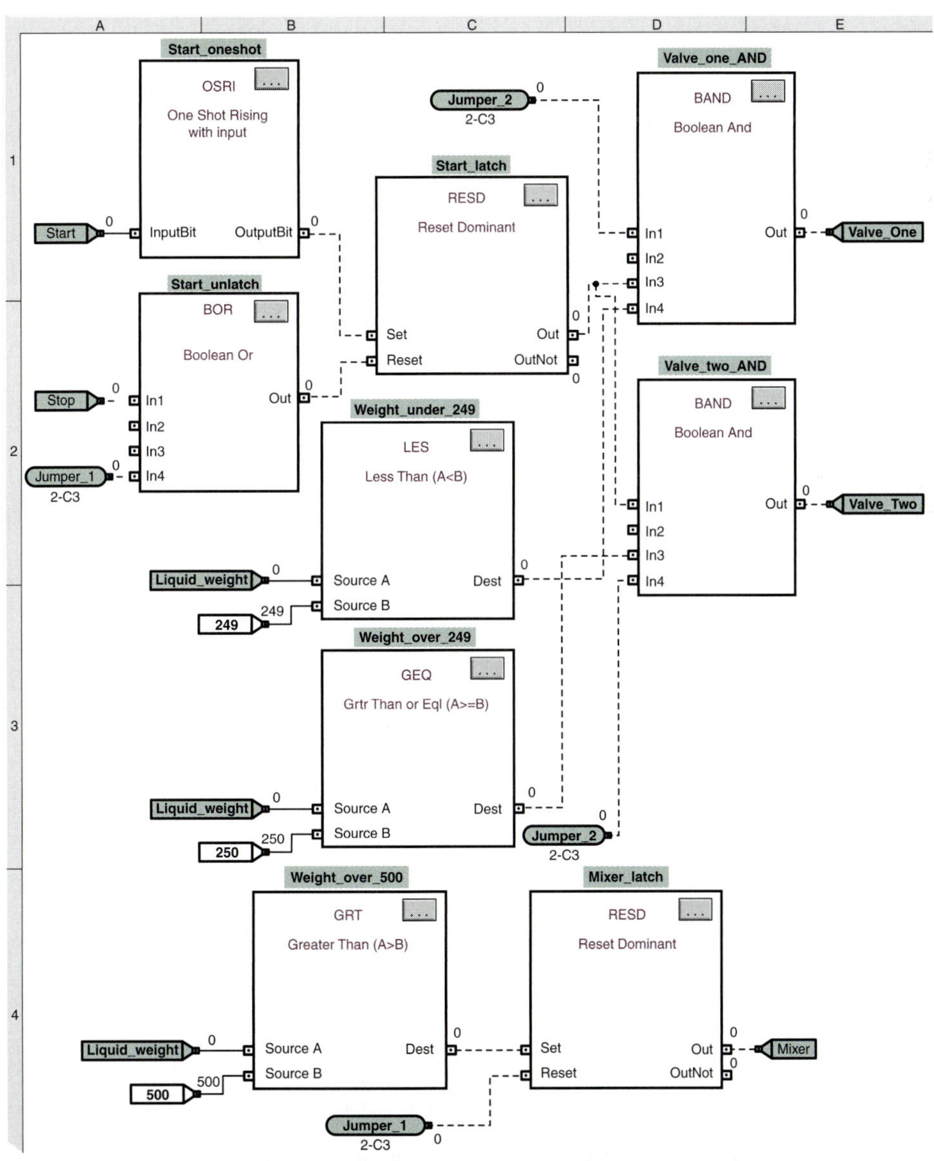

Function block for valve one, valve two, and mixer for Example 13-1

Rehg and Sartori, Industrial Electronics, *1st Edition, © 2006, Reprinted by permission of Pearson Education, Inc., Upper Saddle River, NJ.*

Solution

1. Liquid_weight is less than 500, so the comparator (Weight_over_500) output is not true or 0. The Reset Dominant latch (Mixer_latch) has 0 on the set input and the value generated from Jumper 1. The origin for Jumper 1 (Weight_at_0) is in Figure 13-19; study that routine. Initially Jumper 1 is true or 1 because Liquid_weight is equal to 0. Therefore, the Mixer_latch is 0 on the set and 1 on the reset, so the output is 0 and the mixer is off.
2. When the tank starts to fill, Jumper 1 changes to a 0 since the Liquid_weight is greater than 0. So the input to the Mixer_latch is set 0 and reset 0, the output remains 0, and the mixer is off. When the Liquid_weight exceeds 500, the Weight_over_500 comparator's output is true or 1, so the Mixer_latch set input changes to true or 1. This causes the output to change to true or 1 and the mixer to be energized.
3. The inputs to the Mixer_latch remain set 1 and reset 0 as the tank fills to 1000 and as the tank empties to 500. After the Liquid_weight drops below 500 the Mixer_latch inputs become set 0 and reset 0, the output remains true or 1, and the mixer remains energized. When the tank is empty, Liquid_weight equals 0 and Jumper 1 changes from 0 to true or 1. This causes the Mixer_latch reset input to be true and the output changes from 1 to 0. This turns the mixer off.

FIGURE 13-19: Heat and valve 3 FBD for Example 13-1.

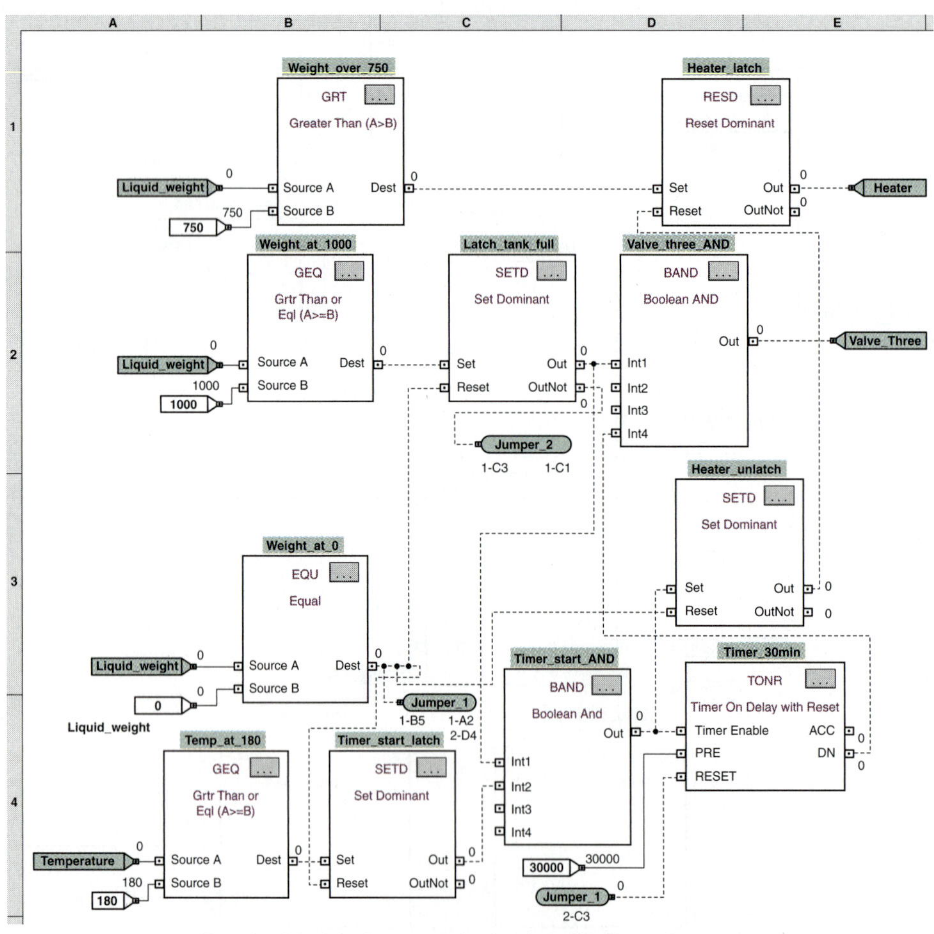

Rehg and Sartori, Industrial Electronics, *1st Edition, © 2006, Reprinted by permission of Pearson Education, Inc., Upper Saddle River, NJ.*

13-5 SITES FOR ALLEN-BRADLEY PRODUCTS AND DEMO SOFTWARE

RSLogix 5000 product information: *http://www.ab.com/catalogs/b113/controllogix/software.html*

RSLogix 5000 demo software ordering and download: *http://www.ab.com/logix/rslogix5000/*

RSLogix 500 demo software download (under Get Software): *http://www.ab.com/plclogic/micrologix/*

Rockwell Automation Logix product information: *http://www.ab.com/logix/*

Rockwell Automation manuals: *http://www.theautomationbookstore.com*; *http://www.ab.com/manuals/*

Chapter 13 questions and problems begin on page 588.

CHAPTER

14

Intermittent and Continuous Process Control

14-1 GOALS AND OBJECTIVES

The goal of this chapter is to present closed-loop feedback control techniques for on/off and continuous processes using PLC as the control element.

After completing this chapter you should be able to:

- Cite the reasons why the on-off control mode is the most popular.
- Define the term *proportional band.*
- Discuss the problem of offset in proportional control, and show why it cannot be eliminated in a proportional controller.
- Explain why the proportional integral control mode overcomes the offset problem.
- Describe the effects of changing the integral time constant (reset rate) in a proportional integral controller.
- Explain the advantage of the proportional integral and derivative control mode over simpler control modes. State the process conditions that require the use of this mode.
- Describe the effects of changing the derivative time constant (rate time) in a proportional integral and derivative controller.
- Describe how digital controllers operate in the control of a process.
- Interpret a table that relates the characteristics of an industrial process to the proper control mode for use with a PLC implementation.

14-2 PROCESS CONTROL

Process control is classified into two categories, *intermittent* and *continuous*, that describe how it tracks the change in the process output. Intermittent control is a discrete or on-off control; therefore, its output usually has just two values—0 *or* 100 percent. In contrast, continuous control has output variability that ranges from 0 *to* 100 percent with an infinite number of values. The control modes for these two categories are as follows:

Intermittent

- On-off control
- Two-position control
- Floating control

Continuous

- Proportional control (P)
- Integral control (I)
- Derivative control (D)
- Proportional integral control (PI)

- Proportional derivative control (PD)
- Proportional integral and derivative control (PID)
- Fuzzy control

The types of controllers that are used to implement these control modes are the pneumatic, analog, and digital controllers. The first six continuous control modes are implemented using either analog or digital controllers, but fuzzy control is implemented only in a digital controller. A PID controller is a three-mode controller that has the characteristics of a proportional (P), integral (I), and derivative (D) mode. PID analog and digital controllers are the most versatile since they can be configured to operate in just proportional mode, proportional plus integral mode, proportional plus derivative mode, or with all three modes present. In general, controllers are not configured for pure integral or derivative mode control.

Process control started with pneumatic instrumentation, pneumatic controllers, and pneumatic final control elements. Today most pneumatic devices have been replaced with digital controllers. Digital controllers were initially developed as the computer implementation of an analog controller, but a modern digital controller is much more versatile than its analog counterpart. A digital controller can be configured to operate in all control modes plus some specialized and novel control algorithms.

Each control mode is addressed in the following sections. Following the sections on intermittent and continuous controllers, a section is presented on how digital control is used to implement the proportional, integral, and derivative modes.

14-3 INTERMITTENT CONTROLLERS

The basic intermittent controller is the *on-off* controller. The operation of the on-off controller is improved with the addition of hysteresis, which is typically called *two-position* control. An extension of two-position control is *floating* control, which provides three or more controller output levels for enhanced process control. These three types of control are discussed in the following subsections.

14-3-1 On-Off Control

On-off control is used in numerous applications such as home appliances, including oven heating, refrigerator cooling, and heating and cooling of all residential and commercial buildings.

In on-off control, the controller output instantly follows the change in the error (E) as shown in Figure 14-1 and where:

$$E = \text{set point} - \text{measured variable}$$

When the set point is less than the measured variable, the controller output is at 0 percent. When the set point is greater than the measured variable, the controller output is at 100 percent. This on-off control method is not as effective because, in the majority of applications, controller output is constantly switching between off and full on. The controller action for this type of control is illustrated in Figure 14-1. Note that negative error produces a 0 percent output response and positive error produces a 100 percent output response.

On-off control is not used as frequently as two-position control because of the frequent change in the control element due to a single trigger point at error equal to zero. The time constant or capacitance and inertia of the process dictate when on-off control can be considered. The time constant, capacitance, or inertia of the process is a measure of how fast the controlled variable will change in the process. For example, the capacitance for an average size house with good insulation would be relatively large, so it would take some time for the temperature to rise and drop. On the other hand, heat placed in a small un-insulated box would not remain at the set level for very long. The box would have a small time constant or low capacitance and inertia. If on-off control is used for processes with small time constants, then the

FIGURE 14-1: On-off control.

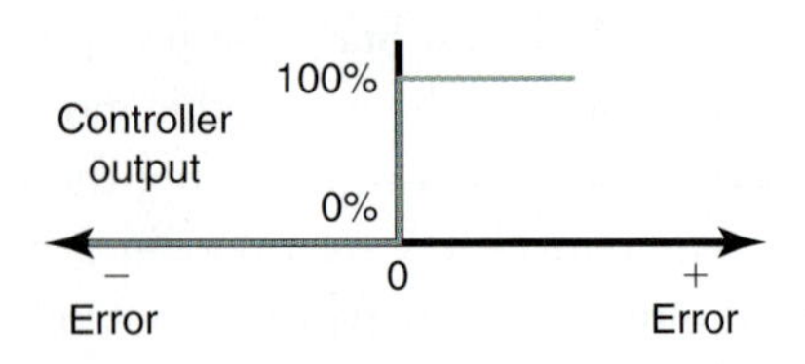

Rehg and Sartori, Industrial Electronics, *1st Edition, © 2006, Reprinted by permission of Pearson Education, Inc., Upper Saddle River, NJ.*

FIGURE 14-2: Digital control for on-off process.

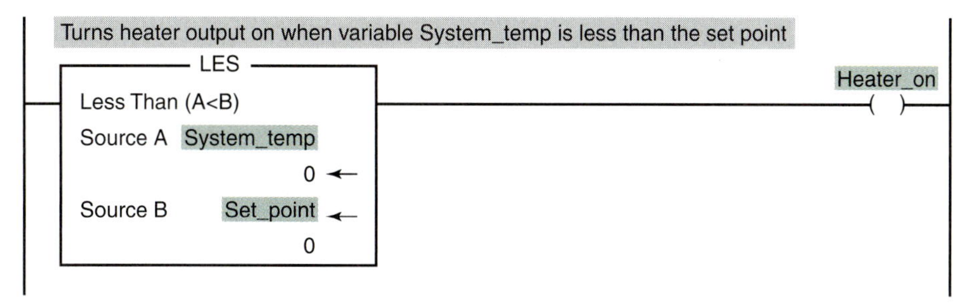

(a) Digital controller: PLC ladder logic

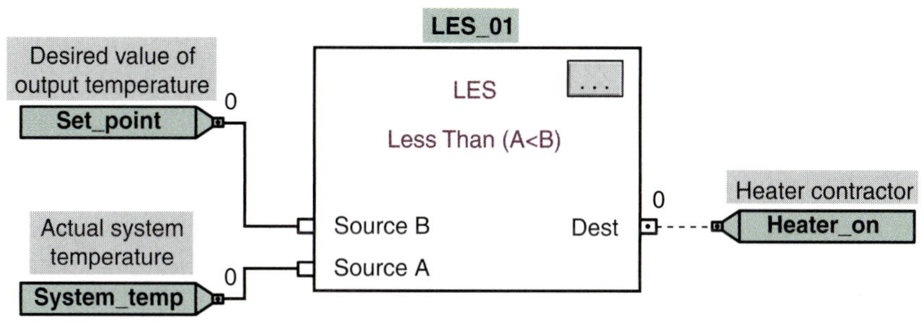

(b) Digital controller: PLC function block diagrams

Rehg and Sartori, Industrial Electronics, *1st Edition, © 2006, Reprinted by permission of Pearson Education, Inc., Upper Saddle River, NJ.*

control contactor is cycling on and off at a rapid rate. However, in processes where the process time constant or inertia is very large compared to the energy supply unit, such as when a large building is heated with a relatively small furnace, then on-off control is a good choice. If the measuring instruments have built-in hysteresis, they will compensate and permit on-off control to be used.

The ladder logic and Function Block Diagram (FBD) PLC solutions for on-off control are illustrated in Figure 14-2. The tag Set_point is the set point for the desired system temperature and System_temp the measured system temperature. The ladder rung compares the measured value to the set point variable and turns on the heater if the set point is above the measured value. The FBD solution performs the same function with one FBD block. Compare the two solutions and see how the ladder instruction is implemented in FBD.

14-3-2 Two-Position Control

Two-position control adds *hysteresis* or a *neutral zone* to the on-off controller operation. The addition of hysteresis causes the turn-on and turn-off error values to be shifted by the width of the neutral zone to the left and right of the zero error value. Study Figure 14-3, which illustrates this concept.

The set point is in the center, so process variable (PV) movement to the right increases the PV and makes the error (SP − PV) value negative, thus the controller output should be *off*. To the left of center, the PV is decreasing and the error is positive and controller output should be *on*. However, the on-off action does not trigger immediately, but is delayed by the value of the error band. This hysteresis or neutral zone causes the controller to drive the *controlled variable (PV)* output value *above* the set point when the controller is *on* (100 percent output), and lets the *PV* fall *below* the set point when the controller is *off* (0 percent output). The addition of this neutral zone or hysteresis region in the control significantly reduces the on and off cycling of the output; however, it causes the PV to vary above and below the desired set point. The value selected for the neutral zone should be large enough to reduce the on-off cycling of the system but within the percentage change in PV allowed by the application. The neutral zone is called the *dead band* in some references. Example 14-1 illustrates this point.

FIGURE 14-3: Two-position controller.

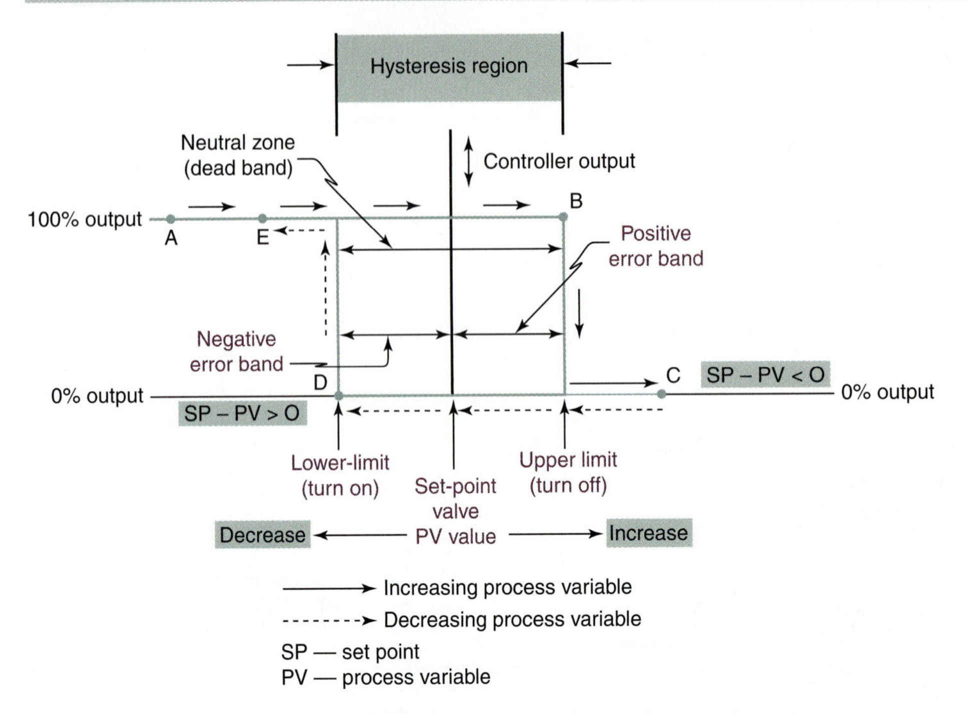

Rehg and Sartori, Industrial Electronics, *1st Edition, © 2006, Reprinted by permission of Pearson Education, Inc., Upper Saddle River, NJ.*

EXAMPLE 14-1

A home is heated with a gas furnace and controlled by a two-position thermostat. The PV is the house temperature and the measured variable (MV) is that temperature at the thermostat. The system specifications are as follows: temperature control range is 0° to 85° F, dead band is 2° F (+/−1° F), room temperature (MV) is 65° F, thermostat setting [set point (SP)] is 70° F, and error when the system is turned on (error = SP − MV = 70 − 65) is 5° F. Describe how the two-position controller will regulate the furnace on and off cycle.

Solution

The control sequence for home heating system is as follows:

1. The two-position thermostat is turned to heat with the measured room temperature (MV) at 65° F and the set point (SP) at 70° F, which produces an error of 5° F that starts the control cycle at point A in Figure 14-3. The furnace turns on and maximum heat is distributed in every room by the furnace blower.
2. The value of MV at the thermostat starts to increase and follows the path marked by the solid arrows in Figure 14-3. As MV reaches and passes the set point (70° F) the furnace continues to generate heat and MV continues to increase.
3. When the MV reaches the upper limit (set point plus positive error band, or 72° F), point B in Figure 14-3, the furnace turns off. The MV moves higher than the upper limit or positive band because the thermostat is a temperature sensor and most have a first-order lag response, which delays turn off slightly. As a result of this delay and heat stored in the heating ducts, the MV continues to rise and reaches point C in Figure 14-3.
4. The MV starts to fall and follows a path marked by the dashed arrows from point C to point D, the lower limit (68° F) in Figure 14-3. Note that the MV is allowed to fall below the set point value of 70° F by a value of 2° F, the negative error band.
5. When the MV reaches the lower limit, point D, the furnace turns on (100 percent output). However, the MV continues to fall to point E because the blower is not turned on until the furnace has heated.
6. This cycle continues until the system set point is changed or the furnace is turned off.

FIGURE 14-4: Process variable output for two-position controller.

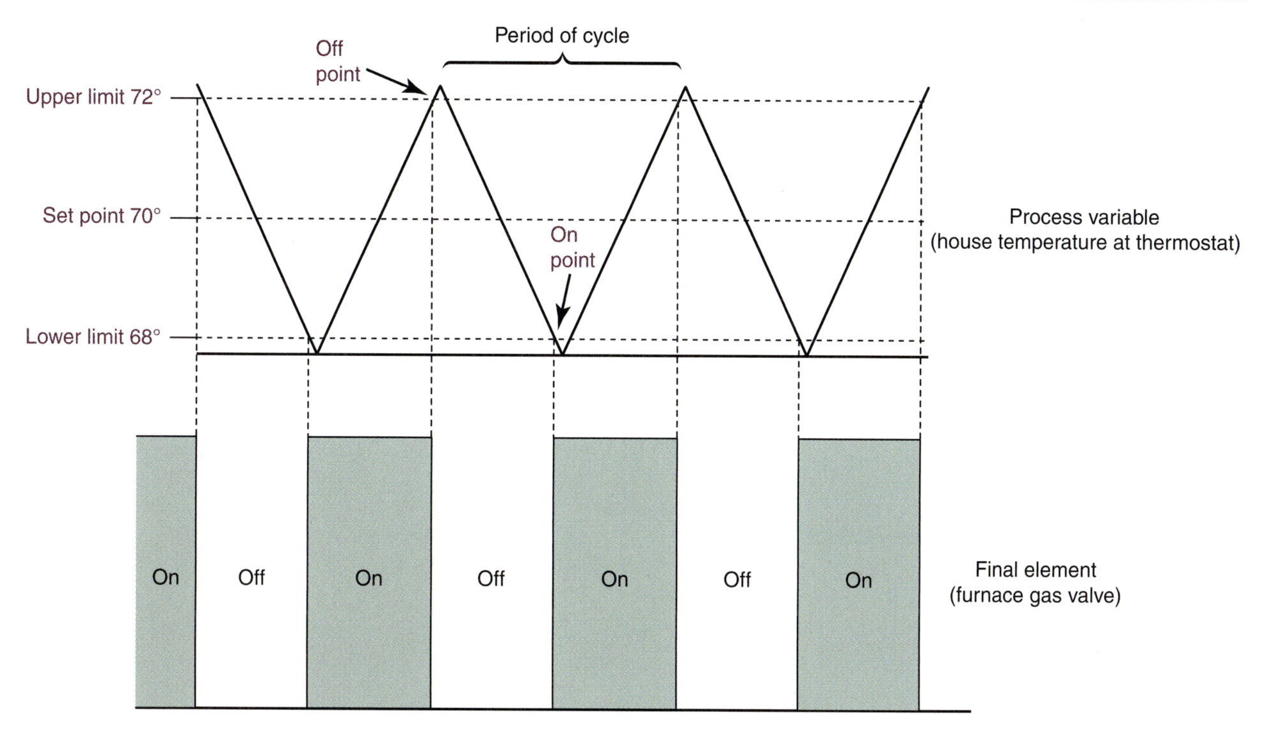

Rehg and Sartori, Industrial Electronics, *1st Edition, © 2006, Reprinted by permission of Pearson Education, Inc., Upper Saddle River, NJ.*

Figure 14-4 illustrates two-position control in a graphical presentation. Note that the process variable is the house temperature and the final element is the furnace gas valve, which is either on or off. The period of the cycle increases as the dead band increases and as the capacitance or inertia of the system increases. Any parameter that increases the heating or cooling part of the cycle will reduce the cycling of the controller. Two-position control can be implemented digitally and mechanically, and each of these implementations is covered in the following discussions.

Digital Two-Position Controller. A digital two-position controller implemented in Function Block Diagrams is illustrated in Figure 14-5. The primary control element is the Set/Reset binary element, SETD (set dominant binary element or flip-flop). SETD Out is true when the Set input is true. The Reset input can be either true or false for this action. SETD Out is false when the Reset input is true and the Set input is false. The digital controller uses the binary element to latch the output on (Set input true) when PV falls below the lower limit. The same latch element is reset as the PV rises above the upper limit (Reset input true and Set input false). In separate addition and subtraction blocks, the limits are calculated using the set point with the positive error band and negative error band constants. Study Figure 14-5 and the preceding description until the FBD program operation is clear.

Mechanical Two-Position Controller. A number of process switches were covered in Chapter 2, including pressure, temperature, and level. These switches and their associated controllers were implemented using mechanically closed electrical contacts. They represent the lowest cost and least complex solution. However, they are limited to some degree in the range of PV values that can be controlled. In addition, a single controller is only usable for a specific measurement span and differential value, which is the hysteresis value.

FIGURE 14-5: Digital two-position controller.

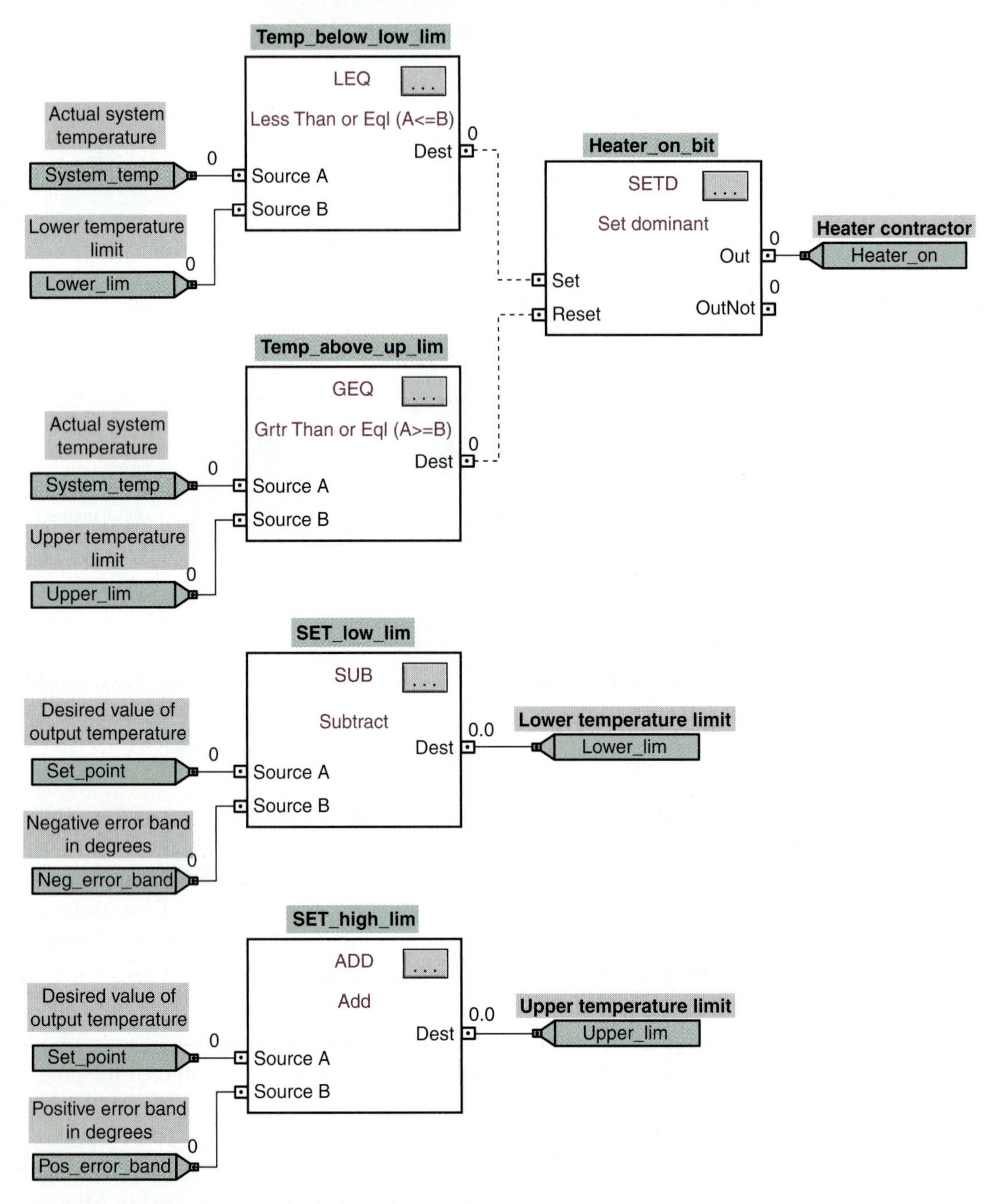

The two-position control pressure switch from Allen-Bradley, shown in Figure 14-6(a), has an adjustable span of 6 to 250 psi and a differential range within that span of 4 to 45 psi. The two-position temperature controller, shown in Figure 14-7, has an adjustable span of 200 to 360 degrees Fahrenheit and a differential range within that span of 8 to 72 degrees. The PV output for the pressure device is illustrated in Figure 14-6(b). Note that the output cycles between the differential values. Definitions of the terms used with the pressure switch and temperature controller are as follows:

Adjustable operating range: total span within which the contacts can be adjusted to trip and reset.

FIGURE 14-6: Two-position pressure controller.

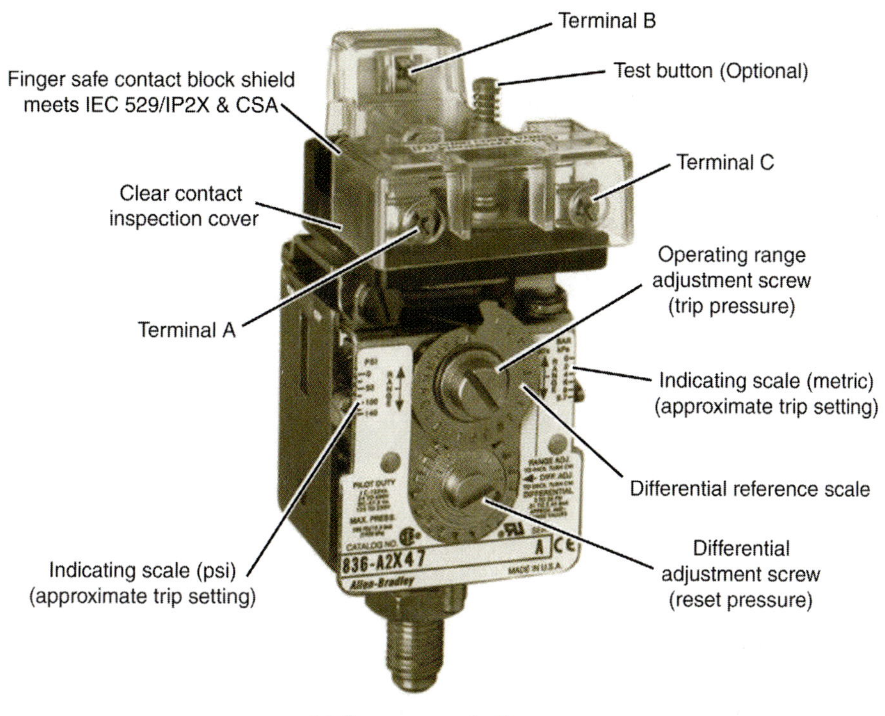

(a) Pressure controller

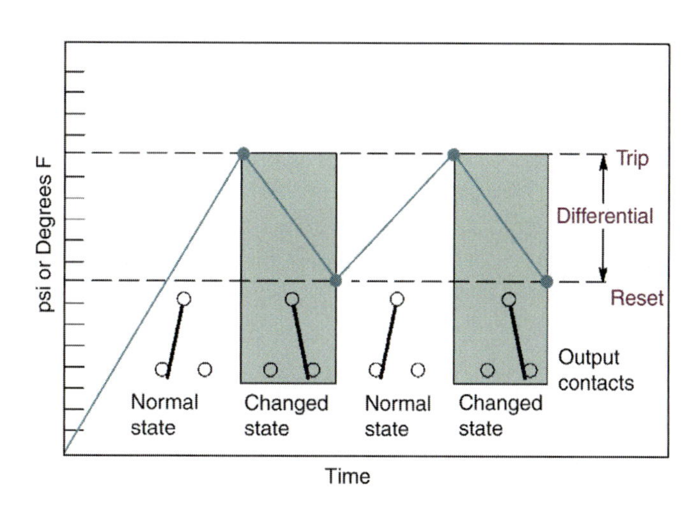

(b) Process variable output

Source: Courtesy of Rockwell Automation, Inc.

Trip setting: higher pressure or temperature setting at which the contacts transfer from their normal state to a changed state.

Reset setting: lower pressure or temperature setting at which the contacts return to their normal state.

Adjustable differential: difference between the trip and reset values.

14-3-3 Floating Control

Floating control is often called multiposition control. It is an extension of two-position control with three (0 percent, 50 percent, and 100 percent) or more controller output levels for control of the process. Figure 14-8 illustrates a multiposition controller with three operating states: 0, 50, and 100 percent, which is used to

FIGURE 14-7: Two-position temperature controller.

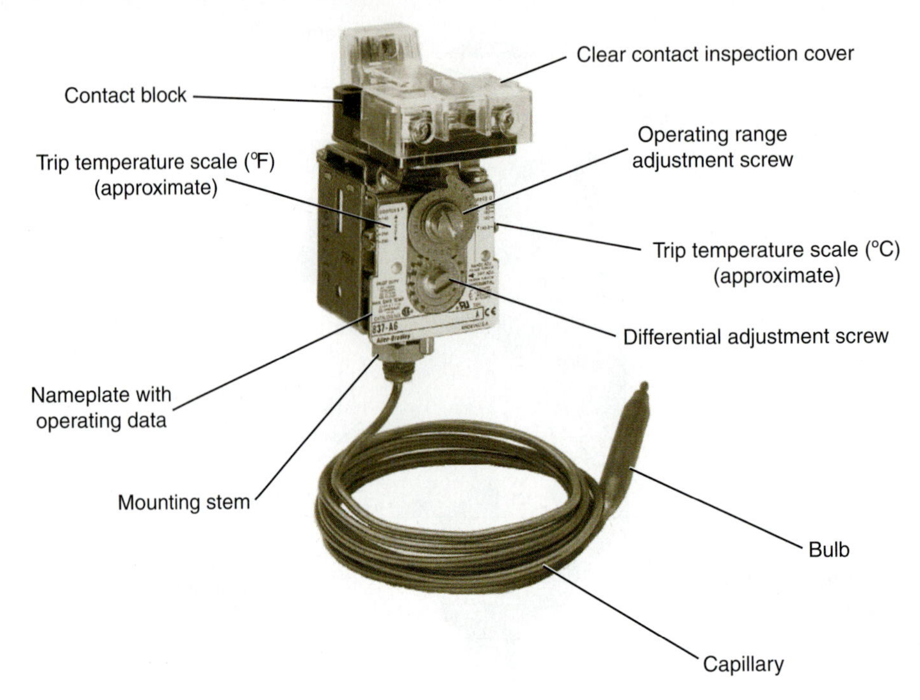

Source: Courtesy of Rockwell Automation, Inc.

FIGURE 14-8: Floating point controller.

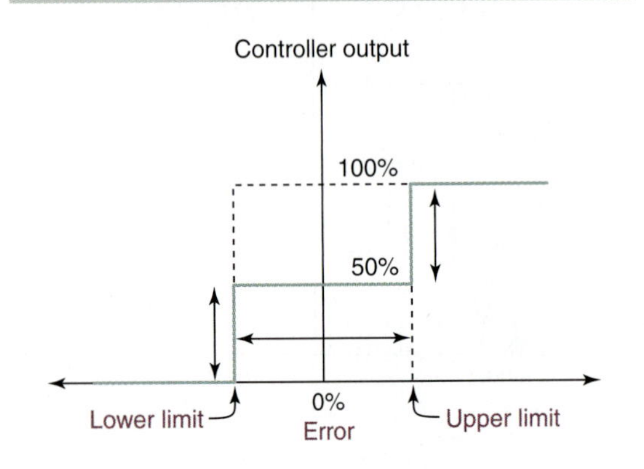

Rehg and Sartori, Industrial Electronics, *1st Edition, © 2006, Reprinted by permission of Pearson Education, Inc., Upper Saddle River, NJ.*

control a cooling system for a process. The controller has 100 percent output when the error value is over a high threshold (process temperature is too high), and the output is 0 percent when the error is less than a minimum threshold (process temperature is too low). If the error value is between the threshold values, then the controller output is 50 percent. Floating control reduces the controller cycling rate compared with on-off and two-position control systems.

Multiposition control using Function Block Diagrams (FBDs) is illustrated in Figure 14-9. Note that three types of comparators (less than, greater than, and limit test) are used to determine where the error is relative to the upper and lower thresholds, and then a selective summer (SSUM) is used to pass the correct PV value. The LES comparator determines if the error is less than the lower limit and the GRT comparator checks for error over the upper limit. The LIM comparator determines if the error falls between the limits. The SSUM takes the integer or floating point values at each select input (In1, In2, etc.) and adds the values for those where the enable input (select1, select2, etc.) is true. The instruction then places the result in Out.

For example, if the error is less than the lower limit, then the select1 input is active and the value for minimum percent output (min_per_out) is applied to the final control element. The select2 and select3 inputs operate in a similar fashion. In this application, the SSUM block does not add the inputs but places one of the three at the out-

FIGURE 14-9: Digital floating point controller.

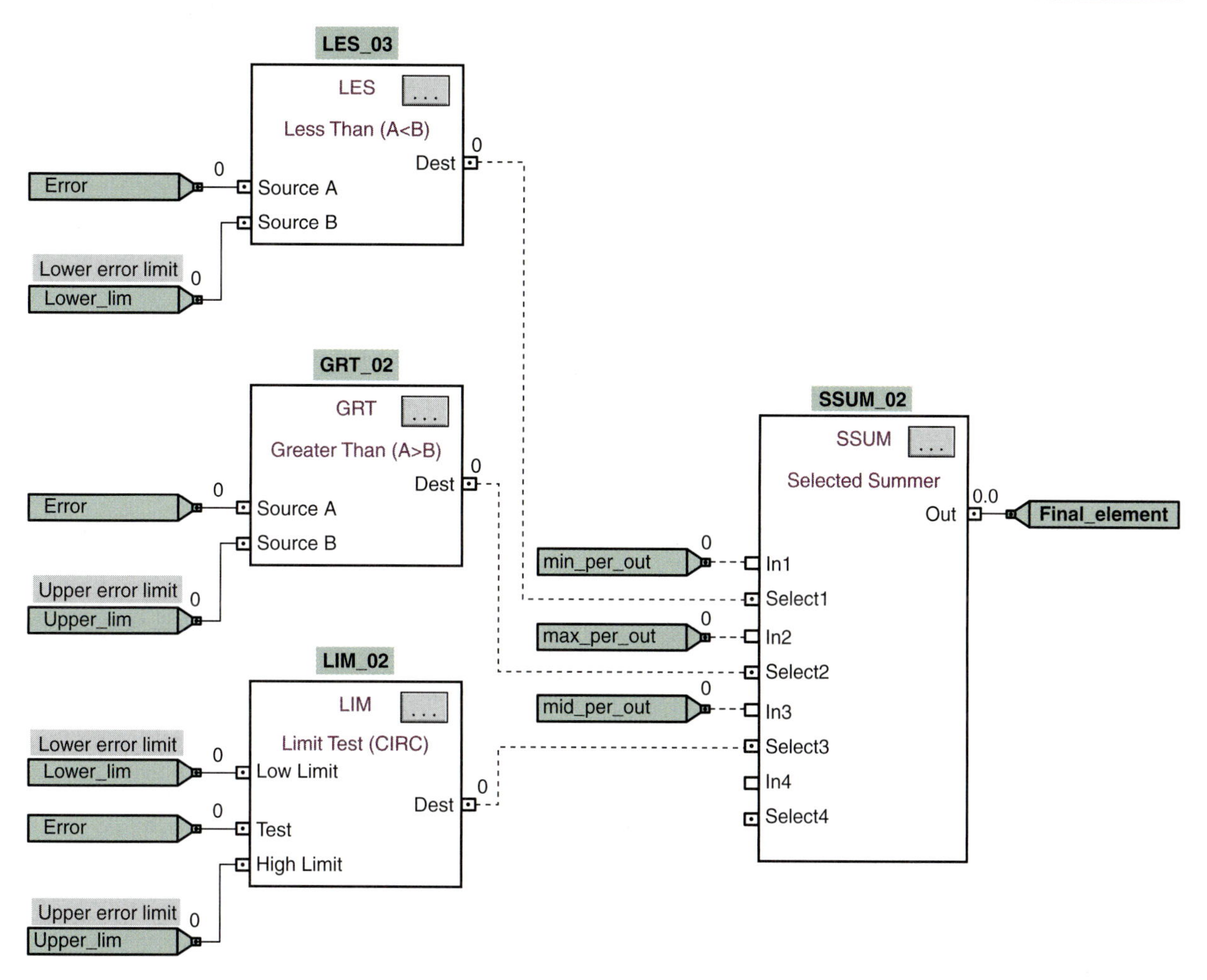

Rehg and Sartori, Industrial Electronics, *1st Edition, © 2006, Reprinted by permission of Pearson Education, Inc., Upper Saddle River, NJ.*

put since only one select input line is true at any one time. Note that not shown in the FBD are blocks to calculate the error (SP − MV) and the lower and upper limit error (Error − negative error band and Error + positive error band, respectively).

14-4 CONTINUOUS CONTROLLERS

Section 14-3 described the operation of intermittent controllers used in control applications where the system inertia or capacitance is large or where limited cycling of the process variable is acceptable. This section addresses continuous controllers. The classification of controllers and control modes is repeated so that you can see the number of continuous control options that are available. They are *proportional control* (P), *integral control* (I), *derivative control* (D), *proportional integral control* (PI), *proportional derivative control* (PD), *proportional integral and derivative control* (PID), and *fuzzy control*. There are few continuous controllers that use analog control in the process industry today. Most use some type of digital control implemented either as a bank of controllers, as a distributed control system (DCS), or with a programmable logic controller (PLC).

14-4-1 Proportional Control

On-off control requires a process PV that can swing between two limits and has sufficient inertia or capacitance that the cycling will not be too severe. However, a proportional controller is necessary when:

- The process must be held at a specific set point value.
- The process has a small capacitance value.
- Disturbances to the controller or process could be present.

Proportional controllers have:

- Analog input and output values, expressed as integers or real numbers that vary over the range necessary for control of the process.
- Final control elements that are continuous devices with outputs from 0 to 100 percent.
- Output values that are the product of the controller gain (K) and the input error (SP − PV or PV − SP).
- An output value that changes the final control element so that the controlled variable moves closer to the desired set point.

The magnitude of the output of the controller has a direct relationship to the input error. If the error increases, then the magnitude of the controller output increases. If the error falls, then so does the output. There are two methods used to change the controller output in relation to the error: *pulse width modulation (PWM)* and *gain-error product.*

Pulse Width Modulation. Although PWM is not used often in process control it should be understood. In PWM the controller output is switched from full *off* to full *on* at a fixed frequency. The magnitude of the error determines the duty cycle of the square wave present at the output. For example, if a small error is present, then the on time for each cycle might be 5 percent of the period and the off time would be 95 percent. This is called a 5 percent duty cycle. The average value of such a square wave would be close to zero. This corresponds to a small error present at the input. Let's look at a larger error value. Such an error might produce a controller output with on time of 80 percent of one period and off time of 20 percent. This pulse train would have a larger average value corresponding to the larger error. Usually in PWM systems the final control element is adjusted to produce a PV value equivalent to the set point when the output from the controller has a 50 percent duty cycle. Thus a zero error into the PWM controller produces a 50 percent duty cycle output. A change in the error from zero causes the duty cycle to increase or decrease based on direct-acting or reverse-acting requirements of the process.

Gain-Error Product. The most frequently used controller configuration is the gain-error product. In this configuration the controller output level is equal to the product of the controller gain (K_P) and the value of the error signal. The analog signal at the output of the controller is either a voltage or a current that is proportional to the size of the error signal. The value of the controller gain (K_P) is dictated by the control requirements of the process. The controller gain can be expressed in two ways: as *proportional gain* or as *proportional band.*

Proportional gain is the ratio of change in output to the change in input, as described mathematically by the following formula:

$$\text{Gain} = K_P = \frac{\text{Percentage output change}}{\text{Percentage input change}}$$

Review the tank level problem introduced in Figure 12-37 and developed further in Figures 12-39 through 12-43.

The system described in those figures and the text supporting them is a *continuous proportional controller*. The diagram in Figure 12-41 indicates that link A is the controller *gain element* for the system, and the gains for different link lengths are calculated in Figure 12-42. Conclusions reached from the analysis of the continuous proportional control system for tank liquid level are:

- A continuous proportional control system has zero steady-state error only at the control variable and set point values established at system startup.
- If a disturbance or set point change occurs in a system with proportional control, then some steady-state error will result.
- The steady-state error and the proportional controller gain have an *inverse* relationship (increase gain and reduce steady-state error).

- Transient response maximum peak and settling time have a *direct* relationship with controller gain (increase gain and increase overshoot and ringing).
- Steady-state error cannot be completely eliminated with increases in controller gain.

Figure 12-41 shows how all of the elements or blocks in the proportional control system are related. Figure 14-10 illustrates how each signal in the system reacts to a load change. Read the description of each process parameter in the figure. Notice how the controller output follows the error value.

FIGURE 14-10: Proportional controller action.

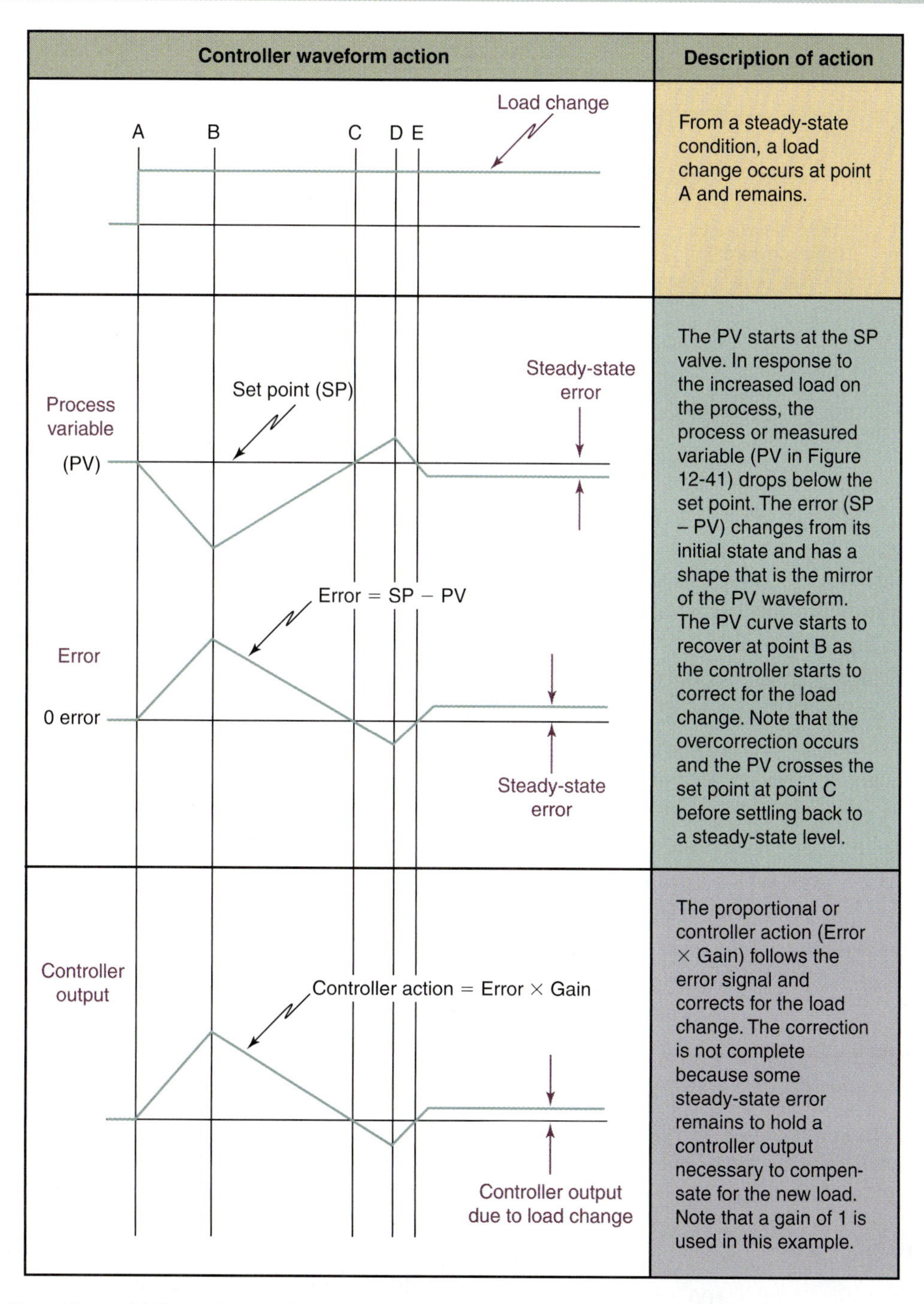

Rehg and Sartori, Industrial Electronics, *1st Edition, © 2006, Reprinted by permission of Pearson Education, Inc., Upper Saddle River, NJ.*

FIGURE 14-11: Proportional controller.

(a) Block representation of a proportional controller

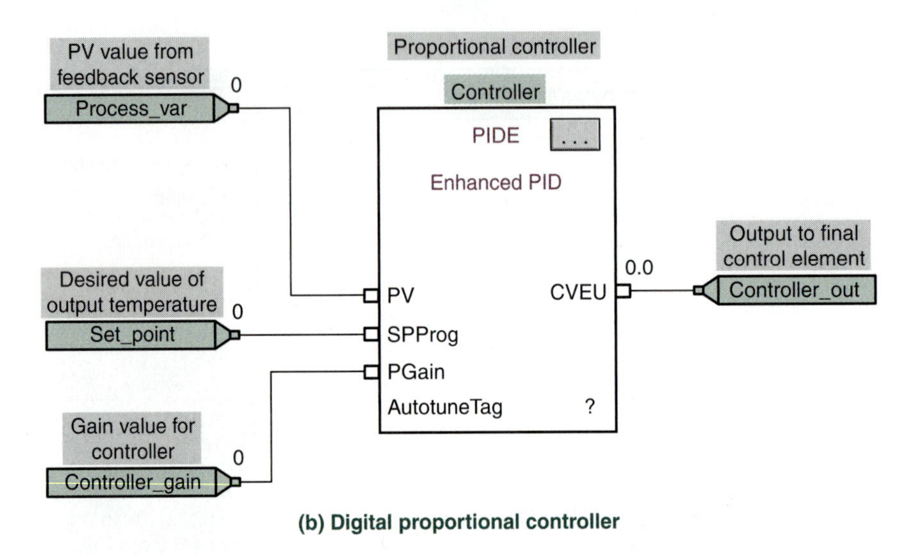

(b) Digital proportional controller

Rehg and Sartori, Industrial Electronics, *1st Edition, © 2006, Reprinted by permission of Pearson Education, Inc., Upper Saddle River, NJ.*

The representation used for a proportional controller in a block format is illustrated in Figure 14-11(a). Note that the proportional gain uses a K_p symbol. The Function Block Diagram PIDE instruction for a proportional controller in a ControlLogix PLC is illustrated in Figure 14-11(b). Note that the error summing junction is inside the controller, so inputs PV and SP are both present. The third input is proportional gain (K_p) and the only output is the controller output (CO) to drive the final control element. This is a simplified view of the PIDE instruction with the integral and derivative gain inputs not shown. Proportional gain can also be expressed as *proportional band* (PB). PB is defined as:

> *The percentage change in the error that causes the final control element to go through its full range.*

Therefore, PB is:

$$PB = \frac{100}{\text{Controller gain}}$$

14-4-2 Proportional Integral (PI) Control

Proportional control is effective for many processes; however, when steady-state error must be reduced to near zero this controller needs another term. Steady-state error is eliminated when *integral action* is added to the proportional action. Integral action is never used alone but is always used in combination with some level of proportional gain.

Let's take another look at the tank level control problem introduced in Figure 12-37 and developed further in Figures 12-39 through 12-43. The load change produced a steady-state error. The change in proportional gain helped to reduce the steady-state error but introduced too much overshoot and ringing after a disturbance or set point change. The balanced output for the tank controller after the initial disturbance, which was shown in Figure 12-39(c), has been redrawn in Figure 14-12 in order to give a closer look at the steady-state error solution. Remember that link A with the fulcrum acts as

FIGURE 14-12: Integral action in tank level control problem.

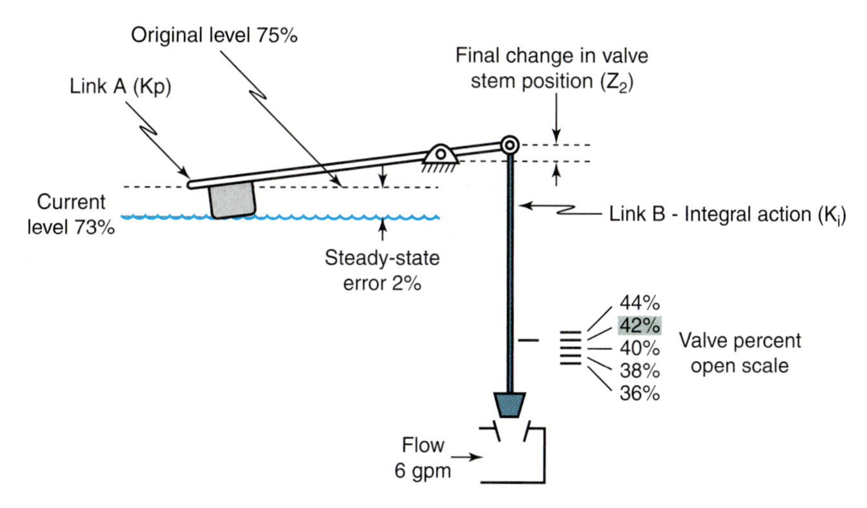

Rehg and Sartori, Industrial Electronics, *1st Edition, © 2006, Reprinted by permission of Pearson Education, Inc., Upper Saddle River, NJ.*

the proportional control element. To eliminate the steady-state error, the float and liquid in Figure 14-12 must be 2 percent higher; however, to correct for the load change the input valve must remain at an input flow of 6 gpm. To satisfy both of these requirements (tank level at 75 percent and the input value at 6 gpm) link B must shorten by the value Z_2 in the figure. Therefore, a link B that is automatically shortened by the value of the steady-state error is the integral controller element. Review this introduction to integral control so that the concept and requirements for the mechanical controller are clear.

The same integral action is implemented in digital controllers through program software. In each system the proportional gain corrects as much of the PV change from the set point or load change as possible, then the remaining steady-state error at the input to the integral controller is used to make the final correction.

The block format for an integral action element is illustrated in Figure 14-13(a). Notice that there are two distinct actions, the proportional and the integral, with the results added to get the final controller output. An FBD PIDE instruction representation for the ControlLogix PLC is illustrated in Figure 14-13(b). A comparison between Figure 14-11(b), the proportional controller, and Figure 14-13(b) indicates that the only change is the addition of the integral gain input.

References to integral action often use the terms *reset, repeat,* and *integral time.* When steady-state error and an integral element are present, the controller output continues to rise after the proportional action has ceased. The rate at which the output increases is determined by the reset value or reset action. The following relationships describe these terms:

Reset = number of repeats per minute or second

Integral time = number of minutes or seconds per repeat

Integral action quickly becomes abstract when you start to use these terms. Reset defines the rate at which the output of the controller is changing relative to the change produced by the proportional control. If the proportional control produced a 10 percent change in the PV after a load change, then the integral element will produce an additional 10 percent change in 1 minute if the reset is 1/min. If the reset was 2/min, then the output adds the additional 10 percent in 30 seconds or adds an additional 20 percent in 1 minute. The higher the reset value is the greater the effect of the integral action. *Integral time* is just the reciprocal of reset action.

The integral waveform is added to the load change plots in Figure 14-10 and illustrated in

FIGURE 14-13: PI digital controller in FBD format.

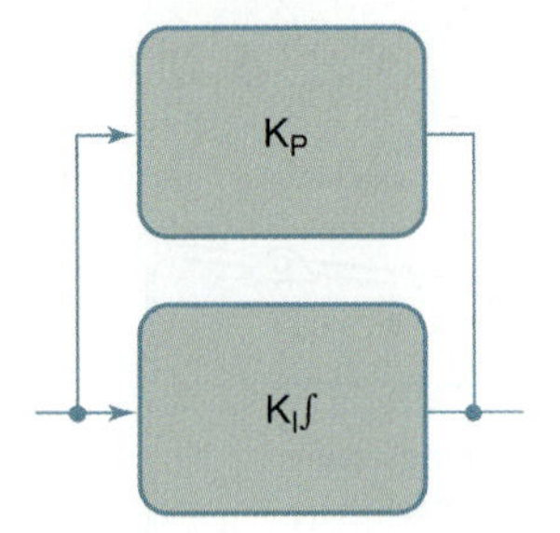

(a) Control system block diagram with an integral action element

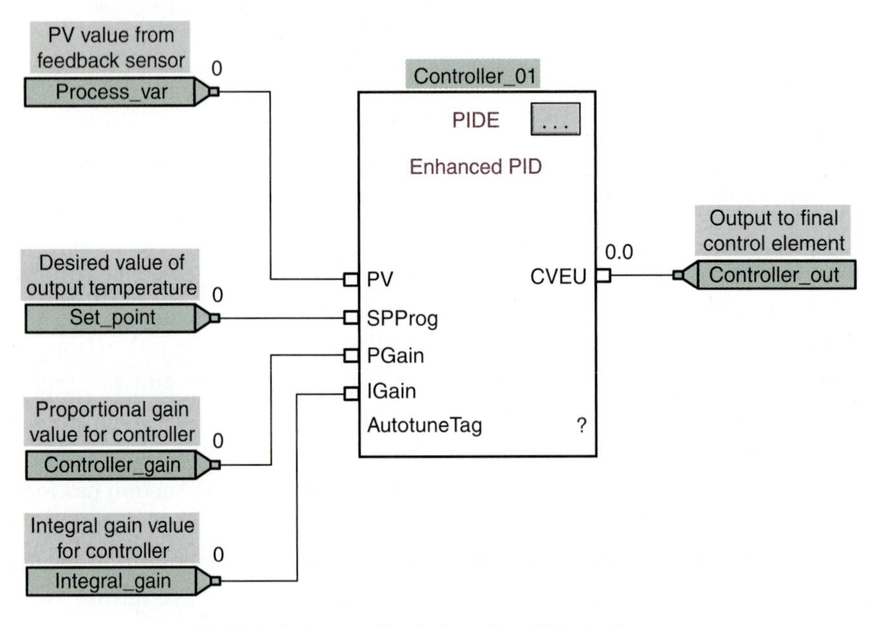

(b) PI digital controller in functional block diagram

Rehg and Sartori, Industrial Electronics, *1st Edition, © 2006, Reprinted by permission of Pearson Education, Inc., Upper Saddle River, NJ.*

the new Figure 14-14. Read the description of the actions in the figure before continuing.

Note the rate of change in the integral output and the effect it has on the steady-state error. The proportional output is superimposed on the integral response to show the effect of the reset action. Also, the integral response has a longer rise time, so the total response for PI usually is not as crisp as for the P response. As a result, integral action eliminates steady-state error but also reduces the quickness of the system response. The addition of derivative action adds quickness back into the system response.

14-4-3 Proportional Derivative (PD) Control

Sudden load or set point changes cause the process variable (PV) to deviate from the set point (SP), and the faster the rate of change of the disturbance, the farther the controlled variable moves away from the set point. Increasing the proportional gain will compensate, but this is likely to cause the PV to overshoot more severely and produce excessive ringing. In some process applications, this situation is undesirable. If the proportional gain is increased to minimize this condition, the controlled variable will likely overshoot and oscillate. The third control mode, *derivative*, is used to correct for these rapidly changing disturbances. The derivative function produces a response whose magnitude is a function of the rate of change of the PV relative to the SP. Therefore, if the error is constant, then the derivative action is zero.

Like the integral, the derivative is not used without some proportional gain present. The

FIGURE 14-14: Proportional plus integral action.

Controller waveform action	Description of action
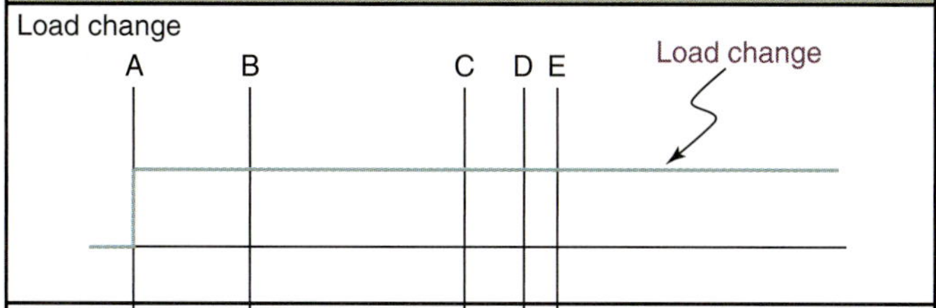 Load change A B C D E Load change	From a steady-state condition, a load change occurs at point A and remains.
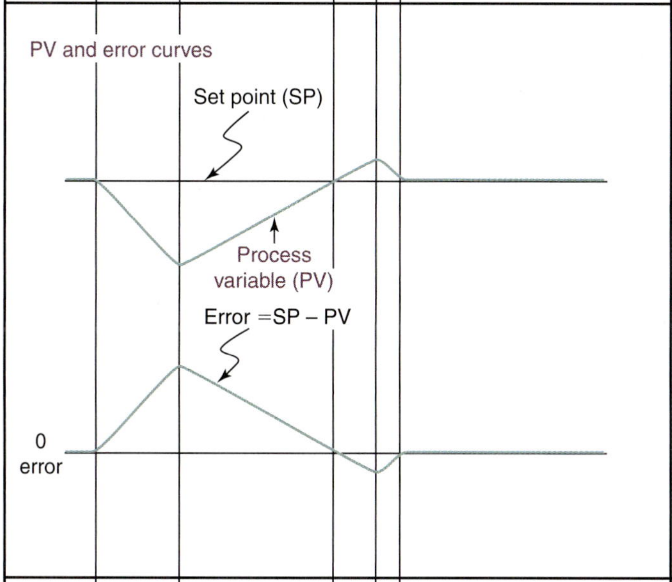 PV and error curves Set point (SP) Process variable (PV) Error =SP – PV 0 error	The PV starts at the SP value. In response to the increased load on the process, the process or measured variable (PV in Figure 12-41) drops below the set point. The error (SP – PV) changes from its initial state and has a shape that is the mirror of the PV waveform. The PV curve starts to recover at point B as the controller starts to correct for the load change. Note that the overcorrection occurs and the PV crosses the set point at point C before settling back to the original PV level.
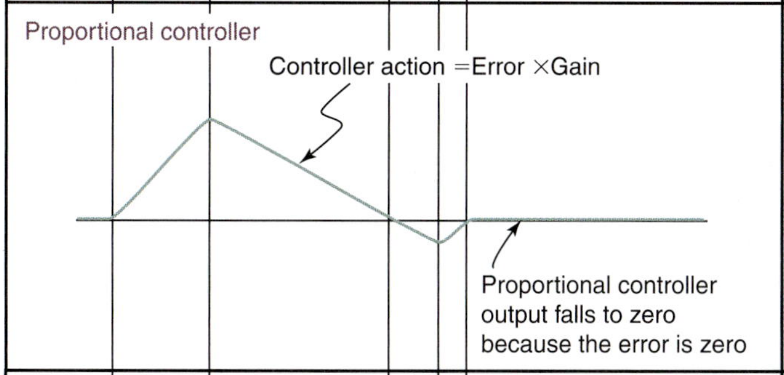 Proportional controller Controller action =Error ×Gain Proportional controller output falls to zero because the error is zero	The proportional or controller action follows the error signal to correct for the load change.
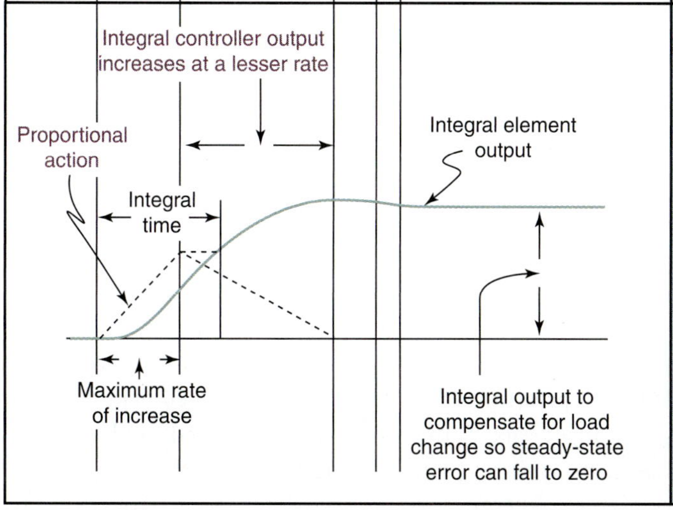 Integral controller output increases at a lesser rate Proportional action Integral time Integral element output Maximum rate of increase Integral output to compensate for load change so steady-state error can fall to zero	Note that steady-state error has been eliminated by the integral output. The rate of integral output is greatest between points A and B, when the error is increasing. It continues to rise between B and C but at a lesser rate because the error is decreasing due to the presence of the integral output. After point E, the integral level remains constant to compensate for the increased load on the system.

derivative element produces an output that combines with the proportional to move the final control element to a position to correct for the disturbance. For example, a large rapidly increasing load change causes a fast rise in the PV and subsequently in the error. The proportional mode element with a gain of 1 goes from 50 to 60 percent output, which matches the percentage change in the error. If only the proportional mode were present, the load change would significantly affect the PV because the proportional mode response is not sufficient to compensate. The derivative mode reacts to the rapid change in the PV by adding a derivative response of 30 additional percent to the controller output. The 90 percent controller output response quickly compensates for the large rapid load change.

The derivative mode parameter, called *rate time,* determines how much the derivative action changes the controller's output. Rate time compares the output duration for PD controllers versus P controllers. For example, a rate time setting of 1 minute will cause a response twice as fast as a setting at 2 minutes.

PD control is unsuitable for systems in noisy environments since noisy signals contain high-frequency components, which are amplified by the derivative action. These amplified signals will appear at the controller output and may cause unwanted changes by the final control element. Derivative control is beneficial for slow response systems that have large and rapid load changes. The derivative mode enables the controller to respond more rapidly and position the final control element more quickly than is possible with only proportional action.

A derivative element provides a short but large change in the magnitude of the controller output in response to a rapidly changing error signal. This prevents the PV from changing severely as the result of a large rapid change from a disturbance or the set point. Figure 14-15 shows that the maximum derivative effect occurs at the step change in the error value.

The block diagram for a derivative action element is illustrated in Figure 14-16(a). Representation using a PIDE instruction from ControlLogix is illustrated in Figure 14-16(b). A comparison

FIGURE 14-15: Derivative gain relationship to a step change in the error.

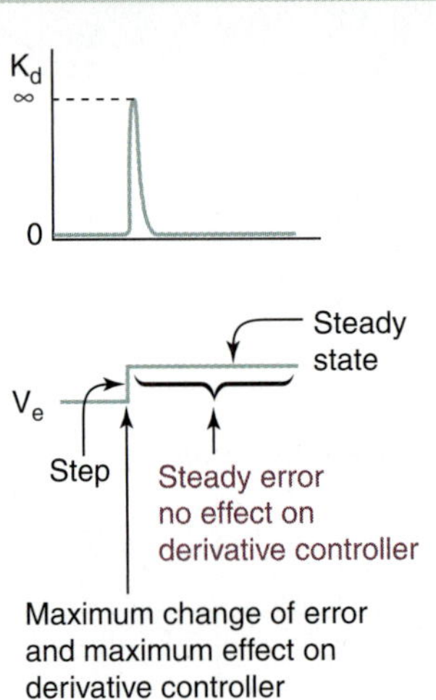

Rehg and Sartori, Industrial Electronics, *1st Edition, © 2006, Reprinted by permission of Pearson Education, Inc., Upper Saddle River, NJ.*

between Figure 14-11(b) and Figure 14-16(b) indicates that the only change is the addition of a derivative gain input to the pure proportional controller.

The derivative waveform is added to the load change plots in Figure 14-10 and illustrated in the new Figure 14-17. The magnitude of the derivative action is related to the slope of the error waveform. The slope or rate of change in the error from point A to B is twice the value from B to C. Therefore, the derivative action for A to B is twice that from B to C, as illustrated with the A and 2A notation in the figure. The response of the system is improved because the controller anticipates a large error and reacts to it. Derivative control cannot be used when a system has noise present since the derivative controller interprets the noise as rapid changes in the error. PD control is effective for changes in set point and load changes, especially those that are rapid and large.

The curves measured from the test system verify this result. In the last several sections control modes designated as proportional (P), proportional integral (PI), and proportional derivative (PD) have been described. The final and one of the most frequently used modes is proportional integral and derivative (PID).

FIGURE 14-16: PD digital controller in FBD format.

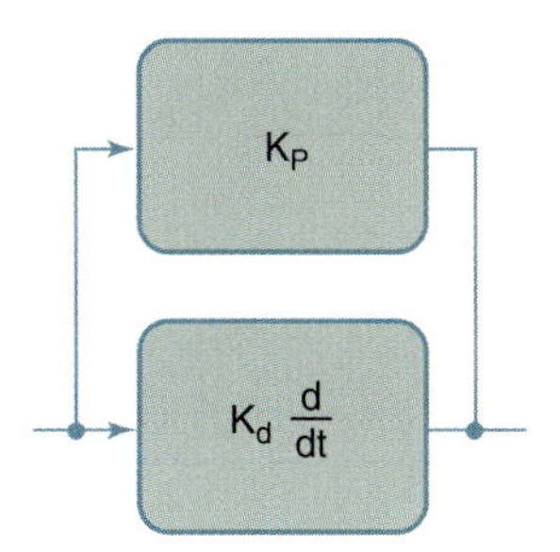

(a) Control system block diagram with a derivative action element

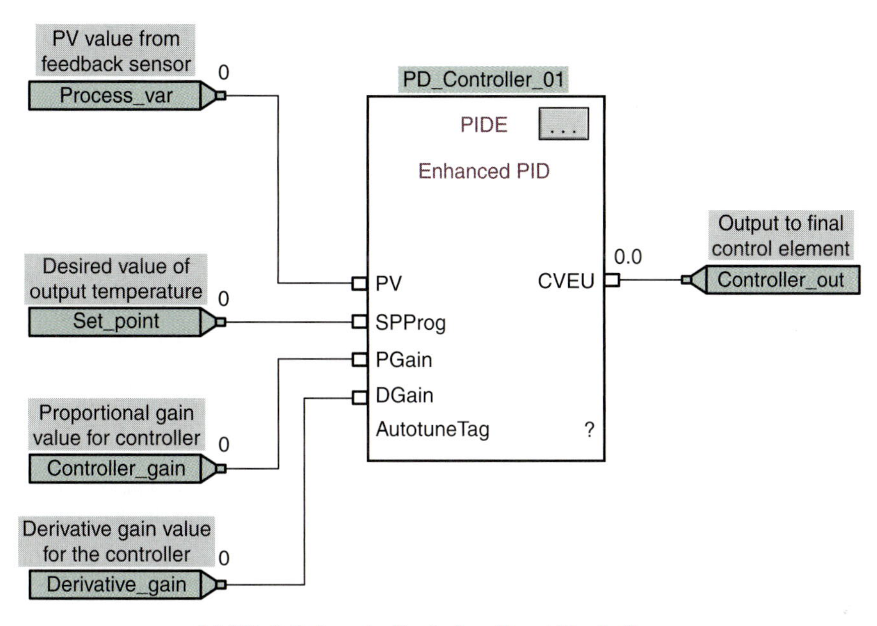

(b) PD digital controller in functional block diagram

Rehg and Sartori, Industrial Electronics, *1st Edition, © 2006, Reprinted by permission of Pearson Education, Inc., Upper Saddle River, NJ.*

14-4-4 Proportional Integral and Derivative (PID) Control

Proportional integral and derivative control is the most common mode used in process control. The PID controller combines all the benefits of each control mode with few of the limitations. When combined into a single control loop the modes complement each other extremely well. Figure 14-18 indicates the parameters that dictate which mode of control would be best for an application. Each control mode in the left column is evaluated versus the process parameters across the top. Note that only one control mode, PID, can tolerate any condition present in the process. Current digital controllers offer all the modes, so the cost to implement PID is no more than a proportional mode controller.

Digital controllers support the P, PI, PD, and PID modes, depending on what control parameters—proportional, integral, or derivative—have gain values greater than zero. A zero in a parameter removes it from the control algorithm. The digital controller implementation of the PID mode is illustrated in Figure 14-19, along with the block diagram symbols used when all three control modes are combined. The PIDE instruction has

FIGURE 14-17: Proportional plus derivative action.

Controller waveform action	Description of action
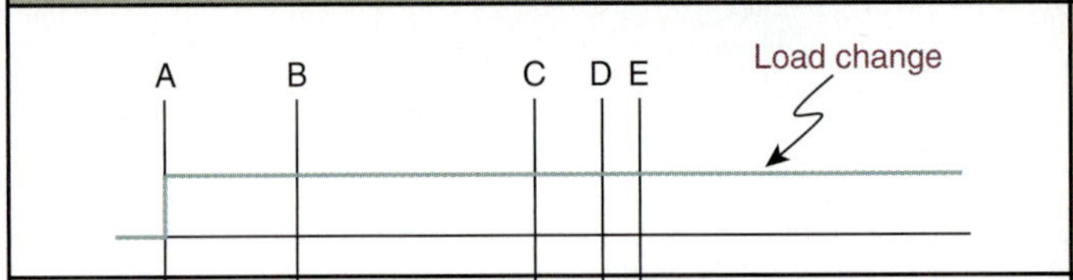 A B C D E Load change	From a steady-state condition, a load change occurs at point A and remains.
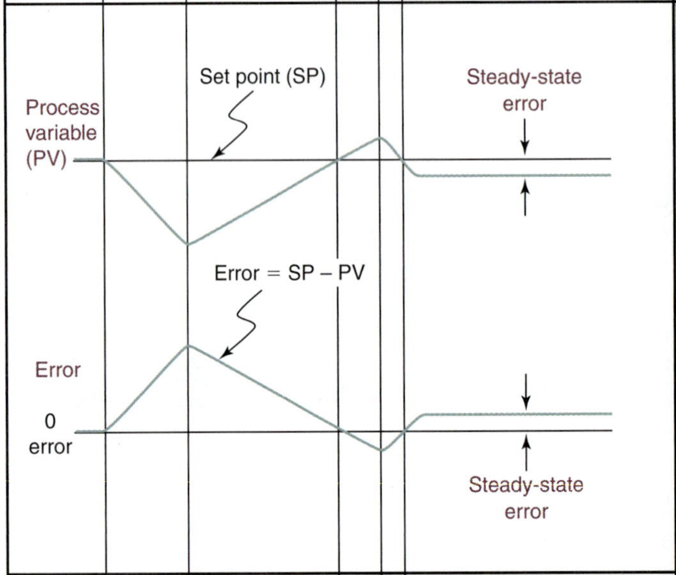 Process variable (PV) Set point (SP) Steady-state error Error Error = SP − PV 0 error Steady-state error	The PV starts at the SP value. In response to the increased load on the process, the process or measured variable (PV in Figure 12-41) drops below the set point. The error (SP − PV) changes from its initial state and has a shape that is the mirror of the PV waveform. The PV curve starts to recover at point B as the controller starts to correct for the load change. Note that the overcorrection occurs and the PV crosses the set point at point C before settling back to a steady-state level.
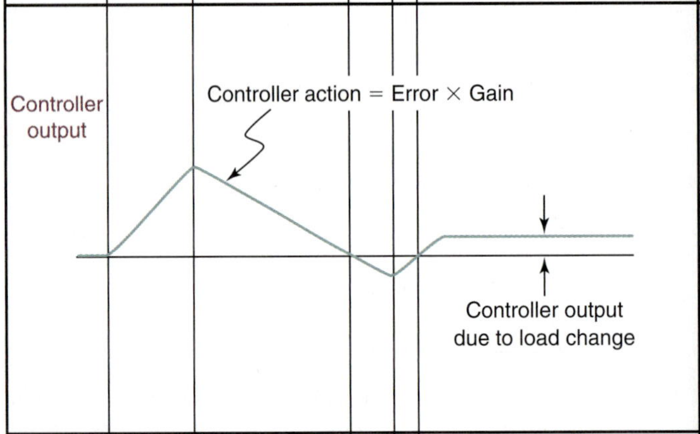 Controller output Controller action = Error × Gain Controller output due to load change	The proportional or controller action (Error × Gain) follows the error signal and corrects for the load change. The correction is not complete because some steady-state error remains to hold a controller output necessary to compensate for the new load. Note the gain is 1 for this example.
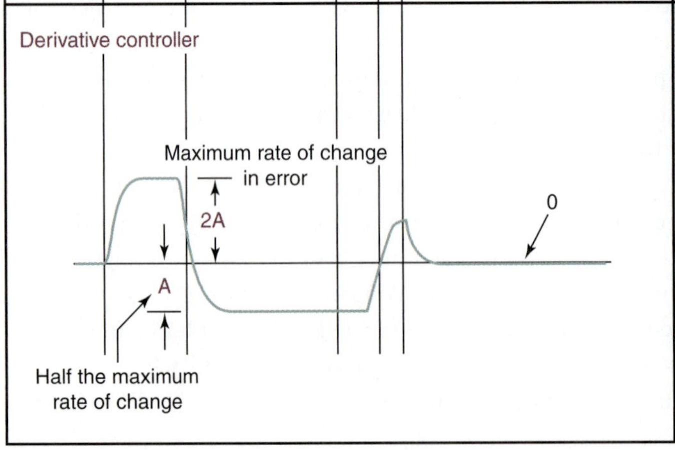 Derivative controller Maximum rate of change in error 2A 0 A Half the maximum rate of change	The digital action responds only to the rate of change in the error signal. Note that the derivative starts at 0 and returns to 0 when the error is not changing. The amount of derivative action between points A and B is twice that between B and C. This results from the fact that the error is changing twice as fast before B than it is changing afterward.

Rehg and Sartori, Industrial Electronics, *1st Edition, © 2006, Reprinted by permission of Pearson Education, Inc., Upper Saddle River, NJ.*

FIGURE 14-18: Control mode selection versus process parameter conditions.

Control mode	Process reaction delay tolerated	Transfer lag tolerated	Dead time tolerated	Size of load disturbance tolerated	Speed of load disturbance tolerated
On-off	Only long never short	Almost none	Almost none	Small	Slow
Proportional (P) only	Moderate or long (cannot be too short)	Moderate	Moderate	Small	Slow
Proportional plus integral (PI)	Any	Moderate	Moderate	Any	Slow
Proportional plus derivative (PD)	Moderate or long	Moderate	Moderate	Small	Any
Proportional plus integral plus derivative (PID)	Any	Any	Any	Any	Any

Rehg and Sartori, Industrial Electronics, *1st Edition, © 2006, Reprinted by permission of Pearson Education, Inc., Upper Saddle River, NJ.*

gain input tags for all three parameters. The full complement of waveforms for a PID controller that has a load change disturbance is illustrated in Figure 14-20. The final waveform shows what the controller output looks like when the process has a set point change or a load change. Study the PID controller output and compare it with the outputs from each mode individually.

14-4-5 Fuzzy Control

Like many topics in this chapter, fuzzy control can fill an entire course. However, this subsection provides you with some basic concepts in fuzzy control and some situations when you may want to consider using it. This type of control uses artificial intelligence to readjust process parameters and is implemented with fuzzy logic algorithms on a digital controller. Rather than using a formula to calculate an output, fuzzy control evaluates rules by following three basic steps:

- First is *fuzzification*. In this step the process errors are changed from continuous variables to linguistic variables such as positive large or negative small.
- Next is *base rules/inference engine*. In this step simple if-then-else rules are evaluated and outputs are inferred, such as if the temperature is cold, then the output is very low.
- Last is *defuzzification*. In this step the output is changed from a linguistic variable to a continuous variable such as a solenoid valve position.

Fuzzy logic controllers (FLCs) implement these three basic steps. Typically, an FLC is placed in front of a PID controller to adjust the set points sent to the PID controller for elimination of overshoot and for improvement of process response to disturbances. In most implementations FLCs are used with conventional PIDs, they do not replace them. Fuzzy control is frequently used in supervisory control problems such as traffic control, quality control, and transportation systems, and is beneficial for the following types of control problems:

- Systems that are complex and difficult or impossible to model.
- Systems that must be controlled by human experts.

FIGURE 14-19: PID digital controller in FBD format.

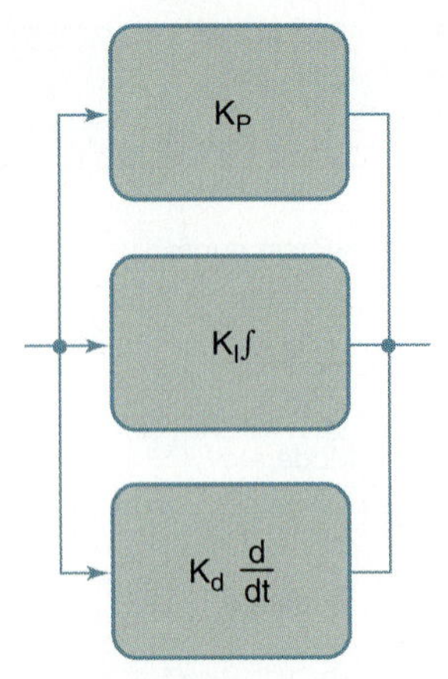

(a) Control system block diagram with all control elements

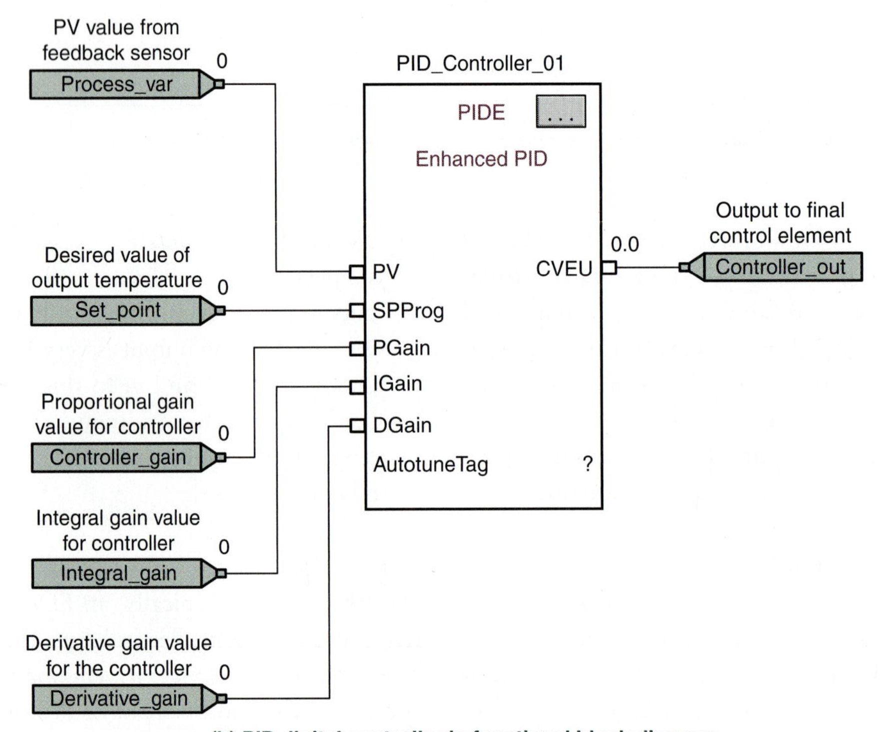

(b) PID digital controller in functional block diagram

- Systems driven by complex and continuous inputs and outputs.
- Systems where inputs or operational rules are based on human observation.
- Naturally vague systems such as in the behavioral and social sciences.

Advantages and Disadvantages of Fuzzy Control. Some of the advantages of fuzzy control include:

- A larger number of system variables can be evaluated.

FIGURE 14-20: PID action.

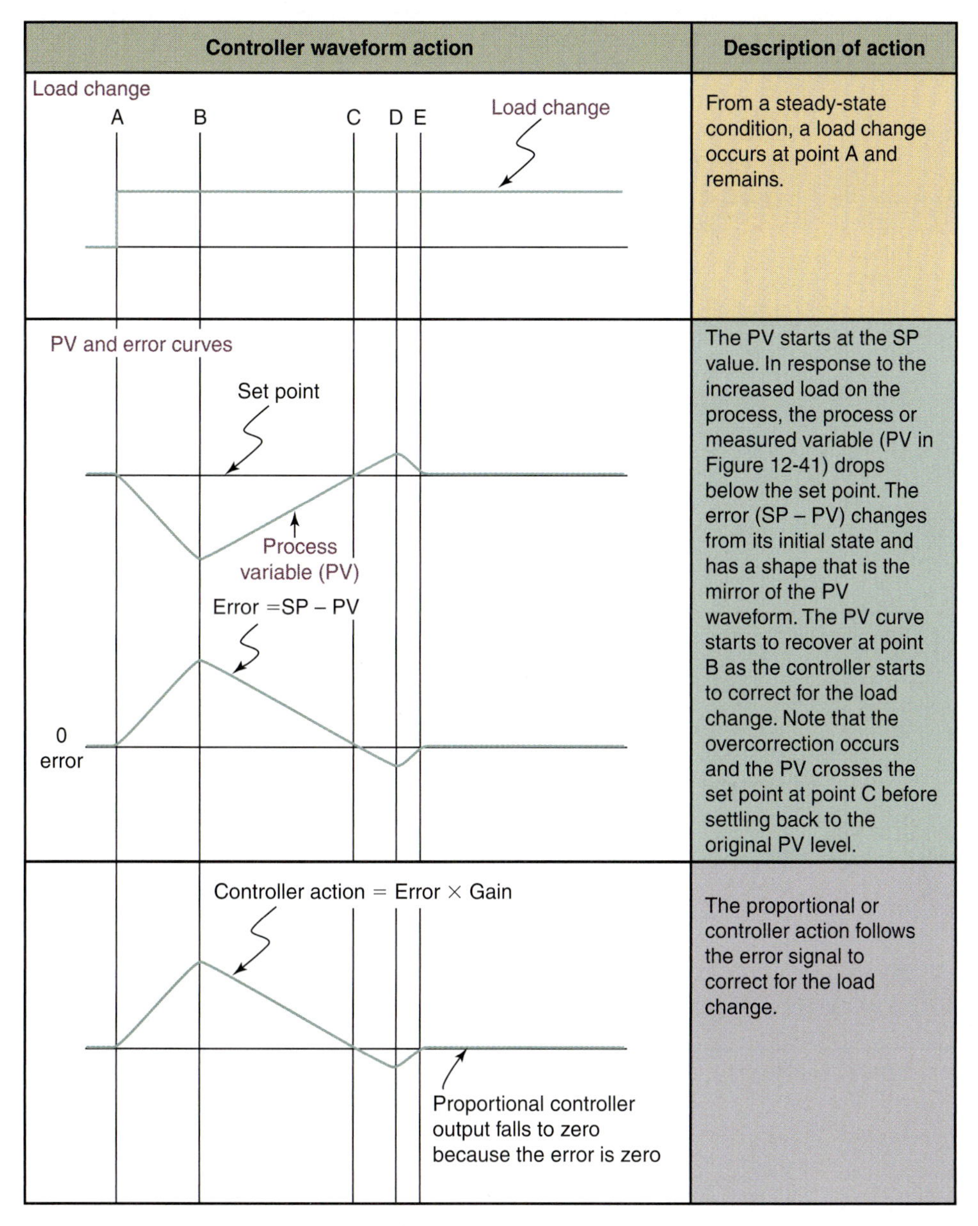

Rehg and Sartori, Industrial Electronics, *1st Edition, © 2006, Reprinted by permission of Pearson Education, Inc., Upper Saddle River, NJ.*

- It is similar to the way humans think because linguistic variables are used.
- It permits a greater degree of uncertainty as the output is related to the input, thus better control is possible for some types of systems.
- It can be used for the solution of previously unsolved problems.
- Systems come together faster because a system designer doesn't have to know everything about the system before starting work.
- Knowledge acquisition and representation is simplified.

The disadvantages are:

- Extracting a model from a fuzzy system is difficult.
- Fuzzy systems require more fine tuning before they're operational.

The availability of low-cost microcomputers has fueled the recent strong gains in the adoption of fuzzy control. However, the cultural bias in the United States for mathematically precise linear models and PID control makes adoption of fuzzy control difficult.

FIGURE 14-20: (Continued).

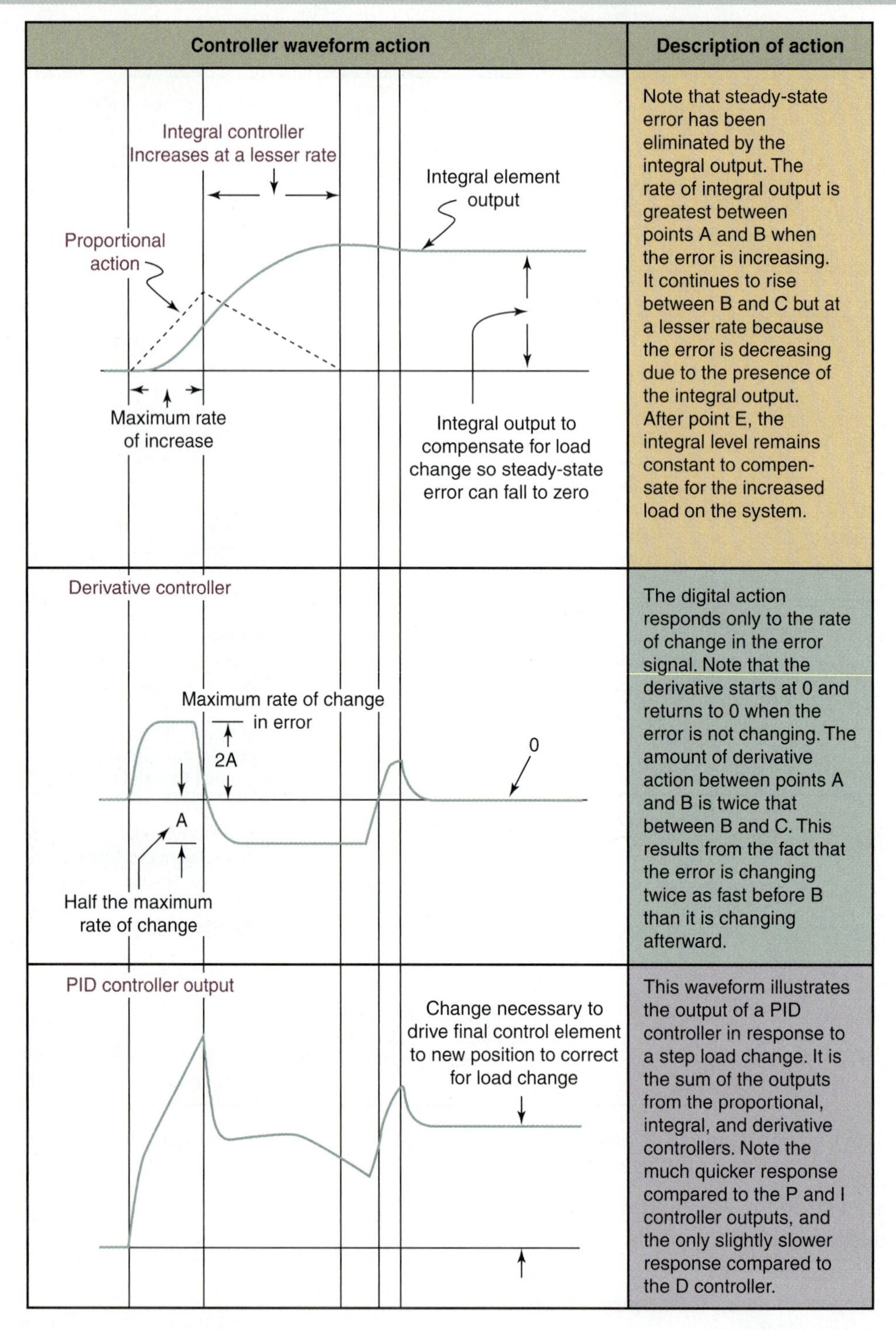

14-5 DIGITAL CONTROL

Let's now take a look at how digital control is used to implement the proportional, integral, and derivative process control modes. Digital controllers have an operational mode that starts from the traditional differential equation for the controller output (CO) but is better expressed as a summation of incremental changes. Both of these expressions for the CO are described in this section. The PID algorithm for the analog controller was given as:

$$CO = K_p e + K_1 \int edt + K_d \; de/dt$$

where CO is the controller output and the error value, e, is the set point minus the process variable (e = SP − PV). Analog controllers are implemented with operational amplifiers that are configured as proportional, integrators, and differentiators. The amplifiers change their output based on a change in the input at the speed or bandwidth of the amplifier. With the high speed present in operational amplifiers, the changes in the output appear to be instantaneous. The term *continuous* is often used to describe analog PID control because there appears to be no delay between the changes in the input and output. There is some very small delay, but the process component change is several magnitudes slower, so the delay is not an issue.

14-5-1 Digital Sample and Hold

A digital control system is not continuous in the sense of an analog amplifier since each action inside a digital system is paced by the speed of the system clock. The digital process control system also operates in this fashion. Figure 14-21 illustrates this concept by showing that the error at the input is read when S1 is closed and the controller output (CO) is changed some time later. The change in S1 and the calculation of the desired output that must occur are done in incremental time slices. The time for that process is called the *sampling* or *system update time*. The update or sampling interval, *T*, generates a series of error values that are available to the controller in the form of an error pulse train. The circuit inside the controller that holds the error values is called a sample and hold circuit. Once new error data is available, the controller calculates an output and makes that output value available until the next update time. The controller calculates and stores a new output only when new error data becomes available; hence, the controller output appears as a set of pulses that change at the same update rate.

FIGURE 14-21: Sampling or system update for a digital controller.

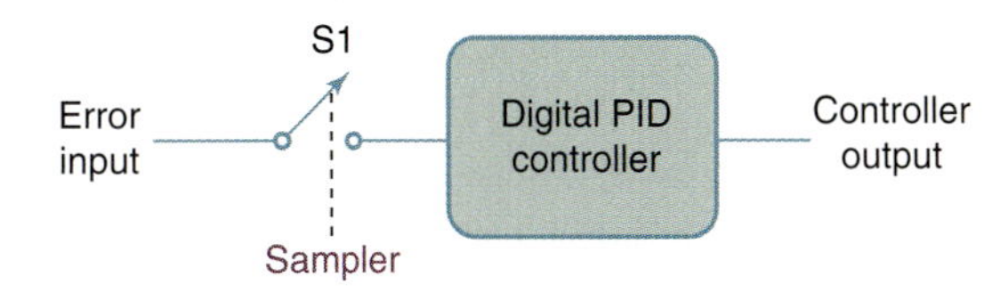

Rehg and Sartori, Industrial Electronics, *1st Edition, © 2006, Reprinted by permission of Pearson Education, Inc., Upper Saddle River, NJ.*

Figure 14-22 illustrates a changing error wave form, the error generated at the system update

FIGURE 14-22: Error input and controller output by update interval.

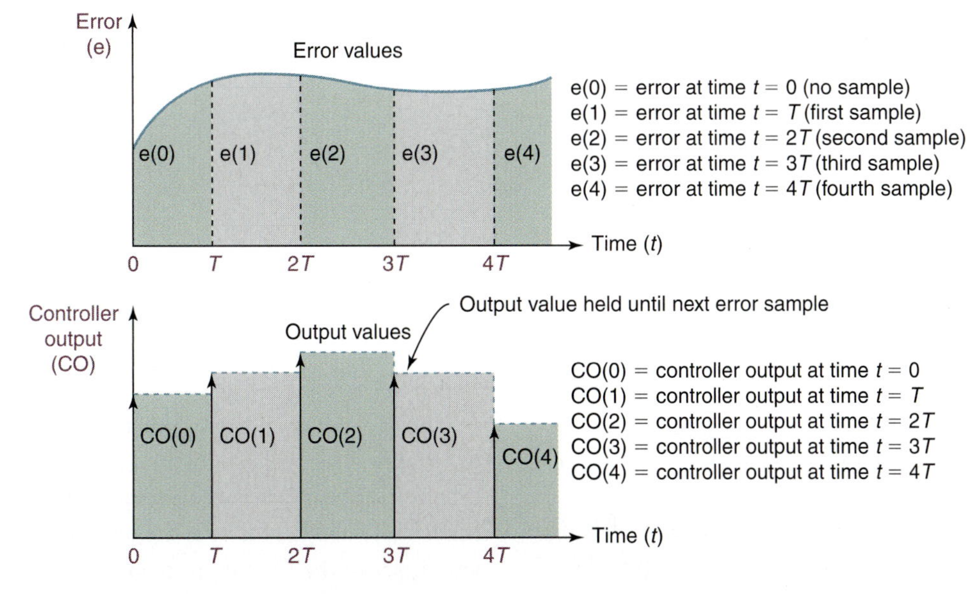

Rehg and Sartori, Industrial Electronics, *1st Edition, © 2006, Reprinted by permission of Pearson Education, Inc., Upper Saddle River, NJ.*

interval, and the corresponding controller output levels. Conventionally, an error pulse train is represented by an e with an n in parentheses, e(n). The n indicates that there are a series of values for the variable e. When the value is specific for one update interval, the n is replaced with the number of the update interval. For example, e(2) represents the error for the second update interval. Study the figure and review this discussion of update intervals until the operation of the digital controller is clear.

14-5-2 Proportional Control Mode

Proportional output is simply the product of current error and the proportional gain coefficient. The expression for the output of an analog controller is:

$$CO = K_p e$$

The analog expression is differentiated to indicate that the error is changing at an interval rate set by the update time. The result is:

$$dCO/dt = K_p \, de/dt$$

The dCO/dt and de/dt terms represent the change in the variable with respect to time. The digital control proportional algorithm says that CO (controller output) changes by a value of K_p (controller gain) times the change in e (input error). Thus the general expression for the digital controller output can be written as:

$$CO(n) = CO(n - 1) + K_p [e(n) - e(n - 1)]$$

where n represents the update interval at which the controller output value is desired, and n − 1 represents values for the previous update interval. Note that the controller output for the previous update interval must be added since the calculation for the current interval is only the change from the previous interval. So at the end of the second update interval (n = 1) or time equal to T in Figure 14-22 the controller output is:

$$CO(1) = CO(0) + K_p [e(1) - e(0)]$$

So the new controller output is the previous controller output plus the proportional gain times the change in the error from interval 0 to interval 1.

EXAMPLE 14-2

Use the algorithm for a digital proportional controller and the value (percent of full scale) of the set point change and the corresponding changes in the process variable for 15 time intervals to complete a table showing how the algorithm generates the controller output.

SOLUTION

The results of the calculations are given in the table in Figure 14-23. The graphs showing set point, process variable, error, and controller output are illustrated as well. Note that the actual controller output cannot change more than 100 percent, so the change goes off of the graph. This means that the controller output is at the 100 percent level.

Example calculation: $CO(5) = 210 = 280 + 7(30 - 40)$

14-5-3 Integral Control Mode

In the integral mode any error present changes the controller output based on the area under the error curve. The algorithm for the integral mode controller is:

$$CO = K_1 \int e \, dt$$

In this analog expression the term $\int e \, dt$ is the area under the error curve in Figure 14-24. Taking the derivative of the analog expressions results in:

$$dCO/dt = K_I e \, dt$$

The dCO/dt term represents the change in the CO value with respect to time and the dt term on the right side is the incremental change in time. This change in time is the update time interval, T, set in the digital controller and shown in

FIGURE 14-23: Solution for Example 14-2.

n—update interval	Set point	Process variable	Controller output CO(n)	K_p	Error e(n)	Previous error e(n − 1)	Previous controller output CO(n − 1)
0	0	0	0	7	0	0	0
1	50	2	336	7	48	0	0
2	50	5	315	7	45	48	336
3	50	10	280	7	40	45	315
4	50	15	245	7	35	40	280
5	50	20	210	7	30	35	245
6	50	25	175	7	25	30	210
7	50	30	140	7	20	25	175
8	50	35	105	7	15	20	140
9	50	39	77	7	11	15	105
10	50	42	56	7	8	11	77
11	50	44	42	7	6	8	56
12	50	45	35	7	5	6	42
13	50	45	35	7	5	5	35
14	50	45	35	7	5	5	35

(a) Steady-state error = 5; final output = 35

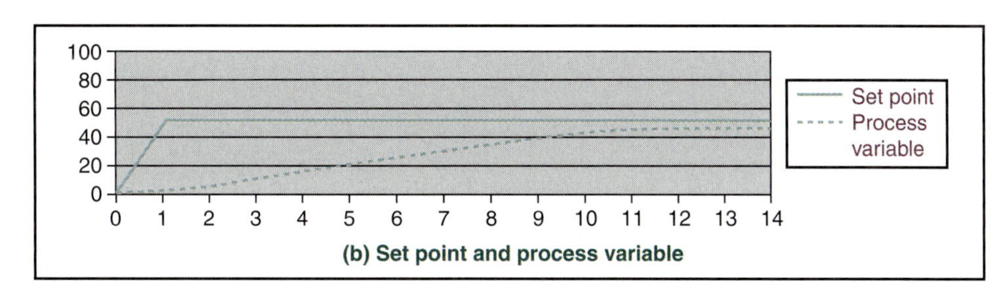

(b) Set point and process variable

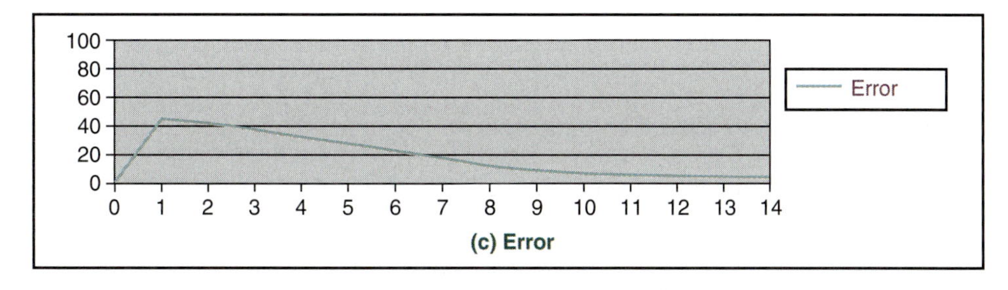

(c) Error

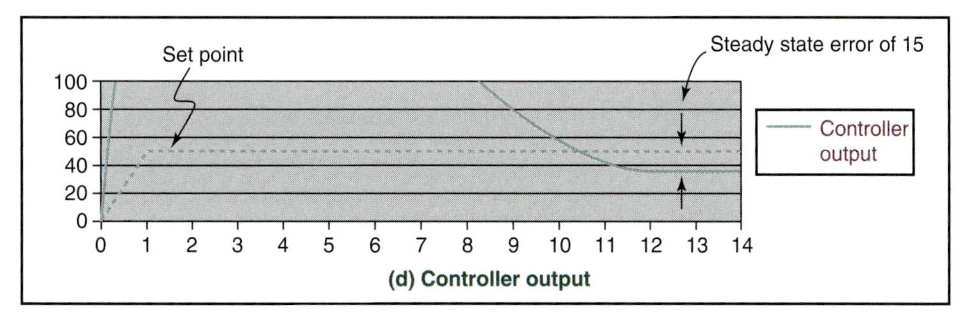

(d) Controller output

FIGURE 14-24: Integral and derivative contributions to digital controller output.

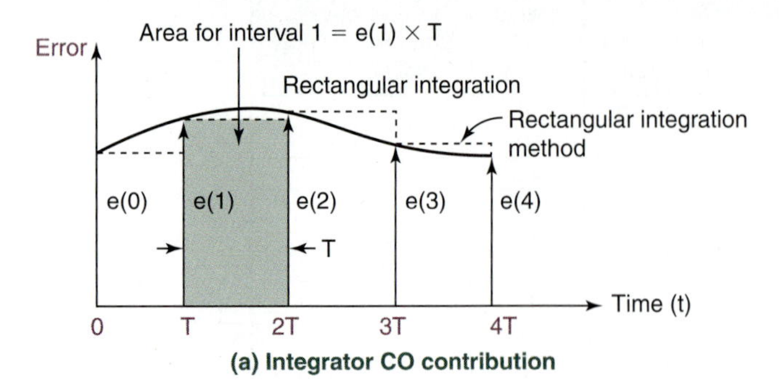

(a) Integrator CO contribution

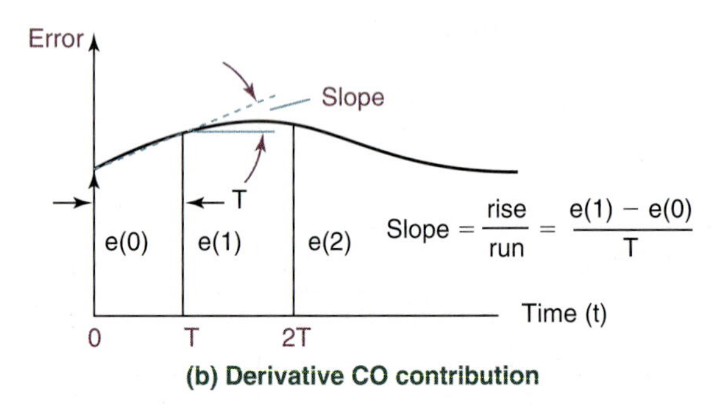

(b) Derivative CO contribution

Rehg and Sartori, Industrial Electronics, *1st Edition, © 2006, Reprinted by permission of Pearson Education, Inc., Upper Saddle River, NJ.*

Figure 14-24. Therefore, the equation can be simplified and converted to single interval as:

$$CO(n) = K_I e(n)\,T$$

The term on the right is the area under the error curve (e value × time duration) times the integral gain, K_I. So the digital control integral algorithm says that CO (controller output) changes by that term on the right as long as an error is present. The value of the integral contribution to the controller output changes at every update interval as indicated in Figure 14-24(a).

The algorithm for a PI digital controller is just the combination of the algorithms for each type individually. The combined terms yield:

$$CO(n) = CO(n-1) + K_P[e(n) - e(n-1)] + K_I e(n)\,T$$

Using this expression, a PI controller can be constructed using any digital controller that has addition, subtraction, and multiplication functions available.

14-5-4 Derivative Control Mode

In the derivative mode any change in the error causes a change in the controller output based on the rate at which the error is changing. The algorithm for the derivative mode controller is:

$$CO = K_D\,de/dt$$

The de/dt term represents the instantaneous slope of the error curve at that point in time. Since it represents the slope it is an indication of the direction in which the error is moving and the rate at which it is changing. The derivative control gives some CO action based on the *tendency* of the error, so a corrective action can be taken even before error has reached a new level. This seems to provide a sort of intelligence to the controller—hence the name *anticipatory control.* The algorithm for the digital controller is found by taking the derivative of the analog CO expression, which results in:

$$dCO/dt = K_D\,d(de)/dt$$

The dCO/dt term represents the change in the CO value with respect to time, and the d(de)/dt term is the rate of change in the error or slope of the error curve over some time interval. The time interval is the PLC update time interval, T, shown in Figure 14-24(b). Therefore, the equation can be simplified and converted to a single interval as:

$$CO(n) = K_D\left[\frac{e(n) - e(n-1)}{T}\right]$$

The bracketed term on the right is the slope from the previous interval to the present and the K_D term is the gain for the derivative component.

The algorithm for a PID digital controller is just the combination of the algorithms for each type individually. The combined terms yield:

$$CO(n) = CO(n-1) + K_p[e(n) - e(n-1)] + K_I e(n)\,T + K_D\left[\frac{e(n) - e(n-1)}{T}\right]$$

Using this expression, a PID controller can be constructed using any digital controller that has addition, subtraction, and multiplication functions available. However, a better delta e estimator is necessary to help smooth out the contribution of the derivative term to the controller output.

Improved Delta e Estimator. Estimator theory offers a number of choices, but a complete discussion is beyond the scope of this chapter. However, an improvement can be added by increasing the number of intervals over which the slope is calculated. A delta e estimator considered is:

$$\Delta e = [e(n) - e(n-1)] - [e(n-1) - e(n-2)]$$

Adding this estimator to the PID digital algorithms results in:

$$CO(n) = CO(n-1) + K_p[e(n) - e(n-1)] + K_I e(n)T + \frac{K_D}{T}[e(n) - 2e(n-1) + e(n-2)]$$

where n is the number of the interval being evaluated, T is the sampling period or loop update time, K is the gain of the respective modes, CO is the controller output, and e is the error, or SP − PV. This form of the equation is often called the textbook PID. However, process engineers frequently make changes to get a better control result. The change made most often is the removal of the set point from the derivative term in the equation. Set point is removed because the derivative action is not necessary for a set point change. In the equation, e equals SP − PV and Δe equals SP − Δ PV. So in situations where the change in error is used it can be replaced with the change in PV since the SP is fixed and does not change. Making that change the PID algorithm becomes:

$$CO(n) = CO(n-1) + K_p[e(n) - e(n-1)] + K_I e(n)T + \frac{K_D}{T}[PV(n) - 2PV(n-1) + PV(n-2)]$$

where PV is the process variable. This is the form of the equation used by many industrial controllers such as the Allen-Bradley ControlLogix System PLC. In some control problems the error term for the proportional part of the equation is changed to PV because SP is removed there as well.

14-6 SCALING IN PROCESS CONTROL

In process control, process input variables typically need scaling, which is the resizing of a signal to meet the requirements of the using system component. Commonly, the input signal to be scaled is from an analog sensor and the scaled value is used by a PID controller as a process variable. The scale instruction (SCL) converts an unscaled input value to a floating point value and is illustrated in Figure 14-25. For example, let's say that the input source is a 12-bit analog input module, which is reading a liquid flow of 0 to 100 gallons per minute (gpm). You typically do not scale the module from 0 to 100 because that limits the resolution of the 12-bit input value (0 to 4095). Instead, you use the SCL instruction and configure the module to input an unscaled (0 to 4095) value, which the SCL instruction converts to 0 to 100 gpm (floating point) without loss of resolution.

The SCL instruction uses the following algorithm to convert unscaled input into a scaled value.

$$Out = (In - InMin) \times [(SclMax - SclMin)/(InMax - InMin)] + SclMin$$

where: InMin and InMax are the minimum and maximum unscaled input values, respectively; and SclMin and SclMax are the minimum and maximum scaled values, respectively.

In simpler terms, the output of the SCL instruction is the input offset times the scaling rate (the terms within the brackets) plus the scaled minimum value.

FIGURE 14-25: Scale instruction.

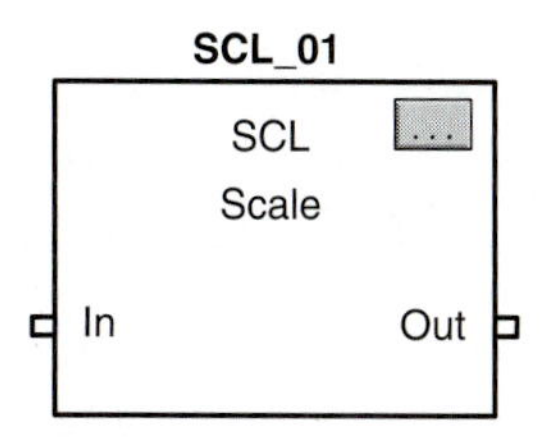

14-7 MANUAL CONTROL MODE AND BUMPLESS TRANSFER

Most industrial process controllers have a *manual control* mode. Operators can switch between automatic and manual mode as illustrated in Figure 14-26. In the manual mode the controller output is manually entered into the controller by the operator and the process final control element responds accordingly. Values for P, I, and D gains are entered, and the PV value is monitored. When the controlled variable reaches a value close to the set point, the process is switched to automatic by the operator. If the PV and SP are close in value the process switches to the automatic mode with the controller taking over control with little or no disturbance in the controlled variable. If the controller has a *bumpless transfer* capability, then the switch can be made with the PV and SP separated by a greater value, and the controller will bring the PV and SP close before switching to automatic.

14-8 LOCATION OF THE INSTRUCTIONS

The location of instructions from this chapter in the Allen-Bradley programming software for Function Block Diagrams is indicated in Figure 14-27.

FIGURE 14-26: Manual control mode.

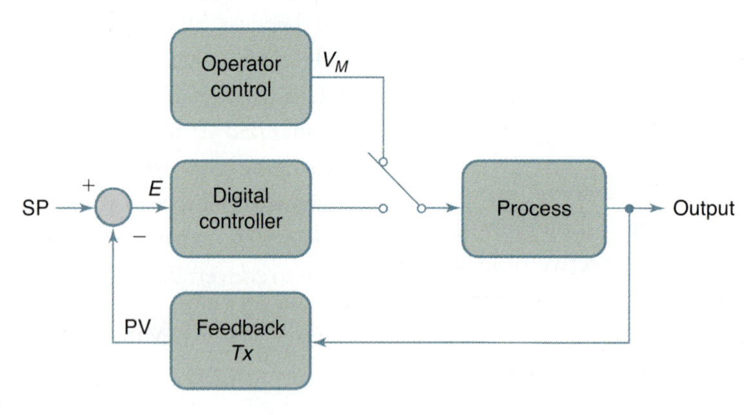

Rehg and Sartori, Industrial Electronics, *1st Edition, © 2006, Reprinted by permission of Pearson Education, Inc., Upper Saddle River, NJ.*

FIGURE 14-27: Location of instructions described in this chapter.

Systems	Instructions	Location
LOGIX	PIDE, SCL	Language Element: ALM, SCL, PIDE, RMPS, POSP, SRTP, LDLG, FGEN, TOT; Favorites, Process, Drives, Filters, Select/Limit
LOGIX	SSUM	Language Element: SEL, ESEL, SSUM, SNEG, MUX; Drives, Filters, Select/Limit, Statistical, Bit, Timer

Chapter 14 questions and problems begin on page 589.

CHAPTER 15

PLC Standard IEC 61131-3—Structured Text Language

15-1 GOALS AND OBJECTIVES

The goals of this chapter are to provide an overview of IEC 61131-3 text languages and to provide the Allen-Bradley implementation of the IEC 61131 Structured Text language.

After completing this chapter you should be able to:

- Describe the difference between the Instruction List and Structured Text languages.
- Define the components of the Allen-Bradley Structured Text language.
- Describe the arithmetic, logical, and relational operators.
- Describe the various constructs of the Structured Text language.
- Create a Structured Text program for automation process problems using the Allen-Bradley implementation.

15-2 OVERVIEW OF IEC 61131-3 TEXT LANGUAGES

There are two text-based languages (Instruction List and Structured Text) that are used to solve automation process problems. Structured Text language is discussed in this chapter. However, let's start with a general description of both languages.

- **Instruction List (IL):** A low-level, text-based language that uses mnemonic instructions like those used in machine code in microprocessors. Consequently, IL is a powerful language, but it has many disadvantages such as a cryptic nature. It is not a good fit for most applications, and as a result it is rarely used. IL should be used only as a last resort. Allen-Bradley does not support this language because of its marginal value in automated process applications.
- **Structured Text (ST):** A high-level, text-based language, such as BASIC, C, or PASCAL, with commands that support a highly structured program development and the ability to evaluate complex mathematical expressions.

15-3 ALLEN-BRADLEY IEC 61131 STRUCTURED TEXT IMPLEMENTATION

The Allen-Bradley version of ST is compliant with the 61131 standard. ST is not case sensitive and uses statements to define what to execute.

The Allen-Bradley ST implements the following components:

- **Assignment:** a statement that assigns values to tags
- **Expression:** part of an assignment or construct statement that evaluates a number or a true-false state
- **Instruction:** a stand-alone statement that contains operands
- **Construct:** a conditional statement used to trigger Structured Text code
- **Comment:** text that explains or clarifies what a section of Structured Text does

15-4 STRUCTURE TEXT PROGRAMMING

The Structured Text language in 61131 is similar to Pascal or other structured programming languages; however, ST has been specifically developed for industrial control applications. The program executes or is scanned on a line-by-line basis, starting at the first line in the program and continuing until the last line is executed. The statements or command syntax are sufficiently robust to avoid jump instructions (GOTO) that are used in some other languages, such as BASIC, to redirect the flow of execution.

ST language is rather straightforward to learn and use, and ST statements are written in a free style where tabs, returns, and comments can be inserted throughout the program. It is particularly useful for complex arithmetic calculations. The ST language is presented using the structure in Allen-Bradley's RSLogix 5000 software as an example.

15-4-1 Assignment Statements

Assignment statements are used to change the value stored within a variable or tag. An assignment has the following syntax:

Tag := expression ;

where:

Component	Description
tag	represents the tag that is getting the new value; the tag must be BOOL, SINT, INT, DINT, or REAL
:=	is the assignment symbol
expression	represents the new value to assign to the tag
If *tag* is this data type:	**Use this type of expression:**
BOOL	BOOL expression
SINT	
INT	
DINT	
REAL	numeric expression
;	ends the assignment

The **tag** retains the assigned value until another assignment changes the value.

The expression can be simple, such as an immediate value like an integer or another tag name, or the expression can be complex and include several mathematical operators and/or functions.

15-4-2 Expressions

Expressions are a part of every assignment statement and are a tag name, equation, or comparison. Expressions are written using the following:

- Tag names that store the value (variable)
- Numbers that you enter directly into the expression (immediate values)

- Functions, such as ABS, TRUNC
- Operators, such as +, −, $<$, $>$, AND, OR

The general rules for expressions include:

- Use any combination of uppercase and lowercase letters.
- For more complex requirements, use parentheses to group expressions within expressions. This makes the expression easier to read and ensures that the expression executes in the desired sequence.

In Structured Text, you use two types of expressions:

- **BOOL expression:** A Boolean expression produces either the Boolean value of 1 (true) or 0 (not true, or false).
  - A Boolean expression uses Boolean tags, relational operators, and logical operators to compare values or check whether conditions are true or false. The Boolean expression tag $>$ 65 is an example.
  - A simple Boolean expression can be a single Boolean tag.
  - Typically, you use Boolean expressions to condition the execution of other logic.
- **Numeric expression:** A numeric expression calculates an integer or floating point value.
  - A numeric expression uses arithmetic operators, arithmetic functions, and bitwise operators. The numeric expression tag1 + 5 is an example.
  - Often, you nest a numeric expression within a Boolean expression. The expression (tag1 + 5) $>$ 65 is an example.

15-4-3 Operators and Functions

Assignment statements also use operators and functions in the development of expressions. You can combine multiple operators and functions in arithmetic expressions to calculate new values. The following arithmetic operators are used in ST.

Arithmetic Operation	Operator	Optimal Data Type
Add	+	DINT, REAL
Subtract/negate	+	DINT, REAL
Multiply	*	DINT, REAL
Exponent (x to the power of y)	**	DINT, REAL
Divide	/	DINT, REAL
Modulo-divide	MOD	DINT, REAL

Arithmetic functions perform a specific mathematical operation on an integer or real constant, a non-Boolean tag, or an expression. The functions supported by ST include:

Function Names	Function Syntax	Optimal Data Type
absolute value	ABS (numeric_expression)	DINT, REAL
arc cosine	ACOS (numeric_expression)	REAL
arc sine	ASIN (numeric_expression)	REAL
arc tangent	ATAN (numeric_expression)	REAL
cosine	COS (numeric_expression)	REAL
radians to degrees	DEG (numeric_expression)	DINT, REAL
natural log	LN (numeric_expression)	REAL
log base 10	LOG (numeric_expression)	REAL

Function Names	Function Syntax	Optimal Data Type
degrees to radians	RAD (numeric_expression)	DINT, REAL
sine	SIN (numeric_expression)	REAL
square root	SQRT (numeric_expression)	DINT, REAL
tangent	TAN (numeric_expression)	REAL
truncate	TRUNC (numeric_expression)	DINT, REAL

15-4-4 Relational Operators

Relational operators, used in expressions, compare two values or strings to provide a true or not true (false) result. The result of a relational operation is a Boolean value. If the result of the comparison is true, then the result is 1; if it is not true, or false, then the result is 0. Relational operators supported by ST include:

Comparison Name	Operator	Optimal Data Type
equal	=	DINT, REAL, string
less than	<	DINT, REAL, string
less than or equal	<=	DINT, REAL, string
greater than	>	DINT, REAL, string
greater than or equal	>=	DINT, REAL, string
not equal	<>	DINT, REAL, string

15-4-5 Logical Operators and Bitwise Operators

Logical operators are used in expressions to perform logical combinations on single bits or Boolean tags, whereas *bitwise operators* perform bitwise logical combinations on bits of DINT tags. If the result tag in the assignment statement is a Boolean data type, then logical operators are used, but if the data type is DINT, then bitwise operators are used. The result of a logical operation is a Boolean data type and the result of a bitwise operation is a DINT data type. Logical and bitwise operators are as follows:

Operator Name	Operator	Logical Operation Data Type	Bitwise Operation Data Type
Logical/bitwise AND	&, AND	BOOL	DINT
Logical/bitwise OR	OR	BOOL	DINT
Logical/bitwise exclusive OR	XOR	BOOL	DINT
Logical/bitwise complement	NOT	BOOL	DINT

The first six examples in the following table illustrate logical operations on Boolean data types and tags. The seventh example is a bitwise operation on DINT data types and tags with a DINT result.

Example	Problem Statement	Required Expression
1	If *fluid_temp* is a BOOL tag and your specification says: "If *fluid_temp_1* is on then . . . "	IF fluid_temp THEN . . .
2	If *fluid_temp* is a BOOL tag and your specification says: "If *fluid_temp* is off then . . . "	IF NOT fluid_temp THEN...
3	If *fluid_temp* is a BOOL tag, *temp* is a DINT tag, and your specification says: "If *fluid_temp* is on and temp is less than 100° then . . . "	IF fluid_temp & (temp <100) THEN...
4	If fluid_temp is a BOOL tag, temp is a DINT tag, and your specification says: "If fluid_temp is on or temp is less than 100° then . . . "	IF fluid_temp OR (temp<100) THEN . . .
5	If *fluid_temp1* and *fluid_temp2* *are BOOL tags and your specification says: "If:* • *fluid_temp1* is on while *fluid_temp2* is off or • fluid_temp1 is off while fluid_temp2 is on then . . . "	*IF fluid_temp1 XOR fluid_temp2 THEN . . .*
6	If fluid_temp1 and *fluid_temp2* are BOOL tags, *open* is a BOOL tag, and your specification says: "If *fluid_temp1* and *fluid_temp2* are both on, set *open* to true."	open := fluid_temp1 & fluid_temp2;
7	If *input1*, *input2*, and *result1* are DINT tags and your specification says: "Calculate the bitwise result of *input1* and *input2*. Store the result in *result1*."	result1 := input1 AND input2;

Note in examples 3 and 4 that a tag or variable was a DINT data type, but it was used in an expression that produced a Boolean result.

15-4-6 Constructs

Constructs are critical building blocks for program development. The structure provided by these elements permits linear program flow from top to bottom without the need for jump instructions. Constructs supported by ST include *IF . . . THEN, FOR . . . DO, WHILE . . . DO, REPEAT . . . UNTIL*, and *CASE . . . OF.* The following table indicates what programming conditions dictate the selection of a specific construct.

If You Want to	Use This Construct
Do something if or when specific conditions occur	IF . . . THEN
Do something a specific number of times before doing anything else	FOR . . . DO
Keep doing something as long as certain conditions are true	WHILE . . . DO
Keep doing something until a condition is true	REPEAT . . . UNTIL
Select what to do based on a numerical value	CASE . . . OF

There are usually a number of variations for the constructs and each has a required syntax that must be followed. Each is addressed in detail in the following sections.

IF . . . THEN. The syntax for the IF . . . THEN construct is:

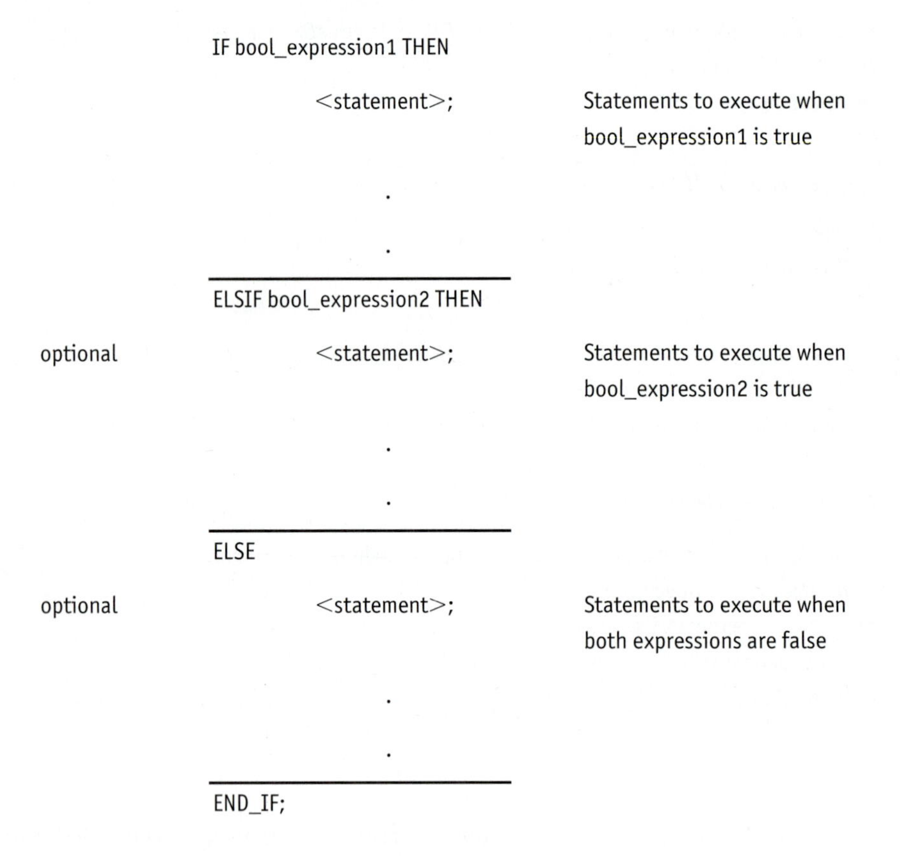

To use ELSIF or ELSE, follow these guidelines:

1. If you have several possible conditions to be evaluated, add one or more ELSIF statements as follows:
   - Each ELSIF represents an alternative path.
   - Specify as many ELSIF paths as you need.
   - The controller executes the first true IF or ELSIF and skips the rest of the ELSIFs and the ELSE.
2. To do something when the entire IF or ELSIF conditions are false, add an ELSE statement.

EXAMPLE 15-1

Turn off a conveyor direction light if the conveyor is moving in the forward direction. Keep the light on if it is not moving in the forward direction. Develop an ST program to solve this automation process problem.

SOLUTION

```
IF conveyor_direction THEN
     light := 0;
ELSE
     Light [:=] 1;
(* The [:=] tells the controller to clear light whenever the controller enters the RUN mode. *)
END_IF
```

Comments beyond a single line are indicated by bracketing them as follows (* comment *). Note the comment added in the previous code.

EXAMPLE 15-2

If the tank temperature is greater than 100 run the pump slow, but if the temperature is over 200 set the pump speed to high. Under all other conditions turn the pump off. Develop an ST program to solve this automation process problem.

SOLUTION

```
IF tank.temp > 200 THEN
     pump.fast :=1; pump.slow :=0; pump.off :=0;
ELSIF tank.temp > 100 THEN
     pump.fast :=0; pump.slow :=1; pump.off :=0;
ELSE
     pump.fast :=0; pump.slow :=0; pump.off :=1;
END_IF
```

In both examples the ELSIF and ELSE are optional and can be eliminated if the problem dictates.

FOR . . . DO. The FOR . . . DO construct performs an operation a specified number of times. The syntax is:

	FOR count := initial_value	
	TO final_value	
optional	BY increment	If you don't specify an increment, the loop increments by 1.
	DO	
	<statement>;	The statement you want to execute a number of times.

	IF bool_expression THEN	If there are conditions when you want to exit the loop early, use other statements, such as an IF . . . THEN construct, to condition an EXIT statement.
optional	EXIT;	
	END_IF;	

END_FOR;

The terms and elements used in the FOR . . . DO syntax are described in the following table.

Operand	Data Type	Format Options	Description
count	SINT INT DINT	tag	Tag to store count position as the FOR . . . DO executes
initial_ value	SINT INT DINT	tag expression immediate	Must evaluate to a number that specifies initial value for count
final_ value	SINT INT DINT	tag expression immediate	Specifies final value for count, which determines when to exit the loop
increment	SINT INT DINT	tag expression immediate	*(optional)* Amount to increment count each time through the loop; if you don't specify an increment, the count increments by 1

EXAMPLE 15-3

The required positions (open = 1 and closed = 0) for 10 valves are stored in an array called Pos [0 – 9]. The tags for the valves are also in an array called Valve starting at element 5. Write an ST routine to set the 10 valves from their database.

SOLUTION

```
FOR count := 0 to 9 BY 1 DO  //count is the index value
    Valve_num := count + 5
//Sets the valve number as 5 higher than the count
    Valve [valve_num] := Pos [count]
//On the first pass - sets the Valve [5] position to the values in Pos [0]
END_FOR
```

Note that single line comments are added to the program by using the // indicators.

WHILE . . . DO. The WHILE . . . DO loop executes for the full time that the Boolean expression is true or 1. The syntax is:

WHILE *bool_expression1* DO		
	<statement> ;	Statements to execute while *bool_expression1* is true.
Optional	IF *bool_expression2* THEN EXIT; END_IF;	If there are conditions when you want to exit the loop early, use other statements, such as an IF . . . THEN construct, to condition an EXIT statement.
	END_WHILE;	

EXAMPLE 15-4

The tank input valve, tag V23, should be open (valve on or power applied) as long as the level switch, tag LS_low, is not true or 0 and the drain valve, tag V10, is closed (valve off or no power applied). Develop an ST program to solve this automation process problem.

SOLUTION

```
WHILE NOT LS_low AND NOT V10 DO           //Low level switch and drain valve
                                          //are not true or 0
          V23 := 1;                       //Turn on input valve
          IF Emerg_stop THEN
               V23 := 0;
               EXIT;
END_WHILE;
```

REPEAT . . . UNTIL. The REPEAT . . . UNTIL loop executes until the Boolean expression is true or 1. This command has a syntax that is like the WHILE . . . DO construct but it operates with complementary logic. WHILE . . . DO executes as long as the condition is true, whereas REPEAT . . . UNTIL operates until the condition is true.

Another difference between WHILE . . . DO and REPEAT . . . UNTIL concerns how execution occurs. The REPEAT . . . UNTIL loop executes the statements in the construct and then determines if the conditions are true before executing the statements again. The statements in a REPEAT . . . UNTIL loop are always executed at least once. The statements in a WHILE . . . DO loop are not executed until the condition is true, and they continue to be executed until the condition is false.

CASE . . . OF. The CASE . . . OF construct uses a numeric value to determine what expression to execute. The CASE construct is similar to a switch statement in the C or C++ programming languages. However, with the CASE construct the controller executes *only* the statements that are associated with the *first matching* selector value. Execution *always breaks after the statements of that selector* and goes to the END_CASE statement. The syntax is:

```
                       CASE numeric_expression OF
                     / selector1 : <statement>; <-- Statements to execute when
                    |              .               numeric_expression = selector1
Specify as          |              .
many                |              .
alternative         <  selector2 : <statement>; <-- Statements to execute when
selector            |              .               numeric_expression = selector2
values              |              .
(paths) as          |              .
you need             \ selector3 : <statement>; <-- Statements to execute when
                                   .               numeric_expression = selector3
                                   .
                                   .
                       ______________________
Optional             / ELSE
                    |          <statement>; <-- Statements to execute when
                    <              .               numeric_expression is not equal to
                    |              .               any selector
                     \             .
                       ______________________
                       END_CASE;
```

EXAMPLE 15-5

Develop an ST program to set the speed of a vehicle and turn on fans and the water pump based on the following switch positions:

Switch position	Speed	Fans	Water pump
1	10 mph		
2	20 mph		
3	40 mph	1	
4 and 5	50 mph	2	
6 through 10	60 mph		1

Solution

```
CASE        speed_setting OF
1    :             speed := 10;
2    :             speed := 20;
3    :             speed := 40; fan1 := 1;
4,5  :             speed := 50; fan2 := 1;
6…10 :             speed := 60; waterpump1 := 1;
ELSE
                   Speed := 0; speed fault := 1;
END_CASE
```

15-5 EMPIRICAL DESIGN WITH STRUCTURED TEXT

The structured text program development constructs are listed in the previous section. After you are familiar with the operation of the commonly used constructs, following these design steps will yield good design results. However, to learn the program structure for an ST solution it is best to study ST equivalents for the ladder logic rungs used in similar control problems.

15-5-1 Standard Structured Text Control Solutions

The following structured text programs indicate how structured text solutions compare to common ladder logic rungs and standard ladder logic diagrams.

ST Standard for Combinational Conditional Logic. The comparison between combinational logic for ladder logic inputs and their ST equivalent is illustrated in Figure 15-1(a) and (b). Read the description of the operation included in the figure. Combinational logic is achieved in ST through the use of Boolean logic functions such as Boolean ANDs, ORs, and NOTs.

ST Standard for Timers. The comparison between timer instructions for standard ladder logic inputs and their ST equivalent is illustrated in Figure 15-2(a) and (b). The ladder logic uses the TON symbol and the ST uses the TONR symbol for a time on timer. Read the description of the operation included in the figures. The timer instructions in ST operate the same as the timers in the ladder logic. However, in ST you can enter operational parameters such as timer preset.

ST Standard for Counters. The comparison between counter instructions for standard ladder logic inputs and their ST equivalent is illustrated in

FIGURE 15-1: ST for standard ladder logic with conditional XIC and XIO logic rungs.

Standard ST Logic for Conditional Control Functions

Turn on an output device for a logical combination of multiple input instructions.

The development of combinational logic in ladder rungs is simply AND/OR or OR/AND instructional groupings as shown in (a). When implementing the same logic in ST as shown in (b), IF...THEN...ELSE statements are used to develop the combinational logic. Note that an XIO instruction in the ladder logic requires that the NOT function is used in the ST to produce the same logical output.

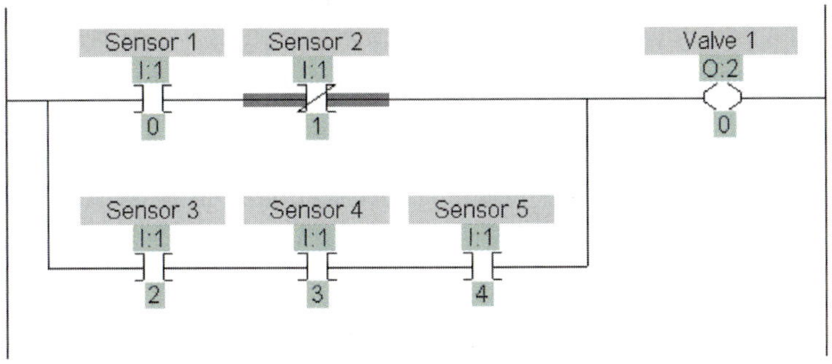

(a)

```
IF Sensor_1 AND NOT Sensor_2 THEN
        Valve_1 := 1;
ELSEIF Sensor_3 AND Sensor_4 AND Sensor_5 THEN
        Valve_1 := 1;
ELSE Valve_1 := 0;
END_IF;
```

(b)

FIGURE 15-2: ST for standard ladder logic for timers.

Standard ST Logic for TONR Timers

Turn on an output device after a set time period.

The standard ladder logic for the TON timer is shown in (a) and the ST equivalent in (b). The input for both is a NO maintain contact selector switch. If the switch is closed, then the ST timer (same notation as the FBD) is enabled. When the accumulator equals the preset value of 30 seconds (30000 x 0.001), the done bit is set and output is true. Note that the timer tag name is inside the parentheses.

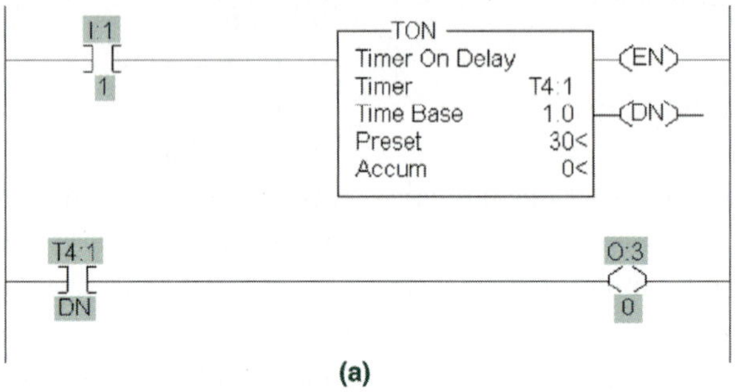

(a)

```
TONR_01.Preset :=30000;
TONR_01.Reset := reset;
TONR_01. TimerEnable := limit_switch1;
TONR(TONR_01);
Timer_state := TONR_01.DN;
Out_3 := Timer_state
```

(b)

Figure 15-3(a) and (b). The ladder logic software has two instructions, CTU and CTD, to handle the count up and count down applications as does the ST implementation. Note that if you want an up counter only, then you eliminate the CU enable, and if you want a down counter only, then you eliminate the CD enable. Read the description of the operation included in the figures.

ST Standard for Latch Outputs. The comparison between output latch instructions for standard ladder logic inputs and their ST equivalent is illustrated in Figure 15-4(a) and (b). The ladder logic software has two instructions, OTL and OTU, to handle applications that require a retentive output. The ST implementation uses set dominant (SETD) and reset dominant (RESD) functions. Read the description of the operation included in the figure.

ST Standard for One-Shots and Math Blocks. The comparison between one-shot and math instructions for standard ladder logic inputs and their ST equivalent is illustrated in Figure 15-5(a) and (b). The one-shot is used in control programs to limit the execution of the output to one scan of the program. This is useful when an instruction executes every scan when the input is true, but the control requires just one execution. This

FIGURE 15-3: ST for standard ladder logic for an up/down counter.

Standard ST Logic for Up/Down Counters

Increment the tag for every false to true transition of an up count signal and decrement the same tag for every false to true transition of a down count signal. Turn on (or turn off) an output field device after the accumulator is equal to or greater than a preset value.

The standard ladder logic for the CTU counter is shown in (a) along with the ST equivalent in (b). The ST instructions have a combination up and down counter, which can be used as an up counter, a down counter, or as a combined up/down counter. Eliminate the CDEnable line for an up counter or eliminate the CUEnable line for a down counter. The output is turned on or off by the counter_state. ST does not have a reset function, but the operation can be handled with an IF...THEN structure. An assignment statement cannot be used since the reset input is a Boolean data type and the counter ACC is a DINT.

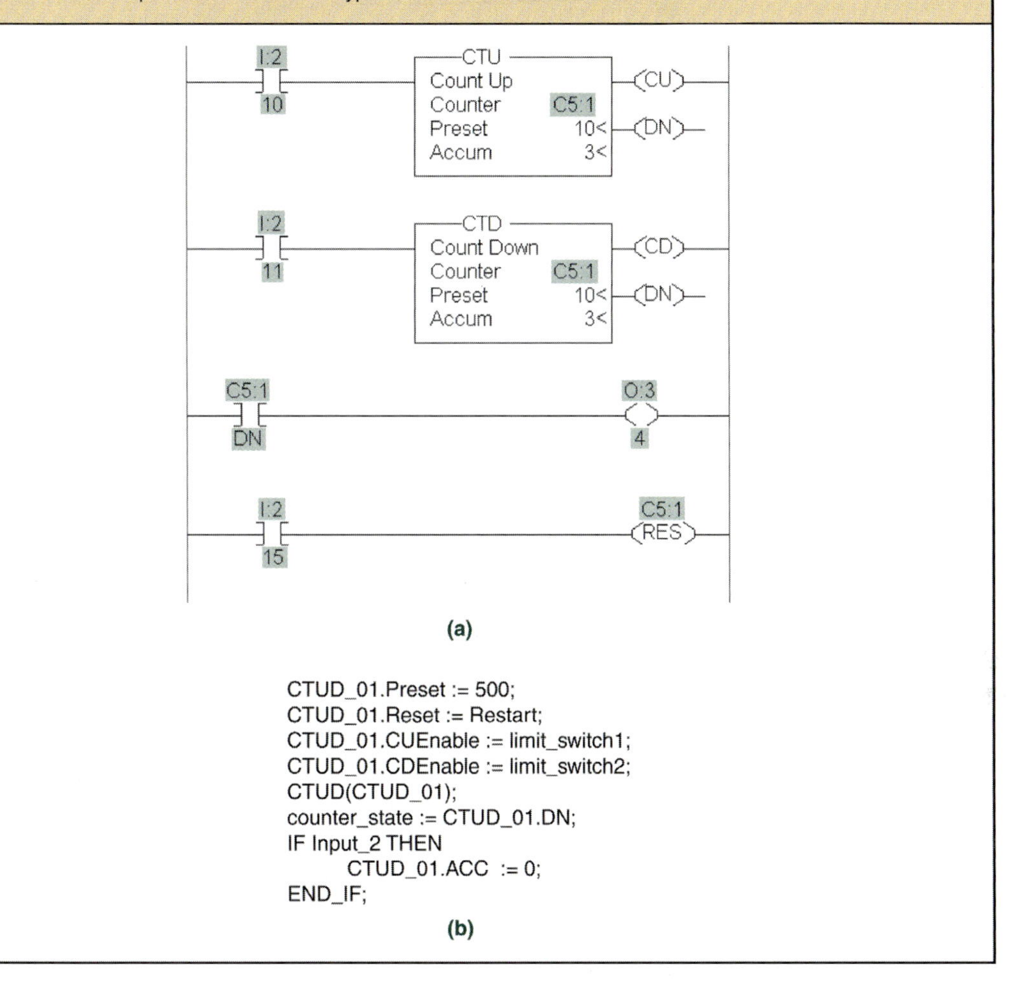

(a)

```
CTUD_01.Preset := 500;
CTUD_01.Reset := Restart;
CTUD_01.CUEnable := limit_switch1;
CTUD_01.CDEnable := limit_switch2;
CTUD(CTUD_01);
counter_state := CTUD_01.DN;
IF Input_2 THEN
        CTUD_01.ACC  := 0;
END_IF;
```

(b)

FIGURE 15-4: ST for standard ladder logic for retentive outputs.

Standard ST Logic for Retentive Outputs

Latch and unlatch an output on with an input signal.

The standard ladder logic for the OTL and OTU instructions is shown in (a) along with the ST equivalent in (b). The ST instructions use a Set Dominant (SETD) instruction, but a Reset Dominant (RESD) could be used as well. The SETD makes the output true when the set input is true regardless of the condition on the reset input.

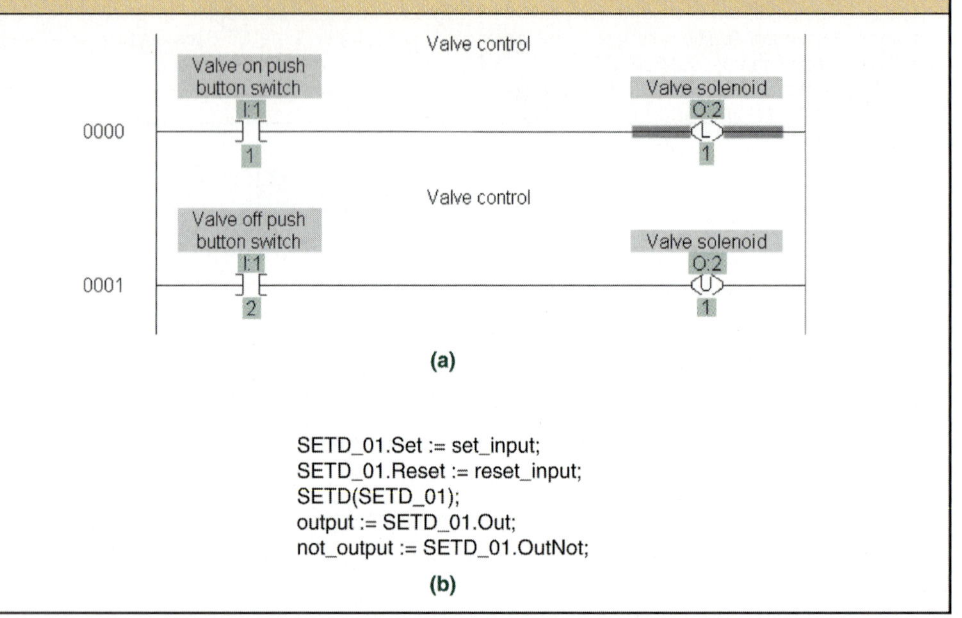

configuration is used most often with math and move instructions and could have a one-shot falling instruction used in place of the one-shot rising. Read the description of the operation included in the figure.

15-5-2 Discrete and Process Implementation

Let's look at an ST implementation for the solution to an automated process example using the process tank as illustrated in Figure 15-6. The process is as follows:

- The tank should have 25 percent of Liquid 1 and 75 percent of Liquid 2 by weight.
- The liquid in the tank is heated when the liquid weight is greater than 750 pounds.
- The heating continues after the tank is full until the liquid temperature reaches 180 degrees.
- The drain cycle starts at 30 seconds after the heater turns off.
- Mixing should start at a liquid weight greater than 500 pounds and continue until the tank has drained.
- Total weight is 1000 pounds.

The ST solution is illustrated in Figure 15-7. Note that the comments imbedded in the program aid the understanding of the solution. Study the ST program until you understand how the constructs and syntax are used for the solution.

FIGURE 15-5: ST for standard ladder logic for one-shots and arithmetic blocks.

Standard ST Logic for One Shots and Arithmetic Blocks

Turn on an output instruction like a math or move instruction for only one scan after an input field device is active.

The standard ladder logic for the one-shot and addition instruction is shown in (a) and the ST equivalent in (b). The input field device(s) determines when the output is active and the one-shot assures that the addition is performed for only one scan. The one-shot makes the instruction look like one that is triggered by a rising edge (false to true) on the input. The State tag enables the ADD function in the IF...THEN statement. While the addition instruction is used in this example, any of the other math instructions (SUB, DIV, MUL, or SQR) or the move instruction (MOV) could be substituted. Also, the one-shot rising is used, but a one-shot falling could be used to trigger on a falling input edge (true to false input change).

```
 I:1      B3      ADD
]  [----[OSR]----Add
  0        3      Source A   N7:1
                                0<
                  Source B   N7:3
                                0<
                  Dest       N7:5
                                0<
```

(a)

```
OSRI_01.InputBit := limit_switch1;
OSRI(OSRI_01);
State := OSRI_O1.OutputBit;
IF State THEN
        sum := value1 + value2;
END_IF
```

(b)

FIGURE 15-6: Process tank.

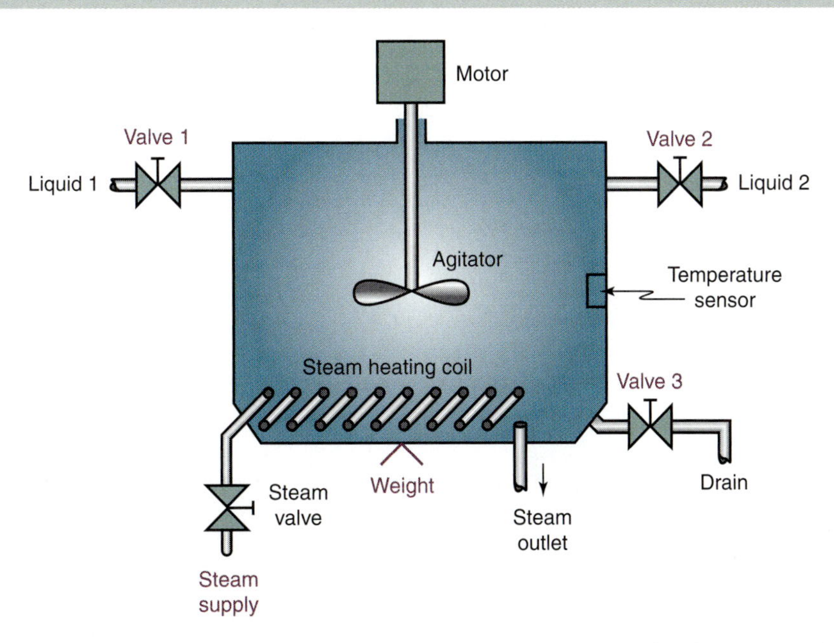

Rehg and Sartori, Industrial Electronics, *1st Edition, © 2006, Reprinted by permission of Pearson Education, Inc., Upper Saddle River, NJ.*

FIGURE 15-7: Structured Text program for the process tank example.

```
//Control for valve one and set control bits Tank_empty and system_reset to 0

IF Start & Liquid_weight < 249 & NOT Latch_tank_full THEN

        (* Note when Booleans, like Start and Tank_full are used in the condition,
        you should not put the relational operator (= 0 or = 1) into the condition.*)

        Valve_One [:=] 1;
        system_reset := 0;
ELSE
        Valve_One := 0;
END_IF;


//Control for valve two

IF Start & Liquid_weight >= 250 & NOT Latch_tank_full THEN
        Valve_Two [:=] 1;
END_IF;


//Control for the mixer

IF Liquid_weight > 500 THEN
        Mixer [:=] 1;
ELSE IF Liquid_weight < 500 & Latch_tank_full THEN
        Mixer := 1;
END_IF;


//Control for the heater

IF Liquid_weight > 750 & Temperature < 179 & NOT Valve_three & NOT Timer_start THEN
        Heater [:=] 1;
ELSE
        Heater := 0;
END_IF;
```

Rehg and Sartori, Industrial Electronics, *1st Edition, © 2006, Reprinted by permission of Pearson Education, Inc., Upper Saddle River, NJ.*

FIGURE 15-7: (Continued).

```
//Latch bit for tank full and turn off valve two

IF Liquid_weight >= 1000 THEN
        Latch_tank_full := 1;
        Valve_Two := 0;
END_IF;


//Full tank and temperature >= 180 timer control

IF Temperature >= 180 & Latch_tank_full THEN
        Timer_start [:=] 1;
END_IF;


//Function Block TONR timer with Preset at 30 min

Timer_30min.Pre := 30000;
Timer_30min.Reset := System_reset;
Timer_30min.TimerEnable := Timer_start;
TONR(Timer_30s);
Timer_done := Timer_30s.DN;

        (* These timer statements are the Structured Text (ST) equivalent for the
        FBD command TONR. Other options here include using a subroutine call to an
        FBD program with the TONR block present, or a subroutine call to a Ladder
        Diagram with a TON timer present.*)


//Tank empty cycle after 30 s time

IF timer_done THEN
        Valve_Three := 1;

END_IF;


//Establish starting condition for control bits

IF Liquid_weight = 0 THEN
        system_reset [:=] 1;
        Mixer := 0;
        Timer_start := 0;
        Valve_Three := 0;
        Latch_tank_full := 0;
END_IF;
```

Chapter 15 questions and problems begin on page 590.

CHAPTER

16

PLC Standard IEC 61131-3—Sequential Function Chart

16-1 GOALS AND OBJECTIVES

The goals for this chapter are to provide an overview of the Sequential Function Chart (SFC) graphic language and to provide an overview of the Allen-Bradley implementation of the SFC language.

After completing this chapter you should be able to:

- Describe the difference between Ladder Logic, Function Block Diagram, Structured Text, and Sequential Function Chart graphic languages.
- Define the components of the Allen-Bradley SFC language.
- Describe the process used to build an SFC program.
- Create an SFC program for automation process problems using the Allen-Bradley implementation.
- Given a problem statement, select the best SFC sequence with the most appropriate action qualifiers.

16-2 IEC 61131-3 STANDARD LANGUAGES

The first three Allen-Bradley 61131 language standards—Ladder Diagram (LD), Function Block Diagram (FBD), and Structured Text (ST)—have been discussed in previous chapters. The final language, Sequential Function Chart, is covered in this chapter. The Sequential Function Chart language is defined as follows:

> **SFC (Sequential Function Chart):** Based on the French standard, GRAFCET, this is a graphical language used most often for sequential control problems. The basic language elements are steps or states that have associated actions that produce the control and transitions with associated conditions used to move from the current state to the next. The actions and conditions can be defined in terms of the four other languages. As a result, the SFC language is often used for the basic sequential program flow with FBD, LD, and ST programs embedded into the SFC structure.

16-3 SEQUENTIAL FUNCTION CHART (SFC)

Sequential Function Chart (SFC) is a powerful graphical technique for describing the sequential behavior of a control program. The definition of the IEC Sequential Function Chart language has been derived from current techniques such as

GRAFCET, a graphical language based on a French national standard.

One of the most important aspects of SFC in the U.S. market is that it displays all the operational states of a system, all the possible *changes* of the states, and the *conditions* that cause those changes to occur. As a result, SFCs are often used to structure a control system solution, and then some combination of Ladder Diagrams, Function Block Diagrams, and Structured Text is used to program the solution. Structured list (SL), a fifth IEC standard language, could also be used with SFC programs, but not with Allen-Bradley SFC implementations since AB does not support the SL language.

16-3-1 Standard SFC Sequences

Because it is a graphical language, SFC is depicted as a series of steps, as illustrated in Figure 16-1. Each step in the SFC represents a *state* that the process exhibits where an *action* is performed. For example, consider Figure 11-9, where a pneumatic cylinder is used to push a part from a magazine parts feeder to a robot pickup point. The cylinder has two states: retracted and extended. The actuator is in each of these states for some period of time during its operational cycle. The signal to change these states, called the *transition condition*, comes from the automation system. In the extended state, the action would be to turn on the pneumatic valve. The transition to the next state, which is retracted, comes from a sensor that indicates a part is in the correct position or from a robot signal indicating that the part was acquired. The actions in an SFC can occur only while a step is active or can continue for more than one step. For example, the mixer in Figure 16-1 is only on during the Mix Solution step, but the action Open Valve 1 is started in the Fill Tank step and ended in the Mix Solution step.

FIGURE 16-1: SFC graphical flow.

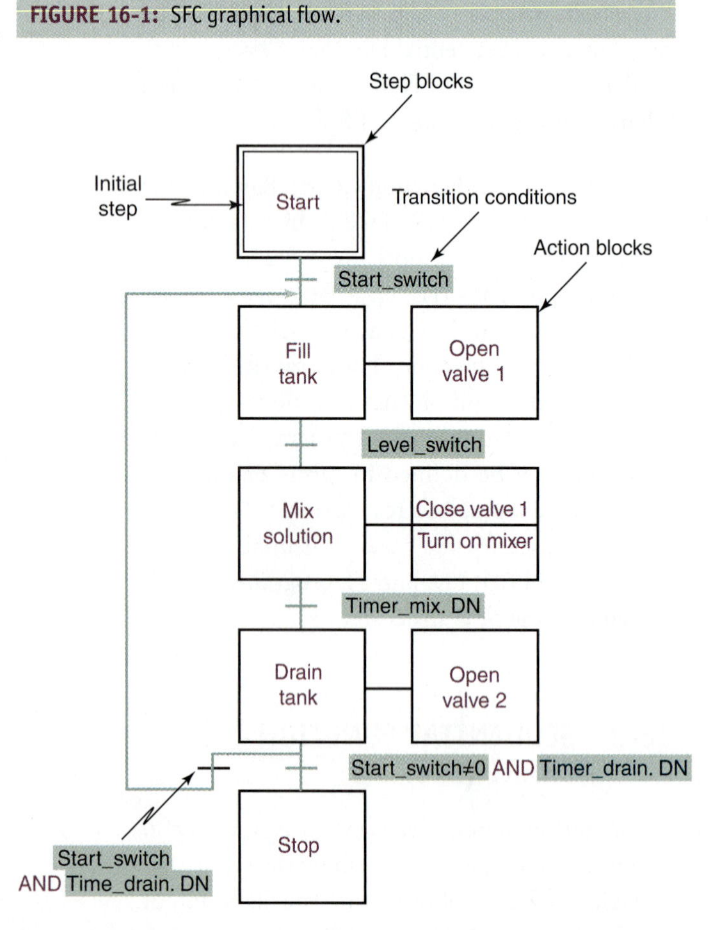

Rehg and Sartori, Industrial Electronics, *1st Edition, © 2006, Reprinted by permission of Pearson Education, Inc., Upper Saddle River, NJ.*

The steps in Figure 16-1 are connected by vertical lines indicating the sequence or flow. The short horizontal line halfway between steps is the *transition condition*. The SFC remains in the current state executing the current action until the transition condition becomes true. At that point the current state becomes inactive and the next state becomes the active state. The flow of control follows the graphically linked rectangles. Study the simple control scheme illustrated in the figure. Each step in the process is listed and the transition between steps links the steps together.

The linear sequence in Figure 16-1 has an initial block outlined with double lines and all remaining blocks outlined with single lines. The steps indicate what action is occurring at the step and the action blocks display the program or a link to the program that produced the action. For example, in the second block the action is Fill tank, and the corresponding action block indicates that a program statement would be placed there, open valve 1. Step 3 indicates that there can be more than one action block and program segment associated with a step, and step 4 indicates that there can be more than one transition condition if a branch is necessary.

Transition conditions are Boolean expressions, which evaluate as either true (move on to the next step) or false (remain in the current step for one more scan). Also, the transition condition can be a logical combination of Boolean tags, as illustrated with the ANDed tags for the fourth step. A simple combination of steps, like that illustrated in Figure 16-1, is called a *sequence*. There are a number of sequence options permitted in SFCs.

16-3-2 SFC Sequences

There are four types of sequences into which all SFCs can be grouped: *single sequence, selection divergence, simultaneous divergence,* and *loop sequence*. Each of these options is illustrated and described in Figure 16-2. You are not limited to the number of steps shown or the number of parallel branches illustrated in the figure.

The single sequence is a linear set of machine states or steps. Single sequences are used to

FIGURE 16-2: SFC sequence options.

To:	Use this structure:	With these considerations:
Execute 1 or more steps in sequence: • One executes repeatedly. • Then the next executes repeatedly.	Single sequence 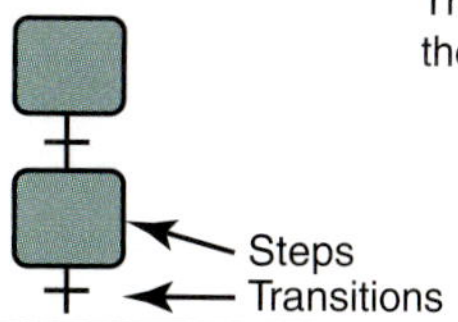 Steps Transitions	The SFC checks the transition at the end of the step: • If true, the SFC goes to the next step. • If false, the SFC repeats the step.
• Choose between alternative steps or groups of steps depending on logic conditions. • Execute a step or steps or skip the step or steps depending on logic conditions.	Selection sequence 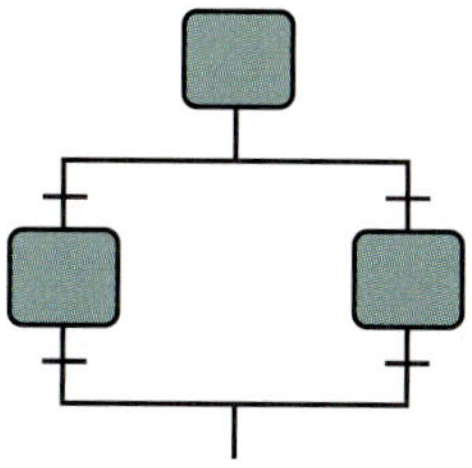	• It is OK for a path to have no steps and only a transition. This lets the SFC skip the selection branch. • By default, the SFC checks from left to right the transitions that start each path. It takes the first true path. • If no transitions are true, the SFC repeats the previous step. • RSLogix 5000 software lets you change the order in which the SFC checks the transitions.
Execute 2 or more steps at the same time. All paths must finish before continuing the SFC.	Simultaneous sequence 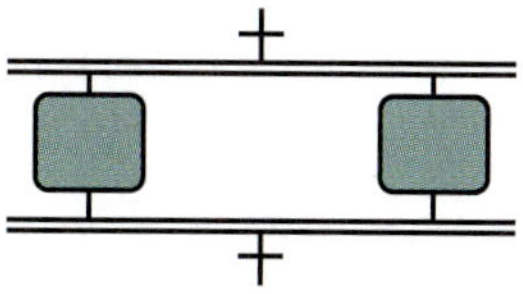	• A single transition ends the branch. • The SFC checks the ending transition after the last step in each path has executed at least once. If the transition is false, the SFC repeats the previous step.
Loop back to a previous step.	Loop sequence 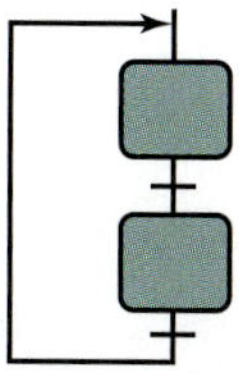	• Connect the wire to the step or simultaneous branch to which you want to go. • *Do not* wire into, out of, or between a simultaneous branch.

Rehg and Sartori, Industrial Electronics, *1st Edition, © 2006, reprinted by permission of Pearson Education, Inc., Upper Saddle River, NJ.*

build the other types of SFC sequence options that follow.

The selection divergence sequence has a transition condition at the top of each of the linear branches. This transition condition is checked in an order specified by the programmer. The first entry transition condition in that order that is true indicates the linear path that is executed. The default condition is to check entry transitions from left to right, starting with the first one on the left. RSLogix 5000 permits the programmer to change the order in which paths are tested to see if their transition condition is true. As a result, you can think of the selection sequence as a priority OR type sequence flow, since the first true linear path out of the total number of linear paths in the order checked is executed.

The simultaneous divergence sequence has a single transition condition that must be true before the simultaneous divergence sequence can be executed. There is also one transition condition at the end of the simultaneous sequence where all the linear paths converge. This final or exit transition condition is not tested by the program until the action blocks for every linear path have finished executing and the action for all steps is satisfied. As a result, this SFC sequence option is an AND type because the exit transition condition cannot be tested until the last step in every parallel path has finished its action(s). The automation control requirements that dictate the use of each sequence are listed in front of each sequence in Figure 16-2, and the operation of each sequence is described in the right column. Review these descriptions if the operation of each sequence is not clear.

16-3-3 SFC Step Programming

After the SFC sequence is established each step must be configured for the solution to the problem. An example sequence from the Allen-Bradley RSLogix 5000 SFC is illustrated in Figure 16-3. Read all the descriptions in the figure before continuing.

A study of the figure indicates:

- Two steps are present.
- The Initialize step has one action block where the tag counter is set equal to zero using a line of program code from structured text (ST).
- The second step, with tag name ServosOn, has three action blocks. The first is an ST program to set the value of the tag counter, and the last two are function calls that turn on the axis 0 and axis 1 servos.
- The four action blocks have default tag names (Action_010, Action_009, Action_000, and Action_008). Action tag names can be changed to any accepted tag name. In addition, each action block has an action qualifier that dictates how that action will be carried out.
- Both the steps and the action blocks have links to their configuration dialog boxes.
- The action description can be just a simple ST assignment statement as in the first action block or a complex series of program statements as in Action_009.
- Transition conditions can be a logical expression or just a Boolean tag.
- A feedback loop enters this sequence between the two steps from another part of the program.
- Step 3 of the sequence is on another page of the program, and the branch to that continuation is shown at the bottom of this sequence.

Configuring the Step. Every step has a tag name to identify the step parameters. The step is configured using the dialog box illustrated in Figure 16-4, and the dialog box is activated from the link in the upper-left corner of the step box (Figure 16-3). Three tabs are visible at the top of the dialog box: *general, action order,* and *tag*. Under the general tab the step type and timers are configured. In addition, time and count data are displayed along with alarm parameters. Action order changes the priority for the actions attached, and tag support the tag name entry. Read the descriptions in Figure 16-4 for the parameters in the general dialog box. The step has a built-in timer that can be used to set the length of time that the step is active. The time value is entered in milliseconds in the preset text box, or an expression can be entered to calculate the desired time. When the preset value is reached, the step timer done bit is set and the step tag ServosOn.DN is true. ServosOn is the tag name for this step.

When more than one action block is attached to a step, the order of execution is set by select-

FIGURE 16-3: SFC example control application.

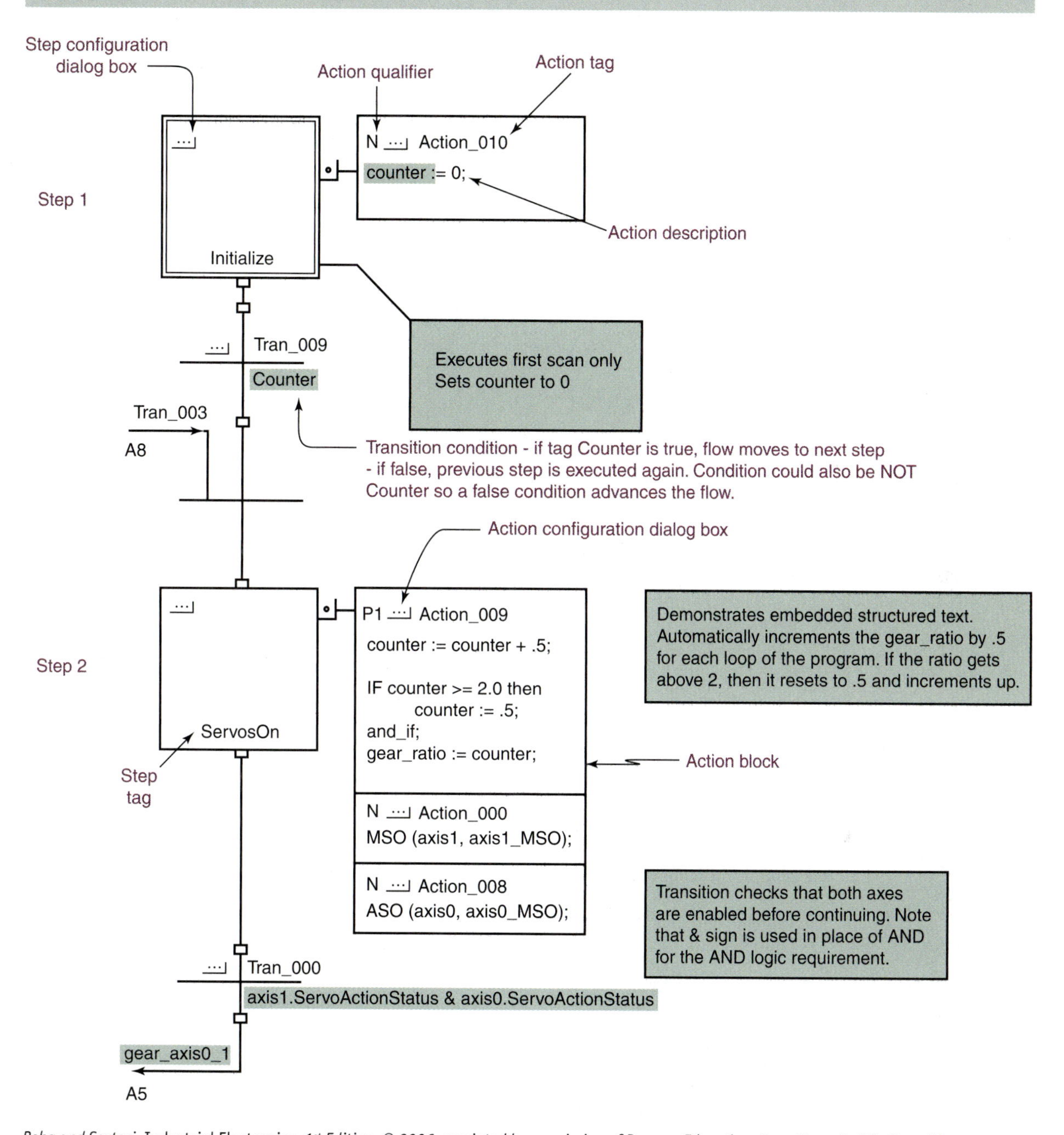

Rehg and Sartori, Industrial Electronics, *1st Edition, © 2006, reprinted by permission of Pearson Education, Inc., Upper Saddle River, NJ.*

ing the action order tab at the top of the step dialog box. This dialog box, Figure 16-5, permits the order or the priority for the actions to be changed. The action at the top of the list has the highest priority. Study Figure 16-5 for the second step in Figure 16-3. The action blocks will execute in the order 009, 000, and finally 008.

Configuring the Action Block. Every step in an SFC is linked to one or more actions, which are described and executed in an *action block*. An action contains a description of one or more behaviors or actions that should occur when the step is active. Action blocks, attached to the step, have three elements that are required and one that is optional. The block (see Figure 16-6) must have an *action qualifier, action tag name,* and *action description* or *assignment statement*. The action description is an executable program that can be entered in any of the supported 61131 languages.

FIGURE 16-4: Step configuration dialog box.

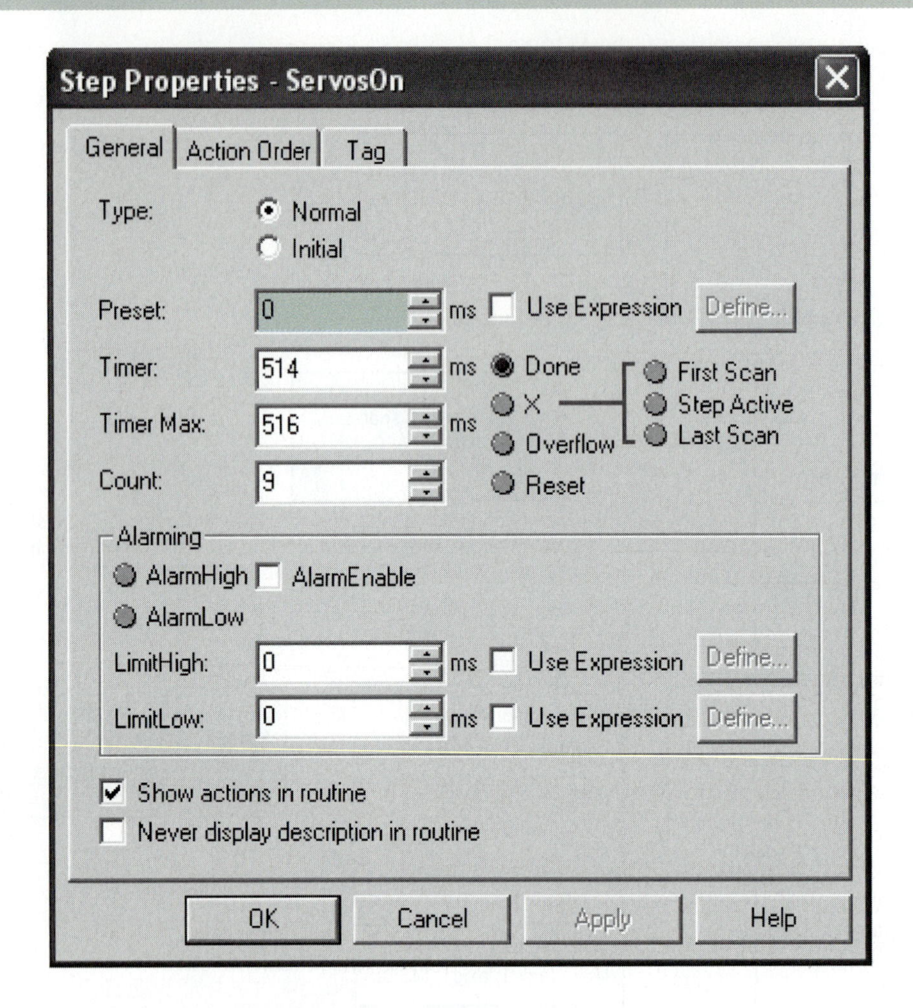

Step Configuration Information

Type: Use the *Type field* to make the first step the Initial step and all other steps Normal steps. Only one step can be an Initial step.

Preset: Each step has a built-in timer. Use the *Preset field* to enter the desired time or check the Expression checkbox and fill in an expression to calculate the timer time. When the accumulator equals the present time, the done bit is set and ServosOn.DN tag is true. ServosOn is the tag name for the step.

Timer: The *Timer field* displays the total time (in milliseconds) that the step is active during the last pass through the SFC sequence.

Time Max: In the *Timer Max field* the highest value ever reached by the Step Timer (T) is stored.

Count: The *Count field* stores the number of times the Step has been activated since the Step Count was reset by the last SFC reset.

Rehg and Sartori, Industrial Electronics, *1st Edition, © 2006, reprinted by permission of Pearson Education, Inc., Upper Saddle River, NJ.*

FIGURE 16-5: Dialog box to set the order of execution for action items.

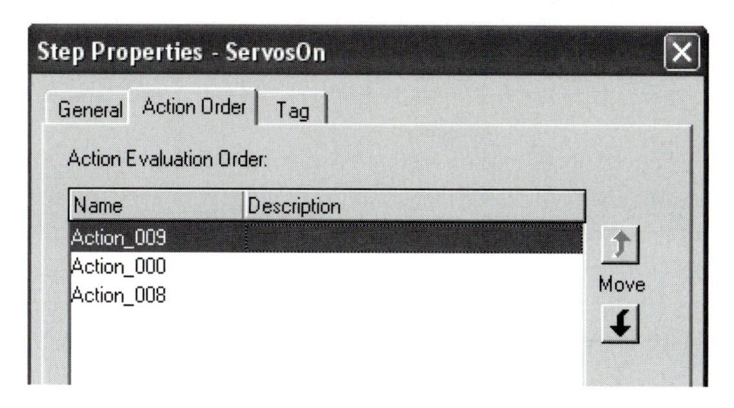

Rehg and Sartori, Industrial Electronics, *1st Edition, © 2006, reprinted by permission of Pearson Education, Inc., Upper Saddle River, NJ.*

FIGURE 16-6: SFC action block configuration.

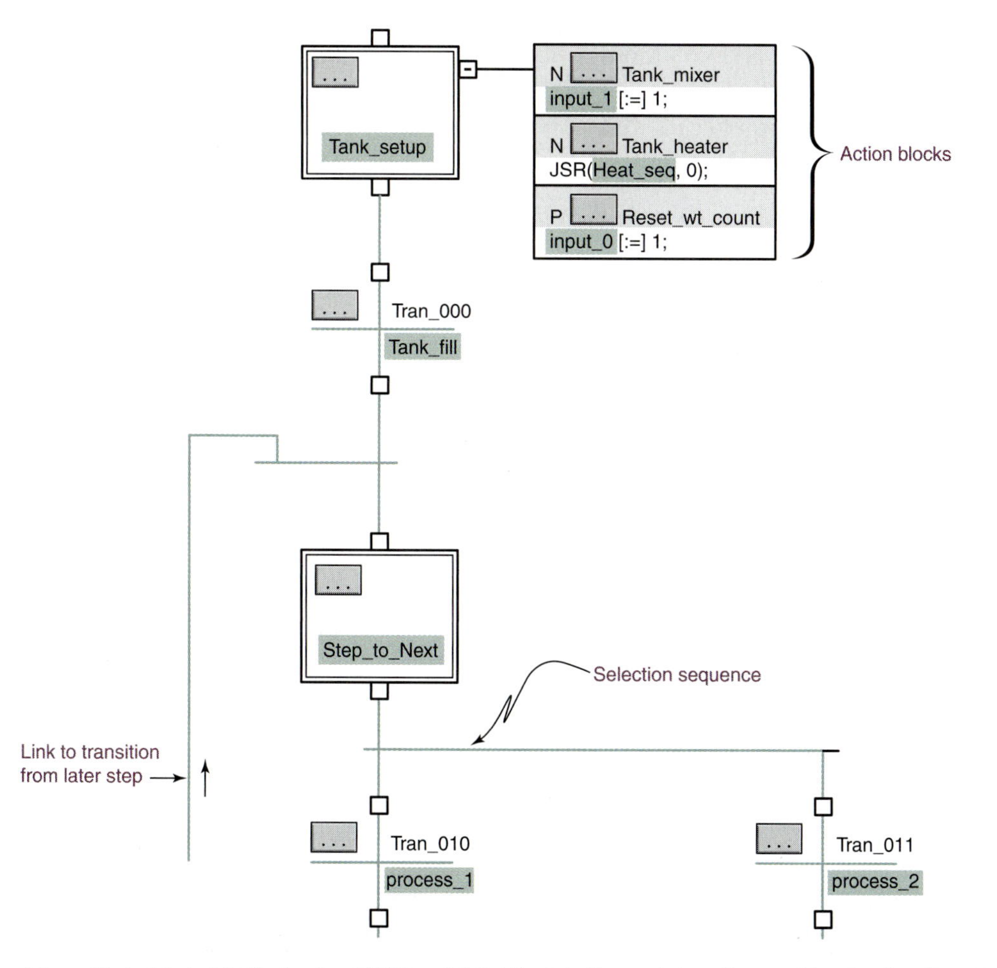

Rehg and Sartori, Industrial Electronics, *1st Edition, © 2006, Reprinted by permission of Pearson Education, Inc., Upper Saddle River, NJ.*

In most SFC programs in the Allen-Bradley system, the actions are programmed in structured text or ladder logic with less use of function block diagrams. Study the SFC in Figure 16-6 and observe the following:

- One step has three action blocks and one has none. The one without an action block is just there so the program can wait for a transition condition (Tran_010 or Tran_011) for one of the selection sequence branches (Process_1 or Process_2) to become active.
- The action qualifiers, described in the next section, are N and P.
- Two action descriptions have ST program assignment statements to turn Input_0 and Input_1 on, and the third has a subroutine call, JSR(Heat_seq,0), which causes the program Heat_seq to be executed with a zero value passed to the subroutine program.
- The two ST assignments use brackets (non-retentive condition) around the colon-equal assignment statement, [:=] to indicate that the tag values should be set to zero when the step becomes inactive.
- The tag names for the three Action boxes are Tank_mixer, Tank_heater, and Reset_wt_count.

Action Qualifier. An action qualifier defines precisely how the action should be executed after the step becomes active. Eleven qualifiers are available and listed in Figure 16-7. The more frequently used qualifiers are described in detail in Figure 16-8. Note that in the timing diagram in the second figure, *Step* represents the state of the step (inactive = low and active = high). *T1 or TN* indicates when the transition condition becomes true relative to the time that the step is active, and *Action1* and *Action2* indicate when the program in the action block is executed relative to the time that the step is active. For example, the timing diagram for an N qualifier indicates that the action program executes the entire time that the step is active. That implies that the program will be scanned more

FIGURE 16-7: SFC action qualifiers.

Qualifier	Description
None	Non-stored, default, same as 'N'.
N	Non-stored, executes while associated step is active.
R	Resets a stored action.
S	Sets or stores an action activity that is later reset with an R qualifier.
L (See Note 1.)	Time-limited action, terminates after a given period.
D (See Note 1.)	Time-delayed action, starts after a given period.
P	A pulse action that only executes once when a step is activated (entered), and once when the step is deactivated (left).
P1	A pulse action that only executes once when a step is activated (entered).
P0	A pulse action that only executes once when a step is deactivated (left).
SD (See Note 1.)	Stored (executed) and time delayed. The action is set active (executed) after a given period, even if the associated step is deactivated before the delay period.
DS (See Note 1.)	Action is time delayed and stored (executed). If the associated step is deactivated (left) before the delay period, the action is not stored or executed.
SL (See Note 1.)	Stored and time limited. The action is started and executes for a given period.

Note: These qualifiers all require a time period.

Rehg and Sartori, Industrial Electronics, *1st Edition, © 2006, reprinted by permission of Pearson Education, Inc., Upper Saddle River, NJ.*

FIGURE 16-8: SFC action qualifiers with timing diagrams.

Qualifier	Description	Timing diagram
N	Normal—the output Action1 executes continuously while the step is active and one more time after the step is inactive.	Step Action1 T1
S	Set (S)—the output Action1 is started when the step becomes active and continues after the step is inactive.	Step T1 Action1
R	Reset (R)—the reset action occurs in a later step where the output Action1 is terminated at the start of the step when the step becomes active and then one more time.	StepN TN
L	Time-Limited Action—the output Action1 is executed for a time duration set in the action configuration and then one time after that time. If the step becomes inactive before the timed duration for the output Action1, then the output Action1 terminates with step termination and then executes one more time.	Step T1 Action1 4s
D	Timer-Delayed Action—the output Action1 execution is delayed for a time duration set in the action configuration and then terminates when the step becomes inactive. One more execution occurs after the step is inactive. If the step time is shorter than the delay timed duration no output Action1 occurs.	Step T1 Action1 4s
P	Pulse—the output Action1 executes once when the step is activated.	Step T1 Action1
P1	Pulse On—the output Action1 executes once when the step is activated.	Step T1 Action1
P0	Pulse Off—the output Action2 executes once when the step is deactivated.	Action2

than once while the step is active, but it will also be executed one more time when the transition condition becomes true and the step is inactive. Study the qualifiers until you understand the range of options available.

In general, the wide range of qualifier types makes the solution to any automation problem possible in an SFC. The N qualifier is used most frequently for actions. The S and R are used when the qualifier program must be scanned for a longer period of time that stretches over more than one step. The P limits the action program to one execution. It is like a single scan of a ladder rung output. Finally, P1 and P2 are used when the action must start and stop during a single step. For example, if the step is timed, the P1 and P2 qualifier could be used in action blocks that turn on a pneumatic valve and then turn it off after a set time.

Transition Condition. The transition condition signals a change in steps or states. When the transition evaluates as true, the current active step becomes inactive and the following inactive step becomes active. It is often desired to have a step become active for specific times. In that case, the transition condition is triggered by the done bit on a timer. If the timer inside the step is used, then the transition condition will be step_tag.DN where step_tag is the tag or name attached to the step. Figure 16-9 shows how a timer in a ladder rung can be called by a subroutine and used to time the end of a step. A study of the figure indicates the following:

- The JSR command in ST calls the LD program called LadderFile with 0 parameters from an action block in the step.
- The LD timer program executes and after 2 seconds sets the DN bit high for one scan.
- The timer.dn Boolean expression in the transition condition location is true after 2 seconds and the SFC transitions to the next step.

This is a good example of IEC program integration where the SFC program uses an ST jump to subroutine function to execute LD rungs to time the step in the SFC.

FIGURE 16-9: SFC transition conditions.

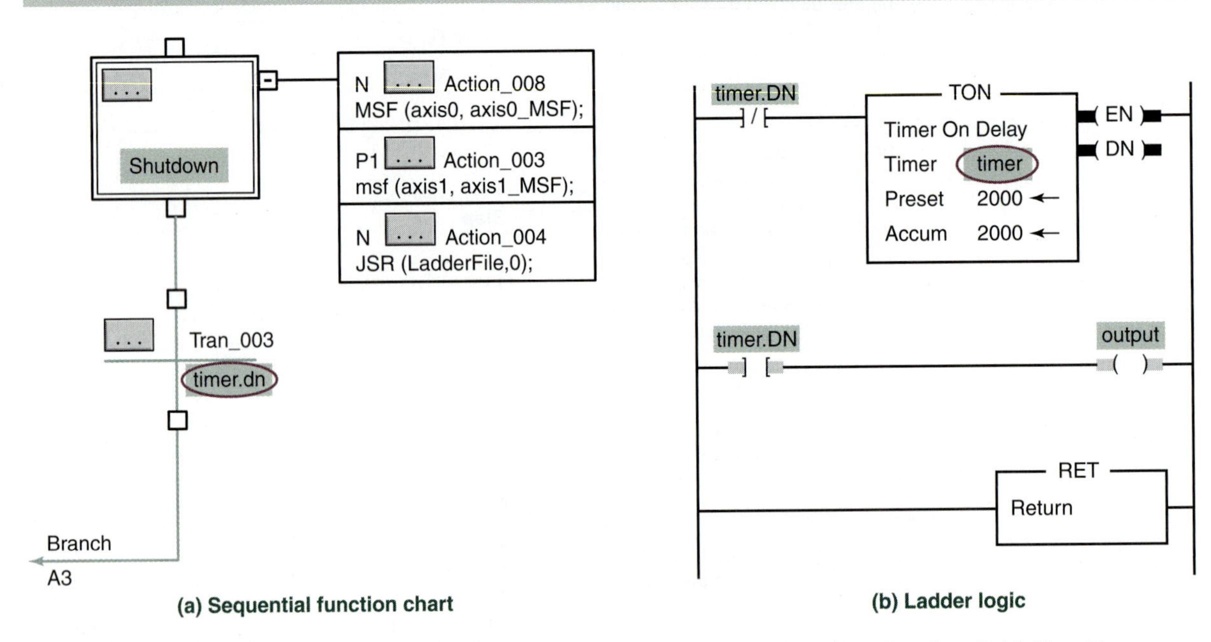

Rehg and Sartori, Industrial Electronics, *1st Edition, © 2006, reprinted by permission of Pearson Education, Inc., Upper Saddle River, NJ.*

EXAMPLE 16-1

Develop an SFC program for execution of the tank problem illustrated in Figure 16-10. The tank should have 25 percent of one ingredient and 75 percent of a second by weight. It should be heated to 180 degrees and then mixed for 30 minutes. Mixing should continue until the tank has drained. Total weight is 1000 pounds, and the control strategy required for the problem is as follows:

- Value 1 = start switch and tank liquid weight less than 250 pounds on fill cycle only
- Value 2 = start switch and tank liquid weight greater than 250 pounds but less than or equal to 1000 pounds on fill cycle only
- Mixer on = tank liquid weight greater than 499 pounds

- Mixer off = tank liquid weight equal to 0 pounds
- Heater on = tank liquid weight equal to or greater than 750 pounds
- Heater off = 30 minutes after tank full and temperature at 180 degrees
- NOT Valve 3 = tank liquid weight equal to 0 pounds on drain cycle only
- Valve 3 = 30 minutes after tank full AND temperature at 180 degrees

SOLUTION

Compare the SFC code for the tank solution in Figure 16-11 with the logic requirements given for the systems.

As you review the SFC solution for the Tank problem, first identify the actions that are produced at each step and compare those with the operation requirements for the process. Next, review the transition logic that takes the program from one step to the next. Note that the transitions are conditions given in the problem for when action should occur in the process. Finally notice that some actions start in one step (S qualifier) but do not end until a later step (R qualifier), so an action can occur during multiple steps. The controller was placed in the Automatic Reset mode, so the non-retentive reset type of equality ([:=]) was used for tags that needed to be reset when the program moved to the next step.

FIGURE 16-10: Process tank.

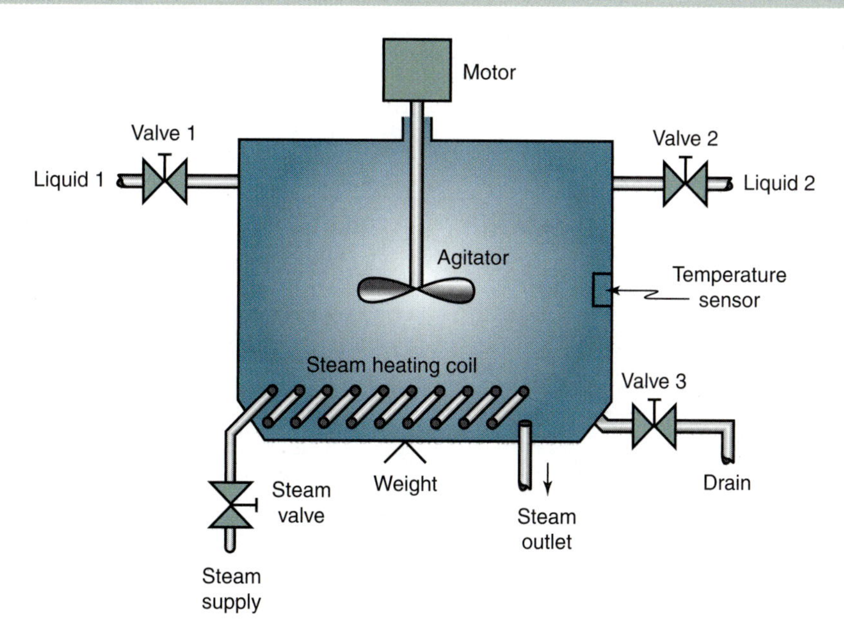

Rehg and Sartori, Industrial Electronics, *1st edition, ©2006, reprinted by permission of Pearson Education, Inc., Upper Saddle River, NJ.*

Note that Temperature_soak.dn in transition Tran_006 is a step timer that is set for 30 minutes.

SFC Program Design. The following design steps describe a process for developing a program for an event-driven sequential process.

1. Study the system operation until you become familiar with every detail of the process.
2. Make a numbered list of every step in the production sequence where an independent action must be executed. Draw SFC steps based on this analysis.
3. Identify and list all outputs for the system and the action required for the output. (These become the SFC action blocks, action descriptions, and output tags.) Group the actions with steps in the process

FIGURE 16-11: SFC solution for Example 16-1.

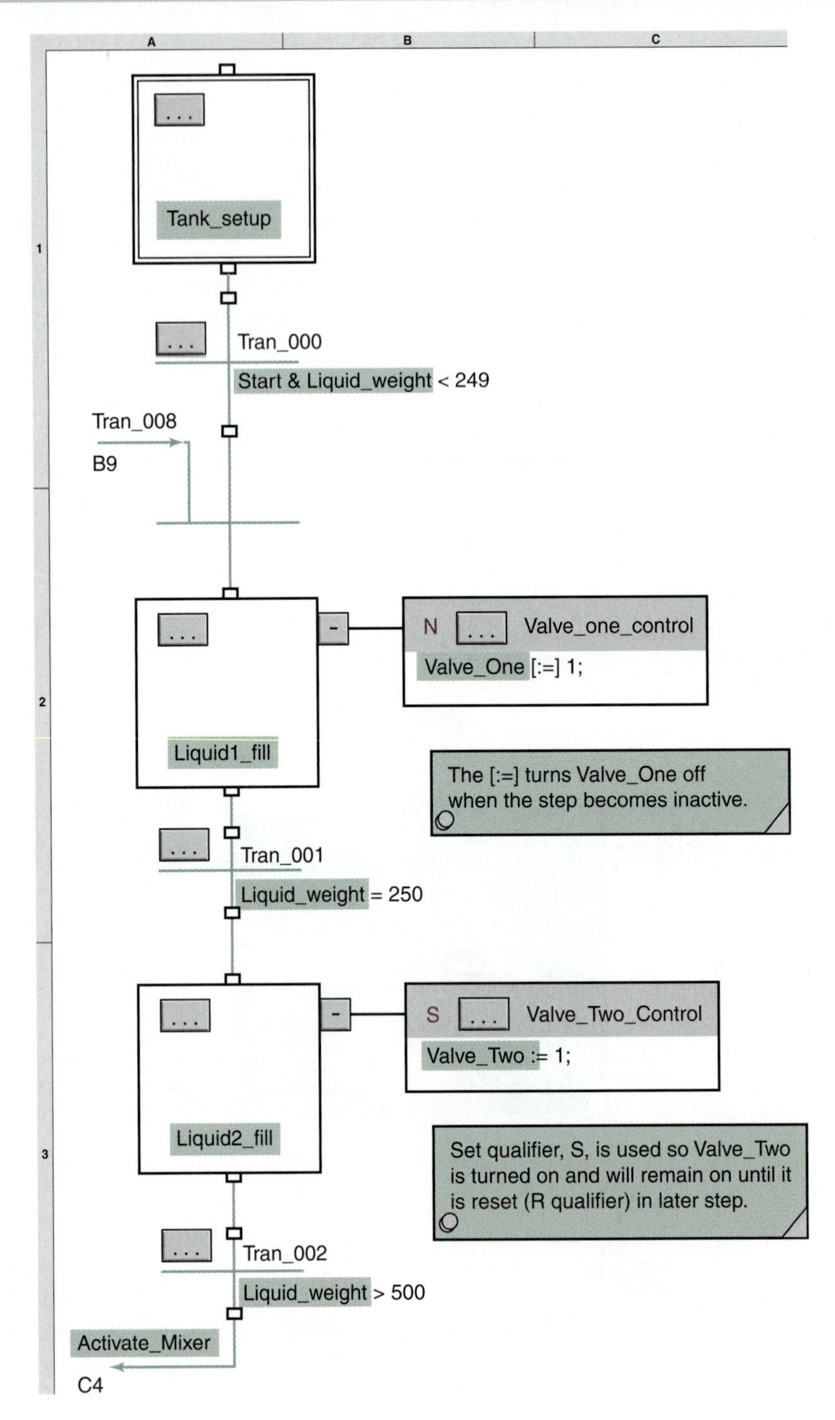

Rehg and Sartori, Industrial Electronics, *1st Edition, © 2006, Reprinted by permission of Pearson Education, Inc., Upper Saddle River, NJ.*

FIGURE 16-11: (Continued).

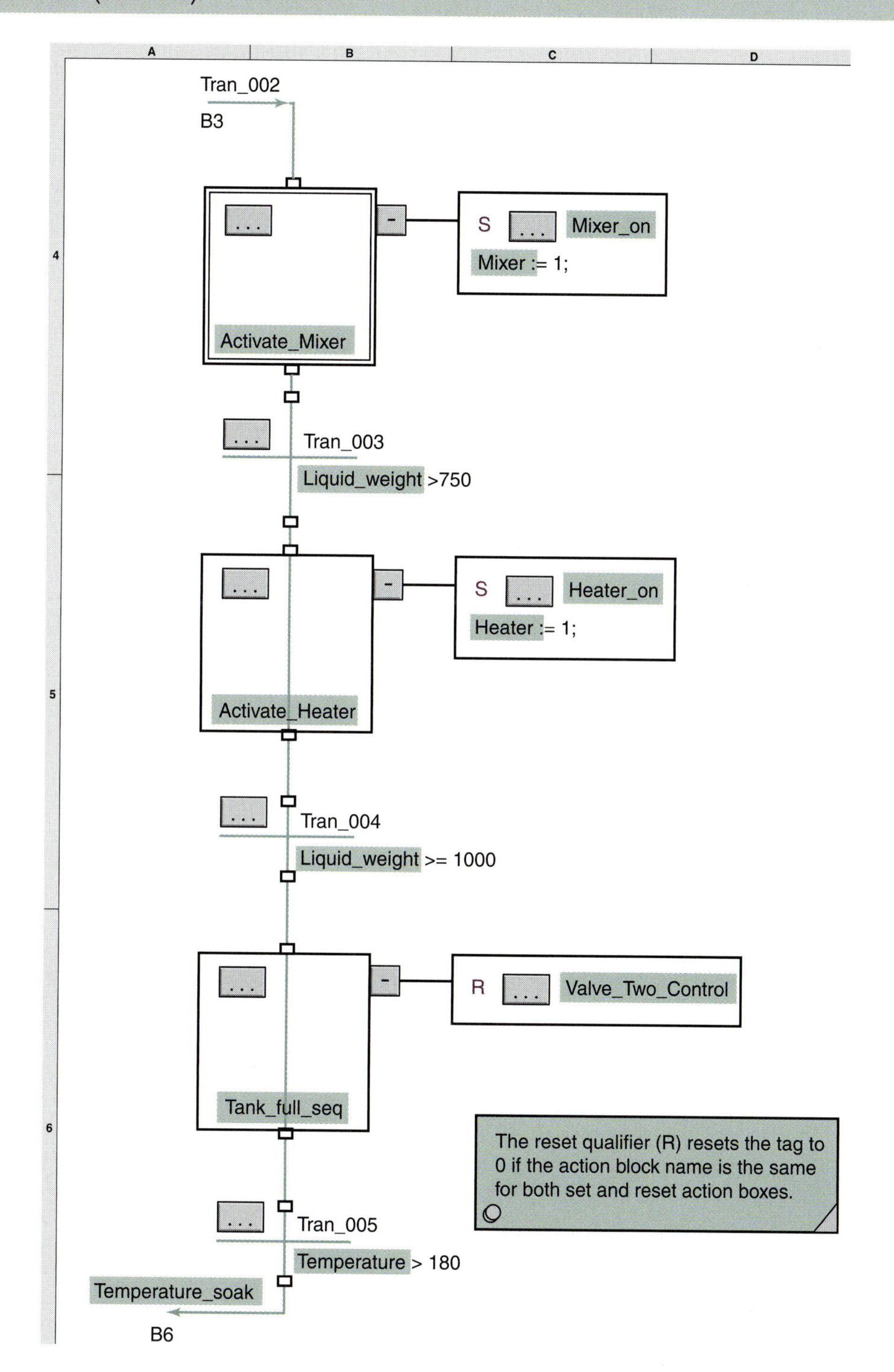

FIGURE 16-11: (Continued).

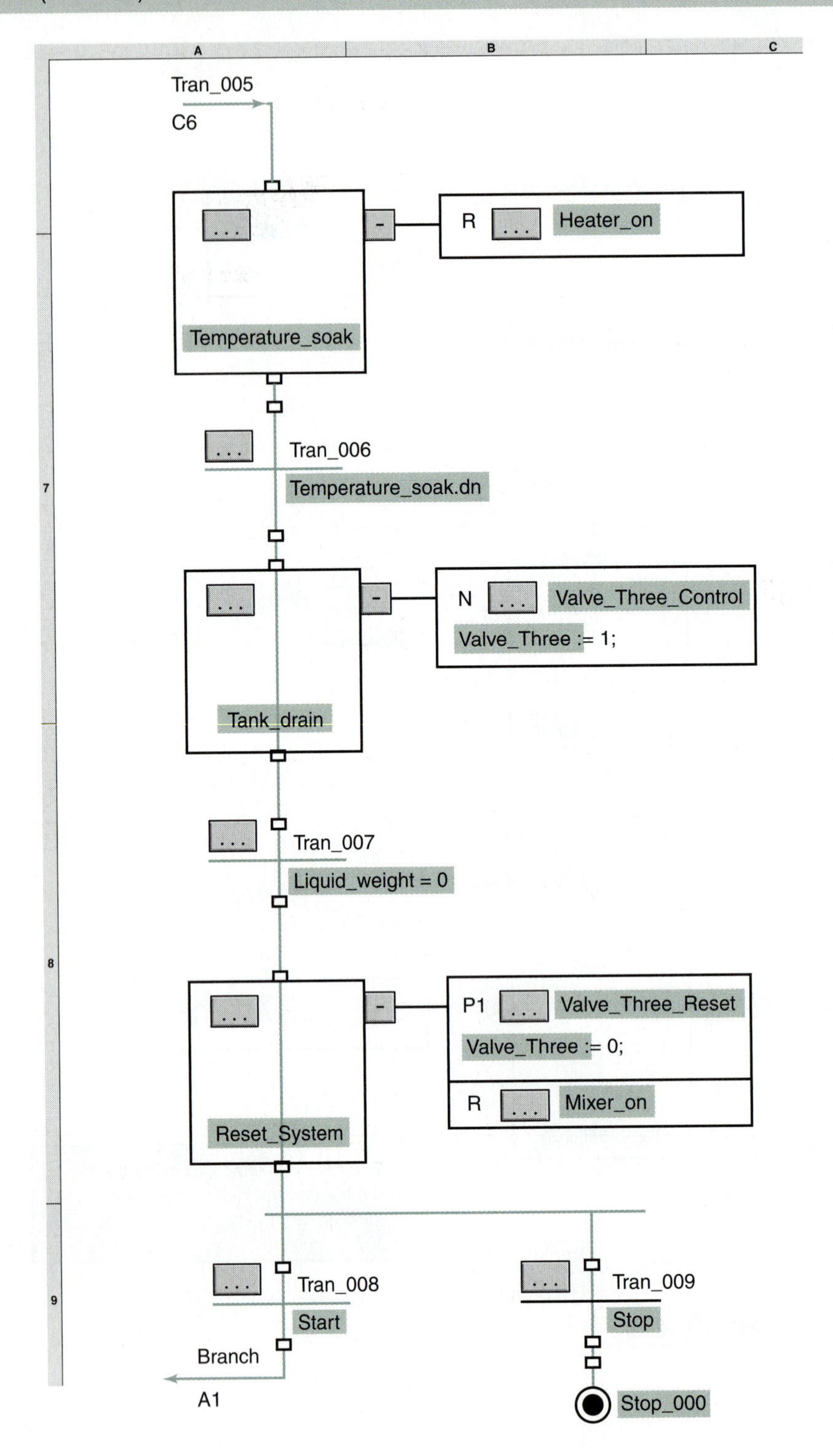

so that the number and type of action blocks for each step is identified. Using the type of action required, assign an action qualifier for each action block. Determine the program language to use for programming each action block. Perform the action programming in the action block or in a subroutine called from the action block.

4. Identify the transition conditions necessary to move from one production step to the next. These become the transition conditions in the SFC to move from one step block to the next. Determine the logic statement for the transition condition and program it in ST, use a function for the transition condition, or use a jump to subroutine for the transition condition. If a subroutine is used, then that program must be developed in one of the IEC languages.
5. Determine the duty cycle for the outputs or actions and determine if the actions must span more than one step, only one step, or just a part of a step. Draw the timing waveforms for the outputs on a process timing diagram if necessary. Verify that the action qualifiers selected earlier support the action timing requirements present in the timing diagrams.
6. List all the sensors in the system. Create input tags based on this analysis.
7. Use the data from steps 1 through 6 to create a Sequential Function Chart in the programming software for the PLC controller.
8. Verify the SFC design.
9. Test the system.

16-4 SITES FOR ALLEN-BRADLEY PRODUCTS AND DEMO SOFTWARE

- RSLogix 5000 Product Information: *http://www.ab.com/catalogs/b113/controllogix/software.html*
- RSLogix 5000 Demo Software Ordering and Download: *http://www.ab.com/logix/rslogix5000/*
- RSLogix 500 Demo Software Download—Under Get Software: *http://www.ab.com/plclogic/micrologix/*
- Rockwell Automation Logix Product Information: *http://www.ab.com/logix/*
- Rockwell Automation Manuals: *http://www.theautomationbookstore.com*, *http://www.ab.com/manuals/*

Chapter 16 questions and problems begin on page 591.

CHAPTER

17

Industrial Networks and Distributive Control

17-1 GOALS AND OBJECTIVES

There are three goals of this chapter. The first goal is to provide a practical introduction to industrial networks such as Ethernet, ControlNet, and DeviceNet, as well as PLC vendor-unique networks. The second goal is to provide an overview of distributive control and distributed I/O. The third goal is to provide basic guidelines for the selection and design of networks.

After completing this chapter you should be able to:

- Describe the purpose and function of Ethernet, ControlNet, and DeviceNet.
- Describe special network interfaces such as SERCOS, smart I/O, remote I/O, serial-data interfaces, and wireless communication.
- Discuss PLC vendor network applications.
- Describe distributive control and distributed I/O.
- Recommend networks based on application requirements.
- Describe network design guidelines.

Note that this chapter provides an overview of networks and is not a substitute for a course in network theory. However, it does give an excellent practical introduction to networks that is essential to achieving a comprehensive study of PLCs.

17-2 PLC NETWORK ARCHITECTURE

The PLC and other intelligent machines are connected into the factory floor network architecture, as shown in Figure 17-1. Note that the network architecture is flattened into three layers—Ethernet, ControlNet, and DeviceNet/FOUNDATION Fieldbus—and that PLCs are interfaced to all three layers. The function and operation of each layer is discussed in some detail in the subsequent sections.

- **Ethernet:** The highest level provides the information layer for data collection and program maintenance.
- **ControlNet:** The middle level supplies the automation and control layer for real-time input/output control, interlocking (coordinating update times between applications), and messaging.
- **DeviceNet/FOUNDATION Fieldbus:** The lowest level provides for cost-effective integration of individual devices—a primary

FIGURE 17-1: Factory floor network architecture with Ethernet, ControlNet, and DeviceNet/FOUNDATION Fieldbus layers.

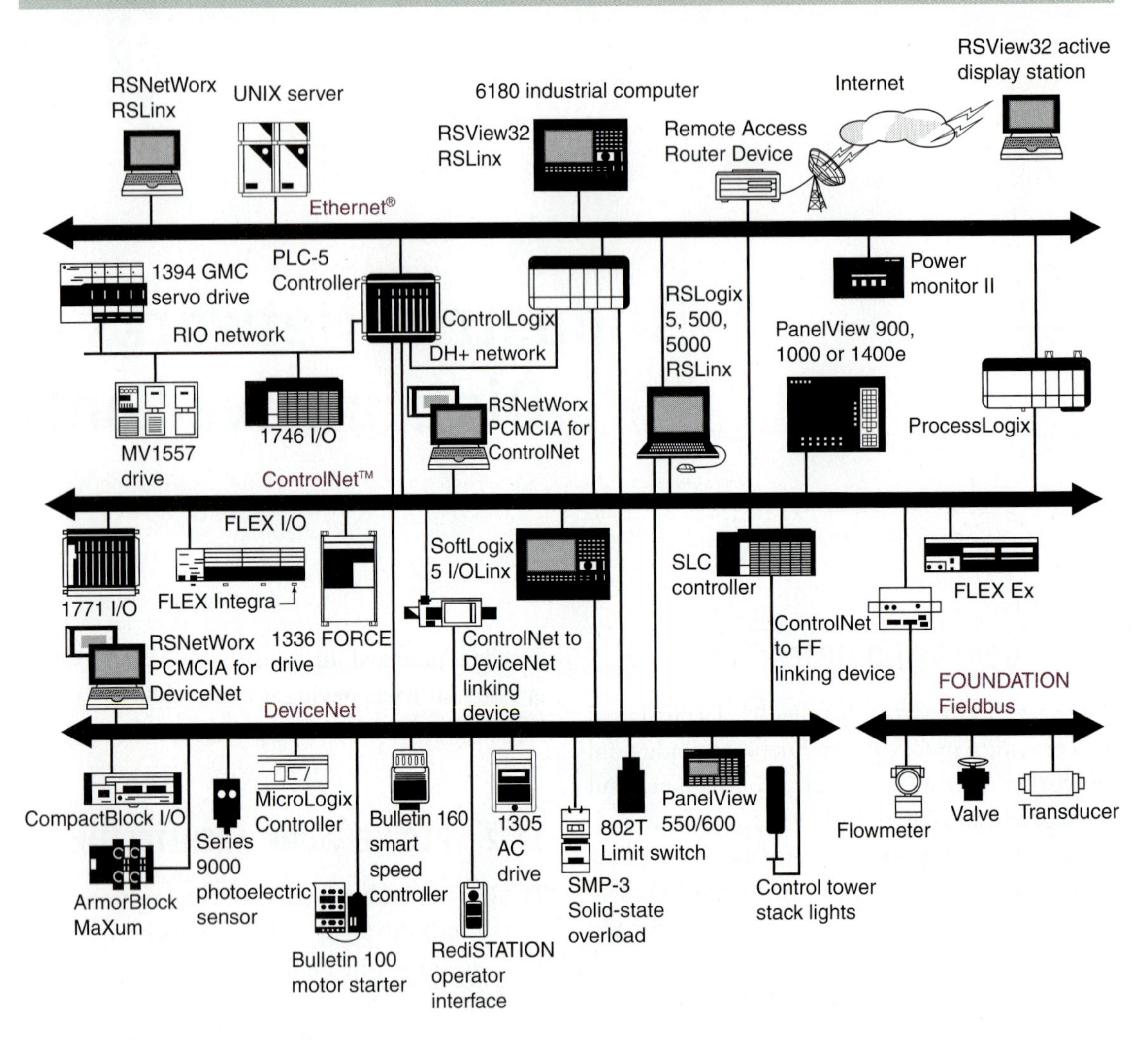

Courtesy of Rockwell Automation, Inc.

discrete interface with DeviceNet and analog control segment in process control with FOUNDATION Fieldbus. DeviceNet and FOUNDATION are two of the many fieldbuses available on the market today. Fieldbus is a term that basically means a network that is optimized to exchange data between small devices and a main larger device(s).

The PLCs and other intelligent machines on the network talk to each other via protocols, which are sets of communication rules. Each network type uses its own protocol. Network protocols handle problems and tasks such as communication line errors, data flow control, multiple device access, failure detection, data translation, and interpretation of messages.

17-3 ETHERNET

Dr. Robert M. Metcalfe invented Ethernet at the Xerox Palo Alto Research Center in the 1970s. It was designed to support research on the Office of the Future project, which included one of the world's first personal workstations, the Xerox Alto. The first Ethernet system ran at approximately 3 megabits per second (Mbps) and was known as experimental Ethernet. In 1980 a multi-vendor consortium turned the experimental Ethernet into an open, production-quality Ethernet system operating at 10 Mbps. A standards committee of the Institute of Electrical and Electronics Engineers (IEEE) then adopted Ethernet technology for standardization with the formal title of IEEE 802.3. Today, the Ethernet standard defines not only the 10-Mbps system, but also the 100-Mbps Fast Ethernet system, the Gigabit Ethernet, and the 10-Gigabit Ethernet. The Ethernet system consists of three basic elements:

1. The physical medium used to carry Ethernet signals between computers.
2. A set of medium access control rules embedded in each Ethernet interface that allows multiple computers to fairly arbitrate access to the shared Ethernet channel.
3. An Ethernet frame that consists of a standardized set of bits used to carry data over the system.

Ethernet and Ethernet Industry Protocol operation are covered in the following subsections.

17-3-1 Ethernet Operation

Each Ethernet-equipped intelligent machine, such as an industrial computer or PLC, operates independently of all other equipment on the network. All equipment or stations attached to an Ethernet are connected to a shared signaling system called the medium. Ethernet signals are transmitted serially, one bit at a time, over the shared signal channel to every attached station. The station listens to the channel and, when the channel is idle, transmits its data in the form of an Ethernet frame or packet. After each frame transmission, all stations on the network must contend equally for the next frame transmission opportunity—that is, when the channel is idle. This operation ensures that access to the network channel is fair, and that no single station can lock out the other stations. Access to the shared channel is determined by the medium access control (MAC) mechanism embedded in the Ethernet interface located in each station.

The Ethernet standard can be implemented in the following ways:

- 10Base5: standard thick coaxial cable
- 10Base2: thin coaxial cable
- 10BaseT: unshielded twisted-pair cable
- 100BaseT: unshielded twisted-pair cable
- 10BaseFL: fiber optic cable

This *x*Base*y* nomenclature is interpreted as follows:

- *x* is the signaling rate in Mbps.
- Base is the term that means the signal uses the cable in a baseband scheme as opposed to a broadband, which is a multi-frequency, multi-channel modulating scheme.
- *y* is some indication of the media type.

17-3-2 Ethernet Industry Protocol

Ethernet Industry Protocol or *Ethernet/IP* is an open industrial networking standard that takes advantage of commercial off-the-shelf Ethernet communication devices and physical media. Unlike many options in industrial Ethernet systems, Ethernet/IP uses an open protocol and is backed by three networking organizations: ControlNet International (CI), the Industrial Ethernet Association (IEA), and the Open DeviceNet Vendor Association (ODVA). Ethernet/IP simplifies interoperability between different vendors' devices on the network by implementing a common DeviceNet application layer over commercial off-the-shelf Ethernet (IEEE 802.3) products. This open standard enables real-time deterministic communications between a wide variety of industrial automation products, including the PLC, robot controllers, input/output adapters, operator interfaces, and supervisory control stations. The term *deterministic* means that the speed of data transfers and the transmission and arrival times are predictable and not determined by conditions present on the network.

Ethernet/IP is constructed from the widely implemented standard used in ControlNet and DeviceNet, which is called Control and Information Protocol (CIP). This protocol organizes networked devices as a collection of objects. It defines the access, object behavior, and extensions, which allow widely disparate devices to be accessed using a common mechanism. Hundreds of vendors currently support CIP with their Ethernet/IP products. Because Ethernet/IP is based on this widely understood and implemented standard, it does not require a new technology learning-curve period.

17-4 CONTROLNET

ControlNet is an open network, which means that any company can develop a ControlNet product without a license fee. It is positioned one level above DeviceNet in the control hierarchy. Its high-speed (5 Mbps) control and data capabilities significantly enhance input/output performance and user-to-user communications. ControlNet uses the producer/consumer network model to efficiently exchange time-critical application information for both processes and manufacturing automation. This model permits all nodes on the network to simultaneously access the same data from a single source. In addition, ControlNet's Media Access Method uses the producer/consumer model to allow multiple controllers to control I/O on the same network segment. This provides a significant advantage over other networks, which allow only one master controller on the network. ControlNet also allows simultaneous broadcast to multiple devices of both inputs and peer-to-peer data, thus reducing traffic on the media and increasing system performance. ControlNet operation and features are covered in the next subsections.

17-4-1 ControlNet Operation

Network access is controlled by a timing algorithm called Concurrent Time Domain Multiple Access (CTDMA), which regulates a node's opportunity to transmit in each network interval. The network is configured by how often the network interval repeats by selecting a network update interval (NUT). The fastest NUT you can specify is 2 milliseconds. Information that is time-critical is sent during the scheduled part of the NUT. Information that can be delivered without time constraints (such as configuration data) is sent during the unscheduled part of the NUT.

17-4-2 ControlNet Features

ControlNet is highly *deterministic* and *repeatable*. Determinism is the ability to reliably predict when data will be delivered, and repeatability ensures that transmitting times are constant and unaffected by devices connecting to or leaving the network. These two critical characteristics ensure dependable, synchronized, and coordinated real-time performance. As a result, ControlNet permits data transfers, such as program uploads/downloads and monitoring of real-time data, in flexible but predictable time segments. Management and configuration of the entire system can be performed from a single location on ControlNet or from one location on an information-level network, such as Ethernet. ControlNet can link a variety of devices, including motor drives, motion controllers, remote input/output modules, PLCs, and operator interfaces. In addition, ControlNet can provide a link to other networks such as DeviceNet and FOUNDATION Fieldbus. The following important features should be noted.

- In most applications, PLCs and industrial microcomputers use ControlNet scanner cards when adding a ControlNet network to an automation system.
- The devices attached to a ControlNet network include PLCs, industrial microcomputers, operator interface terminals, remote I/O modules, motor drives, and personal computers.
- Media used to transmit ControlNet data include both coaxial cable and fiber optic cable.
- ControlNet is a sub-network off of the Ethernet or Ethernet/IP.

ControlNet functions as the integrator of complex control systems such as coordinated motor- and servo-drive systems, weld control, motion control, vision systems, complex batch control systems, process control systems, and systems with multiple controllers and human-machine interfaces. It is ideal for systems with multiple PC-based controllers and PLC-to-PLC communication, allowing multiple controllers—each with their own input/output and shared inputs—to talk to each other with any possible interlocking combination.

17-5 DEVICENET

DeviceNet is an open network standard, which means that any company can develop a DeviceNet product without a license fee. It is based on the reliable Controller Area Networking (CAN) technology, which is used in virtually all industries, including automotive, manufacturing, agricultural, medical, building controls, marine, and aerospace. With many suppliers offering DeviceNet sensors and actuators, designers can select the best combination of devices from multiple suppliers to solve the control problem. As a result, DeviceNet is the fastest growing device network in the world, with over a half million installed devices. DeviceNet operation and features are covered in the next subsections.

17-5-1 DeviceNet Operation

DeviceNet operates on multiple messaging formats, which can be mixed and matched within a network to achieve the most information-rich and time-efficient information from the network at all times: The messaging types are as follows:

- **Polling:** Each device is requested to send or receive an update of its status. This requires an outgoing message and incoming message for each node on the network. This is the most precise but least time efficient way to request information from devices.
- **Strobing:** A request is broadcast to all devices for a status update. Each device responds in turn, with node 1 answering first, then node 2, 3, 4, etc. Node numbers can be assigned to prioritize messages. Polling and strobing are the most common messaging formats used.
- **Cyclic:** Devices are configured to automatically send messages on scheduled intervals. This is sometimes called a heartbeat and is often used in conjunction with change of state messaging to indicate that the device is still functional.
- **Change of state:** Devices only send messages when their status changes. This occupies an absolute minimum of time on the network, and a large network using this type can often outperform a polling network operating at several times the speed. This is the most time-efficient but sometimes least precise way to obtain information from devices because throughput and response time becomes statistical instead of deterministic.
- **Explicit messaging:** The explicit-messaging protocol indicates how a device should interpret a message. It is commonly used on complex devices such as drives and controllers to download parameters that change from time to time but do not change as often as the process data itself. An explicit message supplies a generic, multipurpose communication path between two devices and provides a means for performing request/response functions such as device configuration.
- **Fragmented messaging:** For messages that require more than DeviceNet's maximum 8 bytes of data per node per scan, the data can be broken up into any number of 8-byte segments and re-assembled at the other end. This requires multiple messages to send or receive one complete message.
- **Unconnected message manager:** DeviceNet UCMM interfaces are capable of peer-to-peer communication. Unlike the plain-vanilla master/slave configuration (one device, the master, initiates communication; other devices, the slaves, respond), each UCMM-capable device can communicate with another directly, without having to go through a master. UCMM devices must accept all generic CAN messages, then perform filtering of irrelevant or undesired message types in the upper software layer. This requires more RAM and ROM than ordinary master/slave messaging.

17-5-2 DeviceNet Features

DeviceNet interconnects industrial devices such as limit switches, photoelectric sensors, valve manifolds, motor starters, process sensors, panel displays, and operator interfaces via a single network. Expensive wiring and failure due to the increase in the number of connections is eliminated. It also reduces the cost and time to install industrial automation devices while providing reliable interchangeability of components from multiple vendors. The direct connectivity provides improved communication between devices as well as important device-level diagnostics not easily accessible or available through hardwired input/output interfaces. The following important features should be noted.

- DeviceNet supports individual devices designed with network communications electronics, such as sensors, operator stations, motor starters, motor controllers, pneumatic valves, and microcomputers.
- DeviceNet supports remote I/O ports with a variety of I/O card options. These remote network ports use standard discrete sensors and actuators.
- DeviceNet is generally a sub-network off of a PLC that is connected to a ControlNet, Ethernet, or Ethernet/IP.
- In most applications, PLCs and industrial microcomputers use DeviceNet scanner cards when adding a DeviceNet network to an automation system.
- A large body of sensors and actuators, such as limit switches, proximity and photoelectric sensors, valve blocks, and motor drives, are DeviceNet compatible. This means that these devices are connected directly to the DeviceNet network and no independent power and discrete I/O wiring is necessary.

DeviceNet is a connection-based protocol—all devices are required to establish a network connection prior to exchanging information. It adopts the object modeling approach, in which each information type is structured in different objects. Four basic objects are required to handle these information exchanges:

- **Identity object:** Identification information such as vendor ID, device profile, and revision of a device are stored in this object. Users can identify a particular object by remote access to this object.
- **Message router:** This object handles the explicit messages received by routing them to the proper destination objects.
- **DeviceNet object:** This object stores all DeviceNet-related information such as the device's MAC address and baud rate.
- **Connection object:** This object handles the connection of the messaging module, such as explicit messaging and input/output messaging.

17-6 SPECIAL NETWORK INTERFACES

PLCs connected into the factory floor network architecture often use special network interfaces such as *SERCOS*, *smart I/O*, *remote I/O*, *serial communications*, *wireless communications*, and the *human-machine interface* to further distribute data communications to remote control locations. These special network interfaces are described in the following subsections.

17-6-1 SERCOS Interfaces

This special communications module is a Serial Realtime Communications System, or SERCOS for short. SERCOS is a digital motion control network that interfaces the motion control module in the PLC with the servo motor drive through a fiber optic cable. The fiber permits serial motion data transmissions with improved noise immunity and fast update times. SERCOS allows motion control in velocity, torque, or position modes. Multiple servo drives can be daisy chained on a single SERCOS fiber network.

17-6-2 Smart I/O Interfaces

PLC vendors have proprietary network protocols to allow devices from the same vendor to communicate using that vendor's specific network interface. One common application for a proprietary network is the use of *smart I/O devices*. The term *smart* implies that the device or interface includes a microprocessor so that it can be programmed for network data exchange. PLC vendor-specific network interfaces are discussed in Section 17-7.

Let's look at an operator panel as an aid to understanding the concept of smart I/O interfaces. Operator panels have switches for control of process machines and devices, and the panels have lights to indicate the condition of process equipment. The operator panels are frequently located in a control room away from the process itself. The traditional approach in building an operator panel requires a minimum of one wire per switch and lamp plus several return wires between the operator panel and PLC input and output modules. As a result, the wire bundle between these two devices often has hundreds of wires that must be enclosed in conduit over distances of hundreds of feet. In contrast, the smart operator interface uses the same number of switches and lamps but controls them with a small microcomputer located in the operator panel. This smart I/O interface between the operator panel and the PLC is just a single coaxial network cable that permits the operator panel's microcomputer to communicate with the PLC processor. A large number of smart external devices are available, including motor drives, process controllers, text readout devices, programmable CRT displays supporting full color and graphics, voice input and output devices, and discrete input and output devices.

17-6-3 Remote I/O Interfaces

In an effort to distribute the network control capability across a large automation system, PLC vendors provide *remote I/O or remote rack* capability. Figure 17-2 illustrates a remote I/O link. The rack uses the standard I/O modules for control of machines and processes; however, the processor module is replaced with a remote rack communications module. The processor in the main PLC rack sends control instructions over the single network cable to the communication module in the remote rack and then to the I/O modules included in the remote rack. Use of this technology permits the I/O modules to be located close to the point of control, which eliminates the long wire runs required if the sensors are connected to I/O modules in the main PLC rack.

17-6-4 Serial Communication Interfaces

Serial communication interfaces are either built into the processor module or come as separate modules. In both cases, the interface permits serial data communication using several standard interfaces such as RS-232, RS-422, and RS-485, where RS stands for Recommended Standard, and Firewire, which is an IEEE 1394 implementation. The RS interfaces are used to connect devices such as the smart gages and bar code readers that must transfer quantities of data at a reasonably high rate between the remote device and the PLC. These serial communication interface standards use *data terminal equipment* (DTE) and *data communication equipment* (DCE) terminology. The DTE is the component that communicates

FIGURE 17-2: Remote I/O network.

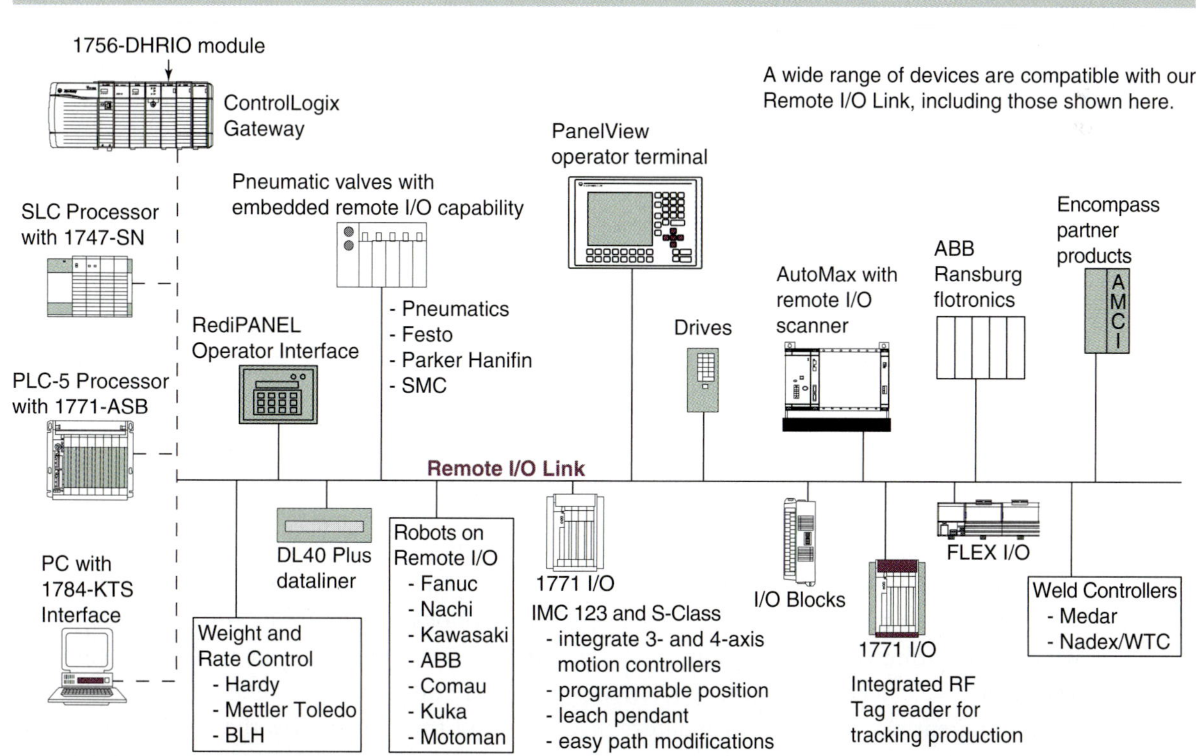

Courtesy of Rockwell Automation, Inc.

TABLE 17-1 Comparison of key characteristics of serial data interfaces

Specifications	RS232	RS422	RS485
Mode of operation	Single-Ended	Differential	Differential
Total number of drivers and receivers on one line (Note 1)	1 Driver 1 Receiver	1 Driver 10 Receiver	1 Driver 32 Receiver
Maximum cable length	50 ft.	4000 ft.	4000 ft.
Maximum data rate (Note 2)	20 Kbps	100 Kbps	100 Kbps

Note 1 – For the RS485, a maximum of 32 drivers can be connected together, providing that only one driver is on at any one time.
Note 2 – For the RS422 and RS485, the maximum data rate for a 40-ft. cable length is 10 Mbps.

with another component somewhere else, such as a PLC communicating with another PLC. The DCE is the component actually doing the communicating or performing the functions of the generator and receiver discussed in the standards, such as a modem. Table 17-1 provides a comparison of some of the key specifications for the RS interface standards.

Firewire Serial Communication. *Firewire* is Apple Computer's implementation of the IEEE 1394 standard high performance serial bus. Firewire is a single plug-and-socket connection on which up to 63 devices can be attached with data transfer speeds up to 400 megabits per second and provides:

- A simple common plug-in serial connector on many types of peripheral devices.
- A thin serial cable rather than the thicker parallel cable used on devices such as a printer or plotter.
- A very high-speed rate of data transfer that will accommodate multimedia applications of up to 400 megabits per second today with much higher rates later.
- *Hot-plug* or plug-and-play capability, meaning that a device can be connected to the bus without powering down the system.
- The ability to chain devices together in a number of different ways without terminators or complicated setup requirements.

Firewire and other IEEE 1394 implementations are expected to replace and consolidate today's serial and parallel interfaces such as the RS-232 and Small Computer System Interface (SCSI), pronounced scuzzy. Products with IEEE 1394 capability include digital cameras, digital video disks (DVDs) and tapes, digital recorders, and music systems. A brief description of the implementation of IEEE 1394 follows.

There are two levels of interface in IEEE 1394, one for the backplane bus within the controller and another for the point-to-point interface between the device and controller on the serial cable. A simple bridge connects the two environments. The backplane bus supports up to 50 Mbps data transfer. The cable interface supports up to 400 Mbps data transfer. The serial bus functions as though devices were in slots within the controller sharing a common memory space. A 64-bit device address allows a great deal of flexibility in configuring devices in chains and trees from a single socket. IEEE 1394 provides two types of data transfer: *asynchronous* and *isochronous*. Asynchronous is for traditional load-and-store applications where data transfer can be initiated and an application interrupted as a given length of data arrives in a buffer. Isochronous data transfer ensures that data flows at a pre-set rate so that an application can handle it in a timed way. For multimedia applications, the data transfer reduces the need for buffering and helps ensure a continuous presentation for the viewer. The IEEE 1394 standard requires that a device be within 15 feet of the bus socket. Up to 16 devices can be connected in a single chain, each with the 15-foot maximum (before signal attenuation begins to occur), so theoretically you could have a device as far away as 325 feet from the computer.

Another approach to connecting devices is the Universal Serial Bus (USB), which provides the same hot-plug capability as the IEEE 1394 standard. It's a less expensive technology, but data transfer is limited to 12 Mbps. Small Computer System Interface (SCSI) offers a high data transfer rate (up to 40 Mbps) but requires address preassignment and a device terminator on the last device in a chain. Firewire can work with the latest internal computer bus standard, Peripheral Component Interconnect (PCI), but higher data transfer rates may require special design considerations to minimize undesired buffering for transfer rate mismatches.

17-6-5 Wireless Interfaces

Wireless communication uses radio frequencies or infrared waves to transmit data between machines or devices where the wireless local area network is located, referred to as WLAN or LAWN. A key component is the wireless hub, or access point, used for signal distribution. To receive the signals from the access point, a machine must have a wireless network interface. Wireless signals are electromagnetic waves that can travel through a medium such as air. Therefore, no physical medium such as cables is necessary in a wireless network. Wireless standards such as IEEE 802.11(WiFi), IEEE 802.16 (WiMAX), Bluetooth, and the European HiperLAN are used in network applications. In general, there are four types of wireless network applications. Each has its own benefit and purpose. Let's examine each:

- **Peer to peer:** This network is one in which each device communicates directly with another device. This is most common in a small and simple network.
- **Point to multipoint:** This is a network where one device talks to many devices at once. The master point broadcasts a message and all devices receive and react to it.
- **Multipoint to point:** In this configuration, remote devices communicate their data back to a central location. This is often used in data collection types of systems.
- **Mesh network:** This type of network is similar to a checkerboard. Each similar colored square on the checkerboard is connected to a neighbor. If we follow the colors we can get to any square on the checkerboard.

Wireless communication is set to invade the factory floor over the next 10 years in a big way. Even as many prepare for a factory floor future with Ethernet, forward thinkers are casting their gaze to a few years beyond. Some foresee a day when sensor-packed machines will talk to host controllers across hundreds of feet without a network cable. At an Intel Developer Forum, a Bluetooth solution was demonstrated. It took control-input commands from a Bluetooth-equipped notebook computer and remotely operated a robotic arm complete with gripper, wrist, elbow, and shoulder. However, many are leery of using Bluetooth protocols for large industrial applications, saying that Bluetooth lacks sufficient broadcasting range and costs too much.

Omron Electronics offers an alternative to the conventional network in its wireless DeviceNet, designed for radio transmission of input and output signals in a factory environment. Figure 17-3 illustrates Omron's WD-30 wireless DeviceNet. The product, based on industrial DeviceNet protocols, uses a unique technology and operates at

FIGURE 17-3: Omron WD-30 wireless DeviceNet network.

Courtesy of Omron Electronics, LLC.

2.4 gigahertz. The company foresees the technology being used to enable controllers to communicate with wireless pressure sensors, temperature sensors, and flow meters in the factory. In many cases, it will make more sense to do it wirelessly as opposed to running wires out to every sensor.

17-6-6 Human-Machine Interfaces

The *human-machine interface* (HMI) is where people and technology meet. This people-technology intercept can be as simple as the grip on an electromechanical hand tool or as complex as the flight deck controls on a commercial jet. Relative to industrial networks the *graphic terminal* is the HMI that is discussed in this section. Graphic terminals offer rugged electronic interface solutions in a variety of sizes and configurations. These robust devices are fully packaged (hardware, software, and communications) and provide HMI operation. You simply download your configured application file, set appropriate communication parameters, and connect the communication cable. Graphic terminals replace traditional wired panels as the input and output mechanism for operator interaction machines. These terminals are capable of providing process information over a variety of industrial networks such as DeviceNet, ControlNet, Ethernet/IP, and Remote I/O.

A popular graphic terminal is the Allen-Bradley PanelView shown in Figure 17-4. The PanelView has a high contrast ratio of 300:1, making this graphic terminal an excellent choice for bright ambient light applications such as environments using halogen or fluorescent lights. PanelView 1400e graphic terminals offer optimum color pixel graphics and high-performance advantages, are available in touch screen or keypad, and are designed to satisfy the most demanding process control applications.

FIGURE 17-4: Allen-Bradley PanelView graphic display.

Courtesy of Rockwell Automation, Inc.

17-7 NETWORK APPLICATIONS

In this section three popular network applications are discussed—the European Profibus, the Allen-Bradley Data Highway, and the Group Schneider (Modicon) Modbus.

17-7-1 Profibus Network

Process Field Bus *(Profibus)* was created in 1989 in Germany by a consortium of factory automation suppliers. It is used primarily in Europe but is gaining worldwide acceptance. Originally developed to enable discrete manufacturing, it has expanded into process automation and enterprise-wide applications. Figure 17-5 shows the architecture of the Profibus, which encompasses several industrial bus standards including:

- *Profibus-DP*, which is a device-level bus that supports both analog and discrete signals.
- *Profibus-PA*, which is a full-function fieldbus that is generally used for process control and process-level instrumentation.
- Prof*ibus-FMS*, which is a control bus generally used for communications between DCS (Distributed Control Systems) and PLC systems.

Profibus-DP, where DP stands for Decentralized Periphery, uses a direct data link mapper, providing access to the user interface with application functions defined in the user interface. It is functionally comparable to DeviceNet. The physical media is defined via the RS-485 or fiber optic transmission technologies. Profibus-DP communicates at speeds up to 12 Mbps over distances up to 1,200 meters.

Profibus-PA, where PA stands for Process Automation, is called the Process Control Network and is an extension of the Profibus-DP for data transmission. It is functionally comparable to FOUNDATION Fieldbus. It uses IEC 1158-2, which provides intrinsic safety for process-level instrumentation, communicates at 30 Kbps, and has a maximum distance of 1,900 meters. These

FIGURE 17-5: Profibus network.

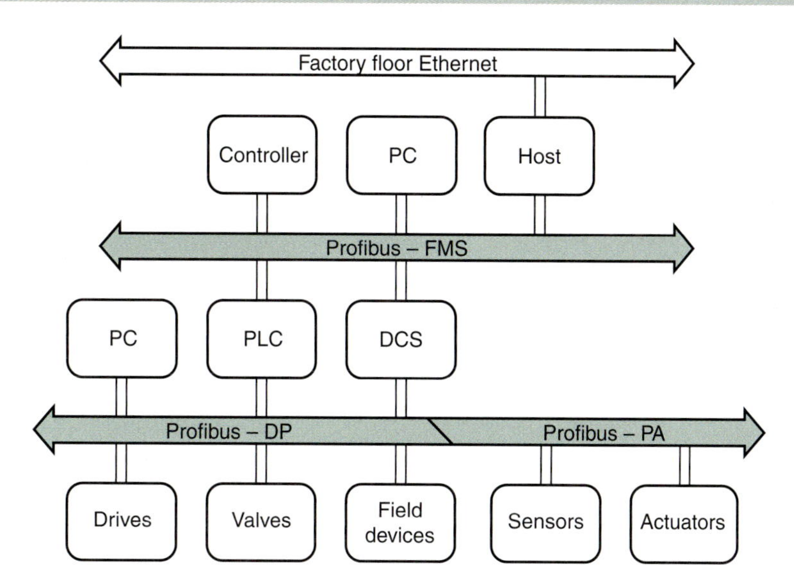

Rehg and Sartori, Industrial Electronics, *1st edition, ©2006, reprinted by permission of Pearson Education, Inc., Upper Saddle River, NJ.*

process control networks are the most advanced fieldbus networks in use today. They provide connectivity of sophisticated process measuring and control equipment and can be easily deployed for new or existing process equipment, and today's engineering tools allow for correct, efficient design. Devices typically connected to process control networks include control valves, temperature and pressure transmitters, level measurement equipment, flow meters, and process analytical instruments.

Profibus-FMS, where FMS stands for Fieldbus Message System, supplies the automation and control for real-time input/output control. It's functionally comparable to ControlNet. It is ideal for systems with multiple PC-based controllers and PLC-to-PLC communication, allowing multiple controllers—each with their own input/output and shared inputs—to talk to each other with any possible interlocking combination.

17-7-2 Data Highway Networks

The Allen-Bradley data highway networks, Data Highway Plus (DH+) and the DH-485, provide simple communication and simple implementation. Figure 17-6(a) illustrates the Data Highway Plus network. DH+ is a bus configuration, token-passing network. A bus configuration network is simply a network with one long cable, a trunk line. To connect your PLCs and other intelligent machines to the network you simply tap into the trunk. In other words, we create a branch off of the trunk line and add our device. So, think of the network as a tree. Token-passing network means that only one intelligent machine on the network can communicate. No token equals no speak. After the device with the token transmits, it electronically passes the token to the next device in line. It's like at a large meeting where only the person with the microphone can speak, and when that person is finished the microphone is passed.

DH+ programming software is available to program the PLC controllers over the DH+ network, which means that a single industrial terminal connected to the network can be used to program all the PLC controllers on all links of the network. Benefits includes switches on each interface that make it easy to reconfigure the network as it changes and incorporate network diagnostics that help to avoid costly downtime and improve network efficiency.

Figure 17–6(b) illustrates the DH-485 network. Note that the DH+ is connected to the DH-485 via a PLC and that a PanelView is connected to the DH-485 network. DH-485 is a local area network designed for factory floor application and allows for the connection of devices such as SLC 500 and MicroLogix 1000 controllers,

FIGURE 17-6: Allen-Bradley Data Highway Plus and DH-485 networks.

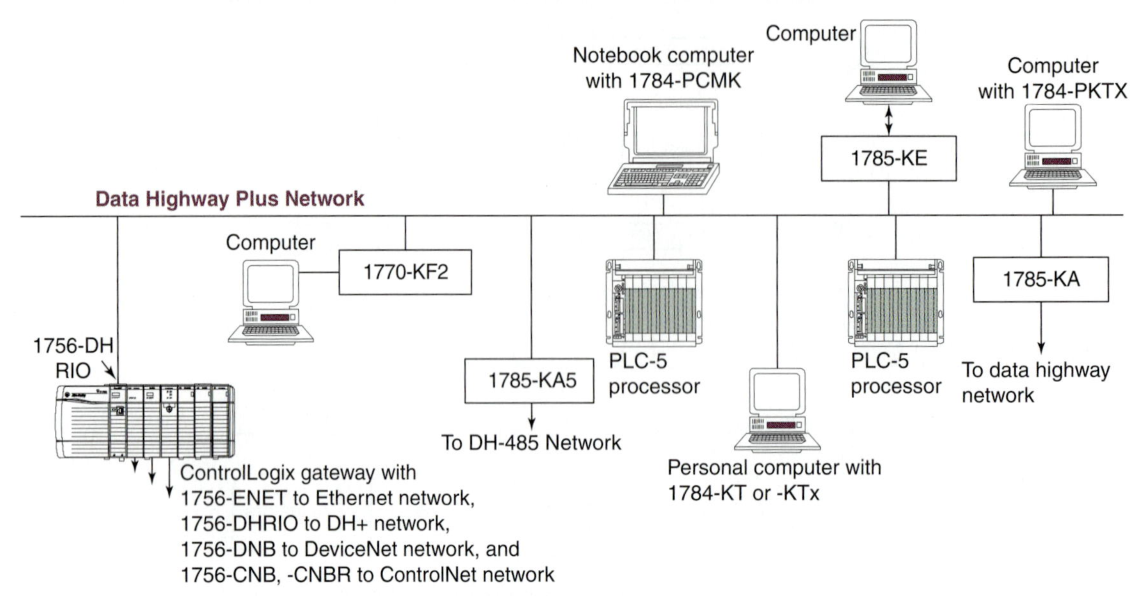

(a) Data highway plus network

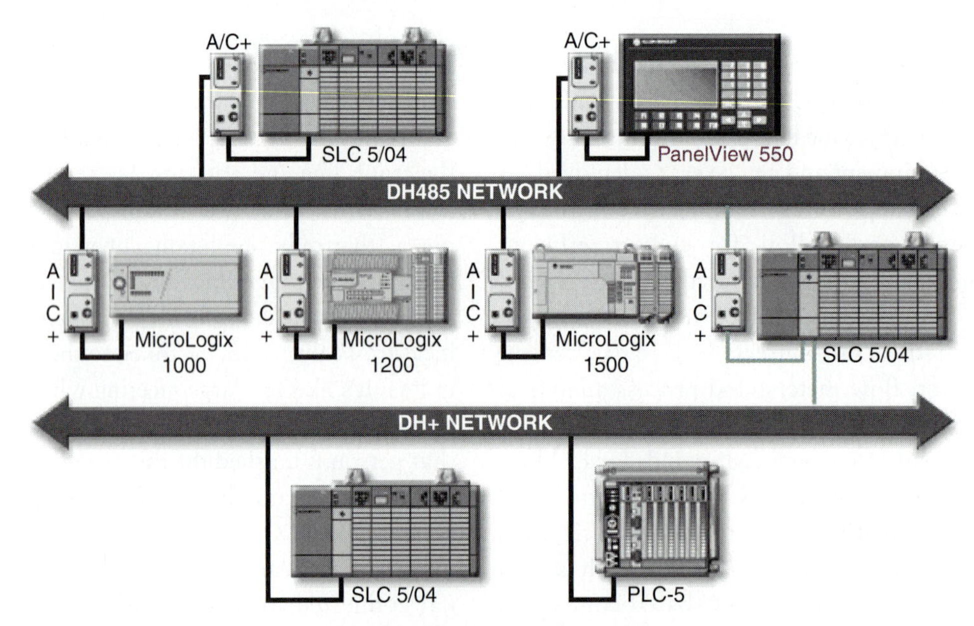

(b) Allen Bradley DH485 network

Courtesy of Rockwell Automation, Inc.

color graphics systems, and personal computers. The DH-485 network offers:

- intercommunication of up to 32 devices
- peer-to-peer capability
- the ability to add or remove nodes without disrupting the network
- a maximum network length of 4000 feet, which can be extended to 8000 feet with two Advanced Interface Converter (AIC+) units

The peer-to-peer network capability means that any controller on a DH-485 network can initiate or respond to communications with any other device on the network. With the ability to initiate communications, any controller can serve as a single initiator (also called a master or parent) on the network. The DH-485 peer-to-peer capability minimizes network traffic by allowing each controller to initiate unsolicited communications. This eliminates the need for a dedicated initiator controller, which can tie up the network with constant polling of the responder (also referred to as slave or child) controllers. You can include other devices running DH-485 as well, for example, operator interface products such as PanelView, bar code products, and drives. Also, a single personal computer can program all the controller units on the network.

17-7-3 Modbus Network

The Group Schneider (Modicon) developed the Modbus network, which uses a master/slave communication technique. This means that only one device (i.e., the master) can initiate communication. The other devices (i.e., slaves) respond to the master's communication messages, sending back the requested data or performing the requested operation. The master communicates with individual slave units or all slave units at one time—a broadcast message. The Modbus protocol establishes the format of the master's query, including a function code defining the required action, any data to be sent, and an error-checking field. The master sends the message, the slave receives it, and replies back in the same format. It's important to note that all messages have a known starting and ending point. This allows the receiving devices to know that a message has arrived, figure out if it's for them or not, and know that the message has been completely received.

Figure 17-7 shows how devices might be interconnected in a hierarchy of networks where, in message transactions, the Modbus protocol is

FIGURE 17-7: Modbus network.

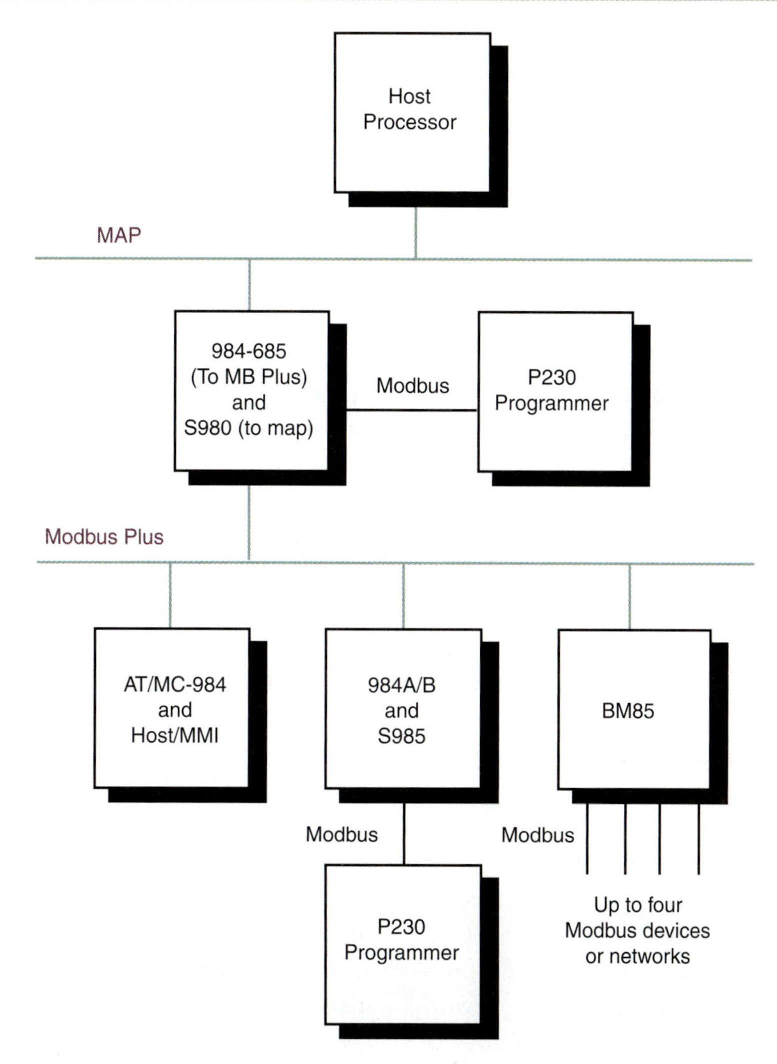

Courtesy of Square D/Schneider Electric.

embedded into each network structure, providing a common language so that all devices can exchange data. Note that Map is a standard network similar to Ethernet and Modbus Plus; it functions similar to ControlNet.

17-8 TROUBLESHOOTING NETWORK SYSTEMS

Troubleshooting is the process of taking a general problem, narrowing it down to one specific component, and then fixing it. The initial step in troubleshooting a network system problem is to collect information from the users as to the nature of the problems that they're experiencing and to collect data from the network. Troubleshooting network systems is especially important because it is rarely done at your leisure—it is strictly reactive. Generally, if something is broken, it needs to be fixed right away because network downtime can cost the company an incredible amount of money.

The *cost of downtime* can be broken into components—repair expenses, business lost, and reduced productivity. The simplest factor of the cost of downtime is the expense of repairing the problem, which is made up of several components, including hardware replacement costs and the salaries of those performing the troubleshooting. Although this is the most tangible component of downtime cost, it is generally the least of the company's concerns. Productivity and revenue lost during periods of downtime can be so great that companies will spend almost unlimited funds to repair a failed network system. Companies that have spent money to build a network often rely on that network for business- and process-critical functions. This is especially true for companies such as Internet service providers, which rely on networking as their primary source of revenue. For these organizations, downtime may translate directly into lost revenue.

Network problems come in many different forms, but generally start with user or operator complaints. One needs an organized troubleshooting process to expeditiously fix the problem process. Start by investigating the network cabling system because many network failures are cable related.

- Make sure that the cable connections are tight. In newly installed networks that's a real problem, and in existing networks machine vibration can cause a loose connection.
- Make sure that the cables are grounded properly. Coaxial cables are troublesome and shielded twisted-pair cables should be grounded at one end of the cable only.
- Make sure that cables run properly. In other words, make sure that the communication cables are not next to power lines. That, of course, will lead to electrical noise issues that tend to give sporadic problems.

When the cabling system proves reliable, troubleshooting using network diagnostics tools such as network analyzers is the next step. Each network has its own unique set of tools, both hardware and software, to evaluate and resolve network problems. Discussion of these tools is beyond the scope of this text. A course on networks, which discusses the Open Systems Interconnect (OSI) model, is a good source. The OSI model systematically defines network architecture in seven layers and is an excellent troubleshooting tool.

17-9 DISTRIBUTIVE CONTROL

Distributive control is the mechanism to access data across a highly distributed network to manage control of field devices. It allows the flexibility to do process control while using the advanced capabilities of network-based process instrumentation. An example of distributive control is the FOUNDATION Fieldbus Linking Device (FFLD). The FFLD includes the unique ability to bridge the Ethernet/IP and the High Speed Ethernet (HSE) to the FOUNDATION Fieldbus network. Bridging these networks facilitates the information flow. This information flow includes device configuration, such as setup, operation, and diagnostic data, and factory floor process information such as temperature and flow data. The advantage of this type of distributive control is that complex control strategies can be developed to best fit an automated application. For complex control scenarios, regulatory loops, which include pumps, can be executed at the controller level and backed up at the device level. For simple control scenarios, the regulatory loop can be run at the device level.

Figure 17-8 illustrates the FFLD networked into a ControlNet network. Note that each of the FFLDs is a ControlNet node that connects to two H1 networks. H1 networks are factory floor local area networks that provide connectivity to field devices. The H1 networks function as DeviceNet networks. The FFLD accesses data across the ControlNet network and distributes the data to the field devices on the H1 networks.

An application of distributive control is shown in Figure 17-9, which illustrates a storage tank system. In this application it is important to control the temperature of the tanks, which are located throughout the facility and where each tank temperature is controlled by a single loop PID controller. Management and maintenance of these distributed sensors and controllers are difficult and require an information technology to inspect them regularly. Through an FFLD, the highly distributed requirements can be maintained

FIGURE 17-8: Distributive control configuration.

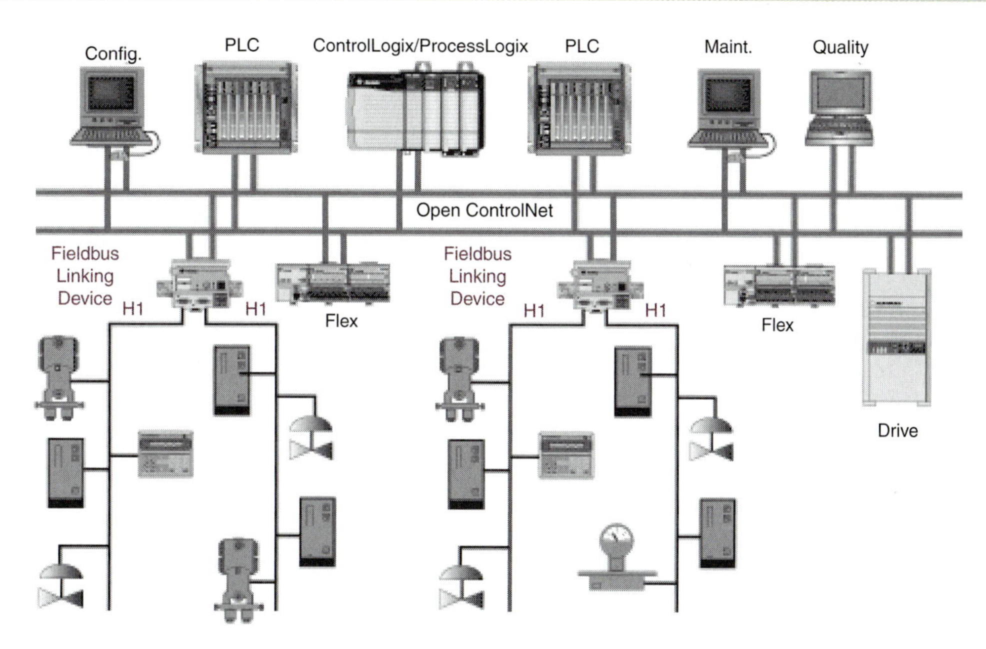

Courtesy of Rockwell Automation, Inc.

FIGURE 17-9: Tank storage system.

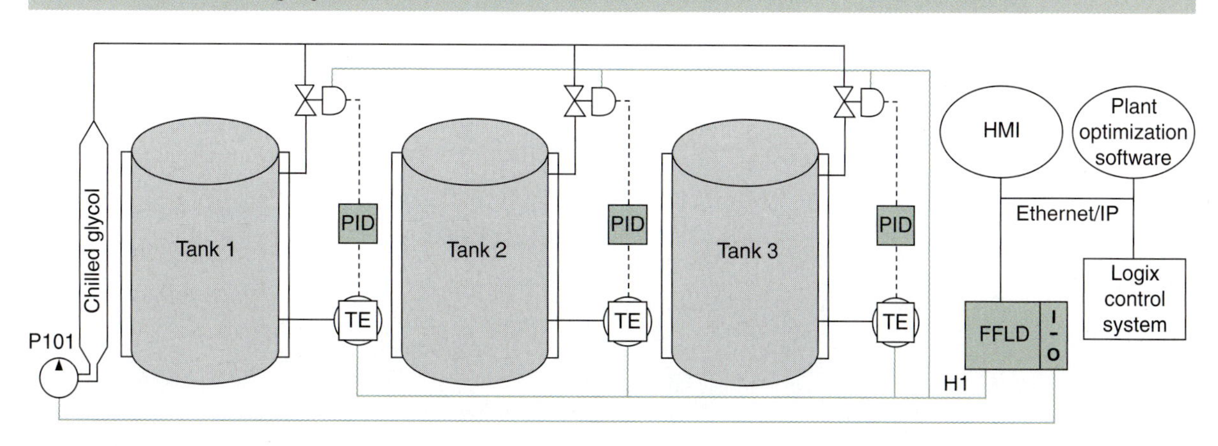

Courtesy of Rockwell Automation, Inc.

while providing remote configuration and monitoring of all loops and devices. Thus critical process or device information can be directly accessed by plant-level systems such as HMI or some plant optimization software.

17-10 DISTRIBUTED I/O

Input/output (I/O) interfaces can be located or distributed around the application or integrated with the PLC. The various PLCs, including the Allen-Bradley PLC-5, SLC 500, and ControlLogix PLCs, offer I/O that can be installed locally in the same chassis as the processor. Additionally, through the use of I/O communication networks, I/O for these platforms can be distributed in locations remote from the processor, closer to the sensors and actuators, which reduces wiring costs. Distributing the I/O provides the freedom to monitor and control I/O across an I/O link, which allows the selection of I/O products from a variety of platforms. There are two types of distributed I/O; these are commonly referred to as in-cabinet I/O and on-machine I/O.

17-10-1 In-cabinet I/O

In-cabinet I/O has the distributed I/O mounted in a central cabinet, thus not requiring an additional enclosure for environmental protection and allowing easier maintenance. In-cabinet I/O is offered in modular and block styles. *Modular I/O* is a system of interface cards and communications adapters that interface directly to the sensors and actuators of the machine/process and communicate their status to the controller via a communication network. It allows the designer to mix and match I/O interfaces and communications adapters. *Block I/O* is a complete assembly of sensor and actuator interface points including a network adapter. It may or may not include a power supply and is available in fixed configurations.

17-10-2 On-machine I/O

On-machine I/O is the placement of automation components directly on a machine rather than housing them in a central cabinet. This is possible with the emergence of more modular, compact devices; plug-and-play connectivity; flexible communication networks; intelligent devices; and a wide array of products with improved environmental ratings. On-machine components can include motor starters, drives, sensors, contactors, network media, and distribution boxes. On-machine I/O provides reduced wiring and system costs, improved mean time to repair (MTTR), enhanced control system reliability, increased productivity, and greater flexibility.

17-11 SELECTING AND DESIGNING NETWORKS

In general, the selection of a network is based on application requirements. Here are some typical application requirements and the network that would best serve the requirements.

If your application requires:	Choose this network
1. High-speed data transfer between information systems and/or a large quantity of controllers 2. Internet/Intranet connections 3. Program maintenance	Ethernet/IP
1. High-speed transfer of time-critical data between controllers and I/O devices 2. Deterministic and repeatable data delivery 3. Program maintenance 4. Media redundancy or intrinsic safety options	ControlNet

If your application requires:	Choose this network
1. Connections of low-level devices directly to controllers without the need of I/O devices 2. More diagnostics for improved data collection 3. Less wiring and reduced startup time than traditional hard-wired systems	DeviceNet
1. Plant-wide and work cell-level data sharing 2. Program maintenance	Data Highway Plus
1. Connections between controllers and I/O adapters 2. Distributed controllers with each having its own I/O communication	Remote I/O
1. Connections to modems 2. Messages that send and receive ASCII characters to and from devices	Serial

In discussing network design guidelines we will concentrate on the Ethernet, ControlNet, and DeviceNet networks because of their popularity. In designing an Ethernet network where many computers, controllers, and other devices communicate over vast distances, it's important to correctly specify the switches and interfaces that are responsible for the high-speed connectivity. Figure 17-10 illustrates an Ethernet network with the hub/switch block and the Ethernet interface (ENI) depicted. When specifying the switches in the hub/switch block in the figure, make sure that the switches have the following design features:

- Full duplex transmit and receive capability on all ports
- High-speed switching design
- Capability to control traffic flow from different systems

FIGURE 17-10: Ethernet network with Ethernet hub/switch block and Ethernet interface module.

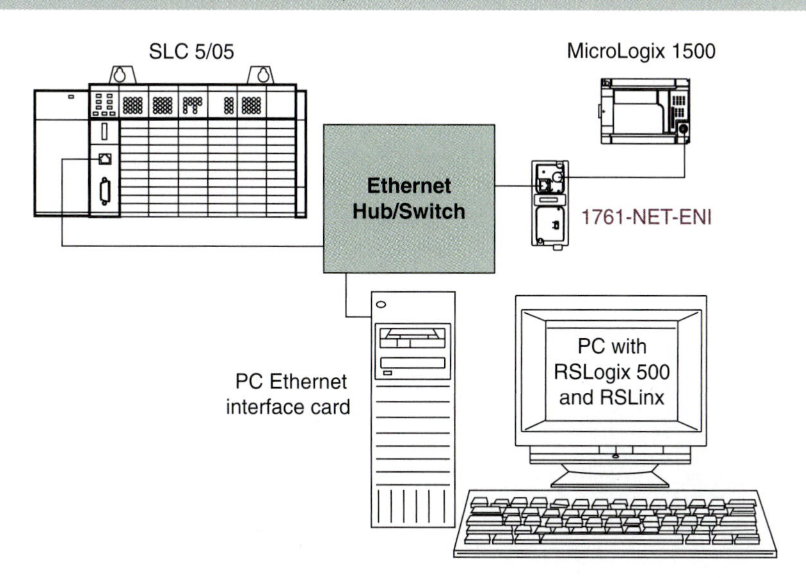

Courtesy of Rockwell Automation, Inc.

- Support redundancy for effective fault isolation
- Frame prioritization
- Address blocking to restrict traffic to a specific range
- Auto-restore of switch configuration capability for replacement
- Method to back up configuration information

Suppliers of switches that encompass these design features include Cisco, Hirshmann, and N-Tron.

In Figure 17-10 the Ethernet interface (ENI) module provides messaging connectivity for all full-duplex devices. The ENI allows network designers to connect controllers onto new or existing Ethernet networks and upload and download programs, communicate between controllers, and generate email messages. A variation of the ENI is the Web-enabled Ethernet interface (ENIW). The ENIW has the same features as the ENI plus the capability to enable the display of data Web pages with user configurable data. Both the ENI and the ENIW are available from the PLC manufacturer.

In designing the ControlNet network that transmits time-critical information and provides real-time messaging services, it's imperative that you consider the following design guidelines:

- Set your design for a maximum of 40 nodes per network, thus providing better performance and having bandwidth available for other communications.
- Design for a minimum reserve of 400 Kbytes of available memory for unscheduled data transfer, thus improving message throughput and workstation response.
- Install serial communication modules in the local chassis, thus avoiding data to be scheduled over the network.
- Save each controller's project file when you change network settings.
- Place each processor and its respective I/O on isolated ControlNet networks, thus reducing the impact of changes.
- Place shared I/O on a common network available to each controller that needs the information.

Figure 17-11 illustrates a ControlNet network depicting two key design components—the scanner module and the adapter module. The scanner module provides scheduled network connections for the controllers, and with scheduled messaging, I/O events can be controlled. The adapter module can enable multiple chassis of I/O modules to produce/consume scheduled I/O on the network. Both the scanner module and the adapter module provide media redundancy and the ability to upgrade firmware. Both of these modules are available from the PLC manufacturer.

In designing the DeviceNet network as a low-level communication link between devices and

FIGURE 17-11: ControlNet network with scanner and adapter.

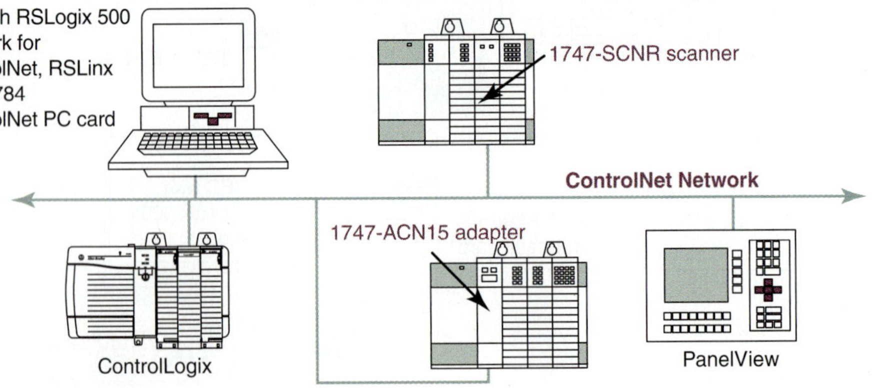

Courtesy of Rockwell Automation, Inc.

controllers, it's imperative that you consider the following design guidelines:

- Verify that the total network data does not exceed the DeviceNet communication module's data table size.
- Place the DeviceNet communication scanner modules in the local chassis to maximize performance.
- Configure a device's parameters before adding the device to the scanner's device list.
- Keep the highest node address open, typically node 63, as a test node so that you can add a new device, then change the address of the new device when it checks out.
- Keep at least one node open, like node address 62, so that a computer can be attached for troubleshooting.

Figure 17-12 illustrates a DeviceNet network depicting two key design components—the scanner module and the DeviceNet Interface (DNI) module. The scanner module is needed to enable communication between the controller and the DeviceNet-compatible I/O devices. The scanner acts as the DeviceNet master, enabling data transfer between the DeviceNet devices—the slaves. Most controllers support multiple scanners installed in a single-processor chassis. The DNI allows connection of compatible devices to the network where the DNI functions as the slave. In addition, the DNI enables the setup of a peer-to-peer communication network on the DeviceNet with other devices using DNIs. Both the scanner and the DNI modules are available from the PLC manufacturer.

In conclusion, the fact is that network design never stops. After the initial requirements analysis, which leads to a basic network design, you buy and deploy equipment. After the network is up and running, the organization discovers unforeseen uses for the network, which leads to design changes, new networking equipment, and new discoveries. The cycle goes on and on.

17-12 WEB SITES FOR INDUSTRIAL NETWORKS

The universal resource locators (URLs) for suppliers of network products and network technology that were discussed in this chapter follow. However, the URLs for supplier sites often change, so if any of the following are not active,

FIGURE 17-12: DeviceNet network with scanner and interface adapter.

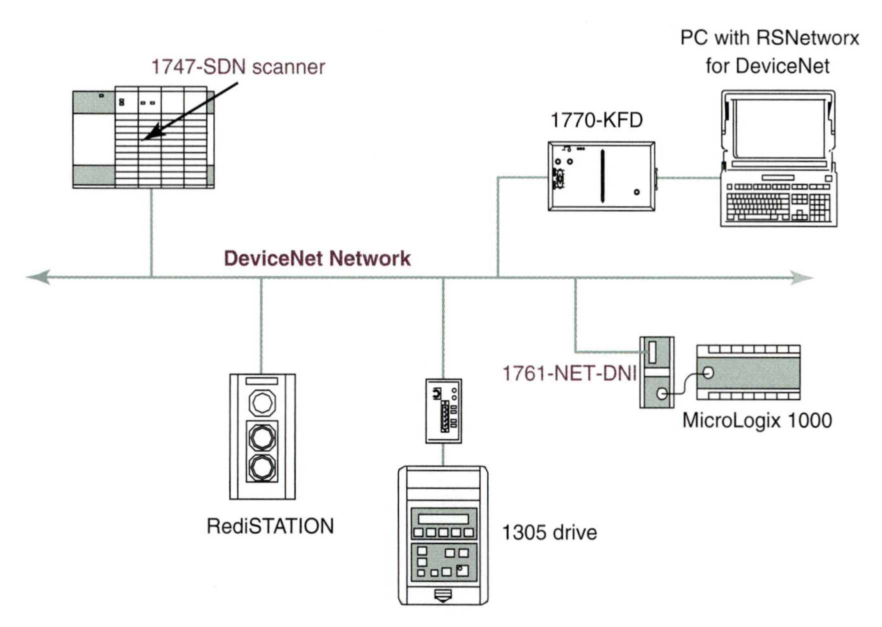

Courtesy of Rockwell Automation, Inc.

use the appropriate network term of interest in a search engine to find on-line material.

- LAN Technology Products at Cisco and Allen-Bradley: http://www.cisco.com, http://www.ab.com
- DeviceNet and Ethernet/IP Technology at ODVA: http://www.odva.org
- ControlNet, DeviceNet, and Ethernet/IP at Allen-Bradley: http://www.ab.com/networks
- Profibus Technology: http://www.profibus.com
- Modbus Technology: http://www.modbus.org
- Fieldbus Technology: http://www.fieldbus.com
- Protocol/Network Analyzers at Acterna: http://www.acterna.com

Chapter 17 questions and problems begin on page 592.

APPENDIX A

Glossary

Absolute pressure	The difference between measured pressure and a perfect vacuum
Accelerometers	Transducers that measure acceleration
Accumulated value	An integer that indicates the number of increments that a timer or counter has accumulated and is generally set to zero
Addressing mode	The means by which a PLC selects the data that is to be used in an instruction
Algorithm	A set of equations or procedures that solve a problem
Alias tag	A tag that references another tag
Ambient temperature	The temperature of the air around a device
ANSI	American National Standards Institute; a government organization that sets electrical standards
Application	The combination of routines, programs, tasks, and I/O configuration used to define the operation of a single controller
ASCII	American Standard Code for Information Interchange; a 7-bit code (with an optional parity bit) that is used to represent alphanumeric characters, punctuation marks, and control-code characters
ASIC	Application-specific integrated circuit used in PLC processors
Assembly language	A low-level programming language that maps into machine language and uses the mnemonic instructions of the CPU
Assignment	A statement that assigns values to tags
Asynchronous	Actions that occur independently of each other and which lack a regular pattern

Backplane	A printed circuit board that delivers power to the plug-in modules and provides a data bus to exchange data between the modules and the CPU
Bandwidth	The frequency range in which a system is designed to operate, expressed in Hertz
Base tag	A tag that represents the memory address where the data is stored
Bellows	A thin, sealed metal cylinder with corrugated sides like the pleats of an accordion
Binary coded decimal (BCD)	Four binary bits that represent decimal numbers 0 through 9
Binary number	Integer values displayed and entered in base 2 where each digit represents a single bit
Bit	The smallest unit of data in the binary numbering system; short for binary digit
BOOL	A data type that stores the state of a single bit, where 0 equals off and 1 equals on
Bourdon tube	A deformed hollow metal tube opened at one end and sealed at the other
Branch	A parallel logic path within a program
Broadcast	A mechanism where an intelligent machine can send data on a network that is simultaneously received by more than one machine
Byte	A group of 8 adjacent bits
Calibration	The procedure of determining the accuracy of a measuring device and replacing or repairing the device if it does not meet a predetermined accuracy
CCD	Charge coupled device
Central processing unit (CPU)	The electronic circuitry that controls all the data activity of the PLC, performs calculations, and makes decisions with its operation controlled by a sequence of instructions
Closed loop	A control system using feedback from a process to maintain outputs at a specific level
CMOS	Complementary metal oxide semiconductor
Comment	Text that explains or clarifies what a section of structured text does
Connection	A communication link between two devices, such as between a controller and an I/O module, PanelView terminal, or another controller
Construct	A conditional statement used to trigger structured text code
Contactor	A relay designed to switch large currents from large voltage sources
Control system load	The value of the manipulated variable required by the process to eliminate the effect of the load

Control system response	The response of any control system to a given input stimulus; this is a measure of how well the control system was designed
Controlled variable (CV)	The actual output from the process system
ControlNet	An automation and control layer for real-time input/output control, interlocking, and messaging
Counter	A device that counts the number of events and controls other devices based on the number of counts recorded
Current sinking	The characteristic associated with an NPN transistor where it is saturated (turned full on) so that it provides a low resistance path to ground
Current sourcing	The characteristic associated with a PNP transistor where it is saturated (turned full on) so that it provides a low resistance path to the power source
Data handling	The movement of data and the manipulation of data by arithmetic and logical operations
Data latching	A technique used to read the value of the input data that will be operated on by the instructions with a function block
Dead time delay	The delay when the feedback sensor is incorrectly located with respect to the location where the process control is occurring
Debouncing	The technique of reducing intermediate noise from a mechanical switch
Debug	The process of locating and fixing software or hardware problems
Decimal number	Integer values displayed and entered in base 10
Decrement	A term indicating that the value of a counter has decreased
Defuzzification	The process that converts fuzzy logic output conclusions to real output data and sends the data to an output device
Determinism	The ability to reliably predict when data will be delivered
DeviceNet	A device layer for cost-effective integration of individual devices
Diagnostics	Software routines that aid in identifying the causes of faults
Diaphragm	A flexible membrane, which is rubber for low-pressure measurements and metal for high-pressure measurements
Differential distance	The difference between the resetting distance and the sensing distance
Differential pressure	The difference between measured pressure and a reference pressure
Dimension	The specification of the size of an array
DINT	A data type that stores a 32-bit (4-byte) signed integer value
Direct addressing	An addressing mode in which the memory address of the data is supplied with the instruction

Discrete inputs	Connections to the PLC that convert an electrical signal from a field device to a binary state (off or on), which is read by the CPU each PLC scan
Discrete outputs	Connections from the PLC that convert an internal ladder program result (0 or 1), which turns an output device off or on
Distributed control	The system organization in which machine control is divided into several subsystems, each managed by a separate PLC
Doppler effect	The frequency of the reflected signal is modified by the velocity and direction of the fluid flow
Double word	Thirty-two bits or four bytes
Down counter	A counter that starts a specific number and decrements to zero
Drum switch	An industrial control switch typically used for motor control
Duplex	Two-way communication
Dwell time	The length of time that a machine such as a PLC pauses between operations
EIA	Electronic Industries Association; an agency that sets electrical/electronic standards
Elapsed time	The total time required for the execution of all operations configured within a single task
Electromagnetic flow sensor	A sensor that measures the electrical charges in flowing fluid, which are proportional to the fluid velocity
Element	An addressable unit of data that is a sub-unit of a larger unit of data
Empirical design	A design approach that develops the ladder solution one rung at a time
Ethernet	An information layer for enterprise-wide data collection and program maintenance
Ethernet/IP	An open industrial networking standard that takes advantage of commercial off-the-shelf Ethernet communication devices and physical media; IP refers to industrial protocol
Examine if closed (XIC)	An instruction that is true if the addressed bit is on and false if the addressed bit is off; XIC refers to open contact instructions
Examine if open (XIO)	An instruction that is true if the addressed bit is off and false if the addressed bit is on; XIO refers to closed contact instructions
Expression	Part of an assignment or construct statement that evaluates a number or a true-false state
Faraday's law	The voltage induced in a conductor moving through a magnetic field is proportional to the velocity of that conductor
Fiber optic cable	Transmits information via light pulses down optical fibers
Fieldbus	A network that is optimized to exchange data between small devices and a main large device(s)

File	A group of words or block of data treated as one unit
Firewire	Serial communication accomplished over a single wire
Flow nozzle	A narrowing spout installed inside a piping system to obtain a pressure differential
Flow rate	The volume of material passing a fixed point per unit of time, where material is a solid, liquid, or gas
Force	A push or a pull
FOUNDATION Fieldbus	A data network capable of handling all the complexities of process management, including process variables, real-time deterministic process control, and diagnostics
FOV	Field of view
Frequency response	The maximum rate at which the output is caused to change states
Full duplex	Data communication in which data can be transmitted and received simultaneously
Function block	Rectangular block with inputs entering from the left and outputs exiting on the right
Functional Block Diagram (FBD)	Graphical language where the basic programming elements appear as blocks
Fuzzy control	The implementation of fuzzy logic algorithms on a digital controller
Gage factor	The ratio of fractional change in electrical resistance to the fractional change in length
Gage pressure	The difference between measured pressure and atmospheric pressure
Gray code	The binary numbering system in which any value can be changed to the next higher value by changing only one bit
Half duplex	Data communication in which data can be transmitted in two directions but in only one direction at a time
Hard wired	Electrical connectivity through physical wiring
Hexadecimal number	Integer values displayed and entered in base 16 where each digit represents four bits
High level language	A programming language tending toward English and mathematical instructions with no implementation details for the CPU operation
Hysteresis	An operational dead band that eliminates false indications such as in a sensor reading
IEC	International Electromechanical Commission; an agency that sets standards, which include PLC programming guidelines
Image table	An area of PLC memory reserved for I/O data where 1s represent an on condition and 0s represent an off condition
Increment	A term indicating that the value of a counter has increased
Index	A reference used to specify an element within an array

Indexed addressing	An addressing mode for referencing an address that is the original address plus the value stored in an index register
Indirect addressing	An addressing mode in which the address of the instruction serves as a reference point instead of the actual address
Instantaneous contacts	Normally open/normally closed contacts that operate independently of a timer's time-delay period
Instruction	The command that causes a PLC to perform one specific operation
Instructional List (IL)	Low-level, text-based language using mnemonic instructions like using machine code in microprocessors
Interrupt	An external request typically from a peripheral device requesting service
Keying	A technique to prevent the insertion of a module or connector into the wrong location or slot
Ladder Diagram (LD)	Graphical language based on traditional ladder logic
Ladder logic	A graphical programming technique that depicts the program control logic on horizontal rungs of a ladder
Laminar flow	Fluid flowing rather smoothly parallel to the walls of the pipe
LAN	Local area network; a group of computers and devices connected to serve a region
Latching relay	A relay that when commanded to a position remains in that position until it's commanded to another position
Least significant bit (LSB)	The bit that represents the minimum bit value of a byte or word
Light emitting diode (LED)	A semiconductor whose junction emits light when current flows through it in the forward direction
Limit switch	An electrical switch that is activated when the motion of a machine or equipment physically contacts the switch
Limit test	A test that determines if a value is inside or outside a specified range
Linear variable differential transformer (LVDT)	A position sensor, consisting of a transformer with a movable core
Load cell	A force transducer that employs a direct application of a bonded strain gage
Machine language	A programming language that consists of 1s and 0s and is used directly by the CPU in the microprocessor
Major fault	A fault condition that is severe enough for the controller to shut down, unless the condition is cleared
Manipulated variable (MV)	The variable regulated by the final control element to achieve the desired value in the controlled variable
Mask	A binary number used to pass and inhibit data, generally where a 1 allows data to pass and a 0 inhibits the data
MCR	Master control reset (PLCs) ; master control relay (relay logic)

Minor fault	A fault condition that is not severe enough for the controller to shut down
Modbus	A network that uses a master/slave communication technique
Most significant bit (MSB)	The bit that represents the greatest bit value of a byte or word
Mnemonic	A small group of letters that assist a person in remembering a word or a phase; e.g., SQR is the mnemonic for the square root instruction
NEMA	National Electrical Manufacturers Association; a standards organization for electrical equipment
Nested subroutine	A subroutine that begins and ends within another subroutine
Nibble	Four bits or one half of a byte
Node	A hardware connection point in a network
Normally closed (NC) contact	A relay contact that provides a conductive path when the relay is de-energized
Normally open (NO) contact	A relay contact that provides a conductive path when the relay is energized
Numeric expression	In structured text, an expression that calculates an integer or floating point
Object	A structure of data that stores status information
Octal number	Integer values displayed and entered in base 8 where each digit represents three bits
Off-delay timer	A timer that changes state some time after power has been removed
On-delay timer	A timer that changes state some time after power has been applied
One shot	A programming technique that sets a bit for only one program scan
One's complement	The inverse of a binary number obtained by changing the 1s to 0s and the 0s to 1s
Open loop	A system that has no feedback or autocorrection
Operands	Symbols in an instruction
Orifice plate	A washer-shaped device that is installed in a piping system; the flow rate is determined from the measurements of pressure in front of and behind the plate
OSHA	Occupational Safety and Health Administration; a government agency that sets and enforces work rules and safety practices
Overflow	An indication that a counter has incremented above its maximum positive number
Overload monitor	A device that typically opens contacts when the monitored object exceeds a specific temperature
Overdamped response	A system response in which the damping coefficient is greater than 1, which causes the response to overshoot the set point before settling to it

Parity bit	An additional bit added to a binary number that makes the sum of the quantity of 1s in the number even or odd
Perfect vacuum	A pressure of zero
Photoelectric device	A device that senses the presence of an object when it either breaks a light beam or reflects a beam of light to a receiver
Piezoelectric effect	When pressure is applied to a crystal, the crystal deforms and produces a small voltage, which is proportional to the deformation
Pilot switches	Various switch types that have a high current rating
Pitot tube	A flow-measuring device, consisting of two tubes placed in the fluid flow that sense two pressures—impact pressure and static pressure
PLC	Programmable Logic Controller; a special-purpose computer designed for single use or one of several controllers on an automation network for the control of a wide variety of manufacturing machines and systems using one of five programming languages
Pole	An internal conductor in the switch that is moved by the switching mechanism
Preset value (PRE)	An integer that indicates the number of increments that a timer or counter accumulates in order to provide some action
Pressure	The amount of force applied to an area, usually expressed in pounds per square inch (psi)
Pressure sensors	Services that detect the force exerted by one object on another
Process	A continuous manufacturing operation
Process variable (PV)	The value of the output that is present at the feedback input of the summing junction
Profibus	European communication standard that was developed to enable discrete manufacturing but has expanded into process automation and enterprise-wide applications
Program	A set of related routines and tags
Proportional band	The percentage change in the error that causes the final control element to go through its full range
Proportional derivative control (PD)	A control used to correct for rapidly changing disturbances in a control system
Proportional integral and derivative control (PID)	A control that combines the benefits of PI and PD controls in a control system
Proportional integral control (PI)	A control used to eliminate the steady-state error component in a control system
Proportional valves	Electromechanical devices, which adjust the flow of fluid over the range of 0 to 100 percent
Protocol	Rules used by machines to communicate with each other

Qualifier	In the action of a sequential function chart (SFC), a qualifier defines when an action starts and stops
Rails	The two uprights in a PLC ladder program
Real	A data type that stores a 32-bit (4-byte) IEEE floating point value
Relay	A remotely operated switch consisting of an electromagnet, a solenoid, and switch contacts
Repeatability	Ensures that transmitting times are constant and unaffected by devices connecting to or leaving the network
Resetting distance	The point at which the output of the sensor changes from on to off as an object is withdrawn from the sensor
Resistance temperature detector (RTD)	A temperature-sensing device that detects a change in resistance in a metal as a function of temperature
Resolution	Indicates how accurately an analog value can be expressed digitally and is specified in bits
Retentive timer	A timer that accumulates time whenever powered and retains the accumulated time when power is removed
Ringing	The oscillations present while the response settles to a final value
Routines	Blocks of code that perform one function such as calculating the sine of an angle or performing an initialization sequence
RS-232, RS-422, RS-423, RS-485	Communication standards that specify electrical, mechanical, and operational characteristics of interfaces for exchanging data between intelligent machines
Safety	The freedom from danger, risk, or injury
Scaling	The resizing of a signal to meet the requirements of the using component of a system
Scan time	The time required to read all inputs, update all outputs, and execute the control program; the scan time is not constant
Selection sequence	In a Sequential Function Chart (SFC), the selection sequence checks each entry transition in a specified order
Sensitivity	The measure of how closely a device can discriminate between levels
Sensors	Devices that are sensitive to a physical condition such as heat, pressure, motion, and light and output an electrical signal proportional to the physical input
Sequence table	A sequential list of operations of a sequencer
Sequencer	A device that is programmed so that a fixed set of actions occurs repeatedly
Sequencing	Predetermined step-by-step process that accomplishes a specific task
Sequential Function Chart (SFC)	Graphical language whose basic language elements are steps or states with associated actions and transitions with associated conditions used to move from the current state to the next

Serial real-time communication system (SERCOS)	An open controller-to-intelligent digital-drive interface specification
Set point (SP)	An integer or real value entered into the control system that indicates the desired value of the process output
Setting distance	The maximum sensing distance for a sensor and the object when worst-case ambient temperature and supply voltage variations are assumed
Shift register	A register that allows the movement of its contents to the right or to the left
Signal conditioning	The converting of a level or type of signal to another level or type of signal to be used by another stage of the system
Simultaneous sequence	In a Sequential Function Chart (SFC), the simultaneous sequence is the AND type—the exit transition cannot be checked until the last step
SINT	A data type that stores an 8-bit (1-byte) signed integer value
Solenoid	An electromechanical device that converts electrical energy into linear mechanical motion
Solenoid value	An electromechanical device that is used to control the flow of air or fluids such as water, inert gas, light oil, and refrigerants
Status indicators	LEDs that indicate the on-off status of an input or output point and are visible on the outside of the PLC
Step	In a Sequential Function Chart (SFC), a step represents a major function of a process
Step response	The time necessary for a signal to go to typically 95 percent of its final value
Strain	The amount of deformation of a body due to an applied pressure or force
Stress	The change in an object when force is applied
String	Group of data types that store ASCII characters
Structured Text (ST)	High-level, text-based language like BASIC, C, or PASCAL with commands that support a highly structured program development and the ability to evaluate complex mathematical expressions
Subroutine	A group of instructions that are outside the main program and executed only when accessed
Syntax	The rules that govern the structure of a language
Tag	A text-based name for an area of the controller's memory where data is stored
Task	It holds the information necessary to schedule the program's execution and sets the execution priority for one or more programs
Temperature	The degree of hotness or coldness of a body or environment

Terminal	An attachment point for field devices on a PLC module
Thermistors	Electronic components that exhibit a large change in resistance with a change in its body temperature
Thermocouples	Temperature-sensing devices that produce a small voltage in the millivolt range as a function of temperature and are constructed using two dissimilar metal wires
Thermopile	The serial connection of several thermocouples to enhance their resolution
Thumbwheel	A mechanical device used to manually enter a number
Timed contacts	Normally open/normally closed contacts that are activated at the end of a timer's time-delay period
Timer	A device that can be preset to a specific number and control the operation interval of other devices
Toggle switch	A small electrical switch with an extended lever
Token ring	A technique that provides an ordered transmission sequence between machines where a binary number or token is circulated among nodes and used by the nodes to gain access to the network
Topology	The shape of the network, in other words, the connection pattern of the network nodes
Transducer	A device that receives one type of energy and converts it to another type of energy and generally includes a sensor and a transmitter
Transistor	A three-terminal device in which the current through two terminals can be controlled by small changes in current or voltage at the third terminal
Transition	In a Sequential Function Chart (SFC), a transition signals a change in steps or states
Transmitters	Devices that convert small signals into larger, more usable signals
Triac	A solid-state device that switches AC current
Troubleshooter	A skilled person employed to locate trouble or make repairs on machinery or technical equipment
Troubleshooting	The intricate process used to solve problems, which is predominantly mental with the use of equipment and including electrical and mechanical manipulation of objects
Troubleshooting problem	A situation in which an answer, solution, or decision is not immediately apparent but may be found with a logical methodology that often has an intuitive component
Truth table	A listing of a set of inputs and the state of an output as a function of the inputs
Turbulent flow	Fluid is flowing down the pipe, but swirling within the flow
Twisted-pair cable	A pair of wires that form a circuit that can transmit data; the wires are twisted to provide protection against crosstalk

Two's complement	The inverse of a binary number obtained by changing the 1s to 0s and the 0s to 1s and then adding 1 to the result
UL	Underwriters Laboratory
Underdamped response	A system response in which the damping coefficient is less than 1, which causes the response to oscillate around the set point before settling to it
Underflow	An indication that a counter has incremented below its maximum negative number
Up counter	A counter that starts a specific number, typically 0, and increments up
Vacuum	A pressure less than atmospheric pressure
Velocity flow sensors	Devices that measure the fluid flow rate based on changes in flow velocity
Venturi	A flow-measuring device that consists of a gradual contraction followed by a gradual expansion within a fluid-carrying pipe
Vision system	A camera and lighting system that is used for part identification, part location, part orientation, part inspection, and range finding
Vortex flow sensor	A sensor that uses a non-streamlined object placed in the fluid flow, which creates vortices in a downstream flow
Word	A group of 16 bits or 2 bytes treated as one unit
Work cell	A group of machines, generally including a robot, working together to produce a particular product
Zero energy state	A system state where all sources of energy are removed to minimize the possibility of personal injury
Zone	The portion of a PLC ladder program that can be enabled or disabled by a control function

APPENDIX B

PLC Module Interface Circuits

B-1 PLC MODULE INTERFACES

There are numerous input and output modules because of the variety of input and output field devices and the voltage and current requirements present. DC modules often have either a current sinking or current sourcing feature, while AC modules are both current sinking and sourcing. This important concept is illustrated in Figures 1-13 and 1-14, and an input module example for current sinking and current sourcing is presented in Figure B-1. A study of this figure indicates that the sourcing sensor normally uses a PNP output transistor for the output circuit, and the sinking sensor uses an NPN transistor for the output. Note how the field device power supply is connected to provide the correct conventional current flow for the sinking and sourcing input modules. Read the description in the figure.

B-1-1 Input Interface Circuits

There are many different types of input circuits used for PLC input modules. Five of the most common (AC/DC, TTL, DC sinking, DC sourcing, and AC) are illustrated and described in Figure B-2. Read the descriptions of the circuits in the figure as you study these Allen-Bradley SLC 500 circuits to find the common elements that are present.

All have opto-isolation between the field device power and the PLC power. Noise and operational problems are eliminated when the power and grounds of the PLC and other work cell systems are isolated and independent. The isolation (diode and transistor without a base at the far right of the circuit) is produced with a light emitting diode (LED) and photo-triggered transistor. Note that the current sinking and sourcing types have *one* diode in the opto-isolator, and the AC types have *two* LEDs in opposite directions to handle the AC current when it changes direction. Circuits (b), (c), and (d) in Figure B-2 have LEDs, which provide an indication on the front of the module when an input signal is present. The input resistor limits input current and the capacitor filters out input noise.

The inverting Schmidt trigger in the TTL interface in Figure B-2(b) produces some hysteresis, so the output will not switch on and off as the input

FIGURE B-1: Current sinking and current sourcing interfaces

Sensor/Module Interface	Input Module/Sensor Schematic
(a) Input Module (Current Sinking) The illustration shows a current sourcing sensor and current sinking input module. Note that a PNP transistor is often used as the final component in the output of the current sourcing type field device. These sensors are called three wire devices. Allen Bradley uses the following color code: brown goes to positive supply, black is the output, and blue is the common.	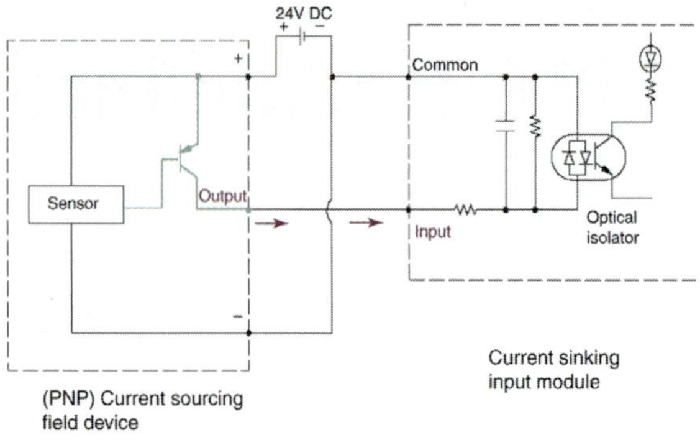 24V DC + − Common Sensor Output Input Optical isolator Current sinking input module (PNP) Current sourcing field device
(b) Input Module (Current Sourcing) The illustration shows a current sinking sensor and current sourcing input module. Note that an NPN transistor is often used as the final component in the output of the current sinking type field device. These sensors are called three wire devices. The same color code from the current sinking model applies here.	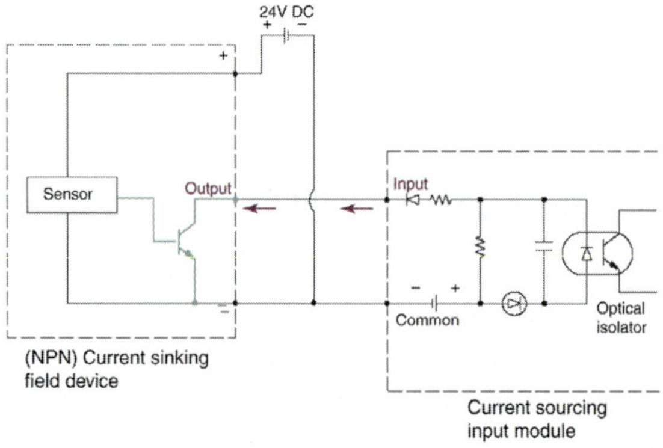 24V DC + − Sensor Output Input Common Optical isolator (NPN) Current sinking field device Current sourcing input module

moves above and below the trigger point. In addition, it inverts the input signal so that a true input produces a 0 or false condition in the input image table. When this module is used the PLC input instruction type must be the opposite of that normally used. For example, if an XIC instruction would normally be used, an XIO instruction must be used when this input module is present. Read the descriptions of all the input circuits in Figure B-2 until you understand how the numerous types of input modules function.

B-1-2 Output Interface Circuits

There are many different types of output circuits used for PLC output modules. Four of the most common (DC, AC, TTL, and relay) are illustrated and described in Figure B-3. Read the descriptions of the circuits in the figure as you study these Allen-Bradley SLC 500 circuits to find the common elements that are present.

Like the input modules just discussed, all have opto-isolation between the field device power and the PLC power. The isolation (diode and transistor at the far left of the circuit) is produced with a light emitting diode and photo-triggered transistor. The circuits operate the same as the input opto-isolators. Many modules also have an LED that indicates on the front of the module when an output signal is present.

The TTL inverter in the output interface in Figure B-3(c) produces the output voltage and drive necessary for the TTL field devices. In

FIGURE B-2: Input module electronic interface circuits

Input Module Description	Input Module Schematic
(a) AC/DC Input Module The AC/DC input modules support both types of input signals. They can be used with all AC field device outputs with compatible voltage levels. In addition, they can be used with sinking or sourcing DC type field devices. Note that an opto-isolator has reversed parallel diodes to handle the AC input and to permit the DC current to flow in either direction for an active input.	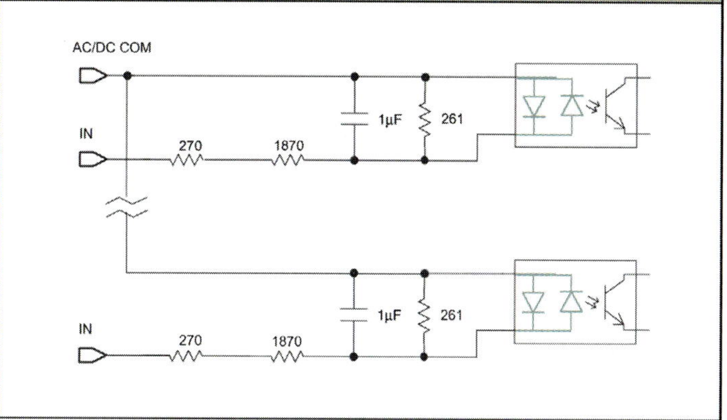 AC/DC COM IN 270 1870 1µF 261 IN 270 1870 1µF 261
(b) TTL Input Module (Current Sinking) The TTL input modules are designed to handle the voltage levels and current drive limits of integrated circuit logic. The inputs are current sinking types so the field devices must be current sourcing models. The modules have a DC voltage source terminal common to all input points and a common also connected to all input ports. The inverting Schmidt trigger is used to add hysteresis to the input and make the switching point crisper. However, the inversion causes the module to place a logic 0 in the input image table when the field device and input are active. This is handled by using the opposite type of logic element in the input ladder.	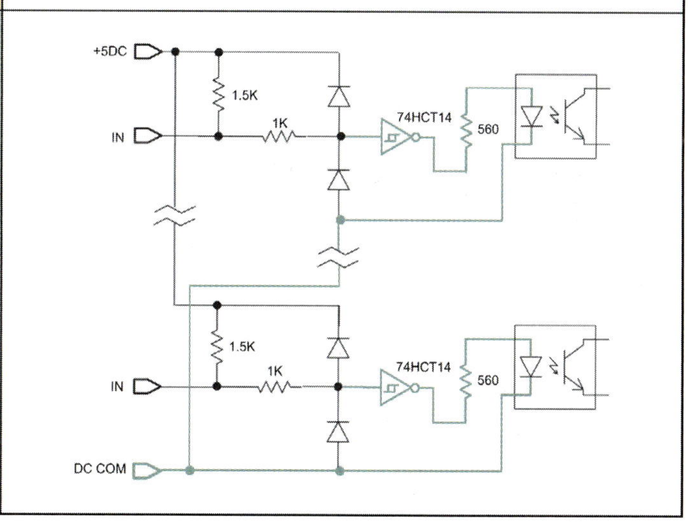 +5DC 1.5K IN 1K 74HCT14 560 1.5K IN 1K 74HCT14 560 DC COM

addition, the inverter inverts the output signal, so a true output or a 1 in the output image table produces a false output condition and an off field device. When this module is used the ladder logic must produce an output state that is the opposite of that normally used. For example, if the output field device must be on for a given condition of the input field devices, then the ladder rung logic must evaluate to a false so that the output is false. This will turn the field device on when the inverter changes the false output to the true state at the terminals of the output module.

The final circuit, a relay output module, uses mechanical contacts, either normally open or normally closed, to switch power on and off to the output field device. Read the descriptions of all the output circuits in Figure B-3 until you understand how the numerous types of output modules function.

FIGURE B-2: (Cotinued)

Input Module Description	Input Module Schematic
(c) DC Input Module (Current Sinking) The sinking input modules have a current flow into the input terminal when the input is active. As a result the signal from the field device at the terminal must be a positive voltage for an active input. To achieve that, the field device must have a sourcing type of output. All of the commons for all input points on the module are connected to the DC common. Note that an opto-isolator is used to separate the power from the sensor from the power for the module.	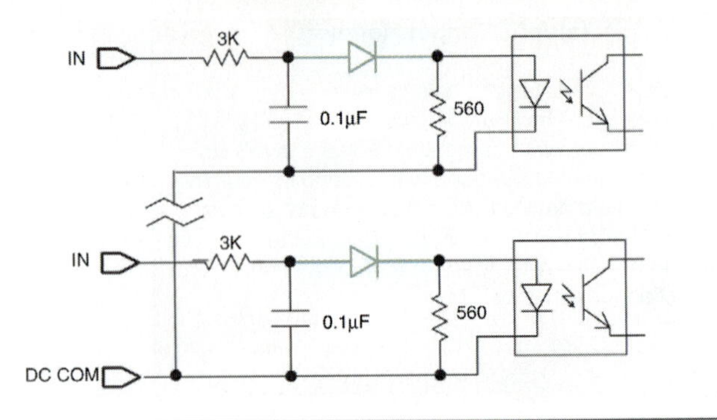
(d) DC Input Module (Current Sourcing) The sourcing input modules have a current flow out of the input terminal when the input is active. As a result the signal from the field device at the terminal must be a ground for an active input. To achieve that, the field device must have a sinking type of output. The other terminals for all input points on the module are connected to the DC power. Again an opto-isolator is used to isolate the two power sources.	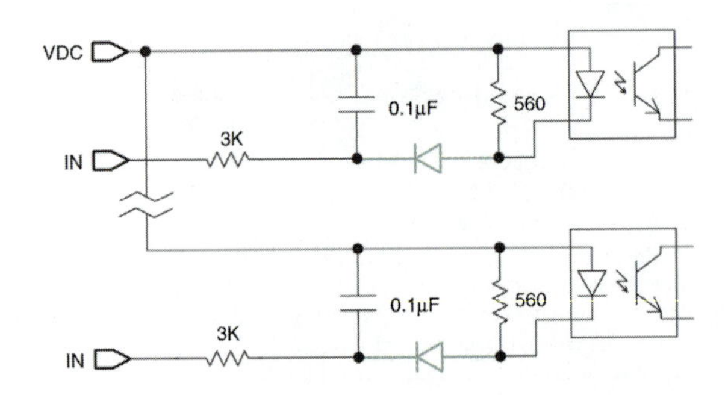
(e) AC Input Module The AC input modules have a current flow that changes direction every half cycle. So the AC modules require two diodes in the opto-isolator each pointing in the opposite direction so that the isolator is conducting on both half cycles. If the field device is a switch contact, then it is placed in series with an AC source. If it is a sensor with an AC ouput it is just connected to the AC module input with the commons aligned.	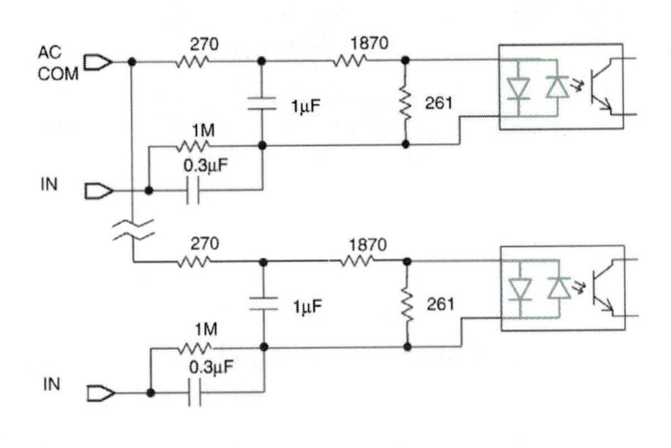

FIGURE B-3: Output module electronic interface circuits

Output Module Description	Output Module Schematic
(a) DC Output Module (Current Sinking) The sinking output modules have a current flow into the output terminal when the output is active. For compatibility, the field device must have a sourcing type of input. All of the commons for all output points on the module are connected to the DC common and a positive DC voltage is applied to each output port as well. Note that an opto-isolator is used to separate the power from the actuator from the PLC power. The sinking output circuit is shown with NPN type transistors for the opto-isolators and drive transistors. A sourcing version of this circuit is also available where the NPN is replaced with a PNP transistor and the power and common lines are reversed.	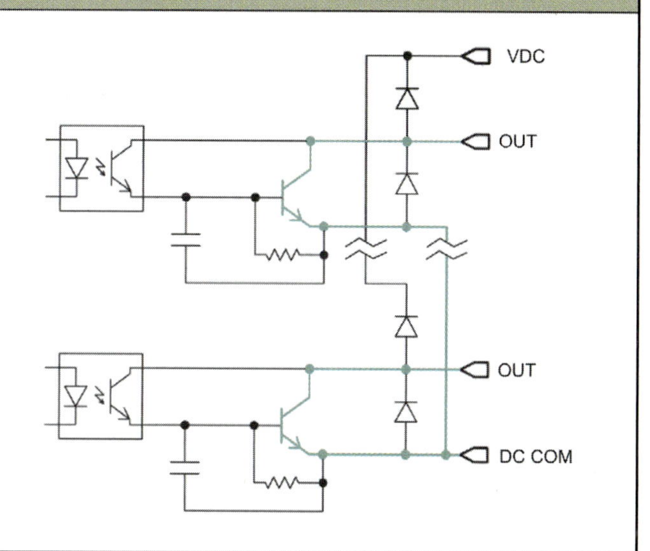
(b) AC Output Module The AC/DC output modules use a Triac to turn on the output field device actuator. An AC source is connected to L1 and the actuator is placed in series with that AC source and connects to the output terminal. They can be used with all AC field device outputs with compatible voltage levels.	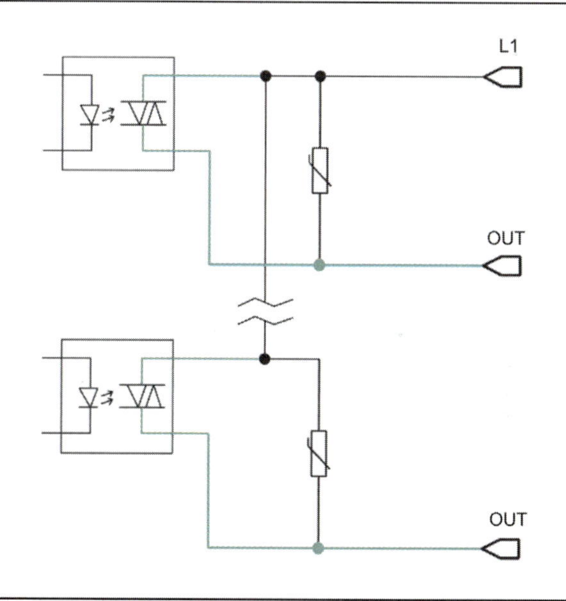

FIGURE B-3: (Cotinued)

Output Module Description	Output Module Schematic
(c) TTL Output Module (Current Sinking) The TTL output modules are designed to handle the voltage levels and current drive limits of integrated circuit logic. The outputs are current sinking types so the field devices must be current sourcing models. The modules have a DC voltage source terminal common to all output points and a common also connected to all output ports. The transistor in the opto-isolator would make the output a sourcing type but the inverting operational amplifier adds inversion to the output signal, making it a sinking output. As a result, the inversion causes the module to output to false or not active when the output image table is true and indicating that the field device should be turned on. This is handled by changing the rung logic so the rung coil is off for an on condition of the field device.	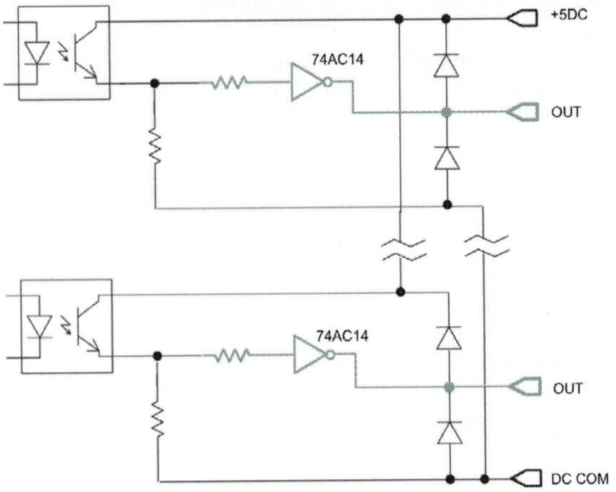 +5DC 74AC14 OUT 74AC14 OUT DC COM
(d) Relay Output Module The relay output modules have a normally open relay contact for each output port. This permits moderately larger load currents and the ability to switch either AC or DC field devices.	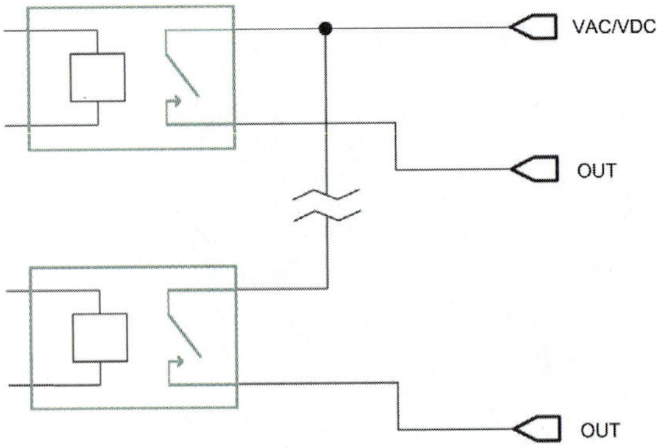 VAC/VDC OUT OUT

APPENDIX C

Programmable Logic Controller History

C-1 INTRODUCTION

Dick Morley conceived the concept of the first programmable controller in the United States on January 1, 1968. He later produced it under the company name Modicon, which is short for MOdular DIgital CONtroller. Morley is pictured with the first PLC in Figure C–1. The first installed Modicon in industry was the model 084. Although the first installation at the Oldsmobile Division of General Motors Corporation and the Landis Company in Landis, Pennsylvania, occurred in 1970, the fledgling company's growth was slowed because of industry concern for replacing relays with computer controlled logic. However, Modicon's growth increased as a result of engineer Michael Greenberg's development of the model 184, a more sophisticated version of the original model. Gradual industry acceptance and the success of this new technology created a global PLC industry in the 1970s. Over time the generic term programmable controller, or PC, became the device designation. The introduction of the personal computer, also shortened to PC, in the late 1970s caused confusion in the technology reference, so the term programmable logic controller, or PLC, was adopted and is still used today.

C-1-1 Relay Replacers

The first PLCs were promoted as "simply relay replacers." This indirect approach concealed the computer nature of the PLC from users who were reluctant to embrace the complexities of computer systems. Reliability was then and continues to be a real concern in manufacturing automation, so early PLC adoptions would have been lost if they were portrayed merely as an industrial computer programmed in ladder logic. Instead the PLC was sold as a new form of mechanical relay, timer, and counter. The present difficulty of moving away from ladder logic programming in the United States is a result of this original promotion strategy.

As the applications for PLCs expanded to a broad range of manufacturing sectors, users demanded more features, including subroutines, complex math functions and data handling, interrupts, analog input/output, proportional integral and derivative process control, distributive control, and the ability to communicate between

FIGURE C-1: Dick Morley pictured with the first PLC—the forerunner to the Modicon 084

Courtesy of Dick Morley and R. Morley, Inc.

PLCs over network cable. However, since each vendor developed its own network technology, the different networks were not compatible. Only systems from the same vendor could share data over the communications network because each was proprietary.

C-1-2 PLCs in the 1980s

A shift away from proprietary, dedicated, and expensive PLC programming terminals occurred when companies outside the PLC manufacturing group provided PLC programming software that would run on standard IBM PCs and compatibles. The PLC and PC became compatible technology, with the PC used for programming and the PLC used for control. Some vendors introduced software that would permit the PC to take over all the functions associated with the traditional PLC.

The 1980s also focused on the initial efforts at PLC standardization in communications with General Motor's standard called Manufacturing Automation Protocol (MAP). The industry also worked on reducing the size of the PLC; as a result, the world's smallest PLC today is about the size of a pack of cigarettes.

C-1-3 PLCs in the 1990s

The 1990s witnessed increased efforts toward standardization of PLC programming languages with the introduction of the International Electrical Commission standard IEC 1131-3 in 1993. This standard was later changed to IEC 61131.

A second major thrust in this decade was the introduction of sub-networks standards, such as ControlNet, DeviceNet, or FOUNDATION Fieldbus, that distribute system control by the PLC over an extended cable network. In addition, the standard assures that devices and controllers from different vendors can coexist on the same sub-network.

C-1-4 PLCs in the 2000s

It is difficult to predict the future in technology due to the current rate of change. However, the Internet is a technology that has had a significant impact in the past and will continue to impact the direction of automation control in the future. Internet/IP is a new technology standard that will open the door for using the Internet as a major network for exchanging automation information between PLCs and the input and output devices, called field devices, used in the processes. An increasing number of wireless networks will be used in automation control in the future. Automation designers in this decade will also witness increased standardization in the programming languages used for PLC automation applications.

APPENDIX D

Questions and Problems

CHAPTER 1 INTRODUCTION TO PROGRAMMABLE LOGIC CONTROLLERS

REFERENCES

Textbook Chapter 1 – Introduction to Programmable Logic Controllers (pp. 3–32)

The following examples and figures are referenced in the questions and problems for this chapter:

- Example 1-1 is on page 15
- Example 1-2 is on page 15
- Figure 1-12 is on page 14
- Figure 1-17 is on page 23
- Figure 1-19 is on page 26
- Figure 1-20 is on page 27

QUESTIONS

1. Write a definition of a programmable logic controller.
2. What international standard governs PLC systems and languages?
3. How big is the PLC industry and how fast is it growing each year?
4. Describe the similarities and differences between PLCs and PCs.
5. Describe the similarities and differences between a PLC solution and a relay ladder logic solution.
6. Name all the major parts of a relay and list four different types.
7. What is the primary difference between a physical relay used in relay ladder logic and the virtual relay used in a PLC ladder logic program?
8. What is the major difference between the input contacts in a relay ladder logic solution and the input instructions used in PLC ladder logic program?
9. Name all the major component parts of a PLC system.
10. Describe each of the major components in a PLC system.
11. What is difference between fixed and modular I/O?
12. Compare and contrast Ethernet/IP, ControlNet, and DeviceNet.
13. Compare and contrast Smart I/O, Remote I/O, and serial interfaces.
14. How does a SERCOS interface differ from a standard serial interface?
15. In what year was the PLC invented and who was the inventor?
16. What was the name of the first PLC company?
17. Why were the first PLCs called "relay replacers" and not industrial computers? (Hint: see Appendix C.)

18. Describe the differences among the three PLC types: rack/slot address based, tag based, and soft PLC.
19. If the Allen-Bradley SLC system input address is I:3/12, describe what each value in the address represents.
20. If an Allen-Bradley SLC system has an output module in slot 5 and an actuator wired to terminal 8, then what is the address?
21. What limits the number of virtual input and output instructions in a PLC ladder logic program?
22. Where are input and output instructions located on a PLC ladder rung?
23. Why do PLC solutions have fewer wiring changes compared to relay logic when changes are made to control systems?
24. List five advantages that PLCs offer over a relay logic solution.
25. How does the body react to electrical shock, and what is muscular tetanus?
26. What is the most dangerous path for an electrical shock in the body?
27. If a woman gets 55 mA of DC current from an electrical machine what body effect is felt?
28. What does the term zero energy state mean?
29. What is the three-step procedure used to verify that a voltage is not present?
30. How should you make your first contact with a bare electrical conductor?
31. What steps are used to respond to a shock victim?

WEB AND DATA SHEET QUESTIONS

Web/CD resource questions should be answered with data acquired from the CD included with the text or from the Web at *http://www.ab.com/*.

1. A vendor plans to use an Allen-Bradley Pico PLC for a machine controller. Determine from the device specifications sheet what model number would provide 6 discrete inputs (24 VDC) and 5 relay outputs.
2. What are the limitations on the load current and voltage for the output field devices connected to the PLC selected in Question 1?
3. A vendor has an SLC 500 system and needs a discrete current sinking input module to handle seven 24 VDC input signals. Determine the model number of the module that should be ordered.
4. The company in Question 3 needs an output module to handle three 120 VAC devices. Two of the devices have a maximum AC current of 1 amp and the third draws 2 amps. The device operates at ambient temperature. Find the model number of the output module that will satisfy the user.
5. Determine what type of motion control modules are supported by the ControlLogix 1756 system.
6. Identify and name the different types of PLC models that are available from Allen-Bradley.
7. Repeat Examples 1-1 and 1-2 for two other PLC manufacturers. Use the URLs provided in Section 1-8, page 32, to access the data at the company websites.

PROBLEMS

1. Select the appropriate Allen-Bradley input module(s) from Figure 1-12 to interface with the following input field devices:
   - Twelve 120 VAC inputs
   - Twenty-four 5 VDC inputs for SPST switches
   - Ten 24 VDC inputs from sourcing-type sensors
2. Select the appropriate Allen-Bradley output module(s) from Figure 1-12 to interface with the following output field devices:
   - Nineteen pneumatic valves powered with 28 VAC that have 0.5 amp continuous and 2 amp surge current
   - Twelve signals for a CNC machine that are 24 VDC current sourcing and require less than 0.2 amp
   - Five 120 VAC motors with 1.2 amp continuous and 4 amp surge current
3. Analyze the control circuit in Figures 1-17(a) and (b), and then answer the following questions.
   a. Which field device is used to control input instruction I1?
   b. What field device condition(s) cause the CR1 instructions to be true?
   c. What input conditions are necessary for the output O2 to be active?

d. The field device switches for the valve and pump 1 are on (the closed position so that voltage is present at the input terminals), the pump 1 motor contactor is on, but no fluid is flowing into the tank. What devices and types of failures could produce this symptom?

4. Analyze the control circuit in Figures 1-19 and 1-20, and then answer the following questions.
   a. What are the conditions necessary for pump 2 to be running?
   b. All three field device switches are on. There is fluid flowing into the tank and pump 1 is operating; however, the pump 2 motor contactor is not active. What devices and types of failures could produce this symptom?
5. A tank is filled with a chemical mixture and heat is used to cause the mixture to react. The control system for this process is illustrated in Figure D-1. Study the figure and describe the conditions necessary for each of the output field devices to be operating.
6. Draw the input and output PLC interface and the ladder logic program to replace the relay ladder logic in the process reaction tank in Figure D-1. The ladder logic is used to control a process reaction tank. Interface the input field devices to the same input module using terminals as follows: selector switch NO to 1, selector switch NC to 2, flow switch to 3, and temperature switch to 4. Interface the 28 VDC output field devices to output module terminals as follows: fill valve to 1 and drain valve to 2. Interface the 120 VAC output field devices to terminals as follows: mixer motor contactor to 1, heater motor contactor to 2, and pilot light to 3.
7. Select a PLC model, input modules, and output modules from Allen-Bradley SLC 500 System Selection Guide on the text CD that would satisfy the requirements in Problem 6. Assume all contactors and the light are 120 VAC and have less than 0.5 amp continuous currents, and valves draw 3 amps.
8. Modifications are necessary for the control system in Figure D-1. A float or level switch (see Section 2-3-3 for the symbol of this switch) is installed in the tank at the 20 percent full level. The level switch is connected to terminal 5 of the input module. Draw the new input interface and change the ladder logic program so that the mixer and heater do not start until the liquid is above the 20 percent liquid level.
9. Determine the level of bodily response for both men and women if a 28 VDC source wire is touched by wet hands. Use minimum resistance values.

FIGURE D-1: Relay ladder logic for Problem 5

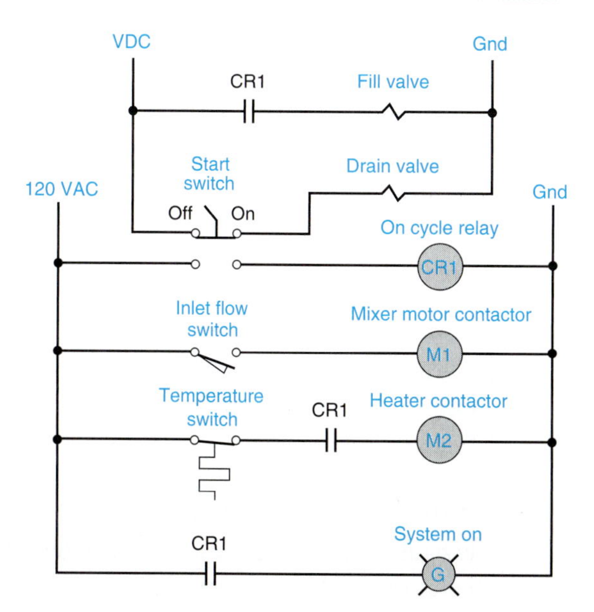

CHAPTER 2 INPUT DEVICES AND OUTPUT ACTUATORS

REFERENCES

Textbook Chapter 2 – Input Devices and Output Actuators (pp. 33–76)
Textbook CD – Allen-Bradley Manuals:

- Capacitive Proximity Sensors
- Inductive Proximity Sensors
- Limit Switches
- Photoelectric Sensors
- Magnetic Voltage Starters Wiring Manual – Bulletins 509 and 609

The following examples and figures are referenced in the questions and problems for this chapter:

- Example 2-2 is on page 35
- Figure 2-38 is on page 60
- Figure 2-59 is on page 72

QUESTIONS

1. What does the switch term *manually operated* imply?
2. Define the following switch terms: (a) pole, (b) NC and NO, and (c) throw.
3. In the schematic representation of switch, what does the dotted line connecting the poles indicate?
4. What switch contact configuration permits different voltages to be switched by the NC and NO contacts?
5. Compare the differences between the following push button switch styles: no guard, full guard, and extended guard.
6. How does a selector switch differ from a toggle switch?
7. What does the term SPDTDB mean when applied to a switch?
8. What does the switch term *mechanically operated* imply?
9. Name four limit switch configurations.
10. What is difference between an open tank level switch and a closed tank level switch?
11. What is the difference in sensing capability between a flow switch and float switch?
12. What is the difference between a contact sensor and a non-contact sensor?
13. When would an analog sensor be selected over a discrete sensor?
14. What are the three basic blocks of an inductive proximity sensor?
15. Contrast the operational principles of the inductive, capacitive, and ultrasonic proximity sensors.
16. What is the solid-state component that is most often used for the light detector in a photoelectric sensor?
17. Compare and contrast the following photoelectric sensor operating modes: diffused, retroreflective, through beam, and polarized retroreflective.
18. Name six operational types of the diffused mode sensor.
19. Define the Dark ON photoelectric sensor function.
20. Describe current sinking and current sourcing relative to an input field device and a PLC input.
21. In a solenoid, what force must the magnetic field overcome to pull the iron core into the coil?
22. Describe the operation of the solenoid-controlled valve.
23. What are the two primary functions of a relay?
24. What is the purpose of a seal-in contact?
25. What is a latching relay?
26. Explain the operation of the two coils in a latching relay.
27. What is the function of the auxiliary contacts in a contactor?
28. List the three main blocks of a motor starter.
29. In general, what are the two types of PLC outputs available to output field devices?
30. Describe current sinking and current sourcing relative to an output field device and a PLC output.
31. What is the problem when you hear a loud humming noise emanating from an AC relay?
32. In troubleshooting proximity sensors, why must you examine the velocity of the parts as they pass the sensor?

WEB AND DATA SHEET QUESTIONS

1. A designer plans to use a limit switch from Omron to control a heater in a package sealer that is rated at 125 VAC and 625 watts. Use the data sheet for the limit switch in Figure D-2 to determine if the switch can handle the load.
2. Identify a control device for the following applications using the Square D and Rockwell Automation Web sites. Indicate the range of measured values available for the devices you select.
   a. A temperature switch to turn on an exhaust fan when internal machine temperature reaches a set level.

FIGURE D-2: Switch voltage and current ratings (Courtesy of Omron Electronics LLC)

Rated voltage	Non-inductive load				Inductive load				Inrush current	
	Resistive load		Lamp load		Inductive load		Motor load			
	NC	NO	NC	NO	NC	NO	NC	NO	NC	NO
125 VAC	10 A		3 A	1.5 A	10 A		5 A	2.5 A	30 A max.	15 A max.
250 VAC	10 A		2.5 A	1.25 A	10 A		3 A	1.5 A		
8 VDC	10 A		3 A	1.5 A	6 A		5 A	2.5 A		
14 VDC	10 A		3 A	1.5 A	6 A		5 A	2.5 A		
30 VDC	6 A		3 A	1.5 A	5 A		5 A	2.5 A		
125 VDC	0.5 A		0.4 A		0.05 A		0.05 A			
250 VDC	0.25 A		0.2 A		0.03 A		0.03 A			

Note: 1. Inductive loads have a power factor of 0.4 min. (AC) and a time constant of 7 ms max. (DC).
2. Lamp load has an inrush current of 10 times the steady-state current.
3. Motor load has an inrush current of 6 times the steady-state current.

b. A pressure switch at the bottom of a tank to start a mixer when the level in the tank reaches the desired level.
c. A temperature switch to turn on a heater when the temperature of the material in a tank drops below the desired level.
d. A pressure switch at the bottom of a tank to open a material inlet valve when the level of material in the tank falls below the desire level.

3. Identify a control relay from Square D and Rockwell Automation capable of switching a 240 VAC lamp that draws 9.5 amps.

PROBLEMS

1. Draw a solution for Example 2-2 using a DPDT (single break) switch contact configuration.
2. Select a sensor type and sensing method or mode for each of the following application situations.
   a. Count boxes (18 inches square) on a 24-inch conveyor belt with cable access from only one side of the conveyor.
   b. Count shiny thermos bottles moving in a production machine. Access is limited to one side of the conveyor, but the sensor can be mounted as close to the bottles as necessary.
   c. Detect plastic clips coming from a bowl feeder on a pair of metal rails.
   d. Detect a small white relay on a dark printed circuit board. Sensor must be located 6 inches above the relay and board.
   e. The level of milk must be verified as the clear plastic milk bottles move down a conveyor. Access to both sides of the conveyor is permitted and there is no limitation on the distance from the bottles.
   f. Detect the leading edge of a square hole in a plate moving down a conveyor. Access above and below the conveyor is permitted.
   g. Detect and count aluminum foil–coated boxes of tea at the entrance to a packaging machine. The sensor must be mounted 2 inches from the box, and a reflector could be mounted on the other side of the object.
   h. Detect the presence of a metal casting at the input to a machining center. There is a single side access restriction, but the sensor can be mounted as close to the slug as necessary.
3. Redraw the circuit in Figure 2-38 using a current sourcing sensor output and indicate the current flow.
4. Redraw the circuit in Figure 2-59 using current sourcing PLC outputs.
5. Draw the control diagram ladder rung that turns on a control relay when the following limit switch conditions are true: LS1 and LS2 are active or LS3 and LS4 are active. A second rung uses a contact from the control relay to make a green indicator lamp active when the control relay is active. All switches and the relay have double pole double throw contacts.
6. Draw the control diagram ladder rung using the correct industrial control symbols for the drilling machine shown in Figure D-3.

FIGURE D-3: Diagram for Problem 6

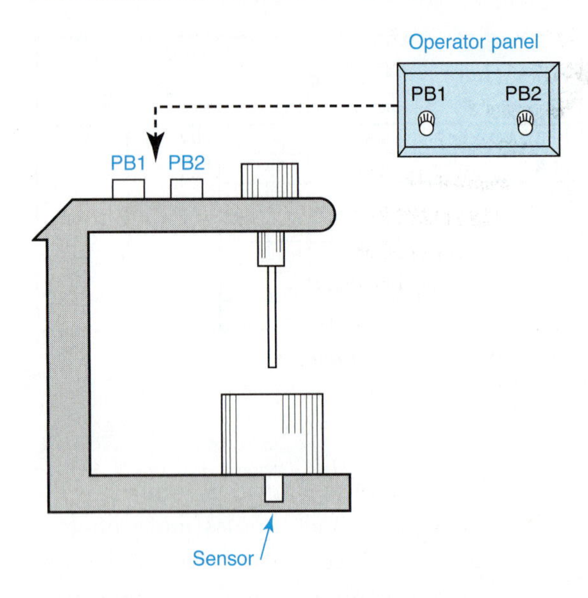

Note that both momentary push buttons (PB1 and PB2) must be depressed, one hand for each push button, and a part to be drilled is in place, centered below the drill, before the drill motor starts.

7. Repeat Problem 6 where a selector switch is used to activate part clamps before the drilling process is started. Make sure the motor cannot be started if clamps are not activated.

CHAPTER 3 INTRODUCTION TO PLC PROGRAMMING

REFERENCES

Textbook Chapter 3 – Introduction to PLC Programming (pp. 77–140)
The following examples and figures are referenced in the questions and problems for this chapter :

- Example 3-14 is on page 130
- Figure 1-17(b) is on page 23
- Figure 3-13 is on page 94
- Figure 3-15 is on page 96
- Figure 3-16 is on page 98
- Figure 3-20 is on page 103
- Figure 3-31 is on page 115
- Figure 3-36 is on page 119
- Figure 3-37 is on page 122
- Figure 3-38 is on page 124
- Figure 3-39 is on page 125
- Figure 3-40 is on page 126
- Figure 3-41 is on page 129
- Figure 3-42 is on page 130
- Figure 3-44 is on page 131

QUESTIONS

1. What are the radix, position digit, and position value in a number system?
2. What are the bases for decimal, octal, and binary number systems?
3. Why do PLCs use the binary number system?
4. Define the following terms: word, byte, nibble, upper byte, lower byte, LSB, and MSB.
5. How is memory in a PLC organized?
6. What is the primary difference between the SLC 500 series and the newer Logix series of PLCs?
7. What is the difference between rack/slot-based memory organization and a tag-based memory system?
8. What is the difference between program files and data files in a PLC system?
9. What is the function of the data files in the PLC 5 and SLC 500 systems?
10. Compare and contrast rack/slot and rack/group memory systems.
11. Define a tag in the ControlLogix PLC system.
12. How do the ControlLogix tag data types compare to those in standard 61131?
13. What advantage(s) do tags have that are not found in rack/slot and rack/group type systems?
14. Define the following ControlLogix terms: project, task, program, main routine, and subroutine.
15. Compare and contrast program tags and controller tags.
16. Compare and contrast terminals, groups, and racks in the PLC 5 addressing system.
17. Describe the addressing format for input and output modules used with a PLC 5 processor.
18. Describe the addressing format for input and output modules used with SLC 500 series processors.
19. Define the following terms in the tag-based addressing system: tag name, alias for, base tag, and type.
20. What are internal control relays or virtual control relays?

21. Describe the format for internal control relay addressing in the PLC 5 and SLC 500 processors.
22. Describe the difference between retentive and non-retentive memory.
23. What safety concern is present when retentive memory is used for an output?
24. Why is it important to be able to address status data in a PLC program?
25. How does the status bit S:1/15 operate, and how is it used?
26. Describe the difference in operation among the following instructions: XIC, XIO, OTE, OTU, and OTL.
27. How do the phases "examine if closed" or "examine on" and "examine if open" or "examine off" relate the field device status to the instruction symbol action?
28. What is the relationship between XIC and XIO and the OTE instruction?
29. What other names are used for XIC and XIO, and which are not good choices?
30. Define continuity and power flow in the ladder logic rung.
31. How can you determine if a ladder instruction is providing power flow?
32. How do OTL and OTU instructions operate and when is it unsafe to use these instructions?
33. What is a sealing instruction and how is it used?
34. Describe all the parts of a PLC scan.
35. What is the relationship between scan time and the size of a PLC program?
36. Describe the problem that the corrected rung in Figure 3-42 fixes.
37. Describe how the correction in Figure 3-42 corrects the problem in the pump control problem in Figure 3-36.
38. Why should all engineers working with PLCs know how to perform program design and relay conversion projects?
39. How does the input instruction selection table in Figure 3-37 function?
40. Describe the operation of the OTE instructions in the PLC 5, SLC 500, and Control Logix PLCs.
41. Why is the ladder logic design for the pump tank problem in Figure 3-41 better than the original design in Figure 3-36?
42. Describe the power flow options present in all of the logic configurations in Figure 3-38.
43. Compare the output format in Figure 3-40(a) with that in Figure 3-40(c).
44. Describe the process used to convert the relay ladder logic to a PLC ladder program.
45. What is a block diagram?
46. How does bracketing help to isolate the problem in a control system?
47. What is the three-step process used in bracketing?
48. Describe each of the following signal flow types: linear, divergent, convergent, feedback, and switched.
49. Why is the divide-and-conquer rule so effective for linear circuit troubleshooting?
50. Why are switched circuits easy to troubleshoot?
51. What is the six-step troubleshooting process?

WEB AND DATA SHEET QUESTIONS

Questions should be answered with data acquired from the CD included with the text or from the Allen-Bradley URL (*http://www.ab.com/*).

1. A vendor plans to use an Allen-Bradley Pico PLC for a machine controller. Determine from the device specifications sheet what model number would provide 6 discrete inputs (24 VDC) and 5 relay outputs.
2. Select input field devices that would satisfy the tank control problem in Figure 3-41.
3. Describe the operation of the PLC 5 when half-slot and two-slot addressing are used.
4. Identify input modules that could be used in a PLC 5 system to take advantage of the half-slot and two-slot addressing.
5. Identify and describe a 32-point input and output module for the SLC 500 that could be used with limit switches, push buttons, and current sourcing sensors.

PROBLEMS

General

1. The input field devices in Figure 3-36 have the following conditions at the start of a scan:
   - The normally closed (NC) high level FS is held open.

- The normally open (NO) low level FS is held closed.
- The temperature switch is not active, so the NO contact is open and the NC contact is closed.
- Start and stop push buttons are all in their normal positions.

Use both the ladder logic program in Figure 3-36 and the given conditions to predict what output field devices will be on at the end of the scan.

2. Write the Boolean equation for the input logic in Figure 3-31 in order for the output to be true.
3. Determine the condition of the input sensors (activated or not activated) for continuity in Figure 3-39(b). The sensor contact configurations are as follows: A – NO; B – NC; C – NO; D – NO; E – NC; F – NO; G – NC; and H – NO.
4. Determine the condition of the inputs sensors (activated or not activated) for power flow in Figure 3-39(c). The sensor contact configurations are as follows: A – NO; B – NC; C – NO; D – NO; E – NC; F – NO; G – NC; and H – NO.
5. In your own words, describe the operation of the process tank in Figure 3-36.

PLC 5

When working these problems, assume that all modules are in rack 2 with single-slot addressing, a 16-point input is in slot 2 (group 2), and a 16-point output is in slot 3 (group 3), unless otherwise noted. Pin numbers are specified in the problem.

6. Using the PLC 5 Selection Guide on the text CD, select the appropriate combination of PLC 5 1771 modules for each combination of the following field devices.
   - Twelve 120 VAC inputs not isolated, twenty-four 5 VDC inputs, and ten 24 VDC isolated inputs for current sourcing sensor.
   - Ten 4 to 20 mA DC analog inputs – five have a common ground and the remaining have individual ground for each signal and for eight type J thermocouple differential inputs.
7. Determine the input and output addresses for the bits with black blocks in Figure 3-13.
8. For the tank problem in Figure 1-17(b), convert the PLC ladder logic solution to a solution using an Allen-Bradley PLC 5 system. Assume that rack 2 is used with single-slot addressing of 16-point input and output modules. The rack is number 2 and the group and terminal numbers are from Figure 3-20. Let bit 5 of word 0 in file B3 be used for the internal relay.
9. Convert the PLC ladder logic solution in Figures 3-41 and 3-42 to a solution using an Allen-Bradley PLC 5 system.

SLC 500

When working these problems, assume that a 16-point input is in slot 2 and a 16-point output is in slot 3 unless otherwise noted. Pin numbers are specified in the problems.

10. Using the SLC 500 Selection Guide on the text CD, select the appropriate combination of SLC 500 1746 modules for each combination of the following field devices.
    - Twelve 120 VAC inputs not isolated where the AC value for the off condition is 40 VAC, twenty-four 5 VDC inputs for current sourcing input field device, and ten 24 VDC electronic protection inputs for current sinking sensor.
    - Ten 4 to 20 mA DC analog inputs with individual ground for each signal and for eight type J thermocouple differential inputs.
11. An Allen-Bradley SLC system uses a rack like the one illustrated in Figure 3-15, with the 16-input DC module in slot 2. Determine the address for a discrete input signal attached to terminal 4 of the module.
12. An Allen-Bradley SLC system uses a rack like the one illustrated in Figure 3-15, with the 24-input DC module in slot 3. Determine the address for a discrete input signal attached to terminal 20 of the module.
13. An Allen-Bradley SLC system uses a rack like the one illustrated in Figures 3-15 and 3-16, with a 32-terminal DC output module in slot 1. Determine the address for the field device attached to terminal 27 of the module.

ControlLogix

When working these problems, assume that a 16-point input is in slot 2 and a 16-point output is in slot 4 unless otherwise noted. Pin numbers are specified in the problems.

14. Using the ControlLogix Selection Guide on the text CD, select the appropriate combination of ControlLogix 1756 modules for each combination of the following field devices.
- Twelve 120 VAC inputs with isolated inputs for five points, twenty-four 5 VDC inputs for sinking sensors, and ten 24 VDC isolated inputs for a combination of sinking and sourcing input field devices.
- Ten 4 to 20 mA DC analog isolated inputs and for eight type J thermocouple inputs.

15. For the tank problem in Figure 1-17(b), convert the PLC ladder logic solution to a solution that uses an Allen-Bradley ControlLogix system. Assume the following: the valve switch and pump switch are in terminals 5 and 6, respectively, and the valve solenoid and pump motor contactor are in pins 3 and 4, respectively.

16. Convert the PLC ladder logic solution in Figures 3-41 and 3-42 to a solution that uses an Allen-Bradley Logix format. Create tag names, data types, alias for, and base tag values for each input and output field device. Redraw the ladder logic with Logix tag names and aliases for addressing.

17. For the two-axis robot control problem described in Figure 3-44, create and list the data for the tag names and data types.

18. For each tag identified in Problem 17, create the alias for and base tag data.

19. Convert the ladder logic in Figure 3-44 to a ControlLogix format.

Challenge

When working these problems, use the PLC processor assigned and the input and output modules specified for PLC 5, SLC 500, and ControlLogix in the previous problem sections.

20. Design a ladder logic control for a single (NO) momentary push button (PB) switch to perform both the start (enable) and stop (disable) functions. The first actuation of the PB turns on an output, and the second causes the output to turn off. Use OTL and OTU output instructions in this design.

21. The tank process in Figure 3-36 is changed as follows:
- Pump 1 (original pump) is on from start until a float switch at the 50 percent level is true (NO contacts are closed).
- Pump 2 (new pump) is on from 50 percent until the high level float switch is active.
- All other tank operations remain unchanged.

Use the empirical design process to create the program for this new tank operation and describe the input and output interface.

22. The process tank like that illustrated in Figure 3-36 has three feed pipes and pumps (pump contactors 1, 2, and 3 are on terminals 1, 2, and 3, respectively) and NO level switches at five levels—0, 25, 50, 75, and 100 percent full (level switches are on terminals 0, 1, 2, 3, and 4, respectively). The process requires that the liquids be added using the following sequence for the feed pumps.
- Pump 1 from 0 to 25 percent full
- Pump 2 from 25 to 50 percent full
- Pumps 1 and 3 from 50 to 75 percent full
- Pumps 2 and 3 from 75 to 100 percent full

Write a program that fills the tank. Start (NO) and stop (NC) push buttons are present on terminals 5 and 6, respectively.

23. Redesign the robot motion control in Example 3-14, using only OTL and OTU instructions for control of the gripper and axes.

24. The robot motion in Figure 3-44(a) must be modified to include a movement of the X-axis to the down position after the Y-axis is extended. Use OTL and OTU instructions for the gripper valve. For a complete cycle, the new motion should be X down, grasp part, X up, Y out, X down, release part, X up, and Y in. Use the empirical design process to create the program for this new motion and describe the input and output interface.

25. The robot illustrated in Figure 3-44(a) (Example 3-14) has a pneumatic horizontal Z-axis actuator added that moves the robot base from left to right. This allows the robot to access two locations when the Z-axis is in the left position (Z pneumatic valve off) and two additional locations when the Z-axis is in the right position (Z pneumatic valve on). The part drop point, position B in Figure 3-44(a), has the Z-axis in the left position. Design a ladder logic system that moves the part to position B when the Z-axis is left and position B when the Z-axis is right. Input terminal 8 true indicates that the Z-axis should be in the right position. The Z-axis left and right sensors are terminals 9 and 10, respectively. The Z-axis valve is connected to terminal 6 of the same output module. Design a program for this robot sequence. A cycle is started when the start selector switch is on and when the pickup sensor indicates that a part is in the pickup location. Use the limit sensors to control actuator and gripper sequences.

CHAPTER 4 PROGRAMMING TIMERS

REFERENCES

Textbook Chapter 4 – Programming Timers (pp. 141–178)
Textbook CD – Allen-Bradley Manuals:

- PLC 5 Instruction Programming Manual – 1675-6.1
- SLC 500 Instruction Set – 1747-RM001D-EN-P
- Logix5000 General Instruction Reference – 1756-RM003G-EN-P
- Logix Advanced Programming Manual – 1756-PM001F-EN-P

The following examples and figures are referenced in the questions and problems for this chapter:

- Example 4-1 is on page 143
- Figure 4-12 is on page 156
- Figure 4-15 is on page 161
- Figure 4-17 is on page 163
- Figure 4-18 is on page 165
- Figure 4-19 is on page 166
- Figure 4-21 is on page 171
- Figure 4-22 is on page 172

QUESTIONS

1. Describe the operation of a mechanical timing relay.
2. What is the difference between the time contacts and the instantaneous contacts of a mechanical timing relay?
3. Describe the differences and commonality between an on-delay relay and an off-delay relay.
4. What do NOTC and NCTO stand for?
5. Define the following timer terms: timer number, time base, preset value, and accumulator value.
6. How do time base values differ among the three Allen-Bradley processor models?
7. Define the following timer bits: timer enable, timer timing, and timer done.
8. What is the relationship between the time base and the preset value of a timer?
9. Compare and contrast the true and false states of the enable bit for the on-delay timer, the off-delay timer, and the retentive timer.
10. Compare and contrast the true and false states of the timer timing bit for the on-delay timer, the off-delay timer, and the retentive timer.
11. Compare and contrast the true and false states of the done bit for the on-delay timer, the off-delay timer, and the retentive timer.
12. What is the difference between a retentive timer and a non-retentive timer?
13. How is the accumulator of a retentive timer reset?
14. What constitutes a cascade timer configuration?
15. What outputs in a PLC timer relate to the timed contact and instantaneous contact in a mechanical or electronic timer?
16. What are the criteria used to select mechanical timing relays?
17. Describe the addressing syntax for timer output bits among the three Allen-Bradley processor models.
18. What types of applications do the four types of TON standard timing circuits satisfy?
19. What types of applications do the two types of TOF standard timing circuits satisfy?
20. Describe how the retentive timer and reset instruction are related.

21. What type of application requires cascade timers?
22. Describe how the empirical design process is amended when timers are required.
23. Why is the two-axis pneumatic robot control solution in Chapter 3 better than the solution in Chapter 4 that uses timers?
24. What may help debug a ladder timer program with timers where the execution time is very fast?
25. How do you use TND to conditionally omit the balance of a program?

WEB AND DATA SHEET QUESTIONS

1. The Allen-Bradley 700-FS timing relay is planned to be used in a process control system, but its reset time must be less than 60 ms. What is the 700-FS reset time, and will it work in the system? Go to the Allen-Bradley Web site at http://ab.com to look up the timing relay's reset time.
2. Determine the maximum integer values for the timer preset number in the three Allen-Bradley PLC system types.

PROBLEMS

General

1. Draw the timing diagram for the active bits and done bits of the traffic light example in Figure 4-15.
2. The operation of traffic lights that controls the traffic flow is specified in Figure D-4.
   a. Draw the ladder diagram to include the control of traffic flow as specified. (Consider using Figure 4-15 as a basis.)
   b. Explain in detail the how the ladder diagram controls the traffic lights.
3. Configure on-delay timers to achieve a time delay of 2006 seconds with each timer having a maximum preset of 999 seconds.
4. Draw the timing diagrams for the two types of standard TON timers in Figure 4-12 (a and b).
5. Draw the timing diagrams for the two types of standard TON timers in Figure 4-12 (c and d).

PLC 5

When working these problems, assume that all modules are in rack 2 with single-slot addressing. A 16-point input is in slot 2 (group 2) and a 16-point output is in slot 3 (group 3), unless otherwise noted. Pin numbers are specified in the problems.

6. Convert the relay ladder logic from Example 4-1 to a solution using the Allen-Bradley PLC 5. Use input pins 0 and 1 and output pin 0.
7. Convert the traffic light solution in Figure 4-15 to a PLC 5 solution. Use the same terminal numbers for the I/O.
8. Convert the the pumping solution in Figure 4-17 to a PLC 5 solution. Use the same terminal numbers for the I/O.
9. Convert the maintenance light and pump control problem in Figure 4-18 to a PLC 5 solution. Use the same terminal numbers for the I/O.
10. Convert the heater control problem in Figure 4-19 to a PLC 5 solution. Use the same terminal numbers for the I/O.
11. Convert the two-handed machine control program in Figure 4-21 to a PLC 5 solution. Use the terminal numbers from the figure and a 1-second time base.

SLC 500

When working these problems, assume that a 16-point input is in slot 2 and a 16-point output is in slot 3, unless otherwise noted. Pin numbers are specified in the problems.

FIGURE D-4: Specifications for Problem 2

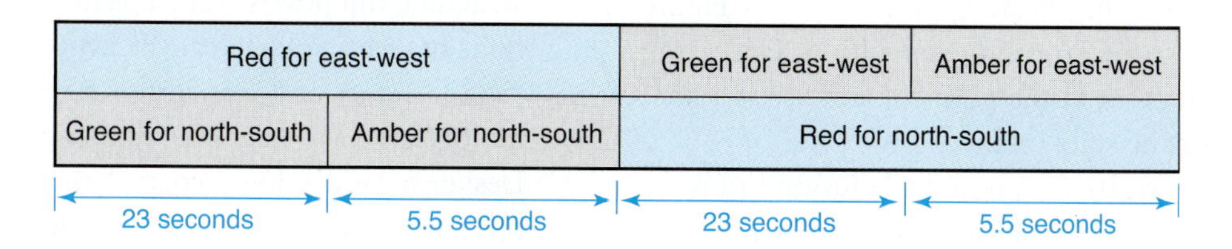

12. Convert the relay ladder logic from Example 4-1 to a solution using the Allen-Bradley SLC 500. Use input pins 0 and 1 and output pin 0.
13. Convert the heater control problem in Figure 4-19 to an SLC 500 solution. Use the same terminal numbers for the I/O.
14. Convert the two-handed machine control program in Figure 4-21 to an SLC 500 solution. Use the terminal numbers from the figure and a 0.01-second time base.
15. Study the ladder logic in Figure 4-17 and answer the following questions.
    a. What timer delays the start of the motor?
    b. What timer prevents the motor from restarting after it is stopped?
    c. Why is the sealing instruction used in rung 0?
    d. Why is the sealing instruction used in rung 2?
    e. What is the total lockout time?
16. Study the ladder logic in Figure 4-18 and answer the following questions.
    a. Why is an RTO timer used in rung 3?
    b. Why is an RES instruction used in rung 5?
17. Study the ladder logic in Figure 4-22 and answer the following questions.
    a. What is the function of rungs 3 and 6?
    b. Why is this solution to the robot control problem not as fast as the one in Chapter 3 that used end-of-travel sensors on the pneumatic actuators?

ControlLogix

When working these problems, assume that a 16-point input is in slot 2 and a 16-point output is in slot 4, unless otherwise specified. Pin numbers are specified in the problems.

18. Convert the relay ladder logic from Example 4-1 to a solution using the Allen-Bradley ControlLogix processor. Use input pins 0 and 1 and output pin 0.
19. Convert the traffic light solution in Figure 4-15 to a ControlLogix solution. Create tag names that fit the problem and use terminal 5 for the start input.
20. Convert the the pumping solution in Figure 4-17 to a ControlLogix solution. Create tag names that fit the problem and use the same terminal numbers for the I/O.
21. Convert the maintenance light and pump control problem in Figure 4-18 to a ControlLogix solution. Create tag names that fit the problem and use the same terminal numbers for the I/O.
22. Convert the cascade timer control program in Figure 4-21 to a ControlLogix solution. Create tag names that fit the problem and use the same terminal numbers for the I/O.

Challenge

When working these problems, use the PLC processor assigned.

23. Modify the pump control problem in Figure 4-17 so that the pump continues to run for 30 seconds after the stop push button is pressed. All other requirements are unchanged.
24. Modify the maintenance light control for the pump control problem in Figure 4-18 so that the light is on continuously for 30 seconds and then starts to flash at a 1-second on–1-second off cycle. The light can be reset at any time while it is on. All other requirements are unchanged.
25. The robot motion in Figure 4-22 must be modified to include a movement of the X-axis to the down position after the Y-axis is extended. The new motion should be X down, grasp part, X up, Y out, X down, release part, X up, and Y in for a complete cycle. Use timers for control of all motion. Use the empirical design process to create the PLC 5 program for this new motion and draw the input and output interface and ladder solution.
26. Modify the solution in Problem 25 to verify that a part is not in the drop location when the robot moves to pick up the next part. The drop part sensor is in terminal 2. If a part is in the drop region, the system will not start the next load cycle.
27. Instead of starting large electric motors by switching full power from a dead stop condition, reduced voltage can be switched for a "softer" start and less inrush current. This motor has two stages in the start routine. Design a ladder logic control that closes a relay when the motor start switch is closed.

FIGURE D-5: Automated conveyor

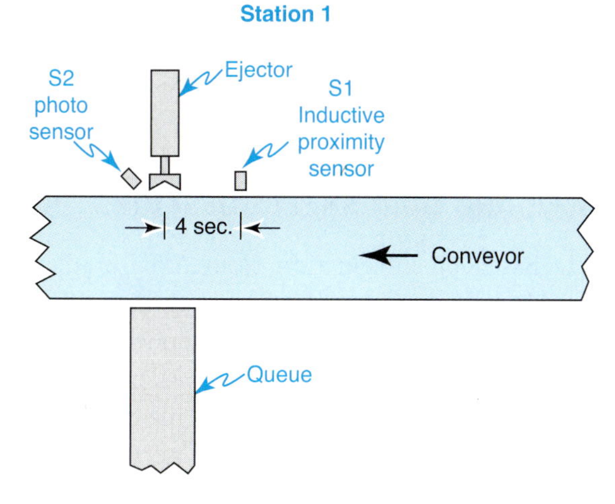

A second relay is closed after 15 seconds, and a third is closed after 40 seconds. Select output terminals assuming that this is the only application for the input and output module.

28. Three conveyor belts are arranged to transport material and the conveyor belts must be started in reverse sequence (the last one first and the first one last) so that the material doesn't get piled on to a stopped or slow-moving conveyor. Each belt takes 45 seconds to reach full speed. Design a ladder logic that would control the start of this three-conveyor system.
29. The conveyor system in Figure D-5 transports a plastic part and a metal part for an assembly. The plastic must be ejected and the metal must pass by the ejector. Design a ladder logic program to provide this control. S1 and S2 are input pins 10 and 12, respectively, and the ejector is on output pin 9.

CHAPTER 5 PROGRAMMING COUNTERS

REFERENCES

Textbook Chapter 5 – Programming Counters (pp. 179–210)
Textbook CD – Allen-Bradley Manuals:

- PLC 5 Instruction Programming Manual – 1675-6.1
- SLC 500 Instruction Set – 1747-RM001D-EN-P
- Logix5000 General Instruction Reference – 1756-RM003G-EN-P
- Logix Advanced Programming Manual – 1756-PM001F-EN-P

The following examples and figures are referenced in the questions and problems for this chapter:

- Example 5-1 is on page 191
- Example 5-2 is on page 193
- Figure 2-24(b) is on page 49
- Figure 5-9 is on page 187
- Figure 5-10 is on page 188
- Figure 5-12 is on page 192
- Figure 5-13 is on page 194
- Figure 5-14 is on page 195
- Figure 5-15 is on page 196
- Figure 5-18 is on page 199
- Figure 5-19 is on page 200
- Figure 5-21 is on page 201
- Figure 5-23 is on page 202
- Figure 5-24 is on page 203
- Figure 5-25 is on page 203
- Figure 5-27 is on page 205
- Figure 5-29 is on page 207

QUESTIONS

1. Describe the operation of a mechanical counter.
2. When does an up counter provide an output?
3. When does a down counter provide an output?
4. Compare and contrast the operation of timers and counters.
5. Explain why the counters in PLC applications are typically retentive.
6. What is the difference between the overflow bit and the underflow bit?
7. Describe the programming process used in cascading two up counters.
8. In an application that requires an up/down counter, why must the counters be located at the same address?
9. Identify the type of counter you would choose for each of the following applications.
   a. Accumulate the total number of components made during the night shift.
   b. Keep track of the number of parts remaining in a bin as parts are extracted. Assume that a full bin contains 24 parts.

c. Keep track of the current number of parts at the first stage of a process as they enter and exit.

10. Compare and contrast the true and false states of the enable bit for the up counter and the down counter.
11. Compare and contrast the true and false states of the overflow and underflow bits for the up counter and the down counter.
12. Compare and contrast the true and false states of the done bit for the up counter and the down counter.
13. What is the default address file for the counter instructions for the PLC 5 and SLC 500 systems?
14. How many counters can the default file 5 support in the PLC 5 and SLC 500 systems?
15. What is the meaning of the letters and numbers in C5:0.0?
16. What data is in files C5:0.0, C5:0.1, and C5:0.2?
17. Describe the HSC in the SLC 500 and MicroLogix processors.
18. In Logix counter instructions, what replaces a counter number like C5:0?
19. Why are one-shot instructions often required in counter applications?
20. What is the difference between the OSR and OSF instructions?
21. What are the data entries in the PLC 5 OSR and OSF instructions?
22. What are the data entries in the ControlLogix PLC OSR and OSF instructions?
23. What is the purpose of cascaded counters?
24. What steps are needed to add counters to the empirical process?
25. What is the purpose of the SUS instruction?
26. Why does process speed create problems when counter values are used to inhibit process actions?

WEB AND DATA SHEET QUESTIONS

1. Based on information from Allen-Bradley, describe how a high-speed counter module operates and how it is programmed.
2. Based on information from Allen-Bradley on the PLC 5 system, describe the difference between an ONS, OSR, and an OSF one-shot instruction.
3. Select an Allen-Bradley through-beam sensor for the bottle counting application in Example 5-1. Verify that the response rate of the sensor is consistent with the needs of the problem.
4. Select a proximity sensor for the can counter in Example 5-2 and Figure 2-24(b). The sensor is located 0.4 inches above the 3 inch cans, the PLC input module is a DC current sinking type, and the response rate must be greater than 300 Hz.

PROBLEMS

General

1. Answer the following questions relative to the ladder diagram in Figure D-6.
   a. What type of counter is used?
   b. When is the red light illuminated?

FIGURE D-6: Ladder diagram for Problem 1

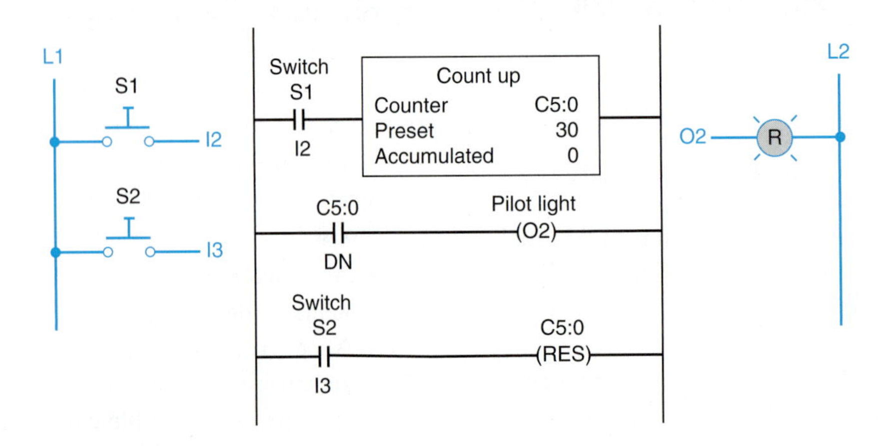

c. When is the counter incremented?
d. What is the accumulated value of the counter when the counter is reset by I3?

2. In an industrial automated process, the quality control requirement is that for every 1000 electronic assemblies that are manufactured, 3 assemblies are to be inspected. Draw the ladder diagram that implements this process using the following operational functions.
   a. A sensor counts completed electronic assemblies as they move along a conveyor.
   b. When 1000 assemblies have passed the sensor, a gate is activated and 3 assemblies are diverted to another conveyor that goes to the quality control station.
   c. After 3 assemblies have been diverted, the gate returns to its original position and the assemblies continue along the conveyor.
   d. The process is repeated for every 1000 assemblies.
3. Explain the sequential operation of the ladder diagram that you generated in Problem 2.
4. Using the ladder diagram that you generated in Problem 2, add a function that turns on a red light if an inspection fails to occur after one thousand units.
5. Draw a ladder diagram using two up counters to count to one million, with each counter having a preset value of one thousand. A green light is illuminated when one million counts have been achieved.

PLC 5

When working these problems, assume that all modules are in rack 2 with single-slot addressing. A 16-point input is in slot 2 (group 2) and a 16-point output is in slot 3 (group 3), unless otherwise noted. Pin numbers are specified in the problem.

6. Convert the counter solution in Figure 5-9 to a PLC 5 solution. Use terminal 5 for the limit switch input.
7. Convert the up counter solution in Figure 5-12 to a PLC 5 solution. Use the same terminal numbers as in the figure.
8. Convert the can packaging counter solution in Figure 5-14 to a PLC 5 solution. Use the same terminal numbers as in the figure.
9. Convert the can jam detection ladder logic in Figure 5-15 to a PLC 5 solution. Use the same terminal numbers as in the figure.
10. Convert the machine queue parts counter ladder logic in Figures 5-23 and 5-24 to a PLC 5 solution. Use the same terminal numbers as in the figure.
11. Convert the cascade counter ladder logic in Figure 5-25 to a PLC 5 solution. Use the same terminal numbers as in the figure.
12. Convert the box sorter ladder logic in Figure 5-27 to a PLC 5 solution. Use the same terminal numbers as in the figure.
13. Convert the queue control ladder logic in Figure 5-29 to a PLC 5 solution. Use the same terminal numbers as in the figure.
14. Describe an application requirement that would dictate the use of the standard ladder logic circuits in Figure 5-19.

SLC 500

When working these problems, assume that a 16-point input is in slot 2 and a 16-point output is in slot 3, unless otherwise noted. Pin numbers are specified in the problems.

15. Convert the counter solution in Figure 5-9 to an SLC 500 solution. Use terminal 5 for the limit switch input.
16. Describe an application requirement that would dictate the use of the standard ladder logic circuits in Figure 5-10.
17. Describe an application requirement that would dictate the use of the standard ladder logic circuits in Figure 5-21.
18. Study the ladder logic in Figure 5-13 and answer the following questions:
    a. What is required for output O:2/3 to be active?
    b. Why does counter C5:1 accumulator never have a count larger than 12?
19. Study the ladder logic in Figure 5-15 and describe why the auxiliary contact from the motor contactor is used to seal in the momentary start switch and not the output O:2/10.
20. Study the ladder logic in Figure 5-15 and describe how the jam indicator system works.
21. Study the ladder logic in Figure 5-15 and descibe how the ladder could be modified to

eliminate the virtual relay B3:0/0 without affecting the current operation.

ControlLogix

When working these problems, assume that a 16-point input is in slot 2 and a 16-point output is in slot 4, unless otherwise noted. Pin numbers are specified in the problems.

22. Convert the up counter solution in Figure 5-12 to a ControlLogix solution. Use the same terminal numbers as in the figure and create tag names consistent with the problem description.
23. Convert the can jam detection ladder logic in Figure 5-15 to a ControlLogix solution. Use the same terminal numbers as in the figure and create tag names consistent with the problem description.
24. Convert the parking lot counter ladder logic in Figure 5-18 to a ControlLogix solution. Use the same terminal numbers in the figure and create tag names consistent with the problem description.
25. Convert the machine queue parts counter ladder logic in Figures 5-23 and 5-24 to a ControlLogix solution. Use the same terminal numbers as in the figure and create tag names consistent with the problem description.
26. Convert the queue control ladder logic in Figure 5-29 to a ControlLogix solution. Use the same terminal numbers as in the figure and create tag names consistent with the problem description.

Challenge

When working these problems, use the PLC processor assigned and the input and output modules specified for the PLC 5, SLC 500, and ControlLogix processors in the previous problem sections.

27. Modify the ladder logic in Figure 5-13 to determine the production rate in cans per minute that are moving through the production system.
28. Modify the jam detection program in Figure 5-15 to include a counter to count the number of jams that occur.
29. A pallet loader, which transfers full packages from the conveyor to a pallet, is added to the can packaging system in Figure 5-13. A sensor detects that a full package (16 cans) is present and the palletizing robot loads the package onto the pallet. A signal is needed to indicate that a package is present and ready to be moved, and another signal is needed that indicates 16 boxes have been moved and a new pallet should be moved into place. Modify the ladder logic so that a 20 to 30 VDC pulse is generated to signal a package is present and that a new pallet is required after the package sensor is true 16 times. The package present output signal is on terminal 4 and the full pallet signal is on terminal 5. The package sensor is connected to input terminal 11. All other requirements are unchanged.
30. This chapter describes one-shot instructions for the Allen-Bradley processors. Design a one-shot instruction using standard ladder logic that produces a true output for just one scan when inputs, whether momentary or continuous contacts, go from false to true. The contacts must become false before a second one-shot output can occur. Hint: Review how inputs (XIO and XIC) and outputs (OTE) respond during a scan.
31. Design a conveyor control ladder logic that delivers 100 boxes to the loading dock whenever the conveyor momentary start switch is pressed. A proximity switch (terminal 0) indicates a box has been delivered. A momentary start (terminal 1 – NO contact) push button and a momentary stop (terminal 2 – NO contact) push button are used to start and interrupt the delivery process. A reset momentary push button (terminal 3 – NC contacts) is used to reset the system, and the conveyor contactor is on output terminal 0. Be careful how you program the NO stop and NC reset push buttons.
32. Timers are used to generate a horn alarm and flashing light alarm with a 1-second on and off time whenever a pressure switch (terminal 5 – NO contact) indicates that the process pressure exceeds a fixed value. Show how ladder logic with a counter can be used to latch the alarm input and operate as follows: alarm present starts both alarm horn (terminal 0) and alarm light (terminal 1); a selector switch (terminal 6 – NO contact) is

switched on to reset the horn but leaves the flashing light until the alarm input condition (process pressure too high) is corrected.

33. Show how an up and down counter could be used to verify that all the parts entering a paint spray booth also leave the paint spray booth. There are five part positions in the booth. Select terminals as necessary.
34. Design a ladder logic control with timers and counters for a 24-hour clock. The seconds are displayed on an RTO timer accumulator, the minutes on a counter accumulator, and the hours on the second counter accumulator.
35. Show how a timer with a large time interval could be used with a counter to produce a time delay of 2,000,000 seconds. Turn on a horn at 2,000,000 seconds.

CHAPTER 6 ARITHMETIC AND MOVE INSTRUCTIONS

REFERENCES

Textbook Chapter 6 – Arithmetic and Move Instructions (pp. 211–240)
Textbook CD – Allen-Bradley Manuals:

- PLC 5 Instruction Programming Manual – 1675-6.1
- SLC 500 Instruction Set – 1747-RM001D-EN-P
- Logix5000 General Instruction Reference – 1756-RM003G-EN-P
- Logix Advanced Programming Manual – 1756-PM001F-EN-P

The following examples and figures are referenced in the questions and problems for this chapter:

- Example 5-4 is on page 193
- Example 6-10 is on page 225
- Example 6-13 is on page 228
- Example 6-14 is on page 232
- Figure 5-18 is on page 199
- Figure 6-1 is on page 215
- Figure 6-7 is on page 220
- Figure 6-8 is on page 221
- Figure 6-10 is on page 223
- Figure 6-11 is on page 224
- Figure 6-12 is on page 225
- Figure 6-13 is on page 227
- Figure 6-14 is on page 228
- Figure 6-15 is on page 229
- Figure 6-16 is on page 230
- Figure 6-17 is on page 231
- Figure 6-18 is on page 234
- Figure 6-19 is on page 236

QUESTIONS

1. What are the four conditions in binary addition?
2. What are the four conditions in binary subtraction?
3. How is the one's complement determined?
4. How is the two's complement determined?
5. Besides the one's and two's complements, what is another method of representing negative binary numbers?
6. Describe double-precision binary arithmetic.
7. What is the mnemonic for the math instruction that sets all bits to zero?
8. What is the mnemonic for the math instruction that raises a number to a power?
9. With Source A equal to 10 and Source B equal to 5, what is the content of the destination for the following arithmetic instructions?
   a. Addition
   b. Subtraction
   c. Multiplication
   d. Division
10. What parameter types are allowed as the source in math instructions?
11. What are the arithmetic status bits found in word 0, bits 0 to 3 in the controller status file?
12. What is the purpose of the status bits in Question 11?
13. Where is the math register located and what does it contain?
14. In the Allen-Bradley PLC 5 and SLC 500 systems, where is the floating point file located and what is its purpose?
15. What process systems and field device types most often use floating point data?
16. Using the dialog boxes in the RSLogix 5 and RSLogix 500 computer-based software for PLC 5 and SLC 500, how is integer data entered?

17. Describe the operation of the MVM instruction.
18. What is the purpose of the NEG instruction?
19. What instruction would you use to ensure that the math or move instruction is active for only one scan?
20. What standard math/move configuration is typically used when it is necessary to have a running total for the values in a register that is changing at regular intervals?
21. Why would you use the standard math/move configuration you selected for Question 20?
22. What steps are used to include math instructions in the empirical design process?
23. What steps are used to include move instructions in the empirical design process?
24. If PLC rungs with math instructions do not produce the correct results, what would you verify first?
25. How would you use the TND and SUS instructions to troubleshoot PLC rungs with math/move instructions?

WEB AND DATA SHEET QUESTIONS

Use the reference manuals on the CD with the text for the following questions.

1. Divide the instructions not covered in Section 6-4-1 into groups to indicate which instruction structure type, Figure 6-1(b), (c), or (d), is used when the instruction is placed in a ladder rung.
2. Describe the application and operation of the following trigonometric function commands: SIN, COS, TAN, ASN, ACS, and ATN.
3. Describe the SLC 500 double divide (DDV) and absolute value (ABS) instructions.
4. Describe the following three PLC 5 instructions: series number sort (SRT), average value (AVE), and standard deviation (STD).

PROBLEMS

General

1. Add in the binary system: (a) 11 plus 3 and (b) 37 plus 13, showing all the steps.
2. Subtract in the binary system: (a) 19 minus 14 and (b) 23 minus 5, showing all the steps.
3. Multiply in the binary system: 11 times 3, showing all the steps.
4. Divide in the binary system: 54 by 3, showing all the steps.
5. Design a ladder logic program for a chemical etching process. Two different products, X and Y, are processed with X having a 10 second etching time and Y having 25 seconds. The etching times are stored in integer registers and selected with a NO two position selector switch with center position off. The process is started with a second NO selector switch. A green light is turned on when the etcher is active.
6. Describe the operation of the program in Problem 5.

PLC 5

When working these problems, assume that all modules are in rack 2 with single-slot addressing. A 16-point input is in slot 2 (group 2) and a 16-point output is in slot 3 (group 3), unless otherwise noted. Pin numbers are specified in the problem.

7. Convert the temperature conversion ladder solution in Figure 6-7 to a PLC 5 solution. Use terminal 3 for the normally open calculate push button.
8. Convert the upper and lower limit temperature chamber ladder solution in Figure 6-8 to a PLC 5 solution. Use terminal 3 for the normally open calculate push button.
9. Convert the right triangle ladder solution in Figure 6-10 to a PLC 5 solution. Use terminal 3 for the normally open calculate push button.
10. Convert the problem in Example 6-10 to a PLC 5 solution. Use the same terminal numbers, and use another input module in slot 4 for the word value.
11. Convert the part color solutions in Figure 6-15 to a PLC 5 solution. Use another input module in slot 4 for the data value.
12. Convert the pallet part total ladder in Figure 6-16 to a PLC 5 solution. Use the same terminal numbers as in the SLC 500 solution.
13. Convert the ball counting and packaging ladder solution in Figure 6-17 to a PLC 5 solution. Use the same terminal numbers as in the SLC 500 solution.

14. Convert the can counting ladder solution in Figure 6-18 to a PLC 5 solution. Use the same terminal numbers as in the ControlLogix solution.
15. Convert the box counting ladder solution in Figure 6-19 to a PLC 5 solution.

SLC 500

When working these problems, assume that a 16-point input is in slot 2 and a 16-point output is in slot 3, unless otherwise noted. Pin numbers are specified in the problems.

16. Convert the box counting and packaging ladder solution in Figure 6-18 to an SLC 500 solution. Use the same terminal numbers as in the ControlLogix solution.
17. Study the move instruction rungs in Figure 6-11 and answer the following questions:
    a. If N7:0 is 423, what is the value in N7:6 after I:1/1 is true?
    b. The scan rate for the rungs is 10 ms and the value in N7:0 increases by a value of 20 every 7 ms. If N7:0 has a value of 100 when I:1/1 becomes true, what is the value in N7:6 if I:1/1 is true for 23 ms?
    c. If you want to capture the value of N7:0 when I:1/1 changes from false to true, how would the ladder rung be changed?
    d. If N7:1 is 25369 and N7:2 is A4D8h, what is N7:3 after I:1/1 is true?
    e. If N7:4 is 578, what is the value in N7:5 after I:1/1 is true?
18. Why are OSR instructions used in the ladder rung in Figure 6-12(c)?
19. Study the ladder logic in Figure 6-13 and determine the value of N7:4 for the following conditions: If the accumulator of counter 3 is 100, I:2.0 has a value of 0000001100100100b, and the mask is 00FC, what is N7:4 after I:1/0 and I:1/1 are pressed?
20. Study the ladder logic in Figure 6-15 and answer the following questions:
    a. What function does the OSR instruction play in rung 0?
    b. What function does the XIO instruction play in rung 1?
21. Why are OSR instructions necessary in the ladder logic in Figure 6-16?

ControlLogix

When working these problems, assume that a 16-point input is in slot 2 and a 16-point output is in slot 4, unless otherwise noted. Pin numbers are specified in the problems.

22. Study the ladder logic in Figure 6-18 and describe why the preset value of the timer is 60,000.
23. Convert the temperature conversion ladder solution in Figure 6-7 to a ControlLogix solution. Use terminal 3 for the normally open calculate push button.
24. Convert the upper and lower limit temperature chamber ladder solution in Figure 6-8 to a ControlLogix solution. Use terminal 3 for the normally open calculate push button.
25. Convert the right triangle ladder solution in Figure 6-10 to a ControlLogix solution. Use terminal 3 for the normally open calculate push button.
26. Convert the problem in Example 6-13 to a ControlLogix solution. Use the same terminal numbers, and use another input module in slot 3 for the word value.
27. Convert the part color solutions in Figure 6-15 to a ControlLogix solution. Use another input module in slot 3 for the data value.
28. Convert the pallet part total ladder in Figure 6-16 to a ControlLogix solution. Use the same terminal numbers as in the SLC 500 solution.
29. Convert the ball counting and packaging ladder solution in Figure 6-17 to a ControlLogix solution. Use the same terminal numbers as in the SLC 500 solution.
30. Convert the box counting ladder solution in Figure 6-19 to a ControlLogix solution.

Challenge

When working these problems, use the PLC processor assigned and the input and output modules specified for the PLC 5, SLC 500, and ControlLogix processors in the previous problem sections.

31. Modify the can packaging problem in Example 6-14 by adding a timer to keep the ejector on for 3 seconds. All other requirements are unchanged.

32. Modify the parking garage car count problem (Example 5-4) solution in Figure 5-18 by adding a push button switch to move the correct value of cars in the lot to the up/down counter accumulator. Use terminal 3 at the input and move the correct car count value from the N7:4 register. All other requirements are unchanged.
33. Design a ladder program with arithmetic instructions that converts degrees Celsius to degrees Fahrenheit.
34. Design a ladder program that moves the value of a count up accumulator to integer register N7:5 every 90 seconds. If the overflow is false the counter accumulator is moved, but if the overflow is true the count is calculated from the current accumulator value and then moved. The counter is reset at the 90-second mark as well.
35. Design a ladder program to move the part color value from a color sensor similar to that illustrated by the timing diagram in Figure 6-14. Input color data is coming from the sensor attached to an additional input module in slot 4, and the value should be placed into N7:0. In this case the color value is valid for 300 milliseconds; the trigger pulse starts 5 milliseconds after the color value is valid and is true for 1.5 seconds. Use modules and terminals from Figure 6-15.
36. The design for Problem 35 must be modified because it is discovered that the trigger pulse starts 200 milliseconds before the color value is valid and is true for 1 second. Create the new design with all other data unchanged.
37. Design a ladder program for counters to count from 0 to 90,000. Use three counters: C5:0, C5:1, and C5:2. When C5:0 nears its maximum allowed value (you decide on the specific number), it maintains its current count and C5:1 starts counting. When C5:1 nears its maximum allowed value (again, you decide on the specific number), it stops counting but maintains its current count and C5:2 starts counting. The system resets itself when the total count is 90,000. Every 20 seconds the current count value is calculated from the three counter accumlator values and moved to a floating point register, F8:1.

CHAPTER 7 CONVERSION AND COMPARISON INSTRUCTIONS

REFERENCES

Textbook Chapter 7 – Conversion and Comparison Instructions (pp. 241–272)
Textbook CD – Allen-Bradley Manuals:

- PLC 5 Instruction Programming Manual – 1675-6.1
- SLC 500 Instruction Set – 1747-RM001D-EN-P
- Logix5000 General Instruction Reference – 1756-RM003G-EN-P
- Logix Advanced Programming Manual – 1756-PM001F-EN-P

The following examples and figures are referenced in the questions and problems for this chapter:

- Example 7-1 is on page 255
- Figure 7-15 is on page 256
- Figure 7-16 is on page 257
- Figure 7-19 is on page 263
- Figure 7-20 is on page 264
- Figure 7-21 is on page 266
- Figure 7-22 is on page 267

QUESTIONS

1. Describe the BCD numbering system.
2. What is the difference between a push switch and a thumbwheel switch?
3. Describe two BCD conversion instructions.
4. What is the advantage of using the hexadecimal numbering system over the binary numbering system?
5. How does the hexadecimal system handle the numbers 10 through 15?
6. Multiples of what number do the following numbering systems use as a weighted value?
   a. Decimal numbering system
   b. Hexadecimal numbering system
7. Describe the compare instruction structure.
8. What is the function of a comparison instruction?
9. When does a comparison instruction provide a true output?
10. What could be an advantage of using a comparison instruction in a ladder program?

FIGURE D-7: Ladder logic for Question 11

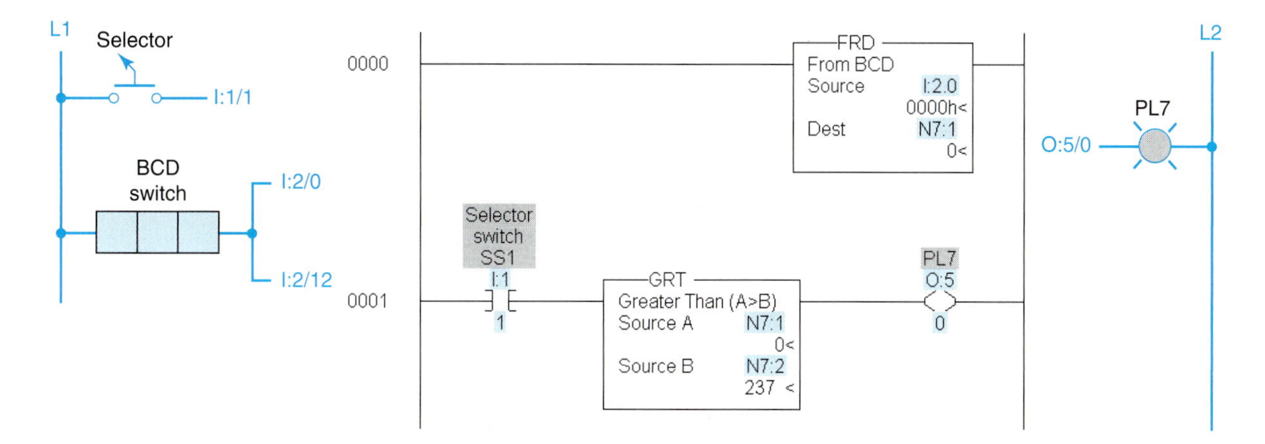

11. With N7:3 set at 237 in Figure D-7, answer the following questions.
    a. Will pilot light PL7 illuminate whenever switch SS1 is closed?
    b. Will the contents of register N7:1 change with SS1 open?
    c. What number(s) must be set in the BCD switch to illuminate PL7?
12. If the compare instruction in Question 11 is a less than instruction, what number(s) must be set in the BCD switch to illuminate PL7?
13. If Source B in Question 11 is the accumulated count of an up counter and SS1 is closed, what is the setting of the BCD switch to illuminate PL7 when the accumulated count is 89?
14. What is the largest number in BCD format that a 16-bit integer register can store?
15. How does the FRD instruction differ from the TOD instruction?
16. For the TOD instruction, what status bit is available in SLC 500 but not in the PLC 5?
17. For the FRD instruction, what status bit is available in SLC 500 but not in the PLC 5?
18. What is the largest decimal value that the ControlLogix TOD instruction can convert to a BCD number?
19. Describe the changes that are required to change the BCD and comparison standard ladder logic from the SLC 500 ladders to PLC 5 ladders.
20. Describe the changes that are required to change the BCD and comparison standard ladder logic from the SLC 500 ladders to ControlLogix ladders.
21. Describe the operation of the OSR instruction with a BCD instruction driving a seven-segment display.
22. List the function of each of the eight comparison instructions discussed in this chapter.
23. What is the difference and similarity between the EQU and MEQ instructions?
24. How does the LIM instruction compare data?
25. How are comparison instructions used to resolve a hysteresis requirement in an automated process?
26. Where is the problem most likely to reside if the field device status, input voltage, and module indicator all agree, but the XIC and XIO ladder instructions do not?
27. Where is the problem most likely to reside if the input voltage, module indicator, and the XIC and XIO ladder instructions agree, but the condition of the field device does not?
28. Where is the problem most likely to reside if the field device status and module indicator agree, but the output coil condition does not?
29. Identify the standard logic ladder that was used to develop each rung in Figures 7-19, 7-20, and 7-21.

WEB AND DATA SHEET QUESTIONS

Use the Allen-Bradley resources from the Web or book CD resources to answer the questions in this section.

1. Example 7-1 has an input from a thermocouple for the oven temperature. Determine

what type of thermocouple would be best for the required temperature range. Find a thermocouple module that would satisfy this application. Write a description of the module explaining its input requirements and data manipulation options.

2. The troubleshooting section indicated that some input and output discrete modules have fuses. Review five input and output modules from two of the three Allen-Bradley PLC types and document how the modules use fuses to protect the module and how you know if a fuse is blown.
3. Use the resources from Allen-Bradley to specify a rack and the necessary input and output modules for the tank control problem in Section 7-6-3.

PROBLEMS

General

If the solution to these general problems requires an input or output address, then use either a generic value or use the PLC processor required for the class.

1. Convert the following decimal numbers to BCD: (a) 257, (b) 490, and (c) 2789.
2. Convert the following hexadecimal numbers to binary numbers: (a) 5D, (b) 3C9, and (c) BCD.
3. Convert the following hexadecimal numbers to decimal numbers: (a) 6C, (b) 1F2, and (c) 92B.
4. Design a ladder rung to illuminate a pilot light when N7:1 has a value greater than 288.
5. Design a ladder program that illuminates a light only when an up counter (C5:0) has a value of 6 or 12.
6. Design a ladder program that illuminates a light only when a timer (T4:0) has a value of less than 12 and greater than 24.
7. Describe how to turn on an output when combinations of two or more bits in registers are equal.
8. Describe how the solution in Problem 7 would change if the goal was to turn on an output when combinations of two or more bits in registers are not equal.
9. Describe how multiple comparison standard ladder logic in Figure 7-15(c) operates.
10. Develop a ladder rung that turns on a binary bit, B3:0/13, when the accumulator in timer T4:10 has a 1101 bit pattern in the least significant byte.
11. Develop a ladder rung that turns on an output, O:2/4, with the limit test instruction when the accumulator in counter C5:12 is between 355 and 1000.
12. Repeat Problem 11 using comparison instructions.
13. Develop a ladder rung using comparison instructions that turns on an output, O:2/8, when register N7:2 is greater than or equal to 1500 and does not turn off the output until the value drops to 1450.
14. Repeat Problem 13 using latched output O:3/0.
15. Develop a ladder rung using comparison instructions that turns on an output, O:2/14, when register N7:2 is not equal to the timer T4:7 accumulator or equal to the second byte of N7:5.
16. Develop a ladder rung using comparison instructions that turns on an output, B3:0/7, when register N7:2 is less than or equal to the counter C5:7 accumulator or greater than or equal to the sum of the counter C5:8 and C5:9 accumulators.

PLC 5

When working these problems, assume that all modules are in rack 2 with single-slot addressing. A 16-point input is in slot 2 (group 2) and a 16-point output is in slot 3 (group 3), unless otherwise noted. Pin numbers are specified in the problem.

17. Convert the process solution in Figure 7-16 from SLC 500 ladder logic to PLC 5 ladder logic. Use the SLC 500 pin numbers in the conversion.
18. Convert the process solution in Figure 7-19 from SLC 500 ladder logic to PLC 5 ladder logic. Use the SLC 500 pin numbers and assume that 16-point input and output modules are placed in the rack as necessary.
19. Convert the process solution in Figure 7-20 from SLC 500 ladder logic to PLC 5 ladder logic. Use the SLC 500 pin numbers and assume that 16-point input and output modules are placed in the rack as necessary.

20. Convert the process solution in Figure 7-21 from SLC 500 ladder logic to PLC 5 ladder logic. Use the SLC 500 pin numbers and assume that 16-point input and output modules are placed in the rack as necessary.
21. Convert the ControlLogix solution in Figure 7-22 to a PLC 5 ladder program. Use the terminal values from the ControlLogix solution.

SLC 500

When working these problems, assume that a 16-point input is in slot 2 and a 16-point output is in slot 3, unless otherwise noted. Pin numbers are specified in the problems.

22. Convert the ControlLogix solution in Figure 7-22 to an SLC 500 ladder program. Use the terminal values from the ControlLogix solution.
23. How do you determine the values for the comparison instructions in Figure 7-22?
24. Study the ladder logic in Figure 7-16 and answer the following questions:
    a. Why does the source address for the FRD instruction I:2.0 have a dot [.] delimiter rather than a forward slash [/] delimiter?
    b. Why are the output registers a combination of N7 and F8 types?
    c. What is the function of the XIC instruction called Heater contactor in the last rung?
25. Study the ladder logic in Figure 7-19 and answer the following questions:
    a. What is the function of the selector switch called Process/Rinse?
    b. What is the function of the timer in rung 1 and sealing instruction from the timer timing bit in rung 0?
    c. What is the function of the two multiply by 2 instructions in rung 3?
26. Study the ladder logic in Figure 7-20 and answer the following questions:
    a. Explain why each input instruction in rung 9 is required.
    b. What is the function of the OSR instruction in rung 10?
    c. What is the function of the sealing instruction in rung 12?
    d. What stops the drain cycle?
27. Study the ladder logic in Figure 7-21 and answer the following questions:
    a. Why is the high level float switch necessary?
    b. Why is the high level float switch sealed in rung 15?
    c. In what two ways is the rinse cycle terminated?

ControlLogix

When working these problems, assume that a 16-point input is in slot 2 and a 16-point output is in slot 4, unless otherwise noted. Pin numbers are specified in the problems.

28. Convert the ladder logic in Figure 7-16 to a ControlLogix ladder. Create tag names as necessary.
29. Convert the process solution in Figures 7-19, 7-20, and 7-21 from SLC 500 ladder logic to ControlLogix ladder logic. Create tag names consistent with the process parameters. Use the integer data types for all variable tags.

Challenge

When working these problems, use the PLC processor assigned and the input and output modules specified for the PLC 5, SLC 500, and ControlLogix processors in the previous problem sections.

30. Modify the ladder solution in Figure 7-16 to input the upper and lower temperature limits in single-degree increments using a two-digit BCD push switch. Use a 16-point input module in slot 5 for assignment of input terminal numbers.
31. Modify the ladder solution in Figure 7-21 to include a single-digit BCD switch for inputting multiple rinse cycles. Add the logic required to make the system execute the multiple wash cycles entered. Use a 16-point input module in slot 8 for asignment of input terminal numbers.
32. Modify the ladder solution in Figure 7-20 to make the system remain at the maximum temperature for 20 minutes before starting the drain cycle.
33. Modify the ladder solution in Figure 7-20 to keep the drain open for 60 seconds after the

low level sensor turns off so that the tank can completely drain.

34. For the process tank problem in Figures 7-19, 7-20, and 7-21, develop the ladder logic to display the following system status: current temperature updated every minute, and a display light for all discrete outputs plus a light to indicate the fill, drain, and rinse cycles. Use the next available slots and add modules for the input and output terminal requirements.
35. For the process tank problem in Figures 7-19, 7-20, and 7-21, develop the ladder logic to display the product weight on three seven-segment readouts with an update time of 120 seconds. The product liquids have a density of 8 pounds per gallon. Use the next available slots and add a module for the output terminal requirement.

CHAPTER 8 PROGRAM CONTROL INSTRUCTIONS

REFERENCES

Textbook Chapter 8 – Program Control Instructions (pp. 273–302)
Textbook CD – Allen-Bradley Manuals:

- PLC 5 Instruction Programming Manual – 1675-6.1
- SLC 500 Instruction Set – 1747-RM001D-EN-P
- Logix5000 General Instruction Reference – 1756-RM003G-EN-P
- Logix Advanced Programming Manual – 1756-PM001F-EN-P

The following examples and figures are referenced in the questions and problems for this chapter:

- Example 6-7 is on page 219
- Example 8-1 is on page 277
- Example 8-3 is on page 283
- Example 8-4 is on page 287
- Example 8-6 is on page 298
- Figure 8-3 on page 276
- Figure 8-4 on page 277
- Figure 8-6 on page 278
- Figure 8-13 on page 286
- Figure 8-15 on page 289
- Figure 8-24 on page 299

QUESTIONS

1. What does the term *fence* mean relative to the MCR instruction?
2. What is the function of the second MCR instruction in a control zone?
3. In Figure 8-3, which of the following are non-retentive: C5:0, O:3/1, O:3/2, and O:3/3
4. In Question 3, which are retentive?
5. What is the purpose of the label instruction?
6. In what situation would two jump instructions have the same address?
7. What are the similarities and differences between the jump instruction and the MCR instruction?
8. Name three cautions the programmer should observe when using the jump instruction.
9. What are two other terms that some PLC manufacturers use for their jump instruction?
10. What is the procedure for selecting and naming the subroutine files in the PLC 5 and SLC 500 PLCs?
11. What is a subroutine?
12. What is parameter passing?
13. Can a conditional return instruction be used and why?
14. What is the advantage of using nested subroutines?
15. What is the caveat in using nested subroutines?
16. How can the caveat in Question 15 be mitigated?
17. Define the terms *global* and *program data values*.
18. What is controller-scoped data?
19. Explain input par and return par.
20. What is the purpose of immediate input and output instructions?
21. What does the address of the IIN and IOT instructions mean?
22. What is the function of a mask with immediate input and output instructions?
23. What is the advantage of IIM and IOM instructions over IIN and IOT instructions?

24. Define the terms *slot*, *mask*, and *length* in an immediate mask instruction.
25. What is the first step in using program control instructions in PLC ladder designs?
26. What is AFI?
27. Describe three suggested methods in troubleshooting a ladder section that contain program control instructions.
28. What happens if excessive backward jumps are performed and the watchdog timer is exceeded?

WEB AND DATA SHEET QUESTIONS

Use the Allen-Bradley resources from the Web or book CD resources to answer the questions in this section.

1. Determine the default value for the watchdog timer for the three Allen-Bradley PLC systems.
2. Use the reference manuals to research the subroutine instructions for the three PLC systems and note any similarities and differences. Which two systems are most alike?

PROBLEMS

General

If the solution to these general problems requires an input or output address, then use either a generic value or use the PLC processor required for the class.

1. Refer to Figure D-8. What is the condition of solenoids A and B for the following input sequences?
   a. All inputs are off, then I:2/2 and I:2/4 are turned on.
   b. Then I:2/1 is turned on.
   c. Then I:2/1 is turned off.
2. Refer to Figure D-8. What is the condition of pilot light 1 for the following input sequences?
   a. All inputs are off, then I:4/2 is turned on.
   b. Then I:4/1 is turned on.
3. Refer to Figure D-9. In what order are the rungs scanned when all the inputs are true?
4. Repeat Problem 3 when all the inputs are false.
5. In Figure D-9, to what rung does the program return after the subroutine has been completed?

FIGURE D-8: Ladder logic for Problems 1 and 2

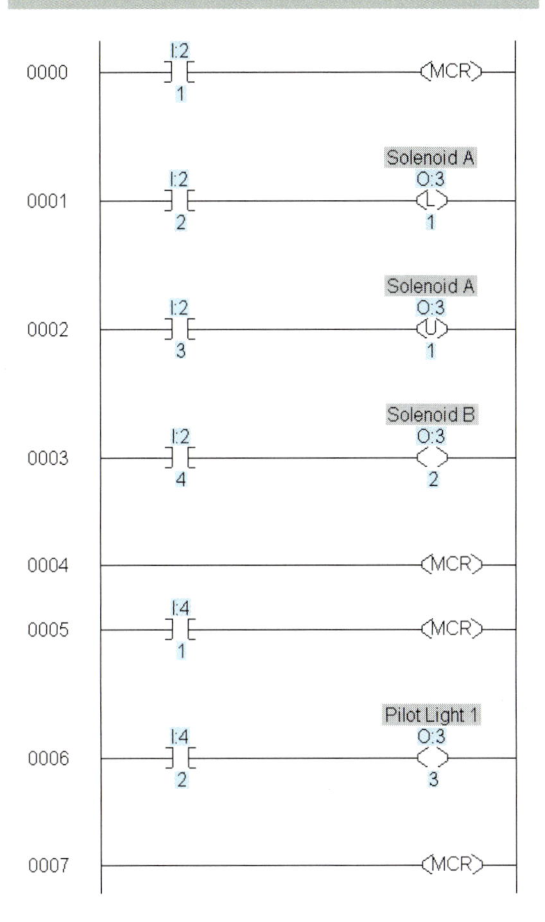

6. Refer to Figure D-9. List only the inputs that must be on to have all the pilot lights on.
7. Repeat Problem 6, listing the inputs that must be on to have all the solenoids on.
8. Refer to Figure D-9. What pilot lights and solenoids are on if only odd number inputs I:4/1, I:4/3, I:4/5, and so forth are on?
9. Repeat Problem 8 if only even number inputs I:4/2, I:4/4, I:4/6, and so forth are on.
10. Refer to Figure D-9. List only the inputs that must be on to have the timer T4:1 start.
11. Repeat Problem 10, listing the inputs that must be on to have the timer T4:2 start.
12. Determine the values for the passed timer values for the following conditions:
    a. Normal 5 seconds on and 2 seconds off, minor fault 3 seconds on and 3 seconds off, and major fault 1 second on and 1 second off.
    b. Normal light on continuously, minor fault 3 seconds on and 1 second off, and major fault 1 second on and 2 seconds off.

FIGURE D-9: Ladder logic for Problems 3 through 11

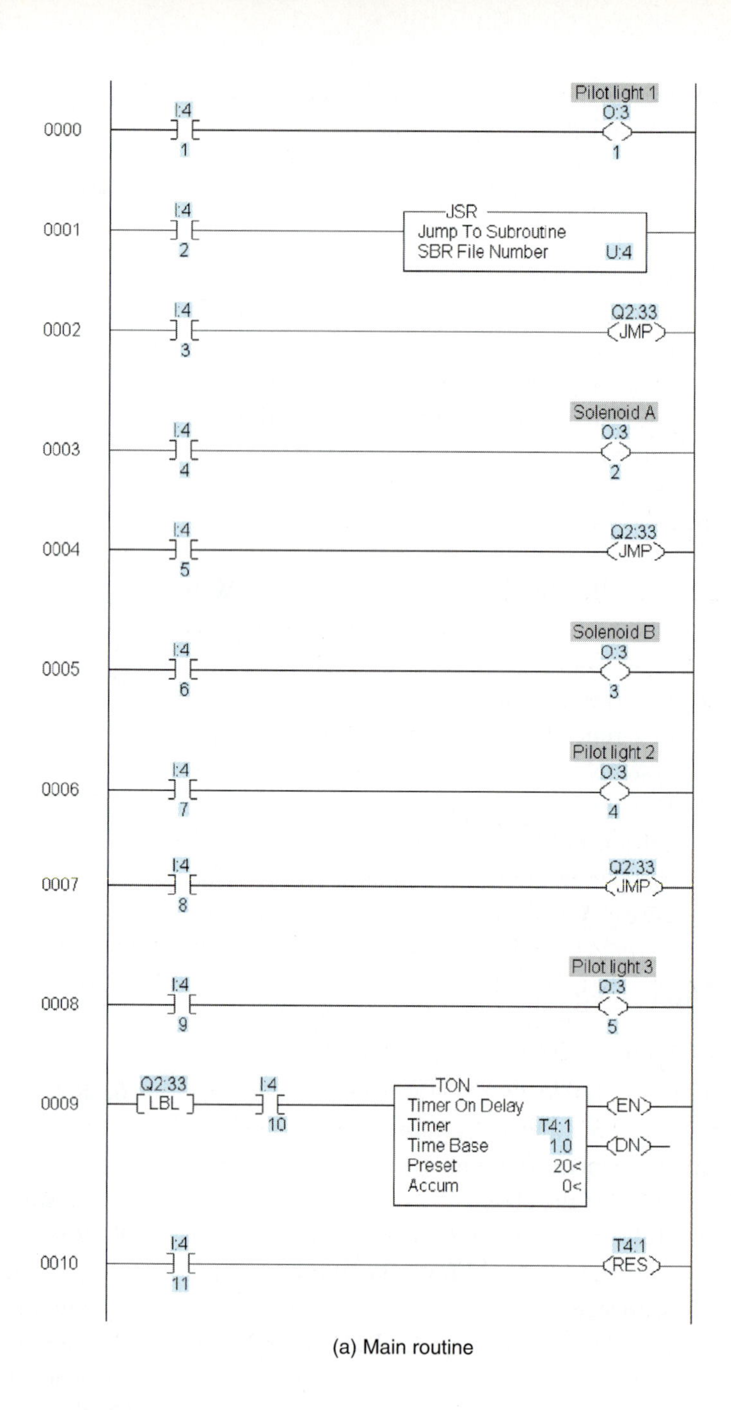

(a) Main routine

13. How many scans are executed in the MCR fence in Figure 8-4? Initialization is momentary push button.
14. In Figure 8-6, what logical input conditions must be present to scan rungs 1 and 2 but not 4 and 5; to scan rungs 4 and 5 but not 1 and 2; and to scan rungs 1, 2, 4, and 5? Is it possible to not scan rung 6?
15. Draw the timing diagram for rungs 0, 1, and 2 in Figure 8-13(b) for Example 8-3.

PLC 5

When working these problems, assume that all modules are in rack 2 with single-slot addressing. A 16-point input is in slot 2 (group 2), and a 16-point output is in slot 3 (group 3), unless otherwise noted. Pin numbers are specified in the problem.

16. Convert the SLC 500 PLC ladder logic with MCR in Figures 8-3 and 8-4 to a solution

FIGURE D-9: (*Continued*)

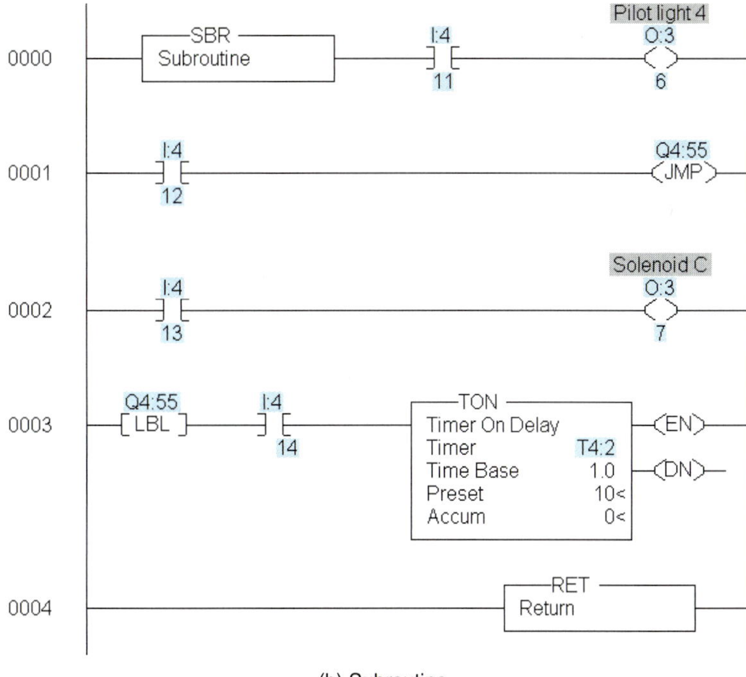

(b) Subroutine

for the PLC 5. Use the terminal numbers in the figures for all input and outputs.

17. Convert the SLC 500 PLC ladder logic with a jump instruction in Figure 8-6 to a solution for the PLC 5. Use the terminal numbers in the figure for all input and outputs.
18. Convert the SLC 500 PLC ladder logic with a subroutine instruction in Figures 8-13 and 8-15 to a solution for the PLC 5. Use the terminal numbers in the figures for all input and outputs.
19. Convert the SLC 500 PLC ladder logic with immediate instructions in Figures 8-15 and 8-24 to a solution for the PLC 5. Use the terminal numbers in the figures for all input and outputs.

SLC 500

Unless specified in the problem, assume that a 16-point input is in slot 2 and a 16-point output is in slot 3 when working the following problems. Pin numbers are specified in the problems.

20. Study the ladder logic in Figure 8-3 and indicate what outputs are true when the following input instruction are true:
    a. I:1/1, I:1/2, and I:1/5
    b. I:1/2, I:1/4, and I:1/5
    c. What is necessary for the solenoid valve in rung 5 to be on?
21. Study the ladder logic in Figure 8-4 and describe the function of binary bit B3/1.
22. Study the ladder logic in Figure 8-6 and answer the following questions:
    a. What input conditions are necessary for the rungs 1 through 6 to be bypassed?
    b. What input conditions are necessary for the rungs 4 through 6 to be bypassed?
23. Study the ladder logic in Figure 8-13 and answer the following questions:
    a. What input conditions are necessary to run the fault routine?
    b. What is the flashing duty cycle of the light in the fault routine?
24. Study the ladder logic in Figure 8-15 and describe how the two levels of faults trigger the lights and the horn.
25. Study the ladder logic in Figure 8-24 and describe why the MOVE instruction in rung 1 moves a $00000000\ 11111111_2$ into O:3.0.

ControlLogix

Unless specified in the problem, assume that a 16-point input is in slot 2 and a 16-point output is in slot 4 when working the following problems. Pin numbers are specified in the problems.

26. Convert the SLC 500 PLC ladder logic with MCR in Figures 8-3 and 8-4 to a solution for the ControlLogix PLC. Create tag

names consistent with the process parameters and use the terminal numbers in the figures for all input and outputs.

27. Convert the SLC 500 PLC ladder logic with a jump instruction in Figure 8-6 to a solution for the ControlLogix PLC. Create tag names consistent with the process parameters and use the terminal numbers in the figure for all input and outputs.
28. Convert the SLC 500 PLC ladder logic with a subroutine instruction in Figures 8-13 and 8-15 to a solution for the ControlLogix PLC. Create tag names consistent with the process parameters and use the terminal numbers in the figures for all input and outputs.
29. Convert the SLC 500 PLC ladder logic with immediate instructions in Figures 8-15 and 8-24 to a solution for the ControlLogix PLC. Create tag names consistent with the process parameters and use the terminal numbers in the figures for all input and outputs.

Challenge

When working these problems, use the PLC processor assigned and the input and output modules specified for PLC 5, SLC 500, and ControlLogix in the previous problem sections.

30. Modify Example 8-1 so that the initialization is performed by the first scan status bit.
31. Modify Example 8-3 so that the light goes from pulsing to continuous after it pulses for 30 seconds. Hint: continue to use one delay timer along with a comparitor.
32. Modify Example 8-4 so that the horn is delayed by 60 seconds for the level 2 fault condition.
33. Modify Example 8-6 by adding an alarm reset using a momentary NO push button.
34. Use ControlLogix ladder instructions to modify Example 6-7 in Chapter 6 by putting the temperature conversion into a subroutine with parameter passing.
35. Use ControlLogix ladder instructions to modify Example 8-3 by using parameter passing to set the time for the horn delay.

CHAPTER 9 INDIRECT AND INDEXED ADDRESSING

REFERENCES

Textbook Chapter 9 – Indirect and Indexed Addressing (pp. 303–314)
Textbook CD – Allen-Bradley Manuals:

- PLC 5 Instruction Programming Manual – 1675-6.1
- SLC 500 Instruction Set – 1747-RM001D-EN-P
- Logix5000 General Instruction Reference – 1756-RM003G-EN-P
- Logix Advanced Programming Manual – 1756-PM001F-EN-P

The following examples and figures are referenced in the questions and problems for this chapter:

- Example 9-1 is on page 311
- Figure 7-20 on page 264
- Figure 9-1 on page 304
- Figure 9-2 on page 304
- Figure 9-3 on page 305
- Figure 9-4 on page 306
- Figure 9-5 on page 307
- Figure 9-12 on page 312

QUESTIONS

1. In addressing modes, explain the meaning of the terms *data* and *operand*.
2. Describe the differences between direct and indirect addressing.
3. In indirect addressing, what is the effective address?
4. Describe indexed addressing.
5. What is a useful application of indexed addressing?
6. Describe the differences and commonalties between pre- and post-indexed addressing methods.
7. How is an indirect address specified in a PLC 5 and SLC 500 instruction?
8. How is an indirect address specified in a ControlLogix PLC instruction?
9. How is an indexed address specified in a PLC 5 and SLC 500 instruction?

10. When does crossing the file boundary occur in indexed addressing?
11. What is the result of crossing the file boundary in indexed addressing?
12. How is an indexed address specified in a Logix PLC instruction?
13. What are two advantages that the ControlLogix PLC has over the PLC 5 and SLC 500 PLCs relative to indexed addressing?

WEB AND DATA SHEET QUESTIONS

1. Use the reference manuals to identify how the PLC 5, SLC 500, and Logix systems respond to crossing a file boundary with indexed and indirect addressing.
2. What options does the SLC 500 offer with respect to crossing file boundary errors?
3. The indexed indirect addressing for the SLC 500 system was not covered in this chapter. Use the reference manuals to describe how that type of addressing works.

PROBLEMS

General

If the solution to these problems requires an input or output address, then use either a generic value or use the PLC processor required for the class.

1. Design a new example solution for Figure 9-4(a) and (c) using an index of 23, resultant data of 58, and a starting base address of 1776. Create your own address numbers and data values as needed to generate this new solution.
2. Find the addressed data for the following addressing problems.
   a. Find the data value using Figure 9-1 for a direct memory address of 2210.
   b. Find the data value using Figure 9-2 for an indirect memory address of 2214.
   c. Find the data value using Figure 9-3 for an indexed base memory address of 2000 and a value in S:24 of 134.
   d. Find the data value using Figure 9-4(a) for a pre-indexed base memory address of 1210 and a value in S:24 of –8.
   e. Find the data value using Figure 9-4(c) for a post-indexed base memory address of 1202 with all other values unchanged.
   f. Find the data value using Figure 9-5 for a Source A of N7:[N7:0] with all other values unchanged.

PLC 5

When working these problems, assume that all modules are in rack 2 with single-slot addressing. A 16-point input is in slot 2 (group 2) and a 16-point output is in slot 3 (group 3), unless otherwise noted. Pin numbers are specified in the problem.

3. Convert the SLC 500 solution in Figure 9-12 for Example 9-1 to a solution for the PLC 5. Use terminal numbers and registers from the figure.

SLC 500

When working these problems, assume that a 16-point input is in slot 2 and a 16-point output is in slot 3, unless otherwise noted. Pin numbers are specified in the problems.

4. Study the ladder logic in Figure 9-12 and describe how the value in register S24 is used in the ladder.
5. How could indexed addressing be used for the solution in Figure 9-12?

ControlLogix

When working these problems, assume that a 16-point input is in slot 2 and a 16-point output is in slot 4, unless otherwise noted. Pin numbers are specified in the problems.

6. Convert the SLC 500 PLC ladder logic indexed solution in Figure 9-1 to an indirect addressing solution with an array for the ControlLogix PLC. Develop tag names consistent with the problem.

Challenge

When working these problems, use the PLC processor assigned and the input and output modules specified for the PLC 5, SLC 500, and ControlLogix processors in the previous problem sections.

7. Modify the temperature/drain cycle for the process tank problem in Figure 7-20 so that a data table of five maximum temperatures can be loaded into the ladder logic using a receipt number from 1 to 5.

8. Integer registers N10:0 through 10 hold data for a manufacturing process. Design a ladder using indirect addressing that moves these values with a single move instruction to N7:0 each time a NO push button is pressed.
9. Repeat Problem 8 using indexed addressing and the S:24 register.
10. An array, Numbers[index], is used to hold 8 data values. Design a ladder that moves the values out to double integer Process_value and keeps a total for the sum of the numbers in Process_sum. The numbers are moved out when Boolean tag Number_out is true and they start with the value in element where index is 0.

CHAPTER 10 DATA HANDLING INSTRUCTIONS AND SHIFT REGISTERS

REFERENCES

Textbook Chapter 10 – Data Handling Instructions and Shift Registers (pp. 317–346)
Textbook CD – Allen-Bradley Manuals:

- PLC 5 Instruction Programming Manual – 1675-6.1
- SLC 500 Instruction Set – 1747-RM001D-EN-P
- Logix5000 General Instruction Reference – 1756-RM003G-EN-P
- Logix Advanced Programming Manual – 1756-PM001F-EN-P

The following examples and figures are referenced in the questions and problems for this chapter:

- Example 10-2 is on page 342
- Example 10-3 is on page 343
- Example 10-4 is on page 344
- Figure 10-5 on page 321
- Figure 10-7 on page 324
- Figure 10-8 on page 325
- Figure 10-15(c) on page 331
- Figure 10-16 on page 332
- Figure 10-26(a) on page 341

QUESTIONS

1. What is data handling?
2. What three terms are used interchangeably with the term *files*?
3. Describe the three types of transfers that are used with words and files.
4. Describe the operation of the AND, OR, and XOR instructions.
5. What is the purpose of the FAL instruction?
6. Which bit is lost during a right shift operation?
7. Which bit is lost during a left shift operation?
8. Describe how you would connect registers together for the three-register shift operation.
9. Describe the rotate function.
10. Explain how a shift register is used in a PLC that is controlling an automated bottle-filling operation on a conveyor system.
11. List and describe the terms within the BSR and BSL instructions.
12. What is the function of the unload bit in a BSL instruction?
13. What is the function of the unload bit in a BSR instruction?
14. When using a sensor as the input to the bit address of a BSL instruction, what is its function?
15. When using a sensor as the input to the bit address of a BSR instruction, what is its function?
16. Why are bit shift registers often used to track parts moving on a conveyor?
17. Describe the FIFO function.
18. List and describe the terms within the FFL and FFU instructions.
19. Why are both the FFL and FFU instructions needed to perform a FIFO function?
20. What is the timing relationship between the FFL and the FFU instructions?
21. Describe the LIFO function.
22. What is the difference between the ladder representation of the FFU instruction and the LFU instruction?
23. What is the difference between the COP and FLL instructions?
24. If the destination file type in a COP instruction is a counter and the source file type is an integer, how many integer words are transferred?
25. List and describe the parameters that are used when programming the COP instruction.
26. List and describe the parameters that are used when programming the FLL instruction.
27. What instructions would you use in applications that require the movement of data

words between registers and arrays without the support of a control data file?

WEB AND DATA SHEET QUESTIONS

1. Identify sensors that would satisfy the control problem in Figure 10-16 for the following conditions: The proximity sensor is 15 cm from the target, the detected material is steel and aluminum, the photo sensors are 45 inches apart with high ambient light, and the boxes are brown cardboard. The PLC input is current sinking.
2. What type of photo sensor is required in Problem 1 when sensor power and wiring is present only on the back side of the conveyor and the PLC input is current sourcing?
3. What type of photo sensor is required in Problem 1 when sensor power and wiring is present only on the back side of the conveyor, the package box is changed to a glossy white, high ambient light is present, and the PLC input is current sinking?
4. Use the Web to search Bimba and Festo valves for a product that could be used for the box ejector. The box weighs approximately 5 pounds and is 10 inches wide and high. Suggest an actuator and valve product from both companies, tooling to hold the actuator and to contact the box, and an output module for each type of Allen-Bradley PLC system that could drive the valve.

PROBLEMS

General

If the solution to these problems requires an input or output address, then use either a generic value or use the PLC processor required for the class.

1. Draw the 1s and 0s in a BSL shift register whose contents are represented by the hexadecimal number 4E3A.
2. Draw the 1s and 0s in the shift register in Problem 1 after two left shifts with a 1 shifted in for both shifts.
3. What is the hexadecimal number that represents the contents of the shift register in Problem 2 after the shifts?
4. Draw the Allen-Bradley FAL instruction to copy 23 words from a data file starting at address N7:10 into a file starting at N10:10.
5. What is the address of the location of the last data word in the destination file in Problem 4?

PLC 5

When working these problems, assume that all modules are in rack 2 with single-slot addressing. A 16-point input is in slot 2 (group 2) and a 16-point output is in slot 3 (group 3), unless otherwise noted. Pin numbers are specified in the problem.

6. Redraw the three FAL instructions in Figure 10-7 with indexed integer registers used in place of the arrays starting from N7:10 and ending at N7:29. Select other integer registers as needed. All other parameters are the same.
7. Convert the SLC 500 ladder logic in Figure 10-15(c) for use on a PLC 5 with the ejector station at position 6 on the conveyor.
8. Convert the SLC 500 solution in Figure 10-16 to a solution for the PLC 5. Use the same pin numbers listed in the current solution.
9. Convert the SLC 500 ladder logic for the problem in Example 10-3 to a PLC 5 solution using the same input and output terminals.
10. Convert the SLC 500 ladder logic for the problem in Example 10-4 to a PLC 5 solution using the same input and output terminals.

SLC 500

When working these problems, assume that a 16-point input is in slot 2 and a 16-point output is in slot 3, unless otherwise noted. Pin numbers are specified in the problems.

11. Using the shift register in Problem 1, draw the SLC 500 BSL instruction when the input is coming from I:3/14 and the shift register is B3:2.
12. Convert the PLC 5 FAL instruction ladder logic in Figure 10-8 to an SLC 500 ladder using the COP and ADD instructions. Use additional instructions and adjust the integer register values as needed for the solution.
13. Design a ladder logic solution for the problem in Example 10-2 using the SLC 500 instruction set.

ControlLogix

When working these problems, assume that a 16-point input is in slot 2 and a 16-point output is in slot 4, unless otherwise noted. Pin numbers are specified in the problems.

14. Convert the PLC 5 FAL instruction ladder logic in Figure 10-8 to a ControlLogix solution. Create tag names for the solution as needed.

15. Convert the SLC 500 solution in Figure 10-16 to a ControlLogix solution. Use the same pin numbers listed in the current solution and develop tags consistent with the problem solution.

16. Convert the SLC 500 ladder logic for the problem in Example 10-3 to a ControlLogix solution. Use the same pin numbers listed in the current solution and develop tags consistent with the problem solution.

17. Convert the SLC 500 ladder logic for the problem in Example 10-4 to a ControlLogix solution. Use the same pin numbers listed in the current solution and develop tags consistent with the problem solution.

18. Draw the AND instruction in Figure 10-5 for ControlLogix that will mask the lower 8 bits of the source tag (current value of 83,843) and move that value to the destination tag. Source and destination tags are In_word and Out_word, respectively. Determine the binary bit and hex output for the destination after the AND operation.

19. Repeat Problem 18 for the XOR instruction using the same source and mask values.

20. Redraw the three FAL instructions in Figure 10-7 when the control register is Control_10, there are 25 elements in the array, the mode is unchanged, and the destination is a three-dimensional array with the values going into the first and third dimensions.

21. Modify the standard ladder logic rungs in Figure 10-26(a) so that both the source and destination are arrays with the starting array elements determined by the tags index1 for the source and index2 for the destination.

22. Identify the source and destination files for Problem 21 if index1 is 10 and index2 is 0. The length parameter is unchanged.

Challenge

When working these problems, use the PLC processor assigned and the input and output modules specified for the PLC 5, SLC 500, and ControlLogix processors in the previous problem sections.

23. Modify Example 10-3 so that the copy instruction is replaced with a FIFO stack with 20 words so that 20 sets of word data are saved.

24. Modify Example 10-4 so that five receipts are saved in memory and each receipt has five ingredients. The starting address for the first receipt is N15:20 and the receipt data is loaded starting at N20:5. Design the ladder logic so that receipt numbers are entered and the corresponding ingredients are placed into process data registers.

25. Modify the ladder logic in Figure 10-15(c) for the ejector station to be at position 6 on the conveyor.

26. Modify the COP instruction in Figure 10-26(a) for a solution with the data table starting at N7:50 and with 10 receipts. Each receipt has three ingredients, and the receipts need to be loaded starting at register N15:25.

CHAPTER 11 PLC SEQUENCER FUNCTIONS

REFERENCES

Textbook Chapter 11 – PLC Sequencer Functions (pp. 347–372)
Textbook CD – Allen-Bradley Manuals:

- PLC 5 Instruction Programming Manual – 1675-6.1
- SLC 500 Instruction Set – 1747-RM001D-EN-P
- Logix5000 General Instruction Reference – 1756-RM003G-EN-P
- Logix Advanced Programming Manual – I756-PM001F-EN-P

The following examples, tables, and figures are referenced in the questions and problems for this chapter:

- Table 11-3 is on page 355
- Table 11-4 on page 357
- Example 11-1 on page 355
- Example 11-3 on page 365

- Figure 1-17 on page 23
- Figure 3-44(b) is on page 132
- Figure 11-4 on page 351
- Figure 11-8 on page 356
- Figure 11-9 on page 358
- Figure 11-10 on page 359
- Figure 11-13 on page 361
- Figure 11-15 on page 364
- Figure 11-16 on page 365
- Figure 11-18 on page 367
- Figure 11-20 on page 368

QUESTIONS

1. What is sequencing?
2. Describe the operation of a drum switch.
3. Define the data that are programmed into a sequencer instruction.
4. What is the function of the file of a sequencer?
5. What is the function of the source and destination addresses of a sequencer instruction?
6. Why are PLC sequencers easier to program than PLC discrete outputs?
7. What is the function of the mask in a sequencer instruction?
8. What is the relationship between the length and the position in a sequencer instruction?
9. What is the difference between the SQI and SQC instructions?
10. How does an SQL instruction operate?
11. In Example 11-1, why are the number of steps set to 12?
12. In what two situations are cascaded sequencers used?
13. How would you modify Table 11-4 if the drying cycle was 330 seconds?
14. How would you use two sequencers in Example 11-1 to set unique times for each step?
15. What guidelines would you use when troubleshooting sequencer rungs?
16. How is the operation of the SQO instruction different during the first scan compared to the rest of the scans?
17. How does the ControlLogix SQO instruction differ from the same instruction in the PLC 5 and SLC 500?
18. What are the two types of triggers used in the SQO instruction to step the sequencer through the output words?
19. What are the two options that are used when timed pulses are the control inputs to the SQO instructions?
20. Describe the operation of the SQO instruction.
21. Describe the operation of the SQI instruction.
22. Describe how the SQI and SQO instructions are used in pairs.
23. When SQO and SQC instructions are used as a pair, what is the function of the FD bit?
24. Describe the function and operation of the control register used in all of the sequencer instructions.
25. Describe the difference between timed and event sequencers.
26. What is the primary application where an SQL instruction is used?

WEB AND DATA SHEET QUESTIONS

1. Search the Allen-Bradley and Square D Web sites for mechanical sequencers and write a brief description describing the product specifications and how they operate.
2. Assume that the parts feeder in Figure 11-9 is plastic and the 1.5-inch-diameter parts are metal. Use Web resources to determine the choice of sensor that could be used in this application.
3. Describe how your answer to Problem 2 would change if the parts feeder is bright aluminum and the parts are plastic.
4. Use the PLC data on the Web or CD to make a table that illustrates what control bits are available with each of the sequencer instructions covered in this chapter.

PROBLEMS

General

If the solution to these problems requires an input or output address, then use either a generic value or use the PLC processor required for the class.

1. Three types of DC motors are wired to a drum switch as illustrated in Figure D-10. Show how the motor windings are connected to the DC power lines for the forward

FIGURE D-10: Illustration for Problem 1

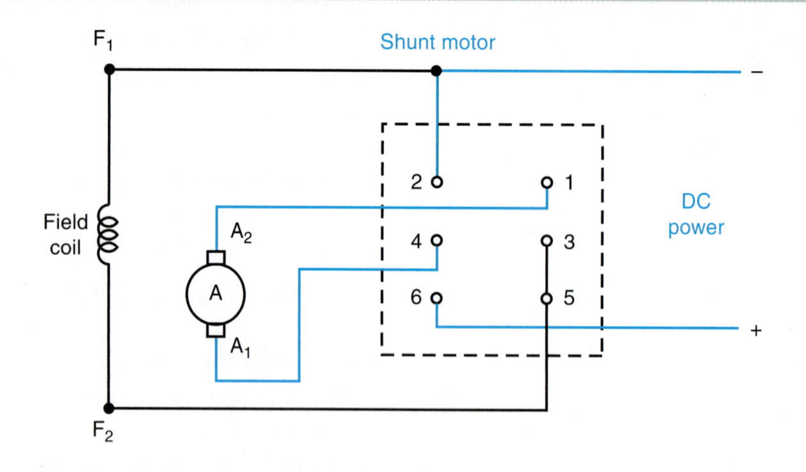

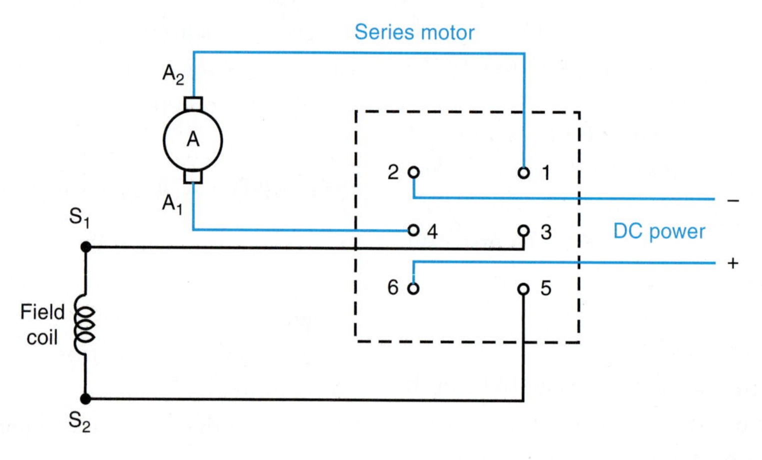

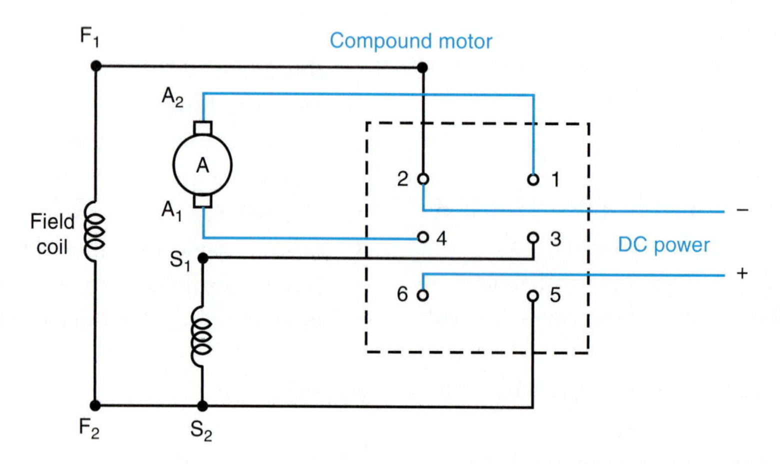

and the reverse positions of the drum switch.

2. List the 1-0 pattern in the destination location when the sequencer program moves the data from position 6 in the file depicted in Figure 11-4.
3. List the 1-0 pattern for Problem 2 if the mask is CCFF.
4. Redo Table 11-4 for the following timing cycles.
   a. A fill cycle and drain cycle of 90 seconds each.
   b. Three wash cycles of 90 seconds, 225 seconds, and 135 seconds each.
   c. A 45-second soap release cycle.
   d. A drying cycle of 270 seconds.

PLC 5

When working these problems, assume that all modules are in rack 2 with single-slot addressing. A 16-point input is in slot 2 (group 2) and a 16-point output is in slot 3 (group 3), unless otherwise noted. Pin numbers are specified in the problem.

5. Develop a PLC 5 standard ladder logic description and ladders for the SLC 500 standard SQO ladders in Figure 11-15.
6. Convert the SLC 500 stop light problem in Figure 11-16 to a PLC 5 solution using the same input and output terminal numbers.
7. Develop a PLC 5 standard ladder logic description and ladders for the ControlLogix standard SQI and SQO ladders in Figure 11-13.
8. Convert the SLC 500 ladder logic in Figure 11-8 to the PLC 5 ladder format. Use the same input terminal numbers.

SLC 500

When working these problems, assume that a 16-point input is in slot 2 and a 16-point output is in slot 3, unless otherwise noted. Pin numbers are specified in the problems.

9. Study the ladder logic in Figure 11-8 and answer the following questions.
   a. Why does N7:10 have a pound (#) sign in front of the N?
   b. What output is on when the sequencer is in position 4 and position 9?
   c. When you compare Tables 11-3 and 11-4, explain why some steps on Table 11-4 are repeated.
10. Study the ladder logic in Figure 11-10 and answer the following questions.
    a. What expression in the OR input logic does not need the one-shot instruction to force the ladder rung back to the false state before cycling to the next sequencer position?
    b. Why is the one-shot instruction necessary in the other rungs?
    c. Why can't the one-shot instruction be placed only in the OR expressions where it is required?
11. Study the ladder logic in Figure 11-16 and answer the following questions.
    a. When O:6.0 has 001100 in the lowest 6 bits, what is the preset value of the timer T4:1?
    b. How long is the E/W yellow light on?
    c. What causes the sequencers to cycle through their positions?
    d. Why is the start selector only in rung 2?
12. Study the ladder logic in Figure 11-18 and describe why the SQC and SQO instructions have the same control register address.

ControlLogix

When working these problems, assume that a 16-point input is in slot 2 and a 16-point output is in slot 4, unless otherwise noted. Pin numbers are specified in the problems.

13. Develop a ControlLogix standard ladder logic description and ladders for the SLC 500 standard SQO ladders in Figure 11-15.
14. Convert the SLC 500 stop light problem in Figure 11-16 to a ControlLogix solution using the same input and output terminal numbers. Use tag names that describe the process.
15. Convert the SLC 500 ladder logic in Figure 11-8 to the ControlLogix ladder format. Use the same input terminal numbers with descriptive tag names.

Challenge

When working these problems, use the PLC processor assigned and the input and output modules specified for PLC 5, SLC 500, and ControlLogix processors in the previous problem sections.

16. Develop a sequential ladder logic solution with one SQO instruction for the robot in Figure 11-20 when the robot does not have end-of-travel sensors on the actuators, and the times for each movement are 2 seconds for the gripper and 2 seconds for each linear actuator movement. Use Figure 3-44 for the input and output pin terminals.
17. Develop a sequential ladder logic solution with two SQO instructions (one for the output conditions and one for the timer times)

for the robot in Figure 11-20 when the robot does not have end-of-travel sensors on the actuators. The times are as follows: gripper close and open is 1 second, X down is 1.5 seconds, X up is 2.5 seconds, and Y out and in are 2 seconds.

18. Develop a ladder solution for the tank system in Figure 1-17 using a single SQO instruction with XIC AND XIO instructions for the input logic.
19. A process has five input states with word values of 0, 1, 2, 4, 8, and 16 (words 0 through 5) and five output steps with word values of 0, 1, 3, 7, 15, and 31 (words 0 through 5). The states must be on or active for the following times in seconds: 33, 150, 3, 45, and 67 for states 1 through 5, respectively. Develop a ladder solution for these conditions.
20. The robot in Figure 11-20 needs the SQI sequencer file or array loaded with input data for each machine state. Develop a ladder solution to perform this load using a manual push button switch. The robot is too large to move manually to each position. Include ladder rungs to manually move it to each position when the data is loaded.
21. Modify the electric signal problem in Example 11-3 so that the E/W lanes have a left turn signal for 10 seconds before the light turns green.

CHAPTER 12 ANALOG SENSORS AND CONTROL SYSTEMS

REFERENCES

Textbook Chapter 12 – Analog Sensors and Control Systems (pp. 373–422)
Textbook CD – Allen-Bradley Manuals:

- Pressure Controls – Bulletin 836
- Temperature Controls – Bulletin 837
- SLC 500 Discrete I/O Modules – 1746-2-35
- SLC 500 I/O Module Wiring – CIG-WD001A-EN-P
- SLC 500 Modular Hardware – 1747-UM011D-EN-P

The following examples and figures are referenced in the questions and problems for this chapter:

- Example 12-2 is on page 404
- Example 12-3 is on page 413
- Figure 12-37 is on page 411

QUESTIONS

1. Define analog sensor.
2. What temperature sensors provide a change in resistance as a function of temperature?
3. What temperature sensors provide a change in voltage as a function of temperature?
4. How is the RTD's change in resistance converted to a change in voltage?
5. How does a thermistor with an NTC differ from a thermistor with a PTC?
6. What is the Seebeck effect?
7. Which temperature sensor is the most linear?
8. What is hydraulic pressure?
9. What is the difference between direct pressure sensors, deflection pressure sensors, and differential pressure sensors?
10. What is a strain gage?
11. What is the purpose of a dummy gage?
12. Describe the three basic segments of a deflection pressure sensor.
13. How is a diaphragm used in a pressure sensor?
14. What is a bellows?
15. Explain the operation of a bourdon tube pressure sensor.
16. Why is the piezoelectric effect applicable to pressure sensing?
17. What is the relationship between the flow rate of a fluid in a pipe and the diameter of the pipe?
18. Compare the terms *laminar flow* and *turbulent flow*.
19. Explain how flow rate is determined using differential pressure.
20. What is a venturi and how is it used as a differential pressure sensor for flow rate measurement?
21. What is the advantage of a venturi over an orifice plate?
22. Explain the operation of a pitot tube relative to flow measurements.
23. How are flow nozzles used in flow measurements?
24. How can an existing elbow in the piping system be used in determining flow rate of the fluid in the system?

25. What do paddlewheel flow sensors and turbine flow sensors have in common?
26. Explain the operation of a vortex flow sensor.
27. On what principle do electromagnetic flow sensors operate?
28. Explain the operation of displacement flow sensors.
29. When would you use an ultrasonic flow sensor instead of other flow sensors?
30. Why is the ultrasonic flow sensor a good troubleshooting tool?
31. Explain the operation of the thermal flow mass sensor.
32. What is the advantage of the Coriolis mass flow sensor over the thermal mass flow sensor?
33. Why is a Coriolis mass flow sensor not practical at high fluid velocities?
34. Why are linear and rotary potentiometers classified as position sensors?
35. Explain the operation of a linear variable differential transformer.
36. What common technique is used in troubleshooting resistance temperature detectors and thermistors?
37. Describe the five application areas that vision systems support.
38. Identify all of the component parts of a vision system.
39. Describe the image capture process in a vision system and the two types of image sensors used.
40. What are the three vision system classifications and configurations?
41. Describe the light sources used to illuminate a vision application.
42. Describe the lighting techniques used in vision applications.
43. What is the function of a PLC analog module?
44. What is the difference between a voltage-sensing and current-sensing analog module?
45. Define unipolar and bipolar analog data.
46. What is backplane current load?
47. What is resolution relative to an analog signal?
48. List and describe the analog modules that are available in PLCs.
49. What are the three major elements of a closed-loop system?
50. What is a transducer?
51. Explain the function of the summing junction in a closed-loop system.
52. List and describe the three major control objectives for an effective control system.
53. Define these three types of transient responses: overdamped, critically damped, and underdamped.
54. How does the cost versus performance trade-off affect control system design?
55. What is system robustness?
56. Write an explanation in your own words that describes why the steady-state error is present in the liquid level problem in Figure 12-37.
57. Assume in Figure 12-37 that the input flow increases to 6 gpm. Describe what happened in the control system and how the system balances after the load change.
58. Define quarter amplitude decay, peak percentage overshoot, minimum integral of absolute error, and critical damping.
59. At what point in the process does the system in Question 57 operate without steady-state error? Describe why this is true.
60. How is bias or offset used to eliminate steady-state error, and what problems are created when that approach is used to reduce steady-state error?
61. Discuss instability relative to a control system.
62. Define proportional gain.
63. What is proportional band?
64. Before troubleshooting a process problem involving a proportional gain controller, what must you first understand?

WEB AND DATA SHEET QUESTIONS

Use the Allen-Bradley Web site at www.ab.com to answer the following questions.

1. Find the SLC 500 RTD analog module type 1746 and list the metals and their resistance values that can be used with the module.
2. What is the part number of the ControlLogix analog I/O module that can accommodate 16 single-ended input channels?
3. Sketch the connections for a three-wire RTD to an RTD analog input module.

PROBLEMS

1. For a temperature-sensing application where the maximum temperature could be 600° Celsius, what RTD metal would you select?

2. What is the flow rate of a fluid flowing through a 6-inch diameter pipe at velocity of 3 feet/second?
3. What is the flow rate using a differential pressure flow sensor if the pressure in front of the restriction is 8.1 psi and the pressure behind the restriction is 8.6 psi, and the restrictor constant is 12?
4. Design a vision application for each of the following situations. Select a vision system type, configuration, performance level, lighting technique, light source and type, and lens. Make all necessary assumptions and justify if necessary.
   a. Count boxes (18 inches square) on a 24-inch conveyor belt with cable access from only one side of the conveyor.
   b. Count shiny thermos bottles moving in a production machine. Access is limited to one side of the conveyor, but the sensor can be mounted as close to the bottles as necessary.
   c. Detect plastic clips coming from a bowl feeder on a pair of metal rails.
   d. Detect a small white relay on a dark printed circuit board. The sensor must be located 6 inches above the relay and board.
   e. The level of milk must be verified as the clear plastic milk bottles move down a conveyor. Access to both sides of the conveyor is permitted and there is no limitation on the distance from the bottles.
   f. Detect the leading edge of a square hole in a plate moving down a conveyor. Access above and below the conveyor is permitted.
   g. Detect and count aluminum foil–coated boxes of tea at the entrance to a packaging machine. The sensor must be mounted 2 inches from the box, and a reflector could be mounted on the other side of the object.
   h. Detect the presence of a metal casting at the input to a machining center. There is a single side access restriction, but the sensor can be mounted as close to the slug as necessary.
5. Describe the conditions of the process tank for the load change described in Example 12-2. Show the changes in fluid level and the drain valve during correction to the load change.
6. Draw the response curve showing the step load change (F_2), response flow (F_1), and controlled variable (tank level) for the problem described in Example 12-3.
7. A level control system has a gain of 6 and a time constant of 0.02 seconds that relate the fluid level in meters to the input flow in cubic meters per second. With the tank empty, the inlet valve is opened with a step change from 0 to 0.25m^3 per second. Determine the level after 0.05 seconds and the final level in the tank.

CHAPTER 13 PLC STANDARD IEC 61131-3 FUNCTION BLOCK DIAGRAMS

REFERENCES

Textbook Chapter 13 – PLC Standard IEC 61131-3 Function Block Diagrams (pp. 423 – 448)
Textbook CD – Allen-Bradley Manuals:

- Logix5000 Process Control and Drives – 1756-RM006C-EN-P
- Logix5000 General Instruction Reference – 1756-RM003G-EN-P
- Logix Advanced Programming Manual – I756-PM001F-EN-P

The following figures are referenced in the questions and problems for this chapter:

- Figure 3-43 is on page 131
- Figure 3-44 is on page 131
- Figure 3-45 is on page 133
- Figure 4-15 is on page 161
- Figure 4-18 is on page 165
- Figure 4-21 is on page 171
- Figure 5-14 is on page 195
- Figure 5-15 is on page 196
- Figure 5-18 is on page 199
- Figure 5-23 is on page 202
- Figure 5-24 is on page 203
- Figure 7-17 is on page 259
- Figure 7-21 is on page 266

QUESTIONS

1. Describe the operation of an FBD program in terms of inputs, outputs, and signal flow.
2. How does the program scan work in the FBD language?

3. Describe how to initiate an FBD program.
4. How does a function block instruction differ from a ladder rung?
5. Compare and contrast the three IEC 61131-3 graphical languages.
6. How are output latches implemented in FBD?
7. Why is the Assume Data Available indicator used in FBD?
8. How is feedback around one function block different from feedback around more than one block?
9. Describe the FBD program design process sequence.
10. Compare the unconditional and conditional ladder logic to the FBD equivalent.
11. Compare the combinational logic for ladder logic inputs to the FBD equivalent.
12. Compare the timer instructions for standard ladder logic inputs and the FBD equivalent.
13. Compare the counter instructions for standard ladder logic inputs and the FBD equivalent.
14. Compare the output latch instructions for standard ladder logic inputs and the FBD equivalent.
15. Compare the one-shot and math instructions for standard ladder logic inputs and the FBD equivalent.

CONTROLLOGIX PROBLEMS

When working these problems, assume that a 16-point input is in slot 2 and a 16-point output is in slot 4, unless otherwise noted. Pin numbers are specified in the problems.

1. Convert the two-handed control ladder logic in Figure 3-43 to an FBD solution. Use tag names and data types that are consistent with the process.
2. Convert the two-axis robot control problem in Figure 3-44 to an FBD solution. Use tag names and data types that are consistent with the process.
3. Convert the relay ladder logic in Figure 3-45 to an FBD solution. Use tag names and data types that are consistent with the process.
4. Convert the traffic light control problem in Figure 4-15 to an FBD solution. Use tag names and data types that are consistent with the process.
5. Convert the pumping system control problem in Figure 4-18 to an FBD solution. Use tag names and data types that are consistent with the process.
6. Convert the two-handed tie down control problem in Figure 4-21 to an FBD solution. Use tag names and data types that are consistent with the process.
7. Convert the packaging system control problem in Figures 5-14 and 5-15 to an FBD solution. Use tag names and data types that are consistent with the process.
8. Convert the parking garage control problem in Figure 5-18 to an FBD solution. Use tag names and data types that are consistent with the process.
9. Convert the machine queue parts control problem in Figures 5-23 and 5-24 to an FBD solution. Use tag names and data types that are consistent with the process.
10. Convert the rinse cycle for the tank control problem in Figures 7-17 and 7-21 to an FBD solution. Use tag names and data types that are consistent with the process.

CHAPTER 14 INTERMITTENT AND CONTINUOUS PROCESS CONTROL

REFERENCES

Textbook Chapter 14 – Intermittent and Continuous Process Control (pp. 449–476)
Textbook CD – Allen-Bradley Manuals:

- Logix5000 Process Control and Drives – 1756-RM006C-EN-P
- Logix5000 Controller Design Considerations – 1756-RM094A-EN-P

The following figures are referenced in the questions and problems for this chapter:

- Figure 14-2 is on page 451
- Figure 14-5 is on page 454
- Figure 14-9 is on page 457

QUESTIONS

1. Describe the two categories of process control.
2. Describe the operation of on-off controllers.

3. How does two-position control enhance an on-off controller?
4. How does floating control enhance an on-off controller?
5. What type of applications would be best suited for on-off control?
6. Describe the operation of a proportional controller.
7. Describe the operation of a proportional integral controller.
8. Describe the operation of a proportional derivative controller.
9. Describe the operation of a proportional integral and derivative controller.
10. What criteria are used to determine the mode of control required for a process system?
11. Describe how integral mode controllers eliminate steady-state error.
12. Describe how derivative mode controllers force the final control element to respond quickly to a large and rapid load change.
13. What is sampling update time?
14. Describe why steady-state error is present in proportional mode controllers after a set point or load change.
15. What are the three basic steps that fuzzy control uses when implemented in a controller?
16. Name two advantages of using fuzzy control.
17. Name two disadvantages of using fuzzy control.
18. Describe how a digital controller functions with the sampling of input data.
19. In what control mode does any error present change the controller output based on the area under the error curve?
20. Why is a delta e estimator needed in derivative control?
21. Why is the set point sometimes removed from derivative control?
22. Why is scaling important in process control?
23. How does a PID controller accomplish bumpless transfer?

WEB AND DATA SHEET QUESTIONS

1. Prepare a list of all of the input and output parameters for the PIDE instruction in the ControlLogix PLC system.
2. Compare the PID instructions for all of the PLC models with the PIDE instruction.
3. Describe what the auto-tune function in the PIDE instruction is and how it operates.
4. Select a mechanical two-point temperature switch and controller from Allen-Bradley for the tank temperature control problem where the process should be held at a temperature of 180° F with a differential of 10 degrees.

CONTROLLOGIX PROBLEMS

When working these problem, assume that a 16-point input is in slot 2 and a 16-point output is in slot 4, unless otherwise noted. Pin numbers are specified in the problems.

1. Modify the ControlLogix FBD solution in Figure 14-2 to a ControlLogix ladder solution. Create appropirate tag names for all parameters. The set point is 180 degrees.
2. Modify the ControlLogix FBD solution in Figure 14-5 to a ControlLogix ladder solution. Create appropirate tag names for all parameters. The set point is 180 degrees, and the dead band is 14 degrees.
3. Modify the ControlLogix FBD solution in Figure 14-9 to a ControlLogix ladder solution. The minimum percent output is 5 percent, the maximum is 95 percent, and the mid-point is 50 percent. Include the ladder rungs needed to determine the upper and lower limits.
4. Design a FBD solution using Figure 14-9 as a model for a controller with four output levels. The levels would be minimum percen, maximum percent, 25 percent, and 75 percent. Create appropriate tag names for all parameters. The set point is 190 degrees and the dead band is 10 degrees.

CHAPTER 15 PLC STANDARD IEC 61131-3–STRUCTURED TEXT LANGUAGE

REFERENCES

Textbook Chapter 15 – PLC Standard IEC 61131-3 – Structured Text Language (pp. 477–494)

Textbook CD – Allen-Bradley Manuals:

- SFC and ST Programming Manual – 1756-PM003G-EN-E
- Logix5000 Process Control and Drives – 1756-RM006C-EN-P

QUESTIONS

1. List the differences between the IL and ST languages.
2. Why is ST a better choice over IL as a programming language for PLCs?
3. Compare ST to another high-level text programming language that you have used, listing how they are different and similar.
4. What is the difference between : = (colon equal sign) and = (equal sign)?
5. What is the difference between an assignment statement and an expression?
6. What are the types that are evaluated in an expression?
7. What is the purpose of the numeric expression?
8. What is the difference between an operator and a function?
9. What are arithmetic operators?
10. What are logical operators?
11. What are relational operators?
12. What are the constructs available in ST?
13. What is the purpose of the ELSIF?
14. If a program has seven ELSIFs and the controller executes the first ELSIF, what happens to the remaining ELSIFs?
15. When would you use the FOR . . . DO construct?
16. Name and describe the four operands of the FOR . . . DO construct.
17. Compare and contrast the WHILE . . . DO and the REPEAT . . . UNTIL constructs.
18. What is the difference between the CASE construct in ST programming and the switch statement in C++ programming?

PROBLEMS

1. Develop an ST program to solve the following automation process problem.
   - If a tank temperature is over 85, run a pump slow.
   - If a temperature is over 170, set the pump speed to high.
   - Under all other conditions turn the pump off.
2. Develop an ST program to solve the following automation process problem.
   - If a liquid flow rate is less than 250 gallons per minute, set the fuel at 100.
   - If a liquid flow is greater than or equal to 250 gallons per minute and the flame is greater than 5 inches, set the fuel to 400, if not, set the fuel to 200.
3. Write an ST routine to set 16 valves from their database. The required positions (open = 1 and closed = 0) for 16 valves are stored in an array called Pst [0 – 15]. The tags for the valves are also in an array called Valve starting at element 8.
4. Develop an ST program to solve the following automation process problem.
   A tank input valve, tag V44, should be open (valve on or power applied) as long as the level switch, tag LS_low, is not true or 0 and the drain valve, tag V15, is closed (valve off or no power applied).
5. Using the IF . . . THEN . . . ELSE construct develop an ST program to set the speed of a water pump based on the following switch positions and set an alarm if the speed is zero.

Switch position	Speed
1	10
2	20
3	30

6. Repeat Problem 5 using the CASE . . . OF construct.

CHAPTER 16 PLC STANDARD IEC 61131-3 – SEQUENTIAL FUNCTION CHART

REFERENCES

Textbook Chapter 16 – PLC Standard IEC 61131-3 – Sequential Function Chart (pp. 495–510)
Textbook CD – Allen-Bradley Manuals:

- SFC and ST Programming Manual – 1756-PM003G-EN-E

- Logix5000 Process Control and Drives – 1756-RM006C-EN-P

The following figures are referenced in the questions and problems for this chapter:

- Figure 3-43 is on page 131
- Figure 3-44 is on page 131
- Figure 3-45 is on page 133
- Figure 4-15 is on page 161
- Figure 4-18 is on page 165
- Figure 4-21 is on page 171
- Figure 5-14 is on page 195
- Figure 5-15 is on page 196
- Figure 5-23 is on page 202
- Figure 5-24 is on page 203
- Figure 6-14 is on page 228
- Figure 6-15 is on page 229

QUESTIONS

1. Why is the Sequential Function Chart language good for developing sequenced-based processes and machines?
2. What current PLC language is SFC based on?
3. Describe the four SFC sequences.
4. Define step blocks, action blocks, and transition conditions.
5. What is an action qualifier?
6. Describe the type of problem for which each action qualifier would be best suited.
7. Describe in your own words the steps used in the development of an SFC program.
8. How is the SFC similar to a PLC sequencer instruction and how are they different?

CONTROLLOGIX PROBLEMS

When working these problems, assume that a 16-point input is in slot 2 and a 16-point output is in slot 4, unless otherwise indicated. Pin numbers are specified in the problems.

1. Convert the two-handed control ladder logic in Figure 3-43 to an SFC solution. Use tag names and data types that are consistent with the process.
2. Convert the two-axis robot control problem in Figure 3-44 to an SFC solution. Use tag names and data types that are consistent with the process.
3. Convert the relay ladder logic in Figure 3-45 to an SFC solution. Use tag names and data types that are consistent with the process.
4. Convert the traffic light control problem in Figure 4-15 to an SFC solution. Use tag names and data types that are consistent with the process.
5. Convert the pumping system control problem in Figure 4-18 to an SFC solution. Use tag names and data types that are consistent with the process.
6. Convert the two-handed tie down control problem in Figure 4-21 to an SFC solution. Use tag names and data types that are consistent with the process.
7. Convert the packaging system control problem in Figures 5-14 and 5-15 to an SFC solution. Use tag names and data types that are consistent with the process.
8. Convert the machine queue parts control problem in Figures 5-23 and 5-24 to an SFC solution. Use tag names and data types that are consistent with the process.
9. Convert the color sensor control problem in Figures 6-14 and 6-15 to an SFC solution. Use tag names and data types that are consistent with the process.

CHAPTER 17 INDUSTRIAL NETWORKS AND DISTRIBUTIVE CONTROL

REFERENCES

Textbook Chapter 17 Industrial Networks and Distributive Control (pp. 511 – 530)
Textbook CD – Allen-Bradley Manuals:

- Logix5000 Controllers Design Considerations – 1756-RM094A-EN-P
- Logix5000 Controllers Import/Export Reference Manual – 1756-RM0841-EN-P

QUESTIONS

1. Describe the three layers in the factory floor network architecture.
2. What are protocols?
3. What does the term *fieldbus* mean?

4. What are the three basic elements of the Ethernet?
5. How would you interpret the *x*Base*y* nomenclature?
6. What advantages does Ethernet/IP have over other industrial Ethernet systems?
7. How does Ethernet/IP simplify the interoperability between different vendors' devices?
8. How is access controlled on the ControlNet?
9. What does deterministic and repeatable mean relative to networks?
10. Describe the following message types available on DeviceNet.
    a. Polling
    b. Strobing
    c. Cyclic
    d. Change of state
    e. Explicit messaging
    f. Fragmented messaging
11. What is connection-based protocol?
12. Why does SERCOS use fiber optic cabling for its connectivity?
13. Compare and contrast the conventional operator panel and a smart I/O interface.
14. What is remote rack capability?
15. What are DTE and DCE?
16. What is the difference between Firewire and other serial communication devices such as RS-485?
17. What is a hot-plug capable device?
18. What is a human-machine interface?
19. What is the function of a graphic terminal on an industrial network?
20. Describe four wireless network applications.
21. Describe the function of Profibus-DP, Profibus-PA, and Profibus-FMS.
22. What is a token-passing network?
23. Describe the communication technique used on the Modbus network.
24. Why is troubleshooting network systems especially important?
25. When investigating the network cabling system, what three steps should a troubleshooter take?
26. What is distributive control?
27. Describe in-cabinet distributed I/O.
28. What is the difference between modular I/O and block I/O?
29. Describe on-machine distributed I/O.
30. What is the function of the Ethernet interface module?
31. Name four features of a hub/switch function in an Ethernet network.
32. What is the function of the scanner in the ControlNet network?
33. In a ControlNet network, why is it important to design in a minimum reserve of available memory?
34. What is the function of the scanner in the DeviceNet network?
35. When designing a DeviceNet network, why is it important to keep at least one node address in reserve?

WEB AND DATA SHEET QUESTIONS

Use the Allen-Bradley Web site, www.ab.com, to answer the following questions.

1. List the transmission characteristics specified for the DH+ network.
2. What is the maximum length in feet for a daisy-chain remote I/O connection?
3. List and describe the four features of the Common Industrial Protocol (CIP).
4. Relative to distributive control, when using an Ethernet/IP to DeviceNet linking device, what is the maximum number of bytes that can be handled?
5. Relative to distributive control, when using a ControlNet to DeviceNet linking device, what is the maximum number of input words that can be handled?

PROBLEMS

1. Your supervisor asks you to recommend a network that can service five bar code readers, five weight scales, and one industrial printer. What network would you recommend and why?
2. An intern shows you her ControlNet network design, proudly commenting that she used all 99 nodes on the network. What is your response and why?
3. You have been asked to connect eight computers for an aircraft propulsion design team so that they can exchange design data. What network would you use and why?

4. As a network designer you have been asked to set up 10 modems in the computer library. What network would you choose and why?
5. You've been asked to set up a network with distributed controllers, each having its own I/O communication. What network would you choose and why?
6. You are a manufacturing engineer on the network design team at an electronics assembly plant. Automated equipment for the new surface-mount facility must be networked with as much fail-safe capability as possible. What network would you choose and why?

Index

Action blocks, 499–502
Action qualifiers, 502–504
Addition instruction (ADD), 217–218
Addressing modes, Allen-Bradley, 303–310
  direct, 303
  indexed, 305
  indirect, 304
  syntax for, 305–310
Alarms, industrial, 71
Allen-Bradley, 4, 4*f*
Analog modules, 399–401
Analog sensors, 373–399
  flow sensors, 380–389
  position, 389–390
  pressure, 377–379
  temperature, 374–377
  troubleshooting, 396–399
  vision, 390–399
  vision systems, 390–399
AND, OR, and XOR instructions, 319–320
  suggested applications for, 341–342
Arithmetic functions, 479–480
Arithmetic instructions, Allen-Bradley, 215–216
  addition (ADD), 217–218
  division (DIV), 219
  move instruction, 223–224
  multiplication (MUL), 219
  operation of, 216–217
  square root (SQR), 222
  subtraction (SUB), 218–219
Arithmetic operators, 479
Armature. *See* Clapper
Arrays, 307–310
ASCII I/O interface, 18, 19–20
Assignment statements, 478
Asynchronous data transfer, 518
Audio output devices, 70–71
  alarms, 71
  horns, 70
Auxiliary contactors, 67

Backplane, 11
BCD. *See under* Binary coded decimal (BCD)
Bimetallic overload method, 68
Binary arithmetic, 211–213
Binary coded decimal (BCD) conversion
  empirical design process and, 258–268
  troubleshooting, 268–272
Binary coded decimal (BCD) system, 241–243
  Allen-Bradley conversion instructions, 243–250
Binary numbers, signed, 213–214
Binary number system, 78–79
  decimal conversion and, 79–80
  notations for, 83
Bit addressing, timer, 153–155
Bit number, 104
Bit operation instructions, empirical design process with, 341–344
Bits, 83
Block diagrams, 134
Boolean expression, 479
Bracketing, 134–137
Business transfer, 476
Bytes, 83

Capacitive proximity sensors, 48
Cascaded counters, 203
Cascaded timers, 167–169
Cascading sequencers, 369
Central processing unit (CPU), 5
Chassis, 86
Clapper, 7
Closed-loop control systems, 401–406
  analysis of, 403–405
  summary of, 420–421
  troubleshooting, 421
Closed tank models, for level switches, 41
Comparison instructions, 250–251
Comparison instructions, Allen-Bradley, 251–258
Comparison ladder logic troubleshooting, 268–272
Constructs, 481–488
Contact devices, 47
Contactors, 10
  auxiliary, 67
Contacts, 7
Contiguous block, 307
Continuous controllers, 457–470
  fuzzy control, 467–469
  proportional control, 457–460
  proportional derivative (PD) control, 462–465
  proportional integral and derivative (PID) control, 465–467
  proportional integral control, 460–463
Control diagrams, 43–45
  alternative names for, 43–45
Controllers
  continuous, 457–470
  intermittent, 450–457

Controller-scoped data, 291
ControlLogix
immediate output instructions, 298
internal control relay bit addressing, 107–108
memory organization, 89–91
sequencer instructions, 252–259
ControlLogix arrays, 307–310
ControlNet, 18, 514, 528
Control relays, 63–64
Control systems
attributes of effective, 406–421
closed-loop, 401–406
Convergent rule, 138
Convergent signal path, 137
Copy (COP) instruction, 338–339
Coriolis mass flow sensor, 388–389
Counter ladder logic, troubleshooting, 207–209
Counters
Allen-Bradley, introduction to, 180–183
Allen-Bradley instructions, 183–202
Allen-Bradley reset instructions, 183–202
cascaded, 203
converting relay ladder logic to PLC logic for, 207
electronic, 179
empirical design process with, 203–207
mechanical, 179
troubleshooting, 207–209
CPU. *See* Central processing unit (CPU)
Current sinking, 14–15
Current sinking devices, wiring for, 60–61
Current sourcing, 14–15
Current sourcing devices, wiring for, 60–61

Data handling, 317–318
troubleshooting, 344–345
Data highway networks, 521–523
Data Highway Plus (DH+), 521
Data latching, 427–428
Data transfer and manipulation instructions, Allen-Bradley, 318–341
AND, OR, and XOR instructions, 319–320, 320*f*
copy (COP) instruction, 338–341
file-arithmetic-logic (FAL) instruction, 320–326
fill (FLL) instruction, 339–341
first in, first out (FIFO) function, 333–336
last in, first out (LIFO) function, 336–338
shift registers, 326–332
Data types
basic, 89
structured, 89
Data values
global, 290–291
program, 290–291
Decimal conversion
binary number system and, 79–80
hex conversion and, 81
Deflection pressure sensors, 377, 378–379
DeviceNet, 515–516, 528–529
DeviceNet Interface (DNI), 529
DH-485 network, 521–523
Differential pressure flow sensors, 381–383
elbow, 383
flow nozzle, 383
orifice plate, 381–382
Pitot tube, 382–383
venturi, 382
Differential pressure sensors, 377, 379
Diffuse mode, 54
Digital control, 470–475
derivative control mode, 474–475
integral control mode, 472–474
proportional control mode, 472
sample and hold, 471–472
Digital two-position control, 453
Direct addressing, 303
Direct pressure sensors, 377–378
Discrete designs, 79
Discrete output instruction, 22
Discrete sensors, 47
Displacement flow sensors, 387
Distributive control, 524–526
Disturbance, 405
Divergent rule, 137–138
Divergent signal path, 137
Divide-and-conquer rule, 137
Division instruction (DIV), 219
Doppler, Christian Johann, 386
Double break contacts, 34
Double words, 83
Dropout current, 74

Elbow flow sensor, 383
Electrical safety, 28–32
Electromagnetic flow sensor, 386
Electromagnetic output actuators, 61–69
contactors, 67–68
control relays, 63–64
latching relays, 67
motor starters, 68–69
solenoid-controlled devices, 61–63
Electromagnets, components of, 77
Electronic counters, 179–180. *See also* Counters
Electronic schematics, 8–9
Electronic timing relays, 145–146
Empirical design process
binary coded decimal conversion and, 258–268
with bit and word operation instructions, 341–344
with Function Block Diagram program, 433–447
with indexed addressing, 310
with indirect addressing, 310
with math/move instructions, 230–235
with PLC counters, 203–207
with PLC timers, 169–174
with program control instructions, 299–301
with sequencer instructions, 369–371
with Structured Text, 488–493
Empirical program design, 126–132
Ethernet, 512–514
operation, 513
Ethernet interface (ENI), 527–528
Ethernet/IP, 18, 19, 513–514
Eutectic alloy overload method, 68
Examine if closed (XIC), 110–115
selecting correct input instructions, 121–123
Examine if open (XIO), 110–115
selecting correct input instructions, 121–123
Execution order, 427
Expansion slots, 5
Expressions, 478–479
Boolean, 479
numeric, 479
Extended guard switches, 36

Feedback rule, 138
Feedback signal flow, 137
Fibrillation, 28
Field device power rails, 220
Field devices, 10, 13, 22
interfacing input, 58–61
interfacing output, 71–73
File-arithmetic-logic (FAL) instruction, 320–326
File name, 104
File number, 104
Fill (FLL) instruction, 339–341
Firewire, 518–519
First in, first out (FIFO) function, 333–336
First pass bit, 109
Fixed I/O interface, 12–13
Floating control, 455–457
Flow nozzle flow sensor, 383

Flow sensors, 380–389
differential pressure, 381–383
displacement, 387
mass, 387–389
measurement methods of, 381
velocity, 384–387
visual, 389
Flow switches, 41
Full guard switches, 36
Function Block Diagram (FBD) language, 425–433
Allen-Bradley RSLogix 5000, 431–433
data latching, 427–428
development sequence, 429–431
empirical design with, 433–447
execution order, 427
feedback loops, 428–429
signal flow types, 427
Fuzzy control, 467–469
Fuzzy logical controllers (FLCs), 467–468

Global data values, 290–291
Graphic terminals, 520

Henry, John, 7
Hexadecimal number systems, 81–83, 250
Holding contacts, 67
Horns, industrial, 70
Human-machine interface (HMI), 520
Hydraulic pressure, 377
Hysteresis, 41

IEC. *See* International Electrotechnical Commission (IEC)
IEC 61131 standard, 423, 424–425
overview of text languages, 477
Immediate input (IIN) instructions
PLC 5, 296
SLC 500, 296–298
Immediate output (IOT) instructions
ControlLogix, 298
PLC 5, 296
SLC 500, 296–298
In-cabinet I/O, 526
Indexed addressing, 304–305
with empirical design process, 310
PLC 5, 306–307
SLC 500, 306–307
troubleshooting, 314
Indirect addressing, 304
empirical design process with, 310
PLC 5, 305–306
SLC 500, 305–306
troubleshooting, 314
Inductive proximity sensors, 47–48
Industrial networks, URLs for, 529–530
Input and output (I/O) addressing
for non-Allen-Bradley vendors, 99–100
PLC 5 rack/group-based, 92–94
SLC 500 Rack/slot, 94–99
tag-based, 100–104
Input current sinking/sourcing circuits, 14–15, 16*f*
Input instructions, Allen-Bradley, 110–116
input label (LBL) instructions, 277–282
Input/output (I/O)
distributed, 526
in-cabinet, 526
on-machine, 526
Input reference symbols (IREFs), 426–427, 428
Input wiring, 59–60
Instantaneous contacts, 143–144
Instruction List (IL) language, 477
Instructions, 23
Integral action, 460–462
Integral time, 461
Interfacing input field devices, 58–61
powering, 58–59
Interfacing output field devices, 71–73
powering, 71
wiring, 71–73
Intermittent controllers, 450–457
floating control, 455–457
on-off control, 450–451
two-position control, 451–455
Internal control relay bit addressing, 104–108
ControlLogix binary bit, 107–108
PLC 5 binary bit, 104–107
SLC 500, 104–107
International Electrotechnical Commission (IEC), 4
I/O. *See* Input and output (I/O)
IREFs. *See* Input reference symbols (IREFs)
Isochronous data transfer, 518

Jump (JUMP) output instructions, 277–282
vs. MCR instructions, 280

Ladder logic, 4–5, 21–28
for Allen-Bradley math/move instructions, 223–229
conversion/comparison, troubleshooting, 268–272
counter, troubleshooting, 207–209
math/move, troubleshooting, 235–238
operation, 244–250
relay, 5–10
Ladder logic programming, 121–133
converting relay logic to PLC solutions, 132–134
empirical program design for, 126–132
multiple inputs, 123–126
multiple outputs, 126
selecting correct XIC and XIO input instructions, 121–123
Laminar flow, 380
Lands, 11
Last in, first out (LIFO) function, 336–338
Latching relays, 67
Left power rail, 10
Lenses, 51–52
Level switches, 41
closed tank models, 41
open tank models, 41
Light detector, 51
Light source, 51
Limit switches, 40–41
Linear potentiometers, 389
Linear rule, 137
Linear signal path, 137
Linear variable differential transformer (LVDT), 389–390
Logic circuit, 52
Logic configurations, 124*f*
Loop sequence, 497

Make before break contacts, 34–36
Manipulation instruction. *See* Transfer/manipulation instructions, Allen-Bradley
Manual control mode, 476
Manually operated switches, 33–39
push button, 36–38
selector, 38–39
toggle, 34
Mass flow sensors, 387–389
Coriolis, 388–389
Master control relays (MCR), 274
jump *vs.*, instructions for, 280
Master control reset instructions, 274–277
Math instructions, Allen-Bradley, 215–216
Math/move instructions, Allen-Bradley
empirical design process, 230–235
ladder logic, 223–229
Math/move ladder logic, troubleshooting for, 235–238
MCR. *See* Master control relays (MCR)
Mechanical counters, 179–180. *See also* Counters
Mechanically operated industrial switches, 39–45
flow, 41
level, 41

Mechanically operated (*Continued*)
limited, 40–41
pressure, 41
temperature, 42–43
Mechanical timing relays, 141–145
instantaneous contacts, 143–144
timed contacts, 142–143
timing relay operation, 144–145
Mechanical two-position control, 453–455
Memory, 5. *See also* PLC
Memory blocks, 84
Memory organization
Allen-Bradley Logix, 89–91
Allen-Bradley PLC 5, 5
Allen-Bradley SLC 500, 86–89
Memory structures, Allen-Bradley
Logix system, 89–91
PLC 5, 86
rack-based memory, 85
SLC 500, 86–89
tag- or variable-based, 86
Mesh networks, 519
Metcalf, Robert, 512
Modbus network, 523–524
Modular I/O interface, 12–13
Morley, Dick, 3
Motherboard, 5
Motor starters, 68–69
Move (MOV) instructions, 222–223. *See also* Math/move instructions, Allen-Bradley
Move with a mask (MVM) instruction, 222
Multiplication instruction (MULT), 219
Multipoint to point networks, 519
Mushroom button guard switches, 36

NC. *See* Normally closed (NC) contacts
Negate (NEG) instruction, 223
Network architecture, PLC, 511–512
Network communications, 10
Networks
designing, 527–529
selecting, 526–527
URLs for industrial, 529–530
Network systems, troubleshooting, 524
NO. *See* Normally open (NO) contacts
No guard switches, 36
Non-contact devices, 47
Normally closed (NC) contacts, 7, 34
Normally open, timed closed (NOTC) contacts, 142
Normally closed, timed open (NCTO) contacts, 142–143
Normally open (NO) contacts, 7, 34
Number systems, 77–81
basics of, 78
binary, 78–80
comparing, 82–83
hexadecimal, 81–83
octal, 80–81
Numeric expression, 479

Octal number system, 80–81
Off-control, 450–451
Off-line programming, 27
Omron Electronics, 419
On-line programming, 27–28
On-machine I/O, 526
Open tank models, for level switches, 41
Operators, 34
arithmetic, 479
bitwise, 480–481
logical, 480–481
relational, 480
Optics, vision system, 396, 398
Orifice plate flow sensor, 381–382
Output coils, Allen-Bradley, 110–116
Output current sinking/sourcing circuits, 15–17, 17*f*
Output device, 52
Output latch, 115–116
vs. sealing instructions, 116
Output reference symbols (OREFs), 426–427
Overloads (OLs), 10
Override instructions, 274

Paddlewheel flow sensor, 384
PanelView, Allen-Bradley, 520
PCs. *See* Program controllers (PCs)
Peer to peer networks, 519
Peripheral Component Interconnect (PCI), 519
Phase-loss sensitivity overload method, 68
Photoelectric sensors, 51
applications, 56, 57*f*
components, 51–52
functions, 55
operating modes, 52–55
troubleshooting, 75
Piezoelectric pressure sensors, 377, 379
Pilot lamps, 69–70
Pitot tube flow sensor, 382–383
PLC 5, Allen-Bradley
immediate input instructions, 296
immediate output instructions, 296
indexed addressing, 306–307
indirect addressing, 305–306
input and output addressing, 92–94
internal control relay bit addressing, 104–107
memory organization, 5
memory structures, 86
sequencer instructions, 349–352, 359–366
status data addressing, 108–110
subroutine instructions, 282–287
PLC industry, growth of, 3–4
PLC logic, converting relay logic timer ladders to, 174–175
PLC manufacturers, URLs for, 32
PLC memory, 83–84
register structure and, 84–92
PLC processor, 11
PLC program design, relay ladder logic conversion and, 121–134
PLCs. *See* Program controllers (PCs); Program logic controllers (PLCs)
PLC solutions, converting relay logic to, 132–134
PLC special-purpose modules, 20
PLC systems
power supply, 11
processor, 11
special-purpose modules, 20
PLC timers, 146
Pneumatic robot design, 265–268
Point to multipoint networks, 519
Polarized retroflective mode, 54
Pole, 34
Position sensors, 389
Post-indexed addressing, 305
Power supply, 11
input/output (I/O) interface, 12
programming devices, 11–12
Pre-indexed addressing, 305
Pressure, defined, 377
Pressure sensors, 377–379
deflection, 377, 378–379
differential, 377, 379
direct, 377–378
piezoelectric, 377
solid-state, 377
Pressure switches, 42
control symbols for, 44*f*
Process control, 449–450. *See also* Controllers
scaling in, 475
Process Field Bus (Profibus), 520–521
Process load change, 405–406
Processor, 5
Profibus (Process Field Bus), 520–521
Program control instructions, 273
Allen-Bradley, 274–295
empirical design process with, 299–301
troubleshooting, 301
Program controllers (PCs), *vs.* PLCs, 5
Program data values, 290–291
Program logic controllers (PLCs)
advantages, 25–28
defined, 4–5

vs. PCs, 5
processors, 10
rack/slot address based, 20
soft, 21
solution, 22–23
standards, 4, 423–425
tag based, 20–21
Program logic controllers (PLCs) systems, block diagram for, 10–11
Programming devices, 11
Proportional control, 457–460
Proportional derivative (PD) control, 462–465
Proportional gain controllers, troubleshooting, 421
Proportional integral and derivative (PID) control, 465–467
Proportional integral (PI) control, 460–463
Proprietary network, 18
Proximity sensors, 47–51
applications, 49–51
capacitive, 48
inductive, 47–48
troubleshooting, 75
ultrasonic, 48–49
Pull-in current, 74
Push button switches, 36–38, 38*f*
extended guard, 36
full guard, 36
mushroom button, 36
no guard, 36

Rack/slot address based PLCs, 20
Reference bloc, 375
Reference junction, 375
Register structure, PLC memory and, 84–92
Relay coil hum, 74
Relay contact ratings, 64–67
rated contact, 65
rated voltage, 64–65
Relay control systems, 8–10
Relay ladder logic, 5, 5*f*
conversion, PLC program design and, 121–134
Relay ladder logic diagrams, 9–10
Relay logic, converting, to PLC solutions, 132–134
Relay logic timer ladders, converting, to PLC logic, 174–175
Relays
control, 63–64
electromagnetic, 7–8
with multiple poles and throws, 7, 8*f*
Square D, 7, 7*f*
troubleshooting, 74
types of, 5*f*
Relay schematic symbols, 66*f*
Remote I/O interfaces, 517
Remote racks, 18
Repeated division method, 83
Resistance temperature detector (RTD), 374
troubleshooting, 396
Response to change, control systems and, 406–410
Retentive timer on-delay (RTO), 146–152
Retentive timers (RTOs), Allen-Bradley, 164–167
Retroreflective mode, 53–54
Right power rail, 10
Robot design, pneumatic, 265–268
Rotary potentiometers, 389

Safety
electrical, 28–32
PLC, 28–32
Sampling update time, 471
Scaling, in process control, 475
Scan, PLC, 116–118
Scan time, 117
Sealing instructions, output latch *vs.*, 116
Selection divergence sequence, 497, 498
Selector switches, 38–39, 38*f*
Sensors, industrial, 45–58
analog, 373–399
discrete, 47
photoelectric, 51–58
proximity, 47–51
troubleshooting, 75
Sequence
defined, 497
loop, 497
selection divergence, 497, 498
simultaneous divergence, 497, 498
single, 497–498
Sequencer compare (SQC), 349
instruction, 366–367
Sequencer function, basic PLC, 349
Sequencer input (SQI), 349
PLC 5 and ControlLogix instructions for, 359–366
Sequencer instructions (SQI), Allen-Bradley, 349–369
ControlLogix, 352–359
empirical design process with, 369–371
for PLC 5 and ControlLogix SQI, 359–366
PLC 5 operation, 350–352
PLC 5 structure, 349–350
SLC 500 operation, 350–352
SLC 500 structure, 349–350
troubleshooting, 371
Sequencer load (SQL), 349
instruction, 367–369
Sequencer output (SQO), 349
Sequencers, cascading, 369
Sequencing, electromechanical, 347–348
Sequential Function Chart (SFC), 495–509
design steps, 505–509
standard sequences, 496–497
step programming, 498–509
SERCOS interface, 18, 19, 516
Serial communication interfaces, 517–519
Shift registers, 326–332
troubleshooting, 344–345
Shock, electrical, 28–30
Signal conditioner, 13–14
Signal flow, 137
Signal flow analysis, 137
Signal flow types, 427
Signed binary numbers, 213–215
Simultaneous divergence, 497, 498
Single pole double throw (SPDT) configuration, 7
Single sequence, 497–498
SLC 500, Allen-Bradley
immediate input instructions, 296–298
immediate output instructions, 296–298
indirect addressing, 305–306
input and output addressing, 94–99
internal control relay bit addressing, 104–107
memory organization, 86–89
sequencer instructions, 249–252
status data addressing, 108–110
subroutine instructions, 282–287
Small Computer Interface (SCSI), 519
Smart I/O interface, 18, 516–517
Soft PLCs, 21
Solenoid valves, 61–63
Solid-state pressure sensors, 377, 379
Solid-state temperature sensors, 376–377
Square D relays, 7, 7*f*
Stability, control systems and, 418–421
Standard diffused mode, 54, 55*f*
Standards, PLC, 423–425
Status data addressing
PLC 5, 108–110
SLC 500, 108–110
Steady-state error, control systems and
correction for, 413–418
correction with bias, 418
Steady-state response, control systems and, 410–431

Strain, 377
Stress, 377
Structured Text (ST) language, 477
  Allen-Bradley version of, 477–478
  arithmetic functions, 479–480
  arithmetic operators, 479
  assignment statements, 478
  bitwise operators, 480–481
  constructs, 481–488
  empirical design with, 488–493
  expressions, 478–479
  logical operators, 480–481
  programming in, 478–486
  relational operators, 480
Sub-network scanner modules, 10
Subroutine instructions, 280–282
  ControlLogix, 287–295
  PLC 5, 282–287
  SLC 500, 282–287
Substraction instruction (SUB), 218–219
Suspend instuction (SUS), 208–209
Switched rule, 138
Switches
  configurations, 34
  manually operated, 33–39
  mechanically operated industrial, 39–45
  terms for, 34
  toggle, 34–36
  troubleshooting, 74
Switch signal flow paths, 137
Syntax, for addressing modes, 305–310
System update time, 471

Tag-based addressing, 100–104
Tag-based memory systems, 86
Tag based PLCs, 20–21
Temperature sensors, 374–377
  comparison of, 377
  solid-state, 376–377
Temperature switches (thermostats), 42–43
  control symbols for, 44*f*
Thermistors, 374–375
  troubleshooting, 396
Thermocouples, 375
Thermostats (temperature switches), 42–43
  control symbols for, 44*f*
Three-wire sensor outputs, 59
Through beam mode, 53
Throws, 7
  double, 34
  single, 34
Timed contacts, 142–143
Timer bit addressing, 153–155
Timer instructions
  Allen-Bradley, 146–152
  PLC, 146
Timer off-delay (TOF) instructions, 146–152
Timer off-delay (TOF) ladder logic, programming, 155–164
Timer on-delay (TON) instructions, 146–152
Timer on-delay (TON) ladder logic, programming, 155–164
Timer parameters, 153–155
Timers
  cascaded, 167–169
  empirical design process with, 169–174
  retentive, 164–167
  troubleshooting, 175–177
Timing relays, 141
  electronic, 145–146
  mechanical, 141–145
Toggle switches, 34–36
Tools, troubleshooting, 134–138
  block diagrams, 134
  bracketing, 134–137
  signal flow, 137
  signal flow analysis, 137–138
Transfer/manipulation instructions, Allen-Bradley, 318–341
  copy/fill instructions, 338–341
  empirical design process with, 341–344
  file-arithmetic-logic (FAL) instruction, 320–326
  first in, first out (FIFO) function, 333–336
  AND instructions, 319–320
  last in, first out (LIFO) function, 336–338
  OR instructions, 319–320
  shift registers, 326–332
  troubleshooting, 344–345
  XOR instructions, 319–320
Transient response, control systems and, 406
Transition condition, 496, 504–505
Troubleshooting
  binary coded decimal conversion, 268–272
  closed-loop control systems, 421
  comparison ladder logic, 268–272
  conversion/comparison ladder logic, 268–272
  counter ladder logic, 207–209
  data handling, 344–345
  indexed addressing and, 314
  indirect addressing and, 314
  input and output modules, 138–140
  input/output devices, 73–74
  ladder logic control systems, 134–140
  ladder rungs with timers, 175–177
  math/move ladder logic, 235–239
  photoelectric sensors, 75
  program control instructions, 301
  proportional gain controllers, 421
  proximity sensors, 75
  relays, 74
  sequence, 138
  sequencer instructions, 371
  shift registers, 344–345
  switches, 74
  tools for, 134–138
Turbine flow sensor, 384–385
Turbulent flow, 380
Two-position control, 451–455
  digital, 453
  mechanical, 453–455
Two-wire diagrams, 9–10
Two-wire sensor outputs, 59

Ultrasonic flow sensor, 386–387
Ultrasonic proximity sensors, 48–49
Universal Serial Bus (USB), 519
URLs
  for PLC manufacturers, 32
URLs for industrial networks, 529–530
User-defined space, 86

Velocity flow sensors
  electromagnetic, 386
  paddle flow, 384
  turbine, 384–385
  vortex, 384–386
Venturi flow sensor, 382
Virtual control relays, 104
Vision systems
  applications for, 390
  classifications, 394
  configurations, 394–395
  lighting and, 395
  lighting techniques, 396–398
  light sources and configurations, 395–396
  operations and applications, 391–394
  optics, 396
  programming methods, 395
Visual and audio output devices, 69–71
  alarms, 71
  horns, 70
  pilot lamps, 69–70
Visual flow sensors, 389
Vortex flow sensor, 384–386

Wireless communication, 519–520
Wireless networks
  mesh, 519
  multipoint to point, 519
  peer to peer, 519
  point to multipoint, 519
Word operation instructions, empirical design process with, 341–344
Words, 83